Engineering Hydrology

Proceedings of the Symposium

Sponsored by the
Hydraulics Division of
American Society of C

Hosted by the
Virginia Section, ASCE

Williamsburg Hilton N
Williamsburg, Virginia
August 3-7, 1987

Edited by Arlen D. Fe

Published by the
American Society of Civil E
345 East 47th Street
New York, New York 10017

DATE DUE

12/7/02

DEMCO 38-297

*TC
5
.E54
1987
589*

ABSTRACT

The Surface Water Hydrology Committee, ASCE Hydraulics Division, sponsored this first Symposium on Engineering Hydrology. The Symposium was held jointly with the ASCE National Conference on Hydraulic Engineering at the Hilton Conference Center in Williamsburg, Virginia, August 3-7, 1987. One hundred and twenty nine papers are contained in these Proceedings. They cover a broad range of hydrologic engineering endeavors, from traditional flood and water supply hydrology, to the many areas which rely on hydrologic analyses. The main session titles are: water quality, reservoir systems, dam safety, sedimentation/erosion, flood forecasting, urban hydrology, groundwater quantity and quality, flood frequency, drought, design storms, microcomputer applications, surface/groundwater systems, and hydrologic modeling and calibration. The objective of the Symposium is to enhance communication of the practical state-of-the-art of engineering hydrology among practicing professionals. The emphasis is on how well the current technology works, rather than new untested theories.

Library of Congress Cataloging-in-Publication Data

Engineering hydrology.

Includes indes.
1. Hydraulic engineering—Congresses. 2. Hydrology—Congresses. I. Feldman, Arlen. II. American Society of Civil Engineers. Hydraulics Divison.
TC5.D54   1987   627   87-19270
ISBN 0-87262-611-3

PREFACE

This is the *first* ASCE Symposium dedicated to *Engineering Hydrology*. The Symposium is a result of the strong interest in Enginering Hydrology expressed in the many ASCE water-related conferences. The Surface Water Hydrology Committee of the Hydraulics Division recognized this need and formed a Task Committee to organize the Symposium. The Symposium was held jointly with the ASCE National Conference on Hydraulic Engineering at the Hilton Conference Center in Williamsburg, Virginia, on August 3–7, 1987.

The Symposium's *objective* is to enhance communication of the practical state-of-the-art of Engineering Hydrology among practicing professionals. The emphasis is on how well the current technology works, not on new untested theories.

The *subject matter* of the Symposium is as far reaching as the definition of Engineering Hydrology. It covers the traditional areas of flood and water supply hydrology, as well as the many areas which rely on hydrologic analyses, including water quality, reservoirs systems, dam safety, sedimentation/erosion, real-time water control and project planning.

Almost all of the papers were *volunteered*—this speaks highly of the interest in and need for more focused ASCE attention on Engineering Hydrology. Each of the papers included in these Proceedings has been accepted for publication in the Proceedings Editor. All papers are eligible for discussion in the Hydraulics Journal. All papers are also eligible for ASCE awards.

The *Table of Contents* shows the order in which the papers were presented. The Symposium began Monday with a keynote paper by Leo R. Beard. Four parallel sessions were then initiated. The entire Symposium rejoined late Monday afternoon to hear about microcomputer applications, followed in the evening with technical exhibits and microcomputer demonstrations. Four parallel sessions were resumed on Tuesday and Wednesday, culminating in one final joint session to end the Symposium. The closing session featured notable leaders of the profession giving their assessment of the current state-of-the-practice and recommendations for the technology of the future.

Many people were responsible for the success of the Symposium. The following persons were responsible for the technical organization of the Symposium:

Symposium Task Committee

Arlen D. Feldman, Chairman
Catalino B. Cecilio
Arthur Cudworth
William Espey, Jr.
Brendan M. Harley

The following persons and groups are *acknowledged* for their invaluable assistance in making the Symposium possible:

Robert M. Ragan, Chairman of The National Conference on Hydraulic Engineering

John C. Peters, Chairman and member of the Surface Water Hydrology Committee, Hydraulics Division

Bill S. Eichert, Director, U.S. Army Hydrologic Engineering Center

Denise Nakaji, Secretary and electronic information system specialist

Gail Ragan, Accommodations, tours, and entertainment arrangements

Harvey Tuvel, ASCE Technical Services Manager

Shiela Menaker, ASCE Book Production Manager

Michael Armenti, ASCE Conference Coordinator

Virginia Section, ASCE

Williamsburg Hilton Staff

The *real thanks* must go to the many dedicated hydrologic engineering professionals who wrote and presented the papers, chaired the sessions, and served on the panel discussions. Their technical abilities and energy are what insures our progress.

Arlen D. Feldman, Editor

CONTENTS

Plenary Session of the Engineering Hydrology Symposium

Session: 01—Welcome and Introduction Speaker
Moderator: Arlen D. Feldman

Session: 02—Flood Forecasting/Warning: Models
Moderator: Gary R. Dyhouse

Session: 03—Water Supply, Part I
Moderator: Chin Kuo

Session: 04—Water Quality, Part I
Moderator: Michael F. Domenica

Session 05—Design Storms
Moderator: Franklin Richards

Session: 06—Flood Forecasting/Warning: Data
Moderator: William Shope

Session: 07—Water Supply, Part II
Moderator: Brendan M. Harley

Session: 08—Water Quality, Part II
Moderator: Robert Ambrose

Session: 09—Urban Hydrology: Modeling I
Moderator: Larry A. Roesner

Session: 10—Microcomputer Applications
Moderator: Bruce A. Tschantz

Session: 15—Flood Forecasting/Warning: Case Examples I
Moderator: Michael W. Burnham

Session: 16—Groundwater
Moderator: Linda S. Weiss

Session: 17—Urban Hydrology: Detention Basins
Moderator: George V. Sabol

xi

Session: 33—Hydrologic Models: Parameter Estimation, Part II
Moderator: Gert Aron

Session: 34—Past, Present, and Future of Hydrologic Engineering
Moderator: William H. Espey, Jr

Panelists: Leo R. Beard, Franklin F. Snyder, Carl F. Izzard and Donald Newton

GENERAL CONSIDERATIONS FOR MANAGING FLOODS AND DROUGHTS

Leo R. Beard, F., ASCE*

Abstract

Work described at this conference and future developmental work of water resource engineers will best contribute to professional improvement if such work is guided by appropriate objectives. Ultimately, these are to promote the general welfare. The civil engineering profession has historically contributed more than any other to the general welfare, but a new approach is necessary if we are to accommodate our work to an increasingly complex social structure and to its impact on the environment and ecology.

When I learned that this talk was to be treated as a keynote for this symposium, I was tempted to change the title to "Why are we here?", but this smacks too much of philosophy. Then I thought of "What are we here for?", but grammarians say that I can't end a sentence with a preposition. Now that's an example of just what I want to talk about. I don't know why I can't end with a preposition, and I think the rule is bad because it was devised without a good objective for determining proper grammatical construction. In anything we do, a good decision requires a good objective, and I want to dwell on the objectives of this conference and of the work we will discuss here.

Some of you may remember the story about the young boy who asked his dad, "Where did I come from?". Immediately the father sat the boy down, saying to himself "This is it." After about 15 minutes about the birds and the bees, the boy responded "But Dad, Jimmy next door came from Chicago, and I want to know where I came from." Sure enough, the father gave a beautiful answer, but the wrong answer, because he misinterpreted the objective.

So why are we here? Some of us need to have technical papers on our resume, some of us like to travel, and some like the professional camaraderie. But I truly believe that most, if not all, of us primarily want to advance the profession or the technology.

---

* Senior Consultant, Espey, Huston & Associates, P.O. Box 519, Austin, Texas 78767, and Professor of Civil Engineering, The University of Texas at Austin.

1

But what do we mean by "advance the profession"? What is our objective? Surely there is merit in developing better procedures for estimating a 100-year flood or designing a dependable water supply; but in doing so, we need to have some measure or criterion for the adjectives, "better" or "dependable". What is our ultimate objective? Is the engineering profession satisfied with designing for a 100-year flood simply because it is popular or legally accepted? Are we satisfied with a municipal water supply that would not fail under the worst historical drought? These are arbitrary criteria, and I would suggest that they are satisfactory in the long run only if the profession is willing to abdicate its responsibility of best serving the needs and desires of society.

We can perform our engineering design in accordance with a technical code, and that makes us handbook engineers. We can compute a 100-year flood in conformance with guidelines or compute a reservoir yield that would show no shortages if the period of hydrologic record were repeated, and that makes us computer operators. What is our mission as engineers?

In the U.S. Constitution, we have a phrase that can be used as our guide -- "To promote the general welfare". There is no profession that has contributed more to the general welfare throughout history than has civil engineering. However, I suggest that we have lost ground in the last 20 years by giving away sound engineering principles in favor of public emotionalism.

If there is any way to overcome the rule of emotionalism in resource management, the engineering profession must find it, and this will require a framework of public values and a set of societal objectives. The emotional stricture that people should move out of the flood plains -- they are just asking for trouble and inviting dependence on society's rescue -- should be countered with good engineering that permits effective use of our most valued lands. The emotional attitude that people who live in water-short areas are not entitled to surplus water from other regions must be countered with engineering demonstrations that water can and must be transferred among river basins in such a way as to contribute to the general welfare, and this means providing perpetually for the use of indigenous waters or their equivalent in the areas of origin when needed. The emotional objection to dam construction must be overcome, where appropriate, by engineering demonstrations that environmental and ecological trade-offs can be effected with a net gain. The environment and ecology are not static and we could not preserve them if we tried. Nature itself is ever-changing, and it certainly can be improved and, to a large degree, protected against the assault of increasing population and industrialization. It takes positive action to protect nature against civilization.

What I am saying is that your deliberations at this symposium and in the future would best be directed toward broad societal objectives. Surely we need better flood frequency and drought analysis techniques, but it is critical that we fit our developing technology into a resources management framework so that it effectively contributes to the general welfare.

The 1936 flood control act stated, in effect, that public works should be undertaken only if the benefits derived exceed the costs. Now this has been interpreted through the years to mean economic benefits that can be measured in dollars as market values. Other social and environmental benefits or costs were not quantified.

About 2 years ago at the ASCE Water Resources Specialty Conference in Buffalo, I discussed how they can and must be quantified and, more generally, how Civil Engineers have relinquished the role of leadership in managing the environment:

We have been burdened with a multi-objective decision mechanism which is scientifically and practically unworkable. Social, environmental and ecological objectives can be quantified and combined with economic objectives into a single objective function if we wish to do so. The key is that overall, we get the most of what we really want for our money. We don't spend big money for small protection or gain if we can do a lot better for our environment in another way for the same money. And this includes fairness to vested interests.

We have been burdened with a discount of future benefits that values one acre-foot of water today more than 1,000 acre-feet 100 years from now, thus stifling investment in engineering works for future generations.

We have capitulated to emotional pressure by a public that loves lakes and hates dams.

We have gone along with a flood insurance program that increases the cost of flooding by administration, profit and overhead costs, yet sets us up for disastrous flooding above the 100-year flood level in 150 to 200 communities per year.

We have occasionally stopped major construction at great costs because of endangered species without regard to the relative importance of the project and the danger to the species.

The profession has failed to demonstrate that inter-basin water tranfers are crucial to our society and that source regions can be fully protected.

We have failed to demonstrate that zero water quality degradation has a social and environmental price that informed society would not normally pay, and many communities are suffering because of this.

And finally we have turned over many engineering decisions to public forums instead of tailoring our engineering to the more general public needs and desires. It is these needs and desires, or social values, or societal objectives that we need to elicit at public meetings, not the detailed design of engineering works.

At that Buffalo ASCE meeting, I was certainly not suggesting that our profession ignore environmental or ecological issues or dictate to the public, but that we need to develop an integrated framework of economic, social, environmental, and ecological values and objectives that can serve as a basis for sound engineering decisions and that will also serve as a basis for public understanding and acceptance of those decisions. Such a framework of public values would be a gigantic undertaking, but, in my view, it is essential to reinstating civil engineering to the status that will enable it to serve an increasingly complex society as it has served a less sophisticated society in the past. In our frustration, we tend to control people instead of serving them. We can serve them only if we have a framework of public values and objectives that we can use to implement our engineering works.

So what does this mean to us in our day-to-day struggle to develop better technology? Let's look at an example.

How do we get a best estimate of a 100-year flood? What is the criterion for "best"? Some of us would readily respond that we get a least-squares unbiased estimate. But we really need to ask, "what is important to our client, to society?" Is streamflow the critical element, or is stage or is damage? In this case, it is obvious that an insurance program or a flood protection program is interested in damages and not simply in flood flows or flood stages. Since damages are a non-linear function of flows, what good does it do to obtain an unbiased estimate of flow if, converting it to damages, results in a biased estimate of damages? So, we have a superb solution to the wrong problem statement and we have failed.

Instead of solving for an unbiased estimate of flow corresponding to a fixed frequency, we can solve for an unbiased estimate of probability or frequency corresponding to a specified flow magnitude, which corresponds to a fixed stage and a fixed damage. Since damage is a linear function of frequency, regardless of stage, an unbiased estimate of probability for each range of magnitude, which we call expected probability, can be converted directly to an unbiased estimate of flood damage, regardless of the flow-damage functional form. Thus, if we can agree that, in the interest of the general welfare, damage reduction is our objective for flood management, we can subscribe to the concept of expected probability that has been debated so often and even rejected for flood insurance purposes by a committee of the prestigious National Research Council of the National Academies of Science and Engineering. Even that body lost sight of our ultimate objective.

Many of our technological debates can be resolved if we can subcribe to a fundamental set of objectives. Remember that for decades there were great arguments in the scientific community about whether the impact of an object on a target is proportional to its velocity or to the square of its velocity. Of course, the argument was eventually resolved by finding that, if the objective (i.e., our concern) is transfer of momentum, then the impact is proportional to the velocity, and, if the objective is loss of energy, then impact is proportional to the square of the velocity.

I can't stress too strongly the importance of going back to our ultimate objective if we are to obtain technical solutions that make sense. Since the science or art of flood hydrology began, the concept of a standard design flood has dominated hydrologic engineering. We need to get away from this and to acquire a comprehensive understanding of flood problems and a more sophisticated solution algorithm.

We sometimes see a statement such as, "I want to be 95 percent sure that the true 100-year flood is not greater than my estimate". So we compute tolerance bounds and add a "safety" factor in our design. We end up with a design flood that is certainly not our best estimate and, in fact, places our design in a region that produces less expected net benefits than would some sort of unbiased estimate. Such a criterion is meaningless, unjustifiably subjective, and constitutes a miscarriage of technology. If we have determined that the optimum design is for the 100-year flood, why should we think that designing for a flood that we estimate to be larger is even "more optimum"? We've got to keep asking ourselves what our real objective is.

Let us relegate the concept of a design flood to simply the magnitude of flood for which a particular feature is designed. It should not be an arbitrary criterion for design. It is not a decision tool, but a result of a complex engineering process. This process is the development of a plan of action that optimizes economically, socially and environmentally the net consequences. It is essential in such a process that all possible magnitudes of flood and associated consequences be considered for each alternative examined, not simply the effects of our project on a design flood. We should end up with a combination of structural and non-structural measures that produce tolerable and hopefully desirable consequences in the event of any reasonably possible flood magnitude. At the very least, any adverse consequences of the project during events exceeding the design flood must be outweighed socially, economically and environmentally by the beneficial consequences during those flooding events that are managed effectively. And we owe it to our clients to point out the various consequences that might occur and their associated probabilities.

We can say the same thing about the design of water supplies and for drought management. We have found that, when pressed, people can get along on substantially less water than is normally used. Some of us think that we should design for that smaller supply. Again, this is controlling the people instead of serving them. What will happen if projected needs are exceeded or if a more severe drought occurs? There is no cushion that could prevent severe suffering. We need to assess all potential demand patterns and all potential degrees of drought severity. We need to demonstrate that our design will not produce intolerable consequences under any anticipated condition or else to make our client fully aware of potential adverse consequences.

When we speak of intolerable consequences, we need to address the sources of water as well as the communities or clients being served. With most of our water in the United States wasting to the oceans over and above estuary needs, it should be unthinkable to deplete our aquifers that may take hundreds and thousands of years to replenish or to deprive the public of water to live where they want when they are willing to pay the economic, social and environmental price. It is equally unthinkable to divert water away from a region without absolute assurance that such water or replacement water will be made available to that region when it is needed. It has got to be feasible to use our renewable resources where and when needed and yet preserve them for use when needed where they exist naturally.

Even more important, we need to protect our ground-water resources against contamination. The disastrous chemical spills on the Rhine River that has upset the ecology for many years are minor compared to potential contamination of some major aquifers that could render them dangerous for thousands of years. Such protection requires well-designed waste management facilities.

Fortunately, this generation has inherited water works for flood protection, water supply, waste management, and hydroelectric and thermal power generation that are absolutely essential to satisfying our needs and desires. The Corps of Engineers, Bureau of Reclamation, TVA, SCS, and many public utilities have constructed reservoirs, diversions, hydroelectric power plants, and other works that would be politically infeasible today. These constitute some of our greatest assets and contribute immeasurably to the strength and welfare of our society. They are the backbone of our industrial and military capabilities and of our good life. We knew the difference between spending and investing, and it seems that we need to rediscover the importance of investing for the future through construction of engineering projects.

The popular media show little appreciation of this, and, as a consequence, we are not constructing works that will be required by the next generation for protecting the environment and ecology while satisfying society's needs.

Civil Engineers have been criticized harshly in the past 20 years as being destroyers of the environment. The bulldozer is a symbol of irresponsible development and lack of environmental awareness. While there might be some justification for this, aquiescence of engineers to the role of anti-environmentalist is not justified in any degree. Civil Engineers have a marvelous record of maintaining a desirable environment in the path of a burgeoning population and explosive industrialization. The problem now is that our social structure has become so complex that the engineer, by default, is leaving important environmental decisions to other disciplines that do not have the capability to do anything about it except restrict our freedom or obstruct progress. We are facing a new challenge.

It is the tradition of the Civil Engineer to serve the public, not to control society. Let us concentrate on goals and objectives that will continue and extend this tradition within a social structure that has become almost incomprehensively complex.

ARE SOIL MOISTURE ACCOUNTING MODELS NEEDED FOR THE REAL TIME FORECASTING OF RIVER FLOWS?

Carlos E. Puente[1] and Rafael L. Bras[2]

The difficulties inherent in using conceptual rainfall-runoff models in conjunction with filtering techniques, for the real time forecasting of river flows, are reviewed. The issue of whether soil moisture accounting models are necessary to guarantee reliable forecasts is investigated by comparing the performance of two alternative models: one with and one without a soil moisture component. Results found in a case study suggest that, given observations of rainfall and discharge only, the soil moisture component could be bypassed and still reasonable flow forecasts could be obtained.

Introduction

This paper is concerned with the problem of forecasting river flows in real time employing conceptual rainfall-runoff models. In particular, this work investigates whether or not a soil moisture accounting model (of the complexity of the Sacramento model, Peck (1976)) needs to be included as one of the components of the rainfall-runoff model. An attempt to try to elucidate such need is presented here by means of a fully developed case study.

Conceptual Models and Framework for Hydrologic Forecasting

Conceptual rainfall-runoff models, with and without a soil moisture component, were employed to obtain flow forecasts. The conceptual rainfall-runoff models used were nonlinear lumped-state and lumped-parameter representations. Such models were variations, obtained by including or excluding components, of a base rainfall-runoff representation. The base model approximates hydrologic response by joining a station precipitation model, a soil moisture accounting procedure, and a channel routing scheme. See Puente and Bras (1984) for a complete mathematical description.

The station precipitation model was developed by Georgakakos and Bras (1984). It uses meterological variables (temperature, pressure, and dew-point temperature) as inputs to produce the precipitation volume rate over the river basin. The soil moisture accounting procedure was a continuous (no thresholds) version of the Sacramento

[1]Department of Land, Air and Water Resources, University of California, Davis, CA 95616.

[2]Department of Civil Engineering, Massachusetts Institute of Technology, Cambridge, MA 02139.

model (Peck, 1976). Such model is driven by the rainfall and evapotranspiration rates (the latter being an input), and produces the volume of water that flows into the river via direct runoff, surface runoff, interflow runoff, and groundwater runoff. The channel router is the conceptual model of Georgakakos and Bras (1982). The river channel is approximated as a cascade of nonlinear reservoirs which are driven by the water produced by the soil, with the outflow from the last reservoir representing runoff from the river basin.

A stochastic nonlinear rainfall-runoff representation was obtained by adding uncorrelated noise terms to the conceptual hydrologic model. Given the random nature of the additive noises, the forecasting problem was solved within the context of stochastic filtering. Specifically, the objective at hand was to find minimum variance estimates of runoff and rainfall based upon the assumed conceptual stochastic model and the fact that noisy measurements of rainfall and runoff become available at regular time intervals. Such nonlinear filtering problem was solved recursively by means of the extended Kalman filter algorithm. This procedure updates current state mean and error covariance estimates using information provided by the current measurements. These estimates are propagated into the future to obtain the forecasts. Recall that the algorithm conserves mass (because the conceptual model does) when in the propagation mode, but it could "create" or "delete" mass, if needed, at times measurement information is incorporated.

Problems Encountered in Practical Applications

The most stringent limitation regarding the applicability of the previously described forecasting methodology, has been the lack of reliable calibration procedures for the parameters of the conceptual rainfall-runoff model and for the parameters that characterize the additive random noise terms. Several attempts to estimate the Sacramento model parameters, while using rainfall records as inputs, have been reported in the literature. They range from using as criteria ordinary least squares, sum of absolute values, and maximum likelihood; see for example Sorooshian et al (1983), Restrepo-Posada and Bras (1985), and Willgoose (1987). Although these works showed the potential of the maximum likelihood and sums of absolute values criteria, the following problems were recognized:

a. existence of non-unique "optimal" parameters,
b. presence of extended objective function valleys which hamper the searching procedure's ability to attain a truly optimum,
c. occurrence of collinearity between some parameters, and
d. existence of parameters which are extremely difficult or impossible to identify.

Puente and Bras (1987a) showed that the quality of the runoff forecasts depends heavily on the noise component characteristics assumed on the stochastic rainfall-runoff representation. An approximate procedure to estimate the dynamics noise parameters (spectral density matrix) was developed by Puente and Bras (1987b). This procedure, which is based on the maximum likelihood principle, estimates a constant in time and diagonal spectral density matrix and assumes all other parameters (model and noise) to be fixed. Later, this paper will illustrate the usefulness of such a procedure.

A Case Study

The river basin considered in this work was the Bird Creek basin (915.6 mi^2) in Oklahoma. Such case study has been employed by several investigators for real-time forecasting studies: Georgakakos and Bras (1982), Restrepo-Posada and Bras (1985), Georgakakos (1986), Puente and Bras (1984, 1987 a,b), Willgoose (1987).

In this case study, it was assumed that the conceptual rainfall-runoff model parameters and the parameters that characterize the measurement errors of the observations (rainfall and runoff) were fixed; for their values refer to Puente and Bras (1984). The spectral density matrix was the only varying parameter besides the different model structures. The results that follow are typical for the whole period of record, irrespective of it being used or not for calibration purposes.

Figures 1a and 1b show respectively the rainfall and runoff[3]. One-step ahead (6-hrs) predictions (dashed lines) and their respective observations (solid lines), for the period comprising April 12 to May 12 of 1960. The spectral density matrix labelled Q-0 (for its values refer to Puente and Bras, 1987a) is used to characterize the dynamics noise term. As can be seen, rainfall forecasts underestimate the actual precipitation, which consequently results in runoff underpredictions.

After applying the approximate maximum likelihood procedure of Puente and Bras (1987b), with Q-0 as initial condition, the spectral density matrix Q-6 is obtained. With such a matrix, the filter is allowed to update more resulting in better forecasts as shown in figures 1c and 1d. Despite the obvious improvements, a detailed sensitivity analysis revealed that drastic changes could be made on the matrix values corresponding to the soil component of the conceptual model without affecting the quality of the runoff forecasts. The implication is that the forecasting capabilities (around Q-6) are primarily due to the channel and to the filtering mechanism: whenever additions or deletions of water are necessary, the filter chooses to update the channel rather than the soil.

To further study the importance of the channel component on runoff forecasting, a two component model was formed by deleting the soil component altogether, i.e. a model with only rainfall and channel components was considered. The runoff forecasting results obtained by using the two component model are shown in Figures 2a and 2b when the respective entries of matrices Q-0 and Q-6 are used to

---

[3]All units in mm/6 hrs

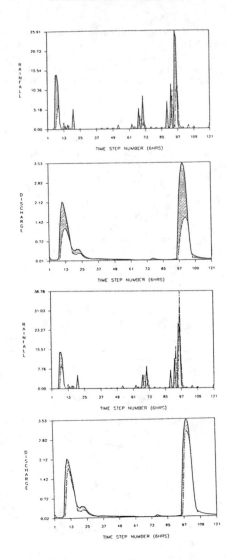

Figure 1. Forecasts for Bird Creek, base model

a) Q-0 6-hrs ahead   b) Q-0 6-hrs ahead
c) Q-6 6-hrs ahead   d) Q-6 6-hrs ahead

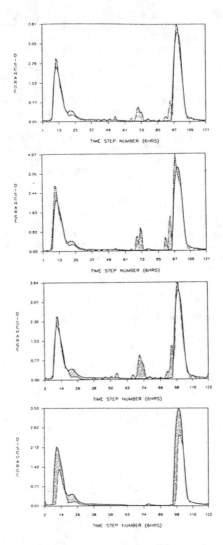

Figure 2. Forecasts for Bird Creek, two component model

a) Q-0 6-hrs ahead   b) Q-0 6-hrs ahead
c) Q-6 12-hrs ahead   d) Q-0 12-hrs ahead
   (Base model)

describe the dynamics noise. The results are clearly excellent and were found consistently throughout the period at hand for Bird Creek (6-hourly data from 1956 to 1961). The goodness of the results is further illustrated by comparing Figures 2c and 2d, which show forecasts two-time steps ahead (12 hours) found using the full model and Q-6, and the two component model and Q-0. Notice that unrealistic runoff peaks are present when the two component model is used; and, therefore, the absence of the soil component is felt. But recall that better forecast are expected when the channel and noise parameters are recalibrated to this new structure. In any case, it seems possible that such unrealistic runoff bursts could be eliminated while maintaining the overall quality of the forecasts if a soil model, much simpler than the Sacramento model, is used as the second component.

An study of the effects the employed rainfall model had on the previously described runoff forecasts was also carried by computing runoff forecasts using observed rainfall values as inputs. The obtained results also showed an extremely fast runoff response when the soil component was not present, indicating a need of a soil component to smooth out such bursts. But as previously found when using the rainfall model, the overall differences on runoff forecasts when the soil is and is not included are not significant, pointing again to the observation that a soil component of the complexity of the Sacramento model seems not necessary.

Conclusions

Results form a case study have shown that a soil moisture accounting model, of the complexity of the Sacramento model, is not necessary to obtain excellent forecasts from rainfall and runoff observations when using a stochastic filtering methodology. Although these results are not readily generalizable, they suggest that closer attention should be paid to the very delicate interactions among model complexity, data requirements, and forecasting methodology.

Acknowledgements

Portions of this research were sponsored by the Hydrologic Research Laboratory of the National Weather Service, U. S. Department of Commerce, under cooperative agreement NA 80AA-H-00D44.

References

Georgakakos, K. P. and R. L. Bras (1982), "A real-time, statistically linearized, flood routing," Water Resources Research 18(3):513-524.

Georgakakos, K. P. and R. L. Bras (1984), "A hydrologically useful station precipitation model, 1. Formulation," Water Resources Research 20(11):1585-1596.

Georgakakos, K. P. (1986), "A generalized stochastic hydrometeorological model for flood and flash-flood forecasting, 2. Case studies," Water Resources Research Vol. 22, No.13, 2096-2107.

Peck, E. L. (1976), "Catchment modelling and initial parameter estimation for the National Weather Service River Forecast System," NWS HYDRO-31.

Puente, C. E. and R. L. Bras (1984), "Nonlinear filtering, parameter estimation and decomposition of large rainfall-runoff models," R. M. Parsons Laboratory for Water Resources and Hydrodynamics, Department of Civil Engineering, Technical Report No. 297, M.I.T., Cambridge, MA.

Puente, C. E. and R. L. Bras (1987a), "Application of nonlinear filtering in the real time forecasting of river flows," Water Resources Research (accepted).

Puente, C. E. and R. L. Bras (1987b), "Error identification and decomposition in large state-space stochastic rainfall-runoff models," Automatica (accepted).

Restrepo-Posada, P. J. and R. L. Bras (1985), "A view of maximum likelihood estimation with large conceptual hydrologic models," Journal of Applied Mathematics and Computation, Modeling the Environment, Vol. 17.

Sorooshian, S., Gupta, V. K. and J. L. Fulton (1983), "Evaluation of maximum likelihood parameter estimation techniques for conceptual rainfall-runoff models: Influence of calibration data variability and length on model credibility, "Water Resources Research 19(1):251-259.

Willgoose, G. (1987), "Automatic Calibration Strategies for Conceptual Rainfall-Runoff Models", Master of Science thesis, Massachusetts Institute of Technology.

The Great Lakes Large Basin Runoff Model

Thomas E. Croley II[*], M. ASCE

The Great Lakes Environmental Research Laboratory (GLERL) developed its Large Basin Runoff Model (LBRM) specifically for modeling runoff and moisture storages in the large river basins (from 100 to 100,000 square kilometers) about the Great Lakes. The LBRM is an interdependent tank-cascade model which employs analytic solutions of physical considerations relevant for large watersheds. The mass balances for snowpack, upper and lower soil zones, groundwater, and surface water are coupled with physically-based concepts of linear reservoir storages, partial-area infiltration, complementary evapotranspiration and heat available (evapotranspiration opportunity) based on available supply, and degree-day determinations of snowmelt and net supply. The 9 model parameters are determined in an automated systematic search of the parameter space to minimize the sum-of-squared errors between actual and model outflow volumes. The physical relevance of the parameters aids in interpreting hydrology and in using hydrologic interpretations to set parameter values. As a conceptual model, the LBRM is useful not only for predicting basin runoff, but for facilitating our understanding of watershed response to natural forces as well. GLERL has used its LBRM in lake level forecast packages for Lake Superior, developed for the U.S. Army Corps of Engineers (Detroit District), and for Lake Champlain, developed for the National Weather Service Northeast River Forecast Center.

Runoff Modeling

The GLERL Large Basin Runoff Model (LBRM), pictured schematically in Fig. 1, consists of several moisture storages arranged as a serial and parallel cascade of tanks (Croley 1983a,b). The main mathematical feature of the LBRM is that it may be described by strictly continuous equations; none of the complexities associated with intertank flow rate dependence on partial filling are introduced. For a sufficiently large watershed, these nuances are not observed, owing to the spatial integration of rainfall, snowmelt, and evapotranspiration processes. Parameters are designed to represent our physical understanding of the watershed hydrology (Croley & Hartmann 1986a). Daily precipitation, temperature, and insolation (the latter available from climatic summaries as a function of location) may be used to determine snowpack accumulations and net supply. Water enters the snowpack (SNW in Fig. 1), if present, and some then is available as net supply to the watershed surface based on degree-day determinations of snowmelt (M); potential snowmelt varies with degree-days as reflected by a melt factor (a_s). Net supply thus occurs only when temperatures are above

[*]Research Hydrologist, Great Lakes Environmental Research Laboratory, 2205 Commonwealth Blvd., Ann Arbor, Michigan 48105.

14

freezing and consists of precipitation and snow-melt. The net supply is divided into infiltration to the upper soil zone and surface runoff in relation to the relative upper soil zone moisture content (USZM); under the partial-area infiltration concept, infiltration is proportional to the net supply rate and to the areal extent of the unsaturated portion of the upper soil zone.

Calculation of outflows from the various storages within the watershed are based on the linear reservoir concept; outflow from a tank is proportional to the moisture in storage as reflected by the linear reservoir coefficient. Percolation to the lower soil zone is dependent on the upper soil zone moisture and the percolation coefficient (α_{per}). Likewise, interflow from the lower soil zone to the surface and deep percolation to the groundwater zone depend on the lower

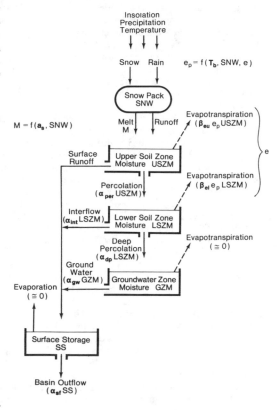

Fig. 1. Runoff Model Schematic

soil zone moisture content (LSZM) and the interflow (α_{int}) and deep percolation (α_{dp}) coefficients, respectively. Groundwater flow depends on the groundwater zone moisture content (GZM) and the groundwater coefficient (α_{gw}), and basin outflow depends on the surface storage (SS) and basin outflow coefficient (α_{sf}).

The evapotranspiration rate from the upper and lower soil zones is proportional to available moisture in the upper and lower soil zone storages, the heat rate available for evapotranspiration (e_p), and the evapotranspiration coefficients for the upper (β_{eu}) and lower zones (β_{el}); it is also based on the concept of complementary evapotranspiration and heat available for evapotranspiration. Over large areas, actual evapotranspiration affects temperatures, wind speeds, and humidities, and hence modifies the evapotranspiration opportunity or capacity; the heat used for evapotranspiration reduces the heat available for additional evapotranspiration. This concept proves superior to classical "potential evapotranspiration" concepts when applied to large areas (Croley 1985). The total amount of heat in a day, to be

split between that used for evapotranspiration and that still available for evapotranspiration, is estimated empirically from the average
air temperature, a long-term heat constant, and a base scaling temperature (T_b). The long-term heat constant is determined from a long-
term heat balance; all absorbed insolation not used for snowmelt
sooner or later appears as other components of the heat balance.
Also, evaporation from the surface storage and groundwater evapotranspiration are neglected here.

Mass continuity yields a first-order linear differential equation
for each of the tanks in Fig. 1; outputs from one tank are used in
lower tanks where their outputs appear as inputs. There are 30 different analytic results (Croley 1982), depending upon the magnitudes
of all inputs, initial storages, and the nine parameters identified
above. Since the inputs and initial storages change from day to day,
the appropriate analytic result, as well as its solution, varies with
time; mathematic continuity between solutions is preserved, however.
Numeric solutions are unnecessary so that approximation errors are
avoided. Furthermore, solutions may proceed for either flow rates or
storage volumes directly without the complication of constraints consideration.

The differential equations for the mass balance can be applied over
any time interval by assuming that the input (precipitation and snowmelt) and the heat available for evapotranspiration are uniform over
the interval. Thus, the resolution of the equations is limited only
by the intervals over which precipitation and temperature data are
available; the mass balance computation interval may be any length
greater than or equal to the interval length for which meteorologic
data are available. The model is applied to daily data with either a
fixed 1-day or a fixed 7-day mass-balance computation interval. Net
supply and the heat available for evapotranspiration are determined on
a daily basis and summed over the time interval as input to the mass-
balance computations. The model is applied to monthly data with a
variable mass-balance computation interval; the interval may represent
28 to 31 days, depending on the month and year, and net supply and the
heat available for evapotranspiration are computed over the same
period.

Parameter Calibration

The LBRM is calibrated to determine the set of parameters with the
smallest sum-of-squared-errors between model and actual daily flow
volumes for the calibration period. The following data are required
to calibrate and apply the model: daily precipitation, daily maximum
and minimum air temperatures, a climatic summary of daily extraterres-
trial solar radiation, and for comparison purposes, daily basin outflows. The area of the watershed is also required. The basin is
first divided into subbasins draining directly to the lake. Meteorologic data from, typically, 150-300 stations about and in the subbasins are combined through Thiessen weighting to produce areally-
averaged daily time series of precipitation and maximum and minimum
air temperatures for each subbasin. Weights are determined for each
day of record, if necessary, since the data collection network changes

frequently as stations are added, dropped, and moved, or fail to
report from time to time. This is now feasible through the use of a
new algorithm for determining the Thiessen area-of-influence about a
station by its edge (Croley & Hartmann 1985a). Records for all "most-
downstream" flow stations are combined by aggregating and extrapolating for ungauged areas to estimate the daily runoff to the lake from
each subbasin.

Snowpack, upper soil moisture, lower soil moisture, groundwater,
and surface storage zones must be initialized prior to modeling.
While the initial snowpack moisture equivalent is easy to determine as
zero during major portions of the year, these variables are generally
difficult to estimate. If the model is to be used in forecasting or
for short simulations, then it is important to determine these variables accurately prior to use of the model. If the model is to be used
for calibration or for long simulations, then the initial values are
generally unimportant. The effect of the initial values diminishes
with the length of the simulation, and after 1 year of simulation, the
effects are nil from a practical point of view. Calibrations are
repeated with initial conditions equal to observed long-term averages
until there is no change in the averages, to avoid arbitrary initial
conditions when their effects do not diminish rapidly.

Applications of the LBRM require the determination of the model's
nine parameters as described elsewhere (Croley & Hartmann 1984).
Parameters are determined in an automated systematic search of the
parameter space to minimize the sum-of-squared-errors between actual
and model outflow volumes. We search each parameter, selected in
rotation, until all parameter values converge to two digits instead of
searching until the sum-of-squared-errors stabilizes. Such an approach is important where synergistic parameter interactions allow the
parameters to change significantly even after the sum-of-squared-
errors has stabilized; this was seen repeatedly in calibrations for
the Ontario, Champlain, and Superior basins (Croley 1983b; Croley &
Hartmann 1983, 1984, 1985b, 1986a).

By combining the meteorologic and hydrologic data for all subbasins
to represent the entire basin, the LBRM may be calibrated in a lumped-
parameter application to the entire basin at one time. Although the
application of lumped-parameter models to very large areas necessarily
fails to represent areal distributions of watershed and meteorologic
characteristics, spatial filtering effects tend to cancel data errors
for small areas as the areas are added together. Distributed-parame-
ter applications, in which the LBRM is calibrated for each subbasin
and model outflows are combined to represent the entire basin, make
use of information that is lost in the lumped-parameter approach; the
integration then filters individual subbasin model errors.

The LBRM captures a "realism" in its structure that has several
advantages over other models. Basin storages, modeled as "tanks," are
automatically removed as respective parameters approach their limits.
Thus, the structure of the model changes within a calibration. This
is achieved without the use of "threshold" parameters in the model
since physical concepts are used which avoid discontinuities in the

goodness-of-fit as a function of the parameters; these concepts appear especially relevant for large-basin modeling. Because the "tanks" relate directly to actual basin storages, initialization of the model corresponds to identifying storages from field conditions that may be measured; interpretations of a basin's hydrology then can aid in setting both initial and boundary conditions. The tanks in Fig. 1 may be initialized to correspond to areal measurements of snowpack and soil moisture water equivalents available from aerial or satellite monitoring (Gauthier et al. 1984).

Studies on the Lake Ontario and Superior Basins (Croley 1982; Croley & Hartmann 1983, 1984, 1986a) show that the simple search algorithm described herein does not give unique optimums for calibrated parameter sets because of synergistic relationships between parameters. However, the calibration procedure does show a high degree of repeatability for recalibrations with different starting values, and consistent parameter values are obtained for subbasins with similar hydrologic characteristics. On the other hand, the nonuniqueness of calibrated parameters was demonstrated by recalibrating for a synthetic data set. The model was calibrated for the entire Lake Superior Basin and then used to simulate outflows to create a new data set for calibration. Subsequent calibration started with a very different initial parameter set and yielded an "optimum" parameter set different from the original with a relatively poor goodness-of-fit. If the original parameter set had been unique, the parameter values produced from the recalibration to the synthetic data set should have been the same as the parameters used to create that data set. This illustrates the nonuniqueness of the parameters, the importance of the starting values used in the search, and the problems inherent in searching the parameter space. Additionally, some components of the LBRM (such as linear reservoirs) are more likely to adequately represent their processes in the real world than others (such as degree-day melting or complementary evapotranspiration). Parameter estimation techniques that properly weight the more accurate parts of the model could improve parameter estimates.

Example Applications

Statistics for the 22 Lake Superior subbasin applications were computed by summing actual flows, extrapolating for ungaged areas, and model flows for all subbasins. The model performs noticeable better for natural flow since diversions and regulations of flows are not represented (Croley & Hartmann 1985b). Comparisons with other runoff models (Croley 1983a,b) show the LBRM to be far superior for estimates of runoff volumes from large basins. In comparisons with a climatic approach (Croley & Hartmann 1984) that use mean daily flow as a predictor of basin runoff, the daily correlation and root mean square error are 0.77 and 0.4 mm, as compared to 0.93 and 0.25 mm for the model application to Lake Superior (Croley & Hartmann 1985b). (Mean runoff is 1.12 mm/day with a standard deviation of 0.67 mm).

GLERL used its LBRM in lake level forecast packages for Lake Superior, developed for the Detroit District U.S. Army Corps of Engineers (Croley & Hartmann 1986b), and for Lake Champlain, developed for the

National Weather Service Northeast River Forecast Center (Croley & Hartmann 1985b). In an evaluation on Lake Superior (Croley & Hartmann 1986b), the forecast package was used with actual meteorology to show how well the runoff model works during a period other than the calibration periods. The monthly correlation and root mean square error in net basin supply (runoff + overlake precipitation - overlake evaporation) for the one-month forecasts were 0.99 and less than 8.7 mm, respectively. (The mean net basin supply was 77.6 mm and the standard deviation was 55.1 mm). The model is judged to do an excellent job in simulating basin runoff to the lake.

References

Croley, T. E., II (1982). "Great Lakes basins runoff modeling." *NOAA Tech. Memo. ERL GLERL-39*, Nat. Tech. Inf. Serv., Springfield, Vir.

Croley, T. E., II (1983a). "Great Lakes basins (U.S.A.-Canada) runoff modeling." *J. Hydrol.*, 64, 135-158.

Croley, T. E., II (1983b). "Lake Ontario Basin (U.S.A.-Canada) runoff modeling." *J. Hydrol.*, 66, 101-121.

Croley, T. E., II (1985). "Evapotranspiration dynamics for large river basins." *Proc. Nat. Conf. Advances in Evapotranspiration*, ASAE, St. Joseph, Mich., 423-430.

Croley, T. E., II, and Hartmann, H. C. (1983). "Lake Ontario Basin runoff modeling." *NOAA Tech. Memo. ERL GLERL-43*, Nat. Tech. Inf. Serv., Springfield, Vir.

Croley, T. E., II, and Hartmann, H. C. (1984). "Lake Superior Basin runoff modeling." *NOAA Tech. Memo. ERL GLERL-50*, Nat. Tech. Inf. Serv., Springfield, Vir.

Croley, T. E., II, and Hartmann, H. C. (1985a). "Resolving Thiessen polygons." *J. Hydrol.*, 76, 363-379.

Croley, T. E., II, and Hartmann, H. C. (1985b). "Lake Champlain water supply forecasting." GLERL Open File Report, Contribution No. 450, Great Lakes Environmental Research Laboratory, Ann Arbor, Mich.

Croley, T. E., II, and Hartmann, H. C. (1986a). "Parameter calibration for the Large Basin Runoff Model." *Multivariate Analysis of Hydrologic Processes*, H. W. Shen, Colo. State Univ., Ft. Collins, Colo., 393-407.

Croley, T. E., II, and Hartmann, H. C. (1986b). "Near real-time forecasting of large lake water supplies." *NOAA Tech. Memo. ERL GLERL-61*, Nat. Tech. Inf. Serv., Springfield, Vir.

Gauthier, R. L., Melloh, R. A., Croley, T. E., II, and Hartmann, H. C. (1984). "Application of multisensor observations to Great Lakes hydrologic forecast models." *Proc. 18th Int. Sym. Remote Sensing of the Environment*, Env. Res. Inst. Mich., Ann Arbor, Mich., 1129-1140.

Probabilistic Forecasts of Continued High Great Lakes Water Levels

Holly C. Hartmann and Thomas E. Croley II[*], M. ASCE

Existing projections of Great Lakes water levels over the next several years do not consider how meteorologic variability affects the hydrologic processes (basin runoff, overlake precipitation, lake evaporation) that control lake levels. We can make such considerations with conceptual model-based techniques developed by the Great Lakes Environmental Research Laboratory (GLERL) for the system-wide simulation of Great Lakes water supplies, connecting channel flows, and lake levels. The package incorporates GLERL's Large Basin Runoff Model (LBRM), which is an interdependent tank-cascade model that employs analytical solutions of climatic considerations relevant for large watersheds. We use meteorologic sequences of daily precipitation and maximum and minimum air temperatures with the runoff model to simulate basin moisture conditions and runoff, and then with empirical estimators of overlake precipitation and lake evaporation to simulate water supplies to each of the Great Lakes. Finally, we use the operational regulation plan for Lake Superior to simulate monthly lake levels and flows through the connecting channels; the regulation plan includes a hydraulic routing model that considers diversions and obstruction of connecting channel flows due to ice jams. The integrated models form a package which enables probabilistic assessments of diversions, regulation plans, or climatic change that reflect the natural long-term variability of the Great Lakes system. The models use meteorologic scenarios based on historical records to determine probabilities of future Great Lakes levels.

Introduction

Continued high precipitation throughout the Great Lakes region since 1970 has created important water management problems associated with high lake levels (Croley, 1986; Quinn, 1986). Lakes Michigan, Huron, and St. Clair experienced record monthly levels from October 1985 through January 1987; during that period, Lake Erie set record levels each month except April 1986. In 1986, each lake reached its highest level recorded this century. In addition, Larsen (1985) suggests that Lake Michigan exceeded its recent high levels by over a meter several times during the last 2000 years and that even higher water levels are possible.

Currently, regulations of Lake Superior use only the historical median or 5%-exceedance-quantile water supplies to regulate water levels over a one- to seven-month planning horizon. Six-month lake level forecasts for each of the Great Lakes, issued each month by the

[*]Research Hydrologists, Great Lakes Environmental Research Laboratory, 2205 Commonwealth Blvd., Ann Arbor, Michigan 48105.

U.S. Army Corps of Engineers (1986), are determined via a trend-and-regression analysis based on past water supplies. Simulations of possible lake levels also have been made by using a hydraulic routing model with low, moderate, and high water supply scenarios (Hartmann, 1987). None of these approaches consider the water stored in the basins about the lakes nor the effect of meteorology on the hydrologic processes contributing to the water supplies (basin runoff, lake precipitation, and lake evaporation). We developed conceptual model-based techniques for generating water supply and lake level forecasts that consider both the existing basin storages and possible meteorology. Outlooks are made for each of the components of a lake's water supplies, which then are used in conjunction with the operational Lake Superior regulation plan and a hydraulic routing model to produce lake level forecasts.

System-wide Model

Our integrated system-wide model package incorporates the GLERL Large Basin Runoff Model (LBRM), which is an interdependent tank-cascade model that employs analytical solutions of climatic considerations relevant for large watersheds (Croley, 1987). The model couples mass balances for snowpack, two soil zones, groundwater, and surface water with physically-based concepts of linear reservoir storages, partial-area infiltration, complementary evapotranspiration, evapotranspiration opportunity based on available supply, and degree day determinations of snowmelt. The snowpack and linear reservoir storages for the soil zones, groundwater, and surface water relate directly to basin moisture storage conditions, which can be used as initial conditions for simulating the basin response to possible meteorologic sequences. The model has been calibrated and applied to each of the 121 subbasins draining directly into the Great Lakes.

The system-wide model also uses estimates of overlake precipitation and lake evaporation. (While the components of water supply are of similar magnitude for each of the Great Lakes, their modeling is not equally sophisticated.) Overlake meteorology is estimated by using overbasin meteorology (described subsequently), since lake effects on near-shore meteorology are more significant than orographic effects throughout the drainage basin. Lake evaporation is estimated by using a mass transfer approach for Lakes St. Clair and Erie, while estimates for Lakes Superior and Michigan-Huron are derived as residuals of a water balance (Quinn and Kelley, 1983).

The system-wide model determines Lake Superior outflows by using "Plan 1977" as implemented by the U.S. Army Corps of Engineers (USACE) for simulation studies (IGLDCUSB, 1981). Plan 1977 attempts to maintain Lakes Superior and Michigan-Huron near their long-term monthly levels; it requires water supply estimates and initial water levels for Lakes Superior, Michigan-Huron, St. Clair, and Erie. For each month during May through November, the Lake Superior outflow structure gates are opened to the average setting required over the remainder of the period through November. Regulation over December through April is accomplished by setting the gate opening to the average required over the 5-month period and leaving it unchanged throughout. However, when simulations begin within the December through April period (e.g., January), the gate setting is averaged over the remainder of the period through April.

Use of the regulation plan requires a hydraulic routing model to determine projected water levels for Lakes Superior and Michigan-Huron, which then affects the balancing of the levels of those lakes through control of the Lake Superior outflows. The hydraulic routing model used in the USACE implementation of Plan 1977 considers net basin supplies, diversions, St. Marys River flows, and ice retardation of flows in the determination of water levels on Lakes Michigan–Huron, St. Clair, and Erie. It also determines flows through the St. Clair, Detroit, and Niagara Rivers. The model uses an iterative approach in level–pool routing to solve a series of stage–fall–discharge equations for each of the connecting channels and continuity for each lake. Future implementations of our system–wide model will instead incorporate the GLERL Hydrologic Response Model (HRM). The HRM uses the same level–pool routing concepts as the USACE hydraulic routing model, but the second–order finite difference solution technique employed provides a 58% savings in computation time.

Lake Level Outlooks

Meteorologic station files for 32 years (1954–1985) of daily data were assembled for all stations in and about the subbasins. Each meteorologic station file contains daily values of minimum air temperature, maximum air temperature, and precipitation for the length of the historical record as well as the location of the station. For each subbasin, for each day, all stations reporting on that day were considered to compute Thiessen weights for the resulting networks for each type of data if those networks had not been considered on previous days. We used an algorithm for determining weights that is especially advantageous for rapidly varying high–density data–observation networks (Croley and Hartmann, 1985); it enables fast computation of daily areally–average meteorology. Meteorology over a lake then is estimated by using the areal–averaged meteorology for the entire basin about the lake. The historical data files were used with the LBRM for each subbasin to determine the basin moisture storage conditions of each subbasin for 31 December 1985. While provisional meteorologic data for 1986 is available for the Lake Superior subbasins from the GLERL experimental water supply forecast system (Croley and Hartmann, 1986a), near real–time data acquisition and reduction procedures (Croley and Hartmann, 1986b) are not yet operational for the remaining lakes (the procedures have been transferred to the USACE and are expected to be implemented soon). The model basin moisture estimates at the end of 1985 thus serve as initial conditions for generating forecasts of basin runoff that can be combined with forecasts of lake precipitation and evaporation as input to the Lake Superior regulation plan for the forecasting of lake levels.

An example probabilistic forecast of lake levels was generated by using an approach similar to the National Weather Service's Extended Streamflow Prediction (ESP) procedure (Day, 1985). Since reliable long–term meteorology forecasts are not available, we assume that all

historical meteorologic sequences are equally likely to recur. Twen-
ty-six years of historical sequences (1954–1979), covering the period
in common between our meteorology and lake evaporation records, are
available for use in generating the water supply and lake level forecasts. To maintain a sample size adequate for subsequent frequency
analysis, only a 1-year lake level forecast is attempted. Because
December Lake Superior regulations require December through April
water supply forecasts, we thus use 25 16-month historical sequences
(January through April of the following year). By using the modeled
31 December 1985 subbasin moisture storages as initial conditions,
each historical meteorology sequence was used with the LBRM to produce
25 sets of subbasin runoff forecasts. The subbasin daily runoff
sequences were then aggregated over each lake basin; the forecast
runoff thus is predicated on the meteorology sequence and the basin
moisture conditions that existed at the beginning of 1986. Outlooks
of overlake precipitation and lake evaporation for each 16-month
sequence were taken directly from the historical records corresponding
to the meteorologic sequence used. In the absence of conceptual
modeling of these elements of water supply, the historical values
provide a simple estimate of possible future conditions; as they are
derived from the same year of record as the subbasin meteorology used
in forecasting runoff, the spatial and temporal interdependencies
between all meteorological processes are preserved. The basin runoff,
lake precipitation, and lake evaporation forecasts were combined to
produce water supply forecasts for each lake. The water supply forecasts were then used as input to the Lake Superior regulation plan and
the hydraulic routing model, along with initial water levels for each
lake on 1 January 1986, to produce lake level forecasts.

The 4% and 96% lake level exceedance quantiles as well as the
median for each lake, as derived from our use of the system-wide model
in an ESP setting, are plotted in Figure 1. Actual lake levels over
1986 are included also for comparison. Although Lakes Michigan–Huron,
St. Clair, and Erie set monthly records throughout 1986, Fig. 1 shows
that each lake could have experienced higher levels, based on past
meteorologic conditions. Lake Michigan–Huron generally remained below
the median level expected, except after the record-breaking October
precipitation over the basin. Lake Erie levels were higher than even
the 4% exceedance level much of the spring and summer; dry autumn
weather over the basin enabled levels to fall below the median during
November, when past meteorology suggested that levels could have been
the highest.

The system-wide model also may be used to assess water management
alternatives. As an example, a long-term deterministic forecast of
lake levels was generated by using the system-wide model as before,
but with only a single historical meteorologic sequence. Many meteorologic futures may be likely; we selected the 25-year historical
sequence of 1954–1979 as one possible scenario. Impacts of changing
diversion rates then were examined by using the system-wide model and
the same 25-year meteorologic sequence, but tripling the Chicago
diversion from Lake Michigan to 283 cms (10,000 cfs) and eliminating
the Long Lac and Ogoki diversions to Lake Superior (presently 159
cms). Due to compensation in the regulated outflows and the connect-

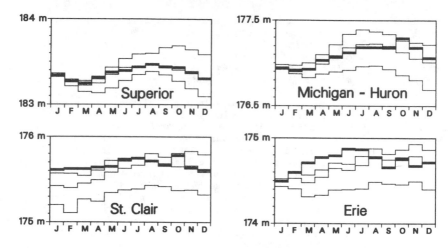

Fig. 1.   4%, 50%, and 96% Monthly Lake Level Exceedance Outlooks (thin
          lines) and Actual Lake Levels (thick lines) for 1986

ing channel flows, the levels practically are unchanged even during
extreme periods.   It appears a change in the regulation plan would be
more significant than these diversion changes in lowering lake levels
during parts of the year.

Summary

   The Great Lakes Environmental Research Laboratory has integrated
its Large Basin Runoff Model, estimates of overlake precipitation and
lake evaporation, the Lake Superior regulation plan, and a hydraulic
routing model for the system-wide simulation of Great Lakes water
supplies, connecting channel flows, and lake levels.   An example
probabilistic forecast made for 1986 illustrates the application of
the methodology and shows that some of the subsequent record levels
set that year were among the highest that could have been expected
based on past meteorology, yet other levels could have been worse than
those actually seen.   The integrated models also enable assessments of
diversions, regulation plan alternatives, or climatic change that
reflect the role of meteorology in controlling lake levels.

References

Croley, T.E., II (1986). "Understanding recent high Great Lakes water
levels." *Proc. Great Lakes Symposium*, Ohio State University, Columbus, Ohio, 60-70.

Croley, T.E., II (1987). "The Great Lakes large basin runoff model."
Proc. Engineering Hydrology Symposium, ASCE, New York, New York, (this
volume).

Croley, T.E., II and Hartmann, H.C. (1985). "Resolving Thiessen polygons." *J. Hydrol.*, 76, 363-379.

Croley, T.E., II and Hartmann, H.C. (1986a). "Areal averaging of point
measurements in near real-time." *Proc. 2nd International Conf. on
Interactive Information and Processing Systems for Meteorology, Ocean-
ography, and Hydrology.* AMS, Boston, Mass., 158-164.

Croley, T.E., II and Hartmann, H.C. (1986b). "Near real-time forecasting of large-lake water supplies; a user's manual." *NOAA Tech. Memo.
ERL GLERL-61*, Nat. Tech. Inf. Serv., Springfield, Vir.

Day, G.N. (1985). "Extended streamflow forecasting using NWSRFS." *J.
Water Res. Plan. and Manag.*, 111, 157-170.

Hartmann, H.C. (1987). "Potential variation of Great Lakes water
levels: a hydrologic response analysis." *NOAA Tech. Memo. ERL GLERL-
xx* (in prep.).

International Great Lakes Diversions and Consumptive Uses Study Board
(1981). "Great Lakes diversions and consumptive uses, appendix B:
computer models - Great Lakes." International Joint Commission, Washington, D.C.

Larsen, C.E. (1985). "A stratigraphic study of beach features on the
southwestern shore of Lake Michigan: new evidence of Holocene lake
level fluctuations." Ill. Dept. of Energy and Nat. Res., Champaign,
Ill.

Quinn, F.H. (1986). "Causes and consequences of the record high 1985
Great Lakes water levels." *Proc. Conf. on Human Consequences of
1985's Climate.* AMS, Boston, Mass., 281-284.

Quinn, F.H. and Kelley, R.N. (1983). "Great Lakes monthly hydrologic
data." *NOAA Data Report ERL GLERL-26*, Nat. Tech. Inf. Serv., Springfield, Vir.

U.S. Army Corps of Engineers (1986). "Monthly bulletin of lake levels
for the Great Lakes." U.S. Army Corps of Engineers, Detroit District,
Detroit, Mich.

WATER SUPPLY ANALYSIS FOR THE MILK RIVER BASIN, MONTANA

Steven K. Sando[1]   and   Richard J. DeVore[2]

ABSTRACT

The U. S. Bureau of Reclamation's Milk River Irrigation Project provides water for 55,000 ha of irrigation in the Milk River basin of Montana. The project includes an interbasin diversion (for the transfer of water from the St. Mary River), three storage reservoirs, and many diversion dams, canals, and lesser structures.

A variable water supply, poor delivery system, and difficulties with water management have led to periodic water shortages for the project. Efforts to make up the shortages are complicated by several factors. The division of water of the Milk and St. Mary rivers between the U. S. and Canada is governed by treaty and Canada is planning to fully develop its water supply. Also, the settlement of reserved water rights of two Indian reservations will affect the available supply in the future.

Planning for the basin has been based on a simulation model that evaluates the effects various strategies have on streamflows, irrigation shortages, or reservoir contents at 35 nodes. The model indicates that shortages are most effectively dealt with through conservation and rehabilitation programs.

INTRODUCTION

Irrigation in the Milk River basin of Montana began in the 1880's and developed rapidly thereafter. Water shortages soon appeared and in 1902 one of the first projects authorized under the Reclamation Act was a storage facility, Sherburne Lake, and a diversion canal from the St. Mary River to augment the Milk River's water supply. The drought of the 1930's again brought shortages and as a result Fresno Reservoir was constructed in 1939. However, periodic and often severe water shortages have continued, primarily due to increased irrigation, continued poor water management and aging of the original diversion and conveyance structures.

The legal setting of the Milk River Irrigation Project does not lend itself to easy solutions for meeting shortages. The flows of the Milk and St. Mary rivers are apportioned between the U. S. and Canada by treaty. Canada is proceeding with plans to construct a reservoir and further develop its share of Milk River water, causing great concern to Milk River irrigators on the U. S. side of the border. Two Indian reservations—the Blackfeet and Fort Belknap—further complicate the question of water supply in the basin. The Winters decision of 1908 established that the Fort Belknap Reservation be given priority use of the natural flow of the Milk River for both present and future needs. Plans to more than double the current irrigated area on that reservation is causing concerns to other irrigators in the basin.

Periodic water shortages coupled with the threat of future depletions have caused the U.S. Bureau of Reclamation (USBR) and the Montana Department of Natural Resources and Conservation (DNRC) to evaluate proposed alternatives for meeting shortages using a simulation model of the Milk River basin. Alternatives include augmenting the supply by interbasin diversion; rehabilitating the conveyance system of the Milk River Irrigation Project; implementing an on-farm conservation program; and increasing storage in the basin.

[1]Hydrologist
  Water Management Bureau
  Montana Department of Natural
    Resources and Conservation
  Helena, Montana

[2]Hydraulic Engineer
  Reservoir Regulation Branch
  Division of Water and Land
  U.S. Bureau of Reclamation
  Billings, Montana

DESCRIPTION OF STUDY AREA AND MILK RIVER IRRIGATION PROJECT

The Milk River is located in northern Montana and southern Alberta, Canada (figure 1). Total drainage area of the basin is 60,350 km^2 of which 40,200 km^2 are in the U. S.

Figure 1. Map Showing Location of Milk River Basin.

The Milk River has a streamflow regime governed primarily by snowmelt. The mean annual flow of the river at the eastern crossing of the international boundary is 10.5 m^3/s, or 329,400 dam^3/yr. Of this, an average of 172,620 dam^3/yr is contributed from the St. Mary River basin. Near the mouth of the Milk River at Nashua, the mean annual flow increases to 19.3 m^3/s or 618,600 dam^3/yr.

The basin climate is semiarid with mean annual precipitation in the U. S. part of the basin ranging from 31.8 cm to 35.6 cm. Topography and soils have been formed primarily through glaciation. The valley floor downstream from Fresno Reservoir is generally flat, mean elevation in this area being about 686 m. Glacial till derived soils are variable in texture ranging from fine sand to heavy clay.

The Milk River Irrigation Project uses three reservoirs to supply eight irrigation districts. Sherburne Lake is located in the St. Mary River basin and releases from the reservoir are diverted via canal into the Milk River near its headwaters. Fresno Reservoir, located 32 km upstream of Havre, is the primary regulating facility for the Milk River Project with a capacity of 127,000 dam^3. The Fort Belknap Indian Reservation is allotted 18,125 dam^3 of Fresno's storage and the remaining 108,875 dam^3 is for use by the non-Indian irrigation districts. Nelson Reservoir is a 71,510 dam^3 offstream storage facility supplying water to some areas in the lower portion of the project.

The eight irrigation districts in the Milk River Project total 55,000 ha. Alfalfa is grown on 40% of this area, native bluejoint hay on 30%, and small grains on 20%. The average seasonal irrigation requirement for the Milk River Project is estimated to be 30.5 cm.

Water is conveyed from Fresno Reservoir to the districts by releases into the Milk River. The water is diverted into unlined canals for distribution to the districts. Condition of these canals is generally poor and often characterized by high seepage rates, slumping of banks, heavy phreatophyte growth and deteriorating check and wasteway structures. The average conveyance efficiency for these canals is estimated to be 64% (SCS, 1978).

Approximately 80% of the Milk River Project area is irrigated with gravity flood systems and 20% with sprinkler systems. On-farm efficiencies of graded border systems, the most common method of flood irrigation in the Milk River Project, average approximately 25% (USBR, et al., 1987). On-farm efficiencies for the sprinkler systems

are approximately 70%. Drainage ditches returning waste water back to the Milk River
tend to be poorly maintained and have heavy phreatophyte growth. Consequently,
non-beneficial consumptive use of irrigation water is high, estimated to be 25-30% of the
total volume of water that would otherwise return to the system.

SIMULATION MODEL OF THE MILK RIVER BASIN

The Milk River simulation model (OPMILK) was developed using the USBR OPSTUDY format
(USBR, 1981). OPMILK employs mass balance computations and a node-link approach to
estimate streamflows, shortages, or reservoir contents at 35 nodes in the Milk River
basin. OPMILK operates on a monthly time-step over a 59 year period of record
(1927-85). A flowchart depicting the general sequence of calculations in OPMILK is shown
in figure 2:

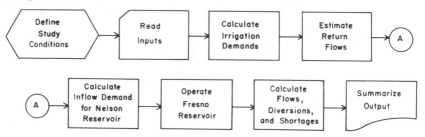

Figure 2. Flow Chart of OPMILK Calculations.

Data input to the model includes historic streamflows, crop irrigation requirements,
irrigation efficiencies, canal capacities, district irrigated acres, municipal water
demands and reservoir capacities, area-capacity tables and evaporation rates. Historic
streamflows for the mainstem of the Milk River at the eastern crossing and 15 major
tributaries to the Milk River are used. Many of the tributary gages have only partial
records for the period 1927-1985, and missing values were reconstituted using the HEC-4
program (U.S. Army Corps of Engineers, 1971). Flows of 26 minor ungaged tributaries were
estimated using drainage area ratios applied to nearby index gaging stations.

Monthly irrigation requirements for alfalfa and small grains were externally
calculated for each year using the USBR CIR77 program (USBR, 1976) based upon the
Jensen-Haise formula. Lacking an appropriate crop coefficient for use in the
Jensen-Haise formula, mean monthly irrigation requirements for native bluejoint grass
were estimated and held constant from year to year. Native bluejoint grass is a cool
season crop grown on marginal soils and is characterized by low yields and relatively low
consumptive use. The annual irrigation requirement for the bluejoint grass was estimated
to be 12.7 cm of which 7.6 cm was assigned to the months of May and June.

After input is read, OPMILK performs initial irrigation system calculations. The
crop irrigation requirements and irrigation efficiencies are used to calculate the
monthly diversion requirements to satisfy the area served by each canal. When total
monthly demand on a canal exceeds the canal capacity, demand is reduced to capacity. A
facility shortage is calculated in this case as the increment of demand in excess of the
canal capacity. Facility shortages are significant in evaluating Milk River irrigation
shortages as they are caused by inadequate canal capacity and cannot be remedied by
increasing water supply.

A preliminary estimate of irrigation return flow is made to determine what part of
the monthly demand can be satisfied by return flows. Monthly return flows for each
district are estimated by a subroutine which calculates the volume of water available for
return by subtracting crop consumptive use and an estimate of irrecoverable loss from the
irrigation diversion. The volume available for return is then lagged over a three month

period, with 70% of the volume returning in the month of the irrigation, 20% in the next month and 10% in the second subsequent month.

After the preliminary irrigation system calculations, OPMILK does initial reservoir computations. To establish demand on the reservoirs, the spatial distribution of demands with respect to tributary inflows and return flows is first analyzed. The Fort Belknap Reservation is given priority use of all tributary inflows upstream of the reservation. Total demand on the reservoirs is equal to the sum of the demands in excess of the tributary inflows and return flows available for use at each demand point. Nelson Reservoir can be dependent on Fresno Reservoir for its inflows, thus a trial operation of Nelson is done each month to establish its demand on Fresno.

Once demands upon Fresno are set, OPMILK does final reservoir computations. Inflows to Fresno, consisting of natural flows and water diverted from the St. Mary River basin are apportioned between the Fort Belknap Reservation and the non-Indian districts. Releases from Fresno satisfy irrigation and municipal demands and then EOM storage volumes are calculated. Distinctions between Indian and non-Indian releases and storage volumes are maintained in all computations and releases are made within the constraints of operating criteria. Reservoir evaporation for each month is calculated from estimates of the mean reservoir surface area for the month and mean monthly evaporation derived from local evaporation data.

With Fresno operations completed, releases are routed through the system. Diversions and return flows at each demand point are simulated in a downstream sequence. Shortages are computed and return flows are recalculated based upon the actual diversion instead of the initial estimate. Accounting of Indian and non-Indian water is maintained through the system. Flows are routed through Nelson Reservoir, where calculation of releases, evaporation, and EOM content are made. Mainstem streamflows of the Milk River are estimated at 17 nodes, six of which have historic gaging records for comparison.

MODELING STUDIES AND RESULTS

To determine the accuracy and applicability of the OPMILK model, a baseline run simulating historic conditions was made. Table 1 shows results of a comparison of simulated streamflows with historic streamflows at two nodes in the basin for the period 1955-1985. The stream gage at Havre is the first gage on the Milk River mainstem below Fresno reservoir. Results at this site are used to demonstrate how well Fresno releases were simulated by OPMILK. Correlation analysis using the Havre gage revealed that simulated and historic values correlated significantly (a 5% significance level is used in all analyses) for both annual flows and May through September irrigation season flows. A paired T-test showed that simulated flows differed significantly from historic for both annual and irrigation season flows.

Table 1. Comparison of Simulated Flows with Historic Flows at Two Locations for the Period 1955-1985

Gaging Site	Historic Mean Annual Flow (1000 dam^3)	Simulated Mean Annual Flow (1000 dam^3)	Annual Flow Correlation Coefficient (r)	Historic Mean Irrigation Season Flow (1000 dam^3)	Simulated Mean Irrigation Season Flow (1000 dam^3)	Irrigation Season Flow Correlation Coefficient (r)
Havre	390	406	0.95	312	329	0.91
Nashua	579	476	0.96	265	195	0.94

The stream gage at Nashua, the last gage on the Milk River mainstem, is 21 river miles upstream of the mouth. Using this gage for comparison indicates how well OPMILK simulates the total operation of the Milk River Irrigation Project. The correlation analysis using the Nashua gage revealed that simulated and historic values correlated significantly for both annual flows and May through September irrigation season flows. A paired T-test showed that simulated flows were significantly different from historic for both annual and irrigation season flows.

Comparison of OPMILK results with historic flows at Havre and Nashua revealed that the model results successfully accounted for the variability of historic flows, but did not accurately duplicate historic flows. Re-inspection of the data showed that the simulated flows consistently deviated from historic flows at both gages during high flow years. Probable reasons for this are: first, a significant part of tributary inflow in the basin is ungaged and had to be estimated. The estimating techniques probably do not accurately represent high flow conditions. Secondly, while certain characteristics of the irrigation systems are factors which vary from year to year depending upon moisture levels and other conditions, they are represented in the OPMILK model by constants. Constant values of such factors as bluejoint grass irrigation requirement and irrigation efficiencies are probably more representative of low rather than abnormally high water supply conditions. Table 2 shows the results of a comparison of simulated flows with historic flows for the 10 driest years during the period 1955-1985. A paired T-test revealed the simulated flows did not differ significantly from historic for either irrigation season or annual flows at the Havre and Nashua gages. Thus, OPMILK model results seem suitable for simulating average and below average water supply conditions, appropriate to the primary study objective of evaluating irrigation shortages in the basin.

Table 2. Comparison of Simulated Flows with Historic Flows at Two Locations for Ten Low Flow Years in the Period 1955-1985

Gaging Site	Historic Mean Annual Flow (1000 dam3)	Simulated Mean Annual Flow (1000 dam3)	Historic Mean Irrigation Season Flow (1000 dam3)	Simulated Mean Irrigation Season Flow (1000 dam3)
Havre	294	291	252	244
Nashua	200	201	73	75

Table 3 presents irrigation shortage estimates for current conditions and future possibilities. The high facility shortage values were substantiated through discussions with ditch riders and irrigators who confirmed that many canals cannot convey enough flow to satisfy all irrigators on the lower parts of the canals.

Table 3. OPMILK Shortage Estimates for Five Scenarios

Scenario	Mean Annual Supply Shortages (1000 dam3)	Mean Annual Facility Shortages (1000 dam3)
1. Current conditions	70	77
2. Future baseline*	65	77
3. Future baseline plus conservation and rehab. program	17	9
4. Future baseline plus Missouri River diversion	32	77
5. Future baseline plus increased storage	43	77

*The future baseline scenario includes rehabilitation of the St. Mary River canal plus the proposed Canadian and Indian developments.

The baseline future scenario envisions three additional developments in the basin. The St. Mary River diversion canal is currently being rehabilitated to its original design capacity, which will result in an increase of 4 m3/s in capacity. The rehabilitation will supply an average of 61,650 dam3/yr to the Milk River Project.

Also included in the baseline future scenario is the proposed reservoir development by the Canadians. To account for this, it is assumed that their entire legal share is consumed. Finally, the baseline future scenario includes the addition of 5,670 ha of irrigation to the Fort Belknap Indian Reservation.

The impact of the Canadians and Indians on demands water shortages are much less than was once envisioned as they are restricted to Milk River natural flow and cannot benefit from present or increased flows received from the St. Mary River diversion. The St. Mary River drainage yields a more firm supply, often providing water when the natural flow of the Milk River is zero. Therefore, in years of short supply, the Canadians and Indian demands have a relatively small impact upon shortages. The Canadians have historically passed an average of 50,000 dam^3/yr of their share but this occurs in years of surplus water supply when the Milk River Project is unable to use or store the water.

Although numerous options were analyzed with OPMILK, three primary alternatives designed to reduce future shortages to acceptable levels are reported. The first alternative involves a comprehensive conservation and rehabilitation program. Rehabilitation and betterment of canals and diversion structures in the Milk River Project has been estimated to result in a 12% increase in the mean conveyance efficiency. Also an on-farm conservation program would result in an 8% increase in the mean on-farm efficiency. This alternative has the benefit of reducing both supply and facility shortages. The second major alternative is a diversion canal which would convey 5.7 m^3/s from the Missouri River to the Milk River at a point just upstream of Havre. While this diversion was once the preferred alternative modeling results have shown that some degree of conservation will be necessary before maximum benefits of increasing the water supply could be realized. The final alternative is increasing storage in the basin through participation with Canada in their proposed reservoir. In this alternative, the Milk River Project is assumed to have available for its use 61,650 dam^3 of storage in the Canadian reservoir.

The conservation and rehabilitation alternative results in the greatest reduction in shortages. Once given only token consideration as a means of reducing Milk River shortages, this strategy is now considered a mandatory initial step to reduce shortages.

SUMMARY AND CONCLUSIONS

The Milk River study is an example of how simulation modeling can clarify a complex hydrologic system and provide direction for the solution of a water shortage problem. Results of the OPMILK model reveal that future developments in the Milk River basin should not increase irrigation shortages to the extent that was once envisioned. Also, the modeling shows the benefits of reducing water demand as opposed to increasing water supply. Currently, a preferred solution to Milk River irrigation shortages is to implement a conservation and rehabilitation program and then assess what additional action, if any, needs to be taken.

APPENDIX I.--REFERENCES
1. U.S. Army Corps of Engineers, 1971. HEC-4 monthly streamflow simulation. Hydrologic Engineering Center. 113 pp.
2. U.S. Bureau of Reclamation, 1976. Methodology to compute crop irrigation requirements using the Jensen-Haise consumptive use formula. Unpublished report. Planning Division, Missouri River Basin, Billings, Montana.
3. U.S. Bureau of Reclamation, 1981. OPSTUDY User's Manual. Central Nebraska Projects Office, Grand Island, Nebraska. 71 pp.
4. U.S. Bureau of Reclamation, U.S. Soil Conservation Service, Montana Department of Natural Resources and Conservation, and Blaine, Phillips and Valley County Conservation Districts. 1987. Milk River Project water measurement data. Planning Division, Missouri River Basin, Billings, Montana.
5. U.S. Soil Conservation Service, 1978. Water Conservation and Salvage report for Montana. River Basin and Watershed Planning Division, Bozeman, Montana. 75 pp.

Estimating Firm Yield and Reliability of Small
Reservoirs on Ungaged Streams in North-Central Montana

by

[1] Stephen R. Holnbeck and [2] Charles Parrett

Abstract

A reservoir-operation algorithm utilizing a sequence of monthly streamflows from a lag-one Markov model is proposed for estimating firm yield and reliability. Streamflow statistics required for the lag-one Markov model are based on regional expressions developed from data at six U.S. Geological Survey streamflow-gaging stations in north-central Montana. The reservoir-operation algorithm for firm yield and reliability was used for hypothetical reservoirs on each of the six gaged streams. The actual streamflow records were first used to determine reservoir capacities required to fully satisfy about 90 different demand levels. Simulated streamflows derived from the regional streamflow statistics and the lag-one Markov model then were used to determine reservoir capacities for the same demand distributions. A regression equation relating capacities determined from the simulated flows to capacities determined from actual flows had a coefficient of determination (r^2) of 0.98 and a standard error of estimate of 33%. Greater streamflow variation and the lack of long-term streamflow records relative to other parts of the State indicate that a worst-case condition was investigated. The proposed algorithm thus appears to be a useful tool for evaluation of existing and proposed reservoirs anywhere in Montana.

Introduction

A method for estimating firm yield and reliability, R_i (percentage of time a given demand is satisfied) of small reservoirs on ungaged streams is currently needed by water-management agencies in Montana. In the past, small reservoirs on ungaged streams were designed with little consideration given to the optimization of water availability, storage capacity, and water demand. Now, however, increased competition for the limited supply of surface water necessitates a more rigorous approach. The purpose of this investigation was to determine if reservoir operation studies could be successfully applied to ungaged systems and thus serve as a tool for evaluation of existing and proposed designs.

[1] Hydrologist, Montana Department of Natural Resources and
Conservation, Helena, Montana 59620

[2] Hydrologist, U.S. Geological Survey, Helena, Montana 59626

Study Area and Data Used

The region studied is in north-central Montana, consists of predominantly plains, and is semiarid. Runoff is variable and is produced by both snowmelt and rainfall. Streams are ephemeral or intermittent (Omang and Parrett, 1984), and few small-stream gaging records are available (Fig. 1).

The procedure used to obtain regional statistical information involved selection of six existing U.S. Geological Survey streamflow-gaging stations and the development of a common base period (1927-80) for those stations. Historic records ranged from 8 to 54 years and drainage areas ranged from about 60 to 283 mi^2. Missing records were reconstructed using the computer program HEC-4 (U.S. Army Corps of Engineers, 1971). Because four of the six sites have historic records that include the drought of the 1930's, it is likely that extreme hydrologic conditions are adequately represented.

Data from the six gaging stations shown in Fig. 1 were used to develop the following regression equations relating mean annual discharge to drainage area and standard deviation of annual mean discharge to mean annual discharge:

$$Q = 0.015 \ A^{1.23} \qquad\qquad (1)$$

$$S = 1.53 \ Q^{0.77} \qquad\qquad (2)$$

where Q = long-term mean annual discharge (ft^3/s),
A = drainage area (mi^2),
S = standard deviation of annual mean discharge (ft^3/s).

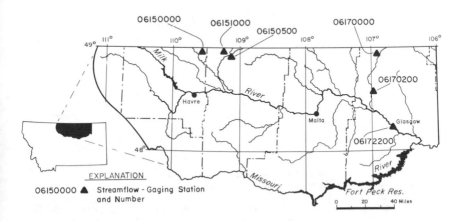

Figure 1.   Location of Study Area and Gaging Stations

Results from Eq. 1 are similar to results reported by Omang and Parrett (1984). Suitable regression relationships between the skewness of annual mean discharge and basin variables and between lag-one correlation coefficient of annual mean discharge and basin variables could not be developed. As discussed by Benson and Matalas (1967), average values of skewness and lag-one correlation coefficients as determined from records of the six gaged streams were considered to be applicable to the study area.

Streamflow Synthesis

The following lag-one Markov generation equation described by Fiering and Jackson (1971) was used to generate a sequence of annual mean streamflows for use in the reservoir-operation algorithm:

$$q_i = Exp \left[\mu + \rho(h_{i-1}-\mu) + t_i \, \sigma \sqrt{1-\rho^2} \right] + a \qquad (3)$$

where q_i = synthesized annual mean streamflow (ft^3/s)
  μ = a logarithmic estimate of sample mean,
  ρ = a logarithmic estimate of sample lag-one
    correlation coefficient,
  h_{i-1} = logarithm of flow,
  t_i = standard normal random sampling deviate,
  σ = a logarithmic estimate of standard deviation
    based on skewness, and
  a = term based on μ and σ.

Eq. 3 is applicable to streamflows having a log-normal distribution, which was confirmed for the study area. Monthly streamflows were generated from the synthetic annual mean streamflows developed using Eq. 3 by a simple percentage distribution of annual mean discharge occurring in each month based on information from the six gaging stations. Anderson and Mereu (1985) demonstrated that the percentage distribution technique is sufficient to generate monthly values from annual streamflow, especially if other more important streamflow uncertainties exist.

Reservoir Operation Algorithm

Data-input requirements for the reservoir-operation algorithm include drainage area upstream from the reservoir, dam height, initial selected storage capacity, and monthly water-demand distribution. Inflows are generated in the program by the lag-one Markov model. A flowchart illustrating the reservoir-operation algorithm is shown in Fig. 2.

The reservoir-operation algorithm assesses reservoir capacity in two ways. If the specified reservoir capacity is more than enough to provide a firm yield equal to the demand, the reservoir capacity is incrementally decreased and the reservoir operation is repeated until the capacity just satisfies demand. If the reservoir cannot fully satisfy demand, the next step depends on whether the reservoir being analyzed is an existing, fixed-size reservoir or a proposed, variable-sized reservoir. If an existing reservoir is being analyzed, the reliability is calculated, and the reservoir size is incrementally decreased. The reliability is calculated again and compared with the previously calculated reliability. When the newly calculated reliability decreases significantly below the last calculated value, the reliability is maximized and the iterations are stopped. If a proposed reservoir is being evaluated, the reservoir size is incrementally increased. So long as reliability is improved, the iterations continue upward. When the reliability is maximized or the firm yield is met, the iterations are stopped.

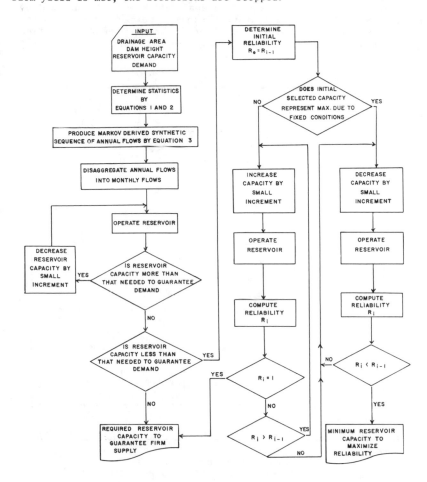

Figure 2. Flowchart of Model

Results

Operation studies were performed for hypothetical reservoirs on each of the six gaged streams in Fig. 1. Demand ranged from about 5 to 125% of mean annual streamflow. Two demand distributions were used. A stockwater demand was based on the specified percentage of mean annual runoff divided equally for 12 months. The second demand was for full service irrigation and was distributed accordingly for May through September. Evaporation, although excluded from this analysis, is included in the model for specific application and is discussed by Holnbeck (1985). About 90 operation studies using different demand levels for stockwater and irrigation were performed. Results comparing required reservoir capacity are illustrated in Fig. 3. Results in

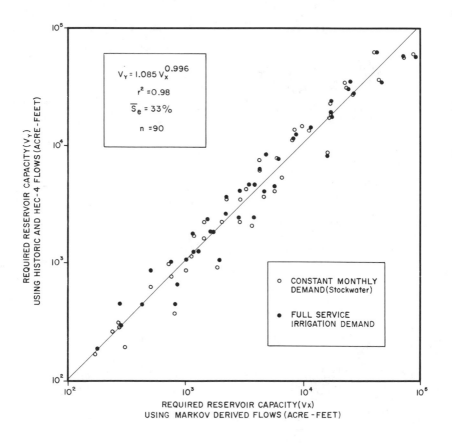

Figure 3.   Comparison of Reservoir Operation Studies

Fig. 3 demonstrate that the operation studies performed using Markov-derived synthetic flows compare favorably with results obtained by using the U.S. Geological Survey and HEC-4 flows. The average standard error of estimate, $\bar{S}_e$, obtained in this study (33%) is approximately equal to that of the mean annual streamflow estimating equation from Omang and Parrett (1984). Thus, the accuracy of the lag-one Markov model apparently is limited to the accuracy of the estimated statistics used in the model. Because of the large variability in streamflow and the general paucity of small-stream data in the study area, the accuracy of the estimated streamflow statistics probably is less than anywhere else in the State. Consequently, application of the proposed methodology in the study area probably represents a worst-case condition. The use of regional streamflow information coupled with a lag-one Markov model thus appears to be a useful tool for estimating reservoir storage requirements at ungaged sites on small streams anywhere in Montana.

Appendix I.--References

1. Anderson, J. E., and Mereu, T. J. (1985). "Suitability of disaggregation methodologies in hydroelectric design studies." Proc. of the Speciality Conf., Computer Applications in Water Resources, ASCE, Buffalo, N. Y., pp. 12.
2. Benson, M. A., and Matalas, N. C. (1967). "Synthetic hydrology based on regional statistical parameters." *Water Resour. Res.,* 3(4) pp. 931-935.
3. Fiering, M. B., and Jackson, B. B. (1971). "Synthetic streamflows." *Water Resour. Monograph 1,* American Geophysical Union, 97 pp.
4. Holnbeck, S. R. (1985). "A proposed methodology to estimate firm yield of small storage reservoirs and ponds in northcentral Montana." Unpublished report, Montana Department of Natural Resources and Conservation, Helena, Mont., 32 pp.
5. Omang, R. J., and Parrett, C. (1984). "A method for estimating mean annual runoff of ungaged streams based on basin characteristics in central and eastern Montana." Water-Resour. Invest. Rep. 84-4143, U.S. Geological Survey, Helena, Mont., 15 pp.
6. U.S. Army Corps of Engineers, (1971)."HEC-4 monthly streamflow simulation." Hydrologic Engineering Center, 113 pp.

Appendix II.--Conversion Factors

The following factors can be used to convert inch-pound units in this report to the International System of units (SI).

Multiply inch-pound unit	By	To obtain SI unit
acre-foot	1,233	cubic meter
cubic foot per second	0.02832	cubic meter per
(ft^3/s)		second
mile (mi)	1.609	kilometer
square mile (mi^2)	2.590	square kilometer

MISSOURI RIVER BASIN WATER ACCOUNTING SYSTEM

by Tsong C. Wei, M. ASCE[1], Warren J. Mellema, M. ASCE[2], and Donald L. Ohnstad[3]

ABSTRACT

A water accounting system was developed for the Missouri River Basin to establish an accurate and acceptable base of information describing water availability and uses, and to provide a system to evaluate the effects and impacts of using additional quantities of water. The system was based on a depletion concept which presumes that the flow at any point in a stream reflects the combination of the surface flow and ground water flow modified by water uses through either man's activities or natural processes upstream of that point in the stream. These water uses are regarded as a streamflow depletion. When the natural depletion remains unchanged, streamflow simply becomes a function of depletions from human activities. By identifying historic depletions and level of human activities or depletors such as population, number of industries, irrigation acreages, and other water consumptive use activities over time, it is possible to evaluate likely effects of additional development on water availability.

I. INTRODUCTION

The Missouri River basin Water Accounting System (WAS) divides the basin, an area of 523,300 square miles, into 93 subbasins. It consists of a database and a number of computer programs to retrieve and manipulate the data. The system database contains monthly data from 1944 through 1978, streamflow at the outlet of each subbasin, 14 water use depletion categories, number of water users, water use rates, narrative legal and institutional information. The system selects the year 1978 as the baseline level, and adjusts the number of users, amount of water used, and streamflow to that development level. The system can retrieve information in the database in tabular or graphic form, perform statistical and trend analysis and provide "what if" type information to evaluate future water uses (7).

II. STREAMFLOW AND BASIN TRANSFER

Historic streamflow data in most instances were directly available from USGS WATSTORE files. Missing data were estimated either from linear regressions, drainage area ratios or total flow ratios by

---

[1] Hydraulic Engineer, Hydraulics and Hydrology Section, Missouri River Division, Corps of Engineers, Omaha, NE.
[2] Chief Hydraulics and Hydrology Section, Missouri River Division, Corps of Engineers, Omaha, NE.
[3] Acting Executive Director, Missouri Basin States Association, Suite 1, 10834 Old Mill Rd., Omaha, NE. 68154

correlating with a nearby station.   The method that resulted in the
least deviation was used to reestablish missing data.   Water transfers
between basins impacts streamflow.   Four imports of water into the Missouri River basin,   two exports out of the basin,   and 13 inter-basin
transfers were considered as having a significant impact.   These data
were used to modify the streamflow records and create an adjusted
streamflow in the database (2).

III. AGRICULTURAL WATER

     Agricultural water uses are the largest depletor of streamflow and
the most complicated and difficult category to quantify.   These uses
were subdivided into irrigation, livestock, conservation measures, farm
ponds,   and forest accretion.   Following is the methodology used for
each category (5).

     Irrigation: Irrigation acreage was determined by using information
available in the U.S. Census of Agriculture, State Crop and Livestock
Reporting Service, and State Permits and Records.   In addition, the
Soil Conservation Service surveyed type, intensity, efficiency of
irrigation and surface and ground water irrigation.   Water requirements
difference of crop patterns was assumed minor, and a single crop
pattern for the year 1978 was adopted.   Sixteen identified irrigation
types were combined to four categories.   The "full service", indicates
that the entire crop irrigation requirement is available in at least
eight out of ten years, "partial service" indicates that the entire
crop irrigation requirement is not available beyond a certain point in
the growing season,   and "waterspreader" diverts flood runoff on
adjacent fields whenever water is available, and "groundwater" pumps
water from a well.

     The crop irrigation requirement (CIR): defined as the amount of
water supplied by irrigation to supplement the effective rainfall to
adequately grow a crop to maturity, was determined either by the
Blaney-Criddle method or by the Jensen-Haise method according to the
preference of each state.   Streamflow depletion (SD) for full and
partial services irrigation was computed by equation below.

$$SD = CIR[1 - NCU(1 - 1/CE*FE)]$$

where FE is the on farm efficiency, CE the conveyance efficiency, NCU
the nonbenificial consumptive use. An annual waterspreader depletion of
9.6 inches was used in Montana, 4 inches in North Dakota, South Dakota
and Nebraska while the CIR method was used in Colorado and Wyoming.

     Livestock: Livestock numbers were compiled from the U.S. Census
of Agriculture and the State Crop and Livestock Reporting Services.
Water requirement of livestock in gallons per head per day are as
follows: milk cows, 28.0; other cattle and calves, 8.0; hogs and pigs,
2.5; sheep and lambs, 2.0; and horses, ponies and mules, 10.0. Annual
livestock water uses were determined from the above data. A monthly
distribution factor was applied to convert to monthly distribution.

     Conservation Measures and Farm Ponds: Contour farming, terraces,
minimum tillage and range pitting that intercept runoff thus deplete

streamflow were also considered. Surveys of the land treatment acreages and the surface area of farm ponds less than 40 acres were conducted by the Soil Conservation Service, Forest Service, and Bureau of Land Management for the year 1978. A distribution curve was used to extend the data back to the year 1944. Annual conservation depletions were computed by USDA method (10) assuming an average climatic condition throughout the period. A monthly distribution factor for March through July was applied to compute the monthly values.

Forest Accretions: Forest management practices affect water yields of a basin through snow distribution and evapotranspiration changes. These influences are such that they can increase or decrease water yield over virgin forest conditions. Information on forest management, forest fires, and road building activities in national forests were furnished by the U.S. Forest Service. Forest accretions were then calculated from Forest Service techniques (5), that utilized water yield study results of Forest Service. Other forested lands were assumed to have an insignificant impact on the water yield.

IV.   MUNICIPAL AND INDUSTRIAL WATER

     Water usage for purposes other than agriculture were grouped into four categories: municipal water use for cities with population exceeding 2,500, self supplied industrial water use, energy water use, and rural domestic water use. Data were collected for years 1944, 1950, 1960, 1970, and 1978. While data sources varied from state to state, municipal water use records and state permitting records were the major data sources. For the rural domestic water uses, per capita consumption rates were provided by each state. The rural population was multiplied by the consumption rate to compute the annual rural domestic water use. Annual water withdrawn and consumed from surface water and ground water were identified. Linear interpolation was used to determine the annual water use for years between survey years. To further breakdown the annual water use figures into monthly figures, two monthly distribution factors were developed. One was for energy and industrial water use, which was evenly distributed for the twelve month, and the other was for municipal and rural domestic water use, which was distributed high in summer months and low in winter months (4).

V.   GROUND WATER

     In a stream-aquifer system where water can freely interchange between the stream and the aquifer, pumping will first withdraw water from ground water storage. The hydraulic head created will gradually move water from the stream and fill the void storage. The replenished water which otherwise is available as a part of the streamflow constitutes the streamflow depletion (3). The Stream Depletion Factor (SDF) method was adopted to quantify the depletion (1). SDF has a dimension of time, and is defined as a^2S/T where a is the distance from a well to the stream, S the aquifer specific yield and T the transmissivity of the aquifer. The dimensionless parameter, t/SDF, where t is the pumping time, was found to be the complementary error function of another dimensionless parameter, v/V, where v is the depletion volume and V the pumping volume and is called SDF function. For a homogeneous and uniform aquifer, SDF value can be determined easily from specific yield and transmissivity of the aquifer and well location. Then applying time and volume of pumping and the SDF value, the streamflow depletion can be computed from the SDF function.

In large areas where homogeneous and uniform assumption do not hold, a SDF function characteristics, that when v/V = 0.28, t/SDF = 1 or the pumping time is the SDF time, was used to determine the SDF value as following. First, transmissivity maps were developed. A 6 by 15 grid, with each grid one mile square, was placed 10 to 20 miles apart along the river. A USGS ground water program (9) was used to each grid system assuming that a well existed at the center of each grid. With continuous pumping from the grid center at a constant rate, a ground water table drawdown was created. The difference of the volume pumped and the volume of drawdown is the volume of streamflow depletion. When the depletion volume became 28% of the total volume pumped, the time of pumping is the SDF value for that point. SDF isograms were constructed to define constant SDF bands. From well permits, the percentage of wells residing in SDF bands in each year in each sub-basin was determined. The ground water uses identified from irrigation and municipal and industrial water use were applied to the well percentage in each band to compute monthly pumping volume. To account for changing pumping rates a technique similar to the unit graph process was adopted. Applying a constant unit pumping rate that creates a positive withdraw and, at an unit time later, adding a constant unit recharging rate that generates a negative withdraw, a unit streamflow depletion histogram was developed for each SDF bands. Assuming linearity, by multiplying the monthly pumped volume to the ordinates of the unit depletion histogram, temporal distribution of stream depletion from that particular pumping operation was obtained. Spatial and temporal summation of the depletion distribution produces the monthly depletion which includes the delayed depletion from previous months.

VI.   RESERVOIR DEPLETION AND MAJOR RESERVOIR OPERATION

Reservoirs with capacities greater than 200,000 acre-feet were defined as major reservoirs, and those with capacities less than 200,000 acre-feet but larger than 40 surface acres were identified as small reservoirs. Using the average monthly net lake evaporation, that is lake evaporation less precipitation, for the 30-year period from 1950 to 1979, and the reservoir surface area provided in the National Dam Safety Inventory, lake evaporation for small reservoirs was determined and used as the small reservoir depletion (2).

Evaporation from major reservoirs also depletes streamflow. In addition, the reservoir operation itself can withhold or release flow to create temporary depletions or accretion to downstream basins. Six Corps of Engineers reservoirs on the Missouri River main stem and 17 Corps of Engineers and Bureau of Reclamation reservoirs on tributaries were designated as major reservoirs. Computer programs for monthly reservoir operation were developed based on generalized reservoir operation criteria for the main stem reservoirs as a system and tributary reservoirs individually or as a system. From the reservoir inflow, reservoir storage, and regulation criteria the reservoir releases and evaporation were computed. The programs were verified using historic inflows to simulate past reservoir operations and were found to be generally satisfactory (6).

VII 1978 DEVELOPMENT LEVEL

Historic streamflow and depletion data reflects the hydrologic condition, water resources development and water use efficiency throughout the period of record in the Missouri River basin. Assuming hydrologic conditions and the water use efficiency remain the same, the WAS selected the year 1978 as the baseline from which to develop a 1978 development level depletion and corresponding depleted streamflow. The 1978 level depletion (D_{78}) is the net depletion calculated by multiplying the historic unit rate of water use (R_h) by the difference in depletors between 1978 (N_{78}) and any years 1944 to 1978 (N_h) or

$$D_{78} = (N_{78} - N_h) * R_h.$$

The 1978 level ground water depletion used the above equation to determine the amount of water pumped in each basin. This value was applied to the unit stream depletion curve to recalculate the delayed impacts on streamflow. Subtracting the 1978 level depletion from the adjusted historic streamflow data, the 1978 level depleted streamflow is obtained. Using the 1978 level depleted reservoir inflow, the reservoir releases were determined from the reservoir operation programs. The difference of historic reservoir release and the 1978 level depleted release was used as the 1978 level reservoir release depletion, which was carried to the downstream node or until another major reservoir was encountered. As the historic streamflow was readjusted to various 1978 level depletion, the depletion might exceeded the streamflow to create a negative value indicating the stream had dried up. In such a case, the streamflow was set to zero and the historic streamflow was used as the actual 1978 level of depletion and the depletion of downstream basins were adjusted accordingly (6).

VIII. MISSOURI RIVER BASIN WATER ACCOUNTING SYSTEM

The Missouri River Basin Water Accounting System consists from a Basic Information Archive System (BIAS), a computer database, a group of utility computer programs and the QUERY system programs. The BIAS contains numerous raw data, computer programs, computational procedures and other information used in developing the depletion and streamflow database. The computer database is the core of the system that contains historic and 1978 development level data on streamflow, 14 categories of depletion, number of depletors, depletion rates, and legal and institutional information. To manipulate, display and exhibit the huge amount of data, a group of utility programs were developed. These programs can present data in various tabular and graphic forms including statistics, regression and trend analysis. Finally, the QUERY program provides the capability to evaluate the impacts of additional water resources development on streamflow. The program assumes various levels of development in the form of depletions and depletors and computes new depleted downstream streamflow values, assuming that hydrologic condition of 1944 to 1978 are repeated (6).

XI. LEGAL AND INSTITUTIONAL INFORMATION

    In addition to the streamflow and depletion data, the database
also includes legal and institutional information, such as general
description of the node outflow gage, length of record, location,
drainage area, overview of the water law and water rights process,
a brief description of compacts and court decrees in effect, and total
water rights appropriated within a basin.

X. CONCLUSIONS

    A comprehensive Water Accounting System was developed for the
Missouri River Basin to establish and serve as an accurate and
acceptable water resources database, to provide information on water
availability, and to serve the ten basin states and federal agencies
for planning and design of future water resources development in the
basin (8). Because of the vast basin size and various water use type,
complex procedures were necessary in determining these depletions.
However, to limit the database size and computation load some
procedures were simplified without unduly sacrificing accuracy and
reliability of the result. The system is operational, and in use
throughout the Missouri River Basin. The WAS represents a step forward
in better management of the basin water resources.

References:

1. Jenkins, C. T. (1968). "Techniques for Computing Rate and Volume
of Stream Depletion by Wells," Vol. 6, no. 2, Groundwater, pp 37–46.
2. "Missouri River Basin Hydrology Study Technical Paper – Surface
Water Supply Including Instream Water Use" (1982). Missouri Basin
States Association, Omaha, NE.
3. "Missouri River Basin Hydrology Study Technical Paper – Ground Water
Depletion" (1982). Missouri Basin States Association, Omaha, NE.
4. "Missouri River Basin Hydrology Study Technical Paper –
Municipal, Industrial, Energy and Rural Domestic Water Use" (1982).
Missouri Basin States Association, Omaha, NE.
5. "Missouri River Basin Hydrology Study Technical Paper –
Agriculture Water Use Including Identification of Irrigation Lands"
(1982). Missouri Basin States Aassociation, Omaha, NE.
6. "Missouri River Basin Hydrology Study Technical Paper – System
Methods and Operation" (1983). Missouri Basin States Association,
Omaha, NE.
7. "Missouri River Basin Hydrology Study – Final Report" (1983).
Missouri Basin States Association, Omaha, NE.
8. Ohnstad, L. D. and Mellema, W. J. (1984). "Organizing for
Development of The Missouri River Basin Water Accounting System,"
Corps of Engineers, Missouri River Division, Omaha, NE.
9. Pinder, G. F. (1970). "A Digital Model for Aquifer Evaluation,"
Chapter C1, Book 7, Techniques of Water Resources Investigations of the
United States Geological Survey, USGS.
10. "Development of A Procedure for Estimating the Effects of Land
and Watershed Treatment on Streamflow" (1966). Tech. Bulletin no. 1352,
Soil Conservation Service, USDA.

OPTIMIZING A MINIMUM INSTREAM FLOWBY

Pamela P. Kenel,[1] A. M. ASCE and Les K. Lampe,[2] M. ASCE

ABSTRACT: The Commonwealth of Virginia is currently investigating methods to establish minimum instream flow requirements which may be applied to flow allocations statewide. Since historical minimum instream flow standards such as maintaining the seven-day, ten-year low flow are not based upon the maintenance and protection of aquatic life, environmental considerations are now being given higher priority when making minimum instream flow decisions. The Crump Creek Reservoir will be one the first major stream diversion projects to be considered in this perspective. The project will consist of the damming of two creeks tributary to the Pamunkey River to create an offline storage reservoir. Water will be diverted from the river into the reservoir when the flow in the river exceeds the minimum instream flow requirement. A method for determining the recommended minimum instream flow for the Pamunkey River is documented in this paper. Key considerations include the identification of critical instream uses and the seasonal period of concern for each. Projected offstream use requirements are identified. The relationship between reservoir yield and minimum instream flow is evaluated for several diversion rates. A measurement of impact for selected minimum instream flow rates is developed for each diversion rate.

Hanover County is located in the rapidly-developing region surrounding Richmond, Virginia. Surpassed only by Henrico and Chesterfield Counties, Hanover is the third-fastest growing County in the Richmond vicinity. Previous studies have predicted that the existing source of water supply, groundwater, will be uneconomical and impractical for use beyond the year 2000. The need to identify an additional water supply of approximately 15 million gallons per day (mgd) resulted in the selection of the Pamunkey River as the best available, developable source. Further studies recommended a surface water reservoir located off-channel, rather than an impoundment of the Pamunkey River, due to serious environmental ramifications and topographic constraints. The broad, flat river bed is not only a documented archeological resource, a superior wetland habitat, and an anadromous fish spawning ground, but the construction cost of a dam on the Pamunkey would not be justified by the small size reservoir required to meet Hanover County's needs.

[1]Water Resources Engineer; Black & Veatch, Inc., Consulting Engineers; Rockville, Maryland.
[2]Director of Water Resources, Black & Veatch, Engineers-Architects; Kansas City, Missouri.

Crump Creek was selected as the site for the off-channel reservoir for several reasons. The area is not currently heavily developed, and land can be obtained with little impact to existing residences. The Crump Creek drainage area is approximately 14.8 square miles, and the reservoir site is buffered by primarily agricultural land uses. Additionally, the site is adjacent to projected local demand centers, lessening water transmission costs.

The proposed Crump Creek Reservoir is an impoundment of two creeks tributary to the Pamunkey River. The proposed dam will be located just downstream of the confluence of the Crump and Pollard Creeks. Reservoir storage will be maintained through pumped diversion of Pamunkey River high flows into the impounded area. The mean annual flow in the ungaged main tributary, Crump Creek, over the 40-year period of synthesized record is estimated at 15 cfs. The USGS maintains a Pamunkey River gage at Hanover, approximately 13.6 miles upstream of the confluence with Crump Creek. Recorded flows at this gage, having a 1,081 square mile drainage area, range from 13 to 39,300 cfs for the period of record from 1941 to 1985. The mean annual flow is 1,035 cfs. Median monthly flows range from 192 to 1,270 cfs, with median flows of 1,270 cfs in March and 192 cfs in September. Flows less than 208 cfs have occurred 20 percent of the time. Selected Pamunkey River flow statistics are provided in Table 1.

From evaluation of flows in Crump Creek, it is evident that an impoundment of Crump Creek cannot independently sustain a reservoir yield of 15 mgd. Peak flows diverted from the Pamunkey will maintain the required yield of the reservoir. The maximum quantity of water which may be diverted from the Pamunkey River will be determined by the minimum quantity of flow required to remain in the river and the capacity of the diversion facilities. In the Commonwealth of Virginia, the Virginia Water Control Board (VWCB) is the agency responsible for setting minimum instream flowby (MIF) values for the waters of the Commonwealth. The VWCB has traditionally required the seven-day, ten-year low flow (7Q10) as an MIF criterion. This criterion was initially established for designing wastewater treatment plants to ensure minimum water quality standards during droughts. A sustained 7Q10 was not intended to maintain or protect the aquatic habitat. Due to potentially negative impacts to environmental resources, care must be taken when setting minimum instream flow objectives. The VWCB recently commissioned a study to evaluate Minimum Instream Flow estimation methodologies and to develop a statewide program for setting and managing minimum instream flow objectives (1). This study concluded that the Tennant Method, which uses fixed percentages of mean annual flow as MIF thresholds, is preferred because of ease in application. However, it recommended that the State maintain flexibility in their program, to account for individual characteristics and critical needs of each location.

Throughout the course of the Crump Creek Reservoir project, meetings were held to identify and clarify issues of concern with the interested state and federal agencies. These agencies included the US Fish and Wildlife Service (USFWS), the US Marine Fisheries Service (USFWS), the US Environmental Protection Agency (USEPA), the US Army Corps of Engineers (USACE), the Virginia Commission of Game and Inland Fisheries (VCGIF) and the Virginia Water Control Board. The minimum instream flowby requirement was pinpointed as an issue of concern by all of the involved agencies.

Initially, existing instream and offstream uses of the Pamunkey River were identified. Instream uses include aquatic habitat for fisheries and wetlands, capacity for domestic waste assimilation, and navigation and recreation. Offstream uses consist of water supply for domestic and industrial users and unregulated irrigation supply for riparian farmers. From the list of existing stream uses, those uses which would potentially be impacted by decreased Pamunkey River flow were identified, as well as the seasonal period of concern for each. Concern was primarily focused on the potential effects that diminished flow would have on the aquatic habitat. Since portions of the Pamunkey River experience tidal action from the downstream York estuary, concerns were raised about increased salinity concentrations during seasonal periods of low flow, July through October. Additionally, the Pamunkey is a well-known spawning ground for anadromous fishes. Concerns about decreased flow during the months of fish migration, March to May, were also raised. From a review of hydrologic conditions in the Pamunkey, it was apparent that during the months of peak river flows, January through April, enough water should be available for all existing instream and offstream uses, including diversions to the proposed reservoir.

The issue of increased salinity concentration under induced low flow conditions has been previously analyzed by the Virginia Institute of Marine Science (VIMS) (2)(3). VIMS developed a model of the York estuary, including its tributaries, the Pamunkey and the Mattaponi Rivers. Analyses of the Pamunkey under induced, low-flow conditions predicted an increased salinity intrusion distance upstream. Previous studies have recommended an MIF no less than 184 cfs, due to potential salinity movement (4). Anticipated impacts due to salinity movement could create a need for a higher instream flowby threshold.

The method to be used for the determination of the Pamunkey River MIF was discussed with the federal and state agencies involved in the project. The USFWS's Incremental Field Instream Methodology (IFIM) was considered, but was not selected. Due to the large Pamunkey flows relative to the amount of anticipated diversion, less than 100 cfs, the benefit gained from uses of the IFIM would not be commensurate with its labor- and cost-intensive requirements. Therefore, the agencies agreed that a range of potential MIF's should be examined with respect to impacts on the yield of the proposed reservoir. Eight separate MIF scenarios would be examined consisting of the following conditions.

1. Seven day, ten-year low flow (57.4 cfs)
2. Tennant 40 (400 cfs) January to June and
    Tennant 20 (200 cfs) July to December.
3. Tennant 30 (300 cfs) January to December
4. Tennant 20 (200 cfs) January to December
5. 50 Percent Exceedance Monthly Flows (Median Monthly)
6. 70 Percent Exceedance Monthly Flows
7. 85 Percent Exceedance Monthly Flows
8. 90 Percent Exceedance Monthly Flows

For the purposes of relating minimum instream flowby values to potential reservoir yield, a computer model was developed to incorporate the necessary system parameters. The variables affecting reservoir yield are listed here. A schematic illustrating the interaction of the model variables is provided as Figure 1.

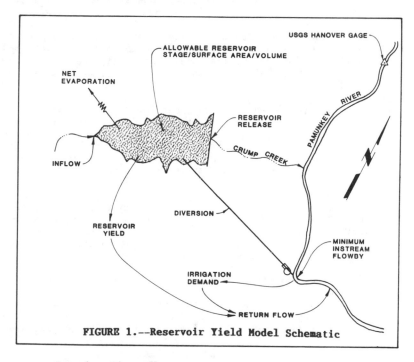

FIGURE 1.--Reservoir Yield Model Schematic

- Pamunkey River flow
- River diversion capacity
- Drainage area return flow
- Minimum instream flowby
- Irrigation demand
- Crump Creek inflow
- Reservoir stage/surface area/volume relationship
- Net reservoir evaporation
- Minimum reservoir release
- Minimum allowable reservoir stage

To directly compare the effects of the various MIF's on the reservoir's yield, only two parameters were varied in the modeling runs. The capacity of the diversion was varied to include 10, 25, 50 and 100 mgd for the eight different MIF rates. Results of the modeling runs are presented in Table 2. Based upon reservoir yield modeling results, preliminary conclusions can be reached concerning optimum diversion size and elimination of several MIF scenarios from further consideration.

1. A 10 mgd diversion is too small to provide 15 mgd water supply and maintain a reasonable pool elevation in the reservoir. Incremental increases in diversion pumping capacity could be useful for the County in the years until the ultimate reservoir yield is needed. However, for purposes of ultimate water supply needs, a 10 mgd diversion is unacceptable.

2. The reservoir's storage capacity becomes a limiting factor at some diversion rate greater than 25 mgd. The 100 mgd diversion provides little additional yield beyond the 50 mgd diversion, and only under the lowest minimum flowby conditions.

3. Although the seven-day, ten-year low flow and monthly 90 percent exceedance values were modeled for reservoir yield comparison, both requirements permit withdrawals to occur during drought conditions. Neither of these flow conditions would be preferable as a minimum flowby requirement due to the resulting stressed state of the aquatic habitat.

4. The median monthly flows were modeled as an upper boundary of potential MIF scenarios. As evidenced by modeling, maintaining median monthly flows in the Pamunkey does not provide enough sustainable yield for the County's projected needs. This MIF is also deemed unacceptable.

5. The extreme MIF scenarios, numbered 1, 4 and 8, should be eliminated from further consideration, reducing the number of conditions to be considered for final recommendation to five. Within this range of MIF's a yield ranging from 19 to nearly 24 mgd could be expected for the 25 mgd diversion rate.

The final selection of an MIF requirement for the Pamunkey River for the purposes of the Crump Creek Reservoir should be based upon several selection criteria. First, the MIF should be based on all known instream and offstream uses of the Pamunkey River. Neither should be sacrificed to support the other. Therefore, the selected MIF should have minimal impact on the water supply needs of the area while maintaining adequate flows to support the instream aquatic habitat. Secondly, the MIF must conform to anticipated state policy. Although the Virginia Water Control Board has ranked Minimum Instream Flowby Regulation as one of its uppermost priorities, presently no final decisions have been made. In order to adhere to the Crump Creek Project's permit application schedule, a decision on the Pamunkey River MIF will be needed imminently by Hanover County. Forthcoming negotiations with the state and federal agencies should result in agreement on an optimum value of minimum instream flowby for all Pamunkey River users.

TABLE 1.—Monthly Comparison of Pamunkey River Flows (cfs)[a]

Month	Median Monthly	70% Exceedance	85% Exceedance	90% Exceedance
JANUARY	919.0	651.0	474.0	343.0
FEBRUARY	1080.0	821.0	586.0	502.0
MARCH	1270.0	928.0	690.0	591.0
APRIL	979.0	722.0	526.0	464.0
MAY	676.0	477.0	346.0	302.0
JUNE	405.0	267.0	193.0	170.0
JULY	252.0	162.0	122.0	110.0
AUGUST	229.0	137.0	96.0	80.0
SEPTEMBER	192.0	122.0	78.0	68.0
OCTOBER	229.0	149.0	85.0	72.0
NOVEMBER	421.0	284.0	173.0	155.0
DECEMBER	675.0	443.0	303.0	228.0

[a] $1 \text{ cfs} = 0.028 \text{ m}^3\text{s}$

TABLE 2.—Comparison of Reservoir Yields (mgd)[d]

MINIMUM INSTREAM FLOWBY SCENARIO	DIVERSION CAPACITY			
	10 mgd	25 mgd	50 mgd	100 mgd
1. 7-day, 10-yr Low Flow[a]	14.4	28.2	50.0	61.4
2. Tennant 40/20[b]	11.7	20.7	30.3	30.3
3. Tennant 30[c]	11.2	19.1	28.2	28.2
4. Tennant 20[b]	12.4	22.4	32.0	33.4
5. Median Monthly Flow	8.5	13.0	14.1	14.1
6. 70% Exceedance Flow	11.1	19.1	25.4	28.2
7. 85% Exceedance Flow	13.7	26.0	43.4	47.3
8. 90% Exceedance Flow	12.9	23.7	35.4	40.8

[a] $7Q10 = 57.4 \text{ cfs} = 1.607 \text{ m}^3\text{s}$
[b] Tennant 40 = 40% Mean Annual Flow = 400 cfs = 11.2 m^3s
Tennant 20 = 20% Mean Annual Flow = 200 cfs = 5.6 m^3s
[c] Tennant 30 = 30% Mean Annual Flow = 300 cfs = 8.4 m^3s
[d] $1 \text{ mgd} = 0.018 \text{ m}^3\text{s}$

APPENDIX.—REFERENCES

1. Camp Dresser & McKee, "Minimum Instream Flow Study: Final Report," for the Commonwealth of Virginia State Water Control Board, 1986.
2. Hyer, P.V., Fang, C.S., Ruzecki, E.P., Hargis, W.J.Jr., "II. Studies of the Distribution of Salinity and Dissolved Oxygen in the Upper York System," Virginia Institute of Marine Science, 1971.
3. Kuo, A.Y. and Fang, C.S., "A Mathematical Model for Salinity Intrusion," Virginia Institute of Marine Science Contrib. No. 475 to Proceedings of the 13th Coastal Engineering Conference, 1972.
4. Camp Dresser & McKee, "Engineering Feasibility Study of a Water Supply Management Program for the Pamunkey River," for the Pamunkey River Water Study Committee, 1984.

Alternative Formulations for Water Quality Management Models

Donald H. Burn*

Various model formulations for water quality management are
examined herein, each of which involves a somewhat different approach
to the problem of determining optimal treatment levels at various
pollution sources. The different model interpretations are of value
to a decision maker in that the decision maker is better able to
choose a solution for implementation which incorporates the pertinent
aspects of the problem.

Introduction

With an ever-increasing burden being imposed on the assimilative
capacity of natural water bodies, the problems of managing water
quality have become more important. It is therefore essential that
due attention be focussed on the implications to a water body of the
implementation of a water quality management scheme.

The role of the water quality modeler is to present to a decision
maker a number of different options or schemes from which the decision maker can choose an alternative for implementation. Since the
decision maker is acting as a representative of society, it is a part
of this individual's task to ensure that the concerns of all members
of society are considered during the decision making process. Thus
the divergent views of environmentalists, industrial dischargers,
government agencies, and the general public must all be given due
consideration. The task of the decision maker can be made easier if
the modeler has provided an array of potential solutions with sufficient diversity to afford the decision maker a degree of latitude in
selecting the final solution. To enhance the effectiveness of the
decision maker, the water quality modeler should reflect the inherent
trade-offs of the waste allocation problem within the model formulations used to generate the water quality plans.

The emphasis in this paper will be on the generation of alternative management options and the quantification of the trade-offs
which are an intrinsic part of the waste allocation problem. The
focus herein will be on determining the optimal releases to allow at
a series of discharge locations in order to maintain an acceptable
level of water quality in a water body. Issues to be resolved are
the definition of an optimal release and the determination of what
constitutes an acceptable level of water quality.

* Assistant Professor, Department of Civil Engineering, University of
Manitoba, Winnipeg, Manitoba, Canada, R3T 2N2.

Modeling Issues

The traditional model formulation used for waste allocation has
been a linear programming (LP) model with the minimization of treatment costs as the objective (see, for example, Loucks et al. [1967]).
The minimization of treatment costs, subject to the attainment of a
minimum water quality standard, can be regarded as a surrogate for
the real objective of the allocation process which should incorporate
a consideration of the benefits resulting from a given plan as well
as the costs associated with the implementation. Ideally, one should
consider the maximization of net societal benefits as a true objective which, however, is probably not attainable. Net social benefits
maximization would consider all the benefits to society of an improvement in the water quality of a water body. Social benefits
include economic benefits such as increased aquaculture and fishing
activity and increased potential for recreational activities as well
as improvements in the aesthetics of the water body and the issue of
the existence value derived from a cleaner environment. The costs
associated with the achievement of the above-noted benefits would be
paid by society either directly or indirectly (for example, through
taxes or increased prices for consumer goods. The difficulties in
quantifying values for the benefits outlined above have often resulted in the use of cost minimization as the objective.

An additional concern for a water quality modeler is the uncertainties which are a part of the water quality process. By including
a measure of uncertainty in the water quality model, not only is the
inherent randomness of the water quality process considered, but by
using other than expected values for water quality parameters, the
risk adverse nature of the decision maker can be incorporated into
the decision making process. Ideally, a water quality planning model
should also be capable of considering equity or fairness in the
allocation of the treatment burden. The goal of equity is to ensure
a sense of fair play and to invoke some degree of sharing of the
total treatment cost. Equity can thus be used as a means of preventing an unreasonable distribution of the treatment costs so that the
competitive balance between industries is not upset.

While all of the above concerns cannot likely be included in a
single waste allocation model, by considering at least some of the
issues in each model selected, it should be possible to illustrate
the trade-offs which exist between the conflicting goals and objectives. The model formulations outlined below are presented with the
intent of elucidating the above-noted trade-offs.

Model Formulations

An allocation model with cost minimization as the objective
results in a linear programming formulation of the form:

minimize $c^T X$ (1)

subject to the constraints

$$\sum_{i} W_i (1 - X_i) \, d_{ij} + D_j \leq S_j \qquad \forall \; j \tag{2}$$

$$X \; \epsilon \; XS \tag{3}$$

where the vector X is the decision vector, and each X_i, $i = 1$, n specifies the required removal at source i, where there are n sources. The vector C contains unit costs associated with a given treatment level; W_i is the pollutant input to the treatment facility at source i; D_j is the pollution arising from the uncontrolled sources at stream location (receptor) j; S_j is the water quality standard at location j; and d_{ij} is a transfer coefficient giving the water quality response at location j as the result of a release of a pollutant unit at location i. The transfer coefficient is thus a function of the river flow, velocity, travel time, and the reaction coefficients which describe the decay of the pollutant as it moves from source i to receptor location j. The constraint equation (3) states that the solution vector must lie within the space given by XS where the limits of this space arise from upper and lower treatment limits, and any equity constraints imposed. With the model formulation given above, the trade-off between treatment cost and water quality can be illustrated through parametric variation of the water quality standards, S_j.

To include uncertainty considerations, it is possible to consider a chance constrained programming (CCP) formulation [Charnes and Cooper, 1963] which after a certain amount of manipulation will give a model of the form [Burn and McBean, 1985]

$$\text{maximize} \; \sum_{j} \sum_{i} W_i \, d_{ij} \, X_{ij} \tag{4}$$

subject to

$$X_{ij} - X_{ip} = 0 \qquad \qquad \forall \; i,j,p \quad p \neq j \tag{5}$$

$$C^T X \leq C_\ell \tag{6}$$

$$X \; \epsilon \; XS \tag{7}$$

where the transfer coefficients, d_{ij}, and the pollutant loadings, W_i, are now interpreted as random variables. The resulting stochastic programming problem can be converted to an equivalent linear programming problem through the use of duality theory and deterministic equivalence relationships (see Burn and McBean [1985] for details). This model formulation also facilitates an elucidation of the cost/quality trade-off through the budgetary constraint right hand side value, C_ℓ.

A second form of stochastic programming can be obtained by considering the objective to be the minimization of the variance of the quality response. The form of the optimization model is then

$$\text{minimize} \; X^T V X \tag{8}$$

subject to the constraints

$$\sum_i W_i \ (1-X_i)d_{ij} + D_j \leqq S_j \qquad \forall \ j \qquad (9)$$

$$X \ \varepsilon \ XS \qquad\qquad\qquad (10)$$

where V is the variance-covariance matrix of the random variables
(i.e., the product of the transfer coefficient and the pollutant
loading) of the problem, and all other symbols are as previously
defined. The problem formulation given by equations (8) to (10)
constitutes a quadratic programming (QP) problem which can be solved
using available algorithms. The trade-off which can be explicitly
illustrated with this model formulation is between the variability of
water quality response and the mean water quality response. An
implicit trade-off between variability and treatment cost can also be
deduced by determining the increase in cost necessary to attain the
same expected water quality response.

Application

The above-noted model formulations were applied to a problem
involving five pollution sources at which treatment plants are
located, and twelve receptor locations at which the water quality is
of concern. The pollutant loadings to each plant were available and
the transfer coefficients describing the response at each receptor
for a unit release from each source have also been obtained. A lower
treatment limit at each source of 80% removal was selected and the
upper bound on treatment was set at 95% removal. The problem facing
the water quality modeler was to determine the pollutant removal to
prescribe at each of the sources in order to maintain an acceptable
level of water quality at the receptor locations. Further details
concerning the problem described herein can be found in Burn and
McBean [1986].

Each of the three model formulations presented above was applied
to the example problem outlined. For each of these formulations, it
was possible to illustrate trade-offs which exist between conflicting
attributes of the management problem (e.g., minimizing treatment cost
versus maximizing the expected water quality response). The result
from the application of each model was thus a transformation surface
depicting a trade-off relationship which can be explicitly elucidated
with that particular model formulation. The transformation surfaces
were different for the three models since each model has a different
objective function as well as differences in the constraint set
included. In the present example, none of the models included equity
constraints although these could have easily been added without
materially affecting the computational burden required to find a
solution. The likely impact of including equity constraints will be
discussed in a subsequent section.

Presented in Table 1 is a summary of the results obtained from the
application of the three models. Results are presented for two
different treatment budgets corresponding to expenditures of 0.40

cost units and 0.23 cost units. In the table, the five pollution
sources are denoted by the letters A–E and the model formulations are
abbreviated as MC for the minimum cost LP model, CCP for the chance
constrained programming model, and MV for the minimize variance QP
problem. The required pollutant removal at each source is indicated
along with the critical dissolved oxygen (D.O.) value at the receptor
with the poorest water quality.

Table 1
Required Treatment for Different Model Formulations

Budget Allocation	Model	Critical D.O. (mg/1)	Removal at Sources				
			A	B	C	D	E
0.40	MC	6.3	.81	.95	.95	.95	.87
	CCP	6.2	.95	.95	.95	.95	.80
	MV	6.0	.95	.95	.94	.95	.92
0.23	MC	4.2	.80	.90	.92	.92	.80
	CCP	4.1	.80	.90	.90	.95	.80
	MV	3.0	.95	.95	.86	.91	.80

Discussion

From the results presented above, it is apparent that a similar
budget allocation (representing the amount spent on pollutant removal
by all sources) can result in different water quality responses as a
function of the model used. As expected, the greatest economic
efficiency is obtained through the use of the minimum cost formulation. When uncertainty is explicitly included in the model, as in the
CCP formulation or the minimize variance model, the effect of this
inclusion is observed directly through a decrease in the mean water
quality response. While the minimize variance model results in the
poorest expected quality of the three formulations, it is also the
model which can guarantee that the variability of the water quality
response will be less than for the other two models. It is further
noted that the differences in the model results are much more pronounced at the lower budgetary level (i.e., lower water quality)
since at the higher budget level, the water quality response at the
limiting receptor is approaching the theoretical upper bound. This
implies that the solution space defined by the constraint set is
comparatively small, as can be deduced from the fact that all sources
are removing at close to the upper limit for treatment.

The implications of the results to the individual sources is also
of interest. Each source can be expected to favour the model which
would result in the lowest removal at that source. This would mean

that source A would favour a minimum cost formulation, or also the CCP model if the budget allocation is at the lower value. This source would definitely not want the MV model implemented. Source B would be indifferent to the model selection except at the lower budget allocation where the MV model would not be favoured. Source C, in contrast, would favour the MV model at the lower budget level and be indifferent otherwise. Source D would like to discourage the use of CCP at the lower budget level but would otherwise be indifferent to the selection of model formulation. Finally, source E would prefer CCP at the higher budget and be indifferent otherwise. Thus if the decision maker is subjected to lobbying, it is possible that groups of sources could form coalitions to try to influence the model choice. An example would be a group consisting of sources A, B, and D, which would advocate the adoption of a minimum cost model.

A final consideration for discussion is the impact that including equity constraints would have on the model results. Adding a constraint to optimization models of the form utilized herein can never result in an increase in the objective function value so it is to be anticipated that equity constraints would decrease the economic efficiency. The decreased efficiency is a penalty incurred in return for the greater fairness in the treatment burden imposed on the individual sources.

Appendix I - References

Burn, D.H., and E.A. McBean, "Optimization modelling of water quality in an uncertain environment", Water Resources Research, 21(7), 934-940, 1985.

Burn, D.H. and E.A. McBean, "Linear stochastic optimization applied to biochemical oxygen demand - dissolved oxygen modelling", Canadian Journal of Civil Engineering, 13(2), 249-254, 1986.

Charnes, A., and W.W. Cooper, "Deterministic equivalents for optimizing and satisficing under chance constraints", Operations Research, 11(1), 18-39, 1963.

Loucks, D.P., C.S. ReVelle, and W.R. Lynn, "Linear programming models for water pollution control", Management Science, 14(4), 8166-8181, 1967.

A Water Quality Model For Urbanized Drainage Basins

C. Y. Kuo[1] , G. V. Loganathan[2] , K. J. Ying[3] and S. P. Shrestha[4]

A Best Management Practices (BMPs) sub-model is incorporated into a continuous version of the Illinois Urban Drainage Area Simulator (ILLUDAS). The model generates pollutographs for three water quality parameters namely, suspended solids, settleable solids, and BOD. Box's complex algorithm is used to find the optimal locations and sizes of the detention devices. Effects of detention ponds and infiltration trenches on water quality parameters can be examined. The model has been applied to a watershed in norhtern Virginia.

Introduction

Non-point source pollution is a major urban-runoff problem faced by many urbanized areas. It has been estimated that the total amount of pollutants contributed to receiving waters by storm-sewer systems is of the same order of magnitude as that released by secondary sewage treatment facilities (Field, 1972). Heaney and Huber (1984) observed significant impacts on DO levels downstream of urban areas during wet weather conditions. The non-point pollution affects the water quality in streams, freshwater impoundments, and coastal waters. This concern has brought in efforts to develop Best Management Practices (BMPs) as pollution control mechanisms. But the implementation of BMPs for stormwater management is complex in developed areas due to space limitations.

The proposed model uses Illinois Urban Drainage Area Simulator (ILLUDAS), developed by Terstriep and Stall (1974), for water quantity simulation. Han and Delleur (1979) extended ILLUDAS for continuous simulation by considering the soil moisture conditions during the intervening dry period of successive storm events. They also modified and added the subroutine DIRT from the model STORM (HEC, 1976) to the original ILLUDAS enabling it to simulate water quality. Wenzel and Voorhees (1980) suggested a procedure for updating the antecedent moisture conditions (AMC) and initial abstractions in a continuous version of ILLUDAS. Noel and Terstriep (1982) extended ILLUDAS to account for the spatial variation of surface runoff and rainfall abstraction. They further extended it to simulate water quality for continuous simulation.

In the present model the focus is towards the effectiveness of different BMPs, their optimal location and design. Modifications have been made within ILLUDAS for computing hydrographs and pollutographs at desired points in a continuous simulation mode. The optimization scheme takes into consideration the limitations in space to al-

---

[1] Prof., Civil Engg. Dept., Va. Tech, Blacksburg, Va. 24061.

[2] Asst. Prof., Civil Engg. Dept., Va. Tech, Blacksburg, Va. 24061.

[3] Grad. Student, Civil Engg. Dept., Va. Tech, Blacksburg, Va. 24061.

[4] Grad. Student, Civil Engg. Dept., Va. Tech, Blacksburg, Va. 24061.

locate required detention volumes resulting from surcharge conditions to the most suitable locations.

Description of the Model

The present model utilizes the "dust and dirt method" of STORM (HEC, 1976) to compute pollutant accumulation on watershed surface since the cessation of previous storm. The information on street sweeping interval, street sweeping efficiency, and the number of dry days is used to compute the total amount of pollutants available for washoff at the beginning of a storm. The amount of pollutant washed off, during any time interval, is then computed based on first order kinetics (Wanielista, 1978). For each computation interval, the pollutograph ordinates are obtained by dividing the amount of pollutant washed-off by the runoff volume for that interval. The pollutographs at the junctions are combined for each time step, and the resulting concentration is computed based on the assumption of complete mixing.

Wet ponds, extended wet ponds, and infiltration trenches are the three types of BMP structures considered in the present study. *Detention basins* (wet ponds and extended wet ponds) are structures that store stormwater and rely upon solids settling processes to remove sediment and suspended pollutant loadings. These are not major impoundments where biological processes play a prominent role in reducing pollutant concentration within the pond. *Wet ponds* are peak-shaving devices which retain stormwater for a short period of time. They retain a permanent pool of water. *Extended wet ponds*, on the other hand, are relatively large facilities that impound water for a time sufficient to achieve high pollutant removal. An *infiltration trench* is a volume control device, having high pollutant removal efficiency (Biggers et. al., 1980). The infiltration trenches have the potential to control surface runoff and replenish groundwater supplies (Cave, 1986).

The modified Puls method is used for flood routing through the wet and extended wet detention ponds. Complete mixing of pollutants is assumed in the pond for the computation of outflow pollutant concentration. Trap efficiency is assumed to be independent of time and sediment re-suspension is neglected. Settling is the the only process by which pollutant removal takes place. Trap efficiencies for wet ponds for suspended solids, settleable solids, and BOD are taken to be 55, 66 and 22 percent of the total load respectively (USEPA, 1983). For extended wet ponds, these values are taken as 91, 92,and 42 percent (Biggers, 1980). The infiltration trench removes the surface runoff till the trench is full. The pollutants in the removed part of runoff are assumed to be completely trapped by the trench and re-suspension is not taken into account.

An optimization scheme based on the complex method (Box, 1965) is used to optimally size and locate the detention structures within the watershed. ILLUDAS is used to check the feasibility of the solutions generated by the Box's method. The algorithm considers the space limitations at various locations within the watershed and flow limitations through the storm-sewer pipes.

Application and Results

The model was applied to a subbasin of the Holmes Run watershed in Virginia. The study area with the existing storm-sewer system is shown in Figure 1. Suspended solids, settleable solids, and BOD were simulated using a storm of 10 hours duration. For continuous simulation, two cases were considered - (i) two storms of durations 6 hours and 16 hours with an inter-event time of 10 hours, and (ii) two storms of the same durations with an inter-event time of 117 hours. To evaluate the effectiveness of BMP structures, five infiltration trenches and an extended wet detention pond are placed at locations shown in Figure 1. Time variation of suspended solids, settleable solids, and BOD concentrations with and without BMP structures (for single storm event) are shown in Figures 2, 3, and 4. Reduction in peak concentrations are shown in Table 1. Hydrographs and suspended solids variation for continuous simulation are presented in Figures 5 and 6 respectively.

Table 1. Effect of BMP Structures on Pollutant Reduction (Single Storm Event)

Pollutant Type	Peak Concentration (ppm)			Percentage Reduction	
	Without BMPs	With Pond only	With Pond and Trenches	With Pond only	With Pond and Trenches
Sus Solids	10.36	1.80	1.18	82.63	88.61
Set Solids	0.34	0.06	0.05	82.35	85.29
BOD	6.49	3.84	2.83	40.83	56.39

Acknowledgement

This study was supported by the Virginia Water Resources Research Center.

References

Biggers, D. J., Hartigan, Jr., J. P., and Bonuccelli, H. A. (1980). "Urban Best Management Practices (BMPs): Transition From Single-Purpose to Multi-Purpose Stormwater Management", International Symposium on Urban Storm Runoff, University of Kentucky, Lexington, KY, pp. 249-274, July.

Box, M. J. (1965). A New Method of Constrained Optimization and a Comparison with Other Methods, The Computer Journal, 8, pp.42-51.

Cave, K. (1986). "Evaluation of the Effectiveness of BMPs for Urban Stormwater Management: Single-Event Simulation", Masters Thesis, Department of Civil Engineering, Virginia Tech, Blacksburg, Virginia, April.

Field, R., and Struzeski, Jr., E. J. (1972). "Management and Control of Combined Sewer Overflows", Journal of Water Pollution Control Federation, 44(7).

Han, J., and Delleur, J. W. (1979). "Development of an Extension of ILLUDAS Model for Continuous Simulation of Urban Runoff Quantity and Discrete Simulation of Runoff Quality", Technical Report Number 109, Purdue Water Resources Research Center, West Lafayette, Indiana, July.

Heaney, J. P., and Huber, W. C. (1984). "Nationwide Assessment of Urban Runoff Impact on Receiving Water Quality", Water Resources Bulletin, 20, February.

Hydrologic Engineering Center (1976). "Storage, Treatment, Overflow, Runoff Model - STORM", Davis, California, July.

Noel, D. C., and Terstriep, M. L. (1982). "Q-ILLUDAS - A Continuous Urban Runoff/Washoff Model", International Symposium on Urban Hydrology, Hydraulics and Sediment Control, University of Kentucky, Lexington, Kentucky, July.

Terstriep, M. L., and Stall, J. B. (1974). "The Illinois Urban Drainage Area Simulator, ILLUDAS", Illinois State Water Survey Bulletin 58.

U.S. Environmental Protection Agency. (1983). "Results of the Nationwide Urban Runoff Program, Executive Summary", Washington, D. C.: Water Planning Division, December.

Wanielista, M. P. (1978). Stormwater Management: Quantity and Quality, Ann Arbor, Michigan.

Wenzel, Jr., H. G., and Voorhees, M. L. (1980). Adaptation of ILLUDAS for Continuous Simulation, *Journal of the Hydraulics Division*, Proceedings of the ASCE, 106(HY11), November.

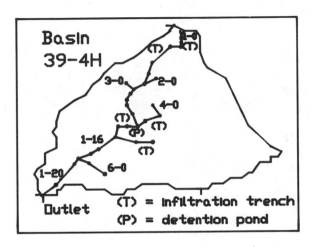

Figure 1. Holmes Run Watershed Sub-basin 39-04H

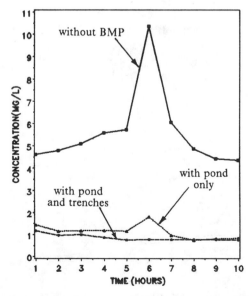

Figure 2. Pollutographs (Suspended Solids) - Single Storm Simulation

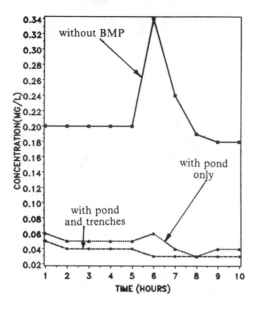

Figure 3. Pollutographs (Settleable Solids) - Single Storm Simulation

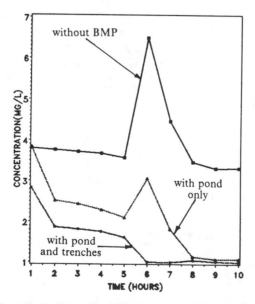

Figure 4. Pollutographs (BOD) - Single Storm Simulation

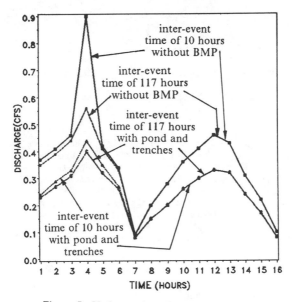

Figure 5. Hydrographs - Continuous Simulation

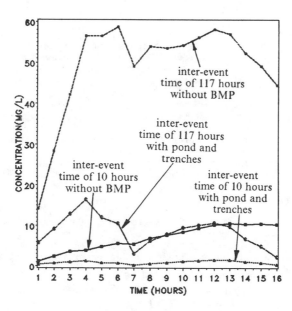

Figure 6. Pollutographs (Suspended Solids) - Continuous Simulation

Stormwater Management Modeling on a Microcomputer

Shaw L. Yu[1], Todd S. Tisdale[2] and Djamel Benelmouffok[3]

Abstract

The Virginia Stormwater (VAST) Model has been developed for use on IBM compatible microcomputers to simulate stormwater runoff and nonpoint source pollution. Hydrologic and nonpoint source pollution modeling techniques employed in VAST are similar to those found in such major models as HEC-1, STORM, and SWMM. VAST is user friendly and has a moderate input data requirement. Computer graphics have been utilized to facilitate presentation of results and to allow interactive model calibration and verification. The model has been successfully tested with data collected for three watersheds near Richmond, Virginia.

Introduction

The VAST model is an event orientated model and can be applied to multiple catchment basins. It combines widely used techniques to (1) compute rainfall abstractions, (2) generate overland flow hydrographs, (3) combine and route outflow from upstream subbasins through the channel downstream, and (4) compute nonpoint source pollution washoff from subbasins.

VAST is composed of three primary and two auxiliary computer programs. The primary programs simulate (1) stormwater runoff, (2) nonpoint source pollutant loadings of suspended solids, settleable solids, BOD, total nitrogen, orthophosphate, and fecal coliforms, and (3) nonpoint source pollution loadings of up to four user specified pollutants. The stormwater runoff program generates basin discharge data that is used by the nonpoint source pollution programs. Auxiliary programs present results generated by the stormwater runoff program and the nonpoint source pollution programs in graphical form on a computer's video monitor. These programs can optionally present field data with simulation results to assist users in the calibration and verification process. Figure 1 illustrates the relationship between programs that comprise the VAST model.

[1] Professor, Department of Civil Engineering, University of Virginia, Charlottesville, VA 22901

[2] Water Resources Engineer, South Florida Water Management District, West Palm Beach, FL 33402

[3] PhD Student, Department of Civil Engineering, University of Virginia, Charlottesville, VA 22901

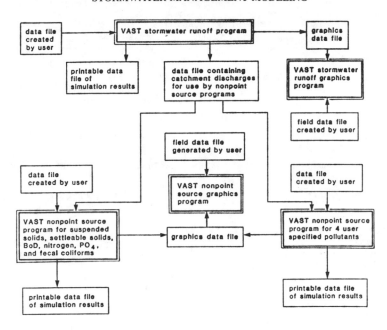

Figure 1.   Relationship Between Programs that Comprise the VAST Model

Modeling Techniques

VAST treats multi-catchments as binary tree structures, in which a downstream subbasin can receive inflow hydrographs from a maximum of two upstream subbasins. By using this basic binary tree structure in a recursive manner, a complex binary tree representation of the entire catchment basin can be developed. PASCAL was used in writing the model because it supports recursion.

Another useful feature of PASCAL is the linked list. This is a data structure whose elements can be created and disposed of during program execution. Therefore, linked lists provide an efficient means of storing data during program execution, and the only data storage limitation is the size of computer memory.

Hydrologic and nonpoint pollution modeling techniques used by VAST are presented in Table 1.

Table 1. MODELING TECHNIQUES USED IN VAST

PROCESS	TECHNIQUE
River Routing	Muskingum Method
Determination of Net Rainfall	(A) Phi Index (B) HEC-1 Rate Loss Function (C) Horton's Formula
Determination of Catchment Runoff	Unit Hydrograph and Net Rainfall
Determination of Unit Hydrograph	(A) Read from Input (B) Clark Method
Pollutant Accumulation on the Land Surface	Technique Used by STORM
Pollutant Washoff	Modification of Technique Technique Used by STORM

Descriptions of the hydrologic techniques can be found, for example, in Viessman et al (1977) and the HEC-1 User's Guide (Hydrologic Engineering Center, 1980). Nonpoint pollution modeling techniques are those described in STORM (Hydrologic Engineering Center, 1977). A procedure was written for each technique, making it easier to incorporate additional, or alternate, techniques into the program.

Running VAST on a Microcomputer

A work diskette for running VAST contains all the VAST object codes and an error message module and support programs for the graphics module.

Once VAST is executed, the user will be prompted to name the input and output files and will begin the hydrologic simulation with given data in the input data file. The program will keep the user informed of its progress by printing messages on the video monitor. While reading from the data file, the program will check for errors in the file, and if it detects any errors, the program will inform the user of the type of error it has found.

When the program finishes executing, it will have created three additional files. The first is the output file which can be printed. The second is a graphics file to be used for producing screen graphics plots. The graphics program is also capable of plotting field data together with simulation results to aid users in the interactive calibration and verification process. The third file is used by the nonpoint source pollution program, which is executed in much the same manner as the runoff program. VAST can accommodate time units of hours, minutes, or seconds.

Although source codes for the various VAST programs are provided on the VAST diskette, they can only be used to generate object codes by using TURBO PASCAL and TURBO GRAPHIX TOLLBOX, both available from Borland International Inc.

VAST Model Testing

The VAST model has been tested using stormwater quantity and quality data collected for three watersheds near Richmond, Virginia. The results are quite satisfactory, as examplified by a flow calibration run for Pocoshock Creek shown in Figure 2.

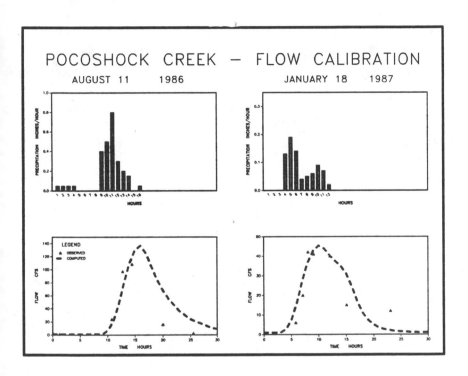

Figure 2

Summary and Conclusions

VAST, written in PASCAL, is an event hydrologic and nonpoint pollution model developed for use on IBM, and IBM compatible, micro-computers. The model uses algorithms found in major models such as HEC-1, STORM, and SWMM. It is user friendly and has graphics capability which permits interactive model calibration and verification. The model has been successfully tested and should prove to be useful for local agencies in their stormwater management studies.

Acknowledgement

Data used for testing the VAST model were collected by the Richmond Regional Planning District Commission under a grant from the Virginia State Water Control Board as part of the EPA 205 (j) program.

Appendix - References

1.   Hydrologic Engineering Center, U.S. Army Corps of Engineers, "HEC-1 Flood Hydrograph Package," Davis, California, 1980.

2.   Ibid, "Storage, Treatment, Overflow, and Runoff Model," Davis, California, 1977.

3.   Viessman, W., J. W. Knapp, G. L. Lewis, and T. E. Harbaugh, "Introduction to Hydrology," Harper and Row, New York, 1977.

WATER QUALITY MANAGEMENT THROUGH PROJECT OPERATION

by Walter M. Linder,* M. ASCE

Abstract

Harry S. Truman Dam, a multipurpose reservoir project, is
located on the Osage River in the headwaters of the Lake of
the Ozarks. Construction of this project created the potential to both improve and degrade the water quality in the
Lake of the Ozarks. This paper describes the features of
the project, the development of a water quality monitoring
program, and operational problems that have occurred since
closure of the embankment in July 1977. Major fish kills
initially occurred in the tailwaters as a result of supersaturation, while minor fish kills have occurred as a result
of low dissolved oxygen levels. Low dissolved oxygen levels
in the downstream gate wells has also been a problem.
Structural modifications and adjustments in project operations have significantly reduced downstream water quality
problems without adversely affecting project performance or
authorized purposes.

Introduction

Harry S. Truman Dam and Reservoir includes hydropower, flood
control, and recreation as project purposes. The Osage
River immediately downstream of Truman Dam forms the headwaters of the Lake of the Ozarks, one of the most highly
developed recreation areas in the midwest. Since completion
of Truman Dam, an excellent warm water fishery has developed
in the tailwaters below the dam. As a result, any downstream water quality problems caused by the operation of
Truman Dam can have serious adverse impacts.

Project Features

Truman Dam controls 11,500 square miles (2,978,500 hectares)
of drainage area, of which approximately 1,600 square miles
(414,000 hectares) are controlled by five upstream reservoirs. The storage capacity of the multipurpose pool is
1,040,000 acre-feet (1.283 x 10⁹ m³) with a surface area of
55,600 acres (22,500 hectares). Approximately 3,918,000
acre-feet (4.834 x 10⁹ m³) of storage are available for
flood control. Depth below the multipurpose pool to the

*Chief, Hydrologic Engineering Branch, U.S. Army Corps of
Engineers, Kansas City District, 700 Federal Building,
Kansas City, Missouri 64106-2896

valley floor at the dam is 46 ft (14 m). Depth to the
invert of the power of the power intakes is 103 ft (31.4 m).

The power plant consists of six reversible slant type pump
turbines. Discharge from power generation can be as high as
65,000 cfs (2,840 m³/s), and 27,500 cfs (780 m³/s) can be
discharged during pumping operations. Only a 2 ft (.6 m)
increment of multipurpose storage was specifically provided
for power generation, as the intent was to rely on pumpback
for firm power capacity. A large fish kill during pumpback
testing resulted in a moratorium on pumping after extensive
studies found no practical available technology for fish
protection. A four-bay gated spillway is available for
flood control releases in excess of power plant capacity.
This spillway capacity varies from 26,000 cfs (736 m³/s) at
multipurpose pool level to 284,000 cfs (8,042 m³/s) at the
top of the surcharge pool.

Supersaturation

Closure of the river channel was accomplished in July 1977,
and flow was diverted through the uncompleted spillway.
Completion of the spillway to its final elevation could not
be started until the closure section embankment had reached
an elevation that would contain the flood of record. A
major fish kill occurred downstream of the dam in the spring
and early summer of 1978 during the passage of moderate
flood flows through the uncompleted spillway. The cause of
the fish kill was determined to be supersaturation. Saturation levels immediately downstream of the dam reached a
maximum of 143 percent and levels above 115 percent were
observed as far as 60 miles (96 Km) downstream in the Lake
of the Ozarks.

Hydraulic model studies at the Waterways Experiment Station
showed that the addition of a 7 ft (2.1 m) wide deflector or
"flip lip" on the downstream face of the spillway a few feet
below the normal level of the Lake of the Ozarks would
provide the best long-term solution. Construction of the
deflectors was completed in November 1978 and work was
started on raising the spillway to its final elevation.
However, severe winter weather and high flows limited the
amount of work that could be accomplished. High flows in
April 1979 resulted in downstream supersaturation levels of
138 percent and another major fish kill occurred. Supersaturation occurred in this instance because the spillway
elevation was still too low for the deflectors to operate
properly. Construction of the spillway was completed in
October 1979 and the gates were closed to initiate impoundment of the reservoir.

High inflows in 1980 and 1981 required significant spillway
releases as none of the power units were as yet available.
The deflectors were not totally effective as they were

designed to operate with tailwater levels produced by combined flows from the power plant and spillway. However, by
limiting spillway discharges between 25,000 and 30,000 cfs
(708 and 850 m³/s) supersaturation levels were held within a
range of 118 to 123 percent, well above the accepted standard of 110 percent. Even though supersaturation remained at
that level for a period of approximately 6 weeks, few if any
fish were killed as a result of supersaturation.

In October of 1986 record rainfall in the Osage River basin
filled the reservoir to within a foot (0.3 m) of the top of
the flood control pool. Combined releases of approximately
70,000 cfs (1,982m³/sec), 30,000 cfs (850 m³/sec) from the
power plant and 40,000 cfs (1,121 m³/sec) through the
spillway, were required for a 2-week period. Downstream
supersaturation levels during this time did not exceed 108
percent, demonstrating the effectiveness of the flip lips in
preventing supersaturation.

Skimming Weir

Thermal simulation studies were conducted in 1972 and again
in 1976 and 1977. These studies were intended to predict
the degree of thermal stratification in Truman Reservoir and
the effect of power releases and pumpback on downstream
water temperatures. The studies concluded that the natural
overbank would act as a broad crested weir if no channel was
excavated between the power intakes and the natural river
channel. Any stratification that occurred would be confined
to the former river channel, and the portion of the reservoir above the flood plain would be essentially homogeneous.
Releases would be somewhat cooler than natural flows in the
spring, nearly the same in the summer, and somewhat warmer
in the fall. No attempt was made to predict dissolved
oxygen (DO) levels in the reservoir.

Temperature and dissolved oxygen data collected in the
reservoir in 1977 after closure of the river channel and
again in 1978 and 1979 showed thermal stratification to be
similar but stronger and more classical than predicted by
the thermal model. A potentially more serious problem was a
very severe oxygen depletion in the mid-levels of the
reservoir with anaerobic conditions near the bottom.

Fill material for the embankment closure section, which was
located at the left abutment, had to be obtained from borrow
areas located upstream of the dam on the right side of the
valley. In order to transport this fill material to the
embankment closure area, a haul road was constructed about
800 ft (244 m) upstream of the spillway and powerhouse.
This haul road was built to an elevation about 3 ft (1 m)
below the multipurpose pool elevation. The opening in the
haul road for the diversion channel to the spillway and
powerhouse was spanned with a bridge.

When it became apparent that water with low DO levels would be released during power operations, a decision was made to remove only the upper portion of the haul road and to fill in the bridge opening. The degraded haul road would act as a skimming weir and provide warmer water with higher oxygen levels for power releases. Observed data suggested an elevation 15 to 20 ft (4.6 to 6.1 m) below the multipurpose pool. However, construction delays and a rapidly rising pool resulted in a final elevation 13 ft (4 m) below the multipurpose pool. Experience has shown the higher weir crest provides better selective withdrawal than would have occurred at the lower elevation.

Skimming Weir Performance

Field data show the skimming weir is effective in limiting the release of water with a low DO content during periods of reservoir stratification. Temperature, DO, and velocity profile data have been collected in the reservoir, the vicinity of the weir, and near the power intakes. Data have been obtained with one through five power units in operation and at reservoir elevations ranging from near the multipurpose pool level to approximately 13 ft (4 m) above the multipurpose pool. During these measurements conditions in the reservoir have ranged from a highly stratified situation with the thermocline and oxycline near the weir crest to a weak stratification with depressed DO levels near the crest and near anoxic conditions 10 to 15 ft (3 to 4.5 m) below the crest. Surface DO levels have been in the range of 6 to 8 mg/l. The data show that some water with reduced DO levels will be drawn over the weir. However, temperature and DO profiles obtained between the weir and the powerhouse intakes show that a significant amount of mixing occurs. As a result the water passing through the power units has an acceptable DO level as indicated by measurements in the downstream channel where DO levels have ranged from slightly below 5 mg/l to nearly 6 mg/l. Had it not been for the skimming weir, releases would likely have been in the 1 to 2 mg/l range on numerous occasions.

Low Dissolved Oxygen in Tailwaters

The first instance of low DO levels in the tailwater area occurred in July 1981 when spillway releases were stopped to reduce flooding on the lower Missouri River. Tailwater DO levels fell rapidly from 8 mg/l to about 1.5 mg/l in 6 to 8 hours. DO levels remained well below the desired 5 mg/l until releases were resumed about 40 hours later. A similar situation occurred in mid-August of 1981 when spillway flows were shut down to start testing a power unit. The power unit was shut down and remained off line about 36 hours. Spillway flows were not resumed because of little or no water in the flood control pool. DO levels again fell to between 2 and 3 mg/l until the power unit was restarted. A minor fish kill occurred in the tailwater area in both

instances. Several days later the power unit was again shut
down. This time a spillway gate was opened to discharge a
small quantity of water. Flows were varied from 500 to 250
cfs (14 to 7 m³/sec), but it was found that a minimum of 500
cfs (14 m³/sec) was required to maintain DO levels above 5
mg/l. No explanation was found for the rapid decline in DO
levels during periods of no flow. It is theorized the fish
population in the tailwater area was so great that the
available oxygen supply was depleted. Since 1981 similar
situations have not been observed.

One of the more persistent water quality problems has been
low DO levels in the downstream gate wells. This situation
frequently develops about mid-summer and continues through
September or until significant reservoir inflow occurs. It
generally develops during an extended period of warm weather
with low reservoir inflows and infrequent power releases.
Fish enter the gate wells through the gate slots in the top
of the draft tubes. After a period of no generation the DO
level in the gate wells drops sufficiently for the fish to
die. On resuming generation, the dead fish are flushed into
the tailwaters and create an adverse public reaction even
though only a small number of fish may be involved. Various
methods have been tried to maintain satisfactory DO levels
in the gate wells. Air bubbling will maintain DO levels,
but will not increase depleted levels. Operating a unit at
speed no load or at partial load will improve DO levels in
adjacent gate wells but only for a brief period of time.
The most satisfactory method has been a continuous spillway
release of about 500 cfs (14 m³/sec). Lessor amounts have
not been adequate to maintain desired DO levels in all of
the gate wells. This has not been entirely satisfactory to
power interests as this water normally comes from the power
pool and does not generate power in the process.

Water Quality Instrumentation and Monitoring Program

An extensive instrumentation and water quality monitoring
program has been developed to detect potential problems and
provide data for making operating decisions relative to
water quality.

Data concerning the quality of water discharged is obtained
automatically by electronic monitoring equipment installed
at a location about 2,000 ft (610 m) downstream of the
powerhouse. This equipment includes (1) an acoustic velocity meter which continuously measures flow velocity and
computes the discharge, (2) a bubble type stage recorder,
(3) water quality monitoring equipment, and (4) a meteorologic station. The water quality monitoring equipment consists of a submerged pump which continuously circulates
water from the discharge channel through sampling chambers
where sensors measure temperature, dissolved oxygen, pH,
conductivity, turbidity, and total gas pressure relative to
atmospheric pressure. The later provides an indication of

saturation levels. All data except velocity and discharge
are collected hourly by a data collection platform and
transmitted every 4 hours via Geostationary Environmental
Satellite to the Missouri River Division downlink facilities
in Omaha, Nebraska. These data are then obtained by the
Kansas City District Water Control computer facilities for
use in making real time operating decisions. DO sensors have
been installed in the downstream gate wells with meter readout in the powerplant control room. The meters are read
hourly when the downstream water temperature reaches and
remains above 21° C. When DO levels approach 4.0 mg/l action
is taken to improve DO levels. This is accomplished by
either a short period of generation or more often by a
continuous small spillway release.

Routine monthly sampling is conducted at three sites in the
vicinity of Truman Dam. These are: (1) the reservoir about
1 mile (1.6 Km) upstream of the dam, (2) the tailwater area,
and (3) in the Lake of the Ozarks about 5 miles (8 Km) below
Truman Dam. At each location, water samples are collected
from the surface and also near the bottom. These samples
are analyzed on site in a mobile laboratory to determine
levels of calcium, chloride, alkalinity, hardness, turbidity, and pH. Other samples are preserved and sent to a
laboratory for a complete chemical analysis. In addition to
obtaining water samples, complete water temperature and DO
profiles are obtained. Weekly water temperature and DO
profiles will be collected at two locations in the reservoir
by on site project personnel during the summer of 1987.
More extensive sampling of the entire reservoir has been
conducted on a periodic basis and during times when unusual
water quality conditions occurred in the reservoir.

Summary and Conclusions

During the 9 years that have elapsed since closure of Truman
Dam several water quality related problems have occurred.
These have been corrected either by structural modification
or by adjustments in operation. Supersaturation has been
significantly reduced by the addition of deflectors on the
downstream face of the spillway and by limiting the amount
of spillway discharges during completion of power facilities. Adaptation of a haul road to serve as a skimming weir
has prevented the release of water with low DO levels during
power generation. Depressed DO levels downstream and in the
gate wells has been corrected by small spillway releases
during extended periods of no power generation. Extensive
instrumentation and periodic monitoring of water quality
conditions provide data needed to make real time operating
decisions whenever necessary.

SPATIAL ATTENUATION OF MAJOR PENNSYLVANIA STORMS

David J. Wall, M. ASCE[1], Gert Aron, M. ASCE[2],
David Kotz, Student Member[3]

ABSTRACT

Design storms are determined by performing frequency analyses on
historic raingage data. These estimates of rainfall, obtained for
specific return periods and durations, can only be considered point
estimates, representative of conditions in the immediate vicinity of
the gage. To utilize these point estimates for more expansive areas,
reduction factors are necessary to account for the attenuation of
storm magnitude with increasing area. The spatial attenuation of
major storms in Pennsylvania was investigated using rainfall data
obtained from the National Climatic Data Center for the statewide
network of raingages and from the Agricultural Research Service for
the dense network of raingages on the Mahantango experimental
watershed in east central Pennsylvania. Individual storms of 1, 2,
3, 6, 12 and 24 hour durations were analyzed using a Thiessen polygon
weighting technique. The results of the individual storm analyses
were then averaged to obtain relative storm magnitude versus area
relationships for two levels of gage network density. The
relationships developed are more representative of rainfall patterns
in Pennsylvania and tend to yield more conservative reductions in
relative storm magnitude than similar curves presented in TP-40.

INTRODUCTION

Reliable estimates of expected rainfall are necessary for engineering
analysis and design. Design storm estimates can be obtained by
performing frequency analyses on raingage data or from published
reports. The most widely used source of design rainfall estimates in
the U.S. is the Rainfall Frequency Atlas of the United States,
Technical Paper No. 40, TP-40 (Hershfield, 1961). TP-40 contains
approximately 50 maps with contours of rainfall depth for durations
of 30 minutes to 24 hours and return periods of 2 to 100 years. The

---

[1]Assistant Professor of Civil Engineering, The Pennsylvania State
University, University Park, PA 16802

[2]Professor of Civil Engineering, The Pennsylvania State University,
University Park, PA 16802

[3]Graduate Student Department of Civil Engineering, The Pennsylvania
State University, University Park, PA 16802

maps, in general, consist of a few widespread contours and lack the resolution needed to identify areas of characteristically high or low rainfall.

TP-40 design rainfalls are considered point estimates, representative of rainfall in the immediate vicinity of the gage. To apply these estimates to larger areas, a set of relative storm magnitude curves, which relate the percent of point rainfall to area, is provided. The curves are duration dependent and show that the percent of point rainfall, relative storm magnitude, decreases as area increases until some terminal value is reached at an area of 400 sq. miles (1036 km^2) or less. Like the rainfall maps, the relative storm magnitude curves are applicable to all parts of the country and suffer the same problem of low resolution.

In this paper, the validity of the TP-40 relative storm magnitude curves for Pennsylvania was investigated. Using a data base of raingage information for Pennsylvania, relative storm magnitude versus area curves were developed and compared to the TP-40 relative storm magnitude curves. This study was part of a general rainfall study conducted for the Pennsylvania Department of Transportation (Aron et al. 1986). The main objective of which was the development of intensity-duration-frequency curves for Pennsylvania.

RAINFALL DATA

The primary source of data was the National Climatic Data Center located in Ashville, North Carolina. Hourly rainfall totals in units of 0.01 inch (0.025 cm) per time interval were obtained for 252 gaging stations with record lengths from 2 to 36 years. This data base represented over three million records. To make the data set more manageable, indpendent storms of significant amounts were extracted. An independent storm was defined as an event separated by at least 24 hours of no precipitation. Each event was then scanned for the maximum 1-, 2-, 3-, 6-, 12- and 24-hour precipitation depths. These amounts were considered significant if the total precipitation for a particular duration was greater than or equal to threshold values which corresponded to a 90% exceedence probability in any one year, according to an earlier study (Reich, et al. 1970).

In addition to the National Climatic Data Center data, rainfall data was obtained for the Mahantango experimental watershed located in east central Pennsylvania from the Agricultural Research Service. The Mahantango basin has a drainage area of 200 square miles (518 km^2) and at one time had up to 45 raingages. Rainfall data to the nearest 0.1 of an inch (0.25 cm) was available for the years 1968 thru 1984 at recording intervals of 5- or 15- minutes. Independent and significant storms were extracted from this data set also.

METHOD OF ANALYSIS

The analysis of the reduction in storm magnitude of the independent, and significant storm events for duration of 1⊢, 2⊢, 3⊢, 6⊢, 12⊢ and 24-hours was conducted as follows:

1. For each duration, the data base was scanned and for each event the storm totals at neighboring gages were extracted.
2. For each event, the point of maximum precipitation in the gage network was located.
3. From the point of maximum precipitation, a set of expanding concentric circles was drawn to encompass all of the raingage.
4. Thiessen polygons were constructed for the network of raingages and the polygon areas corresponding to each gage within each circle determined to obtain the area weights.
5. The area weighted average rainfall within each circle was then calculated.
6. The relative storm magnitudes are calculated by dividing the average rainfall value within each circle by the central maximum value.
7. The relative storm magnitude versue area curves were then plotted, for each duration, and the best fit curves constructed.

The above procedure describes a manual process for which a computer program was written to accomplish what would be an overwhelming manual task.

RESULTS

Mahantango Data Base

The analysis of the Mahantango rainfall data, as described, resulted in the relative storm magnitude versus area curve shown in Figure 1. The 3- and 12-hour curves have been omitted for clarity. In comparison to the curves presented in TP⊢40, the Mahantango curves cover a much smaller area, approximately 60 square miles(155.4 km^2) compared to 400 square miles (644 km^2). In the regions where same comparisons can be made, the Mahantango curves show a greater rate of decrease than do the TP-40 curves. For example, for the 1⊢hour storm and an area of 50 square miles (129.5 km^2), the percentage of point rainfall is 80 percent according to TP-40 and about 70 percent according to the Mahantango curves. The curves developed from the Mahantango data also tend to level off and become horizontally asymptotic at smaller areas than do the TP-40 curves.

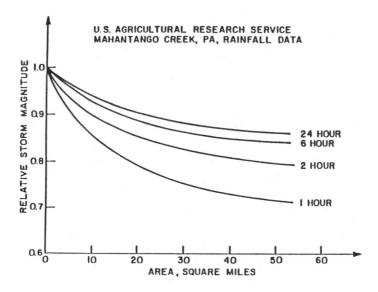

Figure 1   Mahantango Relative Storm Magnitude Curves

National Climatic Data Center Database

The analysis involving the 252 Weather Service gages for Pennsylvania resulted in the relative storm magnitude versus area curve shown in Figure 2. The 2-hour curve fell between the 1- and 3-hour curve and was left off for clarity and the 12- and 24-hour curves were coincident. The Figure 2 curves are flatter and thus more conservative than the TP-40 curves for the first four hundred square miles. For example, for a 6-hour storm and an area of 400 square miles (644 km^2), TP-40 gives a relative storm magnitude of 83 percent while Figure 2 indicates 92%. However, the TP-40 curves start to level off at areas less than 500 square miles (1295 km^2) while the Figure 2 curves continue to decline. At 1000 square miles (2590 km^2) the 24-hour duration value from TP-40 is estimated to be 91 percent, while Figure 2 shows a less conservative 84 percent relative storm magnitude.

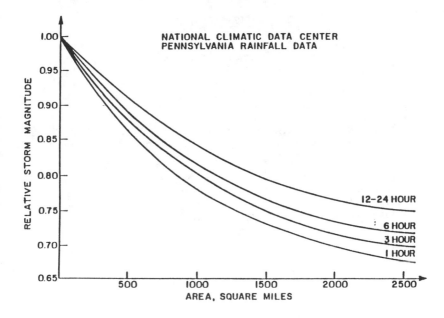

Figure 2   Pennsylvania Relative Storm Magnitude Curves

CONCLUSIONS

The primary objective of this study was to test the validity of the TP-40 relative storm magnitude curves in Pennsylvania. From this study, it appears that TP-40 gives reasonable results. For smaller areas, TP-40 yields more conservative reduction than the curves developed using the Mahantango data base. At areas up to 400 sq. miles (644 km^2), the TP-40 and Pennsylvania rainfall derived curves give similar results. At areas greater than 400 square miles, the TP-40 curves become horizontal and fail to indicate increased reductions with increased area while Pennsylvania curves continue to yield additional reductions up to 2500 square miles (6475 km^2).

The use of threshold values, to reduce the size of the Pennsylvania rainfall data base, has resulted in conservative estimates of relative storm magnitude for Pennsylvania.

REFERENCES

1.   Aron, G., D. J. Wall, E. L. White, C. N. Dunn, D. M. Kotz.,
      (1986). Pennsylvania Department of Transportation Storm
      Intensity-Duration-Frequency Charts (PDT-IDF), Report No.
      FHWA-PA-85-032, Department of Civil Engineering and Institute
      for Research on Land and Water Resources, The Pennsylvania State
      University, University Park, PA.

2.   Hershfield, David M., (1961). Rainfall Frequency Atlas of the
      United State, Technical Paper No. 40, Department of Commerce,
      Washington, D.C., 54 charts.

3.   Reich, B. M., D. F. McGinnis, R. L. Kerr, (1970). Design
      Procedures for Rainfall-Duration-Frequency in Pennsylvania.
      Pub. 65, Inst. for Res. on Land and Water Resources, The
      Pennsylvania State University, University Park, PA.

Design Rainfall for Pennsylvania

Gert Aron, David J. Wall, Elizabeth L. White,
and Christopher N. Dunn[*]

Introduction

Rainfall estimates of acceptable accuracy and reliability are extremely important in any type of hydrologic design, ranging from the simplest rational formula application to the most sophisticated hydrologic modeling project.

The most commonly used source of rainfall estimates is the U.S. Weather Bureau Technical Paper 40 (Hershfield, 1961). This rainfall atlas, which contains roughly 50 maps (see Figure 1) of rainfall contour lines for durations varying between 30 minutes and 24 hours, and return periods between 2 and 100 years, is convenient and simple to use. However, due to their nationwide coverage, the maps necessarily present an image of highly uniform rainfall amounts, with no consideration given to orographic or physiographic features that tend to encourage or discourage intense storm formations.

A supplement to TP-40 was developed jointly between the National Weather Service and the U.S. Soil Conservation Service (NOAA, 1977) in the form of Hydro-35; a rainfall atlas covering the United States east of longitude 105° with 9 maps for durations of 15, 30, and 60 minutes; return periods of 2, 10, and 100 years; and equations to interpolate for intermediate values. These maps suffer from the same lack of resolution as TP-40.

A more detailed study of Pennsylvania storms was conducted in 1970 (Reich, et al., 1970) under the sponsorship of the Pennsylvania Department of Environmental Resources. The report resulting from this study shows some very pronounced areas of large and frequent rainfall amounts. It consists of two principal rainfall contour maps for 2.33-year return period, and durations of 1 and 24 hours, plus a set of graphs and equations to extend the estimates to other durations and return periods.

In 1985, an agreement was reached with the Pennsylvania Department of Transportation to conduct an updated rainfall frequency study, using data up to 1983, and to present the results in the form of charts from which rainfall estimates could be read conveniently for durations between 5 minutes and 24 hours, and return periods between 1 and 100 years. The development of these charts is the main topic of this paper.

---

[*]Department of Civil Engineering, The Pennsylvania State University, **University Park, PA**

Data Sources

   The main source of rainfall data was the National Climatic Data
Center in Ashville, NC. The data, stored on a magnetic tape,
consisted of roughly 3 million hourly records from 153 raingages in
Pennsylvania, spanning the years from 1948 to 1983. This data set was
supplemented by hourly data from 1938 to 1947, acquired by The
Pennsylvania State University Hydrology Laboratory for the 1970 study.
In addition, a tape of 15-minute rainfall records for 45 stations from
1971 to 1984 was also obtained from the national Climatic Data Center.
Finally, rainfall data were supplied by the Agricultural Research
Service, covering the 200 square mile watershed of the Mahantango
Creek, located roughly 30 miles north of Harrisburg, PA. These data
were recorded at 45 closely spaced gages between 1968 and 1984. These
records had been recorded at intervals from 5 to 60 minutes;
unfortunately, the records were reported to the nearest 0.1 inch,
which rendered them of limited usefulness.

Data Processing Problems

   The main problems encountered in processing the rainfall data
were three: the large amount of raw data, the selection of
independent storm events, and the gaps in the data, usually due to
gage malfunction.

   The IBM 3090/200 computer at The Pennsylvania State University
could read tapes with any amount of data, but the disk files for
efficient data analysis were limited to about 16,000 records. The
data on the NCDC tapes contained a large amount of small rainfall
records, obviously of no interest to the rainfall frequency study.
The data were therefore scanned for independent events, within each of
which the largest 1-, 2-, 3-, 6-, 12-, and 24-hour rainfall amounts
were identified. These amounts, sorted by duration, were written to a
file whenever they equaled or exceeded the threshold values listed in
Table 1. As a basis for these threshold values, the 90 percent annual
exceedence probability in the 1970 study was chosen. Any two storm
events were treated as independent if they were separated by at least
24 hours of zero rain. The choice of the interval between independent
storm events was not of critical importance because almost no
consecutive storms above the threshold values were found which were
spaced less than 5 days apart.

Table 3.1. Threshold Values of Significant Storm Totals

Storm Duration, hours	1	2	3	6	12	24
Threshold rainfall, inches	0.6	0.75	0.9	1.1	1.3	1.5

   The gaps, or missing records due to rain gage malfunction, posed
a more serious problem. Such gaps are identified on the tapes by
blank spaces, followed by the rainfall total collected during the
no-record interval, and a special symbol. It would have been easiest
to ignore any gap found; however, it was also recognized that the
malfunction was frequently caused by a storm of high intensity. To
make it possible to use these storm events containing the gap, the

following equation was fit to the average time distribution of storms found in the 1970 study:

$$P(t) = (1 + 0.42 \log \frac{t}{24}) P_{24} \tag{1}$$

in which P_t = storm amount of any duration, in hours
P_{24} = 24-hour storm amount
$\log t$ = base 10 logarithm of any duration t

To illustrate the use of equation (1), let us assume that 4 inches of rain fell over a total gap of 40 hours and that an estimate of the 3-hour storm is needed. According to the equation, $P_{24} = P_{40}/1.09$. For the maximum 3-hour precipitation, the ratio P_3/P_{24} is found to be 0.62. Accordingly, the expected
3-hour rainfall amount is $P_3 = \frac{0.62}{1.09} 4 = 2.27$ inches.

The selection of significant storm events reduced the data sample from 3 million to roughly 26,000 records, stored in two files.

Statistical Analysis

For the storm frequency analysis, the Log Pearson Type III distribution was used. Use of the log-normal distribution was considered, but the investigators did not feel justified to ignore the skew in the data distributions.

Normally, the Log Pearson analysis is applied to annual series of data. Because the records in this case contained many samples of less than 20 years length, and because estimates of the 1-year storm were of interest to the sponsor, the partial series of all events above the threshold values listed in Table 1 were used. The analysis was run like an annual series, but the return periods were multiplied by the ratio of the number of years to the number of records. Separate analyses were run for durations of 1, 2, 3, 6, 12, and 24 hours on the large files of hourly data, and for 15, 30, 45, 60, and 90 minutes on the smaller file of 15-minute records.

Delineation of Storm Regions

At the outset of the project, one of the options considered was to generate one set of rainfall intensity-duration curves for each county. During the project meeting in September 1985, it was decided that county boundaries should have little or no logical correlation with climatic variability. Potter's (1961) physiographic provinces of Allegheny Plateau, Valley Ridge, and Piedmont, which were used in the flood region maps of the PSU-III method for estimating flood peaks on ungaged watershed (Reich, 1970), were suggested for possible regional division. It was recognized, however, that physiographic provinces, while having a logical effect on runoff response, should not affect rainfall intensity or amounts to the extent orographic conditions do.

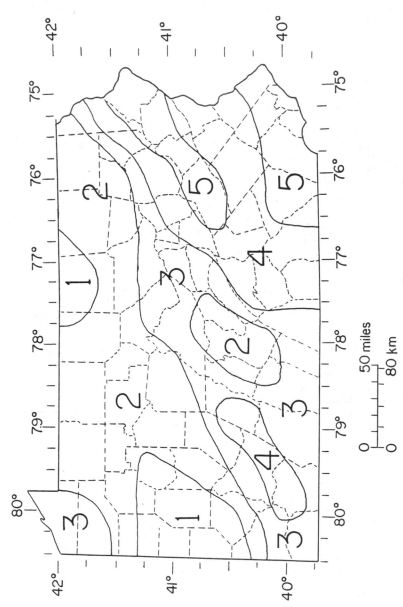

Figure 1. Designated Rainfall Regions

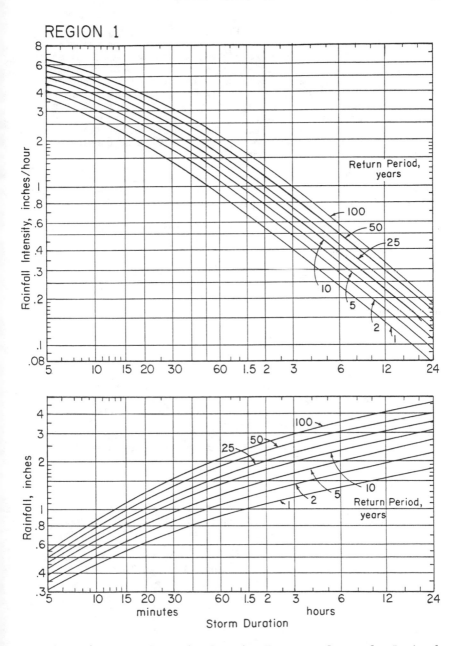

Figure 2.   Storm Intensity-Duration-Frequency Curves for Region 1

To establish regions of similar rainfall intensity, the statistically computed 2- and 100-year storms of 1- and 24-hour duration were plotted on four Pennsylvania raingage maps. By marking these rainfall amounts in colors ranging from yellow to violet in order of increasing rainfall magnitude, and overlaying transparencies of these maps, the regions shown in Figure 1 were delineated. The required boundaries were drawn subjectively by three of the team members drawing individual maps, discussing the differences, and striking a consensus compromise. Some known orographic features, such as the hills near Johnstown and the Poconos were also considered in the final location of the regions.

In each region, average intensity-duration frequency curves were drawn between 1- and 24-hour durations for return periods of 1, 2, 5, 10, 25, and 100 years. Due to the small size of the sample of 15-minute records, only a single set of 15- to 60-minute duration curves was constructed and extended to 5-minute durations with the aid of the standard Yarnell curves (1935). The intensities in these short duration curves were expressed as ratios of their 60-minute intensities, which were then used to extend the five regional sets of curves from 60 to 5 minutes. After some minor smoothing of the composite curves, design storm and intensity curves were constructed for the 5 regions. Figure 2 shows the curves for region 1, the region with the smallest intensities. A full set of design curves for all regions is available from the authors of this paper.

Conclusions

The five sets of charts of rainfall intensities and totals are based on data up to and including 1983. In comparison with the results of the 1970 Reich study, the 1986 charts give lower rainfall intensities and amounts than the Reich report for the shorter durations and return periods, but larger values for storm durations above 6 hours and return periods larger than 10 years. The writers believe that the new charts are the most representative and up-date design rainfall curves for Pennsylvania.

References

1. Hershfield, David M. (1961). Rainfall Frequency Atlas of the United States, Technical Paper No. 40, Dept. Commerce, Washington, DC, 54 charts.

2. NOAA (1977). Five to sixty-minutes precipitation frequency for the eastern and central U.S. National Weather Service Tech. Mem. HYDRO-35.

3. Reich, B. M., McGinnis, D. F., Kerr, F. L. (1970). Design Procedures for Rainfall-Duration-Frequency in Pennsylvania. Pub. 65, Inst. for Res. on Land and Water Resources, The Pennsylvania State University, University Park, PA.

4. Yarnell, D. L. (1935). Rainfall Intensity-Frequency Data. U.S. Dept. Agric., Misc. Publ. No. 204, Washington, DC.

DESIGN STORM SELECTION FOR STORMWATER MANAGEMENT MASTER PLANS

W. Martin Williams, Kelly A. Cave, and Lanchi T. Hoang[1]

The design rainfall event is an important performance standard for a stormwater management master plan. A stormwater master plan for Virginia Beach, Virginia, is used as a case study to illustrate procedures for selecting a design storm event. Statistical analyses of rain gage databases for the 250-square mile study area were used to evaluate seasonal, spatial, and long-term trends within the area. Representative watersheds with different land use patterns were studied with the SWMM model to assess the sensitivity of peak flows to different design storm distributions and durations. The final design storms selected for the stormwater management plan were based on a comprehensive examination of factors influencing peak runoff, thereby overcoming many of the limitations and uncertainties associated with traditional drainage design approaches.

Introduction

The design rainfall event is one of the most important factors in a stormwater management planning program because it serves as one of the major criteria in assessing drainage impacts and establishing performance standards. For example, the design storm is used to evaluate urban runoff, to design stream crossings, and to evaluate and design alternate drainage structures and flood control measures. Conventional approaches to storm drainage design involve peak flow estimates using intensity-duration-frequency (IDF) curves based on average rainfall intensities and methodologies such as the Rational Formula that neglect temporal variations. In reality, the peak discharge at a given location is a function of several factors, including the time of concentration of the drainage system and the duration and distribution of a rainfall event. This is particularly true for watershedwide studies, since peak discharges at downstream locations are highly dependent on the volume and timing of runoff from contributing areas as well as the peak discharge from upstream areas (1). Since the magnitudes, frequencies, and temporal distributions of rainfall events are governed by climatic and geomorphic characteristics of an area, the design storm for a master plan should be based upon local rain gage data.

The City of Virginia Beach (250 sq mi), located in the southeastern corner of Virginia, is characterized by the flat, almost reliefless topography typical of the coastal plain and is underlain primarily by fine-textured, poorly-drained soils. Virginia Beach is currently subject to flooding from both fluvial and tidal sources, which is exacerbated by rapid development. The widely-used Stormwater Management Model (SWMM) (2) is the principal planning tool for the study.

---

[1] Camp Dresser & McKee, Annandale, Virginia 22003

Historical Storms vs. Synthetic Design Storms

Two general approaches used to analyze drainage impacts include the use of historical storms and synthetic design storms. The use of historical storms involves simulating the peak flow responses to either several years of rainfall records with continuous simulation models or to individual historical storm events. A synthetic design storm consists of a hypothetical rainfall hyetograph which is based on the characteristics of a number of historical storms and is generally applicable to a specific area or region. The synthetic design storm is the most widely used approach for master planning studies and has many advantages over the historical storm approach (3). These advantages include: lower time and money requirements for data management and computer model simulation; easy statistical definition in terms of skew, duration, and frequency; their tendency to produce conservative runoff projections rather than underestimates; and their general acceptance in practice, promoting consistency in a particular geographic area. In light of these advantages, the design storm approach was recommended for the Virginia Beach stormwater management master plan.

Selection of Return Periods for Master Planning Studies

Stormwater planning studies have relied upon a range of design storm return periods, depending upon the area and the nature of the stormwater problem and upon adopted criteria of federal, state, local agencies. Alternatives for the Virginia Beach region include design storms with 2-year, 10-year, 25-year, 50-year, and 100-year return periods.

The 2-year event can be used to evaluate measures to manage stream bank erosion potential. Studies have shown that the stream channel shape is governed by the 1.5 to 2-year peak flow and it has become standard practice to adopt the 2-year pre-development peak discharge as the performance standard for minimizing streambank erosion.

Many agencies have adopted the 10-year storm as a basis for designing the minor drainage system including storm sewer appurtenances, inlets, gutters and roadside ditches. Since it exceeds the 2-year streamflow typically associated with bankfull flow conditions, the 10-year flood is an appropriate design event for planning studies of out-of-bank flooding conditions which will occur much more frequently than more extreme events.

The 25-year design storm tends to be the most popular extreme event for stormwater master planning studies. The 25-year event is more conservative than the 10-year design storm typically used for local storm sewer design, but less conservative than the 50- and 100-year events which would require more expensive runoff control measures. A recent U.S. Army Corps of Engineers study of nationwide flood damage concluded that the average annual flood damages within the 25-year floodplain are up to ten times greater than the damages associated with the incremental area between the 25-year and 100-year floodplains (4), which suggests that a 25-year design event may be a reasonable upper limit for stormwater management master planning.

The 50-year design storm was included in the Virginia Beach study because it is the City's current design criterion for drainage structures

for drainage areas greater than 500 acres. In addition, interstate
highway stream crossings are typically designed for a 50-year storm.

The 100-year flood event is the federal flood insurance program
standard and is used by most agencies to establish minimum floor elevations. It is, however, typically too stringent for a stormwater management master plan. Such an extreme event usually cannot be managed with
traditional urban runoff controls. However, runoff control facilities
designed for less extreme rainstorms should be tested with the 100-year
design storm to ensure that the recommended plan does not aggravate the
100-year flood conditions.

Methods for Deriving Design Storm

Most of the approaches for construction of design storm hyetographs
of a specified return period can be grouped into one of two categories:
(a) approaches which rely upon the local rainfall IDF curve; and (b)
approaches which make use of the rainfall characteristics developed from
individual storms.

The first category develops a rainfall pattern which includes the
average rainfall intensity for different durations on the IDF curve for
a selected frequency. The storm developed by this method represents a
variety of storms as one storm. Different methodologies have been proposed in the literature to temporally distribute this rainfall, ranging
from totally advanced (0.0 skew) to totally delayed (1.0 skew) (5). A
more realistic method for temporally distributing the rainfall was developed by Kiefer and Chu (6) in which the IDF curve is used to yield a
distribution such that the ratio of the rainfall falling before the time
of peak intensity to the total rainfall is equal to a predicted ratio
for the area. In a study for the Federal Highway Administration, Chen
(7) suggested some refinements to the Kiefer and Chu method which facilitate design storm development. As another example, the Soil Conservation Service (SCS) has developed standard regional 24-hour storm distributions which can be applied to a specified total rainfall volume (8).

The other category of design storm development uses a distribution
assumed to be typical for the area based on an analysis of a series of
observed rainfall events. One approach is to isolate observed storms of
a given duration and develop a weighted average mass curve (cumulative
volume curve) from the individual mass curve of each event. The resulting peak intensity from the averaged curve can be moderated to values
unrepresentative of a typical storm due to the random fluctuations of
the storm events used in the derivation and to the high variability in
observed storm skews. To overcome this problem, Pilgrim and Cordery (5)
developed a method to determine rainfall distributions using available
rainfall records. This method utilizes the top-ranked storms (with
respect to total rainfall) of a specified duration from the period of
record. This method relies upon ranking statistics on the function of
total rainfall in specified time increments.

Intensity-Duration-Frequency Curves

IDF curves represent statistically defined average rainfall intensities for storms of different durations having different probabilities of
occurrence. IDF data are published by many sources in the form of

curves, tables, and isopluvials. Currently, the most recognized
published sources by the National Weather Service (NWS) for the eastern
United States are TP40 (9) for durations between 1 and 24 hours and
HYDRO 35 (10) for durations less than 1 hour.

Because TP40 was published over 26 years ago, an analysis of local
rainfall data was performed to verify that the TP40 IDF curves were
still applicable for use in the Virginia Beach study area. The analyses
were performed on two NWS recording gages, Williamsburg and Norfolk,
Virginia, both having nearly 40 years of hourly rainfall records. The
analysis was based on an annual extreme series fitted to the Fisher-
Tippett Type I (Gumbel) distribution and adjusted to represent a partial
duration analysis. Comparisons of this analysis with TP40 indicated very
good agreement for the Williamsburg gage, while average intensities were
3% to 20% higher than TP40 at Norfolk with the greater differences associated with the more frequent, shorter duration storms (i.e., 2-year/
3-hour). The differences at the Norfolk gage can be partially attributed
to the regional relationships used in TP40 compared to the point rainfall analysis performed in this study. In lieu of the excellent correlation at Williamsburg and differences that were not significant for critical return periods at Norfolk, TP40 IDF curves were recommended for use
in the Virginia Beach study.

Seasonal And Spatial Variations

To determine if seasonal variations in storm characteristics occur
in the Virginia Beach area, the Norfolk hourly rainfall data were analyzed as three sets or subsets. The first set was the entire period of
record, the second was a seasonal analysis which screened the data to
obtain high-intensity thunderstorm characteristics, and the third was a
seasonal analysis of longer duration storms typifying frontal systems.
The hourly data were screened based on the seasonal period of analysis,
the number of dry hours that must elapse between rainstorms before a new
event is defined, and the maximum intensity and total event rainfall
that will be used in the analysis. Parameters used in the analysis are
specified below:

Input	Period of Record	Type of Analysis Thunderstorm	Frontal Storm
Season	1/1 to 12/31	5/1 to 8/31	7/1 to 10/31
Durations (hrs)	1,3,6,12,24	2,6,12	12,24
Dry hours between events	6	2	12
Maximum Event Length (hrs)	120	8	120
Maximum Intensity (in/hr)	4	4	4
Maximum Event Volume (in)	10	10	10

The seasonal analyses showed that thunderstorms and frontal storms
tend to have different properties, not only in duration and in maximum
intensity, but also in the temporal distribution of the rainfall.
Thunderstorms exhibited a skew of 0.4 on the average, compared to a skew
of 0.5 for the frontal type of storm. The average skew for the complete
annual record was 0.5.

A statistical analysis was performed on daily records at six rain-fall gages near Virginia Beach to determine if spatial variations exist in the study area. The statistical tests, including hypothesis tests on means and standard deviations, indicate there are no significant differences among rainfall records at the six gages.

Design Storm Selection

To investigate the sensitivity of design storm temporal distribu-tions and durations on peak runoff rates, 28 subwatersheds with differ-ent land use plans which range from highly developed to predominately rural were chosen for analysis. Subwatershed drainage areas ranged from 100 to 1,000 acres with times of concentration ranging from 0.3 to 2.3 hours.

Several design storm distributions were examined, including method-ologies by Chen (7), Pilgrim and Cordery (5), and SCS Type II (8). Storms of each type having durations of 24 hours and identical total volumes were simulated and compared to determine the impact of storm distribution on peak rates of runoff. The comparison showed that the Pilgrim and Cordery distribution resulted in significantly lower peak intensities and peak discharges than the SCS Type II and Chen storms. These discrepancies are due to varying skews of the observed data used to constuct the Pilgrim and Cordery distribution, which cause a multi-peaked hyetograph with relatively low peak intensities, compared to the single high-intensity peak of the other two distributions. In general, the SCS Type II distribution resulted in peak discharges ranging from 10 to 20 percent smaller than that of the 24-hour duration Chen storm. A comparison of Chen storms with various skews and durations revealed that the impact that skew and duration have on the peak discharge is depend-ent on the depression storage and infiltration capacity of the basin and on the time of concentration of the contributing drainage area.

The logical approach in examining drainage impacts at a particular location would be to use the design storm corresponding to the maximum peak discharge at the location. This approach is difficult to apply to a regional study because different locations in the basin have different times of concentration, thereby resulting in different critical durations. To standardize the rainfall distribution in the master plan, the SCS Type II 24-hour was selected for all subwatersheds. The Type II rainfall distribution was designed by SCS to contain the intensity of any duration of rainfall for the frequency event being analyzed (8). In other words, the 10-year/6-hour rainfall is contained within the 10-year/24-hour distribution. Also, the 24-hour distribution produces additional total volume on the rising and falling limbs of the distribution, which compensates for losses due to depression storage and infiltration. The skew of the SCS distribution (approximately 0.5) is compatible with the generalized skew of critical storms in the Virginia Beach vicinity. While somewhat higher peak discharges occur for other Chen storm durations, the standardization to the SCS distribution is attractive in that one distribution is required for all points of interest in the watershed for a given return period. In addition, the SCS distribution is universally accepted and used by local engineers in Virginia Beach and throughout the State of Virginia.

Summary and Conclusions

Considerations for the design storm selection include the level of protection or performance standard for design (i.e., return period), the type of event to simulate (i.e., synthetic vs. historical storm), the duration, the distribution of rainfall intensities, and the ability to merge the approach into a workable format understandable to local engineers.

The final design storms selected for the stormwater management plans for Virginia Beach, Virginia were based on a comprehensive examination of factors influencing peak runoff. The limitations and uncertainties of traditional design approaches were overcome by statistical analyses of local hourly rainfall data.

References

1. Hartigan, J.P., "Regional BMP Master Plans," Urban Runoff Quality – Impact and Quality Enhancement Technology, ASCE, New York, NY 1986, pp. 351–365.

2. Huber, W.C., J.P. Heaney, S.J. Nix, R.E. Dickinson, and D.J. Polmann, "Storm Water Management Model User's Manual Version III," Municipal Environmental Research Center, Office of Research and Development, U.S. Environmental Protection Agency, November 1981.

3. Patry, G. and M.B. McPherson, eds., "The Design Storm Concept: Proceedings of a Seminar at Ecole Polytechnique de Montreal: May 23, 1979," December 1979.

4. Johnson, W.K., "Significance of Location in Computing Flood Damage," Journal of Water Resources Planning and Management Division, ASCE, Vol. III, No. 1, January 1985, pp. 65–81.

5. Pilgrim, D.H. and I. Cordery, "Rainfall Temporal Patterns for Design Floods," Journal of Hydraulics Division, ASCE, Vol. 101, No. HY1, January 1975.

6. Kiefer, C.J. and H.H. Chu, "Synthetic Storm Pattern for Drainage Design," Journal of Hydraulics Division, ASCE, Vol. 83, HY4, Aug. 1987

7. Chen, C., "Urban Storm Runoff Inlet Hydrograph Study, Vol. 4: Synthetic Storms for Design of Urban Highway Drainage Facilities," Utah Water Research Laboratory, Utah State University, May 1975.

8. Soil Conservation Service, "Urban Hydrology for Small Watersheds," Technical Release No. 55, U.S.D.A., Washington, D.C., June 1986.

9. Weather Bureau, "Rainfall Frequency Atlas of the United States," Technical Paper No. 40, Jan. 1963.

10. National Weather Service, "Five- to 60-Minute Precipitation Frequency for the Eastern and Central United States," NOAA Technical Memorandum NWS HYDRO-35, June 1977.

11. VA Depart. of Highways and Transportation, Drainage Manual, 1985.

Effect of Rainfall Intensity Distribution on
Excess   Precipitation

Roy W. Koch,[1] A.M. ASCE, and M. F. Kekhia[2]

ABSTRACT

In the development of design storms for input to rainfall runoff models, time variable rainfall intensity must be specified. The resulting hyetograph is used to estimate excess precipitation which is routed to produce a flood hydrograph. Although the routing may mask any effect of the time distribution of rainfall excess on the peak flow, an accurate estimate of the depth of rainfall excess is required if the outflow hydrograph is to have the correct volume.

The effect of the time distribution of rainfall intensity on the prediction of the depth of excess precipitation is evaluated. A comparison is made of the excess precipitation produced by an observed storm, a 10 time increment, variable intensity storm and a constant intensity storm, all with the same depth of rainfall. Excess precipitation is calculated using a physically based model for variable rainfall infiltration for three soil types.

Results indicate that the required complexity of the design storm is dependent on soil type if cumulative excess precipitation depth is used as the criterion. For soil with low infiltration rates, the constant intensity approximation is adequate while for more permeable soils, the variable intensity is required. A bias toward underprediction is noted with more permeable soils under the constant intensity approximation. The 10 increment rainfall model appears adequate for all soil types.

INTRODUCTION

The design of hydraulic structures requires the specification of flowrates for sizing components. Often, only a peak flowrate is necessary but an entire hydrograph is required when storage is being considered such as in the design of detention facilities for urban drainage systems. The design hydrograph is frequently determined.

---

[1]Associate Professor, Department of Civil Engineering, Portland State University, PO Box 751, Portland, OR, 97207.
[2]Research Assistant, Department of Civil Engineering, Portland State University, PO Box 751, Portland, OR, 97207.

using a rainfall-runoff model where excess precipitation is estimated and routed over the watershed and through the channel system.

The hyetrograph of the design storm, as well as those storms used in calibration of the rainfall-runoff model, is approximated as a series with constant intensity over specified time intervals thought appropriate to the particular watershed. the number of time increments into which the storm is divided may affect the estimated excess precipitation either in terms of time distribution, total depth or both. In a study directed at evaluation of a rainfall simulation model, Woolhiser and Osborn (1985) have shown that the representation of a storm can have some affect on the predicted peak outflow from the watershed depending on the routing characteristics. For long times of concetration, the effect of the time distribution of rainfall intensity on peak flow rates is all. However , their results also suggested that there may be effects on the depth of rainfall excess and thus the volume of the outflow hydrograph.

The objective of this study is to evaluate the effect of the time distibution of rainfall intensity on the predicted depth of excess precipitation and determine if there is any bias which results from approximations of the hyetograph. The approach is to first compute excess precipitation from observed storms with known intensity variations. This is accomplished with the application of a physically-based model of variable rainfall infiltration at a point. The storm is then approximated as both a constant intensity event and a variable intensity event with 10 time increments of equal duration. Excess precipitation computed for these approximate storm hyetographs is compared to that calculate from the actual storm to determine any bias resulting from the approximations. Since it is suspected that the soil properties may influence the result, the experiment is carried out for three soils representing a reasonably broad range of hydraulic properties: clay loam, loam and sandy loam.

RAINFALL DATA

In order to evaluate the effect of the time distribution of rainfall intensity, time variable rainfall data are required. Such data are available for selected storms occurring at experimental watersheds operated by the U.S. Department of Agriculture (e.g. Burford et al, 1981). For this study, 20 storms were selected for analysis; 10 each for the Brush Creek watershed in southwestern Virginia and the Walnut Creek watershed near Tombstone Arizona. Criteria for storm selection were a wide range of depths and durations and a reasonably significant runoff event. The selected storms had durations ranging form 0.33 to 10.28 hours and total depths ranging form 0.68 to 6.03 inches. Average intensities were form 0.22 to 2.3 inches per hour.

From the actual hyetograph of rainfall intensity, two approximate storms were developed with the same duration and total depth of precipitation. One storm was represented by constant intensity throughout. The other had variable intensity represented by 10 equal time increments over the storm duration. This variable intensity storm was developed by aggregating the observed storm data as required into the specified time intervals.

INFILTRATION MODEL

For the two watersheds under study, runoff is most likely generated by the Horton mechanism of rainfall excess. This process is characterized, at a point, in a piecewise manner where no excess occurs until the soil surface becomes saturated (the ponding time) afterwhich the excess rainfall is determined by the capacity of the soil for infiltration. Precipitation excess is:

$$p_e(t) = r(t) - f(t) \tag{1}$$

where $p_e(t)$ is the excess precipitation, $r(t)$ is the rainfall intensity and $f(t)$ is the infiltration rate. Rainfall infiltration can be described by (Morel-Seytoux, 1981):

$$f(t) = r(t) \, , \, t < t_p \tag{2}$$

$$f(t) = \frac{S(F_p, h_i)}{2(t - t_p - t_c)^{1/2}} + \frac{K_s}{b} \, , \, t > t_p$$

where t_p is the ponding time, K_s is the saturated hydraulic conductivity, b is the viscous correction factor so that K_s/b is the effective saturated hydraulic conductivity accounting for air effects, $S(F_p, h_i)$ is the rainfall sorptivity, representing the capillary effects on flow in unsaturated soil, F_p is the cumulative infiltration prior to ponding, h_i is the soil water content at the beginning of the storm and t_c is a correction factor so the equation satisfies the condition of $f(t) = r(t)$ at $t = t_p$. Morel-Seytoux (1981) presents the complete expressions for t_p, $S(F_p, h_i)$ and t_c. The total excess precipitation is then the time integral of the rate of rainfall excess over the storm duration.

Application of this expression for infiltration requires specification of parameters describing the hydraulic soil properties. In addition to those already mentioned, the soil water content at natural saturation, h_s, is also required. For this study, three soil types were used representing a broad range of hydraulic behavior: clay loam representing a fairly tight soil, loam representing a soil of intermediate properties,and sandy loam which has a high conductivity. Finally, the initial soil moisture is required. For this study, the field capacity was selected so that the initial conditions reflect a moist soil.

SIMULATION RESULTS

The numerical experiment consisted of calculating the total excess precipitation resulting from each of the 20 observed storms for each soil type with the results taken as the actual excess precipitation. The same calculation was performed for the 10 time

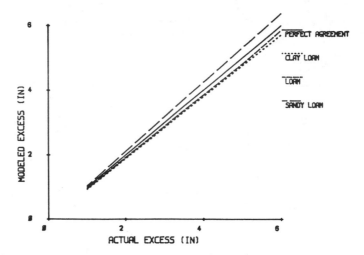

Figure 1.   Comparison of computed excess precipitation from the
             observed hyetograph (actual excess) and the 10 time
             increment, variable intensity hyetograph (modeled
             excess) for three soil types.

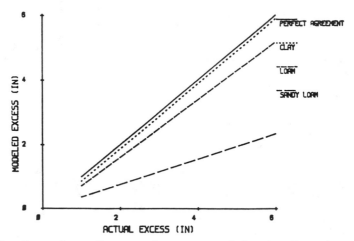

Figure 2.   Comparison of computed excess precipitation from the
             observed hyetograph (actual excess) and the constant
             intensity approximation (modeled excess) for three
             soil types.

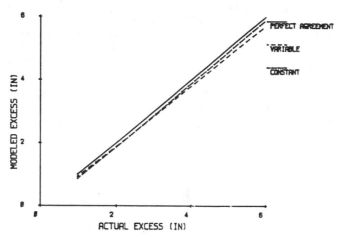

Figure 3.   Comparison of computed excess precipitation for the observed hyetograph (actual excess) and the 10 time increment, variable intensity and constant intensity approximations (modeled excess) for clay loam soil.

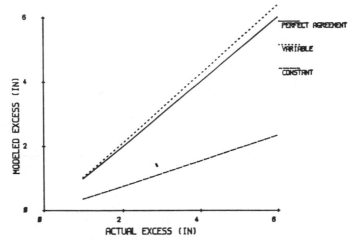

Figure 4.   Comparison of computed excess precipitation from the observed hyetograph (actual excess) and the 10 time increment, variable intensity and constant intensity approximations (modeled excess) for sandy loam soil.

increment, variable intensity and the constant intensity approximations. Figure 1 displays a comparison of the regression line for the rainfall excess computed from the actual rainfall intensity data to that computed using the 10 time increment, variable intensity approximation for each soil type. The line of perfect agreement is also shown. Figure 2 presents similar results for the constant intensity approximation. A comparison of the rainfall intensity approximations for a given soil type is shown in Figure 3 for the clay loam soil, and in Figure 4 for the sandy loam soil. The loam soil results are intermediate to these extremes.

DISCUSSION

Although there is some variation about the regression lines shown in Figures 1 through 4 ($0.94 < r^2 < 0.99$), the trends are indicated quite clearly. From Figure 1, it is apparent that, on average, the 10 time increment approximation is quite reasonable. In contrast, the constant intensity approximation results in a significant underprediction for the more permeable loam and sandy loam soils. However, the constant intensity aproximation appears to be reasonable for the clay loam soil. Figure 3 clearly shows that very little difference occurred in prediction of excess precipitation for the clay loam soil for either intensity approximation, while for the sandy loam soil, the constant intensity approximation resulted in severe underprediction as depicted in Figure 4.

The major implication of this study is that cumulative excess rainfall is only conditionally dependent on the time distribution of rainfall intensity. For soils with fairly low conductivities, the time distribution of the hyetograph is not nearly so important as for situations where the soils are very porous. Further, it appears as if the 10 time increment approximation produces an adequate representation for the prediction of the depth of excess precipitation.

REFERENCES

Burford, J.B., Delashmutt, J.L., and Roberts, R.T. (1981). Hydrologic Data for Experimental Agricultural Watersheds in the United States, 1972. SEA-USDA Misc. Pub. No. 1412.

Morel-Seytoux, H.J. (1981) "Application of Infiltration Theory for the Determination of Excess Rainfall Hyetograph." Water Resources Bulletin. Vol. 17, No. 6, 1012-1022.

Woolhiser, D.A., and Osborn, H.B. (1985) "Point Storm Aggregation - Seasonal and Regional Effects." Multivariate Analysis of Hydrologic Processes. Proceeding of the Fourth International Hydrology Symposium, Ft Collins, CO, July, Water Resources Publications, pp 105-120.

Experiences of the U.S. Army Engineer District, Pittsburgh in Collecting Real-Time Water Control Data

Werner C. Loehlein, P.E., M. ASCE (1)

I. The Need for Real-Time Water Control Management

At present, the U.S. Army Corps of Engineers operates over 400 dams for flood control. Included are many multi-purpose projects which are also operated for low flow augmentation, water quality control, water supply, recreation, fish and wildlife management and hydropower generation. Over the years, the Corps has expended over $10 billion for flood protection while the estimated savings in damages prevented has totaled over $40 billion.

As a result of an ever increasing need to optimize water use, "Reservoir Regulation" has evolved into what is now known as "Water Control Management". In addition to the flood control aspects, national attention has increasingly been focused on the impact that reservoirs have on environmental conditions, water quality, water supply, hydropower and recreation. Reduced manpower resources and funding have been and continues to be a very real problem. These kinds of constraints have caused the management of project purposes to become a very complex task.

In order to effectively meet these demands and improve water control management, it was essential that a systematic approach be taken to enhance the District's capability to collect and manage hydrometeorological data in a Real-Time environment.

II. Real-Time Data Collection

The term Real-Time might best be defined as the time involved in processing data in a sufficiently rapid manner as to influence water control decisions for both normal and emergency conditions. In order to accomplish this task, it is necessary to have sufficient data automatically collected and then transmitted directly to a computer. Thus, several years ago, the Corps began a very active program to have hydrometeorological and water resources data automatically collected and transmitted to Water Control computer systems.

The data collection system chosen for the automation uses the GOES (Geostationary Operational Environmental Satellite)

----

(1) Hydraulic Engineer, U.S. Army Corps of Engineers, 1000 Liberty Ave., Pittsburgh, PA. 15222

system. This system utilizes Data Collection Platforms (DCPs)
in the field which transmit collected data to the GOES, which
in turn transmits the data to an earth station known as a
downlink. This system was determined to be the most cost
effective and reliable way of obtaining Real-Time data.

In mid-1976, the Pittsburgh District installed its first
two DCPs. Since that time, the number has increased until
currently, 141 DCPs are operational. The entire network, when
completed, will have 219 operational sites. The data collected
at DCP stations include various combinations of the following:
river levels, pool levels, precipitation, air temperature and/
or water temperature. All of the stations record and store
hourly values in the memory of the DCP. At selected times,
once every 4 hours, the most recent 8 hours of data is
transmitted by the DCP, through the GOES system and received by
the downlink.

The GOES system consists of two satellites maintained in an
operational status, one at 135° west longitude and the other at
75° west longitude directly above the equator. A spare is
positioned at 105° west longitude. The GOES spacecrafts are
cylindrical in shape, 1.9 meters in diameter, 2.6 meters in
length and weigh 628 kilograms at launch. Their orbit is
circular and about 35,900 kilometers in altitude. The
spacecrafts travel at a speed of approximately 11,000 kilometers
per hour. At this altitude and velocity, the satellites are
stationary above a given point on the earth's surface, hence
the term geostationary.

The Ohio River Division (ORD), U.S. Army Corps of Engineers,
has a downlink, located in Cincinnati, OH. It is being utilized
as the District's primary source of DCP collected data, while
the NOAA/NESS (National Oceanic and Atmospheric Administration/
National Environmental Satellite, Data, and Infornation Service)
downlink at Camp Springs, Maryland, is used as a secondary
source.

A downlink is simply a computer, equipped with an antenna,
demodulators and appropriate software to receive, translate and
store the collected data for dissemination and use in Real-Time.
The ORD downlink, manufactured by Synergetics, Inc., transmits
the data that is collected to a Harris Model 1000 computer
system in the ORD office for processing and high-speed
distribution via X.25 communication lines to the four Corps of
Engineer Ohio River District offices in Nashville, TN,
Louisville, KY, Huntington, WV, and Pittsburgh, PA.

Real-Time Water Control data for the Pittsburgh District is
being received and processed on a Harris Model 800 computer
system.

III. Real-Time Flood Control System Model

With the availability of Real-Time data, considerable
improvements can be made in forecasting and operations work.

In 1984, the Pittsburgh District entered into a contract
with the Hydrologic Engineering Center (HEC), Davis, CA., for
the development of a Real-Time model for operation of the
Monongahela River Basin in the States of Pennsylvania and West
Virginia. This work was completed in 1986. HEC 's continuing
with the development of a Real-Time model for the Allegheny
River Basin. When the Allegheny model is completed, the entire
drainage area above Pittsburgh will be completed.

The completed model primarily utilizes the Real-Time data
collected from the GOES network and other existing data bases
to forecast reservoir inflows and releases. In addition, the
model will forecast streamflows for all downstream control
points.

In order to make Real-Time operational decisions, a rather
comprehensive system of computer software is required. The
capability must exist to allow the following:

a. Entry of data from DCP stations, existing data networks
and manual reporting stations;

b. Storage and retrieval of data;

c. Screening, preliminary processing, editing, archiving,
and transferring of data between users;

d. Forecasting of future watershed conditions;

e. Operation of reservoir systems;

f. Reporting of information in graphical and tabular formats;
and

g. Statistical/economic/utility functions.

The software that is being developed by the HEC addresses
each of these areas in varying degrees. Data may be entered
in several formats to allow incoming data to be stored in the
Data Storage System (DSS). Utilities exist to copy, rename,
edit, purge and transfer data stored in the system.

The forecasting of future conditions in the watershed is
performed by a version of the HEC-1 Flood Hydrograph Package
(HEC-1F) recently adapted to meet Real-Time forecasting
requirements.

Reservoir operation decisions are made with the aid of a
recently adapted Real-Time version of the HEC-5 Simulation of
Flood Control and Conservation Systems Model.

The graphical and tabular display of observed, forecasted
or simulated reservoir operational data is performed by the
DSPLAY routine. Statistical, economic and other utility
routines have been developed to allow forecast and project

operation effectiveness to be evaluated.

The software supervisor routine that provides entry to the various components is entitled Model Control (MODCON). In MODCON, the user specifies the Basin under analysis and the desired time of forecast. From MODCON, the user gains access to any of the above mentioned modules. Control is transferred from one module to another under the direction of the user until the required tasks have been completed.

Capability has been developed to enable Real-Time simulation of snow accumulation and snowmelt. Daily air temperature and snow depth data are used to simulate, on a subbasin basis, snow accumulation, ripening and melt processes. Rainfall attenuation and lag caused by snow on the ground are also accomodated. Rainfall and snowmelt are combined to form a "precipitation equivalent" for input to HEC1F. Future air temperatures and precipitation may be specified for forecasting purposes. The snow depth data that is used for the simulation is observer data received from the National Weather Service (NWS) Automation of Field Operations and Services (AFOS) system or from daily District observer reports.

In the Monongahela Basin, there are competing demands for water quality, navigation and water supply purposes. As a result, a capability for MODCON has been developed to enable evaluation of alternative reservoir operation policies for use during extended low-flow periods. A DSS file was created which contains daily flow data for 14 sites in the Basin for a 50-year period (water year 1930 through water year 1979). An HEC-5 "low-flow" input model for the reservoir system was developed for use with the daily data. During an extended dry period, the current state of the reservoir system can be set in the HEC-5 model, and the model can be run with one or more segments of historical flows to enable evaluation of operational policies.

A primary concern in the development of a Real-Time system is its ease of use. Potential users should not be required to be computer specialists, expert typists, or possess other special skills. The system has been designed to provide human-oriented interfaces between the user and the programs in order to facilitate its use. Commands to the system may be entered through any interactive keyboard terminal. Even though commands have been structured for efficient entry, typing and repetitive entry of data can be a burden on the user. To eliminate this burden, several enhanced user interfaces are provided. As an alternative to keyboard input, the user may select the appropriate command from a command list or menu. The selection is made by simply pointing to the desired entry on a graphics tablet. For example, by pointing at certain areas of menus, data can be displayed, forecasts and reservoir operations can be performed and then reanalyzed with alternative conditions considered. The results of such system operations can be displayed in either tabular or graphical form.

IV. Real-Time Data Collection Equipment

The District has utilized four different manufacturers of Data Collection Platforms (DCPs), i.e., LaBarge, Synergetics, Handar and Sutron. The first 18 units purchased were LeBarge Model 1286. These units required unique programming sets in order to transmit data in pseudo-hexidecimal code. To date, all but 3 of these first generation units have been replaced. The District has installed 75 Synergetics Model 3400A, 52 Handar Model 560A, and 8 Sutron Model 8004D DCPs. These units are easy to program and transmit data in standard ASCII format. In 1986, approximately 330 site visits were made by District field technicians. The most common DCP repairs were the result of dead batteries, drifting frequency of the transmitter and component failures of the master control units.

The District has installed precipitation gages at 74 DCP stations. All of these gages have the capability to record snowfall during the winter months. There are 42 A/C-heated Belfort Model 5-405HA, 26 propane-heated Weathermeasure Model 6041, 5 propane-heated Climatronics Model 2501 and 1 glycometh solution Sierra-Misco Model 5057S tipping bucket rain/snow gages. In 1983, the District's first DCP precipitation gages were placed in the field. Generally, 2-30 pound tanks of propane is enough to allow snowfall recording through the winter months. Most of the operational problems experienced have been the result of plugging from dust, bugs or sometimes leaves. Other stations, even with windshields, recorded noticeably and consistently lesser precipitation amounts than nearby stations. The District has had the graeatest success with A/C-heated gages, however, A/C power is not readily available at most stations. Due to malfunctions at the site the glycometh solution gage could not be fully tested this past winter season.

The District has installed 34 air temperature probes and 13 water temperature probes at its DCP stations. Both the Handar Model 432 and Synergetics Met One temperature sensors have worked very well in the field. The sensors, however, are not interchangeable between DCPs. The only problem with the water temperature sensors has been the need to use divers to securely fasten the sensors underwater.

In 1983, installation was completed on a snow sample station at East Branch Claion River Lake, near Wilcox, PA. With the assistance of the Cold Regions Research and Engineering Laboratory (CRREL), a snow pillow and propane-heated tipping bucket snow/rain gage was interfaced to a LaBarge DCP. After two winter seasons and several malfunctions to the interface of the snow pillow to the DCP, no snow data was collected and the station was discontinued.

V. Conclusion

The benefits of this system include manpower savings at Corps projects and in data handling. The increased amount and

quality of hydrometeorlogical data has improved the water
control management activities, resulting in greater reduced
flood crests at damage centers in the Ohio River Division and
reduced risks of loss of life.

The improvements to Water Control management have been
significant.   The District has immediate access to computed
data, increased data coverage, increased prediction capability
and desireable computer-generated graphics instead of tabular
and manually-drawn plots.

Problems generated with the improvements include: handling
and error checking of the increased volume of data, and the need
to calibrate a range of flood events for forecasting purposes
instead of only analyzing extreme events.   In the District's
current Allegheny River Basin model contract with HEC, mentioned
above, both of these problems will be addressed.

THE ESTIMATION OF RAINFALL FOR FLOOD
FORECASTING USING RADAR AND RAIN GAGE DATA

William J. Charley[*]

Abstract

An inadequate knowledge of the magnitude and spatial distribution of precipitation is often a major limitation in developing accurate river-flow forecasts for use in reservoir operations. Digitized weather radar data can provide useful information regarding the spatial distribution of rainfall, although radar-based estimates of rainfall may be in error due to several factors. The use of radar-rainfall data in combination with rain gage measurements may improve rainfall estimates over those based on either form of measurement alone. This improvement is accomplished by adjusting, or "calibrating", radar-rainfall data with data from rain gages situated within the radar "boundary". A set of rainfall analysis software that incorporates this methodology has been developed by the U.S. Army Corps of Engineers Hydrologic Engineering Center to aid hydrologists in making real-time water control decisions.

The rainfall-analysis software retrieves real-time radar-rainfall data from a National Weather Service RADAP II (Radar Data Processor), and rain gage measurements from data collection platforms via the Geostationary Operational Environmental Satellite (GOES). The radar data from the RADAP II is "calibrated" with the rain gage data using a simple Kriging technique. Subbasin-average rainfall is then computed from the calibrated data and stored in a data base file for subsequent use by a river-flow forecast model. Graphics programs aid in the evaluation of the data. This software system has been implemented for a few pilot watersheds in Oklahoma.

Introduction

A typical rain gage network usually does not provide adequate definition of the spatial distribution of rainfall over a watershed. During a precipitation analysis for river-flow forecasting, a frequent assumption is that averaging or interpolating rain gage data will provide an adequate representation of the average rainfall over a watershed. In many cases this may not be true.

Digitized weather radar data can provide useful information regarding the spatial distribution of rainfall, but this data may contain errors such as the following: 1) the relationship used to compute the rainfall rate from the radar reflectivity assumes standard conditions (e.g., drop size), which may or may not be representative of the actual conditions (Battan, 1973); 2) different types of precipitation (e.g., rain, hail, or snow) have different reflectivities and cannot be represented with the same relationship; 3) atmospheric conditions may cause anomalous propagation of the radar beam and indicate rainfall where there is none; and 4) the radar measures rainfall rates in an elevated volume, not the rate at ground level; evaporation and air currents can significantly alter this rate.

---

*Hydraulic Engineer, U.S. Army Corps of Engineers Hydrologic Engineering Center, Davis, California.

103

Research has shown that radar-rainfall data may be improved by adjusting, or "calibrating", the data with rain gage measurements situated within the radar boundary (e.g., Wilson, 1970). Several algorithms have been proposed to calibrate radar data with rain gage measurements (e.g., Brandes, 1975; Cain and Smith, 1976). However, the rainfall data and the calibration of the data must be carefully evaluated, because improperly calibrated radar data can produce results that are less accurate than would be obtained from either the gage data or radar data alone.

Accumulated digitized radar-rainfall data can be obtained from the National Weather Service's RADAP II radar sites on a real-time basis (Green, et al., 1983; Saffle, 1976). This data is on a grid-cell basis, for which a cell is 3 by 5 nautical miles. The U.S. Army Corps of Engineers Hydrologic Engineering Center (HEC) has developed a set of computer software* for rainfall analysis that can be used to acquire and analyze this data in an attempt to improve estimates of the rainfall over a watershed. No forecasting of rainfall is attempted.

Calibration

The calibration of radar-rainfall data with rain gage measurements proceeds in the following manner. Three hourly accumulated radar-rainfall data for the watershed is automatically retrieved from the NWS RADAP II. This data is decoded and stored in an HEC Data Storage System data base file (HEC, 1985). Concurrently, hourly rain gage measurements are obtained from data collection platforms throughout the watershed via the GOES satellite, and are stored in a similar data base file.

The radar-rainfall data is calibrated by the software component called RADRAN according to the following procedure. The rain gage locations reporting valid data within or near the watershed boundary are identified. The rainfall measured by each gage is compared to the amount measured by the radar at that gage location. If the measured rainfall exceeds a minimum amount (typically 0.1 inch), the ratio of the gage value to the radar value (G/R ratio) is computed. If the ratio is "reasonable" (within user-specified limits), it is used for that site. Otherwise, the algebraic difference between the rainfall measured by the gage and the radar is computed and used for that location.

The radar-measured rainfall for each radar grid cell is adjusted by the G/R ratios and G-R differences from the surrounding gage locations. The area surrounding each cell is divided into quadrants, and the closest two gages within each quadrant are selected. A simplified (linear) Kriging algorithm is applied in order to generate weighting factors, based upon distance, for each of these gage locations. The radar measured rainfall amount for the grid cell is adjusted according to the weighted ratios and differences computed at the selected gages (Charley, 1986).

The adjusted values are averaged over each subbasin in order to compute subbasin average rainfall amounts. The subbasin averages are stored in the data base file for subsequent use by a rainfall-runoff model.

This calibration procedure makes two assumptions that may not always be true. These are that the the rain gage reports the correct amount of rainfall, and that this amount represents the average rainfall in the area corresponding to the radar grid cell. If these assumptions are not valid, the calibrated radar data may produce

---

*This software was developed on a Harris mini-computer, and contains machine dependent code.

results that are worse than those that would be obtained using rain gage data or radar data alone. Therefore, it is important to screen the gage data prior to use, and to evaluate the calibrations.

Several provisions are made to aid in this evaluation. The gage-radar relationship at each gage location is displayed in the output. If the values measured by the gage and the radar are very different, then the calibration may not provide acceptable results. Along with this information, the rainfall measured by each gage is compared to what would have been computed from the calibrated data at that location had that gage not been present. This is accomplished by temporarily removing the gage data from the analysis, then calibrating the radar data at that location. This is repeated for each gage location. The above information provides a quantitative evaluation of the data and the calibration.

The data and the calibration can be evaluated in a qualitative mode by graphical displays. The data can be plotted on a color graphics terminal with rainfall amounts color coded. Outlines of the watershed, rivers, gages and other information can be overlaid on the plot. A similar graphics product may be produced on a dot matrix printer with varying shades of grey.

Results

Models for the RADRAN program were prepared for a few watersheds in Oklahoma, and executed for several storm events. No systematic verification or evaluation procedure was attempted. For most of the events examined, hydrographs computed using rain gage data, and calibrated radar data, were similar. That is the volumes were within about 20 percent and there was little difference in the timing of the runoff peaks. In some of the events, the radar recorded a substantial amount of rainfall that was missed by the rain gages because of the positioning of the storm relative to the gages. An example is presented for the Waurika Lake Basin (located south of Oklahoma City) for the September-October 1986 storms. The subbasin average hyetographs computed for the calibrated radar data, and the gage data only, are presented in Figures 1(a) and 1(b), respectively. For the 30th of September, the calibrated radar hyetograph showed about one inch of rain that was not detected by the rain gages. The calibrated radar data indicates that the storm was situated over the watershed such that the rain gages were located only on the edges of the storm, as depicted in Figure 2(a). The same plot using rain gage data only, presented in Figure 2(b), shows little rainfall over the basin.

It is difficult to evaluate the hyetographs produced by the two procedures, based upon comparing hydrographs computed from these hyetographs against the observed hydrograph, because the basin loss rates necessary for computing the hydrographs are unknown, and cannot easily be determined without bias. However, a relative comparison can be made, as seen in the computed hydrographs depicted in Figure 3. The computed hydrographs were generated using typical loss rates for this area. Unfortunately, the observed hydrograph was computed from changes in reservoir elevation and does not provide an accurate definition of the inflow. The figure does show that, for this event, the hydrograph computed from the calibrated radar data is much closer to the observed hydrograph than that obtained using rain gage data only.

Conclusion

The use of accumulated rainfall data from the National Weather Service's RADAP II radar, adjusted with rain gage data, may give a better spatial estimation of rainfall over a watershed than would be obtained from gage data only. Because of the several

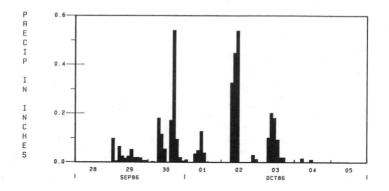

Figure 1(a).   Waurika Lake Basin Hyetograph from Calibrated Radar Data
(total volume: 3.79 inches).

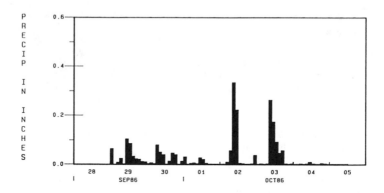

Figure 1(b).   Waurika Lake Basin Hyetograph from Rain Gage Data
(total volume: 2.19 inches).

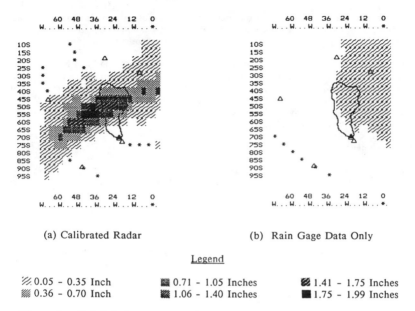

(a) Calibrated Radar   (b) Rain Gage Data Only

Legend

%/ 0.05 - 0.35 Inch   ▦ 0.71 - 1.05 Inches   ▨ 1.41 - 1.75 Inches
%// 0.36 - 0.70 Inch   ▩ 1.06 - 1.40 Inches   ■ 1.75 - 1.99 Inches

Figure 2. Rainfall Over the Waurika Lake Basin for 1500 to 1800 Hours on September 30, 1986 (distances in nautical miles west and south of the radar).

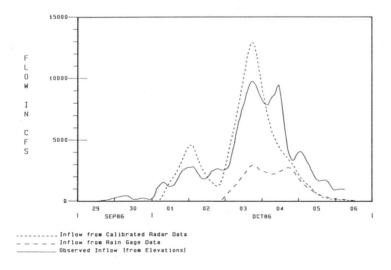

.... Inflow from Calibrated Radar Data
_ _ _ _ _ Inflow from Rain Gage Data
_____ Observed Inflow (from Elevations)

Figure 3. Computed and Observed Inflows (observed inflow calculated from reservoir elevations).

possible errors in the radar-rainfall data, and assumptions made for the data calibration, the data and each calibration must be evaluated by an experienced analyst. Failure to properly evaluate the calibration could lead to erroneous results.

The National Weather Service is currently working on the NEXRAD program which will provide products similar to those obtained from RADAP II, but of superior quality. The NEXRAD program will use algorithms which will attempt to correct for radar errors, and will include a calibration procedure using rain gage information (Ahnert, et al., 1983; Hudlow, et al., 1983). This advancement should provide the hydrologist a valuable tool for river-flow forecasting.

Acknowledgments

The author would like to thank Brian McCormick and the Tulsa District, Corps of Engineers for their help and financial assistance. Appreciation is also expressed to Joe DeVries for his technical assistance and John Peters for his editorial review.

References

Ahnert, P., M. Hudlow, E. Johnson, and D. Green: Proposed "on-site" precipitation processing system for NEXRAD. Preprints, 21st Conference on Radar Meteorology, AMS, Boston, Mass., 1983, 378-384.

Battan, L.: Radar Observations of the Atmosphere. University of Chicago Press, 1973, 324 pp.

Brandes, E.: Optimizing rainfall estimates with the aid of radar. Journal of Applied Meteorology, 14, 1975, 1339-1345,

Cain, D., and P. Smith: Operational adjustment of radar estimated rainfall with rain gage data: A statistical evaluation. Preprints, 17th Conference on Radar Meteorology, AMS, Boston, Mass., 1976, 533-538.

Charley, W.: Weather Radar as an Aid to Real-Time Water Control. Masters of Science Thesis, University of California, Davis, Calif., 1986.

Greene, D., D. Nilsen, R. Saffle, D. Holmes, M. Hudlow, P. Ahnert: RADAP II, an interim radar data processor. Preprints, 21st Conference on Radar Meteorology, AMS, Boston, Mass., 1983, 404-408.

Hudlow, M., D. Greene, P. Ahnert, W. Krajewski, T. Sivaramakrishnan, and E. Johnson: Proposed off-site precipitation processing system for NEXRAD. Preprints, 21st Conference on Radar Meteorology, AMS, Boston, Mass., 1983, 394-403.

Hydrologic Engineering Center: HECDSS User's Guide and Utility Programs Manual. U.S. Army Corps of Engineers, Davis, California, 1985.

Saffle, R.: D/RADEX products and field operation. Preprints, 17th Conference on Radar Meteorology, AMS, Boston, Mass., 1976, 555-559.

Wilson, J.: Integration of radar and rain gage data for improved rainfall measurement. Journal of Applied Meteorology, 9, 1970, 489-497.

MESONETS--A SOLUTION TO IMPROVING FLOOD FORECASTING

by Eugene A. Stallings

Abstract. Floods continue to cause major destruction
annually despite the efforts of Federal water resources
agencies. Improved flood forecasts can reduce flood damages
and corresponding loss of life, but necessary resources are
not available on a nationwide basis.

INTRODUCTION

Each year the United States suffers over $3 billion in
flood damages and 200 lives are lost annually during flood
events. In FY 86 total flood damages approximated $5.5
billion which is the record year. It is interesting to note
that significant events occurred in every region of the
country, with the exception of the Southwest. In October
1985, Hurricane Juan casued eleven lives to be lost, mostly
in the coastal and off-shore waters, and flood loss damages
were over $1 billion. November rains in Central Appalachia
resulted in devastating flash floods that claimed 63 lives
and left damages near $1 billion.

FLOOD DAMAGE

The unfortunate statistics in FY 86 occurred despite
the efforts of the Federal water resources agencies to
reduce flood damages by either structural or non-structural
solutions. The Corps of Engineers, Bureau of Reclamation
and the Tennessee Valley Authority operate numerous flood
control structures throughout the Nation. For example, the
Corps of Engineers flood control projects prevent an average
of $10 billion each year in flood damages. However, the
National Weather Service (NWS) through its nationwide
hydrologic forecasting mandate, is the principal agency
using non-structural means to reduce flood damages.

NATIONAL WEATHER SERVICE HYDROLOGIC OPERATIONS

The NWS meets its hydrologic responsibilities under the
general guidance and direction of the Office of Hydrology in
Silver Spring, Maryland, through the efforts of the thirteen
River Forecast Centers located throughout the United
States.   Each River Forecast Center provides hydrologic
guidance and expertise to a state network of both NWS
Forecast Offices (WSFO's) and NWS service offices (WSOs)
located within each River Forecast Center's area of
hydrologic responsibility. Products generated include flood

Supervisory Hydrologist, National Weather Service, Silver Spring, MD

forecasts, general river forecasts used by navigation,
reservoir inflow forecasts, and water supply outlooks.
Advance warning of an approaching flood permits evacuation
of people, livestock, and property. River and flood
forecasts are primarily concerned with predicting the time
and height of flood crests. There are approximately 3,000
forecast points across the United States and flood forecasts
are issued for about 17,000 flood prone communities.

MESONETS

The National Weather Service is charged with the
mission of providing flood forecasting services to the
nation which can be readily accomplished with unlimited
funds, increased staffing, additional rainfall and stream
gages, and modernized equipment. Obviously, funds and
manpower are limited and the possibility of greatly enhanced
nationwide flood forecasting capability in the next year or
so is not likely. Floods as described earlier continue to
create devastation and certain sections of the populace have
committed themselves to providing improved flood forecasts
now and in the immediate future. Mesonets for the purpose
of flood forecasting are networks of hydrological stations
of sufficient density to sample rainfall and stream
conditions in real time that adequately describe events as
they occur. In the strictest sense, neither the Susquehanna
River Basin flood forecasting improvement project nor the
Integrated Flood Observing and Warning Systems (IFLOWS) are
mesonets. Rather these two innovative enhanced flood
forecasting programs represent a grouping of several
mesonets. For example, the Susquehanna River Basin project
consists of many separate features; IFLOWS, Automated Local
Evaluation in Real Time (ALERT) and improved staffing and
computer capability at the Middle Atlantic River Forecast
Center at Harrisburg, Pennsylvania.

SUSQUEHANNA RIVER BASIN PROJECT

The public demand for improvements in flood warnings, river
forecasts, flash flood guidance and other hydrologic
services prompted the Susquehanna River Basin Commission
(SRBC) to sponsor the proposed flood forecasting system.
The Commission is the single Federal-Interstate water
resources agency with basinwide authority and, working
closely with the Corps of Engineers, U.S. Geological Survey
(USGS) and the National Weather Service, developed a plan.
When completed, the plan will have the capability to
disseminate forecasts within an hour after receipt of data
and to update forecasts at approximately 4-hour intervals.
An effective flood forecast and warning program requires
four basic elements: real-time hydrologic data, hydrologic
forecasting capability, data dissemination capability, and
adequate community preparedness.

The improved forecasting system for the Susquehanna River basin consists of five modules: computer system, data collection network, increased staff, radar component and NOAA weather radios. Each one of the modules will by itself enhance the flood forecasting capability of the National Weather Service for the Susquehanna River, but working in concert with each other, the modules will produce a flood forecasting system that will be a model for any river basin in the Nation.

Improved hydrologic forecasting capability requires increased computer capability for the Middle Atlantic River Forecast Center at Harrisburg, Pennsylvania. When the plan was conceived, the National Weather Service utilized a PRIME 750 in its Headquarters office. In recent months, the PRIME company has marketed a PRIME 9755 (II) computer which has over six times the power of the PRIME 750. The single PRIME will be the principal driving mechanism to develop improved flood forecasts in the Susquehanna River basin. There is no question that the PRIME 9755 (II) will do the job and do it very well.

The staff in the River Forecast Center at Harrisburg is quite competent and professionally very sound, but the improved forecasting program is most unique and extremely sophisticated. The Harrisburg staff must have strong technical support in software development and implementation from the Hydrologic Research Laboratory and the Hydrologic Services Division in Washington and the Prototype Real-Time Operational Test, Evaluation and User Simulation (PROTEUS) Project Office in Kansas City.

The second module in the Susquehanna River Basin project is improved data collection network which needs to be improved in two ways. First, substantially more stream gages and rain gages are needed. The data from these additional sites will minimize the existing uncertainty in estimating runoff from large ungaged areas. Secondly, data transmission from the gage to the MARFC needs to be improved. Normally the data collection platform (DCP) will transmit data at a scheduled time every 3 or 4 hours. During extreme hydrological conditions, it can be programmed to transmit data more frequently (random reporting type). This assures that, when flooding is occurring, current data are available at approximately hourly intervals. Also the data transmission activity does not have to be initiated by people but occurs automatically at the remote data collection site. Naturally, any time savings can be converted into dollars saved through flood damages averted and, most importantly, a potential reduction in flood related deaths.

The data collection network will include 31 new stream gaging stations, 87 new rain gages, and 88 new DCPs. The data from the DCP are received from the satellite by a direct readout ground station at Wallops Island, Virginia. The data are received at Wallops Island and transmitted over dedicated land lines first to the National Oceanic and Atmospheric Administration computer at Suitland, Maryland and then to the MARFC at Harrisburg.

The third module in the Susquehanna River basin project involves the human element. To summarize, the first module provides a sophisticated powerful PRIME computer that can only produce improved hydrologic forecasting with complex software and competent forecasters. Similarly, the enhanced data collection network will produce more data and at a more frequent interval. But a massive powerful computer and an enhanced data base can only produce improved hydrologic forecasting capability for the Susquehanna River basin if there are competent staff at MARFC ready to prepare timely forecasts. Seven people comprised the staff at the MARFC prior to the proposed flood forecasting improvement system. The normal hours of operation are from 6 a.m. to 10 p.m., 7 days a week. Naturally, during flood emergencies, the MARFC operates around the clock every day until the flood waters abate and normal conditions can be resumed. Under the enhanced plan, the staff at the MARFC will increase and forecasts will be available 24 hours a day/7 days a week.

The fourth module is the radar module. The First-Year Plan for the flood forecasting system improvement program proposed a WSR 74-S network radar in the vicinity of Williamsport, Pa., to fill a gap between the existing radars. Three meteorologist technicians will allow Williamsport to be operational 24 hours a day, 7 days a week. At first glance, this appears to be a most appropriate component to providing improved flood forecasting services for the entire Susquehanna River basin. Unfortunately, the Susquehanna River Basin project did not consider the Modernization and Restructuring Plan for the National Weather Service. One major component of the NWS modernization plan is to provide more sophisticated radar throughout the United States, beginning in the early 1990's. The radar system for the future NWS is called NEXRAD (Next Generation Weather Radar). NWS is proposing to install a RADAP II ICRAD at Harrisburg, Pa., and Binghamton, N.Y., and an ICRAD at Williamsport.

The NOAA Weather Radio is the fifth and final module that comprises the flood forecasting improvement plan for the Susquehanna River basin. NOAA Weather Radio is the most direct means to communicate with the general public. The current stations do not reach all areas of the Susquehanna River basin. Two sites are being added at Wellsboro and

Towanda, Pennsylvania, to provide expanded basin coverage. The "Voice of the National Weather Service" provides continuous broadcasts of the latest weather information directly from National Weather Service offices. Taped weather messages are repeated every 4 to 6 minutes and are routinely revised every 1 to 3 hours, or more frequently if needed. Most of the stations operate 24 hours daily. The broadcasts are tailored to weather information needs of people within the receiving area.

MODEL FOR THE COUNTRY

In addition to the technical components associated with the Susquehanna River basin flood forecassting program, there are "people" considerations. The plan to improve flood forecasting is very complicated and is on the "cutting edge of technology." The final system requires a massive interactive capability and will be different from the forecasting configurations of any other National Weather Service Forecast Office. This system will not work without "people." The people are not only NWS hydrologists and river forecasters but USGS staff, Susquehanna River Basin Commission members and volunteer concerned citizens in Williamsport and Harrisburg and other locations in the Susquehanna River basin. When completed, the Susquehanna River basin project will be a model for the rest of the country.

Preliminary studies indicate that improved flood warning, coupled with various preparedness measures, can reduce damages 10 to 15 percent. Assuming that the program proposed here will increase forecast lead times by 6 to 10 hours, thereby achieving reduction associated with the plan would amount to $12.375 million per year. Over a 20-year period, the additional savings would exceed the projected cost of the program by a ratio of 8.6 to 1. Flood damages will be reduced because of the additional lead time and better forecasts. But, most important is the strong possibility that better and more accurate forecasts could reduce the lives lost during flood related events.

IFLOWS

The Integrated Flood Observing and Warning System (IFLOWS) was developed after the creation of the National Flash Flood Program Development Plan in 1978. The goals of the IFLOWS program are to substantially reduce the annual loss of life from flash floods, reduce property damage, and reduce disruption of commerce and human activities.

IFLOWS provides early warnings to local jurisdictions through the use of radio reporting raingages, communications, computers, and specialized software. The central collection point, usually located in the State

Emergency Operations Center, is also called the Central
Processing System (CPS), and is managed by a mini-
computer. It receives, validates, and correlates rainfall,
stream stage, and other meteorological data from automatic
sensors, other National Weather Service systems, and county
centers. The CPS develops information relevant to potential
flash flood threats and quickly disseminates the information
and warning messages to the local jurisdictions. The local
centers (located in the county emergency operations center)
receive the data on their micro-computers. Rainfall data is
received at 15 minutes intervals and cumulative amounts can
be displayed for various time increments up to one month.
Warning messages trigger an alarm and provide a printout of
the message. The local computers can also be used by the
local coordinator to enter volunteer rainfall and stream
gauge observer reports into the system. Local users can
send forward messages to other participating jurisdictions,
Weather Service Offices, and the State Emergency Operation
Center.

     Each jurisdiction receives a micro-computer, radio base
station, automatic reporting rain gauges, and manual
rainfall 11/23 micro-computer with two floppy disk drives.
Normally the computer is located in the sheriff's or police
dispatch office where there is a 24 hour operation. The
radio-base station provides the communications for the
computer and the voice console which interfaces with the
backbone communication system which can be micro, satellite,
VHF or telephone. The radio reporting gauges transmit a
signal when a millimeter of rainfall is measured. The
manual rainfall and stream gauges are to be manned by
volunteers. The startup cost for IFLOWS is borne by the
Federal government through the National Weather Service, and
the state, and the local participating jurisdiction are
responsible to maintain the system once it is installed and
operational.

SUMMARY

     Mesonets are a necessity to improving flood forecasting
services nationwide. A mesonet can vary from a
sophisticated comprehensive project like the Susquehanna
River Basin project or an IFLOWS for a portion of one
community in one state.

LONG-RANGE FORECASTS OF WATER-SUPPLIES
by George N. Newhall, LM ASCE [1]

ABSTRACT

Many engineers and water-users have an urgent need for reliable predictions of future rainfall and water supplies, extending many months or years ahead. Correlation of known historic rainfall patterns with predictable sunspot cycles and meteor-dust dates are shown to assist such predictions, and give evidence of a current multi-year drought.

INTRODUCTION

WATER is essential to all humans, animals and plants for LIFE. How much RAIN and SNOW will there be in FUTURE months and YEARS?

That is a question that all water-users would like to be able to answer better than in the past. What do YOU think the water situation will be in your area in, say, the next five years?

The question of CLIMATIC CHANGES and PATTERNS is of IMPORTANCE not only to water and power companies and agencies, but to YOU. You may not be able to prepare for such VERY-LONG-RANGE changes as the "next Ice Age", discussed by Matthews (4), but you may very well be able to prepare for such possible NEAR-TERM or LONG-RANGE events as DROUGHTS or FLOODS, which may occur in your lifetime or in the next several years.

NEED FOR RESEARCH on CLIMATOLOGY has been mentioned by many engineers and other writers for many years. For example, in 1961, Williams (9) analyzed many historic records of precipitation and river runoff, and their annual variations. He said that correlations between hydrologic data and sunspots have been attempted with varying degrees of success.... In addition to the well-known eleven-year sunspot cycle that affects hydrologic data in an irregular manner, there is evidence of a quasi-100 year cycle that also appears to cause long-term variations in hydrologic data. He concluded "It is hoped that many more studies of this type will be undertaken and that the results will be of assistance in long-range water-resources planning."

---

[1] Consulting Civil Engineer, Retired, P.O. Box 827, Davis, California 95617.

This paper is one result of such further study, including analysis of sunspot records and other factors which can influence precipitation. Hartshorn's 1986 article (3) emphasizes the need for long-range forecasting of rainfall and snowfall in California.

In the past two decades there have been great advances in research and techniques in meteorology and climatology. Matthews (4) described and vividly illustrated several of these in 1976. However, I believe that little or no emphasis has been given to SUNSPOT PATTERNS, or PREDICTABLE METEOR-DUST DAYS, or VOLCANIC DUST, which can affect long-term or short-term patterns of precipitation.

PHYSICAL CAUSE AND EFFECT

A key question that is asked is: What do sunspots and dust have to do with rainfall? What is the PHYSICAL connection?...or logical cause and effect? I see the connection in the following sequence: See Figure 1.
1. Where do rain and snow come from? CLOUDS.
2. Where do clouds come from? EVAPORATION FROM OCEANS.
3. That requires a vast amount of energy to evaporate those millions of tons of water and lift it up to cloud level.
4. Where does all that energy come from? THE SUN.
Sunspots are an indicator of bursts of solar energy.

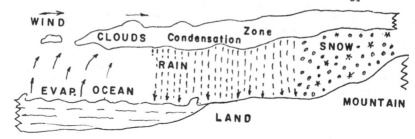

Fig.1. First Part of the Hydrologic Cycle.

5. What causes clouds to drop rain or snow? Temperature below Dew Point, and NUCLEI, such as DUST in cloud. METEOR DUST drops into clouds on predictable dates.

OCEAN WATER BALANCE

Precipitation on ocean minus Evaporation from ocean equals Ocean Water Balance.
$$P - E = OWB$$
A large NEGATIVE OWB is the SOURCE of clouds over land. Those are common over many tropical and temperate oceans.

Precipitation ON OCEANS is largely unused by humans.

SUNSPOTS AND SOLAR RADIATION

Sunspots are dark regions that appear on the sun (Fig. 2),
grow in size, often to tens of thousands of miles across,
and disappear after a life which may vary from days to a
few weeks...or even occasionally longer than a year.   They
often appear in pairs (Fig. 3) or GROUPS, and have strong
magnetic fields. Smith (7) says the fields are as strong
as 3500 gauss;   they are centers of solar activity, from
which extra solar winds radiate, with hydrogen plasma,
electrons, ions, etc.   This can cause AURORAS, radio
interference, and extra solar energy at various wavelengths
in our world.

Fig. 2. Sunspots &
Groups on a fairly
active sun. Sun
rotates toward left.

Fig. 3. Hydrogen whirls in
sun. A pair of sunspots.

Refs. (2),(5),(7), and (8) contain much detail and data on
sunspots.   I present Fig. 4,   from my   analysis of
Waldmeier's data (8) as EVIDENCE of a 178-year long-term
"cycle". Upper curve is 178 years later than lower curve,
particularly in spacing of minima of 11-year cycles.

Annual Means of Relative Sunspot Numbers

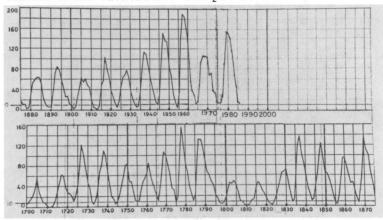

Fig.4.The 178-year repeat pattern of Sunspots.

LATITUDINAL DRIFT.   From start to end of each 11-year cycle, sunspots appear at solar latitudes form about 35 degrees down to about 8 degrees or less.

METEOR DUST AND PRECIPITATION

Do the recurrent showers of meteor dust also increase precipitation?   I believe that there has been enough evidence, world wide, in the last several decades, to indicate that the answer is YES. I suggest that anyone who is interested in METEOROLOGY, CLIMATOLOGY, or WEATHER-MODIFICATION take a look at the theory which Dr. E. G. "Taffy" Bowen, the Australian physicist suggested in 1954...and then test it against his own data, and use it where appropriate.

Dr. Bowen's theory, which was publicized in the Christian Science Monitor Feb. 5, 1954 and later in Reader's Digest and Time magazine, resulted from his study of rainfall records in Australia and many other countries, and of the physical processes involved. He concluded that the showers of meteor dust fall slowly through the earth's atmosphere and arrive down at cloud level about 30 days after each meteor shower. Astronomers have observed those showers of "shooting-stars" to occur on practically the same DATES each year, since the earth passes through masses of "space-dust" regularly in its orbit.   Those millions of dust particles then serve as nuclei around which cloud droplets can coalesce, and can thus cause an increase in the amount and intensity of precipitation wherever in the world it is raining or snowing or hailing ON THOSE DATES.

Following is a list of the approximate dates of the important meteor showers, which are routinely observed by astronomers and are published in astronomy magazines.

Name of Meteor Shower	Dates of Shower	"Meteor Dust Days" (30 Days Later)
Quadrantids	January 2-3	February 1-2
Lyrids	April 20-23	May 20-22
Eta Aquarids	May 4-6	June 3-5
Delta Aquarids	July 28	August 27
Perseids	August 10-13	September 9-12
Giacobinids	October 9	November 8
Orionids	October 18-23	November 17-22
Taurids	November 8-10	December 8-10
Leonids	November 16	December 16
Bielids	November 22-27	December 22-27
Geminids	December 10-14	January 9-13
Ursids	December 22-26	January 21-25

At high elevation stations the effects may show up a day or two sooner. Cloud physicists can study the amount and particle size of such meteor dust in samples taken on those dates.

Two well-known examples of intense heavy precipitation, which caused floods and damage in California occurred in the period December 22-27, in the years of 1955 and 1964, which were about 30 days after the Bielids meteor shower of November 22 to 27.

VOLCANIC DUST

Fine particles of volcanic dust from LARGE, Explosive volcanic eruptions can be carried long distances, even around the world, at or above cloud levels. This can cause lowering of air temperatures and/or coalescing of cloud droplets which may result in increased precipitation, e.g. TAMBORA, East Indies, 1815; & KRAKATAU, Indonesia, 1883.

PRECIPITATION

For a general discussion of precipitation see Barnard (1).

I selected 10 weather stations, worldwide, which had the longest published records (over 100 years) of monthly precipitation, with average annual amounts over 500mm (20 in.). My research included analysis of each of the 10 records, and of all 10 as a group. Below are listed four of the stations:

Name	Record Yrs.	Yr. Start	Mean Ann. Precip.
Milan, Italy	218	1764	1001mm (39.4")
Boston, MA, USA	167	1818	1054mm (41.4")
Adelaide, Australia	146	1839	528mm (20.8")
Portland, OR, USA	114	1871	1082mm (42.6")

Table 1.   Four Selected Precipitation Stations

LONG-RANGE precipitation over continents and worldwide appear to me to be controlled mainly by monthly and annual cumulative amounts of EVAPORATION FROM THE OCEANS, modified in some months by predictable METEOR DUST, and occasional cataclysmic volcanic eruptions, etc.

CORRELATION

Correlations between MONTHLY MEANS of Relative Sunspot Numbers and MONTHLY PRECIPITATION in PERCENT of NORMAL are an important part of the evidence useful in making long-range forecasts of Precipitation, (see Fig. 6). There were eight dry months in that QUIET SUN year in Boston.

Correlations between ANNUAL MEANS of Sunspot Numbers and ANNUAL PRECIPITATION and/or STREAMFLOW in PERCENT of NORMAL have been used by many researchers, including Williams (9), and are also a part of historic evidence useful to long-range forecasting. My analysis of them, as for example in Fig. 5, is that WET YEARS often follow years of Sunspot

Numbers <u>above   40</u>;   and that WET YEARS can also occur near
the end of an 11-year cycle, due to LOW-SOLAR-LATITUDE
spots.   DRY YEARS are usually at or following SOLAR-CYCLE-
MINIMA. DRY MONTHS, often less that 50% of normal, in many
cases follow QUIET SUN months, which always include ZERO
DAYS.

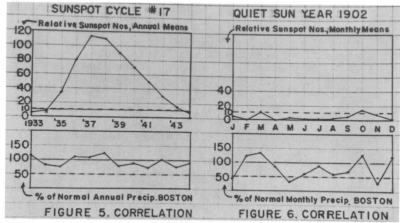

FIGURE 5. CORRELATION   FIGURE 6. CORRELATION

TWO historic examples of POSITIVE CORRELATION I found are:

a)In 1859, a SOLAR MAX year, BOSTON precipitation was 143%.
b)In 1976, a SOLAR MIN year, Portland, OR precip. was 61%.

DROUGHTS

   Researchers have said that 2 or 3 DRY YEARS occur about
every 22 years, i.e. at every other QUIET SUN period: such
as 1931-4, 1953-5, 1976-7.   Why not ALL THE QUIET SUN
PERIODS?   If 1986-89 are QUIET SUN years, then the CURRENT
DROUGHT could be a VERY LONG ONE.   See Fig. 4 for 1808-12
vs. 1986-?

REFERENCES
1.   Barnard, Merrill:Precipitation, in "HYDROLOGY", by
      Meinzer
2.   Bray, RJ & Loughead, RE "SUNSPOTS", 1979, Dover Pub.
3.   Hartshorn JH:Drought or Flood, in WESTERN WATER, Jan'86
4.   Matthews, SW:"What's Happening to our Climate?" in
      NATIONAL GEOGRAPHIC, Nov'76
5.   Schove, JD, ed."SUNSPOT CYCLES", 1983, Hutchinson-Ross
6.   Schulman, E. "DENDRO-CLIMATIC CHANGES IN SEMI-ARID
      AMERICA", 1956, Univ. of Arizona Press
7.   Smith HJ & E. van P.,"SOLAR FLARES",1963, MacMillan
8.   Waldmeier, Prof.Dr.M."THE SUNSPOT ACTIVITY IN THE YEARS
      1610 TO 1960",1961, Swiss Fed. Obs./Schulthess & Co.
9.   Williams, GR "Cyclical Variations in Worldwide Hydr.
      Data" in ASCE Hydraulics Div. Journal, Nov '61

Water Supply Demand Forecast

by

M. Karamouz, M.ASCE, A.S. Goodman F.ASCE
and
L. Bourodimos, M.ASCE
Polytechnic University, Brooklyn, New York

In this study, estimates of location, quantity and timing of
water demands have been investigated for determination of
the adequacy of an existing water supply system, the need
for additional facilities, and desirable changes in
operation. The main objective of this study was to develop
an improved understanding of the traditional methodologies
and modern techniques of mathematical modeling for
estimating water demands. The traditional methods involved
per capita methods and per connection methods and an
estimate of water losses through leaks and unaccounted for
water connections. More advanced techniques involved time
series analysis including the use of autoregressive moving
average models. Past levels of water use (aggregated and/or
disaggregated) were used to estimate the parameters. In the
analysis of water use, it was also necessary to externally
project certain parameters such as population and
employment. Two cases have been studied in this paper.

I - Introduction

Estimates of location, quantity and timing of water
demands are needed to determine the adequacy of existing
water supply systems, the need for additional facilities,
and for better allocation of resources. Despite their
critical importance in the areas of growing demand,
sufficient attention has not been given to water demand
estimates. There has been a generally unquestioning
reliance on traditional engineering approaches such as per
capita values applied to predicted population, per
connection values, and unit use coefficients applied to
households, industrial and commercial employees, or other
variables. These estimates do not adequately reflect the
uncertainty of the variables entering into their
determination.

This study included the testing of both traditional methodologies and time series analysis including the use of auto regressive moving average models

Several significant documents have been prepared in recent years under the sponsorship of the U.S. Army Corps of Engineers. "An Annotated Bibliography on Techniques of Forecasting Demand for Water" (B. Dziegielewski et al, 1981) provides an overview of the literature on this subject. "Forecasting Municipal and Industrial Water Use: A Handbook of Methods" (J.J. Boland et al, 1983) which is intended as a practical guide for planners engaged in performing water use and provide specific suggestions for approaches to common problems. A companion volume "Forecasting Municipal and Industrial Water Use: IWR Main System User's Guide for Interactive Processing and User's Manual" (J.E. Crews and M.A. Miller, 1983) contains computerized procedures for estimating and forecasting water requirements by traditional methods.

II - Case Studies

Two cases have been observed and used in this study to demonstrate conventional and more advanced techniques of forecasting demands:

A) Phase one of a water company in New Jersey
B) A water district system in Long Island, New York.

CASE A

To simulate a real water distribution system, actual demands were placed at various nodes throughout the Phase 1 model. Much preparation and time was spent assigning water demands to four pressure zones that comprise the Phase 1 area. The four pressure zones are divided based on their hydraulic grade line: zone 273 HGL; zone 319 HGL; zone 437 HGL; and zone 453 HGL.

Determination of Demand

The consumption for the Phase 1 system was divided into residential and large users demand. The large users data was obtained from the commercial department billings. The residential demand was based on aerial maps and zoning maps. Consumption values from residential users billing were not used because they were too low and were no identified with a specific consumption date such as maximum day, and were based on a staggered quarterly billing cycle, while large users are billed monthly and it was easier to pro-rate this consumption over a monthly period. After counting the number of nodes for each section the residential demand for each node was calculated by dividing the total residential consumption (total production minus large users billed consumption) by the number of nodes.

Allocation

The demand allocation program developed at the University of Akron, (DMDCNV program, see Sankelli, et al. 1985) allows allocation and re-allocation of water to each pressure zone. As development takes place within a given pressure zone, the consumption increases. It then becomes necessary to reallocate more water to all demand nodes in that pressure zone. The program divides the water (in mgd) to all demand nodes in that zone. Then it adds the large user demands to appropriate nodes.

Other nodes may exist in an area that is non-residential and those will have only large user demand. The increase in demand is a result of requests for new accounts, any major development and/or industrial-commercial expansion, and other identified demands The zoning maps were used to project the extent of developments in an area. The results of the demand program were simulated and checked through the system using the AQUA simulation program (developed at the University of Akron, see Sankelli, et al., 1985). The output of the AQUA program for given hydrant and pipe locations were compared with actual field data involving flow tests and measurements. Agreement within + 10% of actual conditions (for the day being simulated, i.e. Average Day and Maximum Day) was considered satisfactory.

If the simulated results differed significantly from actual data, the demand was re-allocated, the node and pipe arrangement and the friction factor in the database was re-checked, especially in the areas (pressure zones) with unacceptable results. Pressures of less than 20 psi were not acceptable.

For the Maximum Day 1986 model for Phase 1 (which is the model used for the Phase 1 simulations) the pressure zone allocations were as follows:

a)   Zone 273 HGL - 4.028 million gallons
b)   Zone 319 HGL - 12.136 million gallons (this is the Route 1 corridor area)
c)   Zone 437 HGL - 0.924 million gallons
d)   Zone 453 HGL - no allocation yet since little development in this area.

The total consumption for each of the three zones matched the actual available data from the company's billing records and other sources.

CASE B

The total pumpage data for a water district in Long Island from 1971 - 1986 was used to estimate the water demand for years 1987 - 2000. The system loss or unaccounted water for the years 1982 - 1986 was estimated as 10.7%. Two methods were used. The first method was based on the projection of new water accounts, type of land use, the extent of unused land to project the ceiling on the system expansion, and the last 5 years running average of water pumped. The second method was based on the study of the autocorrelation structure of the data and the use of the Autoregressive Integrated Moving-Average (ARIMA) models described by Box and Jenkins (1976).

In order to fit an ARIMA, the following were studied: autocorrelations and partial correlations of historical data and residuals within a 95% confidence interval; chi square test statistic for model adequacy; normality of residuals, and reproduction of a part of historical data using different models. Table 1 shows the historical data, the data estimated using method 1, forecasted data using ARIMA (1,1,0), and forecasted data using ARIMA (4,1,1). For a 16 year historical record, the ARIMA (1,1,0) model was selected because it could closely produce the last six years of historical data using parameters from the first ten years data. The ARIMA (4,1,1) was determined to have better fit, based on other statistics and overall performance of the model. Figure 1 shows the original and forecasted data using the ARIMA (4,1,1) model.

III Summary

Estimates of water demand using different techniques were studied to demonstrate the techniques and the processes involved in making estimates of water demand. This work is part of a comprehensive study that will examine the uncertainty of water supply demand estimates and their effects on water resources management strategies.

**Table 1 - Actual Pumpage Data 1971 - 1986 and
Forecasted Data 1987 - 2000**

Forecasted Data 1987 - 2000 (MG)

Historical data 1971-1986 (MG)	Method 1	ARIMA (1,1,0)	ARIMA(4,1,1)
3303	4134	3919	4285
3120	4185	4131	4125
3143	4236	4146	4464
3075	4288	4250	4476
2730	4339	4314	4289
3284	4390	4396	4539
3293	4441	4470	4628
3434	4492	4548	4458
3402	4543	4624	4621
4143	4595	4701	4760
3701	4644	4777	4624
3757	4691	4854	4717
4270	4734	4930	4877
4044	4764	5006	4786
3955			
4388			

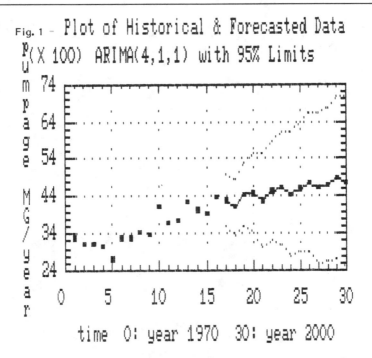

Fig. 1 - Plot of Historical & Forecasted Data
Pumpage (X 100) MG/year ARIMA(4,1,1) with 95% Limits

time 0: year 1970 30: year 2000

Acknowledgment:

The authors wish to thank Andriani Harris for her assistance and constructive comments.

REFERENCES

Boland, J.J., May, W-S, Pacey, J.L., and Steiner, R.C., "Forecasting Municipal and Industrial Water Use: A Handbook of Methods," U.S. Army Corps of Engineers, Institute for Water Resources, July 1983.

Box, G.E., Jenkins, G.M., "Time Series Analysis - Forecasting and Control," Holden-Day, Inc., Oakland, CA 1976.

Crew, J.E., Miller, M.A., "Forecasting Municipal and Industrial Water Use: IWR Main System User's Guide for Interative Processing and User's Manual," U.S. Army Corps of Engineers, Institute for Water Resources, July 1983.

Dziegulewski, B., Boland, J.J., and Baumann, D.D., "An Annotated Bibliography on Techniques of Forecasting Demand for Water," U.S. Army Corps of Engineers, Institute for Water Resources, May 1981.

Sarikeele, S., and Cesario, L., "Advances in Water Distribution Analysis Methods," Proceedings of the AWWR Distribution Symposium, Seattle, WA, September 1985b.

The Northwest Watershed Research Center, 1959-86

Clifton W. Johnson*, M. ASCE, Clayton L. Hanson*, M. ASCE,
and Keith R. Cooley*

Abstract

Hydrologic research has been conducted on the Reynolds Creek
Experimental Watershed by the Agricultural Research Service for nearly
27 years. Some noteworthy instrumentation, developed to solve special
problems in measuring precipitation, soil frost, snow accumulation and
melt, subsurface flow, streamflow, and sediment transport are discussed
in the paper. Results from selected long-term records and special
research projects are briefly summarized as examples of operational
watershed design, operation, and performance.

Introduction

The Northwest Hydrology Research Watershed, headquartered at Boise,
Idaho was established as recommended in U.S. Senate Document 59, 1959.
The major purposes and general area of applicability of the research
watershed are summarized in the following excerpt from the report:

> To gain basic information on runoff characteristics,
> including water yield from plateau and foothill grazing areas
> of the Northwest . . . This watershed center, with satellite
> locations, would give information of value to portions of four
> Western States where little factual information is available.
> A dominant feature of this location is winter precipitation
> (both snow and rainfall), where runoff occurs on frozen soils.

Following investigation of numerous potential rangeland research
watersheds, the 234-km^2 Reynolds Creek Watershed in southwest Idaho was
selected for instrumentation and long-term hydrologic study. Watershed
inventories and instrumentation for measurement of precipitation,
weather, and streamflow began in 1960 (Robins, et al. 1965).

This paper highlights some unique instrumentation, data collection
and processing procedures, and research findings from nearly 27 years
of watershed operation.

Precipitation

The precipitation gage network on the Reynolds Creek Watershed was
established to measure both rain and snowfall as influenced by season
and elevation. The original gage network, established in 1960-61,

*Research Hydrologic Engineer, Agricultural Engineer, and Hydrologist,
USDA-Agricultural Research Service, Northwest Watershed Research
Center, Boise, ID 83705.

consisted of 83 weighing type unshielded gages (Hanson, et al. 1980).
This network was converted to 46 dual-gage sites during 1967-68, because
a considerable portion of the annual precipitation was snow and the
unshielded gages were not measuring snowfall adequately. The network
has since been further reduced to 18 sites after thorough analysis of
the records determined which gage sites best represented different areas
of the watershed. Fewer gage sites significantly reduced costs of
servicing the gages and reducing data.

The dual-gage installations consist of one shielded and one
unshielded weighing-type recording gage. At most sites, the unshielded
gage has a 24-hr per revolution time-scale and the shielded gage has a
192-hr per revolution time-scale. The continuous precipitation record
for the dual-gage sites is computed using a log relationship between
shielded and unshielded gage catch (Hamon 1973).

Average annual precipitation varied from 243 mm at the lower
elevation, 1,184 m, to 1,145 mm at the 2164 m elevation. About 56% of
the annual precipitation fell during November through April at the lower
elevations and 75% at the high elevation sites, which indicates that a
considerable portion of the annual precipitation fell as snow.

The Wyoming shield gage system (Rechard and Wei 1980) was evaluated
to determine how well its precipitation catch compared with dual-gage
values. When the temperature was greater than -2 °C, rainfall and
snowfall catch was essentially the same with both gage systems. Wyoming
shield-gage snowfall catch was less than dual-gage amounts when
temperatures were below -2 °C, with the greatest undercatch associated
with high wind speeds.

Soil Frost

In the fall of 1970, measurements began at eight sites using manual
probing tools to measure frost depth to 12.7 cm. When frost had
penetrated below 12.7 cm, the observation was recorded as 12.7 cm or
greater. The most valuable use of these first data sets was to indicate
when freeze-thaw cycles had occurred at a given location.

During 1975-76, 11 sites were instrumented with soil-moisture
resistance blocks (Burgess and Hanson 1979), and at six of these sites,
frost tubes (Harris 1970) were also installed. The soil moisture blocks
were positioned at 5, 10, 15, 20, and 30 cm. Weekly readings of frost
penetration depths were known to be somewhere between two adjacent
depths and when frost was measured at 30 cm, the depth was recorded as
30 cm or greater. The frost tubes which were installed to a depth of
100 cm were also read weekly. A summary of the weekly soil frost
surveys is in press.

Snow

Snow measurements on the watershed involved: (1) a recording snow
pillow; (2) snow course measurements obtained by hand sampling with
standard snow tube equipment; (3) aerial photographs of snow coverage;
and (4) intensive sampling of a relatively large number of points on
small areas, again using standard hand sampling equipment.

Standard snow course measurements are currently made at seven locations on the watershed, at two week intervals from December through May. The sites were selected in conjunction with the Soil Conservation Service and are sampled according to their procedures. That is, samples are taken near the first and fifteenth of each month at 1.5-m intervals along a 15-m snow course transect. Each sample consists of a core length and weight. These two measurements are then used to make a file of snow depth and density or snow water equivalent with time through the snow accumulation and melt season. Data collection started in 1961 and has continued to the present. A snow pillow at one of the snow course sites provides a continuous record of snow water equivalent using a pressure transducer and an electronic telemetry system.

Aerial photographs of selected snowdrift areas on the watershed have been obtained at critical times to determine the percent of area covered by snow, using a planimeter or digitizing procedures. While this information is important in determining areal depletion relationships, it does not provide data on water storage values. A photogrammetric procedure similar to that used to develop topographic maps was developed to determine differences in snow depth by subtracting ground surface elevations from similar elevations of snow covered sample points (Cooper 1965). Point measurements of snow depth and density were required to compute water volumes.

Water supply forecasting methods using both regression and parametric modeling techniques have been tested and modified to provide improved forecasts (Zuzel and Ondrechen 1975). The parametric modeling techniques require more information, but have the advantage of providing both volumes and hydrographs (peak flows and timing) for use by reservoir managers.

Streamflow

Because of the need for accurate streamflow measurement in steep, gravel-boulder channels, special weirs were developed and modeled in the hydraulic laboratory for Reynolds Creek Watershed site conditions. Results of integrated laboratory modeling and field testing of drop-box weirs and other water-measuring devices, suitable for a wide range in stream discharge and sediment loads, were reported by Johnson, et al. (1966). Drop-box weirs with capacities ranging from 0.6-200 m^3/sec. have operated successfully at 13 sites, 1964-86, under a variety of flow and channel conditions. The streamflow records have been used to develop flow-duration-frequency characteristics for several Reynolds Creek watersheds (Johnson and Gordon 1984).

The maximum flood of record at the Reynolds Creek Outlet station, drainage area 234 km^2, was 108 m^3/sec on December 23, 1984, caused by rain-on-snow with some areas of frozen soil. Peak annual streamflows from Reynolds Creek watersheds above 1,500 m elevation were usually from winter storms with snowmelt and frozen soil (Johnson and McArthur 1973). Thunderstorm runoff was generally limited to small drainage areas below 1,500 m. Water year runoff at the Reynolds Creek outlet station ranged from 9 mm in the 1977 drought year to 200 mm in 1984, the greatest of record.

Erosion and Sediment Transport

Watershed sediment yields have been computed from frequent suspended sediment sample data, taken by a variety of automatic pumping samplers operated at streamflow stations and by occasionally sampling both suspended and bedload sediment by hand (Johnson and Gordon 1986, Johnson and Smith 1978). Also, storm runoff, peak flow, and sediment yield data were analyzed using the Modified Universal Soil Loss Equation, MUSLE (Johnson, et al. 1985). Results of the analysis showed a tendency for the MUSLE to underpredict sediment yields for the largest flood events and to overpredict sediment yields for smallest events, especially where snowmelt was associated with storm runoff.

Rainfall simulators were used at selected sites on the watershed to determine factors in the Universal Soil Loss Equation, USLE, for sagebrush rangelands (Wischmeier and Smith 1978). Soil erodibility, K, factor values were computed using data from rototilled plots and cover-management, C, factor values from clipped and undisturbed plots (Johnson, et al. 1984). Soil losses from grazed and ungrazed plots ranged widely and were sometimes poorly predicted by conventional application of the USLE; therefore, studies are continuing on rangeland areas.

Source Watershed Study

The Upper Sheep Creek source watershed, a 26-ha upstream tributary to Reynolds Creek, was initially instrumented in 1969-71 to collect and process hydrologic data for testing subsurface flow models (Stephenson and Freeze 1974). Because of the limited data base for detailed modeling of the subsurface flow system, the source watershed was more intensely instrumented in 1983-86 by: (1) installing rain gages at three locations; (2) establishing a 300-point snow depth and density sampling grid; (3) installing a series of melt collectors under the drift area and shallow snow zones; (4) installing weather stations; (5) drilling 36 holes to bedrock or a restricting layer and installing piezometers at various depths and hillslope locations; and (6) constructing four additional weirs along the 520-m channel reach to monitor streamflow losses and gains.

Presently, precipitation, air temperature, soil temperature, soil moisture, wind, solar radiation, relative humidity, piezometer pressure, and weir water levels are recorded at hourly intervals on an automatic data acquisition system. The intense snow sampling grid is necessary because of the wide variability in snow depth and density, caused by topography, wind, and alternate periods of freezing and thawing (Cooley and Robertson 1985). The extensive data files are being organized and analyzed for subsurface flow and water balance modeling.

Quality of Watershed Records

Because of the need for high quality data for watershed research, instrument accuracy, proper equipment installation, and rigorous data collection schedules were emphasized on the project (Johnson, et al. 1982). Recording rain gages, water level recorders, pumping samplers, and weather instruments failed most often during periods of subzero temperatures; therefore, recorders were cleaned, serviced, and calibrated each summer and fall to prevent failures during the winter. Also, weekly data collection schedules were printed, showing instrument service dates, names of responsible personnel, and comments from past record examinations. Many unscheduled trips were made to streamflow and sediment sampling stations during severe storm events to observe flood conditions, flush sediment from weirs, mark charts, and manually collect samples for calibration purposes. Specialized instrumentation and procedures were developed or adapted for measuring snow precipitation, snow distribution and melt, soil frost, streamflow heavily laden with sediment, bedload sediment, and subsurface water levels.

Summary

The Reynolds Creek Experimental Watershed, after about 27 years of operation, is an example of how such projects provide hydrologic records suitable for frequency and trend analysis, evaluation of unusual events, and watershed model testing. The success of this project required hydraulic laboratory modeling and instrumentation, long-term data collection and processing, and application of realistic models to watershed conditions. Because of specific site conditions, specialized instrumentation was developed for measuring snow precipitation, snow distribution and melt, soil frost, streamflow heavily laden with sediment, and subsurface flow. Also, the importance of proper installation of instruments, regular calibration, and frequent servicing cannot be overemphasized; especially when the work must sometimes be done under the worst weather conditions.

The Reynolds Creek data base is mature and represents conditions over extensive sagebrush rangelands in the Northwest. Excellent hydrologic records are available for future use in hydrologic research.

Appendix I.—References

Burgess, M. D., and Hanson, C. L. (1979). "Automatic Soil-Frost Measuring System." Agric. Meteor. 20(1979): 313-318.

Cooley, K. R., and Robertson, D. C. (1985). "Determining Variability of Snow Accumulation and Ablation on Western Rangeland." In: Western Snow Conference Proc., Boulder Colorado, pp. 156-159.

Cooper, C. F. (1965). "Snow Cover Measurement." Photogrammetric Engin., 31: 611-619, July.

Hamon, W. R. (1973). "Computing Actual Precipitation." Distribution of Precipitation in Mountainous Areas, Vol I., WMO 326, pp. 159-174.

Hanson, C. L., Morris, R. P., Engleman, R. L., Coon, D. L., and Johnson, C. W. (1980). "Spatial and Seasonal Precipitation Distribution in Southwest Idaho." U.S. Dept. Agric., Science and Education Admin., Agricultural Reviews and Manuals, ARM-W-13, 15 p.

Harris, A. R. (1970). "Direct Reading Frost Gage is Reliable, Inexpensive." U.S. Dept. Agric., U.S. Forest Serv. Res. Note, NC-89. 2 pp.

Johnson, C. W., and Gordon, N. D. (1984). "Streamflow Measurements and Analysis for Northwest Rangelands." Hydraulics Division Specialty Conference, ASCE, Proc., Water for Resource Development, Coeur d'Alene, Idaho, pp. 469-473.

Johnson, C. W., and Gordon, N. D. (1986). "Sagebrush Rangeland Erosion and Sediment Yield." Fourth Interagency Sedimentation Conference Proc., Las Vegas, Nevada, March, pp. 2-132 to 3-141.

Johnson, C. W., and McArthur, R. P. (1973). "Winter Storm and Flood Analysis Northwest Interior." Conference on Hydraulic Engineering and the Environment, ASCE, Proc., Bozeman, Montana, August, pp. 359-369.

Johnson, C. W., and Smith, J. P. (1978). "Sediment Characteristics and Transport from Northwest Rangeland Watersheds." Trans. ASAE, 21(6): 1157-1162, 1168.

Johnson, C. W., Copp, H. D., and Tinney, E. R. (1966). "Drop-Box Weir for Sediment-Laden Flow Measurement." Journal of Hydraulics Div., ASCE, 92(HY5): 165-190.

Johnson, C. W., Hanson, C. L., Stephenson, G. R., and Cooley, K. R. (1982). "Report on Quality of Watershed and Plot Data from the Northwest Watershed Research Center, Boise, Idaho." In: The Quality of Agricultural Research Service Watershed and Plot Data, U.S. Dept. Agric., ARM-W-31, pp. 50-59.

Johnson, C. W., Gordon, N. D., and Hanson, C. L. (1985). "Northwest Sediment Yield Analysis by the MUSLE." Trans. ASAE, 28(6): 1885-1895.

Johnson, C. W., Savabi, M. R., and Loomis, S. A. (1984). "Rangeland Erosion Measurements for the USLE." Trans. ASAE, 27(5): 1313-1320.

Rechard, P. A., and Wei, T. C. (1980). "Performance Assessments of Precipitation Gages for Snow Measurement." Water Resources Series No. 76, Water Resour. Res. Institute, Univ. of Wyoming, Laramie, 195 pp.

Robins, J. S., Kelly, L. L., and Hamon, W. R. (1965). "Reynolds Creek in Southwest Idaho, An Outdoor Hydrologic Laboratory." Water Resour. Res. 1(3): 407-413.

Stephenson, G. R., and Freeze, R. A. (1974). "Mathematical Simulation of Subsurface Flow Contributions to Snowmelt Runoff, Reynolds Creek Watershed, Idaho." Water Resour. Res. 10(2): 284-294.

Wischmeier, W. H., and Smith, D. D. (1978). "Predicting Rainfall Erosion Losses - A Guide to Conservation Planning." U.S. Dept. Agric., Agricultural Handbook No. 537, 58 pp.

Zuzel, J. F., and Ondrechen, W. T. (1975). "Comparing Water Supply Forecast Techniques." In: Watershed Management Symposium Proc., ASCE, Logan, Utah, pp. 327-336.

ENDOW, An Expert System for Screening Environmental
Features for Stream Alterations

F. Douglas Shields, Jr., M. ASCE*, and N. R. Nunnally**

Introduction

Expert systems are knowledge-based systems that replicate some of
the problem-solving ability of human experts. They accept problem
descriptions from users and apply logic to a knowledge base composed
of rules to generate solutions. Expert systems typically are computer
programs and function over very narrow problem domains. In the last
few years, expert systems have developed into the most widespread
application of artificial intelligence, although they are not truly
"intelligent" in the strictest sense. Application of expert systems
to engineering practice is promising. The chief benefits of their use
include saving time for experts, replacement of retired experts,
saving money, and provision of more consistent results than human
experts. The greatest benefit of expert system development, however,
is the codification of a base of knowledge in rigorous, logical terms.
Basic references include Kostem and Maher (1986) and Hayes-Roth,
Waterman, and Lenat (1983).

This paper discusses development of an expert system for Environmental Design of Waterways Projects (ENDOW) by stream alteration specialists who were novices in artificial intelligence. The purposes of
this case study are to aid other novice system builders and to disseminate information about ENDOW to potential users.

Stream Alterations and Environment

Alteration of stream channels for flood control, drainage, or
bank erosion control is widely practiced. Surface mines, bridges or
other crossings, riparian construction, and logging also frequently
require channel modification. Environmental studies of channel
modification have tended to emphasize morphological, ecological or
aesthetic effects rather than to advise the engineer about design
practices and procedures that are compatible with resource protection
and conservation. A body of information on environmental features for
stream alterations does exist, but it is widely scattered among many
types of sources.

In order to concentrate this information, a series of technical
reports were prepared (for example, Nunnally and Shields 1985) that
were organized around either the purposes of the features (for

* Acting Chief, Water Resources Engineering Group, US Army Waterways
  Experiment Station, Vicksburg, MS 39180.
** Professor, Department of Geography and Earth Sciences, University
  of North Carolina-Charlotte 28223.

example, aquatic habitat preservation, aesthetics, recreation) or characteristics of the features (for example, maintenance methods, structures, vegetation). The desirability of a particular environmental feature for a given stream alteration project, however, is controlled by the project setting (hydrograph, sediment load, riparian land use, etc.), overall project design (levees, channel, floodwall, etc.), and the design objectives (habitat conservation, recreation, etc.). Accordingly, a set of tables or matrices were developed that rated the environmental features described in the reports based on stream and basin characteristics, project descriptors, and environmental design goals.

Decision to Produce Expert System Software

Although the tables were an effective summary of knowledge regarding environmental feature feasibility, they had several shortcomings as technology transfer vehicles. They were too long to be reproduced in an easily-used format. In addition, they contained minimal descriptive information. Furthermore, expanding the tables to reflect an growing knowledge base was institutionally and economically difficult.

Development of a set of algorithms (a crude prototype expert system) was attempted next to address the shortcomings of the tables. The algorithms were first expressed as flow charts. Although some progress was made in developing overall logic, the flow charts became unwieldy as more and more alternative design features were added. A decision point was then reached: should resources be devoted to development of some type of software tool, or should efforts be confined to refinement of the existing tables and flowcharts? Advantages and disadvantages of each course of action were weighed as shown in Table 1. Table 1 reflects a particular problem and a particular institutional environment.

Several problems associated with initial use of software on mainframes were avoided by developing user friendly software to run on widely available microcomputers.

Hayes-Roth, Waterman, and Lenat (1983) presented three maxims for determining the suitability of a particular task for expert system development. First, the task must be neither too easy (requires several minutes) nor too difficult (requires several months) for human experts. Second, the task must be well-defined. Third, commitment from a human expert is essential. The contemplated project bordered on violation of the first and third maxims. An expert could use the existing knowledge base (reports, tables, etc.) and generate reasonable solutions in a few hours at most. However, the knowledge base was expected to grow both in the number of environmental features it contains and in the depth of coverage of individual features. The area of inquiry spanned so many disciplines and regional factors that no single human expert exists, although the authors of the aforementioned reports had knowledge of the majority of information resources--both people and documents.

TABLE 1.   Pros and Cons of Software versus Hard Copy System for ENDOW.

	Software	Hard Copy
Expertise for Development	Must acquire	Available
Cost	Unknown	Low, easy to control
Attraction to Users	High	Low
Ease of First Use	Moderately Easy, but requires PC	Very Easy
Dissemination of End Product	Easy, rapid	More difficult
Revisions	Easy, rapid	Slow, expensive
Perform Calculations	Yes	No
Provides Hard Copy Record of Each Session of Use	Yes	No

Conventional wisdom describes expert system development as time-consuming and costly. Two factors tended to mitigate this problem in our case, however: the problem to be solved was not terribly complex, and expert system shells had just become widely available. Shell programs query users for examples and rules and then compose an expert system program based on this input. It seemed that the need for learning a new computer language or hiring a knowledge engineer could be eliminated by buying a shell. This approach was recommended (subsequent to our decision to try a shell) by Ludvigsen, et al. (1986).

Shell Selection

Several reviews of available tools for building expert systems were read before a selection was made. Additional reviews have since been published, and among these are Ludvigsen, et al. (1986), Gilmore and Pulaski (1985), and Mette (1987). Manufacturer's literature was also consulted. The KDS system (KDS Corporation, Wilmette, Illinois) was selected because at the time it offered the largest capacity (number of rules and cases) of the acceptably flexible packages priced less than $2000. The ceiling of $2000 was set by funding constraints for the pilot effort.

ENDOW on KDS

The KDS software consists of a system development module and a
playback module. A user's guide, which was rather difficult to understand, and a demonstration module were also provided. The demonstration module was an expert system built using the development module
for diagnosing problems with a film projector. Expert systems are
constructed using the development module to create knowledge bases, or
"knowledge modules." These knowledge modules are utilized by the
playback module to perform as expert systems.

Using KDS, a knowledge module is built by running the development
module, which queries the user for cases and conditions. Cases are
solutions, such as specific environmental features for stream alteration projects (wing deflectors, recreational trails, artificial wetlands). Conditions are facts or value judgments elicited from the
user (the stream is braided, trees are on one side of the existing
channel, the project will have a design discharge greater than the
ten-year event) which are posed to the system user as questions.
Once several cases and conditions have been entered, the development
module offers several techniques for checking and editing the logic
and wording of cases and conditions. At the end of a session, the
development module generates and stores rules based on the cases and
conditions input by the system composer.

The KDS development module has many attractive features. When a
color display is used, screen colors draw attention to critical information. Detailed instructions are provided to the user on the right
side of a split-screen format. Print-outs of rules, conditions, and
cases are available in several formats. Graphics may be incorporated
into the end product system. Two procedures for generating rules and
two procedures for searching for solutions are available.

The primary disadvantage of the KDS system for this application was
that the underlying logic apparently presupposes that every condition
has a definite impact on every case. Although the development system
can handle situations where the system composer does not know what the
influence of a particular condition is, it always assumes that there
is a definite relationship between each case and each condition. Furthermore, a unique relationship between conditions and cases is
assumed, i.e., a given set of conditions can be associated with only
one case. In situations where there were conditions input that have
no bearing on the desirability of a given environmental feature, KDS
tended to generate nonsensical rules.

PROLOG ENDOW

After considerable effort had been expended trying to adapt the
environmental knowledge base and KDS to each other, the KDS shell was
abandoned in favor of PROLOG, a computer language developed for artificial intelligence applications. A PROLOG program is a list of logical statements in English in a specified format. The PROLOG
interpreter searches a set of solution clauses for all solutions that
satisfy a given set of conditions. An excellent review of PROLOG is
provided by Tello (1985).

Initial composition of the expert system in PROLOG was much easier than for the initial KDS system because the work with KDS had resulted in refinement and organization of the knowledge base. Since PROLOG composition allowed the system developers to write rules directly rather than supplying examples to an inference engine which generated rules, nonsense rule generation was minimized. The current (early 1987) version of ENDOW is a PROLOG code that has separate modules for screening environmental features for streambank erosion control projects and channel modification projects. In addition to performing logical operations, the code also does simple calculations and supplies the user with additional information about selected project features on "help" screens.

ENDOW use

ENDOW runs on any PC with MS/DOS and at least 256K of memory. A user simply enters the command "ENDOW" to run the program. After a series of information screens, the user chooses the module for his structural project type (streambank protection or flood control channel) and provides facts, estimates, and value judgements about the project setting, overall plan, and environmental objectives in response to simple questions from the program. The program then provides a list of candidate environmental features such as "single bank channel modification" and "boatways" for further study. If the user wants additional information regarding any of the suggested environmental features, descriptive information is available on "help" screens.

Refinements planned for future versions of ENDOW include addition of a module for flood control levees, addition of new environmental features for each project type, and inclusion of features to enhance repetitive use including tabulation of previous responses and editing a set of responses.

Summary

The application of expert systems to planning and design of stream alterations holds great promise. A considerable literature exists that describes the environmental effects of channel alterations on specific streams. However, only recently have we begun to codify this information into a general framework for the development of environmental planning guidelines and for assessing potential environmental effects of planned stream alterations. The breadth of knowledge required by this task spans disciplines from engineering, earth sciences, physical sciences, and biological sciences. Expert systems that combine the knowledge of experts from several disciplines can lead to better and more consistent decision making in planning and designing alterations.

Although ENDOW is, at its current stage of development, a rather crude prototype expert system for environmental planning, experiences with its application so far have been favorable. It has been applied to several Corps of Engineer projects that are in the planning stage or that have already been constructed. In nearly every case, ENDOW selected all of the alternatives that were considered or incorporated,

and in some cases, suggested others. With the additional refinements
planned for future versions, ENDOW should become a valuable environmental tool for planning and implementing stream alterations.

Acknowledgements

The work described herein was supported by the Office of the Chief
of Engineers as part of the Water Operations Technical Support Program. Permission was granted by the Chief of Engineers to publish
this information. Mr. David Sinew of the University of North Carolina
-Charlotte, wrote the code for the PROLOG version of ENDOW under an
Intergovernmental Personnel Act assignment with the Corps.
Drs. Thomas Walski and Paul Schroeder and Ms. Anne MacDonald reviewed
a draft of this paper. Mention of trade names herein does not constitute endorsement by the Corps of Engineers.

Appendix I - References

1. Gilmore, J. F. and Pulaski, K. T. 1985. "A Survey of Expert System Tools," A paper presented at the IEEE Applications of Artificial Intelligence Conference, Miami, Florida.

2. Hayes-Roth, Frederick, Waterman, D. A., and Lenat, D. B., eds.
   1983. Building Expert Systems. Addison-Wesley Publishing
   Company, Reading, Mass.

3. Kostem, C. N. and Maher, M. J., eds. 1986. Expert Systems in
   Civil Engineering, ASCE, New York.

4. Ludvigsen, P. J., Grenney, W. J., Dyerson, D., and Ferrara, J. M.
   1986. "Expert System Tools for Civil Engineering Applications,"
   in Kostem, C. N. and Maher, M. L., eds. 1986. Expert Systems in
   Civil Engineering, ASCE, New York.

5. Nunnally, N. R. and Shields, F. D. "Incorporation of Environmen-
   tal Features in Flood Control Channel Projects," Technical
   Report E-85-3, US Army Engineer Waterways Experiment Station,
   Vicksburg, Mississippi.

6. Mette, Stephen. 1987. "Expert Systems' Prices Falling," Lotus,
   Vol 3, No. 1, pp. 17-18.

7. Tello, Ernie. 1985. "The Languages of AI Research," PC Magazine,
   Vol 4, No. 8, pp. 173-189.

EXPERT SYSTEM FOR MIXING ZONE
ANALYSIS OF TOXIC AND CONVENTIONAL DISCHARGES

Gerhard H. Jirka[1], M.ASCE, and Robert L. Doneker[2]

Introduction

U.S. water quality policy - as embodied in Federal and State regulations - includes the concept of a mixing zone, a limited area or volume of water where the initial mixing of an aqueous discharge takes place. Actual water quality standards or site-specific permits apply outside, and at the edge of, the mixing zone. The 1983 Water Quality Standards Handbook [1] gives the rationale and historic develpoment of this policy for conventional pollutants. More recently, special regulations have been proposed for toxic discharges [2] which limit lethal acute concentrations to a spatially highly restricted zone - herein called the toxic dilution zone - typically much smaller than the usual mixing zone. This new policy addresses the particular nature of toxic pollutants for which both acute and chronic limits apply in order to constrain their impact.

The implementation of this policy in the permitting process places the burden on regulators (EPA or State agencies) as well as applicants to predict the initial dilution of a given discharge and the characteristics of its mixing zone. Given the large number of possible combinations of types of pollutant sources, discharge configurations, ambient environments, and mixing zone definitions, the analyst needs considerable training and experience to conduct accurate and reliable mixing zone analyses. The application of predictive models forms the core of such analyses. The unexperienced or moderately experienced analyst faces the following problems:
1) Choice of predictive model(s): Commencing in the early seventies a plenitude of predictive models for a variety of ambient and discharge conditions have been developed. These models range in complexity from simple analytical formulae or nomograms to highly intricate numerical solutions of differential equations. Even though a number of assessment manuals and summaries have been prepared and actual EPA endorsements [e.g.3] for certain models in specific situations have been issued, the average user has no reliable guidance on which model is appropriate or, even more, which is best. Examples abound of "model abuse". Furthermore, there is usually a trend to employ unnecessarily complicated models.
2) Limits of applicability: Even though a particular model may be appropriate for a given discharge that does not mean it will hold over an unlimited range of conditions. Model developers often fail to

1) Professor, 2) Research Assistant, DeFrees Hydraulics Laboratory, School of Civil and Environmental Engineering, Cornell University, Ithaca, NY 14853

specify such limits of applicability or model users may overlook these. For example, a frequent error in the application of EPA "plume models" [3] – all of which assume a theoretical infinite receiving environment – is the consideration of the actual finite depth effects: in reality, the plume may attach to the bottom or may become vertically fully mixed. The model user may fail to recognize these possibilities which can arise due to changes in ambient (e.g. low flow) or discharge conditions. Consequently, analysts have published model "predictions" in which the predicted plume diameter vastly exceeds the actual water depth!

3) Data needs: Once the correct choice of model has been assured, the user is faced with acquiring appropriate data and establishing a design base. For the unsophisticated model user this often becomes a tedious task (especially for highly complex models) as little guidance may be given as to what design base to assume, where to obtain data, which data is crucial and which may be simply estimated.

The case for an expert system

In essence expert systems mimic the way an expert or highly experience person would solve a problem. An expert system is a structured computer program that uses knowledge and inference procedures obtained from experts to enable inexperienced personnel to solve complex problems. The knowledge base includes a set of "objective" or widely accepted facts about a general problem area. It also includes the set of parameters or data an expert would seek in order to characterize a specific problem. The inference procedures are "subjective" rules of judgement which the expert might use when analysing the problem. The inference procedures provide the rules for selecting an appropriate solution to the problem from the knowledge base. The inference procedures allow the system user to search rapidly and systematically through the knowledge base to obtain a solution for the given problem. This element uses structured search techniques based upon mathematical logic.

The development of an expert system for mixing zone analysis promises significant advantages for water pollution control and management:
---it assures the proper choice of model for a given physical situation.
---it assures that the chosen model is applied methodically without skipping essential elements.
---it guides the acquisition or estimation of data for proper model prediction.
---it allows a flexible application of design strategies for a given point source, screening of alternatives and, if necessary, switching to different predictive models thus avoiding rigid adherence to a single model.
---it flags borderline cases for which no predictive model exists suggesting either avoidance of such designs or caution by assigning a degree of uncertainty.
  ---it allows a continuous update of the knowledge base as improved predictive models, experimental data and field experience with particular designs become available.
---it provides a documented analysis listing the knowledge and decision logic that have lead to the problem solution. Thus, unlike

conventional programs or computer algorithims an expert system is not a "black box".

---it provides a common framework which both regulators (Federal or State), applicants and the scientific community can use to arrive at a consensus on the state-of-the-art of impact prediction and control.

---finally, and perhaps most importantly, it provides a <u>teaching environment</u> whereby the initially inexperienced analyst through repeated interactive use gains physical insight and understanding about initial mixing processes.

CORMIX1

The <u>Cor</u>nell <u>Mix</u>ing Zone <u>Ex</u>pert System (CORMIX) is a series of software subsystems for the analysis, prediction and design of aqueous toxic or conventional pollutant discharges into watercourses, with emphasis on the geometry and dilution characteristics of the initial mixing zone. Subsystem CORMIX1 deals with buoyant submerged single port discharges into flowing unstratified water environments, such as rivers, estuaries, and coastal waters. It includes the limiting cases of non-buoyant discharges and of stagnant conditions. The minimal hardware configuration is at the level of IBM-PC/XT with printer for hardcopy output.

Fig.1 shows the system elements of CORMIX1. During system use the elements are loaded sequentially by user prompt or are activated automatically in batch mode:

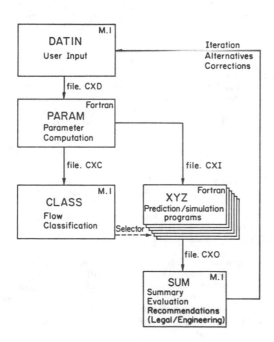

Fig. 1: System elements of CORMIX1. Arrows indicate the data transfer among elements. "File" is the user-specified filename for the specific problem.

1) DATIN is an M.1 (Teknowledge, Inc.) logic program for the entry of relevant data about the discharge situation and for the initialization of the other program elements. The following data groups need to be entered: general identifier information, ambient conditions (geometry and hydrography), discharge conditions (geometry and fluxes), and information desired including legal mixing zone definitions. DATIN performs consistency checks, and provides advice for input parameter selection.

2) PARAM is a Fortran program that computes relevant physical parameters for the given discharge situation. This includes various flux quantities and length scales involving mathematical expressions that are beyond the fundamental operations supported by the M.1 language.

3) CLASS is an M.1 program that classifies the given discharge situation into one of many possible hydrodynamic configurations, e.g. a simple jet or plume, a boundary attached jet, an unstable vertically mixed case, or mixing controlled by the ambient crossflow. Each separate flow configuration has an alphanumeric label, XYZ in general, and a detailed hydrodynamic description is provided.

4) XYZ is a series of separate Fortran programs corresponding to each of the flow configurations XYZ as identified by CLASS. The user chooses the appropriate program. These simulation programs are all based on simple similarity analyses for the turbulent mixing processes that occur within the distinct hydrodynamic zones that comprise a given flow configuation. The programs print out data on geometry (trajectory, width, etc.) and associated mixing (dilution, concentration) following the path of the effluent discharge.

5) SUM is an M.1 program that summarizes the given situation, comments on the mixing characteristics, evaluates how applicable legal requirements are satisfied, and suggests possible design alternatives and improvements.

The purpose of CORMIX is to obviate the need for a detailed hydrodynamic understanding and experience for the novice analyst. A general science or engineering background (at a two or four year college level) appears to be a minimum educational requirement, however, in order to be able to supply relevant data, to interpret the system information, and ultimately to learn and become knowledgeable through interactive system use.

Hydrodynamic Aspects: The Knowledge Base of CORMIX1

CORMIX1 deals with submerged buoyant discharges into a flowing unstratified water body. The system assumes a schematic rectangular cross-section bounded by two banks - or by one bank only for coastal or other laterally unlimited situations. The user receives detailed instructions on how to approximate actual cross-sections that may be quite irregular to fit that rectangular schematization.

Even in this simple schematized ambient geometry there remains a tantalizing amount of geometric and dynamic detail: the discharge location in relation to the bottom or the shoreline; the discharge orientation may be with the flow, against the flow, or vertically upward across the flow, or some other arbitrary angle, the water depth may be deep or shallow; the ambient flow may be stagnant or fast and highly diffusive; and the discharge flow may be non-buoyant or highly buoyant, with high or low efflux velocity. As a consequence there are a myriad of possiblities – as evidenced by available field data and prior laboratory studies – of what the flow in the initial mixing acually does: it may cling to the bottom resulting in high benthic concentations; it may mix vertically due to instabilities in shallow water; it may rise to the surface and the partially intrude upstream against the ambient current; it may interact with the shoreline due to lateral diffusion, etc. etc..

A rigorous flow classification has been developed in order to distinguish among these possibilities and to provide for the appropriate selection of predictive models. All relevant variables are summarized in Fig.2. Dimensional analysis [see 4,5,6] provides the following length scales:

$\ell_Q = Q_0/M_0^{1/2}$ = discharge (geometric) scale

$\ell_m = M_0^{3/4}/J_0^{1/2}$ = jet/plume transition scale

$\ell_m = M_0^{1/2}/u_a$ = jet penetration scale into crossflow

$\ell_b = J_0/u_a^3$ = plume flotation scale into crossflow

These scales when compared to the major length dimension of the near-field, i.e. the depth H, yield the flow classification scheme shown in Fig.3. The classification divides all buoyant discharges into (near-)vertical [V] and into (near-)horizontal [H] cases. Sketches of each flow class are shown on Fig.3. With the exception of flow class H5 that is inherently bottom attached due to dynamic effects in the discharge flow and crossflow Fig.3 does not address possible bottom attachment. Supplemental classifications – on the basis of various ratios formed by ℓ_m, ℓ_b, ℓ_Q and h_0 – distinguish between wake-like and Coanda (dynamic) attachment. These criteria are not shown herein. In total, about 20 major flow classes have been identified for practical relevance. Fig.3 is not concerned with far-field mixing processes, such as buoyant spreading and passive diffusion (caused by the turbulence in the ambient flow with friction factor f). Such processes, including shoreline intertaction, are addressed directly in the simulation models.

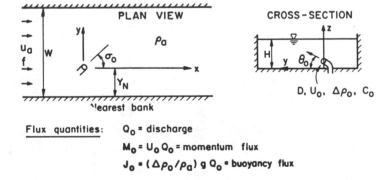

Flux quantities: Q_0 = discharge

$M_0 = U_0 Q_0$ = momentum flux

$J_0 = (\Delta\rho_0/\rho_a) g Q_0$ = buoyancy flux

Fig. 2:   Schematization of discharge configuration assumed in CORMIX1. The rectangular channel may be bounded or unbounded.

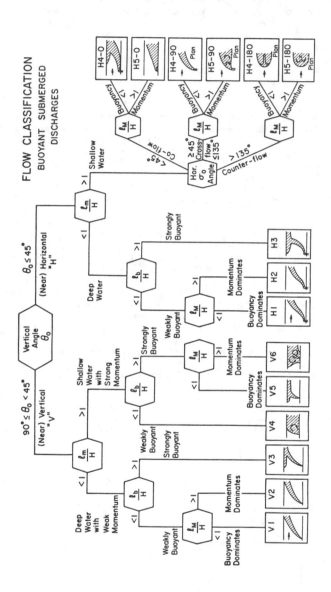

Fig. 3: Flow classification of the near-field of submerged buoyant discharges from a single port outfall in finite water depth H. All sketches are side views of the jet/plume near-field except as noted. The critical values in the length scale criteria are "of order of unity." Exact values are specified from experimental data or detailed numerical models.

The individual simulation programs are all constructed in modular form using simple analytical elements with appropriate transition criteria to proceed from one module to the next. The following modules have been formulated:

01: discharge
12: weakly deflected wall jet
17: strongly defl. wall jet
22: strongly deflected plume
32: surface impingement /
    upstream spreading
34: surface impingement /
    unstable recirculation
61: passive diffusion

11: weakly deflected jet
16: strongly deflected jet
21: weakly deflected plume
31: gradual surface approach
33: surface impingement /
    full vertical mixing
41: buoyant surface spreading
51: wake recirculation

Nine transition rules link these modules and end rules terminate the simulation depending on the specifications of legal mixing zones and/or toxic dilution zones.

The accuracy of the CORMIX1 predictions has been tested with available data and the predictions of detailed buoyant jet predictions (such as the EPA models [3]) which are of course limited to the fully submerged regions (as in modules 11,16,21 and 22). The CORMIX1 prediction agrees well with these reference values.

Conclusions

A working version of CORMIX1 has been developed and is currently being tested in cooperation with various regulatory agencies and design engineers. Initial experience has been very satisfactory. CORMIX1 appears to be a highly flexible tool for the regulatory analysis of diverse environmental situations. It also is adaptive for the rapid evaluation of alternatives in engineering design. The knowledge base and model logic are transparent to the user and thus offer an excellent learning environment.

Acknowledgements
        This work is being conducted as a cooperative effort with the U.S. EPA Environmental Research Laboratory, Athens, Georgia (Dr. Thomas Barnwell, project director).

References
1. U.S.EPA, 1983, Water Quality Standards Handbook, Wash., DC
2. U.S.EPA, 1985, Technical Support Document for Water Quality-based Toxics control, Wash.,DC
3. W.P. Muellenhoff et al., 1985, Initial Mixing Characteristics of Municipal Ocean Discharges (Vol. 1&2), U.S.EPA, Env. Res. Lab., Narragansett, RI
4. R.S. Scorer, 1968, Air Pollution, Pergamon Press
5. S.J. Wright, 1977, Mean Behavior of Buoyant Jets in Crossflow, J. Hydr. Div., ASCE, 103, 499–513.
6. E.R. Holley and G.H. Jirka, 1986, Mixing in Rivers, U.S. Army Corps of Eng.,Tech. Rep. E–86–11, Vicksburg, MS

STOCHASTIC STREAM FLOW ANALYSIS FOR HYDROGRAPH
CONTROLLED WASTE RELEASE

James F. Cruise[1], M.ASCE, and Vijay P. Singh[2], M.ASCE

Abstract

A risk based procedure for the design of hydrograph controlled
release sewage lagoons is derived. The procedure is based on a stochastic analysis of streamflow events above a given threshold level.
Variables of interest include the time intervals between such flow
events and the durations and volumes of these events. A stochastic
model based on representing the streamflow mechanism as a compound
Poisson counting process is derived and tested on various Louisiana
watersheds. This model is shown to adequately represent the stochastic structure of flow events on these streams. Several design procedures based on a combination of maximum time spans and minimum flow
duration and volume are discussed.

Introduction

Many small localities, especially in the southern United States,
cannot afford to construct treatment plants for their domestic waste
products. A feasible solution to the problem may lie in the construction of sewage lagoons to store the waste water during dry periods and
release the excess into nearby streams during periods of high flow in
those streams. In order for these hydrograph controlled release (HCR)
facilities to be properly designed, a stochastic analysis of the flow
characteristics of the receiving stream is necessary. The variables
of primary interest in this regard are shown in Figure 1. The most
important variable is undoubtedly the time interval between successive
periods of high flow, T_s. This period represents the time during
which waste would have to be collected and stored in the lagoon. The
other variables of importance shown in Figure 1 are the duration of
the period of high flow (D_s) and the volume of flow during that
period, V_s. The flow duration represents the length of time during
which waste could be released from the lagoon and the flow volume is
related to the volume of waste which could be released. In order for
a risk based design procedure to be derived, the stochastic structure
of these variables must be investigated.

---

[1] Assistant Professor, Department of Civil Engineering, Louisiana State
University, Baton Rouge, LA 70803.
[2] Professor, Department of Civil Engineering, Louisiana State University, Baton Rouge, LA 70803.

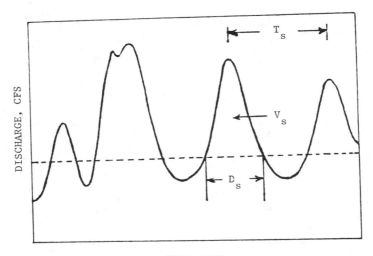

TIME, DAYS

Figure 1.   A typical streamflow time series.

Literature Review

The purpose of this paper is to present a stochastic analysis of
the variables shown in Figure 1 and to discuss some possible design
scenarios for sewage lagoons. The initial work on the design of
hydrograph controlled sewage lagoons was performed by Zitta and Mowry
(1979), Hill and Zitta (1980), and Zitta and Yih (1981). This work
was basically concerned with the design of lagoons based solely on the
time span between high flow events (T_s). Chauhan (1982) attempted to
incorporate an analysis of flow duration and volume into the basic
procedure proposed by Zitta.

The original procedure of Zitta (Zitta and Mowry, 1979; Hill and
Zitta, 1980) was based on a log-normal fit of the observed maximum
annual time spans. Chauhan (1982) altered this procedure to include
an analysis of flow volumes and durations. Log Pearson type III
distributions were fitted to observed data for annual maximum time
spans, volumes and durations. Joint probabilies of annual time
spans and durations were calculated and design recurrence intervals
were selected based upon these joint probabilities. Then, at these
exceedance probabilities, the corresponding volumes of discharge were
read from their probability plots.

The purpose of the research reported in this paper is to expand
the procedure discussed above to include the incorporation of knowledge of the stochastic structure of the streamflow mechanism. The
effects of the threshold value (q_0) used to define the "high" flow
series will be discussed and alternative design scenarios based upon a
combination of maximum time spans and minimum duration and volume will

be analyzed. A physically based algorithm for the design of HCR
sewage lagoons will be derived.

Theory

    The streamflow mechanism was represented by a compound Poisson
counting process (Todorovic, 1978). This model is based upon the
calculation of the extremes (largest values) from a Poisson process.
The calendar year is broken into a number of periods during which the
streamflow series can be considered to be a stationary, homogeneous
Poisson counting process. Seasonal probabilities (summer season,
May – October; winter season, November – April) can be estimated from
this model by utilizing the theory of random number of random variables (Todorovic, 1970). These seasonal probabilities can then be
converted to annual values by the use of joint probabilities; i.e.,
multiplying the seasonal probabilities, assuming the seasons to be
independent.

    It is believed that storms occurring in Louisiana during the
winter season are due to cyclonic climatic conditions, while
thunderstorms are typical of the summer season. Therefore, it seems
reasonable to assume that events in each season would be identically
distributed. Care was taken in data selection to insure that only
independent events were chosen.

    Under the Poisson assumption, the time interval between succes-
sive events is represented by the exponential distribution. Then the
distribution of the time interval between those events is given by:

$$f(t) = \delta \ \exp[- \ \delta(t)] \tag{1}$$

where δ is equal to the inverse of the average time interval. The
distribution of the largest time interval between flow events above a
given threshold value (q_0) for a winter or summer season is given by
Todorovic (1970) as:

$$\phi(t) = \exp[- \ \lambda \ \exp(- \ \delta(t))] \tag{2}$$

where λ is the average sample size for the period of record.

    Equation (2) provides a stochastic model for the maximum time
spans between occurrences of "high" flow events for any season,
regardless of the criteria used to define "high" events. Recent
research (Cruise, 1986; Ashkar and Rousselle, 1981) has shown that the
particular threshold value selected is not critical to the results of
the extreme value distribution calculated from equation (2). Therefore, any convenient value which establishes the serial independence
of the data series can be chosen.

Methodology

    In this research, maximum time spans for HCR sewage lagoon design
were calculated by fitting equation (2) to observed seasonal data from
Louisiana streams. Data were collected for summer and winter series

from various gages in Louisiana representing primarily ephemeral conditions. Ephemeral streams were chosen for this analysis because many streams in the southwest exhibit this condition, and thus it was considered to be the most critical case. All independent events above a convenient threshold value were chosen for the analysis. A preliminary analysis of these data clearly showed that virtually all of the long time spans between events occurred during the summer season. Therefore, it was determined to use only summer series data for the remainder of the analysis since these data represent the most critical conditions. Thus, when probabilities are calculated, they will refer to the summer series data only. A frequency histogram for the summer series time spans for one of the gages used in the analysis is shown in Figure 2. The figure clearly shows that the data exhibit the characteristics of the exponential distribution, as would be expected from the assumed model.

Equation (2) is fitted to the observed summer series time spans by using the maximum likelihood estimate of the Poisson parameter (λ) and the exponential parameter (δ). Once the extreme value distribution (equation (2)) is fitted to observed data, design time spans corresponding to any desired recurrence interval can be calculated.

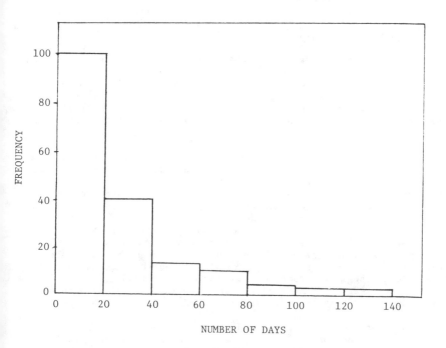

Figure 2.   Frequency histogram of summer series for
            Bogue Lusa, Louisiana (threshold = 100 cfs).

Analysis of the other two variables of interest in HCR sewage
lagoon design are not quite as straightforward. Visual inspection of
the series of flow durations shows that the series possesses a finite
lower bound of one day, and that this bound is encountered several
times during the period of record. Moreover, a specific value of flow
volume is associated with this one day duration, thus establishing a
lower bound for that series also. It can be reasoned that sewage
lagoon design should be based on the concept of maximum anticipated
time spans between flow events and minimum flow duration and volume.
Long time spans during which waste would have to be collected and
stored in the lagoon, followed by one or more short duration periods
during which waste could be released from the lagoon would seem to be
an appropriately conservative design plan. However, a stochastic
analysis of minimum duration and volume is hampered by the occurrence
of the finite lower bound. The reasonable approach would appear to be
to select the lower bound of one day and its associated volume as the
design parameters for those particular variables.

Possible Design Scenarios

The simplest design scenario, the one originally proposed by
Zitta (1980), would be to assume that the design time span is followed
by an event long enough such that the lagoon could be emptied of
waste. Analysis of the data used in this study has shown that this is
not true in all cases. There is some additional risk of functional
failure involved in this assumption.

A more conservative scenario would be to base the design on the
concept of joint probabilities. A design time span would be followed
by an event of one day duration and volume, which would then be
followed by another time span with any desired probability of
exceedance. The total probability of occurrence of this series of
time spans and duration/volume would be the product of the probabilities of the design time spans.

Another design scenario could be based on the joint probability
of the occurrence of the largest and second largest time spans in any
year. The distribution of the second extreme value from a Poisson
process has been derived by Borgman (1961), who provided a convenient
table of probabilities of this function. Under this design, the
largest design time span would be followed, after a one day event, by
the second largest time span with the same exceedance probability.
For instance, if a three year design interval is chosen, the probability of exceedance is $p = 1/3 = .33$. Assuming the independence of
the events, the joint probability of the largest and second largest
event occurring in the same year is given by $p = (.33)(.33) = .10$.
Thus, a ten year design period results from initial use of a three
year period for both the largest and second largest events.

The scenarios discussed above are only a few of the many possible
ones based on the concept of maximum time spans and minimum durations
and volumes. However, they help to establish the concept of a sound
probability based design procedure for HCR lagoons which incorporates
the stochastic structure of streamflow into the analysis.

References

Ashkar, F., El-Jabi, N. and Rousselle, J., "Exploration of an Extreme Value Partial Time Series Model in Hydroscience," Time Series Methods in Hydrosciences, A. H. El-Shaarawi and S. R. Esterby, Editors, Elsevier Scientific Publishing Company, New York, 1982, pp. 76-92.

Ashkar, F. and Rousselle, J., "Some Remarks on the Truncation Used in Partial Flood Series Models," Water Resources Research, Vol. 19, No. 2, April 1983, pp. 477-480.

Borgman, L. E., "The Frequency Distribution of Near Extremes," Journal of Geophysical Research, Vol. 66, NO. 10, October 1961, pp. 3295-3307.

Chauhan, S. S., Statistical Analysis of Stream Events Following Drought Spans with Application in Effluent Storage Design, M.S. Thesis, Mississippi State University, Mississippi State, Miss., December, 1982.

Cruise, J. F., "The Impact of Threshold Selection in Partial Duration Flood Frequency Analysis," Hydrological Science and Technology: Short Papers, American Institute of Hydrology, Vol. 2, No. 2, June 1986, pp. 7-12.

Hill, D. O. and Zitta, V. L., Hydrograph Controlled Release, Engineering and Industrial Research Station, Mississippi State University, Mississippi State, MS, September 1980.

Todorovic, P., "On Some Problems Involving Random Number of Random Variables," Annals of Mathematical Statistics, Vol. 41, No. 3, 1970, pp. 1059-1063.

Todorovic, P. and Zelenhasic, E., "A Stochastic Model for Flood Analysis," Water Resources Research, Vol. 6, No. 6, December 1970, pp. 1641-1648.

Todorovic, P., "Stochastic Models of Floods," Water Resources Research, Vol. 14, No. 2, April 1978, pp. 395-356.

Zitta, V. L. and Mowry, B. S., "Statistical Analysis of Stream Events as Applied to Controlled Release of Lagoon Effluents, Water Resources Research Institute, Mississippi State University, Mississippi State, MS, October 1979.

Zitta, V. L. and Yih, V. J., "Statistical Analysis of Stream Events with Applications in Wastewater Discharge," Statistical Analysis of Rainfall and Runoff, V. P. Singh, Editor, Water Resources Publications, Fort Collins, CO, 1982, pp. 679-699.

Reservoir Water Quality Assessment Using In Situ Microcosms

Richard H. French, M. ASCE[1]
Richard J. Watts, A.M. ASCE[2], James J. Cooper[3]

Introduction

The assessment of reservoir water quality and the design of cost-effective reservoir restoration programs generally require the identification of the limiting nutrient. That is, Liebig's law of the minimum states that the growth of an organism is limited by the nutrient which is available to it in the minimal quantity relative to its needs. It is generally accepted that phytoplankton need a relatively fixed atomic ratio of nitrogen to phosphorus of 16. Thus, when the total forms of nitrogen and phosphorus are used phosphorus is considered limiting when TN:TP > 17; nitrogen is considered limiting when TN:TP < 10; and algal growth is considered to be proportional to phosphorus and/or nitrogen when the TN:TP ratio is between 10 to 17. It is appropriate to note that controlled laboratory experiments have shown that this ratio may actually range from 7 to 27 rather than 10 to 17. When the biologically available forms of these nutrients (TIN and OP) are used, TIN:OP < 8 is generally considered to be indicative of nitrogen limited growth.

Although there are a number of techniques for determining the limiting nutrient, as will be demonstrated, in-reservoir microcosms permit in situ experiments to be conducted under relatively natural conditions on a conveniently small scale and produce additional data and information which can be used to design cost-effective restoration programs. With in situ microcosms the surrounding habitat is protected and the results, with care, can be extrapolated to the whole system. Further, the response of the phytoplankton communities in the microcosms occur quickly with changes in the species composition appearing within days.

Background

Edmondson and Edmondson (1947) first proposed the use of in situ microcosms for ocean trophic studies, and this technique has also been used to study nutrient limiting factors in lakes; see for example, Goldman (1962), and the fertilization requirements of a small lakes; see for example, Kemmerer (1968).

[1]Research Professor, Water Resources Center, Desert Research Institute, 2505 E. Chandler Ave., Suite #1, Las Vegas, Nevada, 89120.

[2]Assistant Professor, Department of Civil Engineering, University of Nevada, Reno, Nevada, 89557.

[3]Aquatic Biologist, Nevada Division of Environmental Protection, Carson City, Nevada, 89710

Lahontan Reservoir is located in west-central Nevada approximately 45 miles (75 km) southeast of Reno. The reservoir, a part of the Newland's Irrigation Project, stores water derived from the Carson River and a transbasin diversion from the Truckee River. Although the primary purpose of this reservoir was and remains to provide irrigation water to the agricultural lands of the Lahontan Valley, the reservoir has become one of the most heavily used recreation areas in northern Nevada.

Water quality problems at Lahontan Reservoir include aesthetics (odor and unsightly conditions), algal toxcity (fish kills), and hypolimnetic oxygen depletion. During the summer months, blue-green algae becomes dominant and often comprises 100% of the phytoplankton community during blooms. Note, green algae are generally considered more desirable than floating blue-green forms because blue-green algae are unplatable to most zooplankton species while green algae are favored. Thus, blue-green algae impede and often block the transfer of energy up the food chain. All of these water quality problems detract from the Lahontan Reservoir both as a fishery and a recreational resource.

On the basis of the results from a reservoir wide sampling program, it was hypothesized that the dominance of blue-green algae in Lahontan Reservoir was due to warm water temperatures, the depletion of biologically available nitrogen in the water column; that is nitrogen is the limiting nutrient, and the ability of the species of blue-green algae present in the reservoir to fix free atmospheric nitrogen.

Although there are several traditional techniques that can be used to identify the limiting nutrient; see for example, Krenkel and Novotny (1980), none of these techniques allow the prediction of the response of the ecosystem to the removal of the limitation. For this reason, microcosm studies were performed to both confirm that the limiting nutrient at Lahontan was nitrogen and examine the response of the reservoir when this limitation is removed. The first microcosm studies were performed on the reservoir in 1982 as a component of the Lahontan Reservoir Water Quality Project, and the second set of microcosm studies were performed in 1986 as a component of a research project sponsored by the U.S. Environmental Protection Agency.

Microcosm Results

In the 1982 study, polyetheylene tubes 0.76 m in diameter and 8 m in length were used as microcosms. Replicate tubes were treated with chemicals which increased the amount of biologically available nitrogen in the water column and with chemicals which increased the biologically available phosphorus in the water column. Replicate tubes with no chemical addition served as a control. In the 1986 study, polyetheylene tubes 0.76 m in diameter and 2 m in length were used as microcosms. Replicate tubes were treated with chemicals which increased the biologically available nitrogen, the biologically available phosphorus, and the biologically available nitrogen and phosphorus.

The results of these studies were as follows. First, the microcosms in which the amount of biologically available phosphorus was

increased had at the end of the experimental period the highest standing crop of phytoplankton and the greatest percentage of blue-greens.
Second, the 1982 control microcosms exhibited phytoplankton concentrations approximately one-third of that in the tubes receiving a phos-
phate addition. Third, the microcosms in which the amount of biologically available nitrogen was increased exhibited a greater initial
algal concentration with a subsequent decline in phytoplankton abun-
dance. The final standing crop of phytoplankton in these microcosms
was only 2.5 to 7.5 percent of that in microcosms in which the available phosphorus had been increased. Furthermore, the final mean composition of blue-greens was about 20 percent compared to a mean of
approximately 90 percent in the other treatments, Cooper and Vigg
(1983).

In summary, increasing the biologically available nitrogen to phosphorus ratio in Lahontan Reservoir generally resulted in an improvement in water quality. There are several methods of increasing the
nitrogen to phosphorus ratio on a reservoir-wide basis. First, phosphorus can be removed from the inflows. Second, the reservoir could
be fertilized with nitrogen. Third, the phosphorus loading to the
reservoir could be decreased while increasing the nitrogen loading.
It should be noted that as demonstrated by the microcosm results the
addition of biologically available nitrogen to the reservoir would increase the primary productivity of the reservoir; however, this
increase in productivity would most likely be non-nitrogen fixing
algae capable of entering the food chain. From the viewpoint of
aquatic biology, the kind of primary productivity is often more important than the quantity when the recreational potential of the aesthetic appearance of the surface waters are considered. However, an
increase in primary productivity could also result in a further decrease in hypolimnetic oxygen concentration.

Conclusions

Although the microcosm method of identifying the limiting nutrient
and studying the response of the ecosystem to the removal of this
limitation is a realistic and cost-effective method of measuring the
cause-effect relationships of nutrient dynamics in a reservoir, water
managers must use caution in interpreting the results and developing
restoration programs. The microcosm while representative is still an
oversimplification of a very complex system. For example, in the
studies described here the potential limiting effect of carbon dioxide
on the system was not considered. Before a restoration is designed
and implemented more detailed ecological measurements such as primary
production diurnal changes in dissolved oxygen and ph, phytoplankton
and zooplankton interactions and toxicity would be required.

Acknowledgements

The work on which this paper is based was sponsored by the Legisla-
ture of the State of Nevada and the U.S. Environmental Protection
Agency under Assistance Agreement R811124-02-0. The authors grate-
fully acknowledge this support.

References

Cooper, J.J. and Vigg, S., "Experimental In Situ Fertilization
    Studies," The Lahontan Reservoir Water Quality Project, Vol. IV,
    Desert Research Institute, Reno, Nevada, 1983.
Edmondson, W.T. and Edmondson, Y.H., "Measurements of Production in
    Fertilized Salt Water," Journal of Marine Research, Vol. 6, 1947,
    pp. 228-246.
Goldman, C.R., "A Method of Studying Nutrient Limiting Factors in situ
    in Water Columns Isolated by Polyethylene Film," Limnology and
    Oceanography, Vol. 7, 1962, pp. 99-101.
Kemmerer, A.J., "A Method to Determine Fertilization Requirements of a
    Small Sport Fishing Lake," Transactions of the American Fisheries
    Society, Vol. 97, 1968, pp. 425-428.
Krenkel, P.A. and Novotny, V., Water Quality Management, Academic
    Press, New York, 1980.

Model Choice and Scale
in Urban Drainage Design

David F. Kibler, George A. Krallis, and Marshall E. Jennings[*]

Abstract

In modelling small urban drainage systems, the engineer has a great
deal of latitude in both the choice of modelling approach as well as
the scale at which he represents the system. This paper addresses
the question of model choice and scale by comparing various model
applications to a 95-acre gaged catchment, known as the Lake Hills
watershed, in Bellevue, Washington. The array of methods and models
available for estimating design floods in the small urban catchment
is significant, perhaps even overwhelming. It includes: physically
based continuous models such as EPA SWMM and USGS DR3M; single-event
models such as the Penn State Runoff Model; various desk-top
microcomputer models based on the Rational formula, SCS TR-55 tabular
hydrograph, and selected unit hydrograph procedures. In addition,
there are several regional flood frequency methods available for the
small ungaged urban site. The main objectives here was to compare
alternative procedures for establishing flood frequency curves in a
small urban watershed, recognizing that in the Lake Hills system, the
authors had access to two years of rainfall-runoff data collected by
the USGS. A secondary issue, which became critical to the primary
objective, was the best method of distributing hourly rainfall
amounts in 5-minute intervals so as to utilize long-term hourly
rainfall records in simulating the small urban catchment.

Characteristics of the Lake Hills Watershed

The Lake Hills watershed is a 95.3 acre (excluding roof tops draining
to dry wells) residential area with single family homes built in the
late 1950's. It is located in the city of Bellevue, Washington,
which lies along the eastern shore of Lake Washington just west of
Seattle. The Lake Hills catchment is fairly steep with slopes of 14
percent. It is approximately 28 percent impervious, with Alderwood
soils throughout the basin. The outfall of the separate storm drain
system discharges to an open channel that flows into Kelsey Creek,
which in turn discharges into Mercer Slough and then into Lake
Washington. There is no detention storage in the present system,
although surcharge and inadvertent ponding have been observed.

[*]Respectively, Professor Department of Civil Engineering, Penn State
Unviersity and ASCE Member; Graduate Research Assistant, Department
of Civil Engineering, Penn State University, University Park; Urban
Studies Coordinator, US Geological Survey, Gulf Coast Hydroscience
Center, NSTL, Mississippi and ASCE Member.

The Lake Hills watershed is one of three Bellevue sites monitored from late 1979 to early 1982 by the US Geological Survey, in cooperation with the City of Bellevue, as part of the US EPA Nationwide Urban Runoff Program [see Prych and Ebbert, 1986]. As a result of the USGS monitoring program, two years of continuous rainfall and runoff data recorded at 5-minute intervals are available. This data base is invaluable from the standpoint of model calibration. It is not usable of course to establish flood frequencies directly and forces the drainage engineer to make a choice of models.

Models and Methods Applicable to Lake Hills

As a first step in evaluating the possible choices of methods/models applicable to the Lake Hills catchment, the authors have assembled flood peak computations from 13 different procedures. In many cases these differ only in the choice of rainfall distribution which often is a governing factor in the runoff computation. This is especially true for the Lake Hills drainage system which has a 15-minute time of concentration and therefore is very sensitive to resolution in the rainfall hyetograph. Table 1 summarizes the six different rainfall distributions applied to the Lake Hills system.

Table 1. Summary of Rainfall Distributions Applied to Lake Hills Catchment (depths in inches)

Time min.	NOAA[a] Atlas	LH[b] No. 1	LH[c] No. 2	LH[d] No. 3	Yarnell[e] 5-min.	Yarnell[e] 10-min.
5	0.030	0.050	0.086	0.085	0.010	0.009
10	0.030	0.050	0.081	0.082	0.018	0.016
15	0.040	0.070	0.084	0.081	0.031	0.027
20	0.070	0.060	0.082	0.080	0.058	0.050
25	0.080	0.140	0.077	0.079	0.128	0.106
30	0.160	0.110	0.087	0.082	0.377	0.292
35	0.290	0.120	0.082	0.081	0.207	0.292
40	0.120	0.120	0.081	0.083	0.084	0.106
45	0.070	0.110	0.085	0.083	0.042	0.050
50	0.040	0.070	0.084	0.090	0.023	0.027
55	0.040	0.060	0.087	0.085	0.014	0.016
60	0.030	0.040	0.084	0.082	0.008	0.009
Totals	1.000	1.000	1.000	0.993	1.000	1.000

Notes:
[a] From NOAA Vol. XI, 1973.
[b] From 15 Lake Hills storms. Distribution is by percent of total depth over most intense 60 minutes.
[c] From 2 years of Lake Hills data. Distribution is by count of occurrence vs. non-occurrence of rain in each 5-minute interval over all storms.
[d] From 2 years of continuous Lake Hills data. Distribution is by depth accumulation in each 5-minute interval over all storms.
[e] Yarnell 5 and 10-minute distributions use the standard form: $i = a/(t+15)^b$, where $a = 84 + 20e^{-P}$, $b = 0.2316 \ln (a/p)$, p = 60-minute rainfall depth for specified return period.

As indicated in Table 1, the Yarnell distributions place between 29 and 37 percent of the total 60-minute rain in the most intense 5-10 minute period and therefore tend to represent sharp short duration events. The LH2 and LH3 distributions, on the other hand, provide a rather uniform distribution with less that 9 percent of the total 60-minute depth falling in any 5-minute period. The NOAA Atlas [1973] and LH1 distributions represent intermediate intensities and have been selected for use in the runoff peak computations which follow.

The runoff peak methods are summarized in Table 2. They range in complexity from Rational formula to discrete storm modelling by the Penn State Runoff Model (PSRM). It is noted that work on continuous modelling by DR3M is still in-progress as of this writing and is not reported here.

Table 2. Summary of Flood Peak Methods
in Lake Hills Catchment

Method No. and Name	Rainfall Distribution	Key Parameters and References
(1) Rational	NOAA Atlas Vol. XI	A = 95.2 acres; C = 0.35; T_c = 15 min.
(2) Rational	LH1	Same as (1)
(3) Rational	Yarnell 5-min.	Same as (1)
(4) USGS 7-parameter urban flood regr.	N/A	A = 0.149; SL = 70; $R12$ = 0.23; ST = 0; BDF = 10; P = 45 RQ_T from Cummans et al., [1975]. See Sauer et al. [1983].
(5) USGS 3-parameter urban flood regr.	N/A	Same as (4)
(6) USGS 7-parameter urban flood regr.		Same as (4), except RQ_T is from area ratio applied to Mercer Creek.
(7) USGS 3-parameter urban flood regr.		Same as (6)
(8) SCS TR-55 graphical method	24-hour Type IA	CN = 80, T_c = .25 hour See SCS TR-55 [1986]
(9) SCS TR-20	24-hour Type IA	CN = 80
(10) PSRM	NOAA Atlas Vol. XI	14 sub-areas; PSRM calibrated on 15 events. See G. Aron [1987] for PSRM ref.
(11) PSRM	LH1	Same as (10)
(12) USGS regression developed from Bellevue sites	MAXR15 from NOAA Atlas Vol. XI	MAXR15 = max 15-min. rainfall rate, in/hr. Reference is Prych and and Ebbert [1986].
(13) USGS regression developed from Bellevue sites	MAXR15 from LH1	Same as (12)

Summary of Flood Peak Results

The results of all flood peak computations for the Lake Hills
catchment are shown in Table 3. The range in discharge peak for any
given return period is substantial, but not unexpected. Several
interesting patterns emerge from Table 3. First, the results
indicate a rough agreement between runoff methods when the same
rainfall distribution is involved. For example, methods (1), (10)
and (12) are similar as are methods (2), (11) and (13) indicating a
very strong dependence on the rainfall distribution method employed.
The NOAA Atlas distributions produce higher flood peaks than LH1 as
expected, but less than the Yarnell 5-minute distributions.
Interestingly, the USGS nationwide regression equations in methods
(4) through (7) produce lower flood peaks than the USGS regression
equations developed from the Bellevue site data in methods (12) and
(13). For reasons not entirely clear, the two SCS methods (8) and
(9) produce flood peaks that are low at the 2 and 5-year levels, but

Table 3. Summary of Lake Hills Flood Peak
Computations

Method No. and Name	Q2 cfs	Q5 cfs	Q10 cfs	Q25 cfs	Q50 cfs	Q100 cfs
(1) Rational, NOAA	33	42	49	58	65	72
(2) Rational, LH1	22	28	33	39	44	48
(3) Rational, Y5	50	58	65	70	77	83
(4) USGS-7	13	15	20	21	24	26
(5) USGS-3	13	15	19	20	23	24
(6) USGS-7 with area ratio	15	20	24	27	31	33
(7) USGS-3 with area ratio	18	23	25	28	31	34
(8) TR-55	7	14	19	28	34	39
(9) TR-20	7	14	19	28	34	40
(10) PSRM, NOAA	26	35	42	50	59	67
(11) PSRM, LH1	17	23	28	35	40	46
(12) USGS Eqn, NOAA	27	37	45	55	65	74
(13) USGS Eqn, LH1	17	23	28	34	40	46

close to the USGS regression results at return periods greater than
10 years. The PSRM results are quite similar to those produced by
the Rational formula and the USGS Bellevue regression equations.

In the absence of long-term flow records at the Lake Hills site, no
firm conclusion on choice of method can be made from the results
presented here. It is very clear, however, that selection of
rainfall distribution is critical to this choice and may even be the
governing factor. To further evaluate this question, the authors
will utilize the capabilities of the continuous distributed USGS DR3M
model to synthesize a long-term runoff trace so that a bench-mark
frequency analysis can be made. Since the nearest available
long-term rainfall record comes from Seattle in hourly intervals, the

approach to segmental rainfall distribution remains crucial to the analysis of urban floods on small responsive systems such as the Lake Hills watershed.

References

Aron, G. "Penn State Runoff Model for IBM PC", Users Manual, Penn State University, Department of Civil Engineering, University Park, January 1987.

Cummans, J. E., M. R. Collings, and E. G. Nassar. "Magnitude and Frequency of Floods in Washington", US Geological Survey Open File Report 74-336, 1975.

Prych, E. A. and J. C. Ebbert. "Quantity and Quality of Storm Runoff from Three Urban Catchments in Bellevue, Washington", US Geological Survey Water Resources Investigations Report 86-4000, Tacoma, Washington, 1986.

Sauer, V. B., W. O. Thomas, Jr., V. A. Strickler, and K. V. Wilson. "Flood Characteristics of Urban Watersheds in the United States", US Geological Survey Water Supply Paper 2207, US Government Printing Office, Washington, D. C., 1983.

US Department of Agriculture, Soil Conservation Service, "Urban Hydrology for Small Watersheds", Technical Release 55, US Government Printing Office, Washington, D. C., 1986.

US Department of Commerce, National Oceanic and Atmospheric Administration, "Precipitation - Frequency Atlas of the Western United States," Vol. XI, Washington, D. C., 1973.

Urban Stormwater Drainage Design By Modified ILLUDAS

G. V. Loganathan[1] , C. Y. Kuo[2] , S. P. Shrestha[3] and K. J. Ying[4]

The Illinois Urban Drainage Area Simulator (ILLUDAS) has been extended for continuous simulation by considering dry periods between storms. The model accepts predetermined hydrographs at desired nodes. Effects of different Best Management Practices (BMPs) measures can be analyzed. Box's complex algorithm is used to determine optimal locations and sizes of detention facilities.

Introduction

Stormwater management programs have been in place in many urban areas across the United States for controlling peak discharges and associated pollutant loadings. The Illinois Urban Drainage Area Simulator (ILLUDAS) is an event model which is used to determine the required pipe size to carry stormwater. It can also be run in an evaluation mode to determine required detention storages at junctions. The model has been modified for continuous simulation which considers the antecedent moisture conditions (AMCs) of soil and the availability of the detention storages in detention facilities prior to a storm. An optimization scheme, based on the complex algorithm (Box, 1965) is used for the Best Management Practice (BMP) structure design. The objective function minimizes the cost of construction, operation, and maintenance subject to the downstream flow constraints and space availability restrictions.

Description of the Model

The ILLUDAS model considers two major rainfall losses, namely surface depression storage and infiltration loss (Terstriep and Stall, 1974). The infiltration loss is determined by considering the soil type and soil water content. The Horton infiltration curve is selected based on AMCs. The model updates the AMC based on inter-event times for storms.

Predetermined hydrographs for subbasins can be input at a given junction. Modifications have been made to route these inflow hydrographs downstream to generate the outlet composite hydrograph. The time intervals of these predetermined hydrographs can be different from the ones generated within the program. An interpolation procedure is applied to adjust the time intervals of the hydrographs. The provision to enter hydrographs generated for upstream subbasins as predetermined hydrographs for the downstream subbasin yields more flexibility to run ILLUDAS and large basins can be easily handled.

---

[1] Asst. Prof., Civil Engg. Dept., Va. Tech, Blacksburg, Va. 24061.

[2] Prof., Civil Engg. Dept., Va. Tech, Blacksburg, Va. 24061.

[3] Grad. Student, Civil Engg. Dept., Va. Tech, Blacksburg, Va. 24061.

[4] Grad. Student, Civil Engg. Dept., Va. Tech, Blacksburg, Va. 24061.

The model can evaluate and design three types of BMP facilities, namely wet detention pond, extended wet detention pond, and infiltration trench. For detention ponds, the modified Puls method is used for flood routing. Extended wet ponds are larger than wet detention ponds and large volume generally provides a longer detention time to enhance the removal of pollutants in the pond. The infiltration trench removes surface runoff till it is full. Runoff entering the full trench is assumed to overflow as outflow in the storage computation for the trench. The infiltration rate through the trench bottom is considered a constant. Figure 1 is the sub-basin of the Holmes Run Watershed in northern Virginia which has been used as the study basin. Figure 2 shows an example hyetograph for a single storm event. The effect of BMP facilities on hydrograph is shown in Figure 3, in which peak flow is reduced by the presence of BMP facilities. The figure includes three stormwater management strategies for this subbasin: without BMP, with an extended wet detention pond located at junction 1-13, and combination of the same pond and infiltration trenches located at junctions 1-2, 1-4, 1-12, 4-1 and 5-0. Figures 4 and 5 show hyetograph and hydrographs for continuous simulation.

Optimization

Location and size of BMP structures play a significant role in the overall cost of the system. It is important to have a basinwide planning and management tool to evaluate and design BMP structures.

Mathematically, the optimization problem can be expressed as:

$$\text{Minimize} \quad \sum_{i=1}^{n} C_i\, S_i$$

Subject to:
  (i)   Hydraulic rules
  (ii) Hydrologic rules
  (iii) Bounds of the type:

$$Q_j \leq Qmax_j$$

$$S_j \leq Smax_j$$

where: C_i = cost/unit volume of storage at site i
        S_i = total volume to be stored at site i
     $Smax_j$ = maximum volume that can be stored at site j
        Q_j = flow through pipe j
     $Qmax_j$ = capacity of pipe j

The storage constraints are due to lack of open space for the construction of BMP structures in urbanized areas. The method uses the complex algorithm (Box, 1965) in conjunction with the modified version of ILLUDAS to obtain an optimal solution to this problem. Details of the Box-Complex method is given in Kuester and Mize (1973). An initial feasible solution is obtained by running the modified ILLUDAS model in the evaluation mode. The algorithm uses the obtained initial solution to generate an initial complex required for the search technique of Box. The complex then moves through the region of search by using expansion and contraction steps. Constraints are checked and adjusted, if necessary, after each expansion or contraction. Once the constrains are satisfied, the cost of the trial design is obtained and compared with the design with the maximum cost within the complex. If the trial design is better than the current most expensive design, the later one is replaced by the new one. Else, the trial vertex is contracted towards the centroid of the remaining design points until an acceptable design is obtained. In the contraction process, if the worst point repeats for several iterations without any improvement in the objective function value, then the worst point is retained and the second worst point is reflected. This process of expansion and contraction continues until a convergence criterion is satisfied or until a specified number of iterations is exceeded. A flow chart illustrating the above steps is given in Figure 6.

Summary

The ILLUDAS model has been modified to study the effects of BMP structures on the basin. The continuous simulation is used to determine the AMC of soil and the detention storage available prior to a storm. The optimization scheme provides the cost, size and location information of BMP structures for the optimal design.

Acknowledgement

This study was supported by the Virginia Water Resources Research Center.

References

Box, M. J. (1965). A New Method of Constrained Optimization and a Comparison with Other Methods, *The Computer Journal*, 8, pp. 42-51.

Han, J., and Delleur, J. W. (1979). "Development of an Extension of ILLUDAS Model for Continuous Simulation of Urban Runoff Quantity and Discrete Simulation of Runoff Quality", *Technical Report Number 109*, Purdue Water Resources Research Center, West Lafayette, Indiana, July.

Kuester, J. L., and Mize, J. H. (1973). *Optimization Techniques with FORTRAN*, McGraw-Hill Book Company, New York.

Terstriep, M. L., and Stall, J. B. (1974). "The Illinois Urban Drainage Area Simulator, ILLUDAS", *Illinois State Water Survey Bulletin 58*.

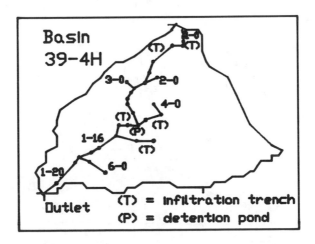

Figure 1. Holmes Run Watershed Sub-basin 39-04H

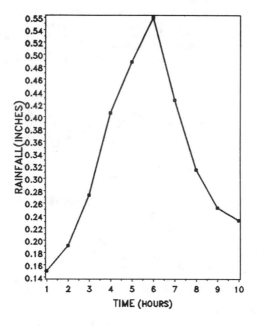

Figure 2. Hyetograph - Single Storm Simulation

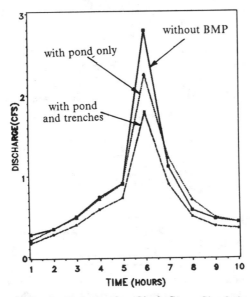

Figure 3. Hydrographs - Single Storm Simulation

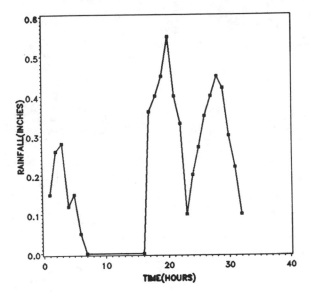

Figure 4. Hyetograph - Continuous Simulation

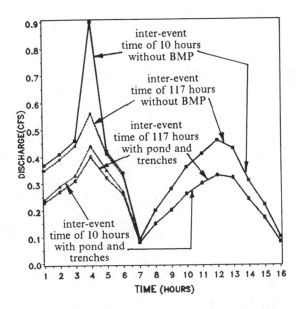

Figure 5. Hydrographs - Continuous Simulation

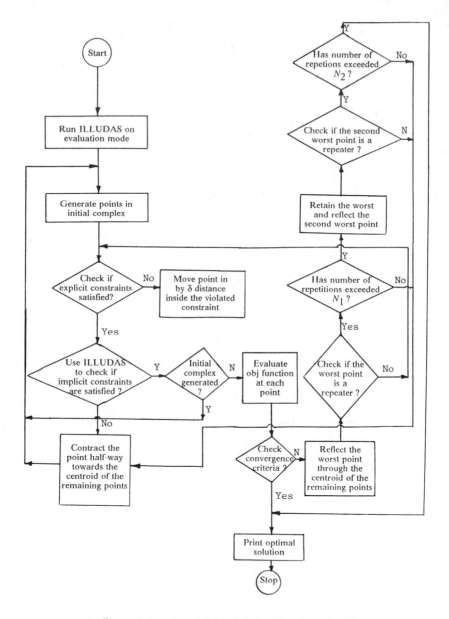

Figure 6. Flow Chart: Modified Box-Complex Algorithm

A PRACTICAL HYDROLOGIC MODEL FOR URBAN WATERSHEDS

Joel W. Toso[1], PhD, A.M. ASCE and Douglas W. Barr[2], P.E., F.ASCE

ABSTRACT

A typical problem in watershed management is determining the peak discharge and runoff volume for numerous and combined watersheds. The methods available to solve this problem are often too complex or, in the other extreme, too empirical and detached from the physical processes involved in rainfall and runoff. This paper will discuss a hydrograph method developed by Barr Engineering Co. in Minneapolis, Minnesota as a practical solution to the problem. An additional and necessary focus is its position amid the crowded array of other urban runoff models.

INTRODUCTION

In keeping with the Symposium theme, the purpose of this paper is to maintain dialog between practicing engineers and researchers. The viewpoint of a practicing engineer will be taken in a discussion of a urban runoff model.

The historical development of urban hydrology is available in texts similar to the ones by Ven Te Chow (1964) and Viessman, Knapp, Lewis, and Harbaugh (1977). A reference to these and a review of the current literature will indicate the voluminous amount of information available on the subject of urban runoff. One is hestitant to add yet one more paper to the group. An effort will be made to raise a discussion.

An effect of the micro-computer on urban hydrology is the numerous micro-computer programs for urban runoff computations on the market. An interesting "shopper's guide" has been provided by Lewis and Gilbert (1985). A dominance is evident in the number of programs using the U.S. Soil Conservation Service (SCS) Curve Number runoff and unit hydrograph methods. There is a concern whether these procedures, developed empirically from rural watershed data, are appropriate as a new "standard" for urban hydrology. The often inadequate documentation of the programs makes it difficult to judge the assumptions and actual procedures used.

---

[1] Hydraulic Engineer, Barr Engineering Co., Minneapolis, Minnesota 55435. Formerly, Research Assistant, St. Anthony Falls Hydraulic Lab.

[2] Principal, Barr Engineering Co., Minneapolis, Minnesota 55435.

There are several models with complex procedures to give refined estimates of urban runoff. In addition to computing water quantity, there are routines to evaluate the water quality produced in the runoff. Continuous simulation of hydrologic processes is also available. These models are limited by the amount of available data and the expertise and experience required to obtain reliable results.

On the other end of the scale of complexity, the Rational Method offers a very quick and easy means of estimating the peak runoff discharge. Experience is required for reliable results from this empirical method. This method is often stretched beyond its limit to arrive at runoff volumes and timing of the peak discharge by applying a triangular hydrograph.

There is a place for the runoff models that apply the basic urban hydrologic principals with readily available data. Models that adhere to simulating each part of the urban runoff process without relying on vague coefficients should be emphasized and made more available. This enables the engineer to visualize and think more reasonably about solutions to runoff problems. The micro-computer has provided a way to apply these models in the same amount of time required with the Rational Method. The Barr Hydrograph Method is discussed below as one example of a practical urban runoff model.

MODEL DESCRIPTION

The Barr Hydrograph Method was developed in the late 1950's, at the same time as the Chicago Hydrograph Method (Tholin and Keifer, 1959). The two methods are similar in approach, being designated as hydrograph methods, but have some significant differences, as will be shown (The reader is referred to Chapter 11 of Viessman, Knapp, Lewis, and Harbaugh, 1977 for the general attributes of the different approaches). The model has been modified and computerized through its years of use in urban watershed management. Recent developments include reprogramming to a structured and modular format, and application to the IBM compatible system.

The model has four basic components: (1) a hyetograph, (2) the abstractions from the hyetograph, interception, infiltration, and depression losses, (3) overland flow routing, and (4) routing to account for flow through swales or ditches, street gutters, and storm sewers to the watershed outlet. There is an additional program to route through a detention pond and/or through the conveyence system to the next watershed that may be added as a routine to the model.

The hyetograph for the Barr Hydrograph Method may be that of an actual storm. If only a rainfall amount and duration are input, the program computes a hyetograph based on a predetermined unit hyetograph. The unit hyetograph currently used is for short duration storms (less than 3 hours) characteristic of summer storms in the Minneapolis, Minnesota area. The distribution of rainfall is based on the studies by Briehan (1940) and the more recent work of Huff (1967) using the second quartile storm. The peak of the hyetograph occurs at the three-eighths point. In comparison to the synthetic storm pattern of Keifer and Chu (1957) the peak is not as pronounced or spiked.

For longer duration storms, serious consideration must be given to including multiple peaks in the synthetic hyetograph. Further study similar to the work by Huff (1967) is desirable to obtain better input for storms between 3 and 12 hour durations.

Abstractions from the rainfall begin with the interception losses. This is taken directly off the first part of the hyetograph until the specified loss in the input data is satisfied. Two values are required, one for impervious areas and one for pervious areas. Typical amounts are 0.05 inches and 0.10 inches, respectively. High rise buildings can increase interception losses significantly.

Infiltration losses to the soil are estimated by the Horton equation. This method of determining infiltration is covered in most hydrology texts and in the ASCE Manual of Engineering Practice No. 28, (1949). The infiltration curve is adjusted during storm periods when rainfall intensity is less than infiltration capacity. Three infiltration capacities are required to define the curve: the initial, the equilibrium and the capacity at some given time, usually one hour. Examples of these three values given in the references above are 2.5, 0.9, and 1.1 inches per hour, respectively. Infiltration losses are not considered for impervious areas.

As the rate of precipitation exceeds the rate of infiltration, some of the water begins to flow overland to the main conveyance system. Part of the water is intercepted by depressions along the route and never reaches the primary collection system. It is reasonable to assume that the many depressions are filled at varying times during the storm and that they are located randomly. An integrated Ogee probability curve is used to distribute the depression losses (See Tholin and Keifer, (1959), or a hydrology text). Typical values for depression losses are 0.10 inches for impervious and 0.15 inches for pervious areas. Depression losses are higher for flatter terrain.

Applying interception, infiltration and depression losses to the hyetograph results in the point runoff hydrograph at any location in the watershed. This point runoff is the supply to the overland flow process. Two point runoff hydrographs are computed, one for the pervious areas which accounts for all the abstractions of the hyetograph, and one for the impervious areas which neglects infiltration losses in the computations.

The overland flow process has the effect of attenuating and lagging the point runoff hydrograph due to detention of the water on the land surface. This detention on impervious surfaces is usually short and can be included or neglected in the computations. For pervious surfaces this detention is significant and depends primarily on the flow length, land slope and surface roughness. A discharge relationship using these factors was formulated from Equation 2b in Horner and Jens (1942). This relationship is included in a storage routing equation to account for the overland flow effects on pervious surfaces. The point runoff hydrograph of the pervious areas is routed using three separate slope and flow length combinations. The surface roughness is considered constant.

The computations up to this point in the Barr Hydrograph Method have generated five runoff hydrographs: one for the impervious areas, three for the pervious areas, which account for three separate choices in the general range of slope and flow lengths in the overland flow process, and one for any water surfaces such as ponds in the watershed. The latter one is the same as the hyetograph assuming there are no significant abstractions and overland flow is not applicable. Given the percent of the total watershed that is included in each of these areas these hydrographs are multiplied by this factor and summed to give a composite runoff hydrograph.

The composite runoff hydrograph includes routing to the the conveyance system of the watershed. This system is made up of swales or ditches, street gutters, and storm sewers. The average velocity weighted by the area and the longest flow length of the conveyance system are used in a storage routing procedure to generate the hydrograph at the outlet of the watershed. The longest flow length would not include a long tributary that is not characteristic of the watershed.

The output of the model gives the outflow hydrograph of the watershed, the peak flow and volume of runoff. This may be input to a separate program for routing through a detention pond and/or through the conveyance system to the next watershed. A routine may be added to the model to perform this task and sum up the hydrographs from a system of watersheds.

INPUT REQUIREMENTS

The inputs required for the model are: (1) precipitation depth and storm duration, or an actual hyetograph, (2) land use breakdown (percent of the total area that is open water, impervious, and pervious; further subdivided by slope and overland flow length factors), (3) interception losses, depression storage losses, infiltration loss rates, and (4) the longest flow length and average velocity of the conveyance system.

These factors would have to be considered even in the Rational Method for an accurate estimate of the peak flow. There would be additional input data requirements for models with greater complexity or refinements.

COMPARISON WITH OTHER MODELS

In the mid-1970's there were several papers published on comparisons of urban runoff models applied to an Oakdale Avenue basin in Chicago, Illinois. An example of this is the paper by Papadakis and Preul (1973). Other discussion is readily available in Chapter 11 of the text by Viessman, Knapp, Lewis, and Harbaugh. During this same period, the Barr Hydrograph Method was applied to the basin with a given storm to illustrate its credibility. The user of the model was

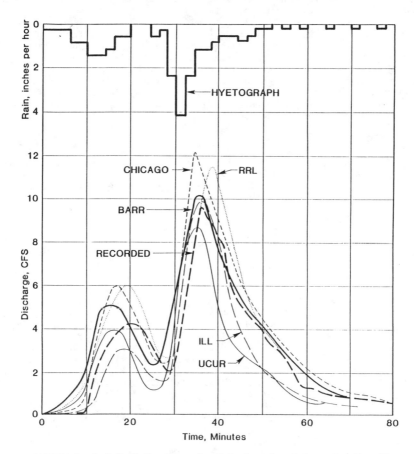

FIGURE 1. Rainfall Hyetograph, Calculated and Recorded Runoff
Hydrographs - Storm of July 7, 1964 - Oakdale Avenue Basin,
Chicago, Ill.

not aware of the recorded results or the results of the other models until after the project was completed.

Figure 1 is a presentation of several model outputs for the July 7, 1964 storm over the Oakdale Avenue basin. The apparently most accepted and complex model, the EPA SWMM, was applied to the basin, but for a different storm (Papadakis and Preul, 1973). It is not the scope of this paper to discuss the attributes of each model. The point is that very acceptable and reliable results are obtainable using basic data and a model with relatively few complexities. The Barr Hydrograph Method is such a model.

CONCLUSION

The Barr Hydrograph Method is an example of a practical urban runoff model that works with each part of the rainfall-runoff process to arrive at a solution. The procedures focus on simulating the storage effects occurring throughout the process of runoff. This method eliminates the necessity of relying on vague coefficients and avoids over complication requiring additional data and more experience. Models such as the Barr Hydrograph Method should receive more emphasis in the current micro software market.

REFERENCES

ASCE Manual No. 28, (1949), Hydrology Handbook, The Hydrology Committee of the Hydraulics Division.
Breihan, E.R., (1940), "Relation of Hourly Mean Rainfall to Actual Intensities," Civil Engineering, Vol. 10, No. 5, pp. 303-305.
Horner, W.W., and Jens, S.W., (1941), "Surface Runoff Determination from Rainfall Without Using Coefficients," Transactions, ASCE, Vol. 107, pp. 1039-1075.
Huff, F.A., (1967), "Time Distribution of Second Quartile Storms," Water Resources Research, 3, No. 4, pp. 1007-1019.
Keifer, C.J., and Chu, H.H., (1957), "Synthetic Storm Pattern for Drainage Design," Journal of the Hydraulics Division, ASCE, Vol. 83, No. Hy4, Aug. 57, p. 1332.
Lewis, G.L., and Gilbert, D.P., (1985), "A Shopper's Guide to Urban Stormwater Micro Software," Proc. ASCE Conference, Florida.
Papadakis, C.N., and Preul, H.C., (1973), "Testing of Methods for Determination of Urban Runoff," Journal of the Hydraulics Division, ASCE, Vol. 99, No. HY9, September, pp. 1319-1335.
Tholin, A.L., and Keifer, C.J., (1950), "Hydrology of Urban Runoff," Transactions, ASCE, Vol. 125, p. 57.
Chow, V.T. (Ed.), (1964), Handbook of Applied Hydrology, New York: McGraw-Hill Book Company.
Viessman, W., Knapp, J.W., Lewis, G.L., and Harbaugh, T.E., (1977) Introduction to Hydrology, Second Edition, Harper & Row.

Design of Seepage Fields for Stormwater Disposal

Stuart D. Wallace [1] and Kenneth W. Potter [2]

Abstract

Stricker Pond is a shallow twenty-acre kettle hole pond in south-central Wisconsin. The pond receives runoff from a closed watershed that is rapidly undergoing low density residential development. A previous University of Wisconsin-Madison study has indicated that under fully developed watershed conditions, the recurrence interval for overflow conditions in the pond will be thirty-two years. This is a significant change from the three hundred year recurrence interval corresponding to pre-development conditions. This study sought to mitigate this situation without transferring the problem to another watershed.

The water budget for the pond suggested that enhancing the seepage capacity of the pond was a means to provide the desired relief. After considering alternatives, an underground seepage field was identified as a possible configuration. Utilizing a hydrologic model developed in the previous University study, a seepage field was designed and implemented. The hydrologic model indicated that the design would substantially mitigate the effects of increased stormwater runoff on the pond water levels, however, initial measurements of the effectiveness of the field indicate higher efficiencies. Guidelines for the design and implementation of similar structures are developed.

Project Background

Stricker Pond is a shallow kettle hole pond, with a surface area of twenty acres, located in Middleton, Wisconsin. The watershed for the pond, approximately nine-tenths of a square mile in area, was an oak savanna and prairie landscape prior to European settlement in the 1800's. Following settlement, the land was used predominately for row crops and pastures. However, in the last twenty years, the watershed has been nearly completely converted to low density

---

[1] Graduate Student, Department of Civil and Environmental Engineering, University of Wisconsin-Madison, Madison, Wisconsin 53706.

[2] Professor, Department of Civil and Environmental Engineering, University of Wisconsin-Madison, Madison, Wisconsin 53706.

residential development.

Sensing that complete development of the watershed would impact the hydrology of the pond, the Department of Civil and Environmental Engineering of the University of Wisconsin-Madison [Mueller, 1984] conducted a study to quantify these effects. During the study a computer simulation model of the water budget for the pond was developed. The model was calibrated with eleven months of field data gathered at the pond and run with twenty years of daily meteorological records from a local station. Analysis of the simulated annual peak pond levels demonstrated that under full development of the watershed, the recurrence interval for the pond exceeding its banks would decrease ten-fold, from three hundred to thirty-two years.

A traditional solution would be to divert excess water during critical periods from the Stricker Pond watershed. The logical diversion in this case would be to another nearby pond. However, that pond also has increased levels due to recent development. Hence, excess water would have to be diverted from that pond to a nearby stream, that is experiencing bank erosion. The diversion of stormwater would only transfer the problem to another location and that was not acceptable to the City of Middleton. The city subsequently commissioned this study to explore alternatives.

Conceptual Development

Alternatives to the traditional approach had to meet criteria of handling the problem within the urbanizing basin and increasing the recurrence interval of the pond stage exceeding the adjacent road levels to that which existed prior to development. Also, the solution should be affordable to construct and maintain.

The hydrologic conditions of the watershed were examined to develop alternatives. It was determined that the watershed is closed, without surface outlets for stormwater under normal conditions. Also, given a low initial pond stage, storage is adequate in the pond to accomadate single large storm events. More specifically, the main factor in long-term regulation of pond levels is the seepage rate of water from the pond between storms, while during acute, large storm events, neither seepage or evaporation play a major role in regulating pond levels. These conditions led to the conclusion that increasing the seepage rate of the pond was the logical point of attack. To pursue this idea, the physical conditions of the landscape surrounding the pond were examined.

Stricker Pond is underlain by a thick layer of impermeable clay. However, there is an area adjacent to the pond, that is covered by one to two feet of a semi-permeable colluvium underlain by extensive sand deposits. In fact, this is the area that provides much of the existing seepage capacity of the pond when flooded.

Given these physical characteristics and the project criteria, a feasible alternative solution was identified. The identified solution was a seepage field, a thin bed of coarse gravel placed in direct contact with a porous substrate and connected by a conduit to an intake structure in the pond (Figure 1). Whenever pond levels exceed the

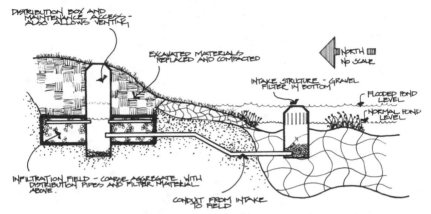

Figure 1. Conceptual Seepage Field Configuration

elevation of the intake, water would be fed to the field by an underground conduit, placed below frost elevation. The configuration of the field would be independent of the ground surface, so material excavated during construction would be replaced on site without alteration of existing grade. Also, water would be taken from the surface of the pond, so sediment loads would be lessened. Within the conduits, sediment could settle and be removed, prolonging the life of the field.

Design Procedure

The steps used to design the seepage field involved first determining the required level of performance for the field and second, determining the required physical dimensions.

The stage-seepage component of the water budget model was modified to simulate the effect of the seepage field. The modification was a series of step increases in the seepage rate for pond stages above a designated intake elevation. Next, for each of the step increase seepage rate cases, twenty year simulation runs were performed, using land use values representing the anticipated fully developed conditions.

Fitting the annual maximum pond levels generated by each
of these runs to an extreme value type I distribution
identified a step increase sufficient to raise the recurrence interval for overflow conditions in the pond from
thirty-two years to one hundred and twenty-five years.

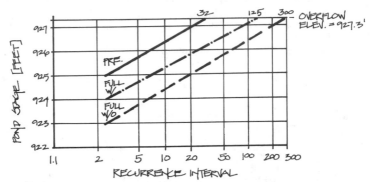

Figure 2. Fitting the peak annual stages for predevelopment;
full development; and full development with the seepage
field to the Gumbel Probability distribution, [n=19]

Using the identified step increase as the desired level of
performance, the physical dimensions of the field were found
through the use of a method for determining the effects of
vertical recharge on unconfined aquifers [Hunt, 1971].
Field data for hydraulic conductivity, depth to the groundwater surface, and the extent of suitable subgrade were used
as physical parameters required in the method. The average
number of days that the pond level would be above the intake
elevation was determined from the model simulation runs and
used as the expectation for the duration of recharge. The
required dimensions were then determined in a trial substitution procedure.

Substitution of a trial seepage field width along with data
for hydraulic conductivity, depth to groundwater, and
thickness of the unconfined aquifer into graphs developed by
Hunt [1971], yielded a value for the average downward
recharge velocity from the field. This velocity, when
multiplied by the trial field width, yielded a trial
volumetric seepage rate per unit length. Varying the trial
field width to enable a match between this trial volumetric
rate and the desired step increase in seepage rate, identified the required field dimensions.

Hunt´s method also enabled the identification of an optimal
configuration for the field. Minimizing the field width,
tended to maximize the downward recharge velocity per unit
length of the field. And minimizing the field width, even
with the required extension of the field length, yielded the

least area necessary to accomplish the desired seepage increase.

Project Construction

The project was approved for construction with $80,000 from the municipal public works budget in the Fall of 1985. The construction documents were produced that winter, bids were let and a contract awarded in the Spring of 1986. The field was constructed in August and September of 1986.

Figure 3. Construction of the seepage field, August 1987. Placing the gravel bed in one-quarter section of the field.

The primary consideration in the determining the construction details was to insure the maximum effective life for the structure. The main source of concern was the prevention of silt and debris from entering and clogging the field. At several points in the system, provisions were made to enable the trapping and removal of sediments. Also, the entire field was covered with a filter material to prevent fine grains in the backfill material from migrating into the field from above.

Results

The simulation of the effect of the seepage field on the recurrence interval (Figure 2) of pond levels indicated that the field is capable of significantly lessening, though not completely mitigating, the influence of urbanization on Stricker Pond. Initial monitoring results from immediately after the construction of the field in August of 1986, indicate that the field may be more effective than predicted. Calculations based on conservative values, methods and assumptions in the design procedure are the likely causes of this discrepancy.

However, to draw conclusions about the application of these results to other situations, it is worth noting that the design method was facilitated by substantial existing data and at the same time constrained by the limits of the suitable area. In addition, the design method assumed that the field would be operating over a sustained groundwater mound. This assumption, while facilitating the modeling of the interaction between the seepage field and the ground-water system, ignores the transition effects between unsaturated and saturated conditions in the zone between the field and the normal groundwater free surface.

Conclusions

To account for the fact that a seepage field is not a system that simply turns on and off but operates in response to characteristics of the physical system, knowledge about four major components is required.

First, the hydrology and conditions of the site or region must be understood in terms of the relationship between rainfall and runoff, the existing landscape characteristics and the extent and location of proposed land uses.

Second, the functional relationships within the seepage field itself must be specified. Determination and location of the intake and field elevations are the most significant aspects of this component.

Third, the extent and physical characteristics of the unsaturated zone, must be determined. Hydraulic conductivi-ty, porosity, horizontal extent, vertical stratigraphy, and depth to the saturated zone are among the needed information for this zone.

Fourth, the extent and physical characteristics of the groundwater system must also be known. The presence or absense of a significant surface gradient, and the depth of the aquifer are most significant.

References

Hunt, B.W., "Vertical Recharge of an Unconfined Aquifer", Journal of Hydraulics Division, ASCE, Vol. 97, No. HY7, July, 1971, pp. 1017-1030.

Mueller, R.H., "Effects of Urbanization on Stricker Pond", M.S. Thesis, University of Wisconsin-Madison, 1984.

LOCALIZED EFFECTS OF SUBSIDENCE
ON FLOODING IN A SMALL URBAN WATERSHED

Martha F. Juch, A.M. ASCE, Steven L. Johnson, M. ASCE,
and David E. Winslow, M. ASCE[*]

Extensive pumpage of groundwater in the Houston, Texas, metropolitan area in the last century has caused compaction of the clay layers in the local aquifers with resulting widespread land surface subsidence. This paper presents the results of a study designed to evaluate localized flooding characteristics of a small urban watershed with respect to subsidence patterns present on a regional scale in the Houston metropolitan area and also those patterns resulting from local groundwater well fields.

Introduction

During the period of record from 1906 to 1978, historical land surface subsidence due to extensive groundwater withdrawal has exceeded 9 feet in the Pasadena and Baytown areas in the eastern metropolitan region while the western portions of the Houston area have experienced nearly 4 feet of subsidence. The severity of the eastern subsidence had caused permanent flooding of some land adjacent to the coast by the early 1970's and substantially increased flooding in areas subject to tidal surges associated with tropical storms. The flooding crisis prompted the City of Houston to dramatically restrict groundwater pumpage in the southeastern portion of the City and initiated the formation of the Coastal Water Authority to develope surface water supplies to the area. In 1975 the Harris-Galveston Coastal Subsidence District was created by the State of Texas to regulate groundwater and control subsidence.

Conversion to surface water in the southeast region of the City near the center of the subsidence pattern began in the late 1970's and dramatically reduced the rate of subsidence in the area. The western portion of the City, however, has continued to experience rapid population growth and related increases in groundwater withdrawal and currently is experiencing subsidence at a rate of one foot per every seven years. The impact of this type of inland subsidence on flooding is not well researched, although increased flooding in areas in the west and southwest Houston metropolitan region in recent years seems to indicate a correlation between the two factors. This paper presents one aspect of a joint venture study implemented to determine the relationship between subsidence and flooding. The study was sponsored by the Harris County Flood Control District, the Harris-Galveston Coastal Subsidence District, the City of Houston, and the Fort Bend County Drainage District.

[*]Winslow & Associates, Inc., 10500 Northwest Freeway Suite 100, Houston, Texas 77092.

Overview

This paper presents the Localized Drainage Analysis phase of the overall "Study of the Relationship Between Subsidence and Flooding" joint venture project performed cooperatively by Winslow & Associates, Inc., Pate Engineers, Inc., and Turner Collie & Braden, Inc. in Houston, Texas. This phase evaluated the effects of subsidence on small localized drainage systems in the Houston metropolitan area with respect to their relationship to street ponding and drainage channel flooding. Street ponding is a regular occurence in the Houston metropolitan area due the flat topography characteristic of the Gulf Coast plain region. The level of street ponding has been shown to have a direct correlation to the capacity of the storm sewer system and lateral channel system to convey runoff into the respective outfall channels. Secondary drainage system surcharge is, therefore, the critical analysis parameter when determining the localized effects of subsidence and was modeled extensively in this phase of the project.

Objectives

The initial goal of the Localized Drainage Analysis was to produce guidelines for the optimum positioning of major water well sites with respect to local drainage patterns. These guidelines were to be developed based upon the impacts of subsidence created by these wells on the local drainage system. This goal was expanded during the project to include projections of localized flooding characteristics resulting from hypothetical regional subsidence gradients. The primary objectives of this analysis were:
  (1) Define the impacts that major water well fields have on small watershed drainage systems;
  (2) Develop generalized drainage standards for locating water wells based on (1) above;
  (3) Determine the significance of projected area-wide subsidence on localized drainage;
  (4) Present a comparison of current City of Houston storm sewer design criteria versus pre-1970's design criteria with respect to the effects on street ponding.

Methodology

A watershed of approximately five square miles was selected as representative of a typical small Houston drainage area and was evaluated on both the regional (macro) level and the local (micro) level. The basin selected encompassed an area amenable to a detailed flooding evaluation of both its primary and secondary drainage systems with computer modeling and had been monitored for many years by a USGS rainfall-streamflow gage, therefore providing extensive historical data for aid in calibrating the model. The drainage system consisted of storm sewers and open ditches constructed well before 1970 and was therefore designed prior to the adoption of current City of Houston and Harris County Flood Control District criteria. This drainage system was analyzed in detail and modeled under various subsidence scenarios with a readily available computer software package. A second input data file was developed which modified the existing drainage system to simulate a theorectical drainage system designed wholly according to current City of Houston and Harris County Flood Control District criteria. This "new" model was then subjected to the same subsidence conditions as the "old" existing drainage system model and a comparison was made of the effectiveness of the current design Criteria.

The localized effects of subsidence on both the "old" existing drainage system and the "new" current-design drainage system were evaluated for two general types of subsidence. The first involved a hypothetical subsidence cone generated from a major water well field (Figure 1), which was simulated in four placement alternatives throughout the watershed. The second involved a general regional subsidence gradient reflecting an overall subsidence gradient occurring across the Houston area (Figure 2), which was simulated across the basin in four opposing directions. The maximum differential subsidence centered in the cone and placed across the basin as a gradient was 1.5 feet.

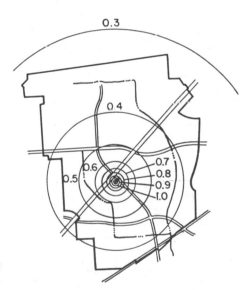

Figure 1   Cone of Subsidence (in feet) due to local well field

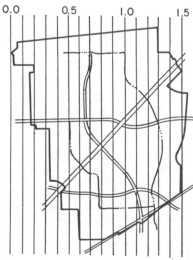

Figure 2   Gradient of Subsidence (in feet) due to regional pumpage

Computer Model

The drainage systems were modeled using the Environmental Protection Agency's (EPA) Storm Water Management Model (SWMM) because of this model's ability to address sheet flow and ponding as well as storm sewer surcharged flow. The model was adapted for use on IBM PC-AT compatible computers by Computational Hydraulics (CHI) and the most current version, PCSWMM3.2, was used in this analysis.

The PCSWMM3.2 model performs overland flow computations and gutter, storm sewer, and channel routing based on rainfall hyetographs, antecedent conditions, land use, topography, channel and storm sewer parameters, and receiving water conditions. The model is also capable of considering backwater conditions, outfall and receiving water levels, and inlet ponding.

In the model representation of the drainage system, conduits and junctions are idealized as links and nodes, respectively. Links transmit flow from node to node and have properties of roughness, length, cross-sectional area, hydraulic radius, and surface width. Nodes correspond to manholes, pipe junctions or channel junctions in the physical system and have variables of volume, head, and surface area. Inflows, such as inlet hydrographs, and outflows, such as weir diversions, take place at the nodes. Model output is described in terms of hydraulic gradient flows and velocities in conduits, and maximum depths of water at conduit ends.

The simulation of the drainage systems required the following assumptions in construction of the computer model:

1. Street system area in each subbasin is simulated by a single trapezoidal open channel with a length approximately equal to one-half the subbasin length in the plane of the major direction of flow and a width equal to the actual street surface area divided by the representative length.

2. Overland flow between subbasins is represented by trapezoidal open conduits leading from the street node with higher ground elevation to the street node with lower ground elevation (the model reversed the flow internally when flooding conditions warranted flow opposite to natural surface grade). Interflow conduits of this type lead to and from nodes at either the upper or lower ends of each street conduit depending on the particular flow characteristics of the two subbasins involved. Interflow conduits were set at elevations three tenths of a foot above the invert elevations of the subbasin street conduit at each node to simulate crests which must be topped before interflow can occur in the street and overland flow systems of the two basins.

3. All runoff was assumed to be contained within the watershed proper so that the drainage basin could be represented as a self-contained system for modeling purposes. Outflow from the system was allowed only at the primary drainage system outfall.

4. Storm sewers were modeled as representative circular conduits leading from the representative trapezoidal open street conduit to the outfall at the receiving channel. The largest diameter storm sewer in each subbasin drainage system was used as the simulated storm sewer conduit. Figure 3 shows a schematic representation of a typical subbasin drainage system.

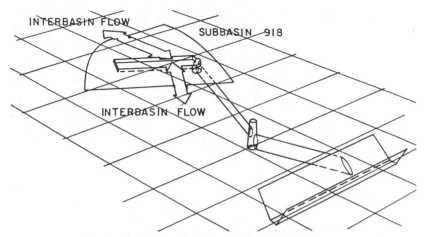

Figure 3   Typical Subbasin Drainage System

Model Calibration

Three steps were completed in the process of calibrating the data sets of PCSWMM3.2 before actual modeling of the subsidence cases were performed.   First the assumptions made in discretizing the watershed into subbasins and in representing the subbasin and routing characteristics with link-node input parameters were verified.   This was accomplished by applying the PCSWMM3.2 model to a smaller, well-documented drainage system where each detail of the simulated network was evaluated independently.   Second, a sensitivity analysis was completed on the input parameters so that the overall computer model for the existing drainage system produced results comparable to historically measured data for the watershed.   Third, both the "new" and the "old" drainage systems were evaluated with standard design storms to verify the consistency of the model between the two systems.

Results of Subsidence Simulations

The impacts of each of the eight subsidence cases and the no-subsidence case on each of the two drainage systems were simulated for the 3-year, 10-year, and 100-year return frequency storm events. The simulation results were then analyzed in terms of depth and duration of street ponding, depth of flooding in the primary drainage channel, and peak flows in the primary drainage channel.
For each of the subsidence conditions analyzed on the "old" existing drainage system, it was noted that the storm sewer capacity is quickly exceeded during all storm events and street ponding occurs.   The depth of ponding increases with storm intensity; however, increases or decreases in street ponding due to the various subsidence cases were very minor.   Street ponding generally varied from 1 foot to 2.5 feet (depending on storm frequency), while the corresponding changes in flood depth due to subsidence were generally less than 0.10 foot.

Modeling of the "new" redesigned storm sewer and ditch system indicated a decrease in street ponding when compared to the "old" system. Street ponding across the majority of the watershed still occurred during all three storm events; however, in several isolated areas no significant ponding occurred during the 3-year under redesigned conditions. While the initial timing of inlet ponding was not significantly impacted by the redesign of the drainage system, the duration of street ponding was reduced from 25 percent to as much as 100 percent over the existing drainage system design. More significantly, no adverse effects on initial timing of inlet ponding or on the duration of ponding occurred due to subsidence based on the simulation results.

The resultant changes of conditions in the primary drainage channel due to the simulated subsidence cases were consistent with the findings of other phases of the study which considered the impacts of regional subsidence on the major bayous in the Houston area. Subsidence conditions that reduced channel slope caused an overall increase in channel water depths and an overall decrease in peak channel flows. Generally, the effect of the local subsidence cone on channel depths and flows is highly localized and is diminished at the outer reaches of the subsidence cone. In addition, changes in slope caused by the subsidence cone or the regional gradient have much less impact on depths and flows in the redesigned drainage system channel than in the "old" existing channel.

Conclusions

The results of this analysis of the localized effects of subsidence on flooding in a small urban watershed indicate that the optimum placement of well fields in the Houston area should be near the drainage divide of small watersheds or on similar areas of high topography within these watersheds. Placement of major well fields adjacent to the primary outfall channel or storm sewer of small basins will tend to increase local flooding near the well field. If unavoidable, placement of well fields adjacent to primary outfall channels or sewers should be combined with drainage system modifications to provide an increase in system capacity based on the predictable effects of subsidence on the channel or sewer gradient. Future predicted regional subsidence patterns which are oriented opposite to the primary drainage system gradient in small basins should be planned for in advance by compensating for potential increased flooding with additional channel freeboard or other design modifications.

MAXIMIZING UTILITY OF INTERFACES FOR HYDROLOGIC MODELS

R. Alex Harrington, A.M.ASCE *
John Y. Ding *

ABSTRACT

From the time a watershed layout has been determined, there is sufficient information for a computer to generate a command sequence for hydrologic simulation. Recognizing this fact, an interface has been designed that relieves the user of many of the tasks associated with data preparation and frees more time for analysis of results. The interface is designed for use on microcomputers and includes a file structure that is geared as much as possible to the way a user naturally organizes data. A variety of hydrologic procedures may be used and they are modularized in a Fortran library that allows routines to be added easily. The output is organized in a variety of summary tables suitable for direct inclusion in a technical report.

INTRODUCTION

The use of numerical models for planning and design of water resources projects has become standard practice since the introduction of 8088 processor-based micro computers in the early 1980s. This development has made computing resources available to the smallest consultant or regulatory agency. While it has been made easier to access and implement various mathematical models as a result of the use of micro computers, it has also become clear that few models take advantage of the full capacity of a computer to relieve the user of the more mundane tasks associated with model application. For example, significant time may be spent in organizing the computational sequence for simulation of a complex watershed. A user will often prepare a schematic outline of a watershed using a link-node or tree-structured layout and from this conceptualized plan, organize data from various sources in the format dictated by the model.

* Conservation Authorities and Water Management Branch, Ontario Ministry of Natural Resources, Room 5620 Whitney Block, Queen's Park, Toronto, Ontario, Canada, M7A 1W3

The Model Development Unit of the Conservation Authorities
and Water Management Branch of the Ontario Ministry of
Natural Resources is responsible for the maintenance and
development of hydrologic, hydraulic and water resources
management models (1,2,3). As the analyses required for
water resources projects become more sophisticated, the
shortcomings of existing models become more evident. Most
hydrologic models in use today were developed on mainframes
to run in batch mode and have been downloaded to
microcomputers with little change in the way in which they
run. As a result, data preparation and entry for such
models are carried out in much the same way as required in
using a mainframe. While some models may employ menus or
other techniques to relieve the user of the monotony of
data entry, this element of model application is usually
dictated by the program input data format, with little or
no thought given to the way a user either prepares or
stores data. In addition, the output from many models is
often a sequential listing, with the model calculations
interspersed with an echo of the input data. Since many
regulatory agencies require consultants to provide some
form of data documentation with reports, considerable time
can be spent organizing output in a format suitable for
inclusion in a technical report. When using a
microcomputer, many users will direct their output to a
file to examine before printing. This can lead to other
irritations because many models use Fortran carriage
control characters in the output which are not recognized
when printed from a file. This results in much time being
wasted in file editing for report preparation.

A user interface has been designed that overcomes many of
these shortcomings. It is a combination of a general
purpose package such as HEC1 (5), a problem oriented
language (6) and a problem oriented library (4) designed to
run on 80286 based microcomputers under DOS. In designing
the interface, three areas were identified as crucial for
meeting user needs. They can be broadly categorized as
data organization, model implementation, and output
generation. The remainder of this paper describes each of
these elements and the way in which they have been
structured to maximize their utility to users.

DATA ORGANIZATION

One of the more tedious aspects of model implementation is
the preparation of input data. This is especially true if
several models are in use at a site since each will require
a different input format and the same data will be reworked
several times. Ideally, data would be organized in a
logical fashion that allows for ease of extraction from a
database of some sort and automated generation of input.

It is also important to be able to restore and possibly
rerun data with minimal effort. Data files have been
divided into several categories reflecting the type of data
they contain. Each data file type has a unique file
extension. For example, watershed characteristics such as
area, main stem length, slope etc. are kept in files with
the extension CHR. The default filename is BASIN.CHR,
although users can change file names (maintaining the
extension). Several basin characteristic files may be set
up, for example, to keep data for particular geographic
regions separate.

Files are also established for rainfall, temperature, snow
data, and observed discharge. By keeping the components
separate, a data library is established which the user can
update as new data become available or as new projects are
undertaken. Each entry in the files is given a unique name
(a watershed name for example) which the program uses to
extract the required data from the library for a particular
simulation. Another file stores the simulation options
chosen by the user that are available for the various
processes included in the model. In addition to measured
or observed data, a parameter file is maintained which
keeps the parameters values to be used for simulating each
watershed.This file structure generally follows the pattern
of a user in data preparation. Each task involved in data
collection has associated with it a data file and the user
does not have to combine different types of data to prepare
to run the model.

MODEL IMPLEMENTATION

In any hydrologic simulation, it is necessary to determine
the watershed layout, with the sub-basins and their
connectivities, and then to organize the data and command
structure to suit both the program requirements and the
watershed characteristics. This can often be a very time
consuming process and requires a great deal of checking of
coding. However, from the time the watershed layout is
established, using for example, a tree structure as shown
in Figure 1, there is sufficient information for the
computer to carry out many of the tasks associated with
model implementation. For example the HYMO model (6),
which has been used extensively in Ontario, requires the
user to code a series of commands and associated data for
each watershed and keep track of hydrograph addition and
routing. The interface described here only requires that
the user code the watershed names and the junctions
associated with the watershed. Referring to Figure 1, the
watershed labelled BAS12 has an upstream junction of 108
and a downstream junction of 109. Headwaters have no
upstream junction. With these data entered for a
watershed, the program finds the basin outlet, and

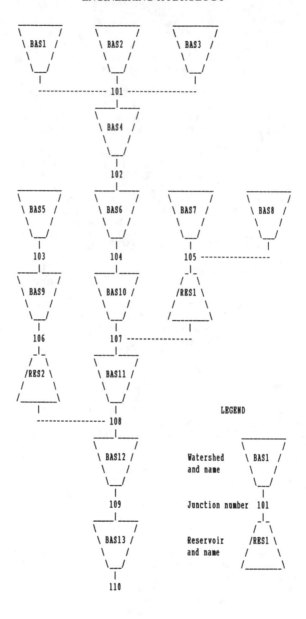

Figure 1
Schematic Watershed Layout

all the paths through the basin. Once the paths have been
established, the program generates the necessary subroutine
calls for computing hydrographs, routing through reaches or
reservoirs and adding hydrographs. A verbal command table
is also generated to allow the user to check that the
sequence generated is correct. Once the computational
sequence has been established, a separate file is created
with the data extracted from the library files to minimize
file access. Any parameter modifications are made to this
file, for optimization of model parameters, for example.
The data for the watershed layout do not have to be entered
in any particular order, making it easy to alter a layout,
for example, to add a reservoir. Basin labels are
character strings so labels having meaning to a particular
job can be used. These labels are used to search the data
files for the appropriate characteristics and parameters to
be used in the simulation.

A number of computational modules are maintained in a
Fortran library to allow a variety of options for a
simulation. Rainfall extractions can be calculated using
SCS curve numbers, Holtan's infiltration model or the
Green-Ampt model. Snowmelt is calculated using the degree
day method or an energy budget method. Hydrographs can be
calculated using either linear or non-linear unit
hydrographs . Channel routing uses either the variable
storage coeffient method or Muskingum Cunge routing.
Reservoir routing uses either storage indication or a
specified release option. These options allow a user
flexibility in tailoring the simulation method to the
particular application.

The model also allows a variety of execution modes. The
program is initially menu driven but this option can be
bypassed. This can be useful when data have been screened
and calibration runs are being made. Therefore, the
program allows users to directly execute a simulation from
a command line in essentially a batch mode. Files may be
edited either through the program or with any text editor.

The model is structured such that new routines for
simulation can be easily added.

OUTPUT GENERATION

Output from the model is organized as much as possible for
user convenience. When calibration runs are being made,
considerable time may be spent viewing model results on the
screen. To minimize fatigue from watching screens of
numbers scroll by, data are written to the screen a page at
a time using system calls. Options allow the user to
choose the level of detail to be displayed. For example,
during initial runs, a user may wish to monitor program

execution to ensure data integrity, but during later runs may only wish to see the results at selected points.

Output is routed to files and is categorized by output type. Plots of hydrographs are kept separate from printed results. A separate file of program execution is maintained, as is a file containing warning messages about conditions that may have been encountered during a run. All output is in tabular form suitable for direct inclusion in a technical report. No special print characters or printer controls are used, to ensure compatibility with most machines.

CONCLUSIONS

A user interface has been designed that attempts to minimize the amount of time a user spends in data entry and editing of output while providing flexibility in the number of computational options available for simulation. The program uses standard hydrologic procedures that are modularized in a Fortran library to allow ease of program expansion or upgrading. This approach to interface design was taken to allow practitioners more time for evaluation of alternatives and analysis of results.

APPENDIX - REFERENCES

1. Ding, J.Y., "Variable Unit Hydrograph", Journal of Hydrology, Vol. 22, 1974, pp. 53-69

2. Ontario Ministry of Natural Resources, "Flood Hydrology-VUH Model User's Manual", April, 1983

3. Ontario Treasury Board, Management Services Division, "Review of Planning for the Grand River Watershed", October, 1971

4. Smith, A.A., "A Problem Oriented Library for Hydrology", in 'System Analysis of Hydrologic Problems', Utah Water Research Library, Utah State University, August, 1970, pp. 305-325

5. U.S. Army Corps of Engineers, Hydrologic Engineering Centre, HEC-1 Flood Hydrograph Package User's Manual, Jan., 1973

6. Williams, J.R. and R.W. Hann Jr., "HYMO - A Problem Oriented Computer Language for Hydrologic Modelling", Agricultural Research Service, USDA, May, 1973

AN URBAN HYDROLOGY MODEL FOR THE PERSONAL COMPUTER

Thomas A. Seybert*, Member and
David F. Kibler**, Member ASCE

ABSTRACT

The use of personal computers to evaluate the effects of urbanization on the hydraulic response characteristics of a watershed is discussed in this paper. Specifically, the use of a group of related hydrologic and hydraulic programs to assist urban planners in estimating runoff rates and volumes is addressed. These programs deal with SCS curve number weighting, design rainfall hyetographs, sub-area hydrographs by the SCS unit hydrograph method, hydrographs by the SCS TR-55 tabular method, Muskingum channel routing, Modified Puls routing, hydrograph combining, hydrograph plotting and multi-stage reservoir outlet analysis. The menu driven computer model allows the selection and interaction of all the programs in an unlimited number of sequences to solve complex hydrologic problems.

INTRODUCTION

In the past ten years state and local governments have come to recognize the need for rules and regulations on land development practices with regard to stormwater management. With this relatively new emphasis on stormwater management, many consulting engineers and urban planners dealing with land development have been required to brush up on current methods in urban hydrology. Because of the demand for engineers to develop skills in urban hydrology and the need for continual updating of these skills, the Department of Civil Engineering and the Office of Continuing Education at The Pennsylvania State University initiated the Penn State Stormwater Short Course and Symposium in October 1979. In addition to presenting methodologies in urban hydrology, the symposium acts as an arena for the exchange of ideas among government officials, professional engineers, planners, as well as academicians engaged in various stormwater management activities. The symposium is held yearly at the University Park Campus of Penn State during the fall semester.

To assist in the presentation of computational methods in the short course, a set of programs for the microcomputer was developed. During the development of the software, the main thrust was practicality and flexibility in providing watershed-wide hydrologic capabilities. In the fall of 1984, the first version of the software package was completed and used in the symposium. The current version is the fourth revision.

* Assistant Professor of Engineering, The Pennsylvania State University, Beaver Campus, Brodhead Road, Monaca, PA 15061.
** Professor of Civil Engineering, The Pennsylvania State University, 212 Sackett Building, University Park, PA 16802.

THE COMPUTER PROGRAM PACKAGE

The package of programs, referred to as the Penn State Urban Hydrology Model (PSUHM), is written in BASIC and solves commonly encountered problems in urban hydrology. The programs are menu driven and written for an IBM compatible computer with 128K memory. Figure 1 shows the menu of the software package as it appears on the computer monitor. A description of each module of the package follows.

Module 0 - Curve Number Weighting: This module allows the user to enter specific watershed data pertaining to soil type and ground cover to weight a total watershed or subarea curve number. For each type of land use within the watershed the user provides a descriptive name, area and curve number. The computer takes this information, performs a simple area based weighting, and reports the weighted curve number.

Module 1 - Design Rainfall Calculations: A central peaking design rainfall with a two hour time base is generated by this program. The user enters the design storm return period and the corresponding one hour rainfall depth. The program computes the hyetograph using standard intensity duration curves and the composite design storm generation method as described in reference [2]. The synthetic storm can be saved as a data file for later use in Module 3.

Module 2 - Manual Entry of a Hydrograph: The user can input a hydrograph (data point by data point) that has been generated by a method outside of the software package. The program will prompt the user to enter the hydrograph starting date and time and data point time interval. Once entered, the hydrograph is saved as a data file for future use in another module.

Module 3 - Hydrograph Generation - SCS Triangular Hydrograph Method: This program uses the SCS unit hydrograph and a design rainfall to create a watershed hydrograph. The computer lists all previously generated design storms (via Module 1) available on the data disk from which a rainfall is chosen. The user enters the hydrograph

```
*****************************************************
*  PENN STATE URBAN HYDROLOGY MODEL - MENU PROGRAM  *
*****************************************************

        HIT          TO RUN
        <0>   CURVE NUMBER WEIGHTING
        <1>   DESIGN RAINFALL CALCULATIONS
        <2>   MANUAL ENTRY OF HYDROGRAPH
        <3>   SUB-AREA HYDROGRAPHS BY SCS U.H.
        <4>   SCS TR-55 TABULAR HYDROGRAPH
        <5>   MUSKINGUM CHANNEL ROUTING
        <6>   MODIFIED PULS ROUTING
        <7>   MSRM - MULTIPLE STAGE ROUTING MODEL
        <8>   HYDROGRAPH COMBINING
        <9>   HYDROGRAPH PLOTTING
```

FIGURE 1 - Menu Screen for PSUHM

starting date and time along with watershed area, curve number and time of concentration. The computer generates the hydrograph using the SCS triangular unit hydrograph method as described in reference [3]. The hydrograph is saved as a data file for later reference.

Module 4 - Subarea Hydrographs by the SCS TR-55 Tabular Method: Module 4 is a computerized version of Chapter 5 of the SCS TR-55 [4]. The program requires the user to enter the storm return period, rainfall depth, number of subareas in the watershed and data for each subarea including drainage area, curve number, time of concentration and time of travel. The program will compute hydrographs for each subarea and the composite hydrograph for the total watershed. The composite hydrograph can be saved as a data file for later use.

Module 5 - Channel Routing by Muskingum Method: The program routes a previously generated hydrograph in a data file through a given reach of channel using the Muskingum method as described in reference [5]. The user must provide a channel wedge coefficient "X", valley storage coefficient "K" and the routing time step. The program checks the time step for routing compatibility and warns the user if the routing interval is unacceptable, or if additional sub-reaches should be used. The routed hydrograph can be saved to a data file for future use.

Module 6 - Routing by Modified-Puls Method: This program routes a flood wave through a reservoir or channel reach and is based on the Modified-Puls method as described in reference [5]. The user must provide a rating curve matrix of elevation, storage and outflow values for the reservoir or channel, obtained from Module 7 or other sources. Once entered, the rating matrix is saved to a data file and can be recalled later in another application. The user selects the hydrograph to be routed from a screen displaying all hydrograph files currently on the disk. The routed hydrograph can be saved to a data file for later use.

Module 7 - Multi-stage Outlet Reservoir Routing: This module routes a flood wave through a reservoir having multi-stage outlets using the Modified-Puls method. The outlet rating curve for the reservoir is modeled by a subroutine program in the module that can handle various discharge types such as orifice flow (rectangular/circular), weir flow (rectangular/proportional), emergency spillways, drop inlets, discharge pipes (headwalls/projecting inlets), and outfall culverts or receiving channels. Specifics on the computation methods of this module are explained in reference [1]. Once the outlet rating curve has been computed, the program will route the inflow hydrograph through the reservoir and generate an outflow hydrograph that can be saved as a data file for later use.

Module 8 - Hydrograph Combining: This program combines any two hydrographs with the same initial time and date. The user selects the two hydrographs from a screen display that lists all of the hydrograph files available on the data disk. The composite hydrograph can be saved to a data file for future use.

Module 9 - Hydrograph Plotting: This program will graphically display on the monitor a single hydrograph or two related hydrographs with the same coordinate axes. If the proper hardware is connected to the computer, a screen dump of the plot can be sent to a printer.

INTERACTION OF MODULES TO SOLVE COMPLEX PROBLEMS

The software package is structured to analyze a variety of hydrologic
design situations by the interaction of the modules through the menu.
Consider the following general example that illustrates the use and
interaction of the software modules.

A simple watershed is divided into two subareas as shown in
Figure 2. For each subarea the following data are available:
* land use classifications with acreage
* hydrologic soil classification(s) and percent coverage
  for each land use classification
* time of concentration
* channel travel time (upper subarea only)
It is desired to establish the runoff hydrograph of the
watershed for a 25 year design storm. Using PSUHM the following
sequence of module selections will generate a solution.

Step #1: Use Module 0 to weight curve numbers for both subareas.
After entering the number of subareas in the watershed the program
will prompt the user to enter a list of descriptive terms of land use
for the first subarea. Land use classifications with more than one
hydrologic soil type should have a descriptive term for each
hydrologic soil type within the land use class. Once this
information is entered, the user must input the specific data of area
and curve number for each land use class. A screen table with curve
number values as a function of land use condition and hydrologic soil
type is provided to assist in the selection of appropriate curve
numbers. After the data for the first subarea have been entered, the
program calculates the weighted value of curve number for the subarea

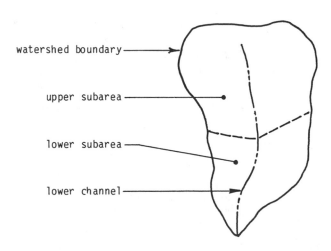

FIGURE 2 - Sketch of Example Watershed

and then proceeds to the second subarea. The final output displays
all of the input data in tabular form plus the final weighted curve
number for each subarea. Upon completion of data output the program
returns the command level to the menu program.

Step #2:   Use Module 1 to generate a synthetic design storm for the
watershed. The module requests two items of information; namely,
return period of the design storm and the corresponding one hour
rainfall depth. With this data the program will generate a central
peaking hyetograph using U.S. Weather Bureau/U.S. Army Corps of
Engineers standard rainfall intensity curves. The design storm must
be saved as a data file on the data disk for later use and thus a
descriptive name for the file, say "P25", is entered when requested
by the program. After the hyetograph is saved on the data disk and
displayed as output on the monitor, the menu will return to the
screen.

Step #3:   Use Module 3 to generate the hydrograph for the upper
subarea. The program will list the names of all the hyetograph files
that exist on the data disk. The descriptive name assigned to the
design storm ("P25") is entered and the computer reads the file into
memory for manipulation. Next, data for the subarea is entered;
namely, area, weighted curve number and time of concentration. The
resulting hydrograph can be displayed on the screen or sent to a
printer. The program asks the user to provide a descriptive file
name for the hydrograph, say "UPPER", then saves this hydrograph to
the data disk. Once the hydrograph data file "UPPER" is saved,
Module 3 sends the user back to the menu program.

Step #4:   Use Module 5 to route the "UPPER" hydrograph through the
lower channel reach. This module immediately lists all of the
hydrograph files that are present on the data disk. The user selects
the hydrograph file "UPPER" and the computer reads the file into
memory. For the channel the user must enter a wedge coefficient, a
valley storage coefficient and a routing time interval. The computer
will display the routed hydrograph, then save it to the disk once the
user specifies the new file name. For this case, the new hydrograph
will get the name of "UPROUTED". Again, the user is returned to the
menu upon completion of the Module 5 task.

Step #5:   Use Module 3 again to generate a hydrograph for the lower
subarea. Referring to Step #3 the required input data are the design
rainfall file ("P25"), area, curve number and time of concentration.
The results of the program calculations are saved as a data file
under a descriptive name, say "LOWER", and the user is returned to
the menu.

Step #6:   Use Module 8 to combine the routed hydrograph of the upper
subarea and the hydrograph of the lower subarea. The user selects
the two hydrograph data files ("UPROUTED" and "LOWER") from a screen
listing of available files on the data disk and also inputs the time
interval between data points for the combined hydrograph. With this
information the combined hydrograph is generated and the solution is
finished. The final downstream hydrograph for the total watershed is
displayed on the monitor or printer and is tabulated along with the
two combining hydrographs. The combined hydrograph could be saved to
the data disk if further manipulation of the output were desired.

OBSERVATIONS

The package of computer programs presented here was written with the hope that a wide variety of hydrologic runoff problems could be solved with a minimum amount of hand calculations. Calculations such as travel time and time of concentration are not well suited for computerized solutions since these calculations are more investigative than iterative in nature.

The software package is used as a teaching aid in the PSU Stormwater Short Course and Symposium and is well received by the participants. In a very short amount of time the users can see a distinct advantage in using the computer to solve their own problems. Several design options for a particular problem can be investigated in a very short time compared to the long-hand desktop methods.

A significant lesson with respect to programming techniques for software versatility was learned as the software package was developed and used. The smaller stand-alone programs networked to each other through a data file manager and menu make the software very easy to learn and very flexible in its use. The user can learn the operation sequence of a particular module without having to worry about the structure of other modules. The modules only interact through the menu and therefore cannot directly affect each other. However, because of this very structure, the package can provide an infinite number of computational sequences to analyze many different situations.

SUMMARY

Because of statewide emphasis on watershed-level planning and local government regulations on methods and procedures of development, stormwater management facilities in civil engineering type work will continue to increase in the future. The use of a personal computer to analyze and design these structures can be a significant time saver for the engineer. In addition, the personal computer can provide the opportunity to economically and efficiently investigate several possible solutions for a particular problem. The concept of stand-alone computing modules networked with a common data file manager and menu program can significantly increase the versatility of the software while reducing its learning curve.

REFERENCES

1. Chamberlain, A. Scott, "A Multiple Stage Detention Basin Analysis and Design," MS Thesis, The Pennsylvania State University, 1986.
2. The Pennsylvania State University - Continuing Education, "Course Notes for Computational Methods in Stormwater Management," Stormwater Short Course and Symposium, October 1986.
3. United States Department of Agriculture - Soil Conservation Service, "Technical Release No. 55, Urban Hydrology for Small Watersheds," January 1975.
4. United States Department of Agriculture - Soil Conservation Service, "National Engineering Handbook - Section 4," August 1972.
5. Viessman, Knapp, Lewis, and Harbaugh, Introduction to Hydrology, 2nd Ed., Harper and Row, 1977.

SITE HYDROLOGY AND BASIN DESIGN USING SPREADSHEETS

Daniel R. Lutz*, Member ASCE
James C. O'Brien**, Member ASCE

ABSTRACT

The Soil Cover Complex Method which was developed by the Soil Conservation Service is a commonly accepted method of evaluating site hydrology. The solution of this method is facilitated by the use of electronic spreadsheets. Closely related is the design of detention basins, a process which can also be facilitated by the use of electronic spreadsheets. With this in mind, several spreadsheet templates have been developed to assist in these procedures. The following paper is a brief description of these procedures and the electronic spreadsheet templates which have been developed.

INTRODUCTION

Municipal land subdivision ordinances in most of the developing townships in Southern Pennsylvania require that the stormwater management system and facilities be designed according to the Soil Cover Complex Method. This method was developed by the Soil Conservation Service, U.S. Department of Agriculture, in the SCS National Engineering Handbook, Section 4 - Hydrology. It also appears in Urban Hydrology for Small Watersheds, Technical Release No. 55.

To facilitate the solution of the site hydrology using this method, templates have been developed for a popular electronic spreadsheet for use on IBM-PC/XT/AT and compatible computers.

SOIL COVER COMPLEX METHOD

The Soil Cover Complex Method is used to compute the peak discharge (cfs) from a particular site for the design storms under consideration. It is based on various parameters which include land use, soil type, total storm

* President, Daniel R. Lutz & Associates, Incorporated, P.O. Box 115, Chadds Ford, PA 19317
** Assistant Professor, Villanova University, Villanova, PA 19085

rainfall, and time of concentration. A detailed topographic site plan map, with contours and soil types plotted, is necessary in order to determine stormwater runoff by this method.

PREDEVELOPMENT CONDITIONS

In order to facilitate the calculations needed for the Soil Cover Complex Method a Predevelopment Condition spreadsheet template was developed. Using the topographic site plan and Urban Hydrology for Small Watersheds, Technical Release No. 55, the land use, area, Cn values, and the time of concentration data are used as input into the spreadsheet template. The template automatically calculates the weighted Cn, the runoff depth Q, the time of concentration, and the predevelopment peak discharge for the 2, 10, 25, and 100 year design storms. This template has the appropriate equations installed in particular cells to do some of the calculations and then uses the results of certain cells as input into other parts of the spreadsheet. A sample printout of the results of the spreadsheet analysis is shown in FIGURE 1 and FIGURE 2.

```
PREDEVELOPMENT CONDITIONS      DRAINAGE AREA #2
-----------------------------------------------------------------------
Predevelopment drainage area #2 contains 14.93 acres of onsite area, 1.62
acres of offsite area to the west of the tract, and 2.73 acres of offsite
area east of the tract.  The total area, therefore, is 19.28 acres.
```

LAND USE	AREA (Acres)	TABLE 2-2 CN	PRODUCT
Buildings, road, paved areas	0.02	98	2.0
1/2 Acre Residential			
Hydrologic group B	2.73	70	191.1
Meadow			
Hydrologic group B	8.90	58	516.2
Hydrologic group C	0.12	71	8.5
Hydrologic group D	2.19	78	170.8
Wooded			
Hydrologic group B	4.14	55	227.7
Hydrologic group C	0.16	70	11.2
Hydrologic group D	1.02	77	78.5
	---------------		---------------
	19.28		1206.04
		WEIGHTED CN =	62.6

```
COMPUTE RUNOFF DEPTH FOR CN =        62.6
```

DESIGN STORM	P (in.) 24 HOUR RAINFALL	Q (in.) RUNOFF DEPTH
2 YEAR	3.2	0.50
10 YEAR	5.0	1.48
25 YEAR	5.6	1.87
100 YEAR	7.2	3.01

$$Q = (P - 0.2S)^2 / P + 0.8S \quad \text{where } S = (1000/CN) - 10$$

FIGURE 1

```
----------------------------------------------------------------------
```

COMPUTE TIME OF CONCENTRATION Tc:

PATH and SURFACE	LENGTH (Ft.)	VELOCITY (Fps.)	TIME (Sec.)
Meadow @ 8.6%	70	1.40	50.0
Woods @ 16.0%	25	1.00	25.0
Meadow @ 12.9%	155	1.70	91.2
Meadow @ 10.0%	200	1.50	133.3
Meadow @ 5.4%	130	1.10	118.2
Woods @ 5.6%	125	0.60	208.3
Woods @ 7.8%	230	0.70	328.6
Meadow @ 6.2%	130	1.20	108.3
Meadow @ 4.6%	240	1.00	240.0

```
                                         ---------------
```

		TOTAL =	1302.9
		Tc in minutes =	21.7
		Tc in hours =	0.36

COMPUTE PEAK DISCHARGE:

DESIGN STORM	qp (csm/in) *	AREA (sq. mi.) **	Q (in.) RUNOFF DEPTH	q (cfs) PEAK DISCHARGE
2 YEAR	620	0.030125	0.50	9.4
10 YEAR	620	0.030125	1.48	27.6
25 YEAR	620	0.030125	1.87	34.8
100 YEAR	620	0.030125	3.01	56.1

* FROM FIG. 5-2 FOR Tc =   0.36

** AREA in acres =   19.28

q = (qp)AQ

FIGURE 2

POSTDEVELOPMENT CONDITIONS INCLUDING DETENTION BASIN ROUTING

The postdevelopment analysis of the site is also based on a detailed graphic plan with contours, soil types, and the post development features such as roads, buildings and stormwater management facilities. The land use, area, Cn values, and the time of concentration data are used as input into a Postdevelopment Condition template. The template automatically calculates the weighted Cn, the runoff depth Q, and the time of concentration in the same manner as was done in the Predevelopment Conditions template.

An Inflow Hydrograph template was developed which computes the inflow hydrograph into a detention basin using values computed by the Postdevelopment Conditions template. This spreadsheet is shown in FIGURE 3.

POSTDEVELOPMENT CONDITIONS

DRAINAGE AREA #2
INFLOW INTO DETENTION BASIN
Inflow per Table 5-3

DESIGN STORM		2 YEAR	10 YEAR	25 YEAR	100 YEAR
DRAINAGE AREA (smi)		0.029922	0.029922	0.029922	0.029922
RUNOFF DEPTH Q (in)		0.75	1.91	2.35	3.62
PRODUCT		0.0224	0.0572	0.0703	0.1083
HYDROGRAPH TIME (hrs.)	TABULAR DISCHARGE (csm/in) Table 5-3 Tc= 0.20 hrs. Tt = 0	INFLOW (cfs)	INFLOW (cfs)	INFLOW (cfs)	INFLOW (cfs)
11.1	23	0.52	1.31	1.62	2.49
11.2	29	0.65	1.66	2.04	3.14
11.3	35	0.79	2.00	2.46	3.79
11.4	41	0.92	2.34	2.88	4.44
11.5	47	1.05	2.69	3.30	5.09
11.6	128	2.87	7.32	9.00	13.86
11.7	208	4.67	11.89	14.63	22.53
11.8	509	11.42	29.09	35.79	55.13
11.9	796	17.86	45.49	55.97	86.22
12.0	641	14.38	36.63	45.07	69.43
12.1	424	9.52	24.23	29.81	45.93
12.2	245	5.50	14.00	17.23	26.54
12.3	170	3.82	9.72	11.95	18.41
12.4	138	3.10	7.89	9.70	14.95
12.5	121	2.72	6.92	8.51	13.11
12.6	104	2.33	5.94	7.31	11.26
12.7	85	1.91	4.86	5.98	9.21
12.8	75	1.68	4.29	5.27	8.12
12.9	71	1.59	4.06	4.99	7.69
13.0	68	1.53	3.89	4.78	7.37
13.1	62	1.39	3.54	4.36	6.72
13.2	56	1.26	3.20	3.94	6.07
13.3	54	1.21	3.09	3.80	5.85
13.4	51	1.14	2.91	3.59	5.52
13.5	49	1.10	2.80	3.45	5.31
13.6	48	1.08	2.74	3.38	5.20
13.7	46	1.03	2.63	3.23	4.98
13.8	44	0.99	2.51	3.09	4.77
13.9	42	0.94	2.40	2.95	4.55
14.0	40	0.90	2.29	2.81	4.33

FIGURE 3

At this stage of the design process the designer must calculate the detention basin volume and design the outlet structure computing the outflow values for selected elevations. Detention basin routing is accomplished using the Flood Routing spreadsheet template. The input data section consists of elevation, storage, and outflow. The inflow hydrograph computed in the Inflow Hydrograph template is also inputed for each design storm.

Detention basin routing is based on solving the continuity equation using the Storage-Indication method. This widely used method of reservoir routing solves the working equation by a process in which interpolations are made in the elevation-discharge and elevation-storage data portion of the template. Spreadsheet macros were written to perform the lookup functions required to completely automate the routing computations. The Flood Routing spreadsheet template shows the inflow and outflow hydrograph, as well as, water surface elevation and storage for each time increment for the various design storms under consideration. The part of this spreadsheet for a ten year design storm is shown in FIGURE 4.

```
---------------------------------------------------------------
            FLOOD ROUTING
            10.06
            (S2/dt + O2/2) = (S1/dt - O1/2) + ((I1 + I2)/2)

            use dt = 6 min.

INPUT DATA
-----------
ELEVATION    STORAGE     OUTFLOW      S/dt + O/2
   (ft.)     (ac.ft.)    (cfs)
             (Input from (Input from
              SHT 18)     SHT 20)
---------------------------------------------------------------
   230.00     0.000        0.00        0.00
   230.60     0.380        7.70       49.83
   231.00     0.634       20.50       86.96
   231.50     0.951       29.90      130.02
   232.00     1.268       49.50      178.18
   233.00     2.014       57.00      272.19
   234.00     2.760       59.00      363.46
     0.00     0.000        0.00        0.00
     0.00     0.000        0.00        0.00
     0.00     0.000        0.00        0.00
     0.00     0.000        0.00        0.00
     0.00     0.000        0.00        0.00
     0.00     0.000        0.00        0.00
     0.00     0.000        0.00        0.00

DESIGN STORM:  10 YEAR
==================================
   TIME      INFLOW     ELEVATION    STORAGE     OUTFLOW
  (min.)     (cfs)      (ft.)        (ac.ft)     (cfs)
---------------------------------------------------------------
     0        0.00       230.00       0.000        0.00
     6        1.31       230.01       0.005        0.10
    12        1.66       230.02       0.016        0.32
    18        2.00       230.04       0.027        0.55
    24        2.34       230.06       0.039        0.80
    30        2.69       230.08       0.053        1.06
    36        7.32       230.13       0.083        1.67
    42       11.89       230.23       0.143        2.90
    48       29.09       230.44       0.277        5.62
    54      |45.49|      230.80       0.504       13.97
    60       36.63       231.09       0.694       22.28
    66       24.23      |231.19|      0.754      |24.06|
    72       14.00       231.13       0.718       22.98
    78        9.72       231.00       0.636       20.55
    84        7.89       230.88       0.555       16.53
    90        6.92       230.78       0.493       13.38
    96        5.94       230.70       0.445       10.99
   102        4.86       230.64       0.407        9.06
   108        4.29       230.59       0.376        7.62
   114        4.06       230.55       0.350        7.08
   120        3.89       230.51       0.326        6.60
   126        3.54       230.48       0.304        6.16
   132        3.20       230.45       0.283        5.73
   138        3.09       230.42       0.263        5.33
```

FIGURE 4

CONCLUSION

The use of electronic spreadsheet templates for the analysis of site hydrology and in detention basin design is a valuable tool for the design engineer. The ease of use and the speed of calculation give the designer the ability to optimize the design in a most cost effective manner.

REFERENCES

Soil Conservation Service National Handbook, Section 4 Hydrology. Soil Conservation Service - United States Department of Agriculture, March 1985

Urban Hydrology for Small Watersheds, Technical Release No. 55. Engineering Division - Soil Conservation Service- United States Department of Agriculture, June 1985

MICRO-ILLUDAS FOR URBAN DRAINAGE DESIGN

Michael L. Terstriep[1], M. ASCE, and Douglas C. Noel[2]

Abstract

The Illinois Urban Drainage Area Simulator (ILLUDAS) was developed at the Illinois State Water Survey in 1969 and has been available for use on mainframe computers since 1974. An enhanced version of the model designed for use on micro computers was released in 1985. The continued popularity of the model is due in part to the simplicity of its internal functions. The user does not feel that he or she is using a black box approach to urban drainage design. The micro version maintains the same theoretical approach of the mainframe version but has additional features. An interactive "front end" has been added to make data entry and modification much easier. Additional options are also available to assist with the incorporation of detention basins into the design. A post processor provides output, including hydrographs, in report format.

Introduction

The Illinois Urban Drainage Area Simulator, ILLUDAS, was first introduced by Terstriep and Stall (1969), and was the result of research done at the Illinois State Water Survey. The original study was intended to investigate the applicability of the British Road Research Laboratory (RRL) methodology (Watkins, 1962) to urban basins in the United States. The result of this study was the computer model ILLUDAS, which provided an objective method for the hydrologic design and evaluation of storm drainage systems in urban areas. This model was made publicly available in 1974 (Terstriep and Stall, 1974).

The RRL method lumped the basin as a whole and accommodated runoff only from the paved areas of the basin that are directly connected to the storm drainage system. The grassed areas were excluded from consideration as were paved areas that were not directly connected. ILLUDAS utilizes the directly connected paved area concept of the RRL method but also recognizes and reproduces runoff from grassed and non-connected paved areas. It allows the user to apply these concepts to any number of sub-catchments.

---

[1]Section Head, Surface Water Section, Illinois State Water Survey, Champaign, Illinois
[2]Associate Hydrologist, Illinois State Water Survey Champaign, Illinois

Surface Runoff

ILLUDAS is applied by first dividing the basin to be studied into sub-basins. A sub-basin is normally a homogeneous portion of the basin tributary to a single inlet or set of inlets that constitute a design point in the drainage network. Separate hydrographs are generated for the contributing paved area and for the contributing grassed area for each sub-basin. These hydrographs are superimposed and routed downstream through the drainage system to the outfall.

The first step in generating the contributing paved area hydrograph for each sub-basin is to determine the paved area directly connected to the storm drainage system. Once this is complete, calculations are made to approximate the time-of-travel from the furthest point on the paved area to the outlet of the sub-basin. The directly connected paved area is then assumed to be linearly distributed with the time-of-travel. Thus for each time increment up to the total travel time there is an equal addition to the contributing area.

Input is the rainfall distribution as a series of intensities of equal duration. The duration is equal to the time increment selected. The rainfall input can be an actual event hyetograph or a design storm. The model has three standard design distributions, a simple uniform distribution, a triangular distribution, and the Huff distribution (Huff, 1967). In addition, as part of a data file which may be edited by the user, up to ten user-defined storm distributions may be stored at any time. The time increment used should be as short as the quality of rainfall data will allow. A longer value may be more convenient to use for very large basins or for very long storms.

Losses representing initial wetting and depression storage are removed from total rainfall before it is applied to the contributing paved area. These losses are combined and treated as an initial mean depth for the paved area abstraction. Since runoff may occur for events having lower total rainfall than the mean depth of the abstraction, a spatial distribution is applied which allows the simulated depth of storage to vary uniformly over the basin from zero to a depth equal to twice the mean value. The total hyetograph less the abstraction is referred to as the paved-area supply rate.

The ordinates of the paved-area hydrograph are computed by a simple convolution of the paved-area supply rate with incremental increases in the contributing paved area. This convolution can be represented as follows:

$$RO_i = \sum_{n=1}^{npa} (PA_n \cdot PASR_{i-n+1}) \Delta t$$

$$
\begin{aligned}
RO_1 &= PA_1 PASR_1 \\
RO_2 &= PA_1 PASR_2 &+ PA_2 PASR_1 \\
RO_3 &= PA_1 PASR_3 &+ PA_2 PASR_2 &+ PA_3 PASR_1 \\
&\;\;\vdots \\
RO_n &= PA_1 PASR_n &+ \ldots\ldots &+ PA_n PASR_1
\end{aligned}
$$

where   RO = surface runoff ordinate
       PASR = paved area supply rate ordinate
          i =current time interval
         PA = contributing area for given time increment
        npa = number of paved area increments
         Δt = time increment

The grassed area hydrograph is determined in a similar process to that of the paved area hydrograph. Runoff from the non-contributing paved area is distributed uniformly in time and added to the rainfall hyetograph. The hyetograph is then subjected to an initial abstraction and infiltration losses. Infiltration curves are provided for four hydrologic soil groups. The user selects the appropriate soil group and specifies an antecedent moisture condition. The antecedent moisture level determines the starting point on the infiltration curve. Infiltration is simulated by continually integrating the Horton equation descripbed by Chow (1964), which allows the model to approximate the current infiltration rate and thereby estimate the maximum potential infiltration for the interval. At any interval in time, the total moisture supply is equal to the sum of the non-contributing paved area runoff, the rainfall for the interval and the depth of moisture in depression storage at the start of the interval. Moisture supply which is in excess of the potential infiltration and available depression storage becomes the grassed-area supply rate. The ordinates of the grassed-area hydrograph are then computed using the convolution described above with the grassed-area supply rate and increments of the contributing grassed area.

The ordinates of the paved-area hydrograph are added to the ordinates of the grassed-area hydrograph to produce the local surface runoff hydrograph, or inlet design hydrograph. These hydrographs are developed for each sub-basin and become an input into the drainage network at a particular point. The input at any location is added to the routed hydrograph from the upstream reach, if any, to become the design hydrograph for the downstream reach.

Channel Routing

Before routing this design hydrograph to the next downstream node, two additional functions of the model are implemented. First, any modifications to flow required by excedence of the reach capacity are performed. This reach capacity may be a user imposed flow limitation or a computed reach capacity in the case of the evaluation of an existing storm drainage system. In these cases, the flow limitation is handled by simply storing all flows in excess of the reach capacity at the upstream end of the reach until such time as the ordinates of the design hydrograph become less than the reach capacity. At this time the surcharged volume is bled back into the hydrograph at the reach capacity until exhausted.

Similarly the user may request that some volume of storage be utilized in a given reach. In this case a procedure is used that gradually increases a limiting discharge until the maximum instantaneous storage volume generated by the procedure described in the previous paragraph equals the requested volume.

The second model feature implemented prior to routing occurs only if a new pipe is being designed for the reach. In this case, the design hydrograph peak after any modification is used to determine which commercially available pipe size would be required to pass the design peak discharge based on the reach slope and the new pipe roughness supplied by the user. Existing drainage systems may also be run in design mode, in which case the model will indicate whether or not the current reach capacity is sufficient to pass the design hydrograph peak or, if not, what size pipe would be needed to avoid surcharging at the upper end of the reach. The pipe diameters are not allowed to decrease in the downstream direction except for an existing reach of sufficient capacity or a reach downstream from a detention basin with a rating curve.

Reach routing to the next downstream node is accomplished in one of three ways. In the first case, a flow through time, Q_t, is computed for the reach based on the design peak and the Manning Equation. The routed hydrograph is then created simply by moving all the ordinates in time by Q_t. This procedure maintains the hydrograph shape, volume, and timing.

The second method attempts to solve the continuity equation for flow through and storage in the reach for each interval. This is an iterative trial and error procedure which utilizes the linear translation as its first estimate of the outflow for any interval.

The third method requires a storage-discharge rating curve with which to route flows in the reach. This option allows the user to perform routing based on a user-supplied rating curve. Dead pool storage is modeled to allow simulation of retention or detention basins. The basic inputs are a storage-discharge rating curve from which the routing is performed. Stage data may optionally be included. A short summary containing the inlet and outlet peaks and peak times is provided. Lag times can be added by specifying the average time of concentration for water entering the inlet end of the reach to impact conditions at the outlet end.

The model allows the user to provide a constant-valued baseflow at any reach in the basin. Some applications require point sources which can not be approximated as a part of a constant-valued baseflow in the receiving stream. MICRO-ILLUDAS allows the user to impose an input hydrograph on the reach. Still other applications require repetitive testing of different alternatives in the lower reaches of an urban catchment. In these cases, the upper reaches of the basin may be simulated once for a given design storm and the resulting outfall hydrograph may be specified as a point source to the remaining catchment area. Observed hydrographs may be included so that the user may easily compare his or her results with the observed data.

The output of the model has been formatted to provide routed and, if so desired, interim hydrographs at the user's discretion. A summary table which describes both the surface runoff and reach characteristics at each node is also printed as part of the output.

Utilities

In order to make the package as complete as possible and to make it "*user friendly*", five programs were developed to be used with the model. The first is an integral part of the model package. It is a data/word processing application through which data sets are created and modified. This program also spots and identifies potential problems with the data set during creation or modification and brings them to the user's attention so that they may be addressed before making a model run with the data set.

A graphics program is included which reads the simulation output and then allows the user to select which, if any, hydrographs to plot for a given reach. Hydrographs may be previewed on a graphics monitor-equipped system prior to the electing to have them plotted.

For the installed base of mainframe ILLUDAS users a data set translation utility is included. After downloading existing ILLUDAS data sets to a microcomputer, this program may be used to reformat the existing data set to be compatible with the micro version of the model.

A program is also supplied with which to create or modify design storm distributions for use in the model. The program creates the Huff and SCS distributions the first time it is run. The user is allowed to list any of the distributions or to specify a duration and depth for a sample event and preview the design storm before using it in the model. This program also generates the endpoints of the triangular distributions for both paved and grassed depression storage and for infiltration if the default uniform (horizontal) distribution is not to be used.

Appendix - References

Chow, Ven Te. 1964. Handbook of Applied Hydrology. McGraw-Hill Book Co., Inc., New York, p. 14-17.

Huff, F.A. 1967. Time Distribution of Rainfall in Heavy Storms. Water Resources Research, V. 3(4):1007-1019.

Terstriep, Michael L., and John B. Stall. 1969. Urban Runoff by the Road Research Laboratory Method. ASCE Journal of the Hydraulics Division, V. 95(HY6):1809-1834.

Terstriep, Michael L., and John B. Stall. 1974. The Illinois Drainage Area Simulator, ILLUDAS. Illinois State Water Survey, Champaign, Illinois, USA. ISWS Bulletin 58, 90 p.

Watkins, L.H. 1962. The Design of Urban Sewer Systems. Department of Scientific and Industrial Research, London, Her Majesty's Stationery Office, Road Research Technical Paper 55.

HEC-1 AND HEC-2 APPLICATIONS ON THE MICROCOMPUTER

Harry W. Dotson*, M.ASCE

ABSTRACT

The Hydrologic Engineering Center (HEC) has developed versions of the HEC-1 Flood Hydrograph Package and the HEC-2 Water Surface Profiles programs for MS/PC-DOS compatible microcomputers (PC). The increased speed, memory and storage capacity of the latest PCs make the use of these large FORTRAN programs highly practical in the PC environment.

Typical tasks that are required when using these batch oriented programs include creating, checking and editing input data; executing the program; and summarizing and displaying the results. The HEC has developed a menu driven user interface or shell program to integrate several application programs, an editor and other utility programs to assist the user in accomplishing these tasks. The interface takes advantage of the unique capabilities and user friendliness found in the PC environment. In addition to the HEC-1 and HEC-2 programs, the integrated package includes a program (SUMPO) for creating summary tables of HEC-2 results, a program (PLOT2) that plots cross-sections and water surface profiles, and a PC version of the Corps of Engineers editor (COED), which features full screen editing and on-line help screens and documentation. The application of these programs is described and demonstrated using the menu driven interface.

INTRODUCTION

The Hydrologic Engineering Center (HEC) has converted several of their engineering application programs to the microcomputer (PC). In order to take advantage of the unique capabilities of the PC and to assist the user in working with these programs, the HEC has developed menu driven, user interfaces for several of their PC application programs. This paper describes the use and features of the menu driven interfaces developed for the Flood Hydrograph Package (HEC-1) and the Water Surface Profiles (HEC-2) programs.

*Hydraulic Engineer, US Army Corps of Engineers, The Hydrologic Engineering Center, 609 Second Street, Davis, CA 95616

HEC-1 APPLICATIONS WITH MENU1

The HEC-1 program menu interface (MENU1) allows the user to define input and output file names, create and edit input data files, run the HEC-1 program and display program output to the printer or console. The opening menu screen for MENU1, listing these choices, is shown on Figure 1.

Figure 1.   MENU1 Opening Screen

Menu options shown are selected either by using the cursor keys or by simply entering the number of the selection desired. The first choice on the menu allows the user to define input and output file names. If this choice is selected a window appears on the screen so that the user can type in the names of the input and output files. An alternative to typing in the names is to enter a question mark (?) and a directory of file names in the current directory will appear in a window on the screen. The user can then select from this directory of file names by using the cursor keys. The second choice on the menu allows the user to create a new or edit an existing input data file using the Corps of Engineers editor (COED). COED has been modified by HEC so that it will perform not only as a powerful line editor but a full screen editor as well. COED also includes extensive, easy to use on-line documentation. Additional enhancements have also been added to allow right justified data entry in 10-8 column fields and program specific variable identification and description. This latter capability will be discussed in a subsequent section. The third choice on the menu is to run the HEC-1 program. If the input and output file names have been selected, the program will begin execution using these file names. If the file names have not been specified, the program will ask for these names prior to execution. Once the HEC-1 program has finished execution, the user is returned to the menu. The fourth choice on the menu allows the user to display the HEC-1 program output file on the screen or send the output to the printer. This choice is made by pressing the space bar to toggle between the console and the printer. A utility program is used to display the output (or any file specified) to the screen and allows scrolling up and down, right and left, searching and many other features. A program called PROUT is used to send output to the printer with carriage control recognition. The last choice on the menu returns the user to DOS.

DATA ENTRY AND EDITING USING COED

A special feature has been added to the COED program that assists the user in preparing input data for specific HEC application programs. This feature includes variable name and field location prompting, automatic tab stops and right justification of input for each variable field, checking for inappropriate (non-numeric) data entry where applicable, and cursor position sensitive input variable definitions. These features are invoked while in the full screen mode by entering the Help Program command and the name of the program for which the input data applies. When the cursor is placed on a line having a two character record ID in the first two columns, the name of the variables for that record are shown in their appropriate position at the bottom of the screen. When the cursor is on a variable position and the Help Variable command is issued, the definition of that variable appears in a window on the screen. Figure 2 illustrates this feature for the variable ITIME in field three of the IT record for the HEC-1 program.

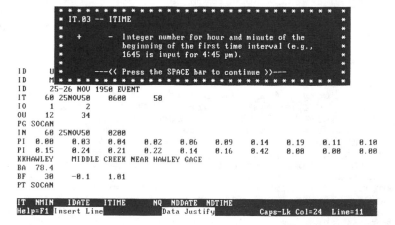

Figure 2.   COED Help Program and Help Program Feature

HEC-2 APPLICATIONS USING MENU2

MENU2 is the menu driven user interface that assists the user when making HEC-2 program applications. MENU2 is similar to the MENU1 interface described previously. It allows the user to define input and output files, create and edit input data files, and display the results. The MENU2 package also uses COED and its associated Help Program and Help Variable capabilities specifically designed for the HEC-2 program. In addition, MENU2 interfaces three supplementary programs, called EDIT2, SUMPO, and PLOT2 that can be used with the HEC-2 program to check input data, create summary tables from HEC-2 results, and plot cross sections and water surface profiles, respectively. The space bar is used from menu item three to toggle the selection of the HEC-2 program or any of these three supplemental programs. Each of these additional features are described in the following paragraphs.

CHECKING HEC-2 INPUT DATA USING EDIT2

The EDIT2 program allows the user to check HEC-2 program input data for errors prior to execution. The program uses the HEC-2 input data file as its input and checks the entire file for fatal errors, as well as possible errors and inconsistencies. The optional ED record can be specified as the first record in the input to set EDIT2 program options for listing input, output format, and change the range of elevation differences that the program checks for in the cross sectional data. An input listing, if not suppressed, and error messages are written to the specified output file. After the results of the EDIT2 run are reviewed, COED can be used to correct the HEC-2 input data prior to execution.

CREATING SUMMARY TABLES OF HEC-2 RESULTS USING SUMPO

The SUMPO program is used to create convenient summary tables using the computed results from HEC-2 multiple water surface profile runs. The user first runs the HEC-2 program, which creates a binary file of results. This binary file becomes the input file for the SUMPO program. The SUMPO program menus, which are similar to those described for MENU2, can be used to select standard, predefined summary tables or to create user defined summary tables by selecting the variables to be displayed. User selected summary table variable lists can be stored for later use. The screen including a window for selecting standard summary tables, which appears after selecting choice 2 from the opening SUMPO menu, is shown on Figure 3.

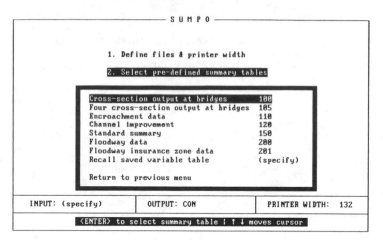

Figure 3. SUMPO Standard Summary Table Selection Menu

PLOTTING CROSS-SECTIONS AND PROFILES USING PLOT2

Graphical interpretations of cross-sectional data and water surface profile computations are an effective way of reviewing HEC-2 program input and results.   The PLOT2 program provides the capability of creating plots of cross-sections from the HEC-2 program input and profiles of water surface elevation or any of the other variables available from HEC-2 program results written to the binary file.   The plots are created by using a series of menus similar to those described for MENU2 and SUMPO.   The opening menu for PLOT2 present a choice of plotting profiles, plotting cross-sections or exiting the program. If plot profiles is selected, a second menu appears giving the user the choice of defining profile options, defining plotting options, plotting the profile or returning to the main menu.   This Plot Profiles Menu also allows the user to specify the names of the HEC-2 program input and binary files.   If define profile options is selected, an additional menu is displayed. This menu allows the user to select the variable to be plotted, whether the invert profile is to be plotted, and the starting and ending cross-section for the profile to be plotted.   If the user selects the define plotting options choice from the Plot Profiles Menu, a menu is displayed that allows the definition of labels and legends, and specification of the plotting device to be used.   Device choices are the color graphics adapter, the enhanced graphics adapter, or the HP 7475 pen plotter. Once the desired profile and plotting options have been selected, the user can select the begin plotting option from the Plot Profiles Menu to plot the defined profile.

   The procedure for plotting cross-sections is similar
to that described for plotting profiles. Choice 2, plot
cross-sections is selected from the PLOT2 opening menu.   A
second menu appears for defining cross-section and plotting
options and specifying file names.   The cross-section
options consist of whether to plot water surface, bridge
geometry, n-values, encroachments or channel improvements
on the cross-section. The plotting option selection is for
defining plot labels, grid and plotting device. An example
cross-section plot is shown on Figure 4.

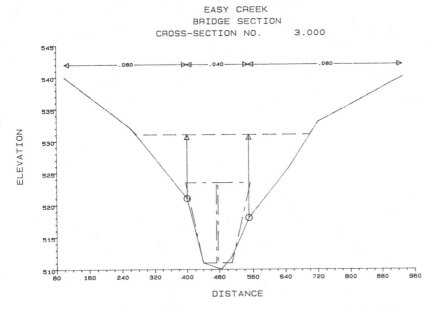

Figure 4. Example Cross-Section Plot Using PLOT2

REFERENCES

   1.   Bonner, Vernon R., 1987, "Computing Water Surface
Profiles with HEC-2 on a Personal Computer," Draft Training
Document No.   26, The Hydrologic Engineering Center, Davis,
CA.

   2.   The Hydrologic Engineering Center, 1985,   "HEC-1,
Flood Hydrograph Package," Users Manual, Davis, CA.

   3.   The Hydrologic Engineering Center, 1987, "COED,
Corps of Engineers Editor," User Document, Davis CA.

   4.   The Hydrologic Engineering Center, 1982,   "HEC-2,
Water Surface Profiles," Users Manual, Davis, CA.

THEORY FOR DEVELOPMENT OF THE
TR-55 TABULAR HYDROGRAPH METHOD

Michael κ. Glazner *

ABSTRACT

Technical Release No. 55 (TR-55), "Urban Hydrology for Small Watersheds," was originally developed by the United States Soil Conservation Service (SCS) in the 1970's, and was revised in June of 1986. TR-55 was developed to provide practical solutions for a wide variety of small watershed hydrology problems including computation of peak discharge, hydrograph generation, reach routing, and detention storage estimates. This paper discusses the applications and theory of TR-55's Chapter 5, "The Tabular Hydrograph Method."

DEVELOPMENT THEORY

The TR-55 Tabular Hydrograph Method was developed by making numerous TR-20 computer runs for different watershed conditions. A one square mile watershed was used to establish unit hydrograph flows per square mile of drainage area.

This section discusses the three major hydrograph shape parameters that were considered in the development of the method. These are: *initial abstraction/precipitation, time of concentration,* and *travel time* (reach routing effects).

Ia/P Ratio

The Ia/P ratio (initial abstraction divided by precipitation) was added to this method in the June 1986 revision to account for the hydrograph shape effects of initial abstraction and precipitation. The Ia/P ratio is inversely proportional to both the CN and precipitation, and hydrograph flow rates are inversely proportional to Ia/P. This ratio can therefore account for variances in both precipitation and CN.

TR-20 runs were made using a CN of 75 with 24-hour precipitation values of 6.67, 2.22, and 1.33 inches to obtain hydrograph tables for Ia/P ratios equal to 0.10, 0.30, and 0.50, respectively. Figure 1 (next page) is a plot of the tabular hydrographs for the three different Ia/P ratios. Linear interpolation between Ia/P table values is allowable since the hydrographs are generally aligned for the three different Ia/P ratios.

Time of Concentration (Tc)

The hydrograph peak is attenuated and shifted to the right as Tc is increased. Tabular hydrographs for various Tc's were developed from TR-20 runs with Tc's ranging from 0.10 to 2.0 hours.

Figure 2 (next page) shows a plot of the tabular hydrographs for various Tc's. Note that the hydrograph peaks do not align for different Tc values, thus making it incorrect to interpolate between Tc tables. Computed subarea Tc's should be rounded to one of the hydrograph table values. Chapter 5 in TR-55 discusses the proper rounding procedure for both Tc and Tt.

* Vice President of Product Development, Haestad Methods, 37 Brookside Rd, Waterbury, CT 06708

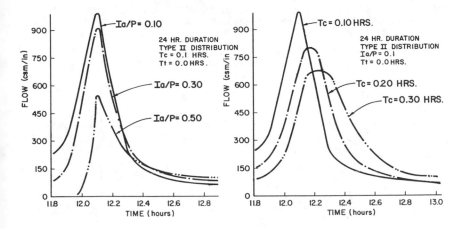

**Plot of Hydrograph Tables
for Different Ia/P Ratios
FIGURE 1**

**Plot of Hydrograph Tables
for Different Tc's
FIGURE 2**

Travel Time (Tt)

This parameter takes into account the reach routing effects on a subarea's hydrograph. The composite travel time (Tt) is the time it takes the water to flow from the outfall of a subarea to the outfall of the entire composite watershed (the sum of the flow times through all of the downstream reaches). Tt should not be confused with Tc, which is discussed in the previous section. Figure 3 shows the flow paths used for computing subarea Tc's and Tt's.

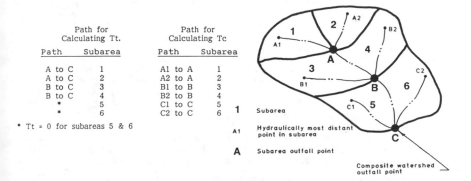

Path for Calculating Tt.		Path for Calculating Tc	
Path	Subarea	Path	Subarea
A to C	1	A1 to A	1
A to C	2	A2 to A	2
B to C	3	B1 to B	3
B to C	4	B2 to B	4
*	5	C1 to C	5
*	6	C2 to C	6

* Tt = 0 for subareas 5 & 6

1   Subarea

A1   Hydraulically most distant
     point in subarea

A   Subarea outfall point

Composite watershed
outfall point

**Flow Paths for Tc and Tt
FIGURE 3**

The hydrograph peak is attenuated and shifted to the right as Tt is increased. Tabular hydrographs for various Tt's were developed from TR-20 runs with Tt's ranging from 0.0 to 3.0 hours.

Figure 4 shows the tabular hydrographs for various Tt's. Note that the hydrograph peaks do not align for different Tt's. Interpolation between Tt tables would result in inaccurate approximations and should not be used. Computed Tt's should be rounded to one of the hydrograph table values using the rounding procedure outlined in Chapter 5 of TR-55.

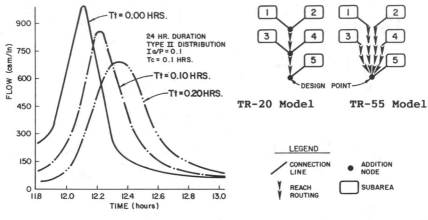

Plot of Hydrograph Tables   Comparison of TR-20 and
    for Different Tt's   TR-55 Reach Routing Models
       FIGURE 4   FIGURE 5

The Tabular Hydrograph Method uses Tt to route a subarea's hydrograph from the subarea outfall down to the composite outfall point (see Figure 3). This method assumes that the stream velocities throughout the reach are approximately the same for all discharge rates. This approach allows each subarea's hydrograph to be routed independently down through the watershed reaches and added directly at the watershed's outfall (see Figure 5).

The SCS took care to develop TR-20 computer runs that properly represented reach routing effects as a function of travel time. Approximately 300 different combinations of reach slopes and cross sectional configurations were analyzed in order to establish a comparison between state-of-the-art dynamic wave routing results and TR-20 routing results. The SCS used the National Weather Service DAMBRK program for approximately 20 runs, and used their own dynamic wave routing program for the remaining runs. These dynamic wave runs were compared to tabular hydrographs resulting from the TR-20 stepped and unstepped routing procedures discussed later in this section and shown in Figure 7.

Figure 6 summarizes the comparisons. Note that using a stepped routing approach with TR-20 resulted in a more accurate comparison with the dynamic wave routing model. Q* is a dimensionless ratio that correlates the reach's routed outflow to its inflow. K* is a factor that accounts for reach length and cross-sectional properties.

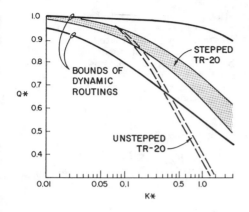

$$Q^* = \frac{Q_o - Q_b}{Q_i - Q_b}$$

Where:

Q^* = Dimensionless flow ratio
Q_b = Base flow
Q_i = Peak Inflow
Q_o = Peak Outflow

$$K^* = \frac{Q_i \ L^m}{X \ V_i^m}$$

Where:

K^* = Reach length factor
L = Reach length (ft)
Q_i = Peak inflow (cfs)
V_i = Runoff volume (ft^3)
X, m = Att-Kin routing coeff. for $Q = XA^m$

**Comparison of Dynamic Wave Routings with the Ranges for
Stepped and Unstepped TR-20 Routings
FIGURE 6**

Since travel time is inversely proportional to stream velocity, the first task for establishing the stepped and unstepped TR-20 models was to make an assumption on stream velocity versus flow characteristics. As a general rule, the flow velocity for a channel increases with an increase in flow rate. However, once flow overtops the channel banks and encounters floodway obstacles such as thick vegetation or man-made structures, the velocity may actually decrease with an increase in flow rate. Hence the assumption of a constant average velocity for all flow rates within a channel was used to establish the Tabular Hydrograph Method.

The TR-20 results displayed in Figure 6 were achieved using a reach velocity of 3.75 ft/sec in combination with various reach lengths. Note that the TR-20 stepped routing approach with constant velocity compares closely to the average of dynamic wave routing runs that were based on an array of computed stream velocities.

The TR-20 modified Att-Kin X and m routing parameters were used to model a constant reach velocity of 3.75 ft/sec for all runs. This was accomplished by setting X equal to 3.75 and m equal to 1.0 for all reaches.

Since the modified Att-Kin X and m coefficients were held constant for all of the TR-20 reaches, different travel times were achieved by using various reach lengths. For example, since the velocity for a reach was 3.75 ft/sec, a 0.10 hour travel time could be modeled by using a reach length of 1350 feet. This can be verified by:

1350 ft / 3.75 ft/sec = 360 seconds = 0.10 hour

Travel time is the main hydrograph shape parameter for the modified Att-Kin routing method in TR-20. Since the m coefficient was held at 1.0, the tabular hydrographs could have been generated from TR-20 runs with any combination of reach length and velocity that yielded the desired travel times. For example, a reach length of 2880 feet and velocity of 8 ft/sec would also yield a 0.10 hour travel time.

Figure 7 is a graphical representation of the stepped and unstepped approaches for a 0.3 hour travel time. To model a 0.3 hour travel time in an unstepped fashion, the inflow hydrograph would be routed through one 4050 ft reach. To model a 0.3 hour travel time in a stepped fashion, three consecutive 1350 ft reaches would be used, with the outflow hydrograph from each reach being the inflow hydrograph for the next downstream reach. Figure 6 indicates that the stepped approach is a better approximation of the dynamic wave routing model.

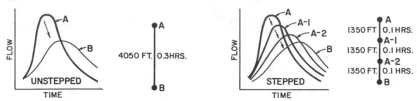

Schematic Representation of Stepped and Unstepped Routings
FIGURE 7

Fine Tuning the Final Model

The previous section on "Development Theory" covered the effects of and considerations behind the three main parameters used to develop the tabular hydrographs. This section will discuss the actual steps that were used with TR-20 to develop the final versions of the tabular hydrographs.

1) Hydrographs were developed for a drainage area of one square mile with Tt=0 (no routing effects), for each distribution type, Tc, and Ia/P ratio listed in the tabular hydrographs.

2) These hydrographs were checked to ensure that their peaks were consistent with the peaks developed for the TR-55 Graphical Peak Discharge Method (GPDM). The hydrograph peaks were modified to match GPDM peaks if the two methods differed.

3) To obtain hydrographs for the various Tt's in the hydrograph tables, the hydrograph for Tt=0 for each Tc, distribution type, and Ia/P combination was routed using the stepped technique previously discussed.

4) Strategic hydrograph ordinate times were selected to effectively represent the hydrograph while significantly reducing the size of the hydrograph tables.

5) In the cases where the hydrograph peak fell between selected hydrograph ordinate times, the peak was included as the flow rate for the nearest selected hydrograph time.

6) Hydrograph peaks for travel times greater than zero were checked (and modified if necessary) to ensure that the hydrograph peak occurred at a time of:

$$T = Tp + Tt$$

Where:

T = Time for subarea hydrograph to peak at the watershed's composite outfall

Tp = Time for subarea hydrograph to peak with Tt = 0 (no routing effects)

Tt = Travel time from subarea outfall to the watershed's composite outfall

HAESTAD METHODS QUICK TR-55 COMPUTER PROGRAM

TR-55 was developed to provide the engineer with a practical, easy-to-learn, hydrologic method for small watersheds. Although the learning time for the methods in TR-55 is relatively short, some of the computations involved can become quite time consuming. Haestad Methods has developed a computer program that greatly reduces the time for TR-55 computations.

Haestad Methods' *Quick TR-55* is now available and provides a friendly environment for quick and easy data entry. It is menu driven and provides a spreadsheet editor for all input data. The program includes the Tabular Hydrograph Method and can plot the resulting hydrographs. It also includes the Graphical Peak Discharge Method and can compute detention storage estimates using the methods outlined in TR-55.

SUMMARY

TR-55 was developed by the SCS in the early 1970's to provide engineers with practical, easy-to-use methods for small watersheds. Since its initial release, TR-55 has become highly popular, and has been approved by reviewing agencies across the United States. The TR-55 document was revised in June of 1986, and included new enhancements such as the Ia/P ratio and Type I, IA, and III distributions.

This paper was written on Chapter 5 of TR-55, "Tabular Hydrograph Method." This method generates pre and post-developed hydrographs for small watersheds. It accounts for initial abstraction and precipitation, time of concentration, and travel time (reach routing effects).

The Tabular Hydrograph Method was developed from numerous TR-20 runs using a 24-hour storm duration and a drainage area of one square mile with different Tc's, Tt's, Ia/P's, and distribution types. The TR-20 reach routings were done in a stepped fashion to approximate the average results from computer runs that used state-of-the-art dynamic wave routing techniques.

REFERENCES

Roger G. Cronshey and William H. Merkel,
United States Soil Conservation Service, Washington, D.C.

Technical Release No. 20 (TR-20),
"Computer Program for Project Formulation Hydrology," Revised, May 1982;
United States Department of Agriculture, Soil Conservation Service.

Technical Release No. 55 (TR-55),
"Urban Hydrology for Small Watersheds," Second Edition, June 1986;
United States Department of Agriculture, Soil Conservation Service.

Quick TR-55 Computer Program
Haestad Methods
37 Brookside Road,
Waterbury, Connecticut 06708
(800) 422-6555

TR-55 Microcomputer Implementation

R. T. Roberts and R. G. Cronshey*, M. ASCE

Abstract

Microcomputer software is now available which implements the Soil Conservation Service's recently revised "Urban Hydrology for Small Watersheds", Technical Release Number 55 (TR-55). This software automates the methods and procedures described in TR-55 for computing runoff volume, peak rates of discharge, hydrographs, and storage retention in small watersheds, especially urbanizing watersheds. Micro-TR55 features interactive, screen-oriented data entry, access via function key to numerous support services, utilization of many special use keys present on IBM-compatible keyboards, extensive use of on-line HELP screens, and the ability to customize the software via user-defined data files. Micro-TR55 is written in compiled BASIC and runs on IBM-compatible microcomputers with 256K of memory.

Introduction

The Soil Conservation Service (SCS) first issued Technical Release Number 55 (TR-55), "Urban Hydrology for Small Watersheds" in January 1975. The TR was developed to provide a set of simplified manual procedures for estimating runoff and peak discharge in small watersheds, with special emphasis given to urban and urbanizing watersheds. In late 1981 an effort was begun to revise the TR to incorporate in its procedures the results of more recent research as well as experience based on use of the original edition. This effort culminated in June 1986 with the publication of a revised edition, Technical Release No. 55 (2nd Edition).

Early in 1985, an SCS/Agricultural Research Service (ARS) programming team joined in a cooperative effort to package the revised TR procedures into microcomputer software to simplify, enhance and standardize their application (Cronshey et al., 1985). An added benefit to both agencies was the experience and knowledge to be gained in effectively applying microcomputer technology to the task of delivering scientific and engineering advances in appropriate and usable forms.

*Computer Specialist, USDA-Agricultural Research Service, Hydrology Laboratory, BARC-West, Bldg. 007, Room 139, Beltsville, MD 20705, and Hydraulic Engineer, Hydrology Unit, USDA-Soil Conservation Service, P.O. Box 2890, Washington, D.C. 20013.

[1]Trade names used in this paper are solely for the purpose of providing information and do not constitute a guarantee or warranty of the product by the U.S. Department of Agriculture over other products not mentioned.

A major goal of this cooperative effort was the development of an effective, generalized "vehicle" for providing an application end user with an environment that included not only automated algorithms, but a complete set of tools for managing all the related processes (e.g., data entry, storage and retrieval of data) implicit in the use of any software system. Traditionally, these activities have been considered separate from an application, an approach which invariably makes them the most error prone and time consuming tasks related to the use of automated systems. This paper discusses the design features incorporated in Micro-TR55 to improve this "man-machine" interface, and which have potential applicability in a wide range of other applications (Rango and Roberts, 1987).

MICRO-TR55 Characteristics

The system software was designed to operate on the IBM[1] personal computer and all 100% PC compatibles. The minimum system requirements for Micro-TR55 include 256K of random access memory, one 5 1/4" floppy disk (hard disk or 2 floppy drives preferred) and standard scan code generation by the keyboard's special purpose and function keys. The software is written in Microsoft BASIC[1] and is designed to run under MS-DOS[1].

System Modules

The development team used a modular design philosophy, which allowed concurrent development and testing of the six interrelated programs that mirror the chapters of the TR, and which facilitate the possible use of each with other applications (see Figure 1).

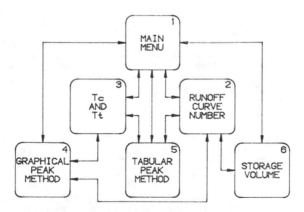

Figure 1. System modules.

The Main Menu (MENU) integrates the five other modules into a unified system. Included in this module are all routines which are applicable system-wide (LOAD, SAVE, PRINT, ZERO), "Help" screen documentation describing system operation, and an automation of the decision process for selecting the TR's methods for computing peak discharge.

Runoff Curve Number (RCN) automates TR-55's cover description/ hydrologic soil group/curve number table and the function it supports, generation of curve number(s) for up to 10 subareas of a watershed, and a resulting composite curve number. Data developed in RCN, drainage area(s) and curve number(s), may be automatically applied by other modules in the system.

The **Time of Concentration/Travel Time** module (TCTT) automates the TR's procedures which calculate time of concentration (T_c) and travel time (T_t) values for overland/sheet, shallow concentrated, and open channel flow types. Time values generated by this module may be automatically applied to the appropriate peak discharge computation module.

The **Graphical Peak Discharge** module (GRAPHIC) computes peak discharge for small, hydrologically homogeneous watersheds. It supports both direct, keyboard data entry and optional system generation of a runoff curve number (CN) from RCN, a time of concentration (T_c) value from TCTT, and rainfall-frequency pairs acquired from "user files", described below.

The **Tabular Peak Discharge** module (TABULAR) computes peak flows and hydrographs for up to ten homogeneous subareas within a watershed, as well as a composite hydrograph for the watershed. Keyboard data entry and CN and T_c value generation, discussed above, is applicable in this module also, and is extended to include T_t through the subarea(s).

The **STORAGE** module automates the procedures described in the TR for computing storage volume for detention basins. It uses one data entry screen, and accepts certain of its required data values either through the keyboard or automatically from GRAPHIC or TABULAR, if available.

Screen Oriented Processing

The TR-55 user interacts with the system through a combination of text and data screens. Program prompts are used in certain situations where additional communication with the user is required, primarily to inform the user of data validation errors. The 23rd line of each display screen is reserved for prompts.

Data Entry Screens

TR-55's data entry screens are simply electronic versions of a standard printed form, consisting of one or more pages of information requests with associated reply spaces or "fields" provided for entering data values. Each data entry screen consists of background information (data item identifiers and associated "fields") overlaid with current data values (See Figure 2). The data fields on a screen are controlled by a numeric pointer array of values which define each field's three attributes: location on the screen, length, and data type (e.g., non-numeric characters invalid in a numeric field). Additional array values define horizontal and vertical movement characteristics between each field and its surrounding fields. This approach for data screen management provides a generalized methodology for developing interactive, maintainable data entry screens.

A program indicates current location on a data screen by displaying a blinking cursor. As data are entered in a field, each valid character replaces the cursor and the cursor moves to the next character position. The PC's cursor movement keys (Arrow up/down/left/right, Enter, Tab, Home, End, PgUp, PgDn), under the direction of the pointer array discussed above, control the user's movement between fields and data entry screens. The keyboard's edit keys (i.e., backspace, insert, delete) are all logically defined by the software, making data editing a simple task.

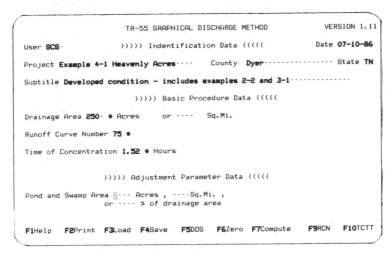

Figure 2. Typical data entry screen.

Text Screens

Each program module contains an introduction screen and various "Help" screens which provide on-line documentation to assist in resolving questions concerning the use of the computer software and to provide limited information about the TR and its data requirements (basic development assumptions included a user familiar with the TR and MS-DOS). Introduction screens are displayed upon module entry. The help screen most appropriate to the user's current location in the system is displayed by pressing Function Key F1. Once in "Help" mode, the user may "PgUp" to a prior help screen, "PgDn" to the next help screen, or "ESCape" to return to the screen from which help was initiated (See Figure 3).

Using the PC's "Special Purpose" Keyboard Keys

Function Keys

Micro-TR55 uses the microcomputer's ten keyboard function keys to provide the user with the ability to interrupt data entry and transfer control to a specific predefined function process. On function termination, control returns to the point of interruption. Each screen display includes information on the functions active at that point in the software. The universal function keys are:

Key	Action	Description
F1	HELP	Display Help Screen(s)
F2	PRINT	Print input data and computation results
F3	LOAD	Read previously entered and SAVEd data
F4	SAVE	Write current input data to a file
F5	DOS	Exit program and return control to MS-DOS
F6	ZERO	Reinitialize system

Function keys F7 through F10 are transient functions, active only in certain portions of the system to support "local" activities.

F7	COMPUT	Run appropriate procedure using current data
F8	SELECT	Display subarea list in CN and TCTT modules
F9	IMPERV	Used in RCN to adjust % of impervious area
F9	RCN	Invoke RCN to generate runoff curve number(s)
F10	TCTT	Invoke TCTT to generate T_c and T_t values

```
    TR-55 GRAPHICAL DISCHARGE METHOD              VERSION 1.11

              ))))) HELP for Basic Procedure Data (((((

  All fields are numeric with maximum field length shown in parentheses.


  DRAINAGE AREA (5) is the watershed area in acres or square miles.

  RUNOFF CURVE NUMBER (2) is the weighted SCS watershed curve number.
       Acceptable values are from 40 to 98.  A value computed by the
       RCN module (F9) is indicated by a '*' after the CN and DRAINAGE AREA
       values.  The RCN procedure requires detailed soils and cover data.

  TIME OF CONCENTRATION (4) is the time in hours it takes for runoff to
       travel from the hydraulically most distant point to the point of
       interest.  A value computed ty TCTT module (F10) is indicated by
       a '*' after the field on the input screen.  THe TCTT procedure
       requires detailed flow path information.



  Esc-exit help    PgUp-go to prior help screen    PgDn-go to next help screen
```

Figure 3. Typical HELP screen.

Cursor Movement Keys

To enter or change an input data value, the user manipulates the screen cursor to the desired position in a screen data field using any combination of the Enter, Tab, Home, End and keyboard cursor, or "arrow", keys. Beginning at the top leftmost field on each data entry screen, the logical sequence for horizontal movement (via the Enter or Tab keys) on the screen is from left to right, top to bottom. Movement vertically is from the current field up/down to the next closest field. Movement between fields initiates data entry for any exited field containing new or altered information.

PgUp and PgDn

PgUp and PgDn control which screen display is presented to the user of Micro-TR55. They are functionally equivalent to turning forward or back one page in a multi-page form. These keys initiate data entry.

ESCape

The ESCape key allows the user to exit some current activity (inspecting help screens, generating a curve number, etc.). After an Escape, control returns to the point at which the process was invoked. An ESCape from the main menu module terminates the application and returns control to MS-DOS.

User-defined support files

"User Files", as defined by Micro-TR55, are support files which an individual TR user may optionally provide to customize the application to a specific geographic area. File COUNTY.RF stores rainfall frequency data by county name. This data can also be organized by individual state (COUNTY.MD, COUNTY.VA, etc.). When these files are used, frequency values are automatically provided during the data entry process. The TYPE.RF file is used to store time of concentration-unit peak discharge equation coefficients for the Graphical Peak Discharge Method. If present, the custom file's name is added to the four standard SCS rainfall distributions. The DISTRIB.RF file stores tabular hydrographs for an additional rainfall distribution for use with the Tabular Peak Discharge Module. The VOLUME.RF file is used by the storage module to obtain non-standard storage volume distribution coefficients.

Program Availability

The technical release and associated computer software (Version 1.11) were disseminated within SCS during October, 1986. At the same time, they were provided to the National Technical Information Service, U.S. Department of Commerce, 5285 Port Royal Road, Springfield, VA 22161, which handles distribution to the general public.

References

Cronshey, R. G., Roberts, R. T., and Miller, N. 1985. Urban hydrology for small watersheds (TR-55 Rev.). Proceedings ASCE Hydraulic Division Specialty Conference, Orlando, FL, August 15-16.

Miller, N. and Cronshey, R. G. 1985. Urban hydrology for small watersheds (Rev. 1985). Proceedings ASCE National Spring Convention, Denver, CO, April 29-May 30.

Rango, A. and Roberts, R. T. 1987. Snowmelt-Runoff Modeling in the microcomputer environment. Proceedings Western Snow Conference, Vancouver, B.C., April 14-16.

Soil Conservation Service. 1986. Urban hydrology for small watersheds, Technical Release 55, Washington, D.C.

Interagency Standards for Automated
Local Flood Warning Systems

Curtis B. Barrett[*]

1. Floods . . . A Growing National Problem

Flood losses continue to increase nationwide. Despite the
various applications of flood mitigation methods . . . flood
damages continue to increase at an average rate of 5% annually.
In FY 1986, over $6 billion in flood damages has been estimated
by NOAA National Oceanic and Atmospheric Administration
National Weather Service (NWS). This represents the highest
(unadjusted for inflation) flood damage loss since records were
accumulated in 1903. There is no similar trend for loss of life.
However, since the 1970's, there has been a significant increase
in flood deaths. The potential for devestation flood losses
continues to increase as more people inhabit floodplains.

According to the Federal Emergency Management Agency
there are 20,000 communities in the nation which are vulnerable
to floods and flash floods. Currently the NOAA NWS provides site
specific flood forecast services to approximately 3,100
locations. Estimates of the current number of community local
flood warning systems is near 1,500. Thus over 15,000 remaining
communities have flood problems but receive only generalized
county flash flood warnings provided by the NWS. Many of these
communities are located along very flashy creeks and rivers
which may crest within minutes to a few hours from the
occurrence of heavy rainfall.

An additional man-made threat of floods are dams. This is
indeed an ironic situation - man has built many dams for flood
control protection, yet these structures also pose a flood threat.

[*] President, River Services, Inc., 3414 Morningwood Drive,
Suite 11, Olney, Maryland 20832.

Encroachment downstream from dams is well documented and heightens the potential for a flood catastrophe.

An inventory of reservoirs conducted in 1978 revealed a count of over 63,000 reservoirs nationwide. Approximately 10,000 reservoirs were categorized as high hazard dams (a failure would result in catastrophic flood losses). Nearly 3,000 dams were declared unsafe.

2. Growth of Local Flood Warning System Concept

Since the NWS cannot now possibly provide site specific flood forecasts for every flash flood prone community, the self help concept of local flood warning was developed by the NWS. A local flood warning system provides community officials with internal self-help flood forecast capability. The NWS and private hydrologic consulting services provide technical support to communities which have flood problems.

A local flood warning system consists of a community coordinator, community volunteers and various types of precipitation stream gages, and a hydrologic forecast procedure. The coordinator collects rainfall and river data either manually or automatically, utilizes a simplified flood forecast procedure to compute a flood crest and informs local response officials of the flood forecast information. Response officials then evoke actions based on an Emergency Operation Plan to save lives and reduce flood damages.

Most local flood warning systems are manual systems as previously described. State-of-the-art automated flood warning systems developed by the NWS have been extremely popular. These systems use event reporting precipitation gages and event river level sensors to instantaneously measure and transmit data to a receiving or base station which evaluates the data continuously.

Event precipitation gages consist of a 1 millimeter tipping bucket mechanism which transmits rainfall data line of site via UHF radio frequency to a base station receiver. Event river level gages send river stage changes (according to pre-set levels) continuously to the base station.

The base station continuously receives precipitation and river data, evaluates the significance of the data, alarms the base station officials if precipitation or river level observations are approaching critical threshold values, and stores the data in a data base for future use. Some automated systems provide a hydrologic forecast model which runs every 10-12 minutes. The model provides local officials the capability to simulate streamflow based on observed and anticipated hydrometeorological conditions.

Automated local flood warning systems have also been applied as monitors to unsafe dams and as "fill and spill" reservoir forecast systems. Locating sensors in the reservoir pool and at the tailwater of the dam provides the capability to detect a dam break as it occurs. Relay of the combined sensor information of a sudden drop in the pool accompanied by a sudden increase in stage at the tailwater gage can activate an alarm warning residents in the floodplain of a dam break.

3.   Automated Local Flood Warning System Development

The NWS River Forecast Center at Sacramento, California developed an automated local flood warning system called Automated Local Evaluation in Real-Time (ALERT). ALERT systems were first established in California and then slowly expanded across the country.

Because of the popularity of these systems and limited resource base of the NWS to meet the increasing demand, private vendors have emerged which are now aggressively marketing ALERT systems (Hardware & Software). Competition amongst these firms to provide expanded capabilities to an ever growing list of community clients has provided some motive for vendors to individualize or create unique system components. This "system uniqueness" is created to attract clients. This possible system uniqueness has also aroused concern by federal agencies. This uniqueness by automated system vendors could possibly isolate systems and prevent adjacent systems from effectively sharing critical hydrometeorological data and information.

Unless guidelines for the collection and transmission of data from a sensor site to a base station were established and agreed upon by the federal agencies and the vendors, the important ability for communities to network data and exchange critical information would be lost.

Additional federal concerns were also for the standardization of processed data transfer from ALERT base station to ALERT base station. This capability was particularly required by the NWS since operational exchange of ALERT base station information with NWS offices is desirable. The NWS and most federal agencies use the Standard Hydrologic Exchange Format (SHEF) to exchange hydrologic data and information.

Other ALERT System concerns for possible guidelines are the use and application of hydrologic models and the standard utilization of data formats archived for future use.

In addition to ALERT, the NWS has separately developed an automated flood warning system in Appalachia. The Integrated Flood Observing and Warning System (IFLOWS) is a cooperative federal, state, and local system. The system is being operated in 7 states eventually covering over 150 counties. This system was designed for the networking of precipitation and river data and does not have the potential problem of data networking that ALERT systems have.

In addition to private vendors marketing automated flood warning system hardware and software, a private hydrologic forecast service is now being offered to communities or industries with flood problems. Using real time data available from the NWS Family of Services data base and local ALERT data hydrologic forecasts are provided for locations other than those forecast by the NWS. Although, NWS hydrologic models are currently used by this company establishment of guidelines for the use of hydrologic models in flood warning systems maybe desirable.

4. Interagency Committee on Local Flood Warning Systems

In September 1935, the Federal Interagency Advisory Committee on Water Data Hydrology Subcommittee established a work group to provide a document which would provide basic technical information on community local flood warning systems and where local officials could seek assistance from various federal and state agencies. In August, 1985, the document entitled "Guidelines on Community Local Flood Warning and Response Systems" (Hydrology Subcommittee, 1985) was published and distributed by the workgroup. In the two years of research and discussion of federal roles and responsibilities, the workgroup concluded the need for an Interagency Standing Committee on flood warning systems was required to provide guidelines and standards to vendors and communities in the following areas: 1) Data transmission formats from sensor site to receiving (base) station, 2) Data transmission format from ALERT base station to any external computer, 3) Guidelines for the use of hydrologic models in automated local flood warning systems, and 4) explore the need for standards for archiving data.

The Interagency standing committee is composed of the following federal agencies: NOAA National Weather Service; U.S. Geological Survey; U.S. Army Corps of Engineers; Bureau of Reclamation; National Park Service; Federal Emergency Management Agency (FEMA); USDA Soil Conservation Service; Tennessee Valley Authority (TVA) and Housing and Urban Development (HUD). Participation from the Association of State Floodplain Managers and the ALERT Users group is encouraged.

The committee has already established standards for data transmission formats from the sensor site to a base station. The committee is now addressing the establishment of additional standards and guidelines mentioned earlier.

This committee has provided many additional benefits to federal, state and local agencies. It provides a mechanism for federal agencies to exchange technical information on agency involvement. It has provided a successful forum for the discussion of agency policies, roles and authorities connected with these systems. This has resulted in changes in agency policy towards LFWS. Finally, it has resulted in the establishment of agency representatives who are knowledgable on LFWS and have become an effective resource to each agency.

Summary

The need exists for establishing guidelines and standards for aspects of local flood warning systems. The demand for increased flood warning systems is expected to increase due to the trend of increasing flood damages. According to FEMA, in the past 50 flood hazard mitigation team flood warning sytems are attractive because of their initial low cost (compared to other flood mitigation methods), the high benefit to cost ratios and the rapid development rate. The proliferation of these systems by various vendors and government agencies could lead to limited benefits unless adequate planning is assured at the federal and state levels. Establishment of agreed upon standards can assure full potential of these systems. Adjacent community automated LFWS established by different vendors will be able to exchange data and information collected by each system. Data and forecast networks of hydrometeorological data can be invaluable in the long run with adequate standards. This foresight in planning was not practiced in the early stages of establishment of GOES Satellite Data Collection Platforms. The result was over 20 different types of data formats that no one agency or data user could obtain.

References

Hydrology Subcommittee. (Workgroup on Local Flood Warning Systems) "Guidelines on Community Local Flood Warning and Response Systems", August 1985.

FEASIBILITY STUDIES OF
FLOOD WARNING-RESPONSE PLANS

Michael W. Burnham and Darryl W. Davis[*], Members ASCE

Introduction

Section 73 of PL 93-251 requires that nonstructural measures be among the alternatives considered in developing flood loss reduction plans by federal agencies. Local flood warning-emergency preparedness plans are nonstructural means for lessening the threat to social, economic, and physical elements of local communities during a flood event. The investigation of flood warning-emergency preparedness plans should be a component of comprehensive flood loss reduction feasibility studies.

Local flood warning-emergency preparedness plans consist of hardware, technical activities, and formal and informal inter- and intra-organizational arrangements and commitments to performance in which the human element is a vital part. The plans are comprised of coordinated actions involving flood detection, warning dissemination, emergency response, post-flood recovery and re-occupation, and continuous plan management (Owen 1976 and the Hydrologic Engineering Center 1979 and 1982). Federal, state, and local governmental agencies and private sector organizations that conduct programs and operations relevant to flood warning-emergency preparedness plans are numerous and diverse.

The type and sophistication of the appropriate measures may vary significantly due to physical characteristics of the stream system, the nature of the problem to the threatened area, resource availability, and institutional factors. The event response and management of a flood disaster, instead of the permanent (long-term) event control or damage potential modification, are unique characteristics of local flood warning-emergency response actions that differentiate them from other flood loss reduction measures.

Investigation Strategy

Flood warning-emergency preparedness plans should be considered in a manner similar to other flood loss reduction measures. That is, flood hazard, flood damage, institutional, and other studies are conducted for existing and future with and without conditions. However, the relatively low cost of implementing the plans, normally $30,000-$70,000 for small communities up to $100,000-$300,000 for

[*]Senior Engineer and Chief respectively, Planning Division, The Hydrologic Engineering Center, 609 Second Street, Davis, CA.

232

for large areas, suggests that the feasibility study be conducted at low cost. The reconnaissance study level of detail seems appropriate in most instances. For larger, more complex basins, increased detail is warranted.

Alternative plans are formulated to address documented needs. They are evaluated in terms of their contribution to resolving the needs, and the best plan is selected and recommended for implementation. The benefits of flood loss reduction plans are determined by comparison of the with and without project conditions. The without project condition is the project setting likely to occur as a result of existing improvements, laws and policies. The with project condition is the most likely project setting expected to exist in the future if the proposed project is undertaken.

Technical Analyses

Four technical analyses areas are discussed which emphasize important aspects of studying flood warning-emergency preparedness systems. They are flood hazard analysis, flood damage evaluations, institutional analysis, and flood scenarios.

Flood Hazard Analysis. Flood hazard analyses are performed to provide information on the flood characteristics of an area. The flood hazard data provides information on the potential for the loss of life, impacts to important community functions, and is required for flood damage evaluations. The analyses include hydrologic and hydraulic studies of existing and future with and without conditions.

Flood hazard analysis provides basic information for defining the existing flood hazard and developing key elements of local flood warning-emergency preparedness plan enhancements. The types of flood hazard information generated include: (1) flood inundation boundary maps; (2) warning times; (3) estimations of depth and velocity relationships; (4) evaluations of the effectiveness and reliability of flood detection systems; and (5) discharge-frequency and discharge-elevation relationships.

Similar hydrologic and hydraulic studies are required for future without conditions and for the plan enhancements analyzed if these conditions effectively alter the hydrology and hydraulics of the study area. The warning time and associated reliability may be significantly enhanced via improvements to the flood process, such as installation of a more automated flood forecasting system, and from better warning dissemination plans.

Flood Damage Evaluation. Flood inundation damage analyses are performed to identify potential damage locations, the type of damage, and the damage reduction associated with implementing temporary flood loss reduction actions such as flood fighting, installation of temporary barriers, and removal or raising of contents.

The existing without condition expected annual damage is calculated by integrating the damage-frequency relationships obtained by combining the hydrologic, hydraulic, and damage data. Future without condition expected annual damage calculations are performed for future time periods throughout the project life as projected changes in conditions warrant. The without conditions equivalent annual damage may then be calculated by discounting the time stream of expected annual damage to the beginning period of the analysis and amortizing over the planning horizon (Hydrologic Engineering Center 1984).

The with condition damage analysis evaluates the residual damage expected to occur if enhancements to existing local flood warning-emergency preparedness plans are implemented. Actions that reduce the flood damage are: enhanced flood detection include forecasting and warning dissemination processes. These can result in more reliable and greater warning times and encourage subsequent enhanced implementation of temporary flood loss reduction measures as a result of better warning, or perhaps development of better planned implementation actions. The calculation of the with plan conditions are performed in a similar manner as the without conditions.

Institutional Analysis. Institutional analysis is the study of formal and informal organizational arrangements for communication, coordination, and conduct of operations required to implement a local flood warning-emergency preparedness plan. It is a principal aspect of the formulation process due to the required interaction of an often large number of organizations needed to successfully implement a successful flood warning-preparedness plan. Specifically, the institution analysis should define the information collection, processing, and dissemination processes of each plan component. The organizational authorities, responsibilities, and general capabilities to implement potential plan enhancements must also be determined. Finally, the cost sharing arrangements to plan, design, and implement the plan enhancements must be negotiated and ultimately defined.

Flood Scenarios. Social scientists performing natural and man related disaster assessments have utilized flood scenarios as a means to describe disaster situations and the social impacts related to the disaster. Scenarios provide descriptive reinactments of past events or may hypothesize situations in order to provide more explicit descriptions of possible catastrophic consequences and actions associated with potential events.

Flood scenarios are at times useful methods for highlighting the value of enhancements to local flood warning-emergency preparedness plans due to the event nature of implementing a sequential set of flood detection, warning dissemination, and emergency response actions. The scenarios can tie together the sequence of institutional operations, flood hazard data and impacts, and flood damage consequences. In addition, risk to life and numerous other social impacts can be described in a realism not possible using technical reporting techniques. The flood scenarios should be developed after the flood hazard analysis, flood damage evaluations, and the institutional analyses are completed.

Benefit-Cost Analysis

The analysis by the Corps of Engineers and other Federal agencies of enhancements to local flood warning-emergency response plans requires that benefit-cost evaluations be performed and the contribution to the National Economic Development account be displayed as an output of the study (U.S. Army Corps of Engineers 1983). The objective of the benefit-cost analysis is to identify, array, and to estimate the cost of actions needed to bring an alternative into operational service and to maintain its viability; and to identify, array and value the output (benefit) of the alternative in commensurate units so that the justification of the investment may be determined. The components of local flood warning-emergency plans are varied and difficult to array and evaluate under the conventional cost and benefit methods. Work is needed to provide a more firm basis for performing the analysis and presenting the results.

Costs. Costs required to implement the enhancements consist of first costs of investigating and implementing the plan enhancements, annual costs of maintaining plan components in a state-of-readiness, and the periodic costs associated with implementing specific actions during flood events. Table 1 summarizes the general cost items associated with implementing a local flood warning-emergency preparedness plan.

It is difficult to distinguish between local flood warning-response plan cost items that fall under the preview of existing agency operations and those not accounted for under existing conditions. The most credible approach is to assign most administrative costs to existing on-going programs and to separate out those items specifically attributed to the proposed enhancements to the existing plans. The cost items listed in Table 1 may be adjusted and tailored to the specific situations and conditions being studied.

Benefits. The benefits from enhancements to local flood warning-emergency preparedness plans are reduction to loss of life, and the reduction of the negative impacts of flood disasters on society in terms of reduced social disruption, business losses, and damage to private and public building and facilities. Table 2 summarizes general benefit categories of enhancements to local flood warning-emergency preparedness plans.

The contribution to increased effectiveness and efficiency of emergency response actions by numerical values is difficult using present benefit analysis procedures. Debates as to valuing the saving of lives and reducing the threats to lives and property have occurred for many years and are continuing. The growing activities in the United States in the implementation of local flood warning-emergency preparedness plans provides evidence that society generally places sufficient value on these activities to support use of public finances and other resources to increase their responsiveness and utility.

TABLE 1

Example Cost Items
Flood Warning - Emergency Preparedness Plan

First Costs

Development of plans
Outfitting/equipping of administrative facilities
Purchase and installation of equipment and hardware
Development/printing brochures, instructions
Stockpiling equipment and materials

Annual Periodic Costs

Updating flood warning-response plans
Updating/printing brochures, instructions, etc.
Operation drills
Supplement/replace stockpiled materials
Equipment operations, maintenance, and replacement

Event Costs

Personnel overtime and emergency hiring
Equipment purchase and rental
Transportation/storage of personnel property
Materials/supplies consumed
Mass care operations

TABLE 2

General Benefit Categories

Category	Contributing Action
Reduced threat to life rescues, public awareness	Barricades, evacuations,
Reduced social disruption	Traffic management, emergency services, public awareness
Reduced health hazards	Evacuations, emergency services
Reduced disruption of services	Utility shutoffs, emergency supplies, inspection, public
Reduced cleanup costs	Flood fighting, self-help loss reduction, efficient resource use
Reduction in inundation damage assistance	Flood fighting, temporary measures, technical

Epilogue

Implementation of a successful flood warning-emergency preparedness plan requires design of formal arrangements for communication and conduct of operations, and the capability to adapt these arrangements to meet possible flooding conditions that would have significant negative effects on the area. The design must be flexible to assure high adaptability to unforeseen flood situations that occur during flood events. The plans must be periodically updated to reflect changed conditions to the flood event, impact area, and to the formal arrangement for conducting the flood warning-emergency preparedness plan.

The major emphasis by federal agencies to date of implementation of local flood warning-emergency preparedness plans has been the installation of flood forecasting hardware, equipment, and enhanced operation procedures of existing federal projects. While these activities are performed to yield earlier recognition of flood threats and more reliable forecasts, greater emphasis on enhancements to the other components and the overall plan of the impacted area is needed. The benefits and subsequent feasibility of any measure is determined from its incrementally enhanced response actions that reduce the potential risk to life, reduce flood damage, and assist in better management of social disruption. Therefore, the consideration of these aspects of local flood warning-emergency preparedness plans is required for determining feasible investments and viable plans.

References Cited

Hydrologic Engineering Center 1984, Expected Annual Flood Computations Users Manual, U.S. Army Corps of Engineers.

Hydrologic Engineering Center 1982, Flood Preparedness Planning: Metropolitan Phoenix Area, U.S. Army Corps of Engineers.

Hydrologic Engineering Center 1979, Physical and Economic Feasibility of Nonstructural Flood Plain Management Measures, U.S. Army Corps of Engineers.

Owen, H. James, 1976, Guide for Flood and Flash Flood Preparedness Planning, prepared for National Oceanic and Atmospheric Administration, Nation Weather Service.

U.S. Army Corps of Engineers 1983, EP 1115-2-1 Digest of Water Resources Policies and Authorities - Chapter 12, Office Chief of Engineers.

OPTIMAL FLOOD WARNING AND PREPAREDNESS PLANS
by
Stuart A. Davis[*]

ABSTRACT: This paper includes a framework for selecting the
optimal investment in warning and preparedness activity. The
primary factors which might influence the magnitude of cost and
damage are described and their interrelationships are
illustrated. An optimization model is presented to illustrate
the minimization of flood damages and flood alleviation costs.
The models expands on previous efforts by: 1) separate
consideration of forecasting and preparedness costs, 2) by
consideration of long term and emergency investments, and, 3) by
the distinction between flood damages that are preventable by
emergency actions and those that are not. Generalized estimates
of variables in the model are critical for economical
application in establishing the optimal levels for the key
decision variables, including investments in the forecasting
system, the response system, and the magnitude of emergency
response.

INTRODUCTION

The following is a framework for selecting the optimal investment
in a warning and preparedness activity. The primary factors which
might influence the magnitude of costs and damages are described and
their interrelationships are illustrated.

The value of the forecast is realized by increased accuracy that
allows for a greater number of accurate responses to actual hazards
and decreased unnecessary response. There is a distinction made
between costs that can be reduced given adequate information, and
costs that cannot be reduced no matter how accurate and timely the
weather forecasts.

Forecasts can only have value as applied to information-sensitive
activities, and then only to the extent that they affect behavior.
Cochrane (1) noted that some activities, such as warnings of heavy
frosts when there is too little time for farmers to take preventive
action, can only add extra time to the farmer's anxiety. When
activity is information-sensitive, behavior can be positively affected
by 1) causing preventive action when a flood is imminent, and 2)
causing individuals not to act when a flood is not going to occur.
Behavior can be negatively affected by 1) inducing needless action in
a false alarm or, 2) discouraging action when a flood is imminent.
These negative actions are type one and type two errors, respectively.

[*]Community Planner, Institute for Water Resources, U.S. Army Corps of
Engineers, Casey Building, Ft. Belvoir, Virginia 22060.

The warning system has benefits to the extent that the errors and the consequences of those errors are reduced. The response system has benefits to the extent that it facilitates evacuation and efficient damage reduction actions.

DECISION RULE

The criteria described above is a very simple relationship. The difficult part is making a complete accounting of the costs and losses. A basic objective for flood damage reduction was established by James and Lee (4). James and Lee stated that the basic objective function is as follows:

$$\text{Minimize } C = C_f + C_s + C_p + C_l \qquad (1)$$

where C is the sum of the total of flood losses and the costs of adjustment. C_f represents the direct cost of flooding, C_s is the costs of structural measure, C_p is the costs of floodproofing, and C_l is the cost of the land enhancement measures.

The model given below expands on James and Lee in the following ways: 1) separate terms are given for forecasting and preparedness cost, 2) long term and short term investments are considered, and, 3) there is a distinction made between flood damages that can be avoided and those that cannot. The model is also flexible to the extent that variables such as flood damage can be broadly defined to incorporate intangible considerations, such as loss of life, psychological trauma, and risk premium.

The variables in the model are defined as follows:

Z = the flood forecast as measured by the number of feet above mean sea level (m.s.l.) or by the stage of water in relation to a designated flood stage.

S = the actual flood event as measured by the height of the flood water above or below a designated flood level.

X_s = the magnitude of investment in structural measures, as measured by the degree of protection offered by the structural measures. This is best given as a reduction in the damage-frequency relationship.

X_f = the magnitude of investment in the forecast system, including a portion of radar forecasting devices, gauging equipment, reporting networks, forecasting information, software and computer hardware, and the costs of labor for individuals involved in the forecasts.

L = the lead time associated with the warning system. For purposes of this model, one level is designated as the minimum amount of time that should be allowing for public safety and economic considerations. The value of L is determined exogenously, with consideration that a minimum lead time is necessary for public safety.

X_p = the magnitude of investment in the preparedness system, as measured by the expected number of people reached by direct, mail, and media warnings.

X_r = the magnitude of investment in permanent floodproofing measures, as measured by the average level of damages reduced for various magnitudes of flooding.

α = the trigger level or the magnitude of the forecasted flood level which must be exceeded before a response will be taken. The trigger level is based on the stipulation that conditions determined by the forecast and stated in the warning must be sufficient to exceed the existing long range investment in flood protection and the degree of risk aversion that might be important in increasing the propensity to respond.

β = the strength of the response, as measured by level of effort made by each individual responding.

R = the magnitude of response to a predicted flood event as measured by the number of people taking action to avoid damage and the number of people evacuating. R is represented by the formula:

$$R = \beta \ (X_p, L) \qquad (2)$$

W = is the publicly issued warning of the forecasted level of flooding.

Two of the primary variables, Z and S, listed above, are stochastic. They are represented in the joint density function:

$$\phi \ (Z, S: X_s, X_f, L) \qquad (3)$$

where the accuracy of the forecast is dependent upon the actual event, the investment in structural flood control measures, including an understanding of how the structural measures will effect the magnitude of flooding with each event, the level of investment in the forecasting system, and the lead time selected.

The total costs of flooding and the alleviating adjustments are as follows:

C_1 (X_s) = the annualized cost of structural flood control. Structural measures include the amortized first cost and the annual operation and maintenance cost of flood barriers and interior drainage, modified channels, and impoundments.

C_2 (X_f) = the annualized cost of the flood forecast system. These costs are assumed to be predominately fixed. They include the first costs of installing equipment, repair, and replacement of equipment, development of computer software, and the continual cost of reporting and recording information.

C_3 (X_p) = the average annual cost of all emergency preparedness measures for investments made regardless of any specific flood event.

Preparedness measures include long term investment in training personnel, public education, and the equipment used at the time of an emergency.

C_4 (X_r) = the average annual cost of permanent floodproofing measures. Floodproofing includes permanent action taken in the construction or remodeling of individual structures, the cost of flood warning and the temporary action taken to alleviate flooding in response to those warnings. It should be noted that long term investment can either be permanent modifications in building structure and use or investments that facilitate emergency flood proofing measures.

C_5 (X_s, X_f, X_p, X_r) = to the long-run fixed cost of flooding, which result regardless of any particular event. These include the foregone land enhancement, modified use of property, and administrative cost of flood insurance that occurs because of the continual flood threat. Land enhancement foregone includes the loss in net return on property or economic rent, from flood-induced modifications in property use.

C_6 $(S: X_r)$ = the annual cost of flooding, which are directly associated with a flood event, but which cannot be avoided by any emergency preparedness action.

C_6 are the total direct costs of flooding, including physical damage to buildings, contents, vehicles, streets, bridges, and utilities that can not be eliminated regardless of the investment in warning and preparedness. Physical damage cannot exceed the depreciated replacement value of the property in question. Damage include the replacement, repair, or the cleanup of property. Also included are direct nonphysical damage, such as temporary lodging, traffic rerouting, and unrecoverable losses in business production and services.

C_7 (R) = the costs of emergency response measures.

C_7 is subject to the following conditions:

$$C_7 = \{ \begin{array}{ll} (Z - X_s, 0) & \text{if } \{Z < X_s \} \\ (B(X_p, L), Z] & \text{if } \{Z \geq X_s \} \end{array} \qquad (4)$$

C_8 (S, R) = the direct cost of flooding which are dependent on the magnitude of the flood event and directly reducible by emergency actions.

C_8 is subject to the following conditions:
$$C_8 = \{ \begin{array}{ll} C_8 (S, 0) & \text{if } Z < X_s \} \\ C_8 (S, R) & \text{if } Z \geq X_s \} \end{array} \qquad (5)$$

The optimal flood policy is determined by the following objective function:

$$\text{Min E (C)} = C_1(X_s) + C_2(X_f) + C_3(X_p) + C_4(X_r)$$

$$+ C_5 (S + X_f + X_p + X_r)$$

$$+ \int_0^\infty \int_0^\infty C_6 (S:X_s) \; \phi \; (Z,S:X_s,X_r,) \; dZdS$$

$$+ \int_0^\infty \int_{\alpha X_s}^\infty C_7 [\beta(X_p,L)Z] \; \phi \; (Z,S:X_s,X_f,L) \; dZdS$$

$$+ \int_0^0 \int_0^{\alpha X_s} C_8 [S,0] \; \phi \; (z,S:X_s,X_f,L) \; dZdS$$

$$+ \int_0^\infty \int_{\alpha X_s}^\infty C_8 [S, \; b(S, \; (X_p,L)Z) \; \phi \; (Z,S:X_s,X_r,X_f,L) \; dZdS$$

$$(6)$$

Interpretation of the Model

There are five decision variables: X_s, X_r, X_f, X_p, and α, indicating the level of investment in structural measures, permanent floodproofing, forecasting, and preparedness, and the level of flooding that must be forecasted before a response would be initiated. These variables have been identified as being critical to the determination of the optimal flood damage alleviation policy.

The total investment in structural measures is an important factor in 1) the total costs of structural measures, 2) the long run fixed costs of flood hazards independent of any flood event, 3) the costs of flooding that cannot be avoided by preparedness measures, 4) the costs of response, and 5) the costs of flooding that can be avoided by preparedness measures. The relationship of X_s^* (optimal level of X_s) is straight-forward for the first three costs listed above. As we increase X_s, the long-run cost of flood hazard and the cost of floods unavoidable with preparedness measures increase proportionately. For C_7 and C_8 there is also a straight-forward relationship for situations where the forecasted flood level exceeds the trigger level times the extent of structural protection, where the optimal level of structural investment increases with the cost of response and decreases with the cost of flooding that can be avoided by preparedness measures. The cost of response and flood damage will not change for flooding that stays below the trigger level.

The level of forecasting is an important factor in 1) the cost of forecasting, 2) long term flood hazard cost, 3) annual cost of response, and, 4) the annual cost avoidable by preparedness. It should be noted that, by increasing the lead time, forecast investment will have an important role in determining the amount of damage that can be avoided by preparedness activities.

It should also be noted without a significant investment in preparedness, the flood forecasting investment may do little to offset the cost of flooding. Mileti (5) noted the difficulties in achieving an effective business and residential response without pre-flood education, an quick warning network, and provisions to reinforce the warning messages.

Floodproofing investment can reduce long term flood costs and event specific flood losses. Flood proofing provides often relatively low cost, and problem specific relief to flooding problems. However the residual risks of floodproofing are often greater than for structural measures. Floodproofing can leave people on islands surrounded by floodwaters.

The trigger level, described by Howe and Cochrane (3), is determined by a hosts of geographically-specific factors, such as risk aversion and a communities recent flood history. The trigger level is optimal when the decision-makers' perception of risks and the value of preventive actions approaches reality.

Several practical considerations that limit the application of this model include: The complexity of the model, particularly with the number and enigmatic nature of the variables involved, prohibit the direct application of the model without the use of generalized estimates of the cost functions and the change in the cost functions with respect to these critical decision variables. Reliable estimates of these variables cannot be made until there is considerably more empirical information with which to calculate or estimate these relationships. Application of any generalized relationships should be with full consideration of how local conditions and the components of any particular alternative may alter the relationships.

APPENDIX.-REFERENCES

1. Cochrane, Harold C., "Modeling the Economic Value of Weather Forecasts," Value and Use of Short Range Mesoscale Weather Information, David George, editor, National Oceanic and Atmospheric Administration, Boulder, Colorado 1982.

2. Day, Harold J., A Study of the Benefits Due to U. S. Weather Bureau River Forecast Service, Pittsburgh, Carnegie Institute of Technology, 1966.

3. Howe, Charles W. and Cochrane, Harold C., "A Decision Model for Adjusting to Natural Hazard Events with Application to Urban Snow Storms," Review of Economics and Statistics, Vol. 58, 1976, pp. 50-58.

4. James, L. Douglas, and Lee, Robert E., Economics of Water Resources Planning, McGraw-Hill, New York, 1972.

5. Mileti, Dennis, D.S., Natural Hazard Warning Systems in the United States: A Research Assessment, University of Colorado, Boulder, 1975.

ROLE OF SURFICIAL AQUIFER IN BARRIER ISLAND
STREAMFLOW GENERATION

Francis M. Nevils, M.ASCE, and Michael E. Meadows, M.ASCE[*]

Abstract

Monitoring results are presented for an experimental
watershed located on a South Carolina barrier island.
Rainfall, streamflow and well data indicate streamflow
occurs due to interaction between the surficial aquifer and
surface streams, and due to surface runoff during periods
of high intensity rainfall. The well data indicate a tidal
influence up to 2000 ft (610 m) inland from the tidal
marsh. A model is proposed for the surficial aquifer-
streamflow interaction.

Introduction

An experimental watershed is being monitored on St
Helena Island, South Carolina, as part of a study to
develop a procedure to delineate those portions of coastal
zone agricultural watersheds which contribute direct storm
runoff during prescribed rainfall events. The purposes for
the experimental watershed and monitoring program are to
obtain basic data about the hydrologic processes involved
in streamflow generation on a barrier island, to test basic
runoff theories, and to provide a basis for developing a
watershed simulation model.

Study Site

The watershed, shown in Figure 1, is approximately 47
acres (19 hectares), nearly rectangular, and bounded on two
sides by roads. A third boundary is a large trench which
drains through a tide gate into a salt marsh that bounds
the site on the fourth side. There is no detectable
surface drainage onto the watershed from offsite areas.
Internal drainage is provided by three parallel ditches
that discharge into the large trench. The site is evenly
graded and relatively flat. Elevations range from 11.5 ft
(3.5 m) MSL in the northwest corner to 6.5 ft (2 m) MSL
between streamflow gages 2 and 3. The site soils consist

----

*Respectively, Graduate Assistant and Associate Professor,
Department of Civil Engineering, University of South
Carolina, Columbia, SC 29208

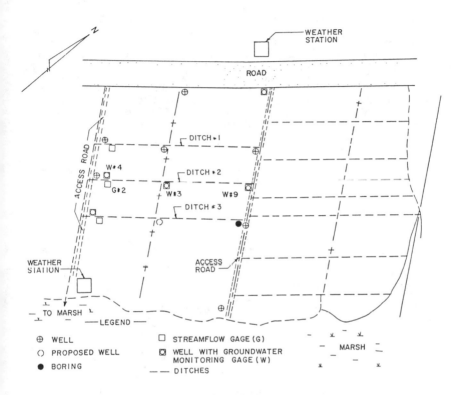

Fig. 1—Watershed Map

of fine to medium sand with little or no clay and silt content.

The watershed is instrumented to continuously monitor rainfall, evaporation, wind direction and magnitude, streamflow and groundwater elevations. There are two rain gages, two evaporation pans, two anemometers, three streamflow gages and twelve shallow groundwater wells. The number of wells with continuous recorders has varied from 2 to 5 depending on the availability of recorders.

Five wells were installed during a geotechnical investigation of the watershed by placing them in the bore holes augered to obtain soil samples and to map the soil stratigraphy. These wells were installed to a depth of 14.5 ft (4.4 m). The remaining wells were installed by jetting them to depths ranging from 10 to 14 ft (3 to 4.3 m).

The three internal drainage ditches were gaged by installing V-notch weirs and float level recorders.

Surficial Aquifer Response to Tides

Monitoring well records indicate tides influence the groundwater system as far as 2000 ft (610 m) inland from the marsh. Well 4 is near the trench and is influenced by the periodic ponding and draining of the trench caused by the closing and opening of the tidal gate during high and low tides, respectively. The groundwater fluctuation averages from 1 to 2 inches (2.54 to 5.1 cm) but a maximum of 4 inches (10 cm) has been recorded. The period of cyclic response closely follows the tidal cycle of 12 hours.

Well 3 is further inland than well 4 and is not influenced by the trench, but still exhibits a tidal influence. The groundwater response is not the rise and fall of the watertable seen at well 4 but a leveling of the natural drawdown curve for a period of time followed by an abrupt drop to a new elevation. As such, the drawdown curve has a characteristic stair-step shape. The time period of the leveling does not match the tidal cycle but is more extended averaging about 24 hours in duration. Well 9, which is the furtherest well inland with a continuous recorder, exhibits similar behavior as well 3 but with a longer time period for the leveling phenomena.

The tidal influence on the groundwater level at each well can be explained as a superposition of the tidal cycle on the natural groundwater drawdown. At well 4 the fluctuations oscillate about the actual drawdown curve. At wells 3 and 9, it is hypothesized the influence is due to the advance and retreat of the saltwater wedge and/or to

the tidally controlled water levels in the salt marsh.
During advance, or high tide, the groundwater gradient is
reduced enough to slow or even stop flow in the surficial
aquifer, but not enough to cause the water level to rise.
During retreat, or low tide, the combined effects of the
reduced tidal influence and the natural drawdown cause a
sharp drop in the groundwater level.

Surficial Aquifer-Streamflow Interaction

     Comparison of the streamflow and well monitoring data
reveal a direct relationship between groundwater elevation
and streamflow. This is clearly evident from the nearly
simultaneous fluctuations in groundwater elevations and
streamflow hydrographs during periods of no precipitation.
This interaction is most important to explaining the role
of the surficial aquifer on streamflow generation during
rainfall events.

     During a rainfall event, typically there is a rapid
rise in the groundwater table several hours after the
beginning of rainfall. The streamflow increases at the
same time or shortly thereafter. This behavior suggests
streamflow largely is a result of increased seepage of
groundwater into the drainage ditch. Table 1 presents the
data from three rainfall events and the resulting rise in
the groundwater and streamflow. Note the groundwater at
wells 3 and 9 begins to rise as early as 1.5 to 2.5 hours
before the streamflow at gage 2.

     The rapid rise in the groundwater may be explained in
one of two ways. First, it is possible that infiltrated
rainfall percolates rapidly through the soil reaching the
water table in only a few hours. Field infiltration and
well tests indicated much of the watershed has surface
infiltration rates ranging from 3 to 6 inches/hr (7.5 to 15
cm/hr) and that the hydraulic conductivity of the subsoil
ranges from 0.75 to 1.25 inches/hr (1.9 to 3.2 cm/hr). The
water table over most of the watershed is high, ranging
from 1 to 8 ft (.3 to 2.4 m) below the ground surface, with
an average of 3.5 ft (1.1 m). A second theory is that the
infiltrated water results in a rapid compression of the
capillary fringe.

Groundwater Maps

     Piezometric maps have been developed from weekly
groundwater level readings taken over a period of several
months. The map for Feb. 18, 1987 is shown in Figure 2.
Several observations (inferences) concerning the
groundwater table and movement can be made from these maps.
The first observation is that the northeast boundary of the
watershed seems to be a groundwater divide. This divide is
created by the influence of two trenches, the one mentioned
previously and a second trench paralleling the first trench

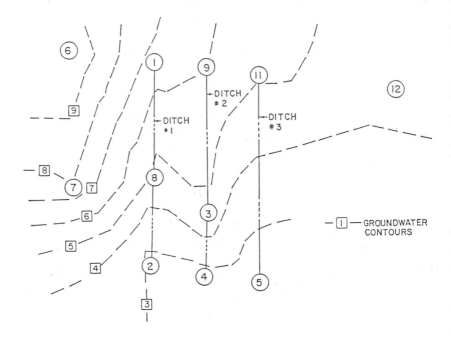

Fig. 2--Piezometric Map

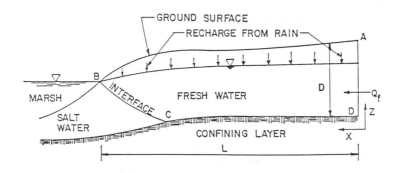

Fig. 3--Conceptual Groundwater Model

but located approximately 1200 ft (366 m) offsite and
northeast of the watershed. A second divide occurs near
ditch 2. This divide is created by groundwater flow to the
first trench and the marsh.

A second and very important observation is that the
lower section of ditch 1 influences the local groundwater
movement. The groundwater contour maps suggest that the
groundwater flows to ditch 1 between wells 8 and 2. The
remaining portion of ditch 1 and ditches 2 and 3 do not
intersect the groundwater table sufficiently to affect the
movement of the groundwater at the site.

The effect of the three ditches on groundwater flow
shown by the contour maps was substantiated by visual
observation and streamflow monitoring. Ditch 1 has had
water flowing almost constantly during the period of study.
Ditches 2 and 3 have had measurable streamflow only during
periods of extremely high groundwater or during high
intensity rainfall events. Apparently, the groundwater
table is below the bottom of ditches 2 and 3 for much of
the year.

The tidal effect can be seen by studying contour maps
generated from readings taken at or near high and low
tides. The tidal influence is most pronounced in those
areas adjacent to the marsh and near the trench.

Surficial Aquifer Model

An event simulation model is being developed which
inputs rainfall, balances infiltration, groundwater seepage
(into or out of the ditches), and soil moisture storage,
and outputs the runoff hydrograph. The most critical
component, and the one which has received the least
treatment in the literature, is the prediction of the time
varying phreatic surface given any tidal and groundwater
inflow conditions. If the position of the phreatic surface
where it intersects the drainage trenches can be predicted,
then the amount of seepage can be determined.

A conceptual model of the groundwater regime at the
experimental watershed is shown in Figure 3. Boundary ABCD
encloses a domain in which the governing field equation is
the Laplace equation

$$\nabla^2 \phi = 0$$

where ϕ is the groundwater potential. The boundary ABCD is
divided into M intervals and a boundary element
discretization scheme is used to write equations for the
location of discrete points on the boundary. A uniform
distribution of all boundary quantities is assumed over
each boundary interval (Q). On the basis of these
assumptions, standard boundary element method techniques

are used for the spatial discretization of the Laplace equation resulting in the following equation (Karabalis 1987).

$$\sum_{i=1}^{M}(A_{ij} - \delta_{ij}\ \beta_i)\ \phi_i = \sum_{i=1}^{M} B_{ij}\ C_i \qquad j = 1,2,\ldots M$$

where

$A_{ij}=\int \partial/\partial n(\ln r_{ij})dQ$

$B_{ij}=\int \ln r_{ij}dQ$

$C_i=[\partial/\partial n(q)]_{Q=Q}$

$\beta_i=(P=P_i)$

r_{ij}=distance from interval i to interval j

q=flux across the boundary

n=outward normal on interval Q.

The boundary conditions are: (a) The flow or flux across boundary CD is zero. (b) Either the flow across boundary AD or the potential on AD is known. (c) The flow or flux across boundary BC is zero. In addition, the time varying slope and position of this boundary depends upon the tidal cycle, the inflows of freshwater and the differences in the densities of salt and freshwater (Bear and Dagan 1964; Vappicha and Nagaraja 1976; Liu, et al. 1981. and (d) The flux on boundary AD is known. The potential and therefore the position are the important unknowns for this problem.

The time variation is handled using a finite difference prediction scheme (Liggett 1977).

Conclusions

The following conclusions can be drawn from the experimental watershed monitoring results:

1.   Streamflow generation is due largely to the surficial aquifer-streamflow interaction. During periods of intense rainfall, surface runoff can occur from those areas with low infiltration rates or surface saturation.
2.   Tides can influence the groundwater system for considerable distances inland from tidal marshes, especially where sandy soils predominate.

The best approach to modeling the tidal effects on the surficial aquifer is to use the boundary element method.

Acknowledgements

This paper is based on research supported in part with funds provided by the South Carolina Sea Grant Consortium. The authors especially wish to acknowledge Seaside Farm, Inc., for providing the experimental watershed site, equipment, and data collection support. The figures were drawn by Ms. Katheryn Olson.

Appendix I.--References

1. Bear, J., and G. Dagan (1964), "Moving Interface in Coastal Aqurfers," J. Hydr. Div., ASCE, 90(4), 193-216.
2. Karabalis, D. L. (1987), "Torsion of Elastic Bars", in D. E. Beskos, Ed., "Boundary Element Methods in Structural Analysis", ASCE, New York (in print).
3. Liggett, J. A. (1977), "Location of Free Surface in Porous Media", J. Hydr. Div., ASCE, 103(4), 353-365.
4. Liu, P. L-F, A. H-D Cheng, and J. A. Liggett (1981), "Boundary Integral Equation Solutions to Moving Interface Between Two Fluids in Porous Media", Water Resources Res., AGU, 17(5), 1445-1452.
5. Vappicha, V. N. and Nagaraja, S. H. (1976), "An Approximate Solution for the Transient Interface in a Coastal Aquifer," J. Hydrology, Elsevier, 31, 161-173.

Table 1.--Groundwater and Streamflow Response to Rain

	Time of Rise (hr)	Starting Elev (ft MSL)	Time To Peak (hr)	Peak Elev (ft MSL)
March 7, 1987				
Rainfall	2300			
Water Level at				
Gage No. 2	0230	2.63	0700	2.75
Groundwater Level		(0.80)		(0.84)
at Well No. 3	0200	4.37	0800	4.74
		(1.33)		(1.44)
February 21, 1987				
Rainfall	0040			
Water Level at				
Gage No. 2	0330	2.61	0930	2.74
Groundwater Level		(0.79)		(0.84)
at Well No. 3	0300	4.32	1100	4.60
Groundwater Level		(1.32)		(1.40)
at Well No. 9	0100	5.46	1000	5.49
		(1.66)		(1.67)
February 22, 1987				
Rainfall	0600			
Water Level at				
Gage No. 2	1030	2.71	1730	2.79
Groundwater Level		(0.83)		(0.85)
at Well No. 3	0930	4.56	1600	4.90
		(1.39)		(1.49)

A LINKED STREAM-AQUIFER SYSTEM MODEL
Anand Prakash[*], F. ASCE and Bahram A. Jafari[*]

ABSTRACT: A linked stream-aquifer system (LISAS) model is described. This model includes a deterministic sequential flow simulation component which generates hydrographs of surface runoff and deep percolation for each sub-watershed in the basin. The hydrograph of deep percolation forms a part of the input to the aquifer-simulation model. The aquifer-simulation model generates hydrographs of ground water flows into or out of the stream at various locations along its course. The surface and ground water flow hydrographs are combined with appropriate lags to develop virgin flow hydrographs at various locations and routed through the stream course using the kinimatic wave method. The application of the proposed model is illustrated by a simplified example.

Introduction

The natural hydrologic budget for an alluvial river basin consists of precipitation in the form of rain or snow, depression storage, soil moisture storage, evapotranspiration, deep percolation, and surface runoff. The deep percolation component enters the ground water storage and may appear as baseflow in the gaining reaches of streams.

To assess the dependability of the water yield of a water resources development plan in a stream-aquifer system, streamflow data extending over several climatic cycles are required. In most cases, only limited observed data are available. To supplement these data using available climatologic information, a linked stream-aquifer system (LISAS) model has been developed. The objective of this paper is to describe the various components of this model and to present an example illustrating its application.

Computational Procedure

The usable stream flows available at any instant at any location along a river channel are given by the summation of the surface and ground water flow hydrographs at that location. The linked model that could accomplish the required computations includes a computer program to simulate the surface runoff component of sequential streamflows and contribution of irrigation and rainfall to the aquifer system by way of infiltration and recharge; a computer program to model the transport of ground water through the stream

* Dames & Moore, 1626 Cole Blvd., Golden, CO 80401

aquifer system which appears as baseflow in the stream; and an algorithm to combine the surface and ground water flow hydrographs and route the same through the stream channel.

A flow chart illustrating the computational procedure of the linked model is shown in Figure 1. The modeling approach consists of the following steps:

(i)   The stream-aquifer system is divided into a number of sub-watersheds based on topography and surface drainage patterns and the stream channel is divided into a number of reaches based on the locations of surface and subsurface tributary inflows.

(ii)   Using the precipitation, temperature, evapotranspiration, and irrigation application data and the hydraulic characteristics of the upper and lower soil zones in each sub-watershed, the hydrographs of surface flows and recharge to the ground water storage are computed using the National Weather Service River Forecast System model (NWS, 1986). A flow chart illustrating the computational steps in the model is shown in Figure 2.

(iii)   Each sub-watershed is subdivided into a number of hydrogeological units based on land-use patterns, i.e., irrigation, dryland farming, etc.

(iv)   The deep percolation or recharge component of the output of the aforementioned model is routed through the stream-aquifer system using a three-dimensional finite-difference ground water flow model (McDonald and Harbaugh, 1984). The algorithms to link the surface and ground water flow models are under development at this time. For the present, the deep percolation or recharge component is routed through the stream-aquifer system using the following analytical equations (Prakash, 1979):

$$
q_{k,1}(t) = K_x\, h_1 \left[\sum_{n=0}^{\infty} e^{-\dfrac{(2n+1)^2 \pi^2 a\, t}{4\, L^2}} \left\{ \frac{2\, h_1}{L} + (-1)^n \cdot \frac{4}{(2n+1)\pi}\, (a+2bL) - \frac{16bL}{(2n+1)^2 \pi^2} \right\} + \right.
$$

$$
\left. \frac{2}{Sy\, L} \sum_{i=1}^{M} \sum_{n=0}^{\infty} e^{-\dfrac{(2n+1)^2 \pi^2 a\, (t-t_i)}{4\, L^2}} \sum_{j=1}^{N} R_{i,j} \left\{ \cos \frac{(2n+1)\pi x_{j-1}}{2L} - \cos \frac{(2n+1)\pi x_j}{2L} \right\} \right]
$$

$$t_i \leq t \leq T$$

(1)

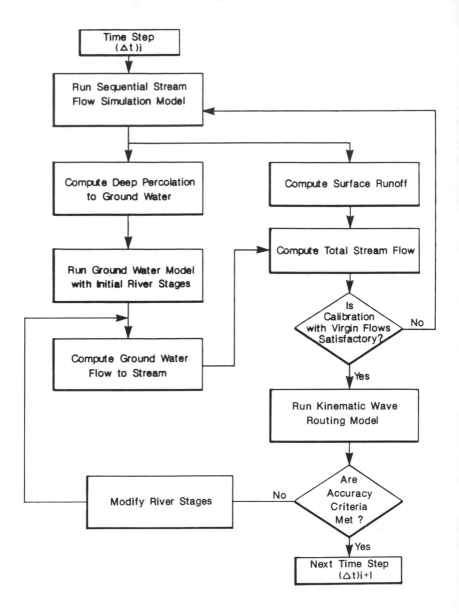

Fig. 1 – Flow Chart for Surface & Groundwater Model Linkage

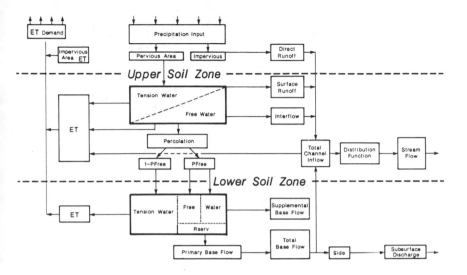

Fig. 2 – Flow Chart for National Weather Service River Forecast System Model

in which $q_{k,1}(t)$ = ground water recharge from 1st (left or right) side of the stream at time t per unit width of hydrogeologic unit k; K_x = average hydraulic conductivity of the aquifer; S_y = average specific yield of the aquifer; h_1 = constant average water surface elevation in the stream reach; $\alpha = K_x h/S_y$; h = average value of saturated aquifer thickness (McWhorter, 1977); t = time; L = length of aquifer on the 1st side of the stream from the bank to the ground water divide measured perpendicular to the stream reach (1 = 1 or 2); a,b = coefficients in the polynomial defining the shape of the initial water table, $h = h_1 + ax + bx^2$; x = distance from stream bank measured perpendicular to the stream reach; M = total number of recharge ordinates occurring at times t_i; $R_{i,j}$ = recharge ordinate occurring at time t_i per unit area of the hydrogeologic unit extending from x_{j-1} to x_j; and T = total time of simulation.

Ground water recharge from the other side (1 = 2) of the stream is calculated using Equation 2.

$$Q(t) = \sum_{k=1}^{K} B_k\, q_{k1}\, (t - r_k) \cdot H(t - r_k) + \sum_{k=1}^{K'} B_k'\, q_{k2}\, (t - r_k') \cdot H(t - r_k')$$

(2)

in which K,K' = total number of hydrogeological units on the two sides of the stream; τ_k, τ'_k = travel times through the stream channel from the midpoints of the kth reach on either side of the stream to the downstream edge of the first (lowermost) reach; B_k, B'_k = widths of the kth hydrogeological units on the two sides of the stream; q_{k1}, q_{k2} = ordinates of subsurface flow hydrographs for the kth reaches on either side of the stream; Q(t) = ordinate of the hydrograph resulting from ground water flows from all reaches upstream of the downstream edge of the first (lowermost) reach; and H(t) = Heaviside's unit step function (Wylie and Barrett, 1982).

(v)   The surface and ground water flow hydrographs at the outlet of each sub-watershed are combined and routed through the next downstream channel reach using the kinematic wave routing approach (USACE, 1981).

Illustrative Example

    To illustrate the computational procedure of the LISAS (linked stream-aquifer system model), the daily surface flow hydrograph has been generated for a 300 sq. mi. sub-watershed using 6-hour precipitation and temperature and daily evapotranspiration data. The subsurface, surface, and combined flow hydrographs at the outlet of the sub-watershed are shown in Figure 3.

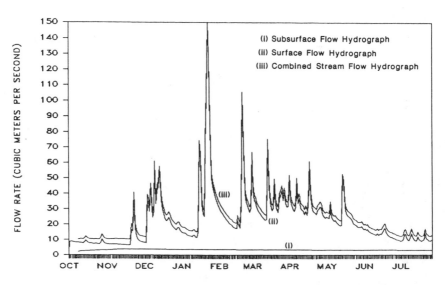

Fig. 3 - Surface, Subsurface, and Combined Stream Flow Hydrographs

Conclusion

A linked stream-aquifer system model has been presented which simulates daily streamflows using climatological, land-use, and hydrogeological data. A sequential flow simulation model is used to generate hydrographs of surface runoff and deep percolation. The deep percolation component is used as input to an analytical ground water flow model which generates and routes the hydrographs of baseflows for the stream. The two hydrographs are combined to develop virgin streamflow hydrographs at the points of interest.

Appendix

McDonald, M.G. and Harbaugh, A.W. (1984). "A Modular Three-Dimensional Finite-Difference Ground Water Flow Model", U.S. Geological Survey, Reston, Virginia.

McWhorter, D.B. (1977). "Drain Spacing Based on Dynamic Equilibrium", Jour. of the Irrigation and Drainage Div., ASCE, Vol. 103, No. I.R.2.

National Weather Service (NWS) (1986). "River Forecast System Model", Silver Spring, MD.

Prakash, A. (1979). "A Deterministic Model to Estimate Low Streamflows", International Symposium on Hydrological Aspects of Droughts, New Delhi, India, Dec. 3-7, 1979.

U.S. Army Corps of Engineers (USACE) (1981). "Flood Hydrograph Package", HEC-1, Hydrologic Engineering Center, Davis, CA.

Wylie, C.R. and Barrett, L.C. (1982). "Advanced Engineering Mathematics", McGraw-Hill Book Company, New York.

SEEPAGE/DRAINAGE INVESTIGATION
COLUSA COUNTY CALIFORNIA

B.E. Manderscheid, M. ASCE *

ABSTRACT: High groundwater in the late winter months saturates the soil profile, causing damage to both crops and soil. Sustained water logging severely damages or destroys orchards. It has generally been accepted that seepage from the Sacramento River is the principal cause of high groundwater within a mile from the river. The study participants gathered, documented and analyzed basic data including soil permeability, the volume, source, movement, and chemical characteristics of the water, and hydraulic gradients. Implementation of the recommended solutions would alleviate seepage and related high groundwater problems that have plagued the Sacramento River area for over 75 years.

INTRODUCTION: The Colusa County Flood and Water Conservation District formed the Zone of Benefit 2 (Zone), consisting of approximately 7,000 acres located adjacent to and west of the Sacramento River between the towns of Colusa and Princeton, near the center of the Sacramento Valley as shown on Figure 1. The Zone, as authorized by state law, authorizes the District authority to provide for conservation of water, water resource development, drainage control, storm, flood, and other water.

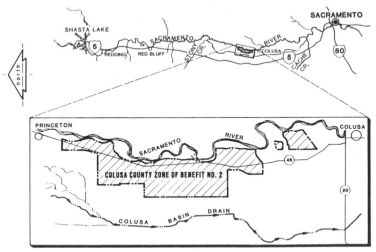

Figure 1. Location of Colusa Zone of Benefit No. 2

* Managing Engineer, Boyle Engineering Corporation, 2277 Fair Oaks Blvd., Suite 305, Sacramento, CA 95825

The Zone is located in the east-central part of the Colusa Basin and is bordered by the Sacramento River on the east and by the Colusa Basin Drain on the west. The river, which bisects the Sacramento Valley, originates near Mount Shasta and flows southerly to the Pacific Ocean. Between Mount Shasta and the Zone, the Sacramento River is fed by numerous small unregulated tributaries originating in the mountains east and west of the valley.

Historically, the Sacramento River, through extensive overflow deposition, has gradually built its riverbed to an elevation approaching that of the surrounding land. During the early part of this century, the Corps of Engineers extensively leveed the river to contain winter and spring floodwaters. The land slopes from the river to Colusa Drain at approximately 15 feet per mile. During periods of high runoff, the surface of the river is frequently higher than the surface of the adjacent lands. Colusa Trough, the low area in the Colusa Basin, parallels the Sacramento River to its junction with the river north of the City of Sacramento. The Colusa Basin Drain, a combination stream and drainage channel located in the Colusa Trough, collects practically all the surface runoff from the foothills and valley areas west of the Sacramento River between Stony Creek to the north and Cache Creek to the south. Above-normal winter precipitation and runoff often cause the Colusa Drain to overflow, flooding the adjacent lands.

Agriculture and related support activities are the economic base of the Zone. In 1985 about 6,300 acres were planted with a variety of orchard, rice, field and row crops.

Basin soils are mostly composed of fine-grained material deposited historically by slowly moving or standing water. Adjacent to the river the soils consist of almost equal parts of clay, silt and sand. Most of the orchards are located on these deposits because of the well-drained soils. Away from the river the soils are principally lean to heavy clays with some sand and silt. Most of the rice is grown in these soils because of the low permeability.

THE PROBLEM: The high groundwater levels and ponded surface water are intermittent but recurring problems. Concurrent with winter precipitation and its accompanying runoff, the water levels rise in the Sacramento River, in the Colusa Basin Drain, and in the soil profile. When water levels in the river rise, water seeps through or under the levees to adjacent lands. In lands away from the river, precipitation percolates through the soil profile. After a period of time, the seepage and percolation cause the soil to become water logged, with ponding occurring when the water table is high. If the river stage remains high and drainage facilities, including Colusa Basin Drain, prove inadequate, water spreads over large areas, damaging both crops and soil. Saturated soil conditions delay or even prevent crop planting and necessary cultural operations. Saturated soil conditions in late spring damage crops already planted, reducing yields. Sustained water logging severely damages or even destroys prune and walnut orchards. Saturated soil conditions also limit the choice of field crops to those less susceptible to damage from soil moisture, which generally are less profitable to grow.

DATA COLLECTION: A field investigation program was developed to determine both the characteristics and the relationships of the area geology, groundwater, precipitation, surface water, river stage and seepage, and water quality. The field investigation, data reduction, and analyses were conducted from June 1985 through June 1986 by the California Department of Water Resources, the U.S. Bureau of Reclamation, the U.S. Soil Conservation Service, and Boyle Engineering Corporation.

During 1985 and 1986, a total of 37 shallow (<18-foot depth), 9 intermediate (35-50 feet), and 2 deep (120 feet) wells were drilled in the study area.

Water levels at all shallow wells were measured by hand from July 1985 to May 1986. The intermediate depth wells were drilled in November 1985 and equipped with continuous water-level recorders which collected measurements from December 1985 to May 1986.

Water quality samples were collected monthly from intermediate depth observation wells during the four-month period February through May 1986. Water samples were analyzed for trace element, general mineral, and pesticide parameters. Pesticides levels were nondetectable or were below safe levels recommended by the California Department of Health Services. All general mineral analyses fell below recommended contaminant levels established by the California Department of Health Services for consumer acceptance limits. At least one of the elements of arsenic, cadmium, chromium, hexavalent chromium, copper, nickel, lead, selenium and zinc exceeded Environmental Protection Agency criteria for drinking water or freshwater aquatic life at one or more of the wells tested at least once during the period of study. Overall, however, water quality does not appear limiting for agricultural production or discharge to the Sacramento River or Colusa Basin Drain.

GROUNDWATER: The configuration and slope of the groundwater table are influenced by precipitation, irrigation, and the river system. The groundwater elevation varies throughout the zone and changes throughout the year, ranging from ground surface to more than 20 feet below the surface.

Underlying the Zone are two aquifer systems confined and unconfined. The confined aquifer consists of the units underlying the Older Basin deposits which act as the confining bed or layer. Results of a pump test conducted to determine if there was leakage between the confined and unconfined aquifers showed no connection. Based upon elevation of the pump test and water-level fluctuations in this aquifer, it was determined that the aquifer had very little effect on the seepage problem.

The unconfined aquifer is hydrologically connected to the river. Therefore, when river stage is low or below adjacent groundwater levels, groundwater flows toward the river, generally during the period between May and October.

When the river stage is high or above adjacent groundwater levels, groundwater flows away from the river and seepage occurs , generally during the period from November to April. The amount of seepage or recharge to the groundwater basin, is dependent on the hydraulic conductivity of the aquifer in that area and the gradient between the river and the water table west of the levee. Hydraulic conductivity and gradient vary widely along the river within the Zone.

SEEPAGE: The term "seepage" as used in this investigation is defined as water attributable to percolation from the Sacramento River which is at or near the ground surface on the landward side of the levee. The timing and ultimate area of seepage are directly related to the depth and slope of the groundwater table. If the water is initially near ground surface and the hydraulic connection to the river is good, it takes little time for a rise in river stage to cause seepage.

Two important factors affecting seepage are the stage or elevation of the water surface in the river above a certain critical base level and the duration of the stage. The higher the river stage, the greater the hydraulic force, and consequently the greater the seepage. The duration of a high river stage determines how far out the water moves onto the adjacent land, and how much soil will become saturated by the seepage.

Topography and levee widths have a very important bearing on seepage and seepage damage. Seepage generally appears first in low spots and depressions, where the difference in head between the river water surface and ground surface is greatest. The widths of the natural levees are highly variable because of the nonuniform method of river deposition due to river meander, variable sediment load, and scours. Thus, the shape of the present levees is somewhat modified from the original form. Generally, the wider the levee the less the rate of seepage flow.

HYDROLOGIC BUDGET: A water balance was obtained by calculating the volumes of recharge and discharge, and the resulting change in storage.

As stated earlier, the groundwater basin underlying the Zone is unconfined above the Older Basin deposits and confined below them.

The specific yield of the deposits that comprise an aquifer determines its storage coefficient. The sands and gravels have a storage coefficient between 0.15 and 0.20; silt and clays, between 0.03 and 0.05. The storage coefficients for the study area were obtained by averaging the specific yields of materials obtained from observation well readings. The change in groundwater storage is determined using the appropriate storage coefficient.

The change in groundwater levels can be computed by dividing the amount of water entering the aquifer by its storage coefficent. For example, one foot of water going in or out of an aquifer with a storage coefficient of 0.20 will change the water level 5 feet; in an aquifer with a storage coefficient of 0.03, the change will be 33 feet.

Recharge: Subsurface inflow from seepage is an important source of recharge. Because sands and gravels have a higher hydraulic conductivity, they respond faster to changes in river stage than do silts and clays. Therefore, areas underlain by sands and gravels experience greater seepage rates during periods of high river stage than areas underlain by silts and clays. As the river recharges the adjacent groundwater basin, causing groundwater levels to rise, the hydraulic gradient declines. This decline will continue until the river stage and the groundwater basin reach hydraulic equilibrium. This rarely happens, however, because the river stage fluctuates faster than the groundwater levels can respond.

Subsurface inflow from the surrounding groundwater basin is generally from the north and west. The hydraulic gradient is very small and the hydraulic conductivity is low, resulting in a low inflow.

Precipitation is a major source of recharge. The ground surface is nearly level so precipitation, instead of running off, stands on the fields and slowly percolates into the groundwater basin. Consequently, standing water has been mistaken for "seepage water".

Discharge: The three major types of discharge are: subsurface outflow, surface runoff, and pumpage. In addition, evaportranspiration removes an unknown amount of water from storage.

Groundwater is removed from the areas by subsurface outflow which is restricted by the hydraulic conductivity and the low water table gradient in the western part of the study area.

Groundwater is also removed by means of surface runoff when the groundwater level rises to the ground surface or intersects a surface drain. Then the water is either evaporated or it leaves the area via the surface drain. The amount of surface runoff depends on the length of time the groundwater table remains at the ground surface and above the drain invert. The ability of the surface drains to remove groundwater from the area during the winter months is greatly influenced by the water-surface elevation of the Colusa Basin Drain.

Groundwater pumpage for irrigation during the spring and summer has little influence on the saturation problem since the irrigation wells pump from the confined aquifer.

Analysis: Saturated soil is the result not only of river stage and duration but also of precipitation, applied water, subsurface inflow, and the elevation of the groundwater table prior to an event.

The elevation of the groundwater table is a result of cumulative effects of recharge, discharge, and aquifer storage. If the long-term recharge is increased while the discharge remains constant, available storage space will decrease. As storage space decreases saturated soil problems could be caused by small amounts of recharge.

The purpose of the hydrologic budget was to quantify the factors affecting groundwater levels. The levels went from the season's low to its high during the period from January 16, 1986, to March 22, 1986. The hydrologic budget was calculated for this time period, dividing the study area into 31 cells. The area of each cell was calculated and water-level data were used to identify groundwater gradients. These were combined with aquifer transmissivities to determine subsurface inflow and outflow. Precipitation data provided additional inflow values. Changes in storage were computed from records showing changes in groundwater level and storage coefficients. These data were entered into a hydrologic budget equation and a balance was achieved.

The hydrologic budget indicates that the recharge rate greatly exceeded the discharge rate for the 2-month period. Seepage and precipitation were the major factors producing the change in storage. Results shows that for the cells adjacent to the river, about 80 percent of the change in storage was from precipitation and 20 percent was from recharge from the river. For the area just west of State Highway 45, about 90 percent of the change in storage was attributed to deep percolation of precipitation. For the western portion of the Zone, almost all (about 98%) of the change in storage was from precipitation.

The analysis showed that the Sacramento River seepage water does not move very far from the river. The river's influence on aquifer storage is limited to approximately 1/2 mile parallel to the river. Seepage is a significant contributor to recharge only in the coarse-grained deposits in the eastern part of the study area, while precipitation is the greater source of recharge in the rest of the area. The hydrologic budget confirmed that discharge from the study area is limited. Increasing the surface and subsurface discharge capacity of the area could greatly reduce groundwater levels. Discharge capacity could be increased by improving surface runoff, improving land surface slopes and providing higher-capacity surface drains, and by installing subsurface drains.

SOLUTIONS: Four alternative drainage plans for alleviating the high groundwater problem were developed. Subsurface drainage systems were provided for each farming unit. The units drained into pipes or ditches and collected in sump areas. Various alternative discharge systems were evaluated. From the sumps drainage was collected into regional discharge lines and pumped to the river or Colusa Drain. Well point collectors were designed for placement adjacent to the river to contain seepage and then pumped back to the river. For some areas surface drainage systems were improved and drained to Colusa Drain. The preferred most economical alternative contained all of these features.

The solutions developed through this effort can be applied to other areas along the Sacramento River with similar high groundwater and seepage problems. Thus, this should be viewed as a model for high groundwater areas along these rivers. State and Federal funding programs and manpower assistance could facilitate the development and implementation of solutions to the seepage and related high groundwater problems that have plagued the Sacramento Valley area for over 75 years.

EVALUATION OF SURFACE-WATER - GROUND-WATER INTERACTION
COEFFICIENTS IN WATER-RESOURCES PLANNING MODEL

Nadeem Shaukat[1]
Manoutch Heidari[2]
Thomas Maddock, III[3]

INTRODUCTION

The planners of water-resources systems are concerned with the identification, selection and scheduling of projects which meet development objectives. A water-resources planning model formulated as a mixed-integer program is used to screen and sequence proposed projects such that the present cost of meeting future water demands is minimized. Surface-water - ground-water interactions are represented in the model by a set of constraints guaranteeing minimum surface-water flow which can be affected by diversions and well-field pumping. When well fields in a planning region are hydraulically connected to streams, consideration must be given to the impact of ground-water pumping on surface-water flows. The streamflow-requirement constraint in the planning model contains surface-water - ground-water interaction coefficients, which relate the quantity of water withdrawn from the stream to the quantity of water pumped by the well fields.

The length of the time period in the planning horizon plays an important role in the simulation of stream-aquifer interaction coefficients. Two approaches for the simulation of these coefficients are investigated. In one approach, long time periods are used, and in the other, shorter time periods are used and the results are aggregated to represent the long time periods. The results obtained from these two approaches are compared.

As expected, the use of stream-aquifer interaction coefficients produced for shorter time periods as compared with those produced for longer time periods showed better results in terms of using the maximum available stream water through well-field pumpage. Furthermore, when the stream-aquifer interaction coefficients were aggregated from smaller time periods of one year to time periods of the planning model, more stream water was shown to be used as compared with the results obtained with non-aggregated stream-aquifer interaction coefficients.

THEORY

When well fields are hydraulically connected with streams in a planning region, a surface-flow-requirement constraint in the water-resources planning model may be required. This constraint contains a surface-water - ground-water interaction coefficient, Ψ, which relates the quantity of water withdrawn from the stream during time

[1]Research Associate, Kansas Geological Survey, Lawrence, KS 66045
[2]M. ASCE, Sen. Sci., Kansas Geological Survey, Lawrence, KS 66045
[3]Professor of Hydrology, University of Arizona, Tucson, AZ 85721

period n and all prior time periods. The stream-aquifer interaction coefficients must be produced for the same time periods as those of the planning periods of the model.

Assume that the aquifer in a planning region possesses the characteristics described below:

1. the saturated thickness of the aquifer is large compared to any drawdown caused by pumpage, i.e. the aquifer's transmissivity is independent of head;

2. water is instantaneously released from storage in the aquifer, i.e. the aquifer's storage coefficient is independent of time;

3. pumping of wells is constant within a given time period, but it may vary from time to time period; and

4. the wells fully penetrate the aquifer, i.e. vertical movement in the aquifer is negligible.

Under these conditions, Maddock (1974) has shown that the stream's cumulative contribution to an aquifer with m pumping wells at time n is

$$F(n) = \sum_{j=1}^{m} \sum_{i=1}^{n} \Psi(j, n-i+1) \, Q(j,i)$$

where

$F(n)$ = cumulative volume of water removed from stream as of the n-th time period

m = number of wells

n = number of time periods

$\Psi(j, n-i+1)$ = fraction of well water supplied by induced flow from the stream to the j-th well in the i-th time period

$Q(j,i)$ = volume of water pumped by the j-th well in the i-th time period.

The Ψ parameters are a function of

1. the river-boundary conditions,
2. the distance of wells to the river,
3. the distance between wells,
4. the well radii,
5. the spatial variability of the aquifer's transmissivity and storage coefficients, and,
6. boundary conditions at points other than the river.

AGGREGATED VS. NON-AGGREGATED INTERACTION COEFFICIENTS

The model shown in Figure 1 is the schematic presentation of a water-resources planning program in Kansas. Its mathematical structure, based on that of Maddock and Moody (1974), is a linear integer programming model seeking to minimize the total cost (capital cost + operation, maintenance and replacement cost) of supplying water to the demand area, D. The numbers in the figure represent the yields, capital costs (CC) and operational, maintenance and replacement costs (OMR) associated with each water resource. The water-planning efforts are underway for the demand area (D), the city of Wichita. The area has a certain demand for treated water which increases with time. The future demands for freshwater were developed from past records of demand and projected forward with consideration of the past and predicted further growth of the area.

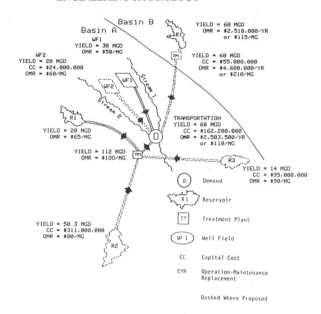

Figure 1:   Schematic Presentation of a Water Planning Project.

Two streams, the Little Arkansas (stream 1) and Arkansas (stream 2) are part of the existing planning model. Four wells with yields of 7.5 million gallons per day (MGD) are located next to stream 1. Well field 2 (WF2), consisting of four wells with yields of 5 MGD, each is also proposed in the model for future demand of the area. The mean streamflows in streams 1 and 2 are 169 MGD and 323 MGD, respectively.

The execution of the model produces optimized pumpages from each water resource based on the demand function for the area, keeping the total project cost for the 20-year planning period at a minimum. A discussion of the planning results is beyond the scope of the paper. Here, only the effects of surface-water - ground-water interaction coefficients produced in two different approaches are compared and reported.

In the first approach, the interaction coefficients are produced for time periods of 5, 4 and 2 years. This was done by setting a time increment, Δt, for calculation of these coefficients equal to 5, 4, and 2 years in a mathematical model designed to calculate these coefficients (see Maddock, 1975; Heidari, 1982 a, b). As expected, the results show that the use of smaller time periods for stream-aquifer interaction coefficients are more responsive than those for longer time periods in terms of producing maximum stream water through well-field pumping. The cumulative water used with time periods of 2 years is 100% as compared to 51.66% for 5-year time periods in WF1 (Figure 2a). In WF2, the cumulative stream water used for 2-year time periods is 41.71%, as compared to 13.80% when time periods of 5 years were used (Figure 3a).

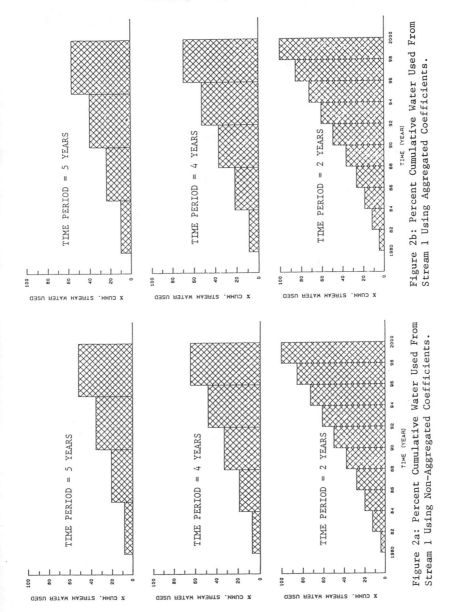

Figure 2b: Percent Cumulative Water Used From Stream 1 Using Aggregated Coefficients.

Figure 2a: Percent Cumulative Water Used From Stream 1 Using Non-Aggregated Coefficients.

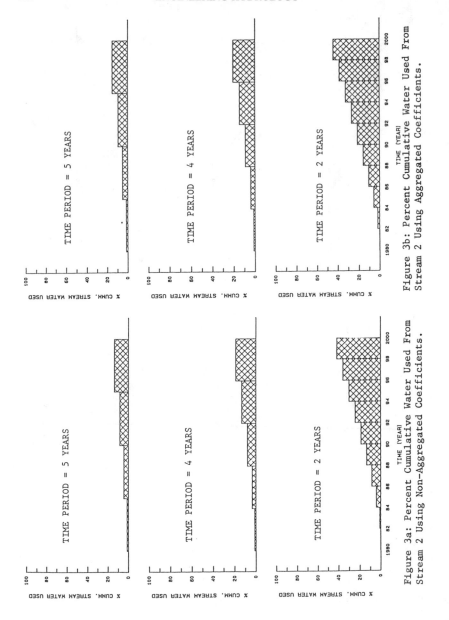

Figure 3b: Percent Cumulative Water Used From Stream 2 Using Aggregated Coefficients.

Figure 3a: Percent Cumulative Water Used From Stream 2 Using Non-Aggregated Coefficients.

In the second approach, the interaction coefficients were first produced for 1-year time periods each and then aggregated to the required time periods of 2, 4 and 5 years for the planning model. This was done following the procedure explained by Maddock (1974) for aggregating the stream-aquifer interaction coefficients. The results obtained using these interaction coefficients show that the stream contribution to pumpages increased up to 5% in certain time periods for WF1, for time periods of 5 and 4 years each. No considerable increase in stream-water use was found with aggregated interaction coefficients for WF1 with a time period of 2 years. This is perhaps because a time period of 2 years is short enough, and aggregation of two 1-year stream-aquifer interaction coefficients does not increase the accuracy to any measureable degree. An increase up to 3% in stream-water use for WF2 also was noticed in time periods of 5, 4 and 2 years.

During this comparison study, each well was treated individually for interaction coefficients. The coefficients for all wells in each well field can also be aggregated depending on the contribution of each well in each time period to reduce the computing time. However, contribution of each well to the well field is hard to assign ahead of time during the planning process.

CONCLUSIONS

1. The stream-aquifer coefficients based on linear system's theory enable us to incorporate the stream contribution to an aquifer into a planning model.

2. The degree of accuracy of these coefficients is directly related to the time period used in the ground-water model for their generation.

3. Experience in this study shows that shorter time periods used in generation of these coefficients produce more accurate stream-aquifer coefficients. However, the short time periods may create a large planning model whose solution may require substantial computing efforts.

4. The stream-aquifer coefficients may be generated for short time periods and aggregated for long time periods. This study shows that such aggregation produces somewhat more accurate stream-aquifer coefficients using aggregated coefficients calculated from short time periods.

APPENDIX - REFERENCES

Heidari, M., "Application of linear system's theory and linear programming to groundwater management in Kansas," Water Resources Bulletin, 18(6), 1003-1012, December, 1982a.

Heidari, M., "Groundwater management options for Pawnee Valley of south-central Kansas": Kansas Geological Survey, Groundwater Series 4, 54 p., 1982b.

Maddock, T., III, "The operation of a stream-aquifer system under stochastic demands": Water Resources Research, 10(1), 1-10, 1974.

Maddock, T., III, and D. W. Moody, "Surface water and ground water interaction phenomena in planning models": presented at the symposium on use of computer techniques and automation for water resources systems, Washington, D. C., March 26-28, 1974.

Maddock, T., III, "A program to compute aquifer response coefficients": U.S. Geological Survey Open-File Report 75-612, 1975.

AN ANALYTICALLY DERIVED DISTRIBUTION FUNCTION FOR FLOOD VOLUME

Van-Thanh-Van Nguyen*, Member, ASCE
Huynh-Ngoc Phien**
Nophadol In-na*

Abstract

A general stochastic model is proposed to determine analytically the distribution function for flood volume. A flood event in this study is defined as an unbroken sequence of consecutive daily flows above a given base level. Accordingly, the volume of a flood could be considered as the sum of all exceedances within its total duration. Analytic expressions for the distribution of flood volume are then derived by assuming that the flow exceedances and flood duration are exponentially distributed random variables. The stochastic model proposed seems to have several advantages over empirical fitting approaches because it could take into account different stochastic properties of the underlying daily stream flow process. Results of a numerical application have demonstrated the adequacy of the analytical approach suggested.

Introduction

Information on the distribution of flood volume is fundamental to successful design and planning of various water resources projects. For instance, when designing storage reservoirs or when considering the water availability for irrigation and water supply purposes, one may wish to know, given the occurrence of a flood, what is the total amount of water that could be expected during the total flood duration. Most previous investigations on flood hydrology involved usually the study of flood occurrences and peak discharges, but very few dealt with flood volume or duration.

In the treatment of flood volume distribution, one common approach is to select a probability distribution function which may appear to fit the observed frequencies of the data. Parameters of such empirical distribution functions are properties of the sample and, usually, have no physical interpretation. The alternative approach is to derive analytically a distribution function for flood volume by taking into account some stochastic properties of the underlying streamflow process. Such analytical procedure is hence more genral and more flexible than the empirical fitting method. The purpose of the present study, therefore, is to suggest an analytical method for determining the probability distribution of flood volume. The method proposed in

*Respectively, Associate Professor and Graduate Research Assistant, McGill University, Montreal, Quebec, Canada H3A 2K6.
**Associate Professor, Asian Institute of Technology, Bangkok, Thailand.

this paper could provide a good fit between observed and theoretical results because it is not necessary to assume that the portion of the hydrograph above a given base level is triangular in shape. This simplistic assumption has often appeared in previous investigations (see, e.g., Todorovic, 1978; Ashkar and Rousselle, 1981).

Theoretical Development

For planning and design purposes, it is usually necessary to consider only those flow discharges whose magnitude exceeds a certain base level. The flows above the given base level will be called "Exceedances". A flood event in this study is defined as an uninterrupted sequence of consecutive exceedances. Accordingly, the total flood duration will be considered as the time between an upcrossing of the selected base level and the next down crossing. Hence, the volume of a flood could be defined as the sum of all exceedances within its total duration.

Let ε_i denote the flow exceedance above a given base level q_o in the i-th day. Consider now an interval of time which consists of n days. In order to consider the volumes of all complete floods that have been observed, all floods are arranged to begin with the first day of the n-day period, and the length of the period considered, must be at least equal to the longest observed flood duration. Let D_n represent the total duration of a flood. The flood volume or the cumulative flow exceedances $V(n)$ during the total duration D_n can be defined as

$$V(n) = \sum_{i=1}^{D_n} \varepsilon_i \qquad (1)$$

Because the flow exceedance ε_i and the flood duration D_n are random variables, the flood volume $V(n)$ is the sum of a random number of random variables. Hence, the distribution function $F_n(x)$ of $V(n)$ can be written as follows (Todorovic and Woolhiser, 1975):

$$F_n(x) = P\{V(n) \leq x\} = \sum_{k=1}^{n} P\{X_k \leq x, D_n = k\} \qquad (2)$$

in which $P\{.\}$ denotes the probability, and $X_k = \sum_{i=1}^{k} \varepsilon_i$ for $k = 1,2,\ldots,n$.

In the above theoretical development, the equation (2) was restricted to daily streamflow process. However, it is readily seen that this restriction is not necessary. The general model, eq. (2), might be used for flow data of any time interval. In the following, for daily streamflow process we consider the following particular cases:

(A) The cumulative exceedance X_k and the flood duration D_n are independent.

In this case we can write:

$$F_n(x) = \sum_{k=1}^{n} P\{X_k \le x\} . P\{D_n = k\} \tag{3}$$

Hence, to determine the distribution function $F_n(x)$, it is necessary to compute the probabilities $P\{X_k \le x\}$ and $P\{D_n = k\}$. In this paper, the flood duration distribution is assumed to be roughly approximated by an exponential distribution. That is, $P\{D_n = k\} = \beta \exp(-\beta k)$. The distribution function $P\{X_k \le x\}$, however, will be analytically derived as will be shown in the following.

If the successive daily flow exceedances $\varepsilon_1, \varepsilon_2, \ldots, \varepsilon_n$ are independent and identically exponentially distributed random variables, it can be readily shown that:

$$F_n(x) = \sum_{k=1}^{n} \left(\frac{\alpha^k}{\Gamma(k)} \int_o^x u^{k-1} e^{-\alpha u} du \right) . (\beta e^{-\beta k}) \tag{4}$$

in which $\Gamma(k) = (k-1)!$ for $k = 1, 2, \ldots, n$, and α is the parameter of the exponential distribution function for daily flow exceedances. However, if there exists a significant correlation between successive values of daily flows, the probability $F_n(x)$ can be written as follows (Nguyen, 1984):

$$F_n(x) = \sum_{k=1}^{n} \left\{ (-1)^{k-1} \sum_{i=1}^{k} \left[\prod_{j \ne 1} (\frac{\lambda_j}{\lambda_i} - 1)^{-1} . (1 - e^{-\frac{\alpha}{\lambda_i} x}) \right] \right\} (\beta e^{-\beta k}) \tag{5}$$

where the λ's are the eigenvalues of the correlation matrix of successive daily flow exceedances, and $\prod (\lambda_j / \lambda_i - 1)^{-1}$ is the product of the $(\lambda_j / \lambda_i - 1)^{-1}$'s for $i, j = 1, 2, \ldots, k$, and $j \ne i$. The parameters α, λ will be estimated from the observed data.

(B)   The cumulative exceedance X_k and the flood duration D_n are dependent

Under this hypothesis we can write

$$F_n(x) = \sum_{k=1}^{n} P\{X_k \le x | D_n = k\} . P\{D_n = k\} \tag{6}$$

In this paper, we assume that X_k and D_n are linearly dependent. That is,

$$X_k = (a D_n + b) \frac{e_k}{100} \tag{7}$$

in which a and b are constant coefficients, and e_k is the random transformed percent residuals. If the distribution of e_k can be adequately approximated by a simple exponential distribution with parameter γ, equation (6) can be written as follows:

$$F_n(x) = \sum_{k=1}^{n} [1-\exp(-\frac{100\gamma}{ak+b} x)].(\beta e^{-\beta k}) \qquad (8)$$

The parameters a,b and γ will be estimated from the observed data.

As mentioned previously, analytical methods for determining flood volume distribution have been used in some previous investigations (Todorovic, 1978; Ashkar and Rousselle, 1981) wherein the hydrograph above a given base flow was assumed to have a triangular shape. In the present study, to compare the calculation of flood volume distribution using the hypothesis of triangular hydrograph with the estimation given by the stochastic models (4), (5) and (8), we assume that the maximum flow exceedance ε_m within a flood can be adequately described by an exponential distribution with parameter θ. Then, assuming ε_m and D_n to be mutually independent we have

$$F_n(x) = 1 - \theta \sqrt{\frac{8x\beta}{\theta}} . K_1(\sqrt{8x\beta\theta}) \qquad (9)$$

where $K_1(.)$ denotes the modified Bessel function of the second kind of order one.

Numerical Application

In the following, to test the descriptive capabilities of the theoretical distribution functions proposed, equations (4), (5), (8) and (9), an illustrative application will be presented using daily streamflow data observed over a period of 28 years (1954-1981) at station 061901 situated on Chamouchouane river in Quebec (Canada). For the moment, only streamflow data in the month of May, representing the flood season in Quebec, are considered. According to the definition of a flood in this study, a total of 29 flood events having durations of from 1 to 60 days have been selected from the 28-year record. The base flow level is 800 m^3/sec.

Figure 1 shows a comparison between observed and theoretical distribution functions for flood volume. In general, except for the model with triangular hydrograph hypothesis, a good agreement between observed and theoretical curves was found for all models (4), (5) and (8). Furthermore, based upon the criterion of maximum deviation between observed and computed distribution functions, the model that could take into account the correlation structure of successive daily flow exceedances, eq. (5), provides the best fit to the observations.

Conclusion

An analytical methodology that can explicityly take into account some observed stochastic properties of an actual streamflow record has been suggested in this study for determining the probability distribution of flood volume. It can be concluded that the proposed

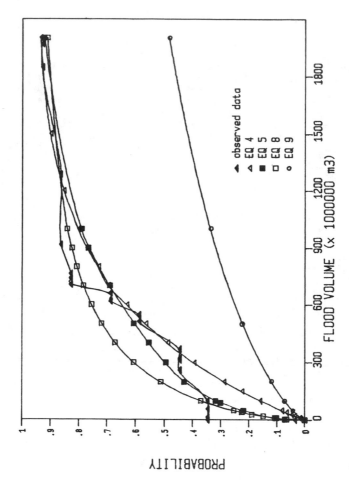

Figure 1 – Observed and theoretical distributions of flood volume
(Chamouchouane River at La Chute aux Saumons, Quebec).

method is more general and more flexible than the empirical fitting approach. Results of a numerical example have indicated the adequacy of the methodology proposed. Furthermore, it has been showed that the simplistic assumption of triangular shape for flood hydrographs above a certain base level, usually appeared in previous investigations, may not be acceptable for estimating the distribution of flood volume.

References

Ashkar, F., and Rousselle, J., A Multivariate Statistical Analysis of Flood Magnitude, Duration and Volume, Statistical Analysis of Rainfall and Runoff, Edited by V.P. Singh, Water Resources Publications, 1981, pp. 651-668.

Nguyen, V-T-V., On Stochastic Characterization of Temporal Storm Patterns, Water Science and Technology, Vol. 16, 1984, pp. 147-153.

Todorovic, P., Stochastic Models of Floods, Water Resources Research, Vol. 14, No. 2, 1978, pp. 345-356.

Todorovic, P., and Woolhiser, D.A., A Stochastic Model of n-Day Precipitation, Journal of Applied Meteorology, Vol. 14, No. 1, 1975, pp. 17-24.

On the Choice Between Annual Flood Series
and Peaks Over Threshold Series
in Flood Frequency Analysis.

Fahim Ashkar(*), Nassir El-Jabi(**) and Bernard Bobée(*)

Statistical techniques are often used to estimate extreme flood events for the design of a flood-control structure or for any other flood-related purpose. Two kinds of models are commonly used in this estimation: (1) Annual maximum (AM) series and (2) peaks over threshold (POT) series. It is not yet clearly known which of these two approaches is more efficient from the statistical point of view. A major drawback of past attempts to compare the statistical efficiency of these two approaches was that these were totally or partially based on large-sample theory. Let Model A be one which focuses directly on AM series and which uses a Gumbel distribution to fit these series. Let Model B be a POT model which uses an exponential distribution for representing "exceedance" magnitude and let the "exceedances" be assumed to be independent. Models A and B are similar in that both lead to a Gumbel distribution for annual maxima, but the estimation of distribution parameters in each model is done in completely different ways. Let X_T be the flood with return period T and let $\hat{X}_T(A)$ and $\hat{X}_T(B)$ be the estimates of X_T under Model A and Model B respectively. Previous studies have compared the variances of $\hat{X}_T(A)$ and $\hat{X}_T(B)$, and have concluded that Model B (POT) is more efficient then Model A (AM) only if the average number of exceedances per year in Model B is higher than about 1.65. We show that this conclusion is not valid for samples of small size, such as those frequently found in hydrology. Results based on small samples are shown to depart considerably from large-sample results.

Introduction

In flood frequency analysis two kinds of flood series are in common use:

(1) Annual maximum (AM) series composed of the largest discharge of each year observed over an N-year period of record;
(2) Peaks over threshold (POT) series, obtained by fixing a base discharge, Q_b, and only observing flood peaks that exceed Q_b. In the POT model, one is interested in the number of flood events that exceed Q_b in a fixed interval of time (o,t) ("flood count"), and also in the magnitude of these flood events, or "exceedances".

(*) Institut national de la recherche scientifique, INRS-Eau, C.P. 7500, Sainte-Foy, Québec, Canada, G1V 4C7.
(**)École de génie, Université de Moncton, New Brunswick, Canada E1A 3E9.

For modeling flood count, the Poisson distribution is very frequently used, whereas for modeling flood magnitude, the exponential distribution is the most commonly employed. The POT model which uses these two distributions shall be called the "POT_{PE}" model, where the "P" in the subscript stands for "Poisson" and the "E" stands for "Exponential".

Although the Poisson and exponential distributions are not the only distributions used in the POT approach, their widespread use warrants some special attention (note for instance that the negative binomial distribution has also been recommended for flood count and the Weibull distribution has been recommended for flood magnitude).

It is well known (Todorovic and Zelenhesic, 1970, Ashkar and El-Jabi, 1987) that under the POT_{PE} model, the maximum annual exceedance follows a truncated Gumbel distribution. The Gumbel distribution can therefore be fitted to annual flood maxima by two ways:

(1) by truncating the flood hydrograph at a certain base level, Q_b, and fitting a POT_{PE} model to the resulting POT series (POT approach);

(2) by considering the untruncated flood hydrograph and fitting the Gumbel distribution directly to the series of annual maxima (AM Approach).

What we will be interested in, is to show which of these two approaches is more "efficient" from the statistical point of view.

The Question of "Efficiency"

Unfortunately, it is very difficult to give a clear-cut definition of the term "efficiency". Some definitions have been given in the literature but none of them is completely satisfactory. Let θ be a certain unknown parameter of a satistical distribution, X, and suppose that θ is estimated by two different methods A and B. Let $\hat{\theta}_A$ and $\hat{\theta}_B$ be the estimates of θ under methods A and B, respectively. Let the variances of $\hat{\theta}_A$ and $\hat{\theta}_B$ be $\sigma_A^2(\hat{\theta})$ and $\sigma_B^2(\hat{\theta})$, respectively. Fisher's definition of relative efficiency of method A with respect to method B is given by

$$Eff(A:B) = \sigma_B^2(\hat{\theta}) \ / \ \sigma_A^2(\hat{\theta}) \qquad (1)$$

Under this definition, method A is said to be more efficient than method B (for estimating θ) if and only if $\sigma_B^2(\hat{\theta}) \ / \ \sigma_A^2(\hat{\theta})$ is greater than 1. It often happens that the two random variables $\hat{\theta}_A$ and $\hat{\theta}_B$ have completely different statistical characteristics. They may for instance follow the same type of distribution but differ in their mean, variance, skewness, etc. The two variables $\hat{\theta}_A$ and $\hat{\theta}_B$ can also differ with respect to the type of distribution they follow. Fisher's definition of relative efficiency (Eq. 1) will be intuitively plausible if the only difference between $\hat{\theta}_A$ and $\hat{\theta}_B$ resides in their variance; i.e. if $\hat{\theta}_A$ and $\hat{\theta}_B$ follow the same type of distribution, with the same skewness, kurtosis, etc. but with different variance. When the form of the distribution of $\hat{\theta}_A$ is different from that of $\hat{\theta}_B$ it becomes very difficult to compare the "efficiency" of the two estimates because in this case comparing $\hat{\theta}_A$ and $\hat{\theta}_B$ will be like comparing apples and oranges. The problem remains, however, that in practice, $\hat{\theta}_A$ and $\hat{\theta}_B$ have to be compared somehow, in order to choose between them.

Traditionally, one of the main uses of estimates like $\hat{\theta}_A$ or $\hat{\theta}_B$ has been to construct confidence intervals for θ, the true but unknown value of the parameter being estimated. These $100\,(1-2\alpha)$ % confidence intervals for θ are often (but not always) put in the form:

$$[L,U] = [\hat{\theta}_A - \kappa_1\sigma_A(\hat{\theta}), \ \hat{\theta}_A + \kappa_1\sigma_A(\hat{\theta})] \qquad \text{under method A} \qquad (2A)$$

$$[L,U] = [\hat{\theta}_B - \kappa_2\sigma_B(\hat{\theta}), \ \hat{\theta}_B + \kappa_2\sigma_B(\hat{\theta})] \qquad \text{under method B} \qquad (2B)$$

where κ_1 and κ_2 are "frequency factors" that depend on the type of distribution that $\hat{\theta}_A$ or $\hat{\theta}_B$ follows.

Let W_A be the width of confidence interval (2A) and let W_B be the width of (2B), so that we have respectively from (2A) and (2B):

$$W_A = 2\kappa_1\sigma_A\,(\hat{\theta}) \qquad (3A)$$

$$W_B = 2\kappa_2\sigma_B(\hat{\theta}) \qquad (3B)$$

Dividing W_B by W_A, we otbain

$$W_B \ / \ W_A = \kappa_1\sigma_B(\hat{\theta}) \ / \ \kappa_2\sigma_A(\hat{\theta}) \qquad (4)$$

In the special case where $\hat{\theta}_A$ and $\hat{\theta}_B$ follow the same type of distribution, but differ only in their variance, we have $\kappa_1 = \kappa_2$, and Eq. 4 becomes:

$$W_B \ / \ W_A = \sigma_B(\hat{\theta}) \ / \ \sigma_A(\hat{\theta} \qquad (5)$$

In this case, a simple relation between Equations 5 and 1 is obtained, namely:

$$(W_B \ / \ W_A)^2 = \text{Eff (A:B)}$$

so that the ratio $(W_B \ / \ W_A)$ can be used as a measure of the relative efficiency of method B with respect to method A in the Fisher sense.

In the present study, we shall use the ratio $W_B \ / \ W_A$ as a measure of relative efficiency, although we shall not necessarily express this ratio in the form of Eq. 4 or Eq. 5. Let us now put the above mathematical formulations into an hydrological context.

Suppose that for designing a flood-control structure, we are interested in estimating a flood event, X_T, corresponding to a return period T. Here the unknown parameter (θ) that interests us is X_T, and its estimate $(\hat{\theta})$ is $\hat{X}_T$. Let $\hat{X}_T$ be obtained by two methods:

Method A :   The Gumbel distribution fitted to annual flood maxima (AM
             approach)
Method B :   The POT$_{PE}$ model applied to the POT series (POT approach)

Confidence intervals for X_T under methods A and B can be constructed
by two different approaches:

(1) Approach based on normal-asymptotic theory (large-sample
    approach);
(2) Approach based on small samples.

   Cunnane (1973) used the first approach to calculate W_A and W_B, and
by computing the ratio (W_B / W_A) [which is equal to $\sigma_B(\hat{X}_T)$ / $\sigma_A(\hat{X}_T)$ in
this case] it was concluded that method B (POT$_{PE}$ model) is more
efficient than method A if and only if the average number of
exceedances per year in the POT model is more than about 1.65.   We
shall see that using small sample procedures can lead to conclusions
quite different from those obtained by Cunnane.   The small-sample
comparison of methods A and B is based on two recent studies:

(1) A study by Ashkar and El-Jabi (1987) which obtains small-sample
    confidence intervals for X_T under the POT$_{PE}$ model;

(2) A study by Bain and Engelhardt (1981) which obtains small-sample
    confidence intervals for X_T under a Gumbel distribution.

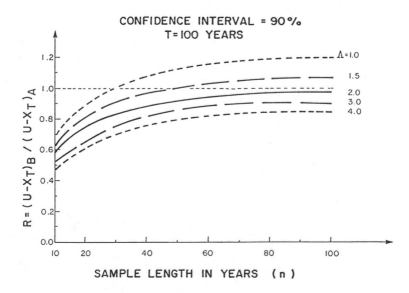

Figure 1. Relative efficiency of the Gumbel distribution with respect
          to the POT$_{PE}$ model.

It was observed from these two studies that the distribution of $\hat{X}_T$ under a Gumbel distribution (Model A) tends to normality considerably slower than under the POT_{PE} model (Model B). For sample sizes typically found in hydrology, upper and lower confidence limits (U and L respectively) for X_T were often found to be not well-centered with respect to X_T ($U-X_T$ typically greater than X_T-L) due to non-normality of $\hat{X}_T$ for these small sample sizes. If attention is restricted to $U-X_T$, which is important for design, and the ratio $R = (U-X_T)_B$ / $(U-\hat{X}_T)_A$ is plotted as in Figure 1, for return period $T = 100$ years, then it can be seen that for small sample size (n < 30 in Fig. 1) the POT model can be more efficient than the AM model (ratio R less than unity) even when the mean number of exceedances per year (Λ) in the POT model is as low as, or even lower than, unity. This result departs from previous results obtained by large-sample theory, which showed that for the POT approach to be more efficient than the AM approach, Λ should be greater than about 1.65. It appears therefore from the results of the present study that the POT approach has probably some favorable characteristics which are not yet evident from previous studies and which should be examined further in future research.

Appendix 1. - References

Ashkar, F. and El-Jabi, N. (1987). "Tables for calculating confidence limits for design flood events under a partial duration series model". Submitted to Journal of Hydrology.

Bain, L.J. and M. Engelhardt (1981)."Simple approximate distributional results for confidence and tolerance limits for the Weibull distribution based on maximum likelihood estimators. "Technometrics, 23(1): 15-20.

Cunnane, C. (1973). "A particular comparison of annual maxima and partial duration series methods of flood frequency prediction." Journal of Hydrology, 18: 257-271.

Todorovic, P. and E. Zelenhasic (1970). "A stochastic model for flood analysis". Water Resource Research, Vol. 6(6): 1641-1648.

A VERSATILE METHODOLOGY FOR FLOOD FREQUENCY ANALYSIS

by Krishan P. Singh,[*] M. ASCE

Abstract

Many observed annual flood series exhibit reverse curvatures when plotted on lognormal probability paper. The occurrence of these curvatures may be attributed to storm and basin factors. A mixed distribution model is needed to analyze such flood series. A versatile flood frequency methodology has been developed which considers power normal, log-Pearson type III, and mixed distribution. Objective detection and modification of outliers/inliers is an integral part of this methodology.

Introduction

Many observed annual flood series exhibit reverse curvatures when plotted on lognormal probability paper. None of the commonly used distributions fits an observed flood series with reverse curvature which may be attributed to seasonal variation in flood-producing storms, dominance of within-the-channel or floodplain flow, and variation in antecedent soil moisture and cover conditions. A mixed distribution model can be used to analyze such flood series. There is no physical rationale for assuming a unique underlying distribution. Distribution parameters can be significantly biased if the flood series has outliers and/or inliers. Such outliers and inliers need to be detected and modified. No detection and modification methodology currently exists except the one developed by Singh and Nakashima (1981).

Mixed Distribution

The magnitude of annual flood peak depends largely on storm and basin factors. Relevant storm factors are the type of storm and the intensity and duration of the storm. Basin factors mainly include relative dominance of within-the-channel or floodplain flow, the antecedent soil moisture condition, and the vegetal cover. Interaction between the distributions of these factors varying form storm to storm and from season to season may produce a flood series resembling a conventional distribution or one exhibiting marked reverse curvature to be dealt with by the mixed-distribution concept.

The mixed distribution model (Singh and Nakashima, 1981) considers the observed annual maximum floods or their logarithms to belong to two populations with means μ_1 and μ_2, variances σ_1^2 and σ_2^2, and relative weights a and $1-a$. Then,

---

[*]Principal Scientist, Illinois State Water Survey, Champaign, IL 61820

$$p(x) = a\, p_1(x) + (1-a)\, p_2(x)$$

$$p_1(x) = \frac{1}{\sigma_1 \sqrt{2\pi}} \int_{-\infty}^{x} \exp\left[-\frac{(x'-\mu_1)^2}{2\sigma_1^2} \right] dx'$$

$$p_2(x) = \frac{1}{\sigma_2 \sqrt{2\pi}} \int_{-\infty}^{x} \exp\left[-\frac{(x'-\mu_2)^2}{2\sigma_2^2} \right] dx'$$

in which p is the probability of exceedance and x=log Q where Q is the annual maximum flood. The parameters μ_1 and μ_2, σ_1, σ_2, and a are linked to mean, standard deviation, and skew of observed flood series by three equations (Cohen, 1967). The parameter values can be obtained with a nonlinear programming algorithm (Singh and Nakashima, 1981).

Outliers and Inliers: Test Statistics

    An outlier in a set of data is defined as an observation or subset of observations which appears to be inconsistent with the remainder of that set of data (Barnett and Lewis, 1978). The inconsistency can be interpreted as the observation being either significantly higher or lower at the high end than the value indicated by the rest of the data, in which case the observation is termed an outlier or an inlier, respectively. If the observation is significantly higher or lower than expected at the low end, it will be termed an inlier or outlier, respectively. Development of suitable statistical tests and methods for objective detection and modification of outliers and inliers is necessary to improve the reliability of flood frequency analyses. Test statistics were developed from extensive Monte Carlo experiments on the basis of departures derived by generating millions of standard normal deviates. The departure is defined as the theoretical standard normal deviate corresponding to a given probability p minus the corresponding generated standard normal deviate.

    The distribution of departures for both the low and high ends is shown in Figure 1. Percent exceedance probability corresponds to percent significance level. For the same level of significance, the outliers have higher departures than the inliers. The outlier can depart considerably from the assumed underlying distribution, but the inlier departs by a lesser amount because the next observation can replace an inlier. For use of the developed departure values, a given flood series has to be transformed to a normally distributed series in order to detect any outliers/inliers.

Detection and Modification of Outliers/Inliers

    The flood series is transformed to resemble a series distributed as $N(\mu, \sigma^2)$ by using power transformation:

$$y_i = (Q_i^{\lambda} - 1) / \lambda; \ \lambda \neq 0, \text{ and } \ y_i = \log Q_i; \ \lambda = 0$$

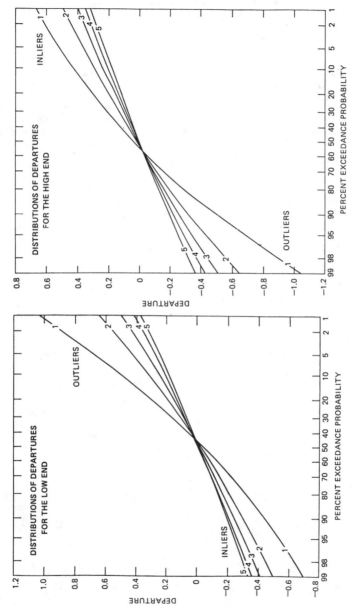

Figure 1. Distribution of departures for both the high and low end

in which Q is the annual flood and λ is the transformation parameter. Value of λ can be obtained with the maximum log-likelihood method. The skew of the y series is very close to zero but the kurtosis may be different from that for a normal distribution. The detection and modification begins at significance or probability level of 0.01. Any outlier/inliers detected are modified at that level. The resulting detransformed series is analyzed to test the departures and the procedure is repeated, if necessary, to ensure that there are no outliers/inliers at the 0.01 significance level. The procedure is followed sequentially from one level to the next and desired distribution statistics and floods are computed before moving to the next level. If no outliers/inliers are detected at a level, no modifications are done at that level. Significance or probability levels of 0.01, 0.05, 0.10, 0.20, and 0.30, or their complements, are used. These levels are designated 1 through 5.

Versatile Flood Frequency Methodology

A computer program has been developed (Singh and Nakashima, 1981) for objective detection and modification of any outliers/inliers from levels 1 through 5 and for computation of various frequency floods with 1) the power transformation method, 2) the log-Pearson type III method, and 3) the mixed distribution method. In the power transformation, the T-yr flood is computed from

$$Q_T = (\lambda \, y_T + 1)^{1/\lambda} \; ; \quad y_T = \overline{y} + Z_T \, s_y$$

in which $\overline{y}$ and s_y are the mean and standard deviation of the y series and Z_T is the standard deviate associated with nonexceedance probability, $1 - 1/T$. The Z_T can be corrected for kurtosis.

With log-Pearson type III, the T-yr flood is obtained from

$$Q_T = (\overline{x} + k_T s)^{10}$$

in which $\overline{x}$ and s are the mean and standard deviation of log-transformed floods and k_T is the frequency factor which depends on both skew and T.

With the mixed distribution, the x for a desired value of p is obtained by finding p for a trial value of x from the equation

$$p(x) = a \, p_1(x) + (1-a) \, p_2(x)$$

and converging to the desired value of x through a fast reiterative process. The program prints any modification of detected outliers/inliers at various levels of significance, various frequency floods, and distribution statistics.

Examples

Annual flood series for the Salt River near Chrysotile, Arizona, and Beetree Creek near Swannanoa, North Carolina are selected to illustrate the results obtained with the proposed versatile flood frequency methodology. The respective drainage areas are 2849 sq mi (7378 sq km) and 5.46 sq mi (14.14 sq km) and the years of record are 60 and 45.

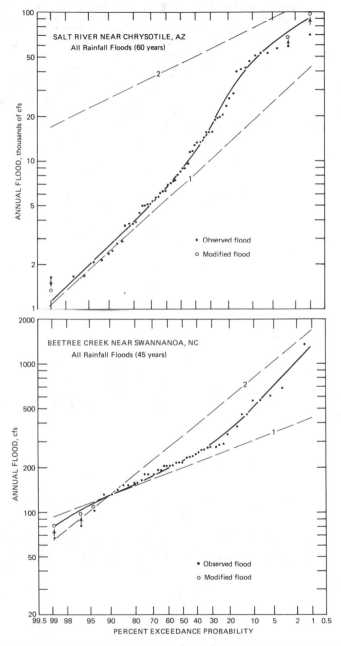

Figure 2.   Observed floods and fitted mixed distributions

Out of 60 annual floods for the Salt River (during the period 1925-1984), 27 occurred in January-March, 11 in August, and none in May, June, or November. Of the 12 highest floods, 10 occurred in December-March and 2 in October; and of the 12 lowest floods, 8 occurred in July-August. At level 4 (significance level of 0.2), the two highest floods were detected as inliers and increased from 70,400 and 58,300 cfs to 97,202 and 67,972 cfs, respectively; but only the lowest flood was detected as an inlier, and was reduced from 1640 to 1394 cfs (1 cfs = 0.0283 m^3/s). The fitted MD curve together with the two component distributions, as well as observed floods and modification of the inliers, are shown in Figure 2. The five parameters a, μ_1, μ_2, σ_1, σ_2 are 0.807, 3.831, 4.658, 0.343, and 0.185, respectively.

A maximum of 9 floods out of a total of 45 for Beetree Creek occurred in March and none in November over the period 1927-1971. High and low floods were distributed over various months. At level 4, the highest flood of 1370 cfs was detected as a minor outlier and modified to 1330 cfs. Three minor outliers and one inlier were detected at the low end and these floods of 67, 81, 102, and 131 cfs were modified to 81, 96, 106, and 121 cfs, respectively (1 cfs = 0.0283 m^3/sec). The fitted MD curve together with the two component distributions, as well as observed floods and modifications of outliers/inliers are shown in Figure 2. The five parameters a, μ_1, μ_2, σ_1, σ_2 are 0.634, 2.305, 2.514, 0.143, and 0.307, respectively.

Conclusions

Mixed populations and mixed distributions can occur because of a host of storm and basin factors. Any outliers and inliers in an observed flood series can seriously bias the distribution parameters. An objective method for detection and modification of any outliers and inliers at various significance levels has been developed and computerized. It is an essential part of the versatile flood frequency methodology developed. Mixed distribution parameters are determined with a nonlinear programming algorithm. The mixed distribution can fit most of the shapes of the observed flood series.

References

Barnett, V. and Lewis, T., 1978, "Outliers in Statistical Data." John Wiley and Sons, New York, NY.

Cohen, A.C., 1967, "Estimation in Mixtures of Normal Distributions." Technometrics, Vol. 9, No. 1, 1967, pp. 15-28.

Singh, K.P., and Nakashima, M., 1981, "A New Methodology for Flood Frequency Analysis with Objective Detection and Modification of Outliers/Inliers." State Water Survey C.R. 272, Champaign, IL.

CONFIDENCE INTERVALS FOR FLOOD CONTROL DESIGN

T. V. Hromadka II, M.ASCE*

Abstract

The line drawn and labeled as the flood frequency curve is seldom identified as to what confidence is associated to the plot. For example while flood control designs are typically based on the flood frequency curve Q_{100} estimate, seldom is it considered that, on the average, and with the other factors and parameters being correct, this Q_{100} estimate has only a 50 percent chance of being greater than the true (unknown) Q_{100} but equivalently has a 50 percent chance of being less than the true Q_{100}.

Consequently, if the goal is to provide protection against Q_{100}, and this level of protection is adopted as the local policy statement for all design purposes, and there is liability should Q_{100} flooding occur, then confidence intervals should be incorporated into the flood control policy statement.

INTRODUCTION

In this paper, the unit hydrograph method (UH) is used to develop estimates of runoff modeling error in the frequently occurring cases where the uncertainty in the rainfall distribution over the catchment dominates all other sources of modeling uncertainty. Indeed, the uncertainty in the precipitation distribution appears to be a limiting factor in the successful development, calibration, and application of all surface runoff hydrologic models (e.g., Loague and Freeze, 1985; Beard and Chang, 1979; Schilling and Fuchs, 1986; Garen and Burges, 1981; Nash and Sutcliffe, 1970; Troutman, 1982).

Schilling and Fuchs (1986) write "that the spatial resolution of rain data input is of paramount importance to the accuracy of the simulated hydrograph" due to "the high spatial variability of storms" and "the amplification of rainfall sampling errors by the nonlinear transformation" of rainfall into runoff. They recommend that a model should employ a simplified surface flow model if there are many sub-basins; a simple runoff coefficient loss rate; and a diffusion (zero inertia) or storage channel routing technique.

In their study, Schilling and Fuchs (1986) reduced the rainfall data set resolution from a grid of 81 gages to a single catchment-centered gage in an 1,800 acre catchment. They noted that variations in runoff volumes and peak flows "are well above 100 percent over the entire range of storms implying that the spatial resolution of rainfall has a dominant influence on the reliability of computed runoff." It is also noted that "errors in the rainfall input are amplified by the rainfall-runoff transformation so that "a rainfall depth error of 30 percent results in a volume error of 60 percent and a peak flow error of 80 percent." They also write that "it is inappropriate to use a sophisticated runoff model to achieve a desired level of modeling accuracy if the spatial resolution of rain input is low" (in their study, the raingage densities considered for the 1,800-acre catchment are 81, 9, and a single centered gage).

Similarly, Beard and Chang (1979) write that in their study of 14 urban catchments, complex models such as continuous simulation typically have 20 to 40 parameters and

---

1   Director of Water Resources Engineering, Williamson and Schmid, Irvine, California and Research Associate, Princeton University, New Jersey

functions that must be derived from recorded rainfall-runoff data. "Inasmuch as rainfall data are for scattered point locations and storm rainfall is highly variable in time and space, available data are generally inadequate for reliably calibrating the various interrelated functions of these complex models."

In the extensive study by Loague and Freeze, (1985), three event-based rainfall-runoff models (a regression model, a unit hydrograph model, and a kinematic wave quasi-physically based model) were used on three data sets of 269 events from three small upland catchments. In that paper, the term "quasi-physically based", or QPB, is used for the kinematic wave model. The three catchments were 25 acres, 2.8 square miles, and 35 acres in size, and were extensively monitored with rain gage, stream gage, neutron probe, and soil parameter site testing. For example, the 25 acre site contained 35 neutron probe access sites, 26 soil parameter sites (all equally spaced), an on-site rain gage, and a stream gage. The QPB model utilized 22 overland flow planes and four channel segments. In comparative tests between the three modeling approaches to measured rainfall-runoff data it was concluded that all models performed poorly and that the QPB performance was only slightly improved by calibration of its most sensitive parameter, hydraulic conductivity. They write that the "conclusion one is forced to draw...is that the QPB model does not represent reality very well; in other words, there is considerable model error present. We suspect this is the case with most, if not all conceptual models currently in use." Additionally, "the fact that simpler, less data intensive models provided as good or better predictions than a QPB is food for thought."

Based on the literature, the main difficulty in the use, calibration, and development, of complex models appears to be the lack of precise rainfall data and the high model sensitivity to (and magnification of) rainfall measurements errors. Nash and Sutcliffe(1970) write that "As there is little point in applying exact laws to approximate boundary conditions, this, and the limited ranges of the variables encountered, suggest the use of simplified empirical relations."

CATCHMENT AND DATA DESCRIPTION

Let R be a free draining catchment with negligible detention effects. R is discretized into m subareas, R_j, each draining to a nodal point which is drained by a channel system. The m-subarea link node model resulting by combining the subarea runoffs for storm i, adding runoff hydrographs at nodal points, and routing through the channel system, is denoted as $Q_m^i(t)$. It is assumed that there is only a single rain gage and stream gage available for data analysis. The rain gage site is monitored for the 'true' effective rainfall distribution, $e_g^i(t)$. The motivation in using a measured $e_g^i(t)$ at the rain gage site is to avoid the necessity of using a multiparameter submodel to approximate $e_g^i(t)$; rather we assume that an accurate value of $e_g^i(t)$ is available, even though this data is measured at the rain gage site which may be located outside of the catchment. The stream gage data represents the entire catchment, R, and is denoted by $Q_g^i(t)$ for storm event i.

LINEAR EFFECTIVE RAINFALLS FOR SUBAREAS

The effective rainfall distribution (rainfall less losses) in R_j is given by $e_j^i(t)$ for storm i where $e_j^i(t)$ is assumed to be linear in $e_g^i(t)$ by:

$$e_j^i(t) = \sum \lambda_{jk}^i e_g^i(t - \theta_{jk}^i), \quad j = 1,2,\cdots,m \qquad (1)$$

where λ_{jk}^i and θ_{jk}^i are coefficients and timing offsets, respectively, for storm i and subarea R_j. In Eq. (1), the variations in the effective rainfall distribution over R due to magnitude and timing are accounted for by the λ_{jk}^i and θ_{jk}^i, respectively. As an alternative to Eq. (1), the $e_g^i(t)$ may be defined as a set of unit effective rainfalls, each unit associated with its own proportion factor; however for simplicity, the use of the entire $e_g^i(t)$ function will be carried forward in the model development.

SUBAREA RUNOFF

The storm i subarea runoff from R_j, $q_j^i(t)$, is given by the linear convolution integral:

$$q_j^i(t) = \int_{s=0}^{t} e_j^i(t - s)\, \phi_j^i(s)\, ds \tag{2}$$

where $\phi_j^i(s)$ is the subarea unit hydrograph (UH) for storm i such that Eq. (2) applies. Combining Eqs. (1) and (2) gives

$$q_j^i(t) = \int_{s=0}^{t} \sum e_g^i(t - \theta_{jk}^i - s)\, \lambda_{jk}^i\, \phi_j^i(s)\, ds \tag{3}$$

Rearranging variables,

$$q_j^i(t) = \int_{s=0}^{t} e_g^i(t - s) \sum \lambda_{jk}^i\, \phi_j^i(s - \theta_{jk}^i)\, ds \tag{4}$$

where throughout this paper, the argument of the arbitrary frunction $F(s - Z)$ is notation that $F(s - Z) = 0$ for $s < Z$.

LINEAR ROUTING

Let $I_1(t)$ be the inflow hydrograph to a channel flow routing link (number 1), and $Q_1(t)$ the outflow hydrograph. A linear routing model of the unsteady flow routing process is given by

$$O_1(t) = \sum_{k_1=1}^{n_1} a_{k_1}\, I_1(t - \alpha_{k_1}) \tag{5}$$

where the a_{k_1} are coefficients which sum to unity; and the α_{k_1} are timing offsets. Again, $I_1(t - \alpha_{k_1}) = 0$ for $t < \alpha_{k_1}$. Given stream gage data for $I_1(t)$ and $O_1(t)$, the best fit values for the a_{k_1} and α_{k_1} can be determined.

Should the above outflow hydrograph, $O_1(t)$, now be routed through another link (number 2), then $I_2(t) = O_1(t)$ and from the above

$$O_2(t) = \sum_{k_2=1}^{n_2} a_{k_2}\, I_2(t - \alpha_{k_2})$$

$$\tag{6}$$

$$= \sum_{k_2=1}^{n_2} a_{k_2} \sum_{k_1=1}^{n_1} a_{k_1}\, I_1(t - \alpha_{k_1} - \alpha_{k_2})$$

For L links, each with their own respective stream gage routing data, the above linear routing technique results in the outflow hydrograph for link number L, $Q_L(t)$, being given by

$$O_L(t) = \sum_{k_L=1}^{n_L} a_{k_L} \sum_{k_{L-1}=1}^{n_{L-1}} a_{k_{L-1}} \cdots \sum_{k_2=1}^{n_2} a_{k_2} \sum_{k_1=1}^{n_1} a_{k_1}\, I_1(t - \alpha_{k_1} - \alpha_{k_2} - \cdots - \alpha_{k_{L-1}} - \alpha_{k_L}) \tag{7}$$

Using the vector notation, the above $O_L(t)$ is written as

$$O_L(t) = \sum_{<k>} a_{<k>} I_1(t - \alpha_{<k>}) \tag{8}$$

For subarea R_j, the runoff hydrograph for storm i, $q_j{}^i(t)$, flows through L_j links before arriving at the stream gage and contributing to the total measured runoff hydrograph, $Q_g{}^i(t)$. All of the constants $a^i{}_{<k>}$ and $\alpha^i{}_{<k>}$ are available on a storm by storm basis. Consequently from the linearity of the routing technique, the m-subarea link node model is given by the sum of the m, $q_j{}^i(t)$ contributions,

$$Q_m{}^i(t) = \sum_{j=1}^{m} \sum_{<k>_j} a^i{}_{<k>_j} \, q_j{}^i(t - \alpha^i{}_{<k>}) \tag{9}$$

where each vector $<k>_j$ is associated to a R_j, and all data is defined for storm i. It is noted that in all cases,

$$\sum_{<k>_j} a^i{}_{<k>_j} = 1 \tag{10}$$

LINK-NODE MODEL, $Q_m{}^i(t)$, AND MODEL REDUCTION

For the above linear approximations for storm i, Eqs. (1), (4), and (9) can be combined to give the final form for the m subarea link-node model, $Q_m{}^i(t)$.

$$Q_m{}^i(t) = \sum_{j=1}^{m} \sum_{<k>_j} a^i{}_{<k>_j} \int_{s=0}^{t} e_g{}^i(t-s) \sum \lambda_{jk}^i \, \phi_j{}^i(s - \theta_{jk}^i - \alpha^i{}_{<k>_j}) \, ds \tag{11}$$

Because the measured effective rainfall distribution, $e_g{}^i(t)$, is independent of the several indices, Eq. (11) is rewritten in the form

$$Q_m{}^i(t) = \int_{s=0}^{t} e_g{}^i(t-s) \sum_{j=1}^{m} \sum_{<k>} a^i{}_{<k>_j} \sum \lambda_{jk}^i \, \phi_j{}^i(s - \theta_{jk}^i - \alpha^i{}_{<k>_j}) \, ds \tag{12}$$

where all parameters are evaluated on a storm by storm basis, i.

Equation (12) described a model which represents the total catchment runoff response based on variable subarea UH's, $\phi_j{}^i(s)$; variable effective rainfall distributions on a subarea-by-subarea basis with differences in magnitude (λ_{jk}^i), timing (θ_{jk}^i), and pattern shape (linear assumption); and channel flow routing translation and storage effects (parameters $a^i{}_{<k>_j}$ and $\alpha^i{}_{<k>_j}$). All parameters employed in Eq. (12) must be evaluated by runoff data where stream gages are supplied to measure runoff from each subarea, R_j, and stream gages are located upstream and downstream of each channel reach (link) used in the model.

The m-subarea model of Eq. (12) is directly reudced to the simple single area UH model (no discretization of R into subareas) given by $Q_1{}^i(t)$ where

$$Q_1{}^i(t) = \int_{s=0}^{t} e_g{}^i(t-s) \, \eta^i(s) \, ds \tag{13}$$

where $n^i(s)$ is the correlation distribution between the data pair $\{Q_g^i(t), e_g^i(t)\}$, for storm event i.

STORM CLASSIFICATION SYSTEM

To proceed with the analysis, the full domain of effective rainfall distributions measured at the rain gage site are categorized into storm classes, $\langle \xi_x \rangle$. Because the storm classifications are based upon effective rainfalls, the measured precipitations, $P_g^i(t)$, may vary considerably yet produce similar effective rainfall distributions. That is, any two elements of a class $\langle \xi_x \rangle$ would result in nearly identical effective rainfall distributions at the rain gage site, and hence one would "expect" nearly identical runoff hydrographs recorded at the stream gage. Typically, however, the resulting runoff hydrographs differ and, therefore, the randomness of the effective rainfall distribution over the catchment, R, results in variations in the modeling "best-fit" parameters (i.e., in $Q_1^i(t)$, the $n^i(s)$ variations) in correlating the available rainfall-runoff data.

More precisely, any element of a specific storm class $\langle \xi_0 \rangle$ has the effective rainfall distribution, $e_g^0(t)$. However, there are several runoffs associated to the single $e_g^0(t)$, and are noted by $Q_g^{0i}(t)$. In correlating $\{Q_g^{0i}(t), e_g^0(t)\}$, a different $n^i(s)$ results due to the variations in the measured $Q_g^{0i}(t)$ with respect to the single known input at the rain gage site, $e_g^0(t)$.

In the predictive mode, where one is given an assumed (or design) effective rainfall distribution, $e_g^D(t)$, to apply at the rain gage site, the storm class of which $e_g^D(t)$ is an element of is identified, $\langle \xi_D \rangle$, and the predictive output for the input, $e_g^D(t)$, must necessarily be the random variable or distribution,

$$[Q_1^D(t)] = \int_{s=0}^{t} e_g^D(t - s) \, [n(s)]_D \, ds \tag{14}$$

where $[n(s)]_D$ is the distribution of $n^i(s)$ distributions associated to storm class $[\xi_D]$.

THE VARIANCE OF HYDROLOGIC MODEL OUTPUT

Consider the $Q_1^i(t)$ model structure in correlating the single rain gage and stream gage. For storm class $\langle \xi_0 \rangle$, there is an associated distribution of correlation distributions, $[n(s)]_0$. Then in the predictive mode, the predicted hydrologic model output is the distribution $[Q_1^0(t)]$ where

$$[Q_1^0(t)] = \int_{s=0}^{t} e_g^0(t - s) \, [n(s)]_0 \, ds \tag{15}$$

For storm time z, the distribution of flow rate values is by $[Q_1^0(z)]$, where

$$[Q_1^0(z)] = \int_{s=0}^{t} e_g^0(z - s) \, [n(s)]_0 \, ds \tag{16}$$

Let t_p be the storm time where the peak flow rate, Q_p, occurs for storm class $\langle \xi_0 \rangle$. Noting that t_p is a function of $[n(s)]_0$, then the distribution of $[Q_p]_0$ is given by

$$[Q_p]_0 = \int_{s=0}^{t_p} e_g^0(t_p - s) \, [n(s)]_0 \, ds \tag{17}$$

Let $\mathcal{D}$ be a single time duration. Of interest is the maximum volume of runoff during duration, $\mathcal{D}$, for storm class $<\xi_0>$. Then the distribution of this estimate is given by

$$[\max_{\mathcal{D}} \int_{\mathcal{D}} Q_1^{\;0}(t)dt] = \max \int_{\mathcal{D}} \int_{s=0}^{t} e_g^{\;0}(t - s)[\eta(s)]_0 \; ds \tag{18}$$

Let A be an operator which represents a hydrologic process algorithm (e.g., detention basin, etc.). Then the output of the operator for storm class $<\xi_0>$ is the distribution

$$[A]_0 = A \left(\int_{s=0}^{t} e_g^{\;0}(t - s) \; [\eta(s)]_0 \; ds \right) \tag{19}$$

The expected value of the hydrologic process A for storm class $<\xi_0>$ is

$$E[A]_0 = \sum_{[\eta(s)]_0} A \left(\int_{s=0}^{t} e_g^{\;0}(t - s) \; \eta(s) \; ds \right) P(\eta(s)) \tag{20}$$

where $P(\eta(s))$ is the frequency of occurrence for distribution $\eta(s)$ in $[\eta(s)]_0$. The variance of predictions of hydrologic process A for storm class $<\xi_0>$ is (for A ( ) being a mapping into the real number line; i.e., giving a single number result),

$$var[A]_0 = \sum_{[\eta(s)]_0} \left(A(\int_{s=0}^{t} e_g^{\;0}(t - s) \; \eta(s) \; ds) - E[A]_0 \right)^2 P(\eta(s)) \tag{21}$$

CONCLUSIONS

A lower bound for estimating the distribution of uncertainty in surface runoff modeling output is advanced. The bound is based on a linear unit hydrograph approach, which utilizes an arbitrary number of catchment subdivisions into subareas, a linear routing technique for channel flow effects, a variable effective rainfall distribution over the catchment, and calibration parameter distributions developed in correlating rainfall-runoff data by the model. Because all hydrologic parameters (e.g., subarea unit hydrographs, channel routing parameters, effective rainfall distribution factors) vary on a storm basis, the unit hydrograph methodology is a reasonable approximation for assessing uncertainty in hydrologic modeling estimates. The uncertainty bound developed reflects the dominating influence of the unknown rainfall distribution over the catchment and is expressed as a distribution function which can be reduced only by supplying additional rainfall-runoff data. It is recommended that this uncertainty distribution be included in flood control design studies in order to incorporate prescribed levels of confidence in flood protection facilities.

APPENDIX I - REFERENCES

1. Beard, L. and Chang, S., Urbanization Impact on Stream Flow, ASCE, Journal of the Hydraulics Division, June, 1979.
2. Doyle, Jr., W. H., Sherman, J. O., Stiltner, G. J. and Krug, W. R., A Digital Model for Streamflow Routing by Convolution Methods, U.S. Geological Survey, Water-Resources Investigation Report 83-4160, 1983.
3. Loague, K. and Freeze, R., A Comparison of Rainfall-Runoff Modeling Techniques on Small Upland Catchments, Water Resources Research, Vol. 21, No. 2, Feb., 1985.
4. Nash, J. and Sutcliffe, J., River Flow Forecasting Through Conceptual Models, Part I - A Discussion of Principles, Journal of Hydrology, Vol. 10, 1970.
5. Schilling, W. and Fuchs, L., Errors in Stormwater Modeling - A Quantitative Assessment, ASCE Journal of Hydraulic Engineering, Vol. 112, No. 2, Feb., 1986.
6. Troutman, B., An Analysis of Input in Precipitation - Runoff Models Using Regression with Errors in the Independent Variables, Water Resources Research, Vol. 18, No. 4, Aug., 1982.

Comparison of Hydrograph Simulation Techniques Used in Structural Design and Watershed Systems Modeling

By
Clarence H. Robbins[1], Andrew J. Reese,[2] M. ASCE and W. Brian Bingham[3]

ABSTRACT

The sizing and design of certain hydraulic structures, watershed model calibration, and other hydrologic design procedures often requires a hydrograph for a large flood with a known peak discharge that may have occurred in past years, or for a design flood of specified peak discharge and return period. At sites where actual hydrograph data are not available, it becomes necessary to estimate a typical or average hydrograph. Traditional methods of unit hydrograph analysis can be used for flood hydrograph estimation, but these require rainfall excess data and convolution techniques to derive the flood hydrograph. Recent studies show that a Inman hydrograph technique can be used to estimate the flood hydrograph directly from peak discharge and basin lagtime. This paper describes a basin study in which traditional unit hydrograph analysis methods were compared to the Inman hydrograph technique. Study results indicate that, where suitable estimates of lagtime and peak discharge can be obtained, the recently developed Inman hydrograph approach can be accurate and cost effective. This is particularly true where time, cost, or data availability constraints preclude the use of other methods and where preliminary flow volume estimates must be made. It was also evident that lagtime is a critical factor and errors in this parameter are directly related to errors in hydrograph width.

INTRODUCTION

The sizing and design of hydraulic structures in rivers, watershed model calibration, and hydrologic design for land development in metropolitan areas often requires a hydrograph for a large flood with a known peak discharge that may have occurred in past years, or for a design flood of specified peak discharge and return period. Typically, the Engineer must select a hydrograph estimation method which will provide the required accuracy, fit the date availability constraints, and be cost effective. Frequently, these criteria are difficult to satisfy because a combination of several hydrograph estimation methods must be employed in order to verify study results.

NOTES:

[1] Senior Hydrologist, MCI Consulting Engineers, Inc., Nashville, TN 37211.

[2] Project Manager, MCI Consulting Engineers, Inc., Nashville, TN 37211.

[3] Engineering Technician, MCI Consulting Engineers, Inc., Nashville, TN 37211.

Several of the most common traditional hydrograph estimation methods include (1) the Snyder synthetic unit hydrograph (U.S. Army Corps of Engineers, 1973); (2) the U.S. Soil Conservation Service (SCS) synthetic unit hydrograph (SCS, 1972); (3) the Clark unit hydrograph method (Clark, 1945); (4) the Nash unit hydrograph method (Nash, 1957); and (5) the Haan unit hydrograph method (Williams and Haan, 1973). Each of these methods have inherent characteristics, data requirements, and basin characteristics or coefficients which must be estimated or calculated. All these methods rely on the unit hydrograph whereby design hydrographs are computed by convolution of the unit hydrograph with rainfall excess. Therefore, rainfall data and methods for estimating rainfall excess are necessary for use of any of the unit hydrograph methods.

Recently, a simple, direct-approach method for estimating hydrographs in ungaged urban or rural watersheds for specified flood peak discharges was developed by Inman (1986) and tested by Robbins (1986). This method utilizes a dimensionless hydrograph derived by harmonic analysis of actual flood hydrographs from 80 basins in Georgia ranging in size from 0.14 to 500 square miles. A design hydrograph is estimated directly from peak discharge and basin lagtime. In Tennessee, T-year peak discharges for rural and urban watersheds can be derived from flood-frequency equations by Randolph and Gamble (1976) and Robbins (1984), respectively. Basin lagtime can be derived from lagtime equations by Robbins (1986).

This paper describes a basin study in which traditional unit hydrograph analysis methods were compared to the Inman hydrograph technique for accuracy and cost effectiveness.

SITE DESCRIPTION

The McCrory Creek Basin was selected as the study basin and is located in the metropolitan area of Nashville, in central Tennessee. McCrory Creek has a total drainage area of 24.3 square kilometers and is a tributary to Stones River. At the study site, the drainage area is 18.6 square kilometers.

The U.S. Geological Survey has maintained a flood-hydrograph station (station number 03430118) at the study site from 1977 to present. An actual flood hydrograph, with a peak discharge recurrence interval of approximately 10 years, and concurrent 15-minute rainfall data was selected from station records for use as the control to which design hydrographs could be compared (figure 1). It was assumed that this hydrograph represents a typical or average 10-year recurrence-interval flood hydrograph at this site.

METHOD OF ANALYSIS

For the purposes of this study, it was assumed that the study basin was ungaged. Therefore, all of the basin characteristics and coefficients were estimated or calculated for the SCS, Clark, Snyder, Nash, and Haan unit hydrograph methods and the Inman hydrograph method. The SCS publication TR-55 (SCS, 1986) served as the basis for the development of the hydrologic data and for the various coefficients used in the analyses.

Watershed and sub basin boundaries, hydrologic soil groups, and land use categories were delineated and digitized on 1:2400 scale topographic maps. The SCS curve number method was used to calculate precipitation losses. Runoff curve numbers were calculated for each sub basin-land use-hydrologic soil group using AMC III conditions and were used to obtain a composit runoff curve number for the entire basin. To estimate travel times, channel lengths and slopes were measured from the topographic maps and FEMA flood study profiles, respectively. Surveyed channel cross sections were used to obtain typical cross sections and hydraulic data. Manning's roughness coefficients were estimated for all channel reaches of the study area during a field reconnaissance of the basin. The basin was modeled as a single entity to allow easier discrimination of method differences.

Using the appropriate basin characteristics and coefficients, unit hydrographs were developed for each of the five unit hydrograph methods. Data from a similar basin near McCrory Creek were used to develop the Clark and Snyder coefficients. A 6-hour synthetic rainfall depth and distribution was generated using TP40 (U. S. Department of Commerce, 1961) and Hydro 35 (NOAA, 1977). A 10-year recurrence-interval flood hydrograph was then synthesized for each of the five unit hydrograph methods.

Finally, a 10-year recurrence-interval flood hydrograph was synthesized for the Inman hydrograph method. The time and discharge ordinates of the hydrograph were generated by multiplying the regression value of basin lagtime (from Robbins, 1986) by the time ratios and the regression value of peak discharge (from Randolph and Gamble, 1976) by the discharge ratios presented in Inman's report (1986).

RESULTS OF ANALYSIS

The observed (control) flood hydrograph and the six design flood hydrographs were plotted for visual comparison and are shown in figure 2. The hydrograph characteristics of peak discharge, volume of runoff, and widths at 50 and 75 percent of peak flow were compiled for all the design hydrographs for comparison to the same characteristics of the observed hydrograph. These characteristics along with percentage differences between the observed and the design hydrographs characteristics are listed in table 1.

As shown in table 1 and figure 2, all of the design flood hydrographs varied considerably from the observed flood hydrograph. However, the Inman hydrograph method produced a design flood hydrograph with a peak discharge, volume, and width that most closely fit the observed flood hydrograph characteristics. In addition, the amount of time used in deriving this design flood hydrograph was only a fraction of the time used for each of the other methods.

SUMMARY

Flood hydrographs generated by several traditional hydrograph simulation methods and a new dimensionless hydrograph simulation method were compared to an observed flood hydrograph to determine which method was most cost effective and accurate for producing a design flood hydrograph in Davidson County, Tennessee. This basin study revealed that the Inman hydrograph method, although originally developed for streams in Georgia, fit observed data better than five other unit hydrograph methods.

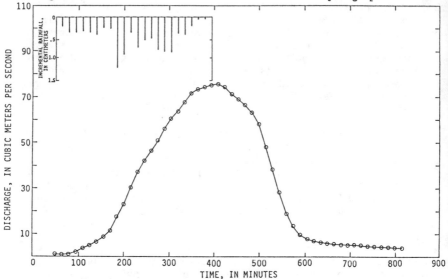

Figure 1.-- Observed flood hydrograph and concurrent incremental rainfall for McCrory Creek, at Nashville, Tennessee, for storm of September 13, 1979.

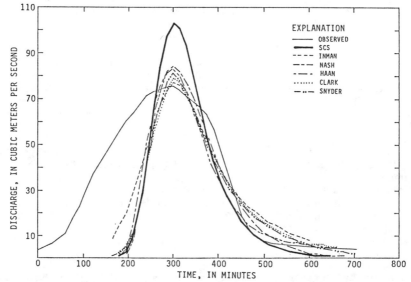

Figure 2.--Comparison of simulated and observed hydrographs for McCrory Creek.

Table 1.— Comparison of the observed hydrograph's characteristics to the characteristics of the synthesized flood hydrographs.
[DA=Drainage Area, Tl=Basin Lagtime, CN=Runoff Curve Number, Tc=Time of Concentration, R=Storage Coefficient, Cp=Peaking Coefficient, N=Nash Constant, K=Recession Constant]

HYDROGRAPH SYNTHESIS METHOD	INPUT PARAMETERS	PEAK DISCHARGE, IN CUBIC METERS PER SECOND	PERCENTAGE DIFFERENCE BETWEEN OBSERVED PEAK AND SYNTHESIZED PEAK	VOLUME OF RUNOFF, IN CENTIMETERS	PERCENTAGE DIFFERENCE BETWEEN OBSERVED VOLUME OF RUNOFF AND SYNTHESIZED VOLUME OF RUNOFF	HYDROGRAPH WIDTH (AT 50 PERCENT (W50) AND 75 PERCENT (W75) OF PEAK FLOW, IN MINUTES	PERCENTAGE DIFFERENCE BETWEEN OBSERVED HYDROGRAPH WIDTHS AND SYNTHESIZED HYDROGRAPH WIDTHS
OBSERVED HYDROGRAPH	-	75.61	-	6.88	-	W75=299 W50=208	-
INNAN DIMENSIONLESS HYDROGRAPH	DA=18.57 Tl=3.41	76.61	+1.31	7.77	+12.92	W50=186 W75=113	W50=-37.79 W75=-45.67
SCS UNIT HYDROGRAPH	DA=18.57, CN=88 Tl=1.51	102.52	+35.58	4.90	-28.78	W50=133 W75=85	W50=-55.52 W75=-59.13
CLARK UNIT HYDROGRAPH	DA=18.57, CN=88 Tc=2.12, R=1.803	79.18	+4.72	4.90	-28.78	W50=163 W75=94	W50=-45.48 W75=-54.81
SNYDER UNIT HYDROGRAPH	DA=18.57, CN=88 Tl=2.05, Cp=0.72	81.31	+7.53	4.90	-28.78	W50=160 W75=92	W50=-46.49 W75=-55.77
NASH UNIT HYDROGRAPH	DA=18.57, CN=88 N=3, Tl=1.51	83.91	+10.97	4.90	-28.78	W50=168 W75=107	W50=-43.81 W75=-48.56
HAAN UNIT HYDROGRAPH	DA=18.57, CN=88 K=1.33, Tl=1.51	82.55	+9.18	4.90	-28.78	W50=147 W75=92	W50=-50.84 W75=-55.77

APPENDIX I. - REFERENCES

1. Clark, C.O., 1945, Storage and the Unit Hydrograph, Am. Soc. of Civ. Eng. Trans, CX, pp. 1419-1446.

2. Hershfield, D.M., 1961, "Rainfall Frequency Atlas of the U.S.", U.S. Dept. of Commerce, Technical Paper 40.

3. Inman, E.J., 1986, Simulation of Flood Hydrographs for Georgia Streams, U.S. Geological Survey Water-Resources Investigations Report 86-4004, 48 p.

4. Nash, J.E., 1957, "The Form of the Instantaneous Unit Hydrograph", Int. Assoc. of Scientific Hydrology, Pub. 45, Vol. 3, pp. 114-121.

5. NOAA, 1977, "Five- to 60-Minute Precipitation Frequency for the Eastern and Central United States", NOAA Tech. Memo. NWS HYDRO-35, June.

6. Randolph, W.J., and Gamble, C.R., 1976, Technique for Estimating Magnitude and Frequency of Floods in Tennessee, Tennessee Dept. of Transportation, 52 p.

7. Robbins, C.H., 1984, Synthesized Flood Frequency for Small Urban Streams in Tennessee, U.S. Geological Survey Water-Resources Investigations Report 84-4182, 24 p.

8. Robbins, C.H., 1986, Techniques for Simulating Flood Hydrographs and Estimating Flood Volumes for Ungaged Basins in Central Tennessee, U. S. Geological Survey Water-Resources Investigations Report 86-4192, 32 p.

9. U. S. Army Corps of Engineers, 1973, Hydrologic Engineering Methods for Water Resources Development-Hydrograph Analysis, Vol. 4.

10. U. S. Department of Agriculture, Soil Conservation Service, 1972, National Engineering Handbook, Sec. 4, pp. 16.1-16.26.

11. U. S. Department of Agriculture, Soil Conservation Service, June, 1986, "Urban Hydrology for Small Watersheds", Tech, Release 55.

12. Williams, J.R., and Haan, R.W., May, 1973, "HYMO, A Problem-Oriented Computer Language for Hydrologic Modelling", User's Manual, ARS-S-0, U. S. Department of Agriculture.

A Possible Definition of the Input Function
for a Rainfall-Runoff Model with Lumped Parameters.

Paolo Bartolini *

Abstract

The spatial variability of rain (and the spatial variability of
the physical characteristics which affect the response of a watershed)
can hardly be taken into account by using rainfall-runoff models with
lumped parameters and lumped input.

It may be useful to introduce, instead of a "mean areal
rainfall", another input function, defined as that input which is
transformed by the model into a known output Q(t). Such a procedure can
be followed when the output is known; the results depend on the choice
of the model.

The relationships between the previously defined input function
and the observed point rainfalls must be examined in order to detect any
regularity. Modeling these regularities makes possible the use of the
definition of the input function even when the output is unknown, thus
allowing the use of the model for the synthetic generation of outputs.

In the following paper, the calibration phase (i.e. the
derivation of the input from known outputs) is performed for a small
creek in northern Italy. A first evaluation of the results is proposed.

Introduction

It is common practice to define the lumped input for a
rainfall-runoff model (RRM) with lumped parameters in terms of more or
less complex elaborations of the records from a selected set of
raingages (point rainfall). This method defines the input as a "mean
areal rainfall"; it is subject to the risk of a poor representation of
the spatial evolution of the rainfall over a given period of time.

Furthermore, such a procedure is strongly affected by the
(incidental) asyncronism between rainfall records.

It may be useful to modify this point of view. More
specifically, it may be useful to introduce a new definition of the
Input Function (IF), as the function which is transformed by a model
(with given parameters) into a known output. In other words, let Q(t) be
the known response of the watershed, $i(x,y,t)$ the real (unknown)
distribution of rain intensity in space and in time, j(t) the function

* Associate Professor; Istituto di Idraulica, Università di Genova.
via Montallegro 1, 16145 Genova, Italy.

which has been defined as the IF of the model. It follows that

$$Q(t) = \Phi(i(x,y,t)) = \psi(j(t)) \tag{1}$$

i.e. the output (the discharge $Q(t)$) is the result of the complex way by which the basin operates on the real distribution of rain ($\Phi(.)$). At the same time, the output can be viewed as the result af the (much less complex) way by which the model works on the IF, ($\psi(j(t))$. In (1) the unknowns are $\Phi(.)$, $i(x,y,t)$, and $j(t)$; $Q(t)$ and $\psi(.)$ are known.

From (1) it follows that

$$j(t) = \psi^{-1}(Q(t)) \tag{2}$$

Once $j(t)$ is derived from (2), the IF for that particular sequence $Q(t)$ and that particular model $\psi(.)$ is fully defined. It is apparent that the problem (2) does not always have a solution which is also physically possible (the $j(t)$ must be non-negative). It may be better to write

$$j(t) = \psi^{-1}(Q(t) + \eta(t)) \tag{3}$$

where $\eta(t)$ is a noise term; the problem may then have an "optimal" (in some sense) approximate solution.

The previous definition of the IF may be very different from that of the "mean areal rainfall". On the other hand, the use of an RRM aims ultimately at estimating, for example, the frequency distribution of floods; this is not in agreement with a procedure which specifies the outputs $Q(t)$ as known functions.

Henceforth, some relationship between the IF and the point rainfalls must be assumed. Once the form of the relationship between the IF and the observed point rainfalls is selected, it must be taken in account also in the calibration phase (when the output is known).

In subsequent sections, after the RRM which was used for testing the procedure is defined, a simplified form of the relationship between the IF and a single point rainfall is proposed.

The Rainfall-Runoff Model

The model used here for the determination of the IF is based on the (modified) Horton model for infiltration, the Wooding model for surface and channel response, one linear reservoir each (in series-parallel) for sub-surface and base flow, and a threshold model for retention. The lack of space does not allow an extensive illustration of the model (herein referred to as HWM), for which the reader may await a forthcoming paper. The model is characterized by the parameter set of Table 1.

Table 1 (Parameter set for the HWM model)

Notation	Description
VOM	Retention capacity (mm)
FO	Dry soil infiltration rate (mm/h)
F1	Saturated " " (mm/h)
V1M	Field capacity of the soil (mm)
A	$V2(t)/Q1(t)$ (1/h)
B	$V2(t)/Q2(t)$ (1/h)
C	$V3(t)/Q3(t)$ (1/h)
ALFAL,ALFAR	Parameters of the kinematic equation $q = \alpha y^m$
ML,MR	to be used in the Wooding model, for the left
ALFAC (AC)	and right side and the channel,respectively,
MC	expressed in S.I. units.

In Table 1, $V2(t)$ and $V3(t)$ are the actual volumes in the first and second sub-surface reservoirs, respectively, and $Q1(t)$, $Q2(t)$ and $Q3(t)$ are the interflow, the deep percolation and the base flow, respectively. The values ML = MR = 2 and MC = 5/3 have been selected, while for the parameters ALFAL and ALFAR the assumption has been made that they are proportional to the square root of the side slopes, by means of the same coefficient RV (1/sec).

It follows that the number of parameters is NP = 9. The geometric characteristics of the catchment are the rectified length of the channel, LC (m); the lengths LL (m) = AL/LC and LR (m) = AR/LC, where AL and AR represent the horizontal projection of the left and right side areas; and the side slopes SL and SR.

Besides the 9 parameters, values must be specified for VO1 (difference between the retention capacity and the initial value of the retained volume); V10 (initial value of the soil moisture); and Q1(0) and Q3(0), from which the initial content of the sub-surface reservoirs can be derived.

Experimental results

The method previously outlined has been applied to Sansobbia Creek (Ligurian Appennini, northern Italy).
The geometric characteristics of the catchment are:
SL = .64, SR = .60, LL = 830 m, LR = 1330 m, LC = 14400 m.
The Input Function has been derived by using a single point rainfall, in order to avoid problems with the syncronism of the records.
The very simple form

$$j(t) = (a + bt) \, i_1(x_1,y_1,t) \tag{4}$$

has been selected to relate the IF $j(t)$ to the point rainfall intensity i_1 at the point (x_1,y_1); 'a' and 'b (1/h)' are the parameters to be estimated in the calibration phase. In using the simple form (4) the goal was that of a gross check of the ability of the model to describe real world situations. The differences between eq.(4) and the definition of "mean areal rainfall" are self-apparent. Furthermore, it is not difficult to recognize in the form (4) the mathematical expression of an hypothesis about the spatial evolution of the rains, during a storm.

The results of the calibration are shown in Fig. 1, where the thick line represents the observed $Q(t)$ (m^3/s) and the thin line represents the calculated output. The results of Fig.1 (time in hours) have been obtained by using the values of the parameters shown in Table 2 (the events are ordered from the top-left). The symbol (*) has been used in those cases for which a specific value for the parameter could not be estimated (the coefficient RV in the absence of surface flow from the sides; the values FO, V1O, V1M, when V1O = V1M, i.e. the soil was initially saturated).

Table 2 (calibration parameters)

Ev.	Month	RV	AC	VO1	FO	F1	V1M	V1O	A	B	C	a	b
1	X	*	1	0	35	20	70	20	.250	.150	.035	1.46	-.017
2	X	20	1	10	35	15	70	20	.120	.115	.020	2.0	-.030
3	XII	20	1	20	*	12	*	*	.080	.115	.023	1.2	.0
4	XI	20	1	15	*	15	*	*	.120	.115	.023	.44	.008
5	X	20	2	6	*	20	*	*	.080	.145	.023	1.09	.0
6	V	20	1	0	35	12	70	25	.080	.250	.025	1.12	.0
7	IV	20	1	0	40	15	70	0	.100	.115	.023	1.1	-.003
8	X	20	1	0	*	15	*	*	.200	.250	.035	1.3	-.049
9	XI	20	1	30	35	10	70	50	.100	.115	.020	.9	.005
10	IV	20	1	0	*	15	*	*	.080	.115	.015	1.4	.001
11	XI	20	1	0	35	23	70	20	.150	.150	.025	2.4	-.041
12	XI	20	1	30	*	25	*	*	.090	.200	.022	1.8	-.015
13	IV	*	1	10	*	25	*	*	.150	.200	.030	1.4	-.010
14	XI	20	1	5	*	5	*	*	.120	.115	.023	1.1	-.005
15	XI	20	1	10	35	22	70	50	.150	.200	.030	1.4	.012
16	II	20	1	0	*	15	*	*	.080	.145	.025	1.2	.0
17	I	20	.5	15	35	8	70	50	.080	.150	.025	1.4	-.003
18	I	*	1	25	45	20	70	40	.150	.145	.025	1.8	-.020
19	X	*	1	10	45	20	70	0	.200	.115	.020	1.3	-.030
20	XI	20	1	30	40	18	70	50	.200	.200	.035	1.15	.003
21	II	20	1	0	20	10	70	30	.080	.115	.020	.4	.011

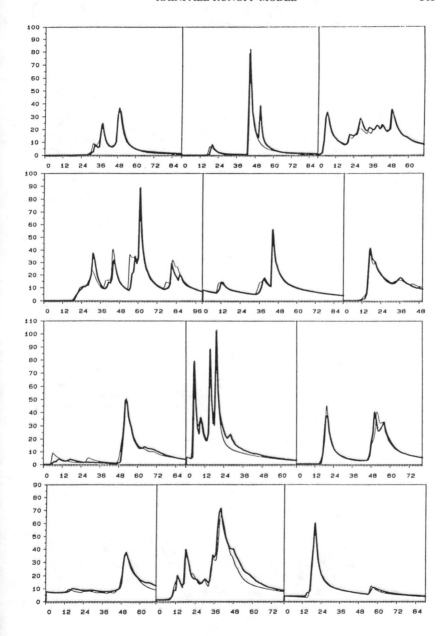

Fig.1   Calibration results

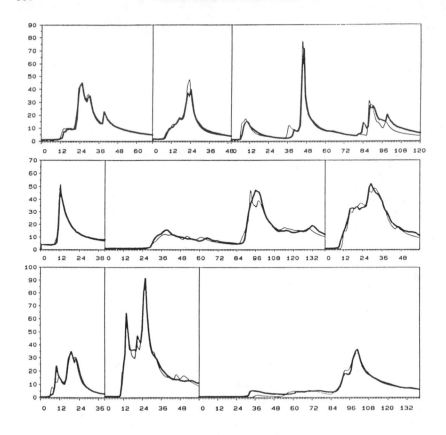

Fig.1 (continued)

Conclusions

It can be said that the procedure (the model and the related IF), has the ability to reproduce real world situations with an acceptable degree of accuracy. The variability of the parameters of the model may be due to the different position of a "storm center", the movement of which is in some way related to the values of the parameters 'a' and 'b'. Some improvement may be expected by using optimization procedures, and more than one raingage record. The simplified form of eq.(4) can be substituted with functions not monotonically varying with time.

It seems possible to find reasonable relationships between the parameters of the IF (in this case, 'a' and 'b') and the characteristics of the point rainfalls (multivariate moments); this allows the procedure to be used for the synthetic generation of runoff.

Kinematic, Diffusion and Dynamic Catchment Modeling

V. Miguel Ponce, M. ASCE, and Donna G. Pipkin[*]

Introduction

This paper reports on a series of tests aimed at developing improved procedures for catchment routing. The objective is to compare kinematic and diffusion models within several scenarios of physical reality and numerical fact. The effects of bed slope and Courant number are specifically isolated herein.

Background

The kinematic wave equation is the basis of most catchment routing models in current use. To derive the kinematic wave equation, the usual statement of conservation of mass in a control volume is coupled with a simplified form of conservation of momentum which accounts only for frictional and body forces. This leads to:

$$\frac{\partial Q}{\partial t} + c \frac{\partial Q}{\partial x} = c\, q_L \dots\dots\dots\dots\dots\dots\dots\dots\dots\dots\dots\dots\dots\dots\dots\dots\dots\dots(1)$$

in which Q= flow rate; c= celerity of kinematic waves; q_L= lateral inflow, per unit of channel length; x= spatial variable; and t= temporal variable.

Equation 1 can be discretized on the x-t plane in several ways. Ponce (3) identified three fully off-centered schemes: (1) Scheme I, forward-in-time/backward-in-space, stable for Courant number less than 1 and convergent as Courant number increases to 1; (2) Scheme II, forward-in-space/backward-in-time, stable for Courant numbers greater than 1 and convergent as Courant number decreases to 1; and (3) Scheme III, unconditionally stable but nonconvergent for all Courant numbers.

These and other similar schemes have been widely referred to as kinematic wave solutions. Schemes I and II are complementary; versions of them are in wide use, for instance in the kinematic wave routing option of HEC-1 (1). Scheme III (and related versions), although nonconvergent, has also been favored because of its feature of unconditional stability (2).

Ponce (3) has applied the concept of matched physical and numerical diffusivity to the distributed catchment routing problem. He compared the solution of scheme III, with its associated "uncontrolled" numerical diffusion, with that of a scheme developed by matching diffusivities (herein referred to as scheme IV), under a wide range of grid resolutions. While scheme III was shown to be highly sensitive to grid size, scheme IV was not. This led Ponce to suggest that scheme IV was altogether better than scheme III.

*Professor and Graduate Student, respectively, Dept. of Civ. Engrg., San Diego State Univ., San Diego, CA 92182.

In scheme IV, the physical diffusivity is given by:

$$\nu = \frac{q}{2S_o} \quad \dots\dots\dots\dots\dots\dots\dots\dots\dots\dots\dots\dots\dots\dots\dots\dots(2)$$

in which $\nu=$ physical diffusivity; $q=$ unit-width discharge; and $S_o=$ bed slope.

Matching physical and numerical diffusivities leads to the cell Reynolds number:

$$D = \frac{q}{S_o c \Delta s} \quad \dots\dots\dots\dots\dots\dots\dots\dots\dots\dots\dots\dots\dots\dots\dots(3)$$

in which $D=$ cell Reynolds number; and $\Delta s=$ spatial interval (either Δx or Δy).

By including the Froude number dependence of the physical diffusivity Ponce (3) extended the concept of matched diffusivities to the realm of dynamic waves. This led to the diffusion-cum-dynamic model (herein referred to as dynamic model), which is the same as scheme IV but with cell Reynolds number defined as follows:

$$D = \frac{q}{S_o c \Delta s} \left[1 - (\beta-1)^2 F^2 \right] \quad \dots\dots\dots\dots\dots\dots\dots\dots\dots\dots(4)$$

in which $\beta=$ exponent in the single-valued rating curve (discharge-area relation); and $F=$ Froude number.

Study Strategy

The sensitivity of catchment response to routing model, bed slope and Courant number is examined herein. Three routing models are chosen: (1) schemes I/II, with scheme I used for Courant numbers less than or equal to 1 and scheme II used for Courant numbers greater than 1, hereafter kinematic wave model; (2) scheme IV, with matched physical and numerical diffusivity, hereafter diffusion wave model; and (3) the diffusion-cum-dynamic model, i.e. scheme IV, but with Froude number dependent diffusivity, hereafter dynamic wave model. In light of Ponce's findings (3), the excessive and uncontrolled numerical diffusion of scheme III precluded its further consideration.

Three bed slopes and three Courant numbers were chosen, encompassing a range of values likely to be encountered in actual practice. The combinations of models, bed slopes and Courant numbers led to a program of 27 runs.

Numerical Experiments

Numerical experiments were designed to test the sensitivity of catchment response to model, bed slope and Courant number. A hypothetical example suited to the overall study objectives was developed. The example consisted of a catchment conceptualized as an open book with two planes adjacent to one channel. The planes are assumed to be impermeable, with rainfall excess (net rainfall) equal to total rainfall. The inflow to the planes is the rainfall excess, and runoff is by overland flow in a direction perpendicular to the channel alignment. Inflow to the channel is ιν lateral contribution from the planes, and outflow from the channel is the catchment response (3).

The dimensions are: plane length (in the direction of plane flow) = 600 ft (183 m); channel length = 1200 ft (366 m). The rainfall is 3 in/hr (76.2 mm/hr), which when multiplied by the catchment area results in a potential maximum peak outflow of 100 cfs (2.83 m^3/s).

The following values of bed slope were selected: (1) mild, 0.0001; (2) average, 0.001; and (3) steep, 0.01. The chosen Courant numbers were: 0.5, 1 and 2. For these bed slopes and Courant numbers, wave celerities and discrete intervals (Δx, Δy and Δt) were chosen to produce a consistent series of 27 runs (Table 1).

Given the catchment geometry, flow path, and wave celerities, rainfall durations were made equal to time-of-concentration, assuming kinematic travel time and neglecting diffusion. Accordingly, for the mild bed slope (0.0001) the rainfall duration was set at 240 minutes, with a runoff volume of 1,440,000 cu ft (40,776 m^3); for the average slope (0.001) it was set at 120 minutes, with a runoff volume of 720,000 cu ft (20,388 m^3); and for the steep slope (0.01) it was set at 60 minutes, with a runoff volume of 360,000 cu ft (10,914 m^3).

For this study the β value for plane flow was set at 3, characteristic of laminar flow, while the value for channel flow was set at 1.5, characteristic of turbulent Chezy friction. Mean flow velocities were calculated by dividing wave celerities by β values. Average flow depths were calculated by assuming laminar flow in the planes and turbulent flow in the channels.

Results

The results of runs 1 through 27, expressed in terms of peak outflow and time-to-peak, are shown in Table 1. These results lead to the following conclusions regarding effect of model, bed slope, and Courant number on catchment response:

Model
1) Kinematic and diffusion models: For the mild slope, the differences in peak outflow between kinematic and diffusion models are very large (see for instance, 99.6529 for run 1 vs 47.2510 for run 2); for the average slope the differences are large (99.2891 for run 4 vs 91.8975 for run 5); and for the steep slope they are small, tending to be masked by the lack of physical diffusion (100.000 for run 16 vs 99.3888 for run 17).
2) Diffusion and dynamic models: The differences in peak outflow between diffusion and dynamic models is quite small, regardless of slope or Courant number (see for instance, 47.2510 for run 2 vs 47.2587 for run 3; also 91.9017 for run 14 vs 91.9095 for run 15).

Bed Slope
3) Kinematic model: Bed slope has generally little effect on peak outflow (for instance, 99.6529 for run 1 vs 99.2891 for run 4, vs 98.5271 for run 7). For Courant number equal to 1, bed slope has no effect on peak outflow (100. for runs 10, 13 and 16). The peak outflow of 100 cfs (2.83 m^3/s) depicts pure kinematic translation, i.e. the complete absence of numerical diffusion and/or dispersion.
4) Diffusion and dynamic models: Bed slope has a substantial effect on peak outflow (see for instance 47.2510 for run 2 vs 91.8975 for run 5, vs 99.2840 for run 8; also 47.2646 for run 12 vs 91.9095 for run

15, vs 99.3920 for run 18). The smaller the bed slope, the greater the diffusion experienced by the hydrograph.

Courant Number

5) Kinematic models: For Courant number equal to 1, there is a complete absence of numerical diffusion and/or dispersion, with peak outflow equal to 100 and time-to-peak equal to rainfall duration

TABLE 1

RUNS, MODELS, BED SLOPES, COURANT NUMBERS AND SUMMARY OF RESULTS

Run	Model	Bed Slope (plane or channel)	Wave Celerity (plane) (fps)	Wave Celerity(ft) (channel) (fps)	Δx (ft)	Δy (ft)	Δt (sec)	C	Peak Outflow (cfs)	Time-to-peak (min)
1	Kinematic	0.0001	0.08333	0.16667	10	20	60	0.5	99.6529	244.
2	Diffusion	0.0001	0.08333	0.16667	10	20	60	0.5	47.2510	304.
3	Dynamic	0.0001	0.08333	0.16667	10	20	60	0.5	47.2587	304.
4	Kinematic	0.001	0.16667	0.33333	20	40	60	0.5	99.2891	123.
5	Diffusion	0.001	0.16667	0.33333	20	40	60	0.5	91.8975	129.
6	Dynamic	0.001	0.16667	0.33333	20	40	60	0.5	91.9049	129.
7	Kinematic	0.01	0.33333	0.66667	40	80	60	0.5	98.5271	62.
8	Diffusion	0.01	0.33333	0.66667	40	80	60	0.5	99.2840	61.
9	Dynamic	0.01	0.33333	0.66667	40	80	60	0.5	99.2875	61.
10	Kinematic	0.0001	0.08333	0.16667	10	20	120	1.0	100.0000	240.
11	Diffusion	0.0001	0.08333	0.16667	10	20	120	1.0	47.2646	304.
12	Dynamic	0.0001	0.08333	0.16667	10	20	120	1.0	47.2646	304.
13	Kinematic	0.001	0.16667	0.33333	20	40	120	1.0	100.0000	120.
14	Diffusion	0.001	0.16667	0.33333	20	40	120	1.0	91.9017	130.
15	Dynamic	0.001	0.16667	0.33333	20	40	120	1.0	91.9095	130.
16	Kinematic	0.01	0.33333	0.66667	40	80	120	1.0	100.0000	60.
17	Diffusion	0.01	0.33333	0.66667	40	80	120	1.0	99.3888	62.
18	Dynamic	0.01	0.33333	0.66667	40	80	120	1.0	99.3920	62.
19	Kinematic	0.0001	0.08333	0.16667	10	20	240	2.0	99.3300	248.
20	Diffusion	0.0001	0.08333	0.16667	10	20	240	2.0	47.2648	304.
21	Dynamic	0.0001	0.08333	0.16667	10	20	240	2.0	47.2734	304.
22	Kinematic	0.001	0.16667	0.33333	20	40	240	2.0	98.6954	124.
23	Diffusion	0.001	0.16667	0.33333	20	40	240	2.0	91.9011	128.
24	Dynamic	0.001	0.16667	0.33333	20	40	240	2.0	91.9099	128.
25	Kinematic	0.01	0.33333	0.66667	40	80	240	2.0	97.3238	64.
26	Diffusion	0.01	0.33333	0.66667	40	80	240	2.0	99.7468	60.
27	Dynamic	0.01	0.33333	0.66667	40	80	240	2.0	99.7521	60.

(see runs 10, 13 and 16). For Courant numbers other than 1 (i.e. 0.5 and 2.0), there is a small numerical diffusion/dispersion effect (99.6529 for run 1, and 99.3300 for run 19), attributed to the high grid resolution used in these runs.

6) Diffusion and dynamic models: Variation in Courant numbers has very little effect on diffusion and dynamic models (see 99.2840 for run 8 vs 99.3888 for run 17, vs 99.7468 for run 26; also 47.2587 for run 3 vs 47.2646 for run 12, vs 47.2734 for run 21). This is attributed to the grid independence of these two models through a wide range of Courant numbers.

Diffusion Effect and Catchment Response

The effect of diffusion on catchment response merits further analysis herein. Test results have clearly shown that bed slope (whether of plane or channel) has a significant influence on the amount of physical diffusion experienced by the outflow hydrograph. Further experience with the diffusion model has led to the following conclusions: (1) For steep bed slopes, the hydrograph diffusion is small; for average bed slopes, it is medium; for mild bed slopes, it is large. Other things being equal, the lesser the bed slope, the greater the diffusion effect experienced by the outflow hydrograph; (2) The diffusion effect causes a delay in peak outflow and an increase in time base. The greater the diffusion effect, the longer it takes the outflow hydrograph to reach its peak value; therefore, the longer the time base of the hydrograph.

In general, catchment response is a function of the following time parameters: (1) rainfall duration t_r; (2) kinematic time-of-concentration t_c; and (3) effective time-of-concentration t_e. For each catchment and rainfall event, there is a potential maximum value of peak outflow, calculated as the product of rainfall intensity and catchment area. Together with rainfall intensity, rainfall duration accounts for the total runoff volume.

Kinematic time-of-concentration is similar to the well established concept of time-of-concentration, i.e. the longest time it takes a parcel of water to travel from catchment divide to outlet, neglecting diffusion effects. It accounts for catchment size and shape, mean velocity and boundary friction. Effective time-of-concentration includes the diffusion effect and is generally greater than kinematic time-of-concentration.

Depending on the relative values of rainfall duration, kinematic time-of-concentration and effective time-of-concentration, catchment flow can be either kinematic or diffusion flow, and can be either subconcentrated, concentrated, or superconcentrated.

Kinematic Flow is that for which effective time-of-concentration is equal to kinematic time-of-concentration, i.e. when there is no measurable diffusion effect. Diffusion Flow is that for which effective time-of-concentration is greater than kinematic time-of-concentration. (Flow with effective time-of-concentration less than kinematic time-of-concentration implies negative diffusivity, clearly a physical impossibility).

Subconcentrated Flow is defined as catchment flow in which rainfall duration is less than effective time-of-concentration; consequently, peak outflow is less than the potential maximum value. Concentrated Flow is defined as catchment flow in which rainfall duration is equal to effective time-of-concentration, with peak outflow equal to the potential maximum value. <u>Superconcentrated Flow</u> is defined as catchment flow in which rainfall duration is greater than effective time-of-concentration; peak outflow is equal to the potential maximum value.

Diffusion vs Dynamic Models

It has been shown that diffusion and dynamic models give practically the same results, with little to be gained by using Froude number dependent diffusivity in catchment modeling. This behavior, characteristic of a wide range of bed slopes--from steep to mild--can be explained as follows: For steep bed slopes, the flow is likely to be kinematic, with both diffusion and dynamic solutions converging to the kinematic solution. For mild bed slopes, the flow is likely to be well below critical, characterized by small Froude numbers. As Froude number approaches zero, the dynamic wave solution converges to the diffusion wave solution (Eqs. 3 and 4). This explains the absence of a substantial difference between diffusion and dynamic wave solutions.

Summary and Conclusions

Discrete catchment models of kinematic, diffusion and dynamic waves are tested under a wide range of bed slopes and Courant numbers. The kinematic model uses an off-centered discretization of the kinematic wave equation. The diffusion model is formulated by matching physical and numerical diffusivities, which gives it grid independence for a wide range of resolution levels. The dynamic model is an extension of the diffusion model to account for Froude number dependent diffusivity. Results show the clear advantages of using the diffusion wave technique instead of the kinematic wave. The kinematic model fails to account for bed slope effect, rendering it unsuitable as a general model of catchment behavior. The diffusion and dynamic models, however, properly account for bed slope effect, producing hydrograph diffusion as required by the physics of the wave phenomena. For Courant number equal to 1, the kinematic model shows a complete absence of numerical diffusion and/or dispersion. The diffusion and dynamic models, however, show grid independence through a wide range of Courant numbers. In light of the present findings, the diffusion model is advocated as a general model of catchment behavior. The kinematic model should be used only in cases where the diffusion effect is negligible. Generally, there is very little to gain by using the dynamic model in lieu of the diffusion model. In practice, uncertainties regarding proper representation of boundary friction would mask the minute increase in accuracy obtained with the dynamic model.

References

1.   "HEC-1, Flood Hydrograph Package, Users Manual," The Hydrologic Engineering Center, U.S. Army Corps of Engineers, Davis, Calif., Sept., 1981, revised Jan., 1985.
2.   Huang, Y. H., "Channel Routing by Finite Difference Method," Journal of the Hydraulics Division, ASCE, Vol. 104, No. HY10, Oct., 1978, pp. 1379-1393.
3.   Ponce, V. M., "Diffusion Wave Modeling of Catchment Dynamics," Journal of Hydraulic Engineering, ASCE, Vol. 112, No. 8, Aug., 1986, pp. 716-727.

A GEOMORPHOLOGICALLY-BASED RUNOFF PREDICTION MODEL: CASE STUDY OF TWO GAUGED WATERSHEDS IN SAUDI ARABIA

By Mohamed N. Allam[1], A.M. ASCE, and Khalid S. Balkhair[2]

ABSTRACT: Using the Geomorphologic Instantaneous Unit Hydrograph (GIUH) and via the convolution transformation, a linear discharge hydrograph prediction model is obtained. Infiltration in streams is represented with a linear function of surface runoff; equals an infiltration coefficient multiplied by the streamflow discharge. Two problems are addressed; (1) estimation of the infiltration coefficient, and (2) estimation of the surface runoff velocity. Analysis of these problems are carried out through applications for two gauged watersheds in Saudi Arabia. A comparison between predicted and observed hydrographs, for representative storms, is presented.

INTRODUCTION

The GIUH, as developed by Rodriguez-Iturbe and Valdes (1979), is the probability density function (Pdf) of the travel time that a drop of water landing anywhere in the watershed takes to reach the outlet. The Pdf of the time of travel in watershed streams is assumed exponential. The travel time of a streamflow is taken equal to the stream length divided by the streamflow velocity. According to Pilgrim (1977), the streamflow velocity at any moment of a given storm may be assumed the same throughout the whole drainage basin. But the velocity changes during the storm, and thence the GIUH. Inclusion of a time-varying velocity in the GIUH would make the approach complicated and impractical to use. But on the other hand, selection of a representative streamflow velocity, for a given storm, is a difficult task. Through case study applications, Allam and Balkhair (1987) found that the wave velocity (V_w) at the peak discharge time is the most representative streamflow velocity for a watershed. This velocity, compared to the other proposed velocities, resulted in the best discharge hydrograph prediction.

Diaz-Granados, Bras and Valdes (1983) proposed another exponential distribution for the time of travel of the streamflow, accounting for the infiltration losses in the streams. Infiltration is expressed as a simple linear function of the surface runoff; equals an infiltration coefficient (K) multiplied by the streamflow discharge. A main drawback of this infiltration expression is that it allows generation of surface-runoffs even for the small rainfall events which most likely, will be

---

[1]Asst.Prof.,Irrigation and Hydraulics Dept.,Faculty of Engrg., Cairo Univ.,Giza,Egypt (Presently at Fac. of Meteorology, Env., and Arid Land Agriculture, King Abdulaziz Univ.,P.O.Box 9034,Jeddah,Saudi Arabia

[2]Graduate Student,Faculty of Meteorology,Env., and Arid Land Agriculture,King Abdulaziz Univ.,P.O.Box 9034,Jeddah,Saudi Arabia

infiltrated completely. But such hydrologic models are usually used for simulating the relatively large rainfall events, and therefore this draw-back may not be of critical importance. A major problem related to this approach, is the estimation of the coefficient K. Another problem is the identification of a representative velocity for the streamflow (V) in this model. Analysis of these two problems, represents the main theme of this study. Procedure for estimating K and V is presented. Verification of this procedure through case study applications for two gauged watersheds in Saudi Arabia is performed. These watersheds are: Wadi Khat in the southwestern region which is a fourth order watershed with an area of 600 Km^2, and Wadi Midhnab in the northern region which is a third order watershed with an area of 20 Km^2. These diversified sizes of the selected watersheds will enable examining the prediction accuracy of the GIUH versus the scale of the watershed.

REVIEW OF THE MATHEMATICAL FORMULATION OF THE GIUH

According to the probabilistic interpretation of the GIUH, stated earlier in the paper, the GIUH (h(t)) can be expressed as:

$$h(t) = f_{TB}(t) = \sum_{s \varepsilon S} f_{Ts}(t) \, P(s) \tag{1}$$

where TB is the travel time to the outlet of a watershed, P(s) is the probability that a water drop will follow path s, Ts is the travel time through path s, S is the set of all possible paths, and $f_{TB}(t)$ and $f_{TS}(t)$ are the Pdfs of TB and Ts, respectively.

Assuming time a drop spends as overland flow is negligible, then the travel time in a particular path is equal to sum of the travel times in the streams of this path. Assuming that the streamflow time of travel is an independent random variable, then the Pdf of Ts is the con-volution of the Pdfs of the times of travel in the various streams of this path. Rodriguez-Iturbe and Valdes (1979) proposed an exponential distribution for the streamflow travelling time:

$$f_{T(i)}(t) = \lambda_i \, e^{-\lambda_i t} \qquad\qquad i<\Omega \tag{2}$$

$$f_{T(\Omega)}(t) = \lambda_\Omega^{*2} t \, e^{-\lambda_\Omega^* t} \tag{3}$$

where $f_{T(i)}$ is the Pdf of the travel time in a stream of order i, Ω is the order of the watershed, $\lambda_i = V_w/\bar{L}_i$, $\bar{L}_i$ is the average length of the streams of order i, and $\lambda_\Omega^* = 2\lambda_\Omega$.

Diaz-Granados, Bras and Valdes (1983) modified $f_{T(i)}$ to account for the channel infiltration losses as:

$$f_{T(i)}(t) = \lambda_i \, e^{-\mu_i t} \qquad\qquad i<\Omega \tag{4}$$

$$f_{T(\Omega)}(t) = \mu_\Omega^* \lambda_\Omega^* t \, e^{-\mu_\Omega^* t} \tag{5}$$

where $\mu_i = V_i/\bar{L}_i(1-I_i)$, I_i is the infiltration portion of the flow in a stream of order i and equals $(1- e^{-KL_i})$, and $\mu_\Omega^* = 2\mu_\Omega$. These values of μ_i and μ_Ω make the area under $f_{T(i)}(t)$ equals $(1-I_i)$.

On the other hand, the probability P(s) can be determined as:

$$P(s) = \Theta_i\, P_{ij}\, \ldots\, P_{\ell\Omega} \qquad\qquad s=1,\ldots,S \qquad\qquad (6)$$

where Θ_i is the probability that a water drop falls in an area draining to a stream of order i and equals the total area draining directly into streams of order i divided by the watershed area, and P_{ij} is the transition probability from streams of order i to streams of order j and equals the number of streams of order i which drain into streams of order j.

ESTIMATION OF THE INFILTRATION COEFFICIENT K

Consider a stream of order i with a unit inflow at the upstream. The net-outflow then at the downstream end (inflow - infiltration) will be $(1-I_i)$. Suppose that this stream drains into a stream of order j. Then the infiltration through the latter stream will be $(1-I_i)(I_j)$ and the net-outflow at the far downstream end will be $(1-I_i)(1-I_j)$. Now consider a surface runoff path s consisting of a stream of order i which drains into a stream of order j which drains into a stream of order ℓ, so forth till reaching the watershed outlet. For a unit inflow at the upstream, the outflow $(1-I_s)$ at the downstream end of this path (watershed outlet) may be computed as:

$$(1-I_s) = (1-I_i)\, (1-I_j)\, (1-I_\ell)\, \ldots\, (1-I_\Omega) \qquad\qquad (7)$$

where I_s is the infiltrated portion of flow through path s. The surface runoff yield (Y) of a watershed of area A_Ω, from a rainfall event of depth d, may then be computed as:

$$Y = A_\Omega d \sum_{s\epsilon S} P(s)\, (1-I_s) \qquad\qquad (8)$$

For a gauged watershed, with known area and drainage pattern, the only unknowns in equations (8) are I_s (s=1,...,S) which are functions of K and stream lengths. By solving this equation for the concerned watershed a value of K corresponding to any rainfall event can be obtained.

ESTIMATION OF THE STREAMFLOW VELOCITY V

Gupta, Waymire and Wang (1980) computed the mean holding time (time lag) of a basin as:

$$K_B = \sum_{s\epsilon S} P(s)\, \bar{T}_s \qquad\qquad (9)$$

where $\bar{T}_s$ is the mean time of travel in path s; equals sum of the mean times of travel in the streams of this path. From equations (2) and (3), $\bar{T}_{(i)}$ can be shown equal to $1/\lambda_i$. In the modified GIUH, as can be shown from equations (4) and (5), $\bar{T}_{(i)}$ is equal to $(1-I_i)^2/\lambda_i$. But the area under $f_{T(i)}$ equals $(1-I_i)$, then $\bar{T}(i)$ of normalized $f_{T(i)}$ (divided by $(1-I_i)$ is equal to $(1-I_i)/\lambda_i$. According to the exponential travel time distribution proposed by Rodriguez-Iturbe and Valdes (1979), K_B can be computed as:

$$K_B = \frac{1}{V_w} \sum_{s\in S} P(s) \; (\bar{L}_i + \bar{L}_j + \ldots \bar{L}_\Omega)_s \tag{10}$$

and according to the exponential distribution proposed by Diaz-Granados, Bras and Valdes (1983), K_B can be computed as:

$$K_B = \frac{1}{V} \sum_{s\in S} P(s) \; (\bar{L}_i \, (1-I_i) + \bar{L}_j \, (1-I_j) + \ldots \bar{L}_\Omega \, (1-I_\Omega))_s \tag{11}$$

Given that K_B for a given storm should be constant for a basin, a relationship between V_w and V; the surface runoff velocities in the GIUH before and after the infiltration modification, respectively; can be derived. From equations (10) and (11), V may be expressed as βV_w where β can be determined via equation (12).

$$\beta = \frac{\sum\limits_{s\in S} P(s) \; (\bar{L}_i(1-I_i) + \bar{L}_j(1-I_j) + \ldots \bar{L}_\Omega(1-I_\Omega))_s}{\sum\limits_{s\in S} P(s) \; (\bar{L}_i + \bar{L}_j + \ldots\ldots\ldots\ldots\ldots\ldots \bar{L}_\Omega)_s} \tag{12}$$

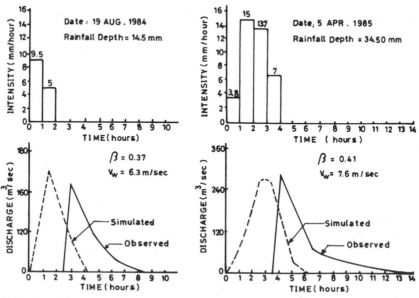

FIG. 1. _ Observed and Simulated Hydrographs for Wadi Khat.

CASE STUDY APPLICATIONS

Rainfall-runoff data on 26 storms for Wadi Khat and 10 storms for Wadi Midhnab was collected from the Ministry of Agriculture and Water. Values of K and β for both Wadis, were computed for all storms. For Wadi Khat, K is found to vary from 0.08 to 0.12 while β is found within the range of 0.34 - 0.45. For Wadi Midhnab, K and β are found to vary from 0.35 to 0.45 and from 0.11 to 0.17, respectively.

The wave velocity is computed here equal to $V_p + \sqrt{gy_p}$, where y_p is the flow depth at the peak discharge time. For both wadis, the velocity V_p for a rainfall-runoff event is taken equal to the peak discharge (Q_p) divided by the cross-sectional area of the main stream at the runoff gauging station. With these values of V_w and β, the velocity V for each storm is determined.

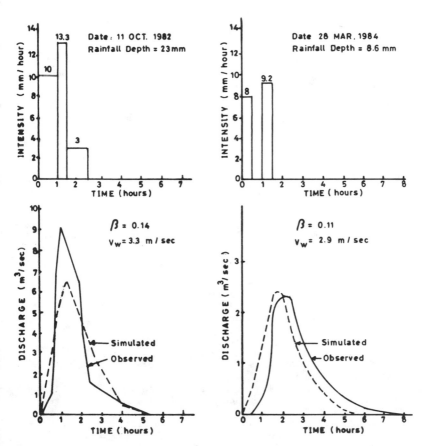

FIG. 2.- Observed and Simulated Hydrographs for Wadi Midhnab.

The surface runoff hydrograph is predicted for each event of the two watersheds. For Wadi Khat, the agreement between simulated and observed hydrographs, in terms of shape and Q_p, is quite good with errors in Q_p being less than 20% in most cases. Large error in the time to peak discharge (T_p) with an average value of 60%, is found. This relatively large error in T_p may be related to the rainfall characteristics in the southwestern region of Saudi Arabia as most rainfall comes in short duration covering small areas probably less than the size of Wadi Khat (Allam and Balkhair, 1987). Figure (1) shows a comparison between the observed and simulated hydrographs for two sample events.

For Wadi Midhnab, the error in the predicted Q_p and T_p is found less than 30% for all events. The error in T_p is probably because of the incorporated linear infiltration expression as the effective rainfall is assumed to start from the beginning of the storm and the effective duration is assumed equal to the storm duration. The observed and simulated hydrographs for two sample events are shown in figure (2).

CONCLUSIONS

1. A methodology for computing the infiltration coefficient K is presented. Comparison of the predicted and observed hydrographs showed that the methodology is efficient.
2. Based on the mean holding time concept, the most representative streamflow velocity V is determined as a function of the wave velocity V_w, geomorphology of the watershed, and the infiltration coefficient K.
3. Incorporation of a physically-based infiltration expression in the GIUH should significantly improve its predictability.
4. Applicability of the GIUH approach for large scale watersheds is questionable particularly for those watersheds which are subjected to short-duration convective storms which cover only small areas.

ACKNOWLEDGEMENTS

This study was sponsored by King Abdulaziz City for Science and Technology in Saudi Arabia by the grant AR7-085.

REFERENCES

1. Allam,M.N., and Balkhair,K.S.1987," Case Study Evaluation of the Geomorphologic IUH," submitted for publication.
2. Rodriguez-Iturbe,I., and Valdes,J.B.,1979,"The Geomorphologic Structure of Hydrologic Response," Water Resources Research,15(5), 1409-1420.
3. Gupta,V.K., Waymire,E. and Wang,C.T.,1980, "A Representation of an Instantaneous Unit Hydrograph from Geomorphology," Water Resources Research, 16(5),855-862.
4. Pilgrim,P.H.,1977, "Isochrones of Travel Time and Distribution of Flood Storage From a Tracer Study on a Small Watershed," Water Resources Research,13(3),587-595.
5. Diaz-Granados,M., Bras,R.L., and Valdes,J.B.,1983, "Incorporation of Channel Losses in the Geomorphologic IUH," TR No.293,Ralph M. Parsons Laboratory,M.I.T., Cambridge,Massachusetts.

OVERVIEW OF A REAL-TIME FLOOD MANAGEMENT MODEL

Richard K. Frithiof,[1] A.M. ASCE and Olcay I. Unver,[2] A.M. ASCE

Abstract:

   A flood management model was developed for the operation of the six reservoirs making up the Highland Lakes System on the Colorado River in Texas. The model allows simulation of over 700 total river miles, which includes three major and one minor tributary along with the mainstem, discretized by some 500 cross-sections. The system, operated by the Lower Colorado River Authority (LCRA), is characterized by sharply rising flood hydrographs. The base model was developed by the University of Texas in Austin. LCRA has further modified the model to include features to better simulate the reservoir system. The model uses data obtained by a hydrometeorological collection system.

Introduction:

   A real-time flood management model was developed at University of Texas at Austin (Unver et al, 1987a,b) to aid in determining reservoir releases from a system of reservoirs under flooding conditions. The model consists of a preprocessor for data control and input, a rainfall-runoff module, a full dynamic wave routing module and a postprocessor for on-screen graphical retrieval of the results. The model was then modified at LCRA to enhance the representation of system dynamics and user features. All model components are interactive with extensive error detection capabilities and sufficient defaults to help the user accomplish the routine inputting and routing tasks. Real-time input data is provided by a remote data acquisition network. Built-in options provide the user with step-by-step control over the simulation and early detection of inadequacies and inefficiencies in the trial reservoir operations.

   Enhancements to the base model were made at three different levels: (i). computational modifications, (ii). user features and screen options, and (iii). output features. The computational procedure employed by the National Weather Service operational dynamic wave model, DWOPER (Fread, 1982), within the routing module was modified to incorporate the actual reservoir components and the hydraulics associated with them. The main iterative loop was tied to the screen, giving the user complete control over the past, current, and future reservoir operations. The enhanced output features provide

[1] Senior Engineer, Lower Colorado River Authority, P.O. Box 220, Austin, TX 78767.
[2] Engineer, Lower Colorado River Authority, P.O. Box 220, Austin, TX 78767.

hard copies of operational decisions and results as well as input data
files for future simulations.

System Overview:

LCRA is an agency of the State of Texas. Created in 1934, it
was given responsibilities to control, store, and sell the waters of
the Colorado River and its tributaries. It has jurisdictional authority in ten counties in central Texas. These counties cover 9794
square miles and, according to the 1980 Census, has a population of
598,856 persons.

LCRA operates six dams, all on the mainstem of the Colorado
River. The total drainage area controlled by the structures is 27,443
square miles. The largest reservoir in the system is Lake Travis with
a conservation storage capacity of 1,171,000 acre-feet. Lake Buchanan
is the next with a storage capacity of 922,000 acre-feet. The total
storage capacity of the remaining reservoirs is 186,000 acre-feet. In
addition to storage, each dam has hydropower generation capacity. The
total available from the six structures is 240 MW.

Drainage Basin:

The modeled basin includes the Colorado River and three major
tributaries: the San Saba River, total drainage area of 3141 square
miles; the Llano River, total drainage area of 4455 square miles; and
the Pedernales River, total drainage area 1281 square miles. These
rivers form the base model.

Both the Colorado and the San Saba Rivers have flood hydrographs that are characterized by flat rising and recession limbs. In
contrast, the Llano and Pedernales Rivers both have steep rising and
recession limbs. In 1952, the gaged discharge at the Pedernales River
near Johnson City went from 9230 cfs at 6 am to 390,000 cfs at 10 am
(the discharge at 9 am was 218,000 cfs). Also during this same event,
Lake Travis rose 56 feet (a net increase of 682,000 acre-feet) over a
seventeen hour period.

LCRA Remote Interrogation System:

The data-acquisition system (EG and G Washington Analytical
Services Center, Inc., 1981), known as the Hydromet, retrieves
discharge, rainfall, and reservoir elevation data from throughout the
LCRA system (Figure 1). The Hydromet, specifically developed for the
LCRA, allows each station to be interrogated both on a user set
interval and independently at any other time.

Currently, LCRA has 20 USGS gaging stations on the Hydromet
system. Each of these stations supply river stage data, accumulated
rainfall, and station status. All information is instantaneous in
nature and therefore requires preprocessing before it is input into
the model.

In addition, there are 13 rainfall gages located in the areas
immediately around the lakes. These gages are used as input into a SCS

rainfall/runoff model. Eventually, these will be used in cooperation
with radar to better define areal coverage of a rainfall event.

       Lastly, the Hydromet allows interrogation of head- and tail-
water gages located at each of the dams. A near term addition to this
part of the system will be metering of both instantaneous and accum-
ulative generation taking place at each structure. Not only is this
information useful during flood operations, it is essential for
maintaining and improving the model's accuracy.

       Communications between the System Operations Control Center
(SOCC) and the gaging stations take place both through microwave and
FM frequency radios. Typically, a signal is sent from the SOCC through
the microwave system to an FM repeater. This repeater then sends the
signal on to the station. Each station responds to a site-specific
code which reverses the above sequence.

Computational Features:

       The base routing module requires an "equivalent" reservoir with
a single gate and a setting-discharge table for partial openings of

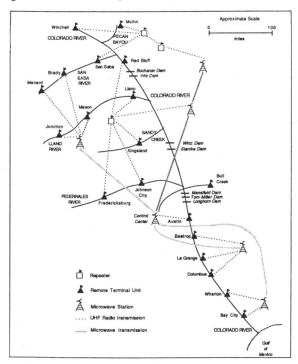

Figure 1. Lower Colorado River Authority Hydrometeorological Data Collection
(Hydromet) Network

the gate. The discharges cannot be specified as a function of elevations. However, in real-time operations, the decision variable is
usually the operation of the different gates and turbines, with
different hydraulic characteristics. Furthermore, the total release
from a reservoir is not only a function of the setting of different
flow structures but also a function of the existing headwater elevation at the reservoir. The DWOPER and GATES submodules of the
original model were modified so inputs to the new model are:

-   number of different gate types,
-   number of gates of each type,
-   location of each gate,
-   partial setting-discharge table for each gate type
    for different headwater elevations,
-   location and rating table for turbines, and
-   rating table for uncontrolled spillway.

The total discharge is computed as the summation of the
releases through each of the individual structures. Changes have been
made to the finite difference equations and partial derivative
computations to allow representation of actual reservoir discharges.
The new decision variables are the settings of the reservoir gates and
turbine operation status.

Screen Features:

The model no longer runs with preset gate operations only.
However, this is still an option. It, by default, pauses at the end of
each iteration for each time step. This enables the user to respond to
the current system status. Thus, the reservoir system is operated on a
step-by step basis, as opposed to after-the-fact operations of the
original model. Originally, a new operation policy could be specified
only after the current simulation was completed. Since a typical time
step requires two to five iterations before the finite difference
computations converge, the user has sufficient time to try and assess
different gate settings. At the end of each iteration, the model
displays the reservoir name, current head elevation, the current (or
default) gate settings, and the resulting release. At this point, the
user chooses one of the following options:

OPTION	Action taken
CHANGE	Change displayed gate setting for current and future time steps
ONLY	Change displayed gate setting for current time
TURBINE	Change turbine status
GO	Go to specified time step without pausing
PRINT	Change model print options
GATE	1. Save current operations in proper format for future run
	2. View/change past operations for future runs
	3. Lag gate operations to start next run from a later time step
SAVE	Save current unsteady flow conditions in proper format for future run

EXIT	1. Stop model execution
	2. Terminate simulation at the end of current time step and re-start with same data
	3. Put execution in hibernation (for input/output inspection etc.)
ADDED	Display storage additions to system and lakes
WATCH	Set watches on specified nodes (for screen display of flow rates and stages at the end of each time step)
BOUNDS	Set bounds on discharges/elevations of specified nodes (if a bound is violated, model pauses and warns user)
LATERAL	Display total lateral runoffs into system
HELP	On-line help

Output Features:

Screen outputs, in addition to the options listed above, include on-screen color graphics of discharge and stage hydrographs, in any number and combination, and the enveloping curves of the maximum water surface elevations and the corresponding discharges and time periods. Specified graphs are saved for plotting on a laser printer. Other output files created by the model are:

File	Contents
1.	Main DWOPER output (discharges, elevations, velocities etc.)
2.	Gate operations specified for current run
3.	Changes made to gate operations during simulation
4.	Storage changes for system and lakes
5.	Reservoir releases
6.	Unsteady flow conditions for specified time steps (in proper format for future DWOPER run)

A typical graphical output is shown in Figure 2. It shows the inflow hydrograph, headwater elevations, and the reservoir releases for Lake Buchanan as a result of the simulation of a historical storm event.

Concluding Remarks:

A user-friendly real-time flood management model has been developed and applied to the Highland Lakes System in Central Texas. The model features include dynamic wave flood routing, rainfall-runoff modeling, interactive input-output, and on-screen color graphics. The model runs on a dedicated Digital Equipment Corporation MicroVax II computer as a part of the LCRA Flood Management System.

Future extensions of the model are:

a). an optimization model to determine the release policy that minimizes flood stages and/or damages in Highland Lakes,

b). a feedback option that will explain and correct any discrepancies between observed and computed flows in the system.

The mathematical development and the corresponding computer models for both extensions have been completed. They will be added to the management model pending calibration and testing.

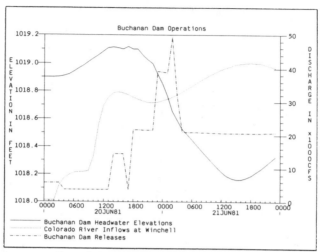

Figure 2. Lake Buchanan Operations, 20-21 June 1981

Appendix A. Unit Conversions

1 cubic foot per second (cfs)=0.02832 cubic meter per second (m^3/sec)
1 foot=0.3048 meter
1 mile=1.609 kilometers

Appendix B. References

1. EG and G Washington Analytical Services Center, Inc., Lower Colorado River Authority Software User's Manual, Albuquerque Operation, 2450 Alamo Ave., S.E., Albuquerque, New Mexico, 1981.

2. Fread, D.L., National Weather Service Operational Dynamic Wave Model, Hydrologic Research Lab., U.S. National Weather Service, NOAA, Silver Springs, Maryland, June 1982.

3. Unver, O.I., L.W. Mays, and K. Lansey, Real-Time Flood Management Model for the Highland Lakes, Journal of Water Resources Planning and Management Division, American Society of Civil Engineers, accepted for publication, 1987a.

4. Unver, O.I., L.W. Mays, and K. Lansey, A Real-Time Flood Management Model for the Highland Lakes System and an Optimization Model for Real-Time Multireservoir Operation, Engineering Hydrology Symposium, American Society of Civil Engineers, Williamsburg, Virginia, August 3-7, 1987b.

EMERSON, IOWA FLOOD WARNING SYSTEM
by WILLIAM P. DOAN*, AM ASCE

ABSTRACT

A flood warning system was installed in 1985 by the Corps of Engineers at Emerson, Iowa as part of Emerson, Iowa Flood Control Project on Indian Creek. The warning system consists of two remote sites for measuring creek levels and rain/snow amounts; one remote site for measuring rain/snow amounts only; one base station; and five small VHF hand-held receivers (beepers) for alarm. This paper presents a discussion of the conception, design, and operation of the Emerson flood warning system. It also describes some of the pitfalls that could have been avoided in the area of customer satifaction in the early part of the operations phase of this project.

INTRODUCTION

On June 15, 1982, the Indian Creek drainage basin in Western Iowa received 6 inches of rain within 4 hours which innundated the central business district of Emerson, Iowa with 6 feet of water and caused millions of dollars in damages and derailed an Amtrack passenger train. This flood prompted the design and construction of a flood control project for Indian Creek at Emerson, Iowa, part of which involved the design and installation of a flood warning system. This paper presents the design considerations and operating history of this system to date.

PROJECT DESCRIPTION

Indian Creek is an elongated drainage basin, approximately 2 miles wide and 20 miles long. At Emerson it drains 43 square miles. A map of the drainage area is shown in Figure 1. A map of the project location is shown in Figure 2. The project consists of a 183-foot levee, which connects the Burlington Northern Railroad embankment with the north approach fill of the U.S. Highway 59 viaduct over Morton Avenue. The town is protected by the railroad embankment to the south and U.S. Highway 59 to the east. Morton Avenue was raised so that the road is just at the Standard Project Storm (SPF) water surface elevation and 2

---

*Hydraulic Engineer, Corps of Engineers, Omaha District
Omaha,NB

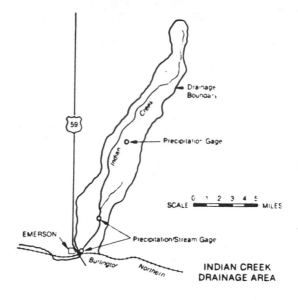

FIGURE 1

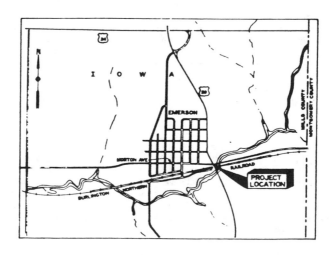

FIGURE 2

feet of freeboard above the 100-year water surface
elevation. The FEMA requirement for freeboard for the 100-
year flood is 3 feet, so the flood warning system is
required to provide enough time to complete the additional
foot of freeboard with a sandbag closure across Morton
Avenue. A map showing the levee profile is shown in Figure
3.

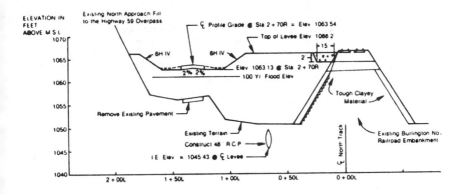

LEVEE PROFILE &
CROSS-SECTION — MORTON AVE. ROAD RAISE (STA. 2 + 70R)
(VIEWED LOOKING EAST)

FIGURE 3

DESIGN

1. Purpose and Need. The main purpose of the flood warning
system is to provide adequate warning time to allow
evacuation of homes not protected by the levee and to
provide enough time to complete the sandbag closure
structure over Morton Avenue during extreme flood events.

2. Hydrology and Hydraulics. Warning times at the project
site for various storm intensities were determined using
the following methodology. A Synder synthetic 1-hour unit
hydrograph was developed for the Indian Creek basin at the
project site based on coefficients for c_p and c_t that were
obtained from a regional study of similar drainage basins.
Hydrographs for the 2, 10, and 100 year storms were
developed by applying runoff coefficients to the unit
hydrograph. A steady state backwater water surface profile
at the project site was developed by use of the HEC-2 Water

Surface Profiles computer program. A plot of the hydrographs at the levee is shown in Figure 4 along with the water surface elevations the flows would reach. The time it would take the flows to reach specified elevations from the three different storms is shown in Figure 5. The time was measured from the centroid of the storm, or at the 1/2 hour mark.

3.Operating Philosophy. The flood warning system selected to give advanced warning of floods involves three remote sensing sites that monitor creek levels and/or precipitation amounts. The data from the remote sites is automatically sent by VHF radio to a base computer which monitors the data and sets off an alarm if threshold criteria have been exceeded. The alarm process involves having a transmitter send an emergency signal to hand held pagers which warn designated Emerson residents that there is a potential flood problem occurring.

OPERATING PROBLEMS

Ideally, the flood warning system is suppose to work by having the alarm system go off when either a 10 percent exceedance rain fall amount is equalled or exceeded (1.9 inches in 30 minutes), or if the stream gages rise above a threshold stage or if the creek level rises 1 foot or higher in 15 minutes, after the creek water level has risen 3 feet above the gage zero elevation. There have been several problems, some minor and some major, which have occurred during the initial one and a half years the system has been in operation.

1.Faulty Stream Gage Readings. The stream gages work by having solid state pressure transducers sense changes in the water pressure due to changes in stage. The transducers work by varying an output voltage from 0 to 5 volts in proportion to the pressure exerted on it. The sensors are calibrated for 256 increments. For example, if the sensor is calibrated to measure up to 10 feet of water, the increment size will be 10/256 or 0.04 feet and 10 feet of water on the sensor will cause it to output 5 volts which is converted to the number 255 at the transmitter. The transmitter will send the binary version of the number 255 to the base computer, which converts it into a reading of 10 feet of water. A series of problems have occurred, in which a sensor would send out a reading indicating that it was under several hundred feet of water, grossly exceeding the threshold criteria and trigger the alarm in the process. The manufactors speculated that somehow water was getting inside the pressure transducer's housing and causing a short in the sensor which would result in an erratic reading. Either the water was leaking directly through the caps at the end of the PVC housing or else it was coming from moisture condensation forming in the plastic conduit leading up to the transmitter and

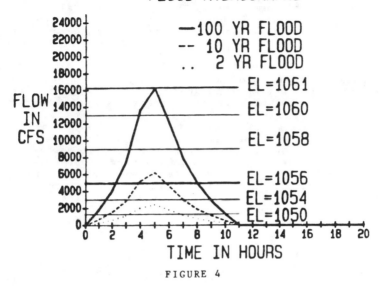

INDIAN CREEK AT MORTON AVE
FLOOD HYDROGRAPHS

FIGURE 4

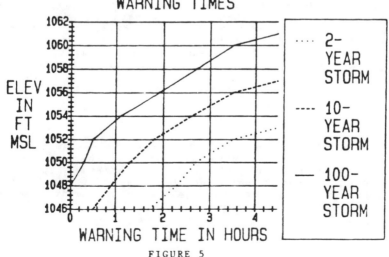

INDIAN CREEK AT MORTON AVE
WARNING TIMES

FIGURE 5

running down into the sensor. The manufactors have since modified the design of the sensor by adding gaskets and tightening the housing cap with more torque to prevent direct leakage into the sensor. The manufactors also added a box filled with desiccant along the conduit line to absorb condensation moisture forming inside of the conduit. There have not been any erratic readings since the modifications were made.

2.Emergency Backup. The flood warning system has been supplied with an emergency backup battery to operate the central base station in case of a power outage. Emerson has experienced two power outages in which the back up system has failed to last more than a few hours before losing power. Both outages happened while the flood warning system was unattended and as the regular power came back on again, it destroyed the self-booting floppy disk in the disk drive on one occasion and burned out the 10 megabyte hard drive on the other occasion. Steps taken to solve this problem have been to add an additional electical surge protector to the computer, add a larger battery to the backup system to provide longer backup time, and to have someone physically shut down the system until the regular power has been fully restored.

3.Remote Site Batteries Run Down Within Two Months. In the early periods of operations, the 12 volt-8 ampere gel cell batteries would run down after 1.5 to 2 months of operation instead of the one year period initally expected. Part of the problem was that due to Indian Creek runoff characteristics of extemely rapid rises, the transmitters were set to transmit at the minimum time interval of 3.6 minutes. The batteries are only good for 20,000 transmissions before they require recharging, which means that at the rate of one transmission every 3.6 minutes, the batteries would only last for approximately 50 days. The transmitters were reset to transmit at 15 minute intervals which translate into a single charge lasting approximately 208 days. It was felt that this 15 minute gap would not seriously jeopardize the effectiveness of the warning system.

CONCLUSION

While the Emerson, Iowa flood warning system has not performed flawlessly, overall, the system has been operational most of the time. What problems there have been, were not major design problems, but rather, initial de-bugging problems that can be solved by a combination of training, experience, and common sense.

The Santa Ana River Flood Forecasting System

Bruce G. Glabau, AM. ASCE and Gregory Peacock *

ABSTRACT: A real-time flood forecasting and reservoir operations simulation system has been developed for the Santa Ana River Basin, located in southern California. The system consists of generalized computer software developed by the U.S. Army Corps of Engineers Hydrologic Engineering Center and the Los Angeles District, and a set of data models describing the characteristics of the Santa Ana River Basin, the flood control system, and the data collection and reporting networks. The water control system includes data retrieval and processing, hydrologic data management, streamflow forecasting and simulation of reservoir operations. The water control system provides a coherent set of tools to increase the effectiveness of the water control manager in identifying future flood problems and evaluating alternative measures for mitigation on the Santa Ana River.

INTRODUCTION

In response to the increasing need to more effectively operate the existing flood control system, a real-time flood forecasting and reservoir operations simulation system has been developed for the Santa Ana River Basin, located in southern California. The purpose of the Santa Ana River Real-Time Water Control System is to aid the water control manager in operating the Santa Ana River flood control system by (a) the acquisition and display of real-time data that reflects the current status of the watershed and water control facilities, (b) the production of forecasts of runoff for the entire basin, with or without the addition of future precipitation, and (c) the derivation of an optimized reservoir system operation plan.

MODELING DESIGN

The Santa Ana River Real-Time Water Control System consists of computer software developed by the U.S. Army Corps of Engineers Hydrologic Engineering Center(HEC) and the Los Angeles District(LAD), and a set of data models describing the characteristics of the Santa Ana River Basin, the flood control system, and the data collection and reporting networks. In general the modeling system operates in the following manner:

* Hydraulic Engineers, Los Angeles District, Corps of Engineers, P.O. Box 2711, Los Angeles, Ca 90053-2325.

1. Data Acquisition and Preparation - IAD programs (1) collect real-time data, convert it to engineering units, check for and report alarm conditions, load all data into a HEC Data Storage System (2,3) master database, and prepare a summary report of the real-time information. Additionally, quantitative precipitation forecasts can be entered into the master database using a IAD interactive program. A forecast specific subset database is created using HEC software (4) by extracting data from the master database which covers a selected forecast time window.

2. Watershed Modeling - The watershed modeling consists of generating subbasin average precipitation hyetographs, computing streamflow forecasts for subbasin and streamflow control points, and simulating reservoir operations. Regular interval time-series subbasin average hyetographs are developed from the irregular interval real-time precipitation gage data stored in the master database by the HEC program PRECIP (4) in conjunction with the IAD data model (1). These hyetographs are stored in the master database or the forecast subset database. PRECIP also calculates a single antecedent moisture index based on subbasin precipitation and mean monthly evaporation.

Streamflow forecasting is performed by the HEC program HEC-1F (4) in conjunction with IAD data models (1). HEC-1F is a modified version of the HEC program HEC-1 Flood Hydrograph Package (5) and has the capability of calculating runoff from a complex multi-subbasin watershed using a Snyder unit hydrograph approach and Muskingum hydrologic routing methods. For streamflow forecasting, two HEC-1F data models are developed. The first is a parameter estimation model, termed the E-MODEL, which calculates discharge hydrographs for gaged headwater subbasins by estimating "best fit" loss rate, unit hydrograph and baseflow parameter values using real-time information. The second HEC-1F data model, termed the F-MODEL, is the streamflow forecasting model which calculates discharge hydrographs for ungaged subbasins and routes, combines and blends all the discharge hydrographs throughout the basin. Blending consists of replacing the calculated hydrograph ordinates up to the time of forecast with observed real-time gage data, and providing a smooth transition to the calculated hydrograph. The forecasted hydrographs are stored in the forecast subset database for use in reservoir simulation and output display.

The HEC program HEC-5 (6), in conjunction with IAD data models (1), is used to simulate the sequential operation of the reservoir system. Reservoir releases are determined by HEC-5 in accordance with constraints at downstream control points while keeping the reservoirs in the system "in balance". The forecasted reservoir discharges, elevation and storages are stored in the forecast subset database for subsequent display and analysis.

3. Model Control - The watershed modeling is controlled by MODCON, a HEC interactive execution program designed to facilitate use of the components of the modeling software system (4). The program can be used to designate user-specified rainfall-runoff parameter values (loss rate, baseflow, and future precipitation) on a zonal (multi-subbasin) basis for the generation of streamflow forecasts. For reservoir simulation, MODCON can be used to specify future reservoir releases to test various alternative operation plans. MODCON also has

the capability to create and send off (for execution) batch jobs that involve the sequential execution of a series of modeling programs. Additionally, summary displays of data status, input-parameter status, job status and program output can be viewed. HEC utility programs (2) can be accessed to create graphical plots and tabular displays of variables of interest.

BASIN DESCRIPTION

The Santa Ana River Basin is located in southern California near the City of Los Angeles and drains approximately 2,450 square miles to the Pacific Ocean. Approximately 37 percent of the basin is mountainous with the remaining area consisting of broad alluvial valleys. The climate is mild with warm, dry summers and cool, wet winters. The mean annual precipitation varies from 10 inches in the low valley areas to 45 inches in the higher mountains. Streamflow is perennial in the headwaters of the Santa Ana River and is generally intermittent in the valley sections. Streamflows increase rapidly in response to high-intensity precipitation. Unregulated flows in the lower reaches of the Santa Ana River have an annual average flow near 100 cubic feet per second with an extreme peak of record estimated at 320,000 cubic feet per second.

FLOOD CONTROL SYSTEM

Four flood control dams are located in the Santa Ana River Basin. Three of these structures, Prado Dam, San Antonio Dam, and Carbon Canyon Dam, are operated by the Los Angeles District of the Corps of Engineers. The fourth, Villa Park Dam, is a multipurpose reservoir with a seasonally varied flood control pool operated by the Orange County Environmental Management Agency. In addition to these flood control facilities, eight water supply structures which affect floodflows are located in the Santa Ana River Basin.

DATA NETWORK

The real-time data collection network located in the Santa Ana River Basin consists of the Corps' Los Angeles Telemetry System and the Orange County, San Bernardino County, and Riverside County event reporting stations sponsored by the National Weather Service, California-Nevada River Forecasting Center. Both information systems are used to develop irregular interval time series data sets. The Los Angeles Telemetry System stations report on a user-specified time interval and the county event reporting stations automatically report when a pre-determined change in stage or accumulated precipitation is reached. There are 13 river sites, 6 reservoir water surface elevation sites, and 50 precipitation sites in the combined network of real-time gages.

PRECIPITATION MODEL

The Santa Ana River precipitation data model is used in conjunction with the HEC program PRECIP to generate regular interval average hyetographs for each subbasin in the watershed. The data model contains areal weighting criteria used in calculating the subbasin hyetographs. Precipitation zones used for designating estimates of future precipitation in the watershed, through the model control program MODCON, are specified. Eight relatively homogenous precipitation zones were identified in the Santa Ana River Basin. Additionally, evaporation rates used to develop the antecedent moisture index for each precipitation zone are contained in the data model.

STREAMFLOW FORECASTING MODELS

The HEC-1F parameter estimation data model (E-MODEL) developed for the Santa Ana River Basin consists of eight gaged headwater subbasins. The HEC-1F streamflow forecasting data model (F-MODEL) consists of 21 ungaged subbasins and 19 streamflow concentration points. These data models were developed in the following manner.

1. Representative rainfall-runoff parameters for gaged subbasins were derived using data from four historical flood events and information from previous hydrologic studies (7). For the ungaged subbasins, the Snyder unit hydrograph parameters were then established by regression analysis. Loss rate and baseflow parameters were established by regionalizing the gaged values over ungaged subbasins with similar physical characteristics. Preliminary data models were then constructed with these parameter estimates and streamflow routing criteria obtained from previous hydrologic studies (7).

2. The preliminary data models were then calibrated using historical information from two additional flood events. The basin wide calibration was accomplished by adjusting the initial estimates of the rainfall-runoff parameters and streamflow routing criteria until a "best fit" was achieved with observed data at concentration points throughout the watershed.

3. Loss rate and baseflow zones were developed using information from the calibration results. These regional zones represent areas of relative homogenous loss rates and baseflows. They are applied during forecast operations and allow parameter changes in multiple subbasins with changes in zonal parameters designated through the modeling control program MODCON. Four loss rate and four baseflow zones were identified in the basin.

4. The forecasting capabilities of the calibrated HEC-1F data models were verified using historical data from an additional flood event that occurred in the basin. The forecast models were tested by simulating forecast conditions and developing "real-time" streamflow forecasts.

RESERVOIR OPERATIONS MODEL

The Santa Ana River HEC-5 data model which describes the

reservoir configuration and contains the regulation criteria used to simulate reservoir operations was developed in the following manner:

1. Reservoir project data describing the flood control system was compiled and a initial HEC-5 data model were constructed.

2. The initial data model was calibrated with historical data from a recent runoff event.

3. The calibrated HEC-5 data model was verified by simulating a real-time forecasting application using data from an additional historical flood event.

The resulting HEC-5 data model describes seven of the twelve reservoirs modeled in the Santa Ana River Basin. The remaining reservoirs are ungaged but are adequately simulated by the HEC-1F streamflow forecast process using direct storage-discharge relationships. A LAD interactive program (1) was developed to facilitate updating the starting storage conditions at the time of forecast.

CONCLUSION

A real-time flood forecasting and reservoir operations simulation system has been developed for the Santa Ana River Basin. The water control system includes data retrieval and processing, hydrologic data management, streamflow forecasting and simulation of reservoir operations. Streamflow forecasts are based on observed precipitation and streamflows, estimates of future precipitation, selected subbasin loss rates and baseflows, and expected reservoir releases. Future regulated flows at control points and future storage levels can be anticipated using the reservoir operations simulation model. These tools increase the effectiveness of the water control manager in identifying future flood problems and evaluating alternative measures for mitigation on the Santa Ana River.

APPENDIX 1. REFERENCES

1. Los Angeles District, U.S. Army Corps of Engineers, February 1987. Santa Ana River Real-Time Water Control System. Los Angeles, Ca.

2. Hydrologic Engineering Center, U.S. Army Corps of Engineers, November 1985. HECDSS-User's Guide and Utility Program Manual. Davis, Ca.

3. Hydrologic Engineering Center, U.S. Army Corps of Engineers, November 1985. Water Control Software - Data Acquisition. Davis, Ca.

4. Hydrologic Engineering Center, U.S. Army Corps of Engineers, August 1985. Water Control Software - Forecast and Operations. Davis, Ca.

5. Hydrologic Engineering Center, U.S. Army Corps of Engineers, 1985.

HEC-1, Flood Hydrograph Package, User's Manual. Davis, Ca.

6. Hydrologic Engineering Center, U.S. Army Corps of Engineers, 1985. HEC-5, Simulation of Flood Control and Conservation Systems, User's Manual. Davis, Ca.

7. Los Angeles District, U.S. Army Corps of Engineers, 1975. Review Report on the Santa Ana River Main Stem - Including Santiago Creek and Oak Street Drain, for Flood Control and Allied Purposes. Los Angeles, Ca.

DENVER AREA REAL TIME SYSTEM

Doug J. Clemetson*, A.M. ASCE

ABSTRACT

Cherry Creek, Chatfield and Bear Creek Dam and Reservoir projects were designed by the Corps of Engineers with the primary purpose of providing flood protection for the highly urbanized Denver, Colorado area. Drainage basins upstream and downstream from these projects are subject to extreme thunderstorm activity, resulting in severe flash-flooding conditions. In recent years, water control management practices have advanced rapidly due to the installation of automated data acquisition systems and the development of computer software to aid in the analysis of real-time hydrologic and meteorologic information.

This paper presents a discussion of the development of the Denver Area Real Time System which will be used to regulate the Denver reservoir system on a real-time basis. Its main components are: an automated data acquisition system, the SSARR model for predicting runoff from rainfall and snowmelt, the HEC-5 model to simulate the operation of the Corps' reservoirs, and the Hydrologic Engineering Center's Data Storage System and associated Water Control Software to assist in the analysis and display of real-time information and forecasts.

INTRODUCTION

Perhaps one of the most severe flash-flood regions in the United States lies in the Front Range region along the eastern slope of the Continental Divide. This region, with steep slopes and sparsely vegetated land covers allows for rapid runoff from violent orographic-intensified thunderstorms. Floods in this region can also be generated or compounded by rapid melting of deep snowpack from the high elevations of the Rocky Mountains. In Colorado, the Front Range region has been rapidly urbanizing, despite the severe flood threat posed by its geographic setting.

Background Information. Settlement of Denver, Colorado first began in 1858 as gold seekers rejected the warnings from friendly Indians of great floods and established their encampments near the confluence of Cherry Creek with the South Platte River in what is now downtown Denver (Follansbee and Sawyer,1948). Six years past before the first major flood occurred during 1864, in which nineteen lives were lost. Four major floods later, the mayor of Denver recognized the flood threat to the rapidly growing metro area and established the Cherry Creek Flood Commission following the flood of 1912 when two lives were lost.

Nearly one-hundred years after the first settlers arrived in Denver and its population had grown to about 600,000, the Corps of Engineers constructed Cherry Creek Dam to control flooding along Cherry Creek. In the mid-1970's, the population of the Denver metropolitan area exceeded

---

* Hydraulic Engineer, Hydrology and Meteorology Section, Omaha District, U.S. Army Corps of Engineers, Omaha, Nebraska

one million and the Corps of Engineers constructed Chatfield Dam and
Bear Creek Dam to control floods originating in the South Platte River
and Bear Creek basins, respectively. These projects, referred to as the
Denver Tri-Lakes projects, provide flood protection for the metropolitan
Denver area which has a present population exceeding 1.5 million and is
expected to double within the next fifty years (Corps,1986a).

Two major devastating floods have occurred in this region recently,
which include the South Platte River flood of June 1965 in which
thirteen lives were lost (Corps,1967) and the Big Thompson Canyon flash
flood of 1976 in which at least 135 persons were killed (Dept. of
Commerce,1976). Combining the heavily urbanized Front Range with the
potential for life-threatening flooding hazards creates a serious
situation that warrants close monitoring of the hydrologic and
meteorologic conditions.

Basin Description. Above Henderson, Colorado, the South Platte
River basin drains a total of 4,713 square miles along the eastern slope
of the Continental Divide as shown on figure 1. The shape of the basin
is somewhat circular with the eastern one-quarter covering the foothills
and high plains and the western three-quarters covering the Rocky
Mountains. Elevations throughout the basin range from 5,000 feet above
mean sea level (m.s.l.) at Henderson to over 14,000 feet above m.s.l. at
several mountain peaks along the Continental Divide. Approximately 60
percent of the basin is above 8,000 feet m.s.l., while about 20 percent
is above 10,000 feet. Several water supply reservoirs, transmountain
diversions from the Colorado River basin, and diversions for municipal
and agricultural water supplies are located throughout the basin.

Chatfield Reservoir, located just downstream from the confluence of
Plum Creek with the South Platte River, contains about 200,000 acre-feet
of flood control storage and about 27,000 acre-feet of storage for water

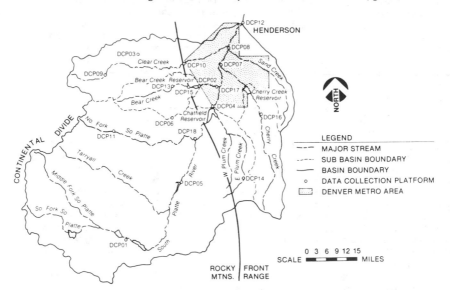

Figure 1. South Platte River Basin Above Henderson, Colorado

supply and recreational purposes. Cherry Creek Reservoir, located
eleven miles upstream from the mouth of Cherry Creek, has about 80,000
acre-feet of flood control storage and 13,000 acre-feet of conservation
storage.   Bear Creek Reservoir is located about eight miles upstream
from its confluence and contains about 53,000 acre-feet of controlled
flood storage and 2,000 acre-feet of storage for sediment.
    Existing Flood Control Operations. Regulation of the Denver Tri-
Lakes projects is provided by the Omaha District Reservoir Regulation
Section.   Reservoir design releases for Chatfield and Cherry Creek
reservoirs are 5,000 c.f.s. and for Bear Creek Reservoir, 2,000 c.f.s.
The South Platte River does not have sufficient capacity to handle the
reservoir design releases as channel capacities in the reach downstream
from the confluence of Clear Creek are less than 5,000 c.f.s.   This is
well below the combined design releases for the Tri-Lakes system;
consequently, it often takes several weeks to evacuate flood storage
from the reservoirs. Evacuation is made at a rate near the downstream
channel capacity.   This is a major concern because fast peaking
uncontrolled runoff from the plains and mountain tributaries downstream
from the projects can rapidly cause any remaining channel capacity to be
exceeded, resulting in flood damages. Travel time from Chatfield Dam to
Henderson is approximately 12 hours. Peaking times of tributaries above
and below each project are rapid and are generally less than one day.
Therefore, early detection of heavy rainfall and runoff is very
important in this area to make rapid forecasts of streamflow so the Tri-
Lakes system can be operated to help prevent damages from occurring.

REAL TIME SYSTEM COMPONENTS
    To allow close monitoring of the hydrologic and meteorologic
conditions and assist in the Corps flood control operations of the
Denver Tri-Lakes projects, the Denver Area Real Time System (DARTS) was
developed.   DARTS is a real-time streamflow forecasting and reservoir
operation system for the South Platte River basin upstream from
Henderson, Colorado.   Components of DARTS include a data acquisition
system, a streamflow forecasting model, a reservoir simulation model,
and associated software to assist in analysis of results (Corps,1984b).
A master control program is used to facilitate the operation of DARTS,
which is depicted by the flow chart on Figure 2.   Computer programs
utilized by the system were developed by the Hydrologic Engineering
Center (HEC), the North Pacific Division, and the Omaha District.
    Data Acquisition System. Eighteen Data Collection Platforms (DCP)
were installed throughout the South Platte River basin above Henderson
to collect and transmit hydrologic and meteorologic data which include
precipitation, river stage, air temperature, and reservoir pool
elevation.   These data are transmitted to the Geostationary Operational
Environmental Satellite (GOES) where the signal is amplified and relayed
to the downlink at the Federal Building in Omaha, Nebraska, where the
Omaha District offices are located. A satellite dish located on the
roof of the office building receives the signal from the GOES and
conveys the raw data to a VAX 11/730 minicomputer for preliminary
decoding. After preliminary decoding, the VAX transfers the data to the
North Pacific Division's AMDAHL computer located at Portland, Oregon.
Final decoding is completed in the AMDAHL computer before the data are
stored in the Missouri River Automated Data System (MRADS) database
(Corps,1982b). Once in the MRADS database, the DATA conversion package
is used to retrieve the data, check for errors, estimate missing values,

allow user input of additional data, and prepare input data files for
the streamflow forecasting model.

   Streamflow Forecasting System. In order to forecast streamflow in
the South Platte River basin above Henderson, the Streamflow Synthesis
and Reservoir Regulation (SSARR) model was selected as the primary tool.
Development of the SSARR model was completed by the North Pacific
Division of the Corps of Engineers (Corps,1986b).   The SSARR is a
numeric model of the physical hydrologic processes of a river basin
system.   Throughout a river basin, streamflow can be synthesized by
simulating the effects of rainfall, snow accumulation and melt, channel
routing, diversions, and reservoir regulation and storage.   Algorithms
are included for modelling the physical processes of   snowmelt,
interception, evapotranspiration, soil moisture, baseflow infiltration,
routing of runoff into the stream and to downstream locations through
channel and lake storage, and simulation of reservoirs under controlled
or uncontrolled modes of operation. Calibration of the SSARR model to
the South Platte River basin was completed by following the basic

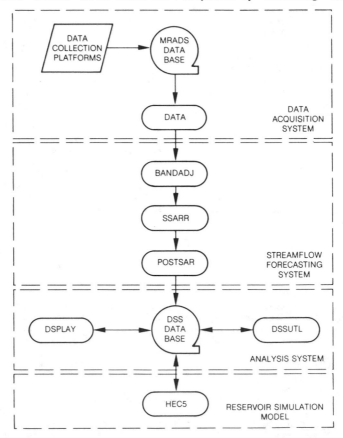

Figure 2.   Denver Area Real Time System Flow Chart

procedures outlined by Cundy and Brooks (1981). In order to improve the reliability of the SSARR model for real-time operational forecasting, it was given the capability to adjust from the computed initial conditions to observed conditions at the time the forecast is made. To automatically set up the necessary input data to use this option, a program called BANDADJ is used. Currently, the South Platte River basin SSARR model is configured to synthesize five-day forecasts of inflows to Chatfield, Cherry Creek and Bear Creek reservoirs, and to forecast the runoff from the uncontrolled downstream area between the Tri-Lakes projects and the Henderson gage. It can also be used to provide long-term forecasts based on historic or normalized meteorologic data. Computed forecast results from the SSARR model are stored in the HEC Data Storage System (DSS) database using the POSTSAR program, which was developed by the HEC (Corps,1984a). Once the data are stored in the DSS database, utility programs can be used to analyze and display the results and the HEC-5 model can access the data to simulate the operation of the Tri-Lakes system.

Reservoir Simulation Model. The HEC-5 program was originally developed by the Hydrologic Engineering Center to assist in planning studies for evaluating and sizing proposed reservoir systems and for evaluating the performance of existing reservoir systems (Corps,1982a). As demonstrated by Pabst and Peters (1983), recent modifications to the HEC-5 program have allowed application of the model to a system of reservoirs for real-time simulation. Two modes of real-time operation can be used with the HEC-5 model. Under the first type of operation, reservoir releases are determined by the program in accordance with specified operating criteria for each reservoir and downstream control point, keeping the flood control storage in each reservoir balanced. For the other mode of operation, the user can specify releases from each reservoir to analyze the effects on downstream flow conditions. Inflow hydrographs to each reservoir and hydrographs for the downstream uncontrolled areas are obtained from preceding SSARR model forecasts through the DSS database. Output results from the HEC-5 analysis are stored in the DSS database for subsequent use.

Information Analysis System. Both the SSARR and HEC-5 models have the capability to generate a tremendous amount of output data. Since, in a real-time environment, it is not practical to review such volumes of data, a database and associated software are used to store and retrieve necessary information quickly. The database selected for use with the DARTS is the DSS database, developed by the HEC to provide for efficient storage and retrieval of hydrologic data (Corps,1983). Associated software used in conjunction with the DSS include DSPLAY and DSSUTL, which were also developed by the HEC. The DSPLAY program provides graphical or tabular displays of data stored in the DSS database. DSSUTL is a utility program to provide maintenance functions on the DSS database. Results of the SSARR model forecasts and HEC-5 simulations can be quickly displayed as hydrograph plots or tabular information to allow the water control manager important information in making real-time water management decisions.

SUMMARY

A real-time water control system has been developed to forecast streamflow and simulate reservoir operations in the South Platte River basin near Denver, Colorado. Hydrologic and meteorologic data are collected automatically through a network of data collection platforms.

Additional data may be collected and input manually to improve the model performance. Streamflow forecasts may be made for either short-term (less than 5 days) or long-term (more than 5 days). A complete short-term forecast including summary tables and plotted hydrographs takes less than 30 minutes. Additional forecasts with alternative precipitation forecasts or different reservoir operation schemes may be completed in less than one hour. DARTS has been tested on both the Control Data Corporation computer system and on a HARRIS-500 mini-computer. Currently, DARTS is being implemented on the North Pacific Division's AMDAHL computer as it is planned to begin using the system in 1988 for day-to-day operation of the Tri-Lakes reservoirs.

It should be recognized that, although the DARTS is a complex system of sophisticated computer software and equipment, it is only a "tool" available to the water control manager to assist in making real-time water control decisions. As new information and experience in using the system become available it will be refined and updated as necessary to improve performance and reliability.

REFERENCES

1. Cundy, Terrance W. and Brooks, Kenneth N., 1981. "Calibrating and Verifying the SSARR Model - Missouri River Watersheds Study", Water Resources Bulletin Vol. 17 No. 5 pp. 775-782.

2. Follansbee, Robert and Sawyer, Leon R., 1948. "Floods in Colorado", USGS Water-Supply Paper 997, Washington D.C.

3. Pabst, Arthur F. and Peters, John C., 1983. "A Software System to Aid in Making Real Time Water Control Decisions", Technical Paper No. 89, Hydrologic Engineering Center, Davis, California.

4. U.S. Army Corps of Engineers, 1967. "Report on the Floods of June 1965 - South Platte River Basin, Colorado and Nebraska", Omaha District, Omaha, Nebraska.

5. U.S. Army Corps of Engineers, 1982. "HEC-5 Simulation of Flood Control and Conservation Systems, Users Manual", Hydrologic Engineering Center, Davis, California.

6. U.S. Army Corps of Engineers, 1982. "Water Control Data Collection and Management System", Missouri River Division, Omaha, Nebraska.

7. U.S. Army Corps of Engineers, 1983. "HECDSS User's Guide and Utility Program Manuals", Hydrologic Engineering Center, Davis, California.

8. U.S. Army Corps of Engineers, 1984. "Data Management Software for Real-Time Water Control of the South Platte River Basin", Special Projects Memo No. 84-4, Hydrologic Engineering Center, Davis, California.

9. U.S. Army Corps of Engineers, 1984. "Denver Area Real Time System", Volume 1 through 3, Omaha District, Omaha, Nebraska.

10. U.S. Army Corps of Engineers, 1986. "Metropolitan Denver Water Supply Draft Environmental Impact Statement", Omaha District, Omaha, Nebraska.

11. U.S. Army Corps of Engineers, 1986. "Users Manual - SSARR Model, Streamflow Synthesis and Reservoir Regulation", North Pacific Division, Portland, Oregon.

12. U.S. Department of Commerce, 1976. "Big Thompson Canyon Flash Flood of July 31-August 1, 1976", Natural Disaster Survey Report 76-1, National Oceanic and Atmospheric Administration, Rockville, Maryland.

GROUNDWATER MODELING - A MANAGEMENT TOOL

Daniel J. Whalen* and Gregory A. Johnson**

ABSTRACT

A hydrogeologic investigation within the South Mountain Complex of south central Pennsylvania was performed to develop a groundwater management program.

The investigation had four main objectives:

1. Determine the accrued volume of groundwater at the site on an annual basis.

2. Calculate the safe yield of groundwater that may be removed from the site.

3. Delineate regions with a high probability of producing high yield production wells.

4. Select potential high yield production well sites.

To ensure that there was an adequate volume of groundwater available in the watershed areas, a hydrologic budget was developed. A background review of the existing published literature, well logs, pump test results, and other pertinent data was conducted. Geological, hydrological and climatological data were compiled to interpret the watershed characteristics.

While background information was being compiled, an investigation of the fractured bedrock environment was being conducted. Using LandSat imagery, low altitude aerial photography, topographic maps and direct field measurements, the fracture fabric of the bedrock was defined. By interpreting the above information, a hydrologic concept was developed.

A finite difference groundwater model was developed and used to simulate conditions at the site. After the model was calibrated, sensitivity tests were performed, then predictive modeling was conducted. The model was used to simulate different groundwater pumping schemes and to predict the impact to the hydrologic cycle.

*Geological Engineer, Chemistry and Earth Sciences Division, Buchart-Horn, Inc., P.O. Box M-55, York, PA 17405

**Geological Engineer, Chemistry and Earth Sciences Division, Buchart-Horn, Inc., P.O. Box M-55, York, PA 17405

The results of this investigation included the calculation of safe yields of groundwater from the aquifer and the delineation of sites with the potential for high yield production wells.

GEOLOGY

The topography at the site consists of moderately high and steep rolling hills and valleys. These hills form the topographic boundaries to a "V-shaped" catchment basin (watershed). The approximate area encompassed by these boundaries is 6.7×10^6 m^2.

The soils overlying the watershed are composed of weathered bedrock material. These soils are derivatives of coarse angular blocky rock fragments resulting from fracturing, weathering and colluvial deposition. Consequently they are poorly sorted and consist of fine sand, silt, clay and abundant rock fragments. Field observations indicate that the rock fragments appear to increase in size with increasing depth. This gradational zone may extend vertically in excess of 60 meters. The actual interface between the unconsolidated soil and the weathered bedrock material is not clearly defined at the site. Field investigations and a review of core borings at the site indicate a gradual transition from unconsolidated soil (decomposed bedrock) to a highly fractured, and weathered bedrock material. Therefore a distinct soil/bedrock interface is not believed to exist.

The bedrock formations in the area are medium to coarse grained interbedded sandstones and quartzites. Bedding planes are visible in most of these units, some display a parting between each interface.

On a regional scale, the watershed is on the southern tip of a large northeasterly trending anticline. Beds strike predominantly northeast except on the ends of the structure where they vary according to the limits of the anticline. Consequently, bedding planes follow the shape of the anticline. Since the watershed is located at the southern edge of the anticline, strike and dip of the beds vary widely.

Smaller scale folds or ripples are present over the watershed area which influence the drainage and erosion pattern. The wavelengths of some folds encompass the entire watershed, exerting a complex structural influence on both the surface hydrology and the hydrogeological environment.

An extensive study of fractures and jointing was conducted in order to describe the secondary structural fabric of the bedrock.

LandSat imagery (1:1000000), Pennsylvania Geological Survey 7.5 min. geologic/topographic maps (Fauth, 1968) and low altitude aerial photographs (1:2400) were examined to investigate the region for large scale structural features. Lineaments and fracture traces were mapped in the area with three primary trends; east-west, northeast and north-northwest.

Direct field measurements were incorporated into the fracture analysis to define a correlation between the macro and local scale. Structural measurements of bedding planes, joint sets and fractures were taken at the site and in the surrounding area.

The results of the direct structural investigation indicate that the fracturing is predominantly near vertical and strikes northeast, north-northwest and east-west. The bedding planes in the area surrounding the watershed had a predominate north to northeast strike and a dip of approximately 45 degrees to the south.

Jointing was found to penetrate multiple bedding planes. Well developed joint sets are present, each with a unique orientation. With increasing depth jointing becomes less developed until, at an estimated depth of 60 meters, joints are closed and the rock becomes competent.

This field information was projected in combination with the fracture and lineament analysis onto topographic maps of the site.

HYDROGEOLOGY

The aquifer at the site exists in the fractured weathered bedrock material. The weathering process has enhanced the primary porosity and permeability of the bedrock. The secondary porosity and permeability created by fractures and open bedding planes has also increased the aquifer's ability to transmit and store groundwater. The combination of primary and secondary features within the aquifer have created excellent conditions for the development of ground water resources within the watershed.

Beneath the aquifer is the unweathered bedrock which has a lower primary porosity and permeability due to the nature of the consolidated sandstone and quartzite. The upper weathered bedrock acts as the aquifer and the lower unweathered bedrock acts as an aquitard. Transmission of groundwater through the aquitard is believed to be limited. This interface is used in the numerical ground water flow model to determine the vertical extent of the aquifer.

The hydrologic budget was calculated by using the estimated volume of precipitation that fell within the watershed. Average precipitation values from the surrounding areas were used in the calculation. To calculate the inflow of groundwater to the watershed, the following annual budget was used; Precipitation (P), .99 meters; Infiltration (I), .33 meters; Runoff (R), .13 meters; Evapotranspiration (ET), .53 meters.

Where; P = R + I + ET   or
       .99 m = .33 + .13 + .53

Taking the total area of the watershed and multiplying by the estimated infiltration rate, an average annual accrued volume of groundwater entering the site was calculated. The total area within the watershed is approximately 6.7 x 10^6 square meters. At .33 meters of infiltration per year, the aquifer will receive approximately 2.2 x 10^6 m^3/yr or 6000 m^3/day. If an average of 6000 m^3/day enters the system then 6000 m^3/day must leave the system. For the purpose of modeling, precipitation does not occur on a daily basis but rather it is accumulated continuously with time over the year.

Aquifer storage is important to consider in determining the management of ground water resources. Assuming that the following conditions exist at the site: 6.7 x 10^6 m^2 for recharge, 60 m thick aquifer throughout the site, and an average 35% porosity. The volume of groundwater that will be stored beneath the site is therefore approximately 1.4 x 10^8 m^3 of water. Therefore, the volume of ground water in storage would allow for removal of large quantities from the aquifer during drought conditions. If the aquifer is allowed to recharge during non-drought conditions, the balance within the hydrologic cycle would not be upset.

The safe yield of the basin is calculated by using 80 percent of the total volume of recharge to the system. This is 6000 m^3 per day or a safe yield of 4800 m^3 per day. The factor of safety is included due to possible leakage from the system, and error in estimated hydrologic parameters.

To determine the capabilities of the aquifer to perform under different hydrologic conditions and various groundwater withdrawal schemes, a numerical ground water flow model was developed. The hydrogeologic concept was used to define the site parameters necessary to design the numerical groundwater flow model to simulate the site.

NUMERICAL GROUND WATER FLOW MODEL

A Modular Three Dimensional Finite Difference Groundwater Flow Model (McDonald and Harbaugh, 1984), a widely used and well documented flow model, was applied to the site in its two dimensional form for steady state flow conditions. To use the model successfully the hydrogeologic concept of the site was translated into the model. The hydrogeologic concept defines the hydrogeologic parameters as follows. The aquifer is under watertable conditions in a highly weathered and fractured bedrock media with overlying unconsolidated soil. The hydraulic conductivity values range from 9.2 x 10^{-7} to 2.4 x 10^{-8} m/sec. The thickness of the aquifer is 60 meters. Infiltration was calculated as 33 percent of the average annual precipitation. Because the bedrock was highly fractured, the aquifer was modeled as an equivalent continuum. That is, due to the high fracture density at the site, discrete fracture modeling was unnecessary. The aquifer was treated as an equivalent porous medium by spatial averaging of the system.

Boundary conditions define the extent of influence simulated by the computer model. The topographic highs which act as groundwater flow divides are considered no flow boundaries for the computer

model.   Conditions   or events outside the boundary have no effect on
the ground water system.   Internal boundaries of the system consist
of stream boundaries which behave as discharge zones.

To   input   the hydrogeologic parameters into the model a   finite
difference grid was established.   The study area   was discretized
using a 19 x 15 cell grid with each cell representing a 203 x 203
meter area.

Calibration of the model was achieved when the calculated   head
values   (watertable) matched observed head values,   and responses   to
yield tests performed on existing wells.   The sensitivity of   the
model was investigated by modifying aquifer parameters and   observing
those parameters affecting the model solution.   It was determined
that large cell sizes negated aquifer heterogeneities, and changes in
aquifer thicknesses disturbed the desired flow field.

Using the hydrologic budget and the hydrogeologic concept,
different pumping schemes and well positions were simulated to
achieve optimal ground water withdrawal.   Several important factors
were determined from the results of the pumping simulations; 1)
pumping wells located outside discharge zones tend to have excessive
drawdown; 2)   production wells located along discharge zones could
produce large quantities of groundwater with minimal drawdown of the
water table; 3) Wells spaced in excess of 400 meters had minimal
interference with each other; 4) wells located in zones of high
fracture density and at the intersection of fracture traces and
lineaments produce large quantities of groundwater with minimal
drawdown.

CONCLUSIONS

The results from this study exemplify the importance of a
hydrogeologic concept when designing a groundwater flow   model.   All
factors of the hydrologic and geologic system must be investigated
and combined to aid in the perception of aquifer behavior.

The hydrologic cycle was examined and reduced to its primary
components of   precipitation,   runoff,   infiltration   and
evapotranspiration. Infiltration and runoff rates were assumed by
observing the nature and vertical extent of the soils.

The geologic environment was examined because of its influence
on groundwater flow.   Without background literature and actual field
information, the structural fabric of the bedrock could not be
defined.   The use of LandSat imagery and low altitude aerial
photography further enhanced the perception of the geological
environment.

Next the hydrologic and geologic characteristics were merged   to
define a hydrogeologic concept.   Aquifer characteristics from the
concept were then translated to the model.   During calibration,
careful attention was placed on ensuring model input would represent

actual field conditions. Finally the model could be used for simulation with reasonable predictive capabilities.

This study demonstrates the importance of an understanding of the hydrologic and geologic systems before model simulation or prediction should be performed. Comprehension of a hydrogeologic concept will enhance model input and output and therefore closely represent actual field conditions.

REFERENCES

Fauth, J.L., "Geology of the Caledonia Park Quadrangle Area, South Mountain, Pennsylvania," Commonwealth of Pennsylvania, Department of Internal Affairs, 1968.

McDonald, M.G., and Harbaugh, A.W., "A Modular Three Dimensional Finite Difference Ground Water Flow Model", U.S. Department of the Interior, U.S. Geological Survey, 1984.

Effect of Length of
Precipitation Record on Recharge Estimates

Richard H. French, M. ASCE*

The effect of the length of precipitation record on groundwater recharge estimates is examined using data from eleven pretipitation gages located on the Nevada Test Site. It is concluded that the length of precipitation record can have a significant effect on estimates of natural recharge.

Introduction

An understanding of the temporal and spatial distribution of precipitation is crucial in arid and semi-arid environments where the available water resources must be carefully assessed and managed. In the Southwestern regions of the U.S. a portion of the precipitation which falls recharges the groundwater system while the remainder either directly evaporates or becomes surface runoff. Since one of the primary sources of water in these regions is groundwater, rapid regional growth has increased the importance of being able to accurately estimate the groundwater in storage and the rate at which natural recharge of the groundwater system is taking place.

Although there are a number of methods that can be used to estimate regional groundwater recharge, many of these methods assume a priori a detailed knowledge of annual and seasonal precipitation. For example, in Nevada potential groundwater recharge is estimated by a technique developed by Maxey and Eakin between 1947 and 1951, in Watson et al. (1976). The Maxey-Eakin method for potential recharge estimation has as its basis: 1) the assumption that recharge must equal groundwater basin discharge, 2) an annual precipitation map, Houghton et al. (1975), asserts that annual precipitation is an unspecified but known function of elevation; and 3) the assumption that the amount of potential recharge is proportional to the amount of annual precipitation, Col. (2) of Table 2.

Although the Maxey-Eakin methodology is empirical and has never been completely justified, it has and continues to be used in Nevada to manage the groundwater resources of the state; that is, groundwater cannot be withdrawn at a rate that exceeds the rate of recharge. One of the primary assumptions, as noted above, is that the amount of annual precipitation is known. However, in most arid and semi-arid regions, the number of precipitation gaging stations is limited, the period of record available for each station is often short, and the annual and seasonal precipitation is highly variable. Finally, it should be noted that the scientific basis for the development of the annual precipitation map which is a primary component of the Maxey-

*Research Professor, Desert Research Institute, Water Resources Center, Las Vegas, NV 89120.

Eakin technique for estimating potential recharge is unknown.

    As a part of its continuing research program with the U.S. Depart-
ment of Energy, the Water Resources Center of the Desert Research Institute has developed and continues to maintain and analyze a precipitation data base for the over 1,300 mi^2 (3,370 km^2) Nevada Test Site
(NTS). This data base consists of the records of 40 gaging sites with
periods of record ranging from less than a year to 27 years. In examining the effects of a recent update of the data base, French (1986)
noted that rather small changes in the length of the period of record
resulted in rather large changes in the calculated annual precipitation. It is the purpose of this paper to examine the effect of
these shifts in calculated annual precipitation on potential recharge
estimates. In the following analysis, only the 11 stations on the NTS
with a reliable period of record of 9 or more years are used. Further, potential recharge for the total NTS area is estimated without
regard to the individual hydrologic basins which compose this area.

Estimation of Annual Precipitation

    French (1983) hypothesized that for Southern Nevada a satisfactory
relationship between annual precipitation and elevation was

$$\log(p) = a + bE \tag{1}$$

where p = estimated annual precipitation, E = elevation above mean sea
level, and a and b are coefficients. Models similar to Eq. (1) have
also been developed for other areas of the Southwest where precipitation is strongly affected by topographical relief; see for example,
Osborn (1984). The accuracy of empirical precipitation models such as
Eq. (1) in estimating precipitation depends on the both the validity
of the hypothesized functional relationship and the quality and quantity of the data available to calibrate the coefficients.

    In Table 1, the 11 precipitation gaging stations used in this study
are identified and located. An examination of the horizontal and vertical distribution of these stations demonstrates that the long-term
precipitation monitoring coverage of the NTS is not uniform. However,
the precipitation coverage of the NTS is much better than that generally available in the Southwestern U.S. both in terms of the number
of gages per unit area and the length of record associated with each
gage. In Cols. (6) and (7) of Table 1, the calculated average annual
precipitation and the associated standard deviation for the period of
record complete through calendar year 1979 are summarized, and in
Cols. (9) and (10) the analogous data for the period of record complete through calendar year 1985 are summarized. In Col. (11) of
Table 1 the percent changes in calculated annual precipitation due to
the addition of approximately 6 years of record are summarized. While
the percentage changes in the mean values of annual precipitation are
reasonably large, it must be noted that the standard deviations are in
many cases 50% of the average values and thus such shifts in the mean
values are not unexpected. In all cases the change in the average
annual precipitation is positive; that is, the values in Col. (9) are
larger than the corresponding values in Col. (6). These two aspects
of the data summarized in Table 1 are discussed in French (1987).

TABLE 1. CALCULATED ANNUAL PRECIPITATION AT ELEVEN LOCATIONS ON THE NEVADA TEST SITE, NEVADA (1 ft. = 0.305 m, 1 in. = 2.54 cm)

Station	Elevation ft.	Location Nevada Coordinate System		Annual Record thru 1979			Annual Record thru 1985			% Change in $\bar{p}$
		East ft.	North ft.	Years of Record	$\bar{p}$ (in.)	σ (in.)	Years of Record	$\bar{p}$ (in.)	σ (in.)	
(1)	(2)	(3)	(4)	(5)	(6)	(7)	(8)	(9)	(10)	(11)
Desert Rock	3298	688,100	681,750	16	5.34	2.28	22	6.07	2.64	+13.6
Mercury	3770	696,738	695,045	9	5.83	2.03	15	6.45	2.37	+10.6
4JA	3422	610,605	740,840	22	4.56	1.77	27	5.01	2.73	+9.8
Cane Springs	4000	663,600	751,000	15	7.75	3.93	21	8.19	3.71	+5.7
Well 5B	3080	705,200	747,600	16	4.60	2.00	22	5.06	2.19	+11.3
Mid Valley	4660	644,500	809,250	13	9.11	4.18	19	9.67	4.04	+6.1
Yucca	3920	680,875	803,600	20	6.52	3.56	26	6.77	3.48	+3.8
40 MN	4820	610,600	837,100	17	7.10	2.92	23	7.93	3.47	+11.7
Tippi- pah Springs #2	4980	638,650	838,600	15	9.47	4.37	21	9.68	2.06	+2.2
BJY	4070	679,100	842,300	19	6.03	2.94	25	6.55	3.18	+8.6
Area 12 Mesa	7490	631,400	888,900	20	12.46	5.99	26	12.96	5.96	+4.0

Finally, it should be mentioned that seasonal precipitation exhibits
the same trends as annual precipitation; that is, the addition of approximately 6 years of data also results in an increase in the calculated values and summer and winter season precipitation.

If the logarithms of the annual precipitation data in Col. (6) are
regressed against the elevations in Col. (2), Eq. (1) is

$$\log(p) = 0.40 + 0.00010E \tag{2}$$

with r = 0.90; and if the precipitation data from Col. (9) are regressed against the elevations in Col. (2), Eq. (1) is

$$\log(p) = 0.46 + 0.000094E \tag{3}$$

with r = 0.90. In Eqs. (2) and (3) p is in inches and E in feet. Although there is a only a small difference between the coefficients in
Eqs. (2) and (3) this small difference results in a significant difference in potential recharge estimates as will be demonstrated in the
next section. It is appropriate to note that whether there is a statistically significant difference between Eqs. (2) and (3) is not relevant. The question of recharge is examined from the viewpoint of an
administrator who has a set of data and must make a decision regarding
potential recharge and hence allowable pumpage of an aquifer system.
From the viewpoint of this administrator whether there is a statistical difference between Eqs. (2) and (3) is irrelevant since he would
not be aware of the other equation. Thus, we are inquiring into the
magnitude of the error that is made given limited data on which a
water resources management decision must be made.

Potential Recharge Estimate

Eqs. (2) and (3) can be used with the Maxey-Eakin methodology to
estimate the potential recharge on the NTS, Table 2. In Cols. (1) and
(2) of Table 2, the essentials of the Maxey-Eakin methodology are defined. For example, in an area where the annual precipitation exceeds
20 in (51 cm) the potential recharge is 25% of the annual volume of
precipitation. Elevations on the NTS range from 2,700 ft (820 m) to
7,500 ft (2,290 m). In Cols. (3) and (4) the elevation ranges on the
NTS obtained from Eqs. (2) and (3) respectively corresponding to the
precipitation ranges in Col. (1) are summmarized. In Cols. (5) and
(6) the areas corresponding to the elevation ranges in Cols. (3) and
(4), respectively, are summarized. In Cols. (7) and (8) the estimated
annual and spatially averaged precipitation for the elevation ranges
in Cols. (3) and (4), respectively, are summarized. Finally, in Cols.
(9) and (10) the potential recharge for the NTS area is estimated as
the product of Cols. (2); (5) or (6); (7) or (8) and appropriate conversion factors. A comparison of the total potential recharge numbers
in Cols. (7) and (8) indicates that the effect of adding approximately
6 years of record was to increase the estimated total potential
groundwater recharge for the NTS area by approximately 25%.

It must be realized that the 25% change in potential recharge is
the result of two factors. First, as noted in Table 1, the addition
of 6 years of data to the period of record resulted in an increase in

TABLE 2. COMPARISON OF ESTIMATED POTENTIAL RECHARGE ON THE NEVADA TEST SITE USING THE MAXEY-EAKIN TECHNIQUE AND TWO PRECIPITATION DATA BASES (1 ft. = 0.305 m, 1 in. = 2.54 cm)

Annual Precipitation in.	Maxey-Eakin Recharge %	NTS elevations corresponding to precipitation in Col. (1)		NTS areas corresponding to elevation ranges		Estimated annual precipitation corresponding to elevation ranges		Potential Recharge	
		Data thru 1979 ft.	Data thru 1985 ft.	Data thru 1979 (mi)²	Data thru 1985 (mi)²	Data thru 1979 in.	Data thru 1985 in.	Data thru 1979 ac-ft.	Data thru 1985 ac-ft.
(1)	(2)	(3)	(4)	(5)	(6)	(7)	(8)	(9)	(10)
>20	25	none	none	--	--	--	--	--	--
15-20	15	none	none	--	--	--	--	--	--
12-15	7	6790-7500	6590-7500	160	200	13.03	13.27	7780	9910
8-12	3	5030-6790	4710-6590	380	460	9.86	9.86	5990	7260
<8	0	2700-5030	2700-4710	800	680	6.19	6.48	0	0
Total potential recharge								13,800	17,200

annual precipitation at each of the 11 stations examined. This upward movement of the annual values of precipitation resulted in a change in the values of the regression coefficients, Eqs. (2) and (3). These are changes that are the direct result of lengthening the period of record analyzed; however, these changes then directly affect the Maxey-Eakin method of potential recharge estimate; for example, compare the elevation ranges in Cols. (3) and (4) of Table 2 or the NTS areas corresponding to these elevation ranges in Cols. (5) and (6). Thus, moderate upward changes in annual precipitation combined with this technique of potential recharge estimation combine to yield large shifts in the estimated potential recharge.

Discussion

The foregoing abbreviated analysis serves to emphasize several important aspects of the precipitation-recharge problem in arid and semi-arid regions. The social/legislative purpose of estimating recharge is to prevent the issuance of water rights that will result in the overdraft or mining of the groundwater available in a basin. Without discussing the validity of the Maxey-Eakin methodology or comparing it to other techniques, it is clear that the problem of determining the spatial and temporal distribution of precipitation must be addressed first. Finally, there is increasing evidence that winter precipitation at higher elevations is responsible for much of the natural recharge of the groundwater system at the NTS; Russell et al. (1986).

Acknowledgement

This paper is partially based on work performed for the U.S. Dept. of Energy under contract DE-AC08-85NV10384.

References

French, R.H., "The Effect of a Lengthening Period of Record on Annual and Seasonal Estimates of Precipitation at the Nevada Test Site, Nevada," Water Resources Center, Desert Research Institute, Las Vegas, Nevada, in press.

French, R.H., "Daily, Seasonal, and Annual Precipitation at the Nevada Test Site, Nevada," Pub. No. 45042, Water Resources Center, Desert Research Institute, Las Vegas, Nevada, 1986.

French, R.H., "Precipitation in Southern Nevada," ASCE, Journal of Hydraulic Engineering, Vol. 109, No. 7, July 1983, pp. 1023-1036.

Houghton, J.G., Sakamoto, C.M., and Gifford, R.O., "Nevada's Weather and Climate," Sp. Pub. 2, Nevada Bureau of Mines and Geology, University of Nevada, Reno, 1983.

Osborn, H.B., "Estimating Precipitation in Mountainous Regions," ASCE, Journal of Hydraulic Engineering, Vol. 110, No. 12, December, 1984, pp. 1859-1863.

Russell, C.E., "Hydrogeologic Investigations of Flow in Fractured Tuffs, Rainier Mesa, Nevada Test Site," thesis presented to the University of Nevada, Las Vegas, Nevada, 1987.

Watson, P., Sinclair, P., and Waggoner, R., "Quantitative Evaluation of a Method for Estimating Recharge to the Desert Basins of Nevada," Journal of Hydrology, Vol. 31, 1976, pp. 335-357.

WELL FIELD MANAGEMENT DESIGNED TO
MINIMIZE IMPACT ON SURFACE-WATER FLOW

William K. Beckman[*], A.M. ASCE

Abstract

Restrictions placed on the operation of a well field used for public water supply initiated an investigation to develop management techniques that would maximize utilization of the well field. Management alternatives were identified and then evaluated with the aid of a computer model. Insight to the relationship between the well field, aquifer and nearby river was gained. Two alternatives were found to have potential for extending well field operation while meeting the imposed restrictions.

Introduction

The Ramapo Valley well field consists of 10 production wells screened in sand and gravel along a 2-mile stretch of the Ramapo River in southeastern New York. The well field delivers an average of 0.35 to 0.44 m^3/s and a maximum of 0.61 m^3/s to the public supply system of the Spring Valley Water Company (SVWC). A large portion of the water pumped by the wells is derived from infiltration from the Ramapo River.

To the south, the Village of Suffern operates four wells, completed in the same aquifer, having an average total yield of about 0.066 m^3/s. Further south, the village discharges treated sewage effluent to the Ramapo River. The village is restricted from discharging effluent when river flow falls below the 7-day low-flow having a 10-year recurrence interval (MA_7CD_{10}), a flow of about 0.35 m^3/s at the Mahwah, New Jersey gage just downstream of the waste treatment plant.

Since low flows of the Ramapo River falls below 0.35 m^3/s naturally, a large well field dependent on induced infiltration would have the potential to aggravate the effluent assimilation capacity of the river. The New York State Department of Environmental Protection recognized this situation and included operational restrictions in the water supply permit for the Ramapo Valley well field. The restrictions require the well field to be shut down whenever the river flow leaving the well field area falls below 0.35 m^3/s as measured at the "Regulatory Weir" located between the SVWC and Suffern well fields.

This restriction on the operation of the Ramapo Valley well field prompted an investigation to determine management guidelines that would maximize pumpage during low river flow situations. The alternatives identified for well field management were no action, long-term

*Associate, Leggette, Brashears & Graham, Inc., 72 Danbury Road, Wilton, Connecticut 06897

pumping rotation, short-term pumping rotation, sequential pumping
reduction and river-flow augmentation. Each alternative was evaluated
on the premise that the maximum yield was needed from the well field
while minimizing the impact on the river and adhering to the operational restrictions. The computer model of the aquifer and well
field, that was developed to fulfill another condition of the permit,
was used to evaluate the management alternatives before the entire
well field was put online.

No Action Alternative

Well-field operation under this alternative would depend on the
system demand. River flow would not be a consideration until the
low-flow operational restriction took affect. With a low total
withdrawal rate (less than $0.13 \text{ m}^3/\text{s}$), and not anticipating an
increased or extended demand, this alternative is satisfactory until
incoming river flow falls to 0.39 to $0.44 \text{ m}^3/\text{s}$. River flow entering
the well field area is measured at the "Monitoring Weir"; located
upstream and beyond the influence of the well field. This alternative
no longer meets the requirements if a larger yield is required, the
well field has already been operating at a large rate, an increase in
pumping rate is expected, or extended pumping at low rates is desired
after incoming river flow as measured at the Monitoring Weir falls
below 0.35 or $0.39 \text{ m}^3/\text{s}$.

In general, the no action alternative is limited to very specific
conditions which depend not only on the river flow but also the
existing and anticipated operation of the well field. As a result,
little potential exists for well field operation and flexibility in
management during low-flow situations when using the no-action alternative.

Long-Term Pumping Rotation Alternative

The seasonal shift of centers of pumping attempts to operate
wells near the river during times when river flow exceeds $1.1 \text{ m}^3/\text{s}$,
taking advantage of induced infiltration. In contrast, during low-
flow periods, pumpage would be shifted to wells more distant from the
river in order to utilize aquifer storage. Model results indicated
that because the aquifer is narrow, even the well farthest from the
river derives a large portion of its water from the river and that no
significant benefit is derived from this alternative. Therefore, this
alternative was judged as inapplicable.

Short-Term Pumping Rotation Alternative

The sources of water pumped by wells in the Ramapo Valley are
aquifer storage, river infiltration and recharge from precipitation.
This rotation technique attempts to take advantage of the relatively
large amount of pumped water derived from aquifer storage before
increasingly larger portions of water are induced from the river.

Two groups of wells were selected so that the total withdrawal
rate from each group was $0.17 \text{ m}^3/\text{s}$ and the areas influenced by pumping
would not overlap. Three situations were simulated. Two simulations

started from stable pumping conditions and cycled pumpage between the
well groups at 2 or 5-day intervals. The third situation started from
non-pumping conditions and cycled pumpage at a 5-day interval.

In each case, the loss of river flow from infiltration was
greater than the infiltration developed without rotating pumpage
between the well groups. These larger river flow reductions were due
to residual cones-of-depression that continued to induce infiltration
after pumpage had moved to another location. This phenomenon often
caused the reduction of river flow by infiltration to be larger than
the total pumping rate. As a result, short-term rotation of pumping
was concluded to be an unsuitable alternative for well-field management during low river flows.

Sequential Pumping Reduction Alternative

This alternative was analyzed to determine whether the shut down
of one or more wells in various sequences would have different effects
on river flow to the degree that consideration as a management option
was worthwhile. Since the effects of recovery closely depend on the
conditions under which pumping has been conducted, it was necessary to
define the river-aquifer relationship under pumping conditions prior
to examining recovery from pumping. Declining river flow and stabilized pumping conditions were used as the starting points for this
analysis.

After some period of pumping, the percentage of pumped water
derived from induced infiltration stabilizes. Each situation stabilizes differently and the amount of infiltration depends on the
simulated pumping rate, the river flow or stage, and the array of
operating wells. At pumping rates less than about 0.22 m^3/s, the
total pumpage can be derived from infiltration. However, at rates
more than about 0.39 m^3/s, a maximum induced-infiltration situation is
created due to the hydraulic separation between the river bed and
aquifer water level. In the latter cases, regardless of the amount of
pumpage, the largest possible amount of infiltration is equal to the
maximum infiltration that could be developed under the existing
hydraulic conditions. As a result, the ratio of leakage to total
pumpage ranges from 55 percent at high pumping rates to 100 percent at
low rates. Similarly, the total amount of recovery of river flow that
could be expected when wells are turned off may range from 55 percent
to 100 percent of the rate being pumped.

Another factor to be considered is the time required after the
termination of pumping for full recovery to occur, if it occurs. Just
as 100 percent of the water is not derived from induced infiltration
upon start-up of pumping, neither can river flow be expected to
immediately recover in an amount equal to the pumping rate upon
shutdown of the well. To simplify comparison of results, recovery was
initiated after pumping periods of equal duration and expressed in
terms of residual impact of pumping on river flow.

Applying this information to well-field management would require
forecasting the rate of decline of river flow far enough in advance so
that, after consideration of the factors involved, the proper wells

could be shut down in time for their recovery to meet the decline in
river flow. The factors which influence recovery and should be
considered are the amount and duration of pumping prior to recovery,
the river flow rate and the combination of pumping wells. Seasonal
factors include sedimentation, scouring and river-water temperature.
Though all these factors can be accounted for, projecting river flow
one to three weeks in advance is particularly speculative, especially
when runoff from a small rain storm could significantly alter stream
flow and make unnecessary the prior shutdown of a well and the accompanying loss of water to the system. Potentially, more water could be
lost from the system than gained if pumpage reductions were frequently
negated due to precipitation.

In general, the use of sequential pumping reduction for well-
field management will permit well-field operation in limited short-
term situations which without management would have been shut down.
Hydrogeologic factors did not indicate any particular advantage to
shutting down specific wells prior to others. Therefore, the suggested order of shutdown was from the largest capacity well to the smallest. This was believed to be a site-specific phenomenon and should
not to be generally applied to other sites without first examining the
hydrogeologic factors. Additionally, this technique does not extend
well field operation below river inflows of 0.39 to 0.44 m^3/s.

River Augmentation Alternative

The goal of the river augmentation analysis was to determine
whether or not discharging one or more wells into the river to meet
low-flow restrictions could extend well-field operation under low
river-flow conditions. During these low-flow conditions, it was
assumed that there is little or no rainfall so that natural recharge
to the ground-water system is nonexistent. Pumpage would be derived,
therefore, solely from aquifer storage and induced infiltration from
the river.

The ratio of water derived from these two sources is the critical
factor in determining the success of river augmentation procedures.
The portion of water a pump-back well obtains from induced infiltration is simply recycled back into the river for neither a gain nor
loss of river flow. Any net increase in river flow which may result
when a well is discharged into the river is directly equal to the
portion of pumped water which was derived from aquifer storage. As
with the other cases, the proportion is also pumping-time dependent.

Numerous simulations were made to define the ratio of pumped
water derived from aquifer storage to that from induced infiltration.
All the factors described in the analysis of Sequential Pumping
Reduction as affecting infiltration (amount, location and duration of
pumping and river flow/stage) also affect river augmentation schemes.
Because each of these factors may vary, there are a very large number
of situations that are possible. For ease of analysis and to present
a few examples of the potential of river-flow augmentation, some
simplifying assumptions were made.

1. Wells used for augmentation of river flow started with the southernmost well, with additional wells being located progressively to the north.

2. Wells operated to the system started with the northernmost well, with additional wells being located progressively to the south.

3. Infiltration rates were based on simulations of several withdrawal rates from two river stages and flows (0.44 and 0.66 m^3/s). Adjustments were made from these base values for other pumping rates and river flows.

The first test case postulated that all ten wells were in operation so that maximum infiltration would be occurring. Assumed demand on the system also required that maximum possible water be pumped. Initially, river flow at the Regulatory Weir was presumed to be adequate to meet all restrictions. However, the incoming river flow at the north end of the well field was assumed to decline, forcing more and more of the well production to be diverted from the distribution system to the river. Since maximum infiltration had been in effect, the smallest possible amount of water was being taken from aquifer storage for this pumping scheme, making this a worst-case situation in terms of impact on river flow.

If, for example, while all wells are pumping to the system, the flow in the Ramapo River as measured at the Monitoring Weir declines from 0.70 to 0.66 m^3/s, the southernmost well would have to be shut down so that flow measured at the Regulatory Weir would be more than 0.35 m^3/s. Eliminating the impact of that one well on river flow would be sufficient to meet restrictions at the Regulatory Weir for as long as conditions remain stable. Should conditions change, there would be no problem if demand on the system were reduced or river flow increased. However, should flow at the Monitoring Weir fall to 0.61 m^3/s, a deficit of 0.022 m^3/s (below 0.35 m^3/s) would be realized at the Regulatory Weir, resulting in shutdown of the well field unless the southernmost well were then discharged to the river. In this situation, 0.52 m^3/s would be pumped to the system and 0.088 m^3/s to the Ramapo River. Again, this condition would remain stable for at least two weeks if all other factors remained constant.

In contrast to this first case, the second test attempted to define the optimum effects of river augmentation. The major difference between this and the first test case was in the initial assumptions. For the second test case, the situation required that pumping would be needed for only one additional day. Therefore, augmentation to allow this pumping would be needed for one day. The other changed assumption was that wells to be used for augmentation had been idle and were, therefore, being started from static conditions. The last assumption remained unchanged so that whatever wells happened to be operating, would have been operating long enough to establish maximum pumping impact on the river for that particular array of wells.

The model results indicated that the first two factors, the duration of river-flow augmentation and the starting condition of the augmentation wells, drastically alter the conditions under which augmentation may be possible. Due to the changed initial assumptions for the second case, a larger portion of pumped water is derived from aquifer storage. This results in larger net increases in river flow when a well is pumped to the river. Therefore, fewer wells are required than for the first case to maintain pumpage under the same flow condition. Also, pumpage to the system can continue at lower river flows than in the first case for similar rates of pumping.

In each of the two cases, the procedure would continue by diverting well discharge to the river as needed to match decreases of increasing river flow. These two cases were presented as extremes to illustrate the potential of augmentation. There are many situations in between, to which augmentation may be used to carry pumping over a short-term situation or provide temporary relief during a long-term event.

Generally, augmentation of river flow is the only method of those analyzed that permits some well-field operation when river flow entering the well field at the Monitoring Weir is less than 0.35 m^3/s. This management alternative has its greatest potential for short periods of time (1 or 2 days). Although possible for longer durations, it should not be considered a definite or practical solution for all low-flow situations. Augmentation also provides the greatest flexibility of well-field operation when pumpage to the system is less than 0.22 m^3/s and incoming river flow is between 0.26 to 0.44 m^3/s.

Summary

There is not one rule for operation of the Ramapo Valley well field that will cover all situations all of the time. Instead, each situation should be evaluated with consideration of system demand, hydrologic conditions and well-field operation in terms of the recent past, present and near future. When incoming river flow at the Monitoring Weir reaches 0.88 m^3/s, the evaluation process should be initiated, alternative courses of action outlined and well-field operation modified in preparation of further action. When flow at the Monitoring Weir reaches 0.70 m^3/s, the best alternative should be chosen and put into action.

The permit condition of maintaining a minimum flow at the outlet of the well field during operation complicates management due to a time lag between certain remedial actions and their eventual effects. The sequential reduction of pumping and river-flow augmentation alternatives both have potential for extending well-field operation during low-flow conditions. River-flow augmentation was shown to have the greatest potential and flexibility for extending well field operation during low-flow periods. Based on experience, however, sequential reduction of pumpage is the easier to implement and has proved to be satisfactory thus far.

UNDERGROUND DRAINAGE OF URBAN RUNOFF
FROM NATURAL DEPRESSIONS

By Jobaid Kabir[1], A.M. ASCE

Areas of natural depressions with improper drainage usually poses
potential of frequent flooding of the surrounding areas, various health
hazards, and unacceptable asthetic scenes. Due to the unfavorable
topographic conditions, drainage of such natural depressions calls for
pump driven systems of high capital expenditures and expensive and
troublesome maintenance procedures. A unique subsurface system was designed and implemented for draining such a natural depression in the
City of West Lake Hills, Texas. This subsurface drainage system uses
a part of the natural depression as a detention basin for storms of
design frequency of 25 years or less. A filter system was installed in
the natural depression to remove undesireable elements of the surface
drainage. Filtered surface runoff is then collected by a system of
perforated pipes and delivered to a drain well located at the lowest
point of the natural depression. This system of managing storm runoff
has proved to be very cost-effective, trouble-free, and efficient for
the City of West Lake Hills, which experiences several high intensity
rainfalls annually.

Background

The natural low lying area on the east side of Revelle Road in the
City of West Lake Hills, Texas, works as a retention pond in its present condition. Storm water drainage from an area of approximately
twenty acres reaches this retention pond. Part of this water reaching
the site evaporates and the remainder infiltrates into the soil. Close
observation of this retention pond indicates that the water level drops
fairly fast, even on a cool and cloudy day. This indicates that infiltration is higher than that amount evaporated from the surface.

It was observed by City of West Lake Hills' officials that runoff
from high intensity storms causes frequent overflow affecting the
surrounding neighborhood.

Project Objective

The purpose of the proposed project is to reduce overflowing of the
natural retention pond located east of Revelle Road in the City of West
Lake Hills, Texas.

---

[1]Hydraulic Engineer, ARE Inc - Engineering Consultants, 2600 Dellana
Lane, Austin, Texas 78746

Area Geology

The site for the drain well is located adjacent to the Edwards Aquifer Zone. No data was available to show any ground water use in the local area and no water wells exist surrounding the proposed project site (5).

Table 1 of Reference No. 4 summarizes the water-bearing properties of the Edwards and associated limestones aquifer. The Edwards and associated limestones represent the upper portion of the Fredericksburg Group and the lower portion of the Washita Group of the Cretaceous System. They lie above the Walnut Formation and below the Del Rio Clay. Collectively, these limestones are considered the principal aquifer in Travis County and include, in ascending order, the Comanche Peak Limestone, Edwards Limestone, Kiamichi Formation, and Georgetown Formation.

Ground water in the Edwards and associated limestones aquifer may be described as a calcium carbonate, and sometimes magnesium carbonate water, generally becoming a sodium sulfate water downdip (4). Still further downdip, it becomes a sodium chloride water. It is very hard, usually fresh, and normally neutral. Its quality decreases rapidly downdip or to the southeast (Figure 12, Reference No. 4). Decreasing water circulation through faults, increasing temperature as the depth of the aquifer increases, and solution of the rocks cause the ground water to become more highly mineralized downdip.

Records of chemical analyses from selected wells in Travis and adjacent counties are given in Table 7 of Reference No. 4. The chloride, sulfate, and dissolved soils from selected wells completed in the Edwards and associated limestones aquifer are shown in Figure 12 of Reference No. 4. The water is generally within the recommended limits for drinking water as established by the Texas Department of Health, except near its downdip limit of fresh to slightly saline water where higher concentration of dissolved minerals occur. In most areas, except in the extreme downdip area, water from the Edwards and associated limestones aquifer is suitable for public supply, irrigation, and industrial use.

The source, significance, and range in concentrations of chemical constituents for ground water collected from the Edwards and associated limestones aquifer are given in Table 2 of Reference No. 4. The recommended primary and secondary constituent levels should be considered when evaluating the quality of water for public and domestic use. It should be noted that these concentration limits will apply except where suitable public water supplies are not available or cannot be made available at a reasonable cost.

Iron content in the Edwards and associated limestones aquifer range from 0 to 13 mg/1 in 32 samples with 31 percent exceeding the recommended 0.3 mg/1. Sulfate content range from 4 to 2,750 mg/1 in 182 samples with less than 9 percent exceeding 300 mg/1. Chloride content range from 4 to 6,050 mg/1. Fluoride content range from 0 to 4.8 mg/1 in 147 samples with 35 percent of the samples collected in Travis County exceeding the recommended upper limit of 1.4 mg/1. The range in

nitrate content was 0 to 88 mg/1 in 163 samples with less than 4 percent of the samples exceeding 45 mg/1. Dissolved solids content range
from 173 to 10,190 mg/1 in 185 samples with only 11 percent exceeding
1,000 mg/1.

Water Quality of Storm Runoff

In 1983, the City of Austin published a report (1) showing the water
quality of storm runoff from several areas in and around the city limits. This extensive study included a site in the City of Rollingwood
representative of low density residential areas. A total of sixteen
events were monitored for the Rollingwood watershed in this study as
summarized in Table 14 of Reference No. 1. Due to the proximity of the
proposed project site to the Rollingwood watershed, and similar nature
of topography, geology, and population, this data was assumed to be
representative of the water quality of the runoff reaching the retention pond. A summary of storm water runoff pollution concentration for
the Rollingwood watershed is shown in Table 5 of Reference No. 2. More
details of these pollution data are shown in Tables 37 and 39 of Reference No. 1.

Review of the pollution data mentioned above indicates that, generally, the level of pollution is low in the Rollingwood watershed. On
the basis of this, it was assumed that the pollution level of the storm
water runoff reaching the retention pond will also be low. However,
water quality data from the Rollingwood watershed showed a significantly high level of lead and methylene chloride in the storm water runoff.
At this point, it was decided that water quality analyses of the storm
runoff reaching the retention pond should be made.

Two water samples were collected from the retention basin and analyzed by the Environmental Laboratory of the Lower Colorado River Authority. Results of these laboratory analyses are shown in Table 1.
Water sample 1 was collected at 5:00 p.m. on November 14, 1985, at
which time the water level was very low and quite some time had passed
following the last rain storm. Water sample 2 was collected from the
"first flush" storm water runoff after a rainfall event in the evening
of November 14 and morning of November 15, 1985. This second sample
was collected at 12:00 noon on November 15, 1985.

Table 1. Results of laboratory analyses

Sample No.	Parameter	Results	Units
1	Lead, Total	0.005	mg/1
	COD	20	mg/1
	TOC	7	mg/1
	Methylene Chloride	5	mg/1
2	Lead, Total	0.005	mg/1
	COD	14	mg/1
	TOC	4	mg/1
	Methylene Choride	5	mg/1

Results of laboratory analyses in Table 1 indicate that levels of lead, methylene chloride, COD, and TOC are very low. This also shows that a significantly high level of lead and methylene chloride in the Rollingwood watershed is not representative of the storm water runoff reaching the proposed project site and, therefore, should be of no concern.

Pollutant Removal

The City of Austin is currently developing guidelines for designing filter beds for removing pollutants from surface runoff. According to these guidelines (3), sand bed filters of a certain combination can be used for the removal of most of the detrimental pollutants from surface runoff. Pollutant removal efficiencies of sand bed filters are given in Reference No. 1. The City of Austin developed this guideline using preliminary findings of its Storm Water Monitoring Program. Following this guideline, a sand bed filter and collection system for the filtered water was designed for the proposed site. It was proposed that the filtered water be collected by a system of 4" diameter PVC pipes and then delivered to the well by a 6" main. Considering the low level of pollution in the water reading, the retention basin, and pollutant removal by sand bed filters, it may be concluded that the quality of the water reaching the aquifer is equal to that of the aquifer, if not better.

Filterbed Maintenance

Due to the gentle slope of the area discharging runoff to the proposed site and extensive vegetation at the ground surface, it is expected that the maintenance requirements for the proposed filterbed will be minimal. However, to ensure proper functioning of the retention basin and filterbed, the following maintenance and inspection schedule was proposed.

Retention Basin

Removal of silt when accumulation exceeds 6 inches
Removal of accumulated paper, trash, and debris every 7 months
Vegetation growing within the basin will not be allowed to exceed 18 inches in height at any time
Annual inspection and repair of the structure
Any time a detention pond does not drain completely, corrective maintenance is required

Filtration Basin

Removal of silt when accumulation exceeds one-half inch
Removal of accumulated paper, trash, and debris every 6 months
Vegetation growing within the basin is not allowed to exceed 6 inches in height
Corrective maintenance is required any time draindown does not occur within 36 hours
Annual inspection and repair of the structure

REFERENCES

1. The City of Austin, "Final Report of Nationwide Urban Runoff Program in Austin, Texas", January 1983.

2. City of Austin, "Interim Water Quality Report - Hydrologic and Water Quality Data for Barton Creek Square Mall and Alta Vista PUD", Fall 1985.

3. City of Austin, "Water Quality Design Guidelines for Filtration and Detention/Retention Basins", currently being prepared.

4. Texas Department of Water Resources, "Occurrence, Availability, and Quality of Ground Water in Travis County, Texas", Report No. 276, June 1983.

5. Hargarten, Richard A., City Administrator, City of West Lake Hills, Telephone Interview, September 19, 1985.

EVALUATION OF STORMWATER MANAGEMENT STRUCTURES IN MARYLAND
PROPORTIONED BY SCS TR-55

By Donald Woodward,[1] M. ASCE and Helen Fox Moody[2]

Abstract

Various state and local agencies in Maryland are requiring that the Soil Conservation Service Technical Release 55 be used to evaluate impact of urbanization on peak rates of discharge. In 1981, these agencies began to wonder if the procedures in TR-55 (SCS, 1975) were oversizing the outlet of stormwater management structures. Thus, the desired regulation of frequent events was not happening. In 1983, SCS in Maryland began a study to evaluate the concern of these agencies. The results of this study are presented.

Introduction

Various state and local agencies in Maryland are requiring that the impact of urbanization on peak rates of discharge be evaluated using Soil Conservation Service Technical Release 55, "Urban Hydrology for Small Watersheds" (TR-55) (SCS, 1975). TR-55 is also being used to determine the amount of storage needed to mitigate the increase in peak discharge due to urbanization.

In 1981, these same local agencies began to wonder if the procedures in TR-55 (SCS, 1975) were oversizing the outlet works of stormwater management structures and thus the desired regulation of frequent rainfall events was not happening. It had been reported that installed stormwater management structures had no high water marks associated with significant amounts of storage.

---

[1] Head, Hydrology Unit, Soil Conservation Service, P.O. Box 2890, Washington, D.C. 20013
[2] Agricultural Engineer, Soil Conservation Service, Hartwick Building, Room 522, 4321 Hartwick Road, College Park, Maryland 20740

In 1983, SCS in Maryland initiated a program of measuring the rainfall and maximum reservoir stage at four existing stormwater management structures. The selected structures are representative of urban or urbanizing watersheds in Montgomery County. Table 1 gives the watershed characteristics for the selected stormwater management structures. Selection was based on availability of original calculations, ease of access to outlet works and general condition of the structure. Montgomery County Department of Environmental Resources provided copies of the original calculations.

It was decided to utilize an existing recording rain gage network in Montgomery County rather than install a standard rain gage at each site. In all cases, there were two or more recording rain gages near each site.

A standard plastic pipe crest staff gage was installed by SCS at each site. Stage was determined by observing the height of regranulated cork on a measuring stick. Initially, the county provided observers for the crest staff gages. SCS provided the observer for the last year of the study.

Watershed Characteristics

The characteristics of the four watersheds are shown in Table 1.

The stormwater management structures were proportioned using standard storage routing techniques. The original stage-storage information was checked by comparing limited field surveys with original site maps. There were minor difference between design and as-built stage-storage information. In three of the four sites the stage-discharge information was checked. The as-built condition varied from the original design for the fourth site. The tabular hydrograph procedure in TR-55 (SCS, 1975) was used to develop the needed inflow hydrographs.

Rainfall Data

The National Weather Service Technical Paper No. 40 (Hershfield, 1961) indicates the 2-year, 24-hour precipitation for Montgomery County is 81.3 mm. A review of the available records for seven standard rain gages used in this study indicated that during a six year period (1980-85), only three 24-hour periods at two gages had precipitation exceeding this amount. The maximum annual 24-hour record precipitation and in Table 2.

Table 1.--Watershed Characteristics

Site ID	Drainage Area	Hydrologic Soils	Present Runoff Curve Number	Present T_c	Outlet Dia.
	-ha-			-hrs-	-cm-
Horizon Hills	33.3	B & D	79	.13	76.2
Bethesda Bus Park	3.66	B	90	.21	25.4
Rock Creek Manor	2.63	B	84	.04	20.3
North Farm	.49	B	75	.13	45.7

Table 2.--Precipitation
Largest (24-hour) Amount
(mm)

| Year | ------------------ Rain gage Number ------------------ | | | | | | |
	243	235	244	145	138	346A	335
1980	48.0	25.4	55.9	86.4	38.1	26.7	27.9
1981	28.4	38.1	27.9	82.6	61.7	44.5	61.0
1982	26.9	88.9	26.7	26.7	27.2	21.8	23.6
1983	51.8	36.8	73.4	81.3	68.6	50.8	78.7
1984	63.5	50.8	68.6	59.7	68.6	66.0	53.3
1985	79.0	80.0	81.3	78.7	77.5	72.4	72.4

Storm Data

Table 3 lists the date, the amount, and the duration for the selected events. The precipitation is the arithmetic average of storm total at the two nearest rain gages. The three highest reservoir stages at each site for period 1983-1986 were selected for detailed study. It is interesting to note that an event larger than the 2-year 24-hour amount was not in the sample set.

Table 3.--Selected Events

Watershed	------------- Events --------------			
Horizon Hills	5/16/83 78.7 15	8/1/84 68.8 2	9/27/85 78.7 15	date amount (mm) duration (hrs)
Bethesda Bus Park	3/28-29/84 65.6 34	7/13/85 53.3 2	9/27/85 72.4 15	date amount (mm) duration (hrs)
Rock Creek Manor	5/16/83 78.7 15	3/28-29/84 61.7 34	9/27/85 78.7 15	date amount (mm) duration (hrs)
North Farm	5/22/83 24.1 15	8/1/84 44.4 2	7/20/86 39.4 4	date amount (mm) duration (hrs)

Detailed Studies

Table 4 indicates the results of the detailed studies. The Soil Conservation Service DAMS2 computer program (SCS, 1982) was used to develop the inflow hydrograph and route it through the site using the original stage storage-discharge information. Actual recorded precipitation amounts and temporal distributions were used as input data.

The curve number was adjusted to represent the moisture condition of the soil profile at the time of the storm. The amount of precipitation that occurred in the prior 5-day was used as an index of the watershed conditions. With no prior precipitation it was assumed that dry conditions (AMC I) existed and with 50-80 mm of prior precipitation it was assumed that wet conditions (AMC III) existed. The amount of curve number adjustment was based on information in Table 10.1, National Engineering Handbook, Section 4 - Hydrology (NEH-4) (SCS, 1985) and the percentage of impervious area in the watershed. The adjustment for Bethesda Bus Park was less than North Farm because there is more impervious area in Bethesda Bus Park watershed than in the North Farm watershed.

Observer notes indicated that on two visits the low stage orifice on Rock Creek Manor structures was partially plugged and some debris was removed. It was assumed that 50% of the low stage orifice area was plugged for the May 1983 and September 1985 events.

Table 4.--Results of Simulation Studies

	Horizon Hills	Bethesda Bus Park	Rock Creek Manor	North Farm
Date	5/16/83	3/28-29/84	5/16/83*	5/22/83
Recorded Stage (m)	111.2	76.6	93.0	94.7
Simulated Stage (m)	111.1	76.7	92.0	94.7
Reservoir Depth (m)	.5	.5	.9	.3
CN	91	96	90	88
Soil Condition	Wet	Wet	Wet	Wet
Date	8/1/84	7/13/86	2/28-29/84	7/20/86
Recorded Stage (m)	111.5	77.0	92.7	94.6
Simulated Stage (m)	111.5	77.0	92.6	94.6
Reservoir Depth (m)	.8	.8	.6	.3
CN	62	90	90	75
Soil Condition	Dry	Average	Wet	Average
Date	9/27/85	9/27/85	9/27/85*	8/1/84
Recorded Stage (m)	111.0	76.6	93.2	94.8
Simulated Stage (m)	111.0	76.6	93.2	94.8
Reservoir Depth (m)	.4	.4	1.1	.5
CN	62	82	80	75
Soil Condition	Dry	Dry	Dry	Average

* For these events it was assumed that some plugging of low stage orifice had occurred.

Conclusions

As a result of this study it was concluded that:

(1) Because precipitation greater than the 2-year 24-hour event
     was not measured, a significant amount of storage did not
     occur. The maximum water depth measured was 1.1 meters.

(2) The maximum recorded reservoir stage was readily simulated in
     10 of the 12 selected events using existing watershed
     conditions, actual precipitation data and standard SCS
     hydrologic techniques. In the other two events, some
     plugging of the low stage orifice was assumed. While
     simulation of a recorded hydrograph is the best index of
     accuracy, simulation of a reservoir stage is a useful index
     of accuracy of a hydrology procedure.

(3) The standard SCS hydrologic techniques in TR-55 worked well
     in these small urban watersheds.

(4) For the four selected sites, the low stage orifice and
     principal spillway size would not change using the June 1986
     version of TR-55 (SCS, 1986).

References

Hershfield, D.M. 1961. "Rainfall Frequency Atlas of the United
State for Durations from 30-Minutes to 24-Hours and Return Period
From 1 to 100 Years." U.S. Department of Commerce, Weather Bureau
Technical Paper No. 40. Washington, D.C.

Soil Conservation Service. 1975. "Urban Hydrology for Small
Watersheds," Technical Release 55. Washington, D.C.

_____. 1982. "Structure Site Analysis Computer Program - DAMS2
(Interim Version)." Technical Release 48. Washington, D.C.

_____. 1985. National Engineering Handbook, Section 4 -
Hydrology, Washington, D.C.

_____. 1986. "Urban Hydrology for Small Watersheds," Technical
Release 55. Washington, D.C.

Design Criteria for Detention Basins in Urban Drainage

Abbas A. Fiuzat, M., ASCE and Scott E. Sonnenberg[*]

Abstract

Many communities require the use of detention basins in order to manage the increased runoff discharge in all newly developing watersheds. The authors have shown earlier that the typical design criteria for detention basins can cause increased peak discharges from watersheds. In this paper the effect of design rainfall frequency on peak discharges has been reported, and design criteria for detention basins have been proposed to maintain the watershed discharges at peak developed or peak undeveloped discharges.

Introduction

Rapid rate of urbanization and growth of cities into rural areas has been resulting in increased runoff flows, reduced flow times, and frequent floodings. In an effort to encounter the problem and protect the welfare and safety of their residents, many communities now require the construction of private detention facilities on all future developments. These installations constitute a small portion of the cost of development and hence encounter little opposition from the residents. Using an idealized watershed model, the authors have shown that the typical design criteria now in use for the design of detention basins can result in increased peak discharges from the watershed (Sonnenberg, 1986; Sonnenberg and Fiuzat, 1987). In this paper the effect of the recurrence interval of design rainfall on peak watershed discharges with detention basins has been reported, and criteria have been compared in order to keep the watershed discharge below the peak discharge prior to development, or below the peak discharge for a developed watershed that utilizes no detention basins.

Model Development

Three models were needed for use in a computational analysis of watersheds using detention basins, namely, a watershed model, a runoff model, and a detention basin model. In order to select a watershed model that was realistic, the city of Greenville in South Carolina, which has been suffering from problems of rapid development, was used as a guide. Comparing 11 watersheds constituting the metropolitan city area, an idealized watershed was selected with a total area of 6000 acres (2500 hectares), an elevation drop of 300 ft (100 m) and a concentration time of 135 minutes. For developing runoff hydrographs

*Assistant Professor, Department of Civil Engineering, Clemson University, Clemson, South Carolina and Landscape Architect, William Renninger Associates, Greenville, South Carolina, respectively.

from the watershed, the area was divided into four equal subbasins, flow passing sequentially from one to the next.

The runoff model used was the Modified Rational Method (ASCE, 1969), which states
    Q = CIA
where Q = the runoff flow rate; C = runoff coefficient; I = design rainfall intensity; and A = area. For the area in acres and intensity in inches per hour, the flow rate will be in cfs (28 cfs = 1 m^3/s). A runoff coefficient of 0.3 for the undeveloped watershed and 0.7 for the developed watershed were used. For the rainfall intensity, the intensity-duration-frequency curves for Greenville, South Carolina were used with a 10-year recurrence interval (City of Greenville, 1979). Other tested frequencies are given in the next section.

The Modified Rational Method gives good estimates of peak discharge, but is not suitable for the recession curve of the flow hydrograph. Because the basin was subdivided into subbasins, the SCS dimensionless synthetic hydrograph was used to find the recession curves of the flow hydrographs from subbasins (Mockus, 1957).

The detention model constituted a facility that would release water at a constant rate equal to the peak runoff rate before development. The Design Time of Concentration (DTc) for such basins is typically 5 to 10 minutes, the latter value being used in this work. The facility holds water until the total volume of outflow equals the total volume of inflow, after which time the facility is no longer holding or storing water. This time interval is called the design holding time (DHT).

The Results

Calculations of the outflow hydrograph from a watershed with detention facilities are complicated by the fact that several detention facilities release water at staggered times and varying distances from the outlet. The simple models explained above reduce the complexity of calculations, but the procedure is still too lengthy to be presented in this space. Application of the selected models has been explained in sufficient detail by Sonnenberg (1986). The results showed that using the present criteria for the design of detention basins, the peak watershed discharges can actually increase beyond the levels with no detention. Figure 1 shows a typical result, where development causes increased runoff flow and reduced time to peak, and the use of typical detention facilities increases the peak discharge further. The design time of concentration of the detention basin for the case of Fig. 1 was 57 minutes, beyond which the discharges with detention are more than the discharge without detention.

The effect of using 5-year and 2-year recurrence intervals for the design of detention basins is shown in Fig. 2. Although the peaks are smaller, durations of floodings do not seem to be appreciably different for these cases. Also, the differences between the peak discharges of watershed with no detention and watershed with detention are very similar. This indicates that both the intensity and duration

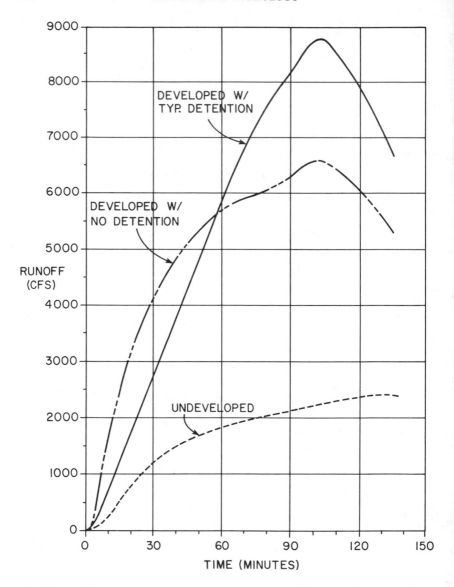

Fig. 1.   Outflow hydrographs for undeveloped and developed watersheds
          and with the use of typical detention basins
          (1 cfs = 0.036 m^3/s)

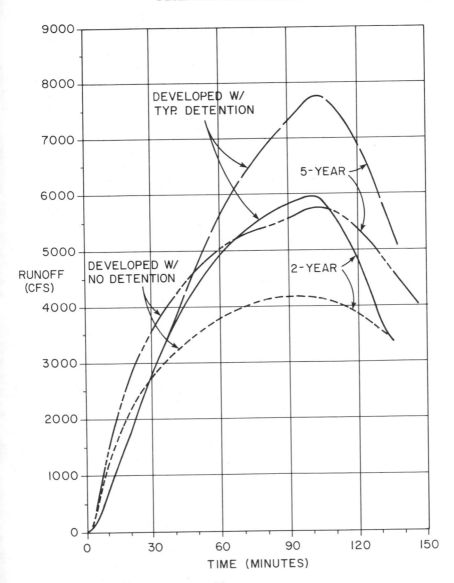

Fig. 2. Outflow hydrographs for 2-year and 5-year recurrence interval rainfall events (1 cfs = 0.036 m³/s)

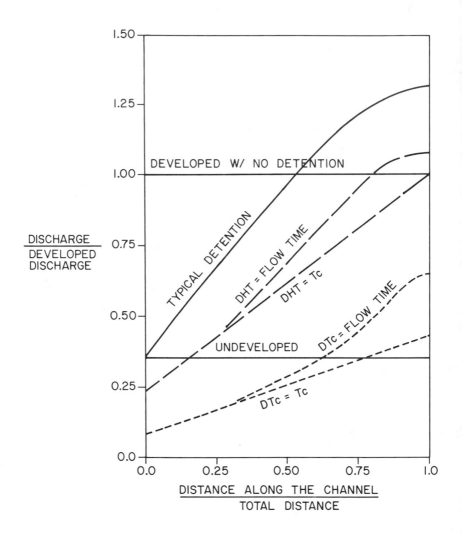

Fig. 3. The effect of different design criteria on watershed outflow discharges

of flooding will be about the same. If the detention basins are
designed for smaller recurrence interval rains, the flooding problem
will recur more often, but not less severely. The problem can be
avoided by changing the design criteria for the detention basins.

If it is desired to keep the peak watershed discharge at the
undeveloped peak flow, the DTc could be set equal to the total
watershed time of concentration. For a constant release rate
detention facility, this will keep the peak watershed outflow at the
level of the undeveloped peak flow. But before the watershed time of
concentration is reached, the outflows will be much less than those
for undeveloped watershed, resulting in extremely large detention
basins. A more relaxed criterion would be to set the DTc equal to the
flow time from each individual detention basin to the outlet of the
watershed. Thus, the largest basins would be the ones farthest from
the outlet. However, both of these criteria are very restrictive and
will result in unduly large detention facilities. It may be more
practical to design detention basins such that the peak watershed
discharge with detention is equal to the peak discharge of the
developed watershed without the use of detention basins. In that
case, the DHT can be set equal to the total watershed time of
concentration. Again, a somewhat more relaxed criterion would be to
set the DHT equal to the flow time from each individual detention
basin to the watershed outlet. The effects of the four criteria given
above have been show in Fig. 3, on a non-dimensional plot.

Conclusions

With the present design criteria, the use of detention facilities
on newly developing watersheds can result in peak runoff flows higher
than the peak flows without detention basins. This situation would
exist irrespective of design rainfall recurrence interval. Depending
on how strictly the situation needs to be controlled, the design
criteria of the detention basins could be changed in several possible
ways, in order to keep peak discharges below specified rates.

Appendix - References

ASCE, Design and Construction of Sanitary and Storm Sewers, Manual on
    Engineering Practice, No. 37, ASCE, 1969.

City of Greenville, South Carolina, Design Criteria and Standard
    Drawings, October 1979.

Mockus, V. "Use of Storm Watershed Characteristics in Synthetic
    Hydrograph Analysis," U.S. Soil Conservation Service, 1957.

Sonnenberg, S.E. "The Effect of Detention Basins on Peak Watershed
    Discharges," Masters thesis, Clemson University, May 1986.

Sonnenberg, S.E., and Fiuzat, A.A. "The Effect of Detention Basins
    on Peak Watershed Discharges," Computational Hydrology '87
    Conference, Anaheim, California, July 12-16, 1987.

DETENTION BASIN DESIGN USING CONTINUOUS SIMULATION

*By Lindell E. Ormsbee, M. ASCE

Traditionally, stormwater detention basins have been sized using the design storm approach. One of the most important limitations of this approach is that antecedent conditions (e.g. soil moisture, initial detention storage levels, etc.) are typically neglected. The main reason that the design storm approach continues to be employed is due to its simplicity and the perceived lack of a better alternative.

As an alternative to the design storm approach, a continuous simulation model is proposed for use in the design of detention basins. The model accepts standard NWS hourly rainfall data as input. The hourly rainfall data may then be disaggregated into smaller time intervals using a built in empirical rainfall distribution model. A continuous runoff series for both the developed and pre-developed conditions is then generated using a continuous soil moisture accounting strategy and a nonlinear reservoir runoff model. For each individual storm a maximum peak discharge constraint is obtained based on the pre-development runoff series. The runoff series for the developed condition is then routed through a generic reservoir from which a series of required storages may be obtained. The final output from the model includes both peak discharge and required storage frequency curves from which design decisions can be made. The model can be used to design individual basins or to construct general frequency curves for a wide range of possible design conditions.

Introduction

Several approaches have been proposed for use in the planning and design of stormwater detention basins. These include: (1) the design storm approach, (2) the derived distribution approach, and (3) the use of continuous simulation. By far, the most popular approach is the design storm approach. Construction of a typical design storm requires a specific rainfall frequency and duration along with an assumed temporal distribution. Through the years, numerous design methodologies have been developed for use with the design storm approach (ASCE, 1983).

In recent years, the design storm approach has been criticized for several reasons (James and Robinson, 1982). Probably the most important deficiency of the method is that antecedent moisture conditions are typically not considered. As a result, the frequency of the stormwater runoff is assumed to be equal to the frequency of the design storm. Such frequencies will only be equal if the watershed is completely impervious.

* Asst. Professor of Civil Engineering, University of Kentucky, Lexington, KY 40506.

376

A more recent approach to stormwater modeling, especially from a water quality standpoint, is the derived distribution approach. This method is based on the statistical distributions of different storm variables. Using different hydrologic relationships, distributions are derived for the dependent variables such as runoff, storage volume, and overflow (Loganathan, et al., 1985). Due to the hydrologic simplifications involved in most derived distribution approaches, the developed methodologies are generally applicable on a macro planning scale.

A third approach for use in the planning and design of stormwater detention basins is the use of continuous simulation. Such an approach is beneficial in that the effect of antecedent conditions on storage structure performance can be determined. In addition, the use of continuous simulation allows the development of storage frequency curves for use in the design of detention facilities.

One of the most commonly stated disadvantages of continuous simulation is that such an approach is extremely time consuming and expensive. With the advent of inexpensive and computationally efficiency microcomputer technology this complaint is generally no longer valid. In those cases where computation time is excessive substantial reductions may be made through the use of simulation models which employ variable time steps (Smith and Alley, 1982). Alternatively, continuous simulation models may be used to screen critical design events that can then be analyzed in more detail (Walesh, et al., 1979; Ellis, et al., 1981).

Continuous Simulation Model

In the current study a continuous simulation model was developed for generating both peak discharge and required storage volumes statistics for small urbanized watersheds. The program is based on a modified version of the Synoptic Rainfall Data Analysis Program - SYNOP (Hydroscience, 1979). SYNOP was originally developed for use in generating statistics for four different storm parameters: storm intensity, storm duration, unit volume and time between storms. For the current study the program was modified to yield statistics on peak discharge and required storage volumes.

Peak discharge statistics are obtained from an analysis of a continuous runoff series. The continuous runoff series is generated using a continuous soil moisture accounting strategy and a nonlinear reservoir runoff model. Rainfall excess is generated using the SCS runoff equation. The soil moisture used in the runoff equation is determined as a function of the 5 day antecendent rainfall volume and the hydrologic curve number.

Required storage statistics are obtained by routing the continuous runoff series through a generic detention basin. For each individual storm a different required storage may be obtained. The required storage for each storm will be a function of the inflow hydrograph and the maximum peak discharge. The most common design criteria for stormwater detention basins is that the peak discharge after development does not exceed the peak discharge which occurred before development. As a result, the pre-development peak discharge is usually a design constraint when routing the post-development hydrograph.

When using continuous simulation two different approaches may be used in determing the required storage. The first approach is to use the same peak discharge for each detention basin routing as shown in Figure 1. This discharge may be determined from a frequency analysis of pre-development discharges or set by some external design consideration. It should be recognized that the storage statistics resulting from this approach are a function of the frequency of the selected peak discharge. Frequency curves developed using this approach would thus represent the frequency of a storage volume for a detention basin with a design frequency corresponding to the the frequency of the selected peak discharge.

Instead of using a constant release rate, different release rates can be set for each individual storm. These release rates can be determined using the peak discharges corresponding to the hydrographs for the pre-development conditions as shown in Figure 2. The storage statistics resulting from this approach are thus independent of a single design frequency. Instead, they are a function of the frequency of the pre-development peak discharge resulting from the same corresponding rainfall event.

In the proposed model, both the single and multiple release rate approaches may be used. When using the multiple release rate approach the model generates both a pre development and post-development runoff series. In general, the multiple release rate approach is more appropriate for a design situation while the single release rate approach is more appropriate for an analysis application.

Detention Basin Routing

In order to determine the required storage statistics, each hydrograph is routed through a generic detention basin. This routing can be accomplished using several approaches. The simplest approach is to assume a constant release rate as shown in Figure 3. Flows above the release rate are considered detained until the recession limb of the inflow hydrograph drops below the specified release rate. Although such an approach is easy to model it is not very realistic.

In the proposed model, hydrographs are routed through the generic detention basin using a linear reservoir model. In this case the storage (S) is assumed to be a linear function of the discharge (O):

$$S = K * O \qquad (1)$$

Combining equation (1) with the continuity equation:

$$I - O = \frac{dS}{dt} \qquad (2)$$

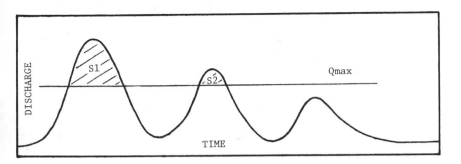

FIGURE 1.   Required Storages With a Single Discharge Constraint

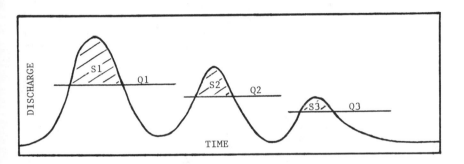

FIGURE 2.   Required Storages With Multiple Discharge Constraints

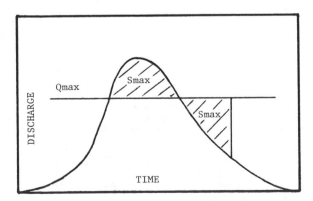

FIGURE 3.   Constant Release Rate Routing

yields:

$$I - O = K * \frac{dO}{dt} \tag{3}$$

If equation (3) is expressed in finite difference form then the following linear routing equation may be obtained:

$$O2 = C1 \ (\ I1 + I2 \ ) - C2*O1 \tag{4}$$

where:

$$C1 = \frac{\Delta t}{2*K + \Delta t} \tag{5}$$

and

$$C2 = \frac{2*K - \Delta t}{2*K + \Delta t} \tag{6}$$

and Δt is the time interval in seconds and K is the storage coefficient. For a given time interval Δt, Q2 is a function of the storage coeffient. For each reservoir routing K may be obtained from equation (1):

$$K = \frac{Smax}{Omax} \tag{7}$$

where Omax is the maximum discharge and Smax is the maximum storage that occurs when the hydrograph is routed through the detention basin.

In the proposed model, each post-development hydrograph is routed through the generic detention basin by applying equations (4-7) in an iterative manner. To begin the analysis, an intial estimate of Smax is obtained. The simplest way to estimate Smax is to assume a constant outflow rate as shown in Figure (3). Once Smax is determined it is used with Omax to obtain an initial estimate of K. Using this value of K along with equations (4-6) an outflow hydrograph is generated. Once the outflow hydrograph is obtained an improved estimate of Smax is obtained (and thus K) and the routing is repeated. This process is continued until the storage constant converges.

Rainfall Model

The proposed model accepts standard NWS hourly rainfall data as input. If desired, the hourly rainfall data may be disaggregated into smaller time intervals using a built in empirical rainfall distribution model. The model creates a unique rainfall frequency distribution for each individual hour using two linear segments. The slope and time of intersection of these two segments is a function of the ratio of the volumes of precipitation in the hour of interest to the volumes of precipitation in adjacent hours. Once the frequency distribution is generated, the total rainfall volume for a particular hour is distributed over the hour using a monte carlo simulation approach.

Summary

A continuous simulation model has been developed for use in designing
stormwater detention basins. The final output from the model includes
both peak discharge and required storage frequency curves from which
design decision can be made.   As a result of the use of continuous
simulation, antecedent conditions are considered.

APPENDIX I.   REFERENCES

ASCE. (1983). Annotated Bibliography on Urban Design Storms.

Ellis, F. W., Blood, W. H., and Baig, N. (1981). "A Continuous
Hydrologic Simulation Approach to Watershed Management," 1981
International Symposium on Urban Hydrology, Hydraulics, and Sediment
Control, University of Kentucky, Lexington, Kentucky, July 1981, pp. 1-
7.

Hydroscience Inc. (1979). "A Statistical Method for the Assesment of
Urban Stormwater Loads - Impacts - Controls," EPA - 440/3 - 79 - 023,
Office of Water Planning and Standards, Washington, D. C.

James, W. and Robinson, M. A. (1982). "Continuous Modeling Essential for
Detention Design," Proceedings of the ASCE Conference on Stormwater
Detention Facilities, Henniker, New Hampshire, pp. 163-175.

Loganathan, G. V., Delleur, J. W., and Segarra, R. I., (1985). "Planning
Detention Storage for Stormwater Management," ASCE Journal of Water
Resources Planning and Management, Vol. 111, No. 4, pp. 382-398.

Smith, P. E., and Alley, W. M., (1982). "Rainfall - Runoff - Quality
Model for Urban Watersheds," Applied Modeling in Catchment Hydrology,
V. P. Singh Editor (Proceedings Intl. Symp. on Rainfall - Runoff
Modeling, Mississippi State Univ. 1981) Water Resources Publications.

Walesh, S. G., Lao, D. H., and Liebman, M. D. (1979). "Statistically
Based Use of Event Models," Proceedings of the International Symposium
on Urban Storm Runoff, University of Kentucky, Lexington, Kentucky, pp.
75-82.

APPENDIX II.   NOTATIONS

The following symbols are used in this paper.

I = Inflow
O = Outflow
S = Storage
K = Storage coefficient
Δt = Time interval (secs)
$C1$ = Routing coefficient
$C2$ = Routing coefficient
$Omax$ = Maximum discharge
$Smax$ = Maximum storage

THE STORAGE MATCHING TECHNIQUE OF DETENTION BASIN DESIGN

Thomas H. Price - Kenneth Potter*

ABSTRACT

Detention basis are in widespread use for runoff control in urbanizing areas and are often constructed to control a specific, often rare design event (25 to 100 year flood). While the rare events are important to control, the more frequent events are also important. The ideal detention facility would be one that controls the events of all frequencies or preserves the pre-development distribution of floods after development.

In recent years, use of continuous rainfall - runoff simulation with historic rainfall data has become much more popular for the design of detention basins and reservoirs. While it may be common for detention facilities to be designed to preserve the peak flow of a few historic events, it is the distribution of floods that should be preserved.

Unlike using design rainfall events of critical duration to develop design hydrographs, continuous simulation of historic data does not result in hydrographs with the same frequency as the rainfall (or runoff) that produced them. To avoid the complexities of continuous simulation, the Storage Matching Technique of detention basin design has been developed.

INTRODUCTION

This paper presents the Storage Matching Technique (SMT) of detention basin design developed by the authors. The SMT relates peak flow and volume frequencies to develop a storage-outflow curve that preserves the pre-development distribution of peak flows after development. A detailed procedure for using the technique and an explanation of why it works is presented. A case study is presented where the technique was used to design a detention basin with good results·.

DETENTION BASIN DESIGN FOR A RANGE OF FLOOD FREQUENCIES

Detention basins that control urban runoff may be designed for large rare events, for small frequent events, or for the full range of events. Typically, detention basins designed to control the large events do nothing to control the small events. Although, in the past it was more common to design for the large rare events, more recently there has been a shift in design philosophy to control the full range of runoff events.

* Respectively, water Resources Engineer, Donohue & Associates, Inc., Itasca, Illinois, Professor of Civil Engineering, U.W. Madison, Wisconsin.

The smaller, more frequent events (2-10 year floods) may actually be just as important as the larger events. While each individual small event may not cause as much damage as a large event, it occurs much more frequently and can cause more total damage over the life of the structure. Recently, the trend in some areas has been to design such that the post-development distribution of floods is attenuated down to the pre-development distribution of floods in hopes that the effects of urbanization will be offset by the detention basin.

RAINFALL-RUNOFF MODELS FOR DETENTION BASIN DESIGN

Typically, rainfall-runoff models are used in the design of detention basins. Some of these models are based on continuous simulation using historic rainfall while other models use design rainfall events to develop design hydrographs. Design of detention basins using design rainfall events is fairly straight forward. The designer finds the volume required to reduce the peak flow of the post-development hydrograph of critical duration to the pre-development level and provides that amount of storage when the detention basin is releasing the pre-development flow rate. However, when using continuous simulation, designing a detention basin to control all recurrence intervals is not straightforward because the design flow or volumes are not directly developed. When using continuous simulation and historic rainfall, it cannot be assumed that the recurrence interval of each peak flow is the same recurrence interval as the rainfall that produced it. In addition, the recurrence interval of a particular hydrograph may vary depending on whether it is examined in terms of peak flow or volume.

One approach to develop design parameters when using continuous simulation would be to attempt choosing a few design hydrographs from the simulated group. However, because of the variation in recurrence interval between hydrograph peaks and volumes, selection of hydrographs with the design peak flows and volumes would require considerable effort. If one considers that it is really the distribution of floods that should be considered and not individual storms, a more systematic, less subjective approach would be more appropriate.

THE STORAGE MATCHING TECHNIQUE (SMT)

The Storage Matching Technique (Price, 1986) was developed to make use of peak flow and runoff volume distributions to develop storage-outflow curves for detention basins or reservoirs. The storage-outflow curve is the combined stage-storage and stage-discharge relationships of a reservoir.

The SMT is used with continuous simulation models or with models that can be used to develop event hydrographs from many years of rainfall data.

Development

The SMT is based on the following three relationships. First there is a probability distribution for the peak storage used in a reservoir. Second, there is a desired or target probability distribution of the peak outflows from the reservoir (typically the pre-development distri-

bution of peak flows).   Third,   there is a unique relationship between
the peak storage and peak outflow from a reservoir (the storage-outflow
curve).   The third   relationship   relates   the first two relationships.
Hence, the three   relationships   are   not   independent.   The amount of
storage used by any particular   event   is dependent on the storage-outflow relationship of the reservoir.   The SMT alleviates the problem of
interdependence between relationships by assuming that, on the average,
the outflow from the reservoir   for a given post-development hydrograph
is the pre-development peak   flow   that   results from the same rainfall
event.   By assuming a triangular   shaped outflow hydrograph with a peak
flow equal   to   the   pre-development   flow,   the   approximate   storage
requirement for that event   can   be   computed   as shown in Exhibit 1. A
probability distribution can be   fit   to   the approximate storages from
all the hydrographs in a flood series.

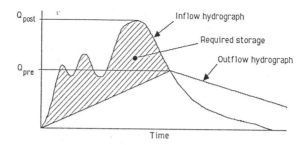

EXHIBIT 1 - REQUIRED DETENTION STORAGE

Now from this distribution of peak storages, the peak outflow distribution can be derived from   the storage-outflow curve. Conversely, if one
knows   the   desired   distribution   of   outflows   (the   pre-development
distribution), the two   distributions   can   be   combined   to obtain the
storage-outflow curve.   The storages and pre-development flows of equal
recurrence interval are matched, for   the range of recurrence intervals
that are to be controlled,   to   obtain the storage-outflow curve of the
reservoir.

Step-by-Step Procedures

The following steps   are   used   in   the   Storage   Matching Technique to
obtain a storage-outflow curve for a reservoir.

1.   Develop   pre-development   and   post-development   rainfall-runoff
     models of the watershed.

2.   Using   historic   rainfall   and   the   above   models,   develop   pre-
     development and   post-development   hydrographs   for   the period of
     record.

3.   Using the post-development hydrographs   and the corresponding pre-
     development peak   flows,   develop   storage   requirements   for each
     hydrograph assuming triangular shaped outflow hydrographs.

4.   Develop the pre-development peak flow distribution from the flows
     in (2) and develop the peak storage distribution from the storages
     in (3).

5.   Match storages and pre-development peak flows of equal recurrence
     interval to obtain the design storage-outflow curve of the
     reservoir.

6.   Route the post-development hydrographs through the reservoir with
     the storage-outflow curve obtained in (5).

7.   Compute the distribution of peak reservoir outflows and compare it
     to the distribution of pre-development flows to evaluate the
     performance of the reservoir.

8.   If the two distributions are insufficiently close, the procedure
     can be rerun as shown in the case study below of Pheasant Branch
     Creek in Madison, Wisconsin.

Case Study

The following example is taken from a branch of the South Fork of
Pheasant Branch Creek in Madison, Wisconsin. The creek has been experiencing increased overbank flooding and some severe streambank erosion
problems. To alleviate these problems, a regional detention basin was
considered for reducing flow rates.   The SMT was used for preliminary
design of the detention basin.   The rainfall-runoff model "DR3M",
(Alley and Smith) obtained from the U.S.G.S., was used to estimate both
the pre- and post-development flows.   Sixty-eight years of rainfall
data obtained from a nearby airport were used for the simulation. The
model uses a variable time step.   Daily rainfall was used between
events for moisture accounting and five minute rainfall data is used
during events. A couple of the most significant rainfall events per
year were used to develop hydrographs.   The Log Pearson Type III (LP3)
distribution was used when fitting probability distributions to the
flow data.

The watershed is expected to develop from its current, mostly agricultural land use to mostly medium density residential land use. Exhibit 2 shows the peak flow distributions for the existing and expected,
ultimate land use. The approximate storage required to reduce each
post-development peak to the pre-development level was estimated by
assuming a triangular shaped outflow hydrograph, as shown in Exhibit 1.
The area under a triangle is equal to 0.5 times the height times the
base. The assumed shape of the outflow hydrograph can be varied by
using a fraction other than 0.5 when calculating the area under it.
The LP3 distribution was fit to the required storages.

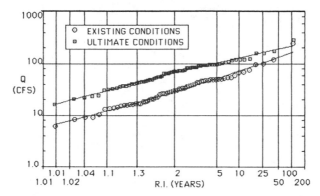

EXHIBIT 2 - PEAK FLOW DISTRIBUTION FOR EXISTING CONDITIONS
AND FUTURE CONDITIONS WITHOUT DETENTION

Then storages and pre-development flows with recurrence intervals of
1.3, 2, 5, 10, 25, 50, and 100 years were matched to obtain the
storage-outflow relationship for the detention basin. Note that it was
assumed that there was no outflow when the reservoir was empty. When
the post-development hydrographs were routed through the reservoir, the
distribution of pre-development flows and distribution of routed post-
development flows were close, but the routed post-development flows
were consistently greater.

Since the routed flows were too high, additional storage was required.
Thus, the shape of the outflow hydrograph was assumed to be something
narrower than a triangle by using a fraction of 0.45 when computing its
volume. After repeating the steps of the SMT with the new outflow
hydrograph shape, the pre-development and routed post-development distributions were in very good agreement. The storage-outflow curves
from the two trials are shown in Exhibit 3. The 2, 5, 10, 25, 50, and
100 year peak flows for pre-development, post-development without
detention, and post-development with detention for the two trials are
shown in Exhibit 4. Finally, Exhibit 5 shows the plotted peak flow
distributions for the pre-development condition and the post-development condition routed through the final storage-outflow curve.

EXHIBIT 3   STORAGE-OUTFLOW CURVES

Outflow (cfs) (m³/s)	Trial 1 Storage (acre-feet) (m³)	Trial 2 Storage (acre-feet) (m³)
0 (0.00)	0.0 (0)	0.0 (0)
18 (0.50)	1.8 (2200)	1.9 (2300)
29 (0.82)	2.6 (3200)	2.8 (3500)
52 (1.48)	3.9 (4800)	4.2 (5200)
72 (2.05)	4.7 (5800)	5.1 (6300)
104 (2.95)	5.7 (7000)	6.3 (7800)
132 (3.75)	6.3 (7800)	7.1 (8800)
165 (4.68)	7.0 (8600)	7.9 (9800)

EXHIBIT 4   PEAK FLOW DISTRIBUTIONS

Recurrence Interval (year)	Pre-development Flows (cfs)(m³/s)	Post-Development Flows (cfs)(m³/s)	Routed Post-Development Flow First Trial (cfs)(m³/s)	Routed Post-Development Flow Second Trial (cfs)(m³/s)
2	29 (0.82)	66 (1.87)	30 (0.85)	28 (0.79)
5	53 (1.50)	104 (2.95)	54 (1.53)	50 (1.42)
10	73 (2.07)	130 (3.68)	75 (2.12)	70 (1.98)
25	105 (2.97)	165 (4.67)	109 (3.07)	102 (2.89)
50	133 (3.77)	191 (5.41)	141 (3.99)	133 (3.77)
100	166 (4.70)	218 (6.17)	179 (5.07)	170 (4.81)

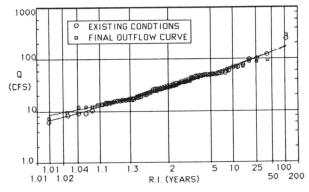

EXHIBIT 5 - PEAK FLOW DISTRIBUTIONS FOR EXISTING CONDITION
AND FUTURE CONDITIONS WITH THE FINAL STORAGE-OUTFLOW CURVE

Exhibit 5 shows that the peak flow distribution before and after
development with the reservoir in place are virtually identical. The
SMT caused the post-development distribution to slide down the graph to
virtually match the pre-development distribution.

VERIFICATION

The SMT has also been tried on other, hypothetical watersheds with
equally good results. It appears, as might be expected, the more years
of data used, the better the technique works. A good match between the
pre-development peak flows and the routed post-development peak flow is
generally obtained for recurrence intervals up to the plotting position
of the largest event.

SUMMARY

The SMT provides a useful tool for designing detention basins or reservoirs when continuous or near continuous simulation is used. While it
is computationally intensive to fit the flood distributions and to

compute the approximate required storages, this work can be easily and
quickly performed by computer.   The method gives good results that can
be easily checked by routing the post-development hydrographs through
the computed storage-outflow curve.

REFERENCES

Alley, W.M. and Smith P.E., "Distributed Routing Rainfall-Runoff Model--
Version II"; U.S.G.S Open-File Report 82-344, no date given.

Price, T.P., "An Investigation of Two Rainfall-Runoff Models for Detention Basin Design", Masters Project, May, 1986.

SEASONAL RULE CURVE RESERVOIR OPERATION

Ralph A. Wurbs,[1] M.ASCE, and Michael N. Tibbets[2]

Abstract: This paper addresses the potential for increasing the beneficial use of existing reservoirs in the state of Texas through seasonal rule curve release policies. Seasonally varying operating procedures have been adopted to only a very limited extent in the state.
Although seasonal variations in risk of flooding, flood damage susceptibility, water supply demands, and streamflow availability occur, the
time periods when flood control and conservation storage are most
needed significantly overlap. However, seasonal rule curve operation
is concluded to offer significant potential for improving surface water
management in the state in those situations in which needs for flood
control and conservation purposes are severely taxing the available
storage capacity.

Introduction

   Many multi-purpose reservoirs contain storage capacity for both
flood control and conservation purposes. Operation is based on the
conflicting objectives of maximizing the amount of water available for
conservation purposes, such as water supply, hydroelectric power, and
recreation, and maximizing the amount of empty space available for
storing flood waters to reduce downstream damages. Vertical zones, or
pools, are typically allocated to flood control and conservation purposes by a designated top of conservation pool elevation. The top of
conservation pool may be set at a constant elevation, or a seasonal
rule curve operating policy may be adopted to reflect seasonally varying conditions. A rule curve specifies the top of conservation pool
elevation as a function of time of the year. Storage capacity is re-
allocated between flood control and conservation purposes in a set
annual cycle.

   This paper is based on the results of a university research effort,
presently underway, directed toward expanding capabilities for optimizing the beneficial use of existing reservoirs in Texas. Permanent
and seasonal reallocation of storage capacity between flood control
and conservation purposes is a major focus of the study. Wurbs,
Cabezas, and Tibbets (1985) and Tibbets (1986) document work completed
to date on evaluating the feasibility of seasonal rule curve operations. The present paper is a discussion of the potential for improving the effectiveness of existing reservoirs by adopting seasonal
variations in storage capacity allocated to flood control and water
supply purposes.

[1]Associate Professor, Civil Engineering Department, Texas A&M
University, College Station, TX 77843.
[2]Staff Civil Engineer, City of Paris, Paris, Texas 75460.

Seasonal Rule Curve Operation in Texas

Texas has 187 major reservoirs with total controlled storage capacities greater than 6.17 million m^3 (5,000 acre-feet), including 182 operational and five projects under construction (Wurbs 1985). Conservation, flood control, and total controlled capacities of 49.3 billion m^3, 22.8 billion m^3, and 72.1 billion m^3, respectively, are contained in the major reservoirs. Three reservoirs are used only for flood control, 152 are conservation only, and 32 contain both flood control and conservation capacity. About three-fourths of the total conservation storage capacity is designated for municipal and industrial water supply, and most of the remainder is for irrigation. Most of the flow through the state's 21 hydroelectric power plants is from releases for municipal, industrial, or agricultural water supply. Recreation is also a major consideration in reservoir operation in the state.

Institutional arrangements for developing and managing reservoirs are based on project purposes. With one exception, all the reservoirs in Texas containing controlled flood control storage were constructed by the federal agencies. The one exception also happens to be the smallest and oldest of the major flood control projects. The U.S. Army Corps of Engineers operates the flood control storage in all the federal reservoirs except for the two projects on the Rio Grande River owned and operated by the International Boundary and Water Commission. Municipal and industrial water supply has traditionally been a local responsibility, with the federal government confining itself to a secondary role. However, municipal and industrial water supply storage is included in all but two federal reservoirs in Texas. Costs allocated to water supply are reimbursed by the nonfederal sponsors which have contracted for the conservation storage capacity. Releases from conservation storage are made at the discretion of the nonfederal sponsors.

Although seasonal rule curves are fairly common in many parts of the United States, this type of operating policy has not been widely adopted in Texas. The top of conservation pool has been varied seasonally at four reservoirs in the state.

Two Corps of Engineers projects, Lake O' the Pines and Wright Patman Lake in the northeast corner of the state, are operated in accordance with seasonal rule curves. The operating curve for Lake O' the Pines provides for raising the top of conservation pool 0.46 m from mid-May through mid-September for recreation purposes. The rule curve for Wright Patman varies significantly during the year in response to an interim operating agreement with the conservation storage sponsor to provide additional municipal and industrial water supply. The top of conservation pool is constant from November through March and varies with date from April through October. The top of conservation pool peaks on June 1 at a level 2.1 m above the winter pool level. A permanent reallocation of flood control storage capacity to conservation is planned for Wright Patman Reservoir upon completion of construction of Cooper Reservoir upstream.

The top of conservation pool elevations for Falcon and Amistad Reservoirs on the Rio Grande River can be, at the discretion of the International Boundary and Water Commission, temporarily raised for seasonal rule curve operation. However, the optional encroachment into the flood control pool does not routinely occur each year, and the magnitude of encroachment can vary in those years in which the seasonal pool raise is implemented.

Reallocating flood control storage capacity to conservation purposes is a management strategy for responding to increasing demands and changing conditions. Permanent and/or seasonal reallocations to enhance conservation purposes could receive significant attention in the state in the future. The role of a seasonal rule curve would be to minimize adverse impacts on flood protection while increasing water supply or hydroelectric power yield and/or enhancing recreation. Another common situation is a reservoir operated for only conservation purposes. Adoption of a seasonal flood control pool could provide some flood protection while minimizing adverse impacts on conservation operations.

Seasonally Varying Factors Affecting Reservoir Operation

Dugas (1983) performed a statistical analysis of daily precipitation data from 36 stations located throughout the state. The probabilities of various amounts of precipitation being equalled or exceeded during each of the 52 weeks of the year were estimated for each station. Griffiths and Ainsworth (1981) describe Texas weather, including major flooding events, during the period 1880 through 1979.

Patterns of seasonal precipitation vary for different areas of the state. However, rains generally occur most frequently in late spring as a result of squall-line thunderstorms, with most areas showing a peak in May. Tropical storms and hurricanes are a perennial threat to the Texas Gulf coastal region during the summer and autumn, occurring most frequently in August and September. Forty-two of the most severe floods in the state were distributed among months of the year as follows: January, 1 flood; February, 1; March, 2; April, 5; May, 7; June, 6; July, 4; August, 2; September, 10; October, 1; November, 1; and December, 2 floods.

Agricultural and municipal water demands, hydroelectric power demands, recreational use, and evaporation are highest during the hot summer months concurrently with relatively low reservoir inflows. Consequently, raising the top of conservation pool elevation during the wet spring months to capture additional rainfall for use during the summer should enhance conservation operations. However, extreme flood events tend to occur during the period from April through October. September is a particularly critical month for flooding. The majority of the flood control benefits attributed to reservoirs in Texas are related to agriculture, for which damage susceptibility is highest during the summer. Consequently, the time periods during which flood control capacity and water supply capacity are most needed significantly overlap.

A seasonal encroachment into the flood control pool during the spring and summer still has an advantage over a permanent reallocation from the perspective of minimizing the reduction in flood protection. Since significant time will likely be required to fill the additional conservation capacity after the designated top of conservation pool elevation is raised, the reservoir storage level may still be relatively low when a spring flood occurs. Also, additional protection is provided for the relatively infrequent yet possible winter flood. The hot dry months of July and August tend to naturally deplete the storage prior to September floods.

Case Study Analysis

A 12-reservoir system in the Brazos River Basin operated by the Corps of Engineers and Brazos River Authority is presently being modeled. However, initial work has focused on a single reservoir in this system. Waco Reservoir was selected for the initial case study analysis partially because the Corps of Engineers, at the request of the Brazos River Authority and City of Waco, recently studied and recommended a permanent reallocation of storage from flood control to water supply. Waco Reservoir also provided a good case study because it is located in central Texas and has physical and hydrologic characteristics and operating procedures representative of typical multipurpose reservoirs in the state.

Waco Reservoir has flood control, conservation, and sediment reserve capacities of 680 million m^3, 128 million m^3, and 85 million m^3, respectively. All the conservation capacity is committed for providing municipal and industrial water supply for the City of Waco and its suburbs. The Corps of Engineers and Brazos River Authority plan to permanently reallocate 58.6 million m^3 of flood control capacity to water supply by raising the top of conservation pool 2.13 m. The reallocation will increase the firm yield from 2.29 m^3/s to 2.89 m^3/s. The flood control capacity will be reduced from a 100-year to about an 80-year recurrence interval design storm (USACE 1982).

The present study (Tibbets 1986) focused on seasonal rule curve operations. Firm yield and expected annual flood damages were computed for alternative constant and seasonally varying operating plans using the HEC-5 computer program (USACE 1983).

Seasonal rule curve operations were found to achieve essentially the same increases in firm yield as corresponding permanent reallocations. Raising the top of conservation pool elevation 2.13 m either seasonally or permanently results in increasing the firm yield from 2.29 m^3/s to 2.89 m^3/s, as long as the seasonal pool is raised no later than early May and lowered no earlier than late September.

An analysis was also performed in which total demands were divided into the categories of firm and secondary. Firm yield represents a demand level met continuously during a period-of-record simulation. Secondary yield represents a level of additional demand that can be met a portion of the time. Essentially the same secondary yield versus reliability relationship can be achieved with either seasonal or permanent reallocations of the same magnitude.

Expected annual flood damages, computed using the damage-frequency method, were used to compare seasonal and constant operating plans. A seasonal reallocation of flood control capacity to conservation in Waco Reservoir results in slightly smaller increases in expected annual damages than a permanent reallocation of the same magnitude, assuming perfect forecasting of future flood flows. However, average annual damages computed assuming 24 hours of foresight are slightly less for a seasonal rule curve than for the corresponding constant operating policies at the lower as well as higher top of conservation pool elevation. Most of the flood damages occur in reaches which are 96 to 130 hours travel time below the dam, which is not unusual for flood control reservoir systems in Texas. Imperfect streamflow forecasting reflected in the model results in premature releases from the flood control pool which would contribute to downstream flooding several days later. The rule curve operation resulted in spring flood waters being stored in a yet unfilled seasonal conservation pool instead of the flood control pool, and thus provided an incidental flood control benefit. A seasonal encroachment into the flood control pool increases the risk of an extreme event overtopping the flood control pool, but could possibly reduce the risk of premature releases, due to imperfect streamflow forecasting, contributing to flooding during less severe, more frequent events.

Summary and Conclusions

In Texas, as elsewhere, population and economic growth and depleting groundwater reserves are resulting in intensified demands on surface water resources. Management strategies for optimizing the beneficial use of limited reservoir storage capacity are becoming increasingly important.

Many factors affecting reservoir operation are seasonal in nature and should be considered in evaluating the feasibility of rule curve operations. Risk of flooding, flood damage susceptibility, water supply demands, and streamflow availability in Texas, like most other places, vary greatly during the year. Since the time periods when flood control and conservation storage capacity are most needed overlap in Texas, seasonal rule curve operation involves trade-offs between project purposes. However, seasonal rule curves can still be beneficial in enhancing one purpose while minimizing adverse impacts on the other. Distinct seasons of high flood threat and high water demands conveniently occurring at separate times of the year are not necessarily required for seasonal rule curve operation to be useful.

Acknowledgements

The research on which this paper is based was financed in part by the United States Department of the Interior through the Texas Water Resources Institute as authorized by the Water Research and Development Act of 1984 (P.L. 98-242). Contents of this publication do not necessarily reflect the views and policies of the Department of the Interior, nor does mention of trade names or commercial products constitute their endorsement by the U.S. Government.

Appendix 1 - References

Dugus, W.A. (1983). "Agroclimatic Atlas of Texas, Part 5, Precipita-
    tation and Dry Period Probabilities at Texas Research and Extension
    Centers," MP1513, Texas Agricultural Experiment Station, Texas A&M
    University.

Griffiths, J.F., and Ainsworth, G. (1981). "One Hundred Years of Texas
    Weather, 1880-1979," Monograph Series 1, Office of the State
    Climatologist, Texas A&M University.

Tibbets, M.N. (1986). "Feasibility of Seasonal Multipurpose Reservoir
    Operation in Texas," Master of Science Thesis, Civil Engineering
    Department, Texas A&M University.

U.S. Army Corps of Engineers (1982). "Waco Lake Storage Reallocation
    Study, Recommendation Report," Fort Worth District.

U.S. Army Corps of Engineers (1983). "HEC-5 Simulation of Flood and
    Conservation Systems, Users Manual," Hydrologic Engineering Center.

Wurbs, R.A. (1985). "Reservoir Operation in Texas," Technical Report
    135, Texas Water Resources Institute, Texas A&M University.

Wurbs, R.A., Cabezas, L.M. and Tibbets, M.N. (1985). "Optimum
    Reservoir Operation for Flood Control and Conservation Purposes,"
    Technical Report 137, Texas Water Resources Institute, Texas A&M
    University.

HYDRAULICS OF ENLARGING CANNONSVILLE RESERVOIR

A. F. Monaco, M ASCE*

ABSTRACT: This paper describes the hydraulic studies performed for enlarging the Cannonsville Reservoir with a gate system which could raise the normal storage level without having an adverse effect on structures in the reservoir or on the downstream environment. Without special measures, the higher initial level could result in floods encroaching on bridges over the reservoir, and increased flood damage downstream. Measures were devised to avoid these impacts, including a gaging network and flood prediction system to determine appropriate spillway release rates during floods; and conceptual design of a multiple gate system with independent operation to provide the necessary flexibility. To avoid flood releases higher than those under present spillway conditions without encroaching on the bridges, it was necessary to devise three alternative schedules of lowering and raising the gates for use in different ranges of predicted magnitude of floods.

1. INTRODUCTION

Ebasco Services was contracted by N.Y.S. Department of Environmental Conservation to perform a feasibility study for the enlargement of their Cannonsville Water Supply Reservoir. The principal scope of services included an evaluation of the hydrology and hydraulics of installing gates on the spillway and subsequent impacts on both the downstream and upstream environment; a determination of the modifications required to the embankment dam, emergency weir structure, bridges, and roads to mitigate those impacts; an evaluation of the costs and benefits to be derived from the proposed reservoir enlargement; and a determination of the feasibility of raising the reservoir level. The project would enlarge the storage capacity of New York City's existing Cannonsville Reservoir in Delaware County by 13 billion gallons (4922 Ha-m) by installing gates on the spillway, thereby raising the maximum pool elevation 8 ft (2.4m).

HYDRAULIC STUDY

The drainage area of the West Branch of the Delaware River above the Cannonsville Dam is 456 sq miles (1182 sq km). Elevations in the basin range from 1150 ft (351m) to 3343 ft (1091m) msl. The riverbed has a very uniform slope of 8-9 ft (2.4-2.7m) per mile. Due to the large size of the Cannonsville Reservoir and potential

*Senior Consulting Civil Engineer - Ebasco Services Incorporated, 2 World Trade Center, New York, NY 10048

downstream hazard, it was necessary to determine the Probable Maximum Precipitation and corresponding flood, and route it through the reservoir to check the adequacy of the spillway.

PMP depths for a range of storm sizes and for durations from 6 to 72 hours were determined from Hydrometeorological Report No. 51. The 6-hour increments of the PMP were arranged to obtain the highest PMF peak discharge.

A comparison of seasonal PMP indicated that for the project site, the maximum flood will occur in the summer since the snow melt contribution in winter cannot make up the difference between summer and winter PMP.

The transformation of a known precipitation pattern over a drainage basin to an outflow hydrograph depends on the characteristics of the basin. The preferred method of determining these characteristics is to calibrate a hydrologic model by reproducing historical flood hydrographs from known rainfall events. One of the most comprehensive and widely-used models for this purpose is HEC-1, developed by the U.S. Army Corps of Engineers.

A model of the Upper Delaware River Basin using the HEC-1 program was developed for the New York District of the Corps of Engineers. The model was calibrated using the precipitation data and flood hydrographs of historical storms occurring in November 1950; August 1955; and June 1983.

Initial loss and infiltration rate were determined by matching the rainfall to the historical storm hydrographs for the gaged sub-basins within the Cannonsville drainage basin. The values of basin parameters determined from the 1955 storm were used for all flood determinations since it is the largest of the three storms and therefore the most suitable for modeling major hypothetical storms. A check was made of the validity of these basin parameters by reproducing the flood hydrograph of the June 1973 storm from information available regarding precipitation at various stations throughout the basin. Runoff hydrographs were recorded at the towns of Walton and Stilesville, upstream and downstream of the reservoir respectively. Using the basin parameters determined from the 1955 storm, the discharge hydrographs for the 1973 storm were reproduced with the HEC-1 program, and found to agree reasonably well with the observed hydrographs.

The PMP distribution was then applied to the basin to determine the Probable Maximum Flood. Base flow and loss conditions were assumed to be the same as those determined by the Corps of Engineers for the 1955 storm. The resulting PMF peak inflow was 237,900 cfs (6737 cms). Assuming an initial water level of 1150 ft (351m) msl and the present spillway conditions, the peak outflow was calculated to be 203,700 cfs (5769 cms) and the peak reservoir stage is 1172.51 ft (357m) msl. Since the crest of the dam is at 1175 ft (358m) msl, the maximum freeboard is 2.49 ft (0.8m).

To evaluate the safety of the dam against overtopping waves during the PMF a 55 mph (88 kmph) wind with a minimum duration of 14 minutes was applied to the reservoir surface at the peak of the flood. This produced a significant wave height of 2.2 ft (0.7m) with a period of 2.7 seconds, and a maximum wave height of 3.7 ft (1.1m). Wave runup will occur on the 2:1 slope near the crest of the dam. The significant wave is calculated to run up 2.5 ft (0.8m) and the maximum wave 3.7 ft (1.1m) above the still water level on the riprapped embankment. Thus both of these waves could overtop the dam under present conditions.

SPILLWAY MODIFICATION

The Cannonsville Reservoir has a storage capacity of 300,000 acft (37000 Ha-m) at its present normal pool level of 1150 ft (351m) msl.

The present spillway consists of an ogee shaped "low weir" 240 ft (73m) long. The crest of the low weir is at elevation 1150 ft (351m) msl. Adjacent to this is a broad crested emergency spillway 560 ft (171m) long at elevation 1158 ft (353m) msl as shown in the photograph below.

CANNONSVILLE DAM SPILLWAY

In consideration of the spillway modifications required, the principal objectives were to avoid costly relocations and modifications to bridges and other structures while satisfying the State and City criteria for passage of the 100-yr flood.

Under the present spillway conditions the town of Deposit would be partially flooded during a 100-yr flood, and the peak discharge from Cannonsville reservoir would be 31,023 cfs (879 cms). In order to restrict outflow to 31,000 cfs (878 cms) for the enlarged reservoir, it is necessary to throttle the discharge. Therefore, an evaluation of the required modification to the spillway was made to determine the number and size of gates that should be installed in order to provide the controls to both water level and outflow and still have confidence that the embankment dam will not be inadvertently jeopardized. These considerations resulted in a decision to recommend 4 - Pelican type gates, each 110 ft (34m) wide by 8 ft (2.4m) high separated by 5 ft (1.5m) wide piers. This resulted in a 465 ft (142m) long gated spillway, and a 335 ft (102m) long emergency spillway as shown on Figure 1 below.

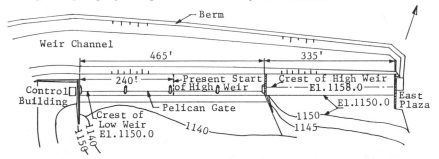

Figure 1. Plan of Spillway With Recommended Modifications

The same PMF inflow hydrograph was routed through the reservoir with the modified spillway. The flood was assumed to begin with the gates upright and the reservoir at its new normal pool level at 1158 ft (353m) msl. With the spillway modification and all four gates operating, the maximum stage would be 1171.0 ft (357m) and peak outflow would be 204,100 cfs (5780 cms). If one of the four gates were stuck in the upright position, the maximum reservoir stage would be 1172.53 ft (357.5m) msl and the peak outflow would be 205,200 cfs (5811 cms). The maximum level with two gates stuck would be 1173.9 ft (357.9m) msl and the dam would still not be overtopped.

OPERATING SCHEDULE

The objective of the spillway operating schedule is to obtain the desired storage for augmenting downstream releases and to mitigate adverse upstream and downstream effects from the higher operating level.

Under normal conditions, the gates would be in the upright position ponding water to 1158 ft (353m) except during the winter months: December, January, February and March, when it is recommended that two of the four gates be lowered completely so as to avoid increased build-up of ice in the reservoir and to simulate spillway conditions close to the present conditions. Following the above winter period, during April when spring runoffs normally begin, the gates would be raised to store water after the flood peaks and begins to recede. This operating procedure will avoid any incremental adverse effects from ice jams either upstream or downstream, since the first flushing of the reservoir of any ice formed during the winter months would be done under reservoir conditions similar to those which exist presently.

A review made of the historical records indicates there was only one year where storing water during the cold months could have provided the additional volume needed to fill the reservoir to the 1158 ft (353m) level. Therefore, there is no significant benefit derived from raising the gates prior to the month of April.

The operation of the gates during passage of floods subsequent to the initial filling of the storage volume between elevation 1150 to 1158 ft (351-352.7m) would be integrated with an "Early Warning System" which would predict the inflow flood hydrograph and establish the limiting outflow discharge consistent with maximum attenuation of the flood peak and with minimum upstream impact due to pondage of excess inflow over outflow.

For floods in excess of the 500-yr return period flood, up to and including the PMF, the procedure would be as follows:

1. It is assumed that the reservoir is full to elevation 1158 ft (353m) at the start of the storm.

2. At the start of the storm lower the reservoir down to 1157.0 ft (353.7m).

3. Operate the gates so as to maintain elevation 1157 ft (353.7m) by letting outflow equal to inflow until the gates are fully seated on the crest of the dam.

4. Keep that position until the flood begins to recede and the water level drops to 1158 ft (353m) again.

5. Operate the gates to maintain the reservoir level at 1158 ft (353m) elevation until they are in the fully upright position.

For estimated peak inflow below the 500-yr return period flood and down to the downstream no-damage limit flow, the gate operating procedure would be as follows:

1. At the start of the storm, release water to lower the level to elevation 1157.0 ft (353.7m).

2. Maintain the reservoir water level at 1157.0 ft (353.7m) until the limiting outflow discharge is reached, and hold this outflow constant by throttling the gates, and storing excess inflow over outflow. For the 100-yr flood, this limit would be 31000 cfs (878 cms) and for a 50-yr flood the limit would be 23400 cfs

(663 cms) which compare favorably with the outflow peaks with the existing spillway for the same return period flood.

3. After the flood crests, hold gate position until the water level drops to elevation 1158.0 ft (353m).

4. When the level drops to elevation 1158.0 ft (353m) gates are operated to hold that level until they are in the fully upright position.

For estimated peak outflows below the no-damage limit, maintain the reservoir level just below 1158.0 ft (353m) by letting outflow equal inflow and ride out the storm without overtopping the emergency spillway.

It is proposed to operate the gates either remotely from the Rondout Central Station (using automatic or manual control) or manually at the dam site. A description of the proposed control system follows.

The control and instrumentation system will provide the following:

1. Local manual control of the gates by means of switches and indicators located on a small control panel in the gate control building at the spillway.

2. Supervisory Control and Data Acquisition system (SCADA) with the master station at Rondout. This will permit control of the gates by manual means via the supervisory control from Rondout. It will also provide information on pool level, and gate position.

3. A precipitation and river level reporting system using several precipitation/river level or precipitation only units to report via space radio to a central receiving station at Rondout via a repeater as required by the topography. These units are battery powered and require an annual battery replacement. This system will provide continuous transmission of updated information on rainfall, snowpack conditions, temperature, reservoir water level and stream flow throughout the drainage basin to a central receiving station at Rondout.

4. Receiver/decoder station and computer with enhanced alert software at Rondout to provide a precipitation monitoring system using the data from the remote gauging stations, make flood peak predictions and establish gate operation criteria. This will provide present and predicted hydrographs and generate gate commands based on operating strategies incorporated in the program.

FLOOD OPERATION STUDIES OF THE VALDESIA RESERVOIR SYSTEM IN THE DOMINICAN REPUBLIC

By J.T.B.Obeysekera[1], J.D.Salas[2],M. ASCE, H.W.Shen[3],M. ASCE and G.Q.Tabios[4]

ABSTRACT   The Valdesia reservoir system, located on the Nizao River in the Dominican Republic, is designed to provide irrigation water to the Nizao project areas and hydroelectric energy to the national power network. The system consists of a main reservoir, concrete dam and a spillway, a diversion and a spillway system a short distance downstream. This paper presents the application of state-of-the-art hydrologic techniques to investigate the flood control operation of the Valdesia system which experiences floods due to local weather phenomena and also extreme floods due to Atlantic tropical cyclones.

Introduction

The Valdesia Reservoir System is located in the Nizao watershed in the Dominican Republic. Due to its proximity, Nizao watershed has a potential for experiencing a direct strike from a severe tropical cyclones which originate in the Atlantic Ocean. The hurricane David which occurred in late August of 1979 passed over Nizao watershed and caused extensive damage to the regulating structures of the Valdesia system.

The reservoir system consists of a main reservoir(Valdesia), dam and spillway, a power plant and outflow regulating works, together with an afterbay(Las Barias), diversion and spillway system a short distance downstream. The Valdesia reservoir is impounded by a concrete dam designed for maximum storage of 153 million cubic meters at an elevation of 150 m.a.s.l. The spillway runs the entire length of the top of the dam and is controlled by five radial gates. The power plant which consists of two Francis turbines rated at 30 MW each is located near the dam and has an intake with a maximum capacity of 90 m³/s. The Las Barias afterbay located about 15 km downstream of Valdesia is of much smaller size(3.0 MCM capacity at 77 m.a.s.l) consists of a concrete dam and a regulated spillway with seven radial gates. The primary purpose of the Las Barias reservoir is to regulate daily peak period power releases from Valdesia to provide stable discharges to the irrigation canals.

The material presented in this paper is a part of a larger investigation to study the operational management and safety of the Valdesia system in the Dominican Republic(Salas et al., 1986a,b; Obeysekera et al.,1986a,b,c,d; Labadie et al.,1986; Shen et al.,1986) The overall study entailed many areas

[1]Water Resour. Eng.,South Florida Water Management District, West Palm Beach Florida.
[2]Professor, Dept. of Civil Engg.,Colorado State University, Fort Collins, Colorado.
[3]Professor, Dept. of Civil Engg.,University of California,Berkeley, California.
[4]Post Doctoral Fellow,Dept. of Civil Engg.,Colorado State University, Fort Collins, Colorado.

including development of methods of operation for energy, irrigation and flood control, analysis of the safety of the dams including maintenance and inspection, and the development of an organizational plan for implementing the operational procedures. This paper presents the techniques and products employed to investigate the operation of the Valdesia system for flood control. Emphasis is given to the discussion of techniques used rather than specific results obtained which can be found in the series reports listed in the references.

Flood Operation Studies

The primary concern in flood operation studies for Valdesia system is dam safety. In developing operating rules however, some consideration was given to control releases from the Valdesia system. Unnecessary release of water from the system which otherwise may be available for irrigation and power generation was also another criteria. Forecasting is an important element not only for providing advance warning in case of catastrophic floods particularly those due to Atlantic hurricanes but also for efficient operation during small floods.

The elements of flood operation studies is illustrated in Figure 1. The proposed system for real time operation of floods is conceptualized in Figure 2. Shown in these two figures are also the various computer models developed and used for the study. A basic requirement of the entire study is that all computer models must be developed to run on a small microcomputer.

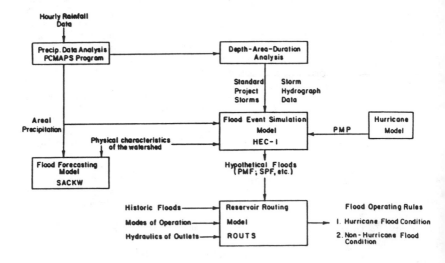

Figure 1. Techniques and models used for flood operation studies

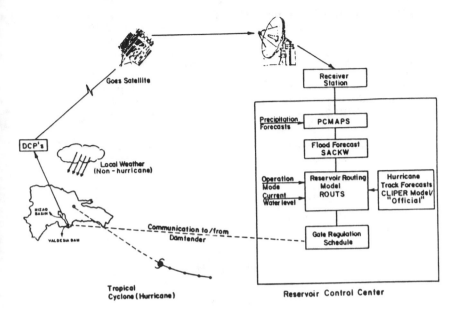

Figure 2. Proposed system for real time operation for flood control

Hypothetical floods: For developing hypothetical floods two types of hypothetical design storms were considered:(a) Standard Project Storms(SPS); and (b) Probable Maximum Precipitation(PMP). Both types of hypothetical storms require depth-area-duration(DAD) curves. These were developed from about 25 observed historic storms using a versatile precipitation data analysis program named PCMAPS. It is an easy to use software which has many capabilities including:(a) Interpolation of a precipitation field using data available at gaging stations in and around a basin;(b) Computation of areal precipitation over a given area defined by a set of coordinates using one of several techniques including Thiessen, Multiquadric, Optimal, and Kriging interpolation; (c) Printer display of isohyetal maps of areal precipitation; and (d)Derivation of depth-area data for computing DAD curves. The program is designed in a modular form such that different options may be combined according to the needs of the user. A sample display of isohyetal maps obtained by PCMAPS for Nizao watershed is illustrated in Figure 3.

Storms associated with Atlantic hurricanes and severe tropical storms were separated from those resulting from local weather phenomena. Consequently, a SPS was developed for each storm type. The temporal distribution for SPS was derived from the observed storms. Given the location of Nizao watershed, a hurricane is most likely to produce the PMP. A hurricane model developed by U.S. Weather Bureau(1961) and subsequently modified by local engineers in the Dominican Republic was used to compute the PMP. A word of caution is that this model produces extremely high precipitation values and sufficient care must be exercised to interpret its results.

The HEC-1 model of U.S. Army Corps of Engineers was used for computing hypothetical floods from design storms. In view of steep slopes in Nizao watershed and the lack of sufficient observed hydrograph data to calibrate other methods, the kinematic wave routing is used for both overland flow and channel routing. The SCS method is chosen for computing loss rates.

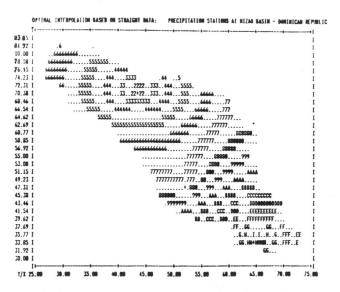

Figure 3. Isohyetal map computed by the program PCMAPS

Flood routing: A computer model named ROUTS to simulate the hydraulic characteristics of spillways and various outflow regulating works of the Valdesia - Las Barias system was developed. A gate regulation schedule incorporated into the model computes the necessary gate openings during the passage of the flood and consequently it can be used not only for design flood routing studies but also for real time operation as shown in Figure 2. Three modes of operation for flood routing was included in the model: (a) operation by induced surcharge method; (b) existing operation procedure which essentially attempts to balance inflow and outflow; and (c) hurricane operation in which all gates are left fully opened from the beginning of the flood.

Flood forecasting: The Sacramento Soil Moisture Accounting model of the U.S. National Weather Service River Forecast System was combined with the kinematic wave routing routines of the HEC-1 model to develop a microcomputer version of a continuous streamflow model for flood forecasting. This model entitled SACKW has a modular structure which is flexible to simulate the soil moisture accounting and streamflow routing in watersheds partitioned into several subbasins. The partitioning is accomplished in two levels. The first level, partitions the watershed homogeneous subwatersheds for soil moisture accounting whereas a second level of partitioning is required for kinematic wave flow routing. The model has two operational modes: the calibration mode for parameter estimation, and the forecasting mode for simulating basin hydrology under some specified or known set of model parameters. In calibration mode, the parameters of the soil moisture accounting model may be calibrated manually or automatically using the constrained Rosenbrock optimization technique. The time scale of simulation is at the least on an hourly basis and can be at longer time intervals but as multiples of one hour.

A total of four years(1972-1975) of data was used for model calibration. Hourly rainfall data at nine stations were averaged over three subbasins using optimal interpolation in PCMAPS to compute three time series of average rainfall over each subbasin. In spite of poor quality of some available data a reasonably successful calibration was obtained for the entire Nizao watershed.

Hurricane forecasting: For flood control operation of the Valdesia system forecasting of both track and precipitation amount of hurricanes is important. After a review of several models currently in use for hurricane track forecasting a simple regression model named CLIPER which is demonstrated to predict hurricane tracks in the south zone of North Atlantic(those initially at, or south of 24.5° N) with lesser errors than more sophisticated models was chosen to aid in real time operation. This model is used as a preliminary to the "official" forecasts provided by the National Hurricane Center in Miami. The CLIPER which stands for CLImatology and PERsistence(Neumann and Pelissier, 1981) is a system of regression equations fitted to several predictors available from past observations on the motions of tropical cyclones in the North Atlantic. The predictors required to forecast the track basically relate to the position(latitude and longitude), velocity at the current position and at the position 12 hrs ago, maximum wind speed, and the day number. The model provides forecasts of position up to a lead time of 72 hours.

Results

It was found that the temporary flood control storage above the normal pool level of 145 m.a.s.l in Valdesia is small and advance drawdown of the reservoir to levels below 145 m is time consuming due to capacity constraints. The spillway at Valdesia does not have sufficient capacity to pass the hurricane SPF and the PMF. An upstream reservoir at Jiguey can effectively control SPF although it will not provide sufficient control for PMF. The above conclusions must be considered in the light of uncertainties that may be present in PMP estimates due to gross assumptions associated with the hurricane model.

Acknowledgment

This paper is an outgrowth of the project entitled "Operational and Safety Studies of the Valdesia Reservoir System" funded by the World Bank and the Dominican Republic government. The project was carried out through the cooperative agreement CSU/IICA. Thanks are due to INDRHI and IICA staff in the Dominican Republic for their assistance throughout the entire project.

References

Labadie, J.W., V. Floris, N. Chou, D.G. Fontane, and W. Shaner,Operational and safety studies of the Valdesia Reservoir,Volume II:Normal Operation, Colorado State University, Fort Collins, Colorado, July, 1986.

Neumann, C.J., and J.M. Pelissier, An analysis of Atlantic Tropical Cyclone Forecast Errors, 1970-1979, Mon. Wea. Rev.,109,1248-1266,1981.

Obeysekera, J.T.B., G.Q. Tabios, F.A. Pons, J.D. Salas, and H.W. Shen, Operational and safety studies of the Valdesia Reservoir,Volume I:Hydrologic Studies, Colorado State University, Fort Collins, Colorado, July, 1986a.

Obeysekera, J.T.B., K.L. Hiew, G.Q. Tabios, J.D. Salas, and H.W. Shen, Operational and safety studies of the Valdesia Reservoir,Volume III:Flood Operation Studies, Colorado State University, Fort Collins, Colorado, July, 1986b.

Obeysekera, J.T.B., J.D.Salas, H.W.Shen, J.W.Labadie, and G.Q.Tabios, Operational and safety studies of the Valdesia Reservoir,Volume VI:Transfer of Technology and Training, Colorado State University, Fort Collins, Colorado, July, 1986c.

Obeysekera, J.T.B., G.Q.Tabios, J.D.Salas, H.W.Shen, Hydrologic Modeling System(CSU-HMS), Users Manual, Colorado State University, Fort Collins, Colorado, June, 1986d.

Salas, J.D., H.W.Shen, J.W.Labadie, and J.T.B.Obeysekera, Operational and safety studies of the Valdesia Reservoir,Summary Report, Colorado State University, Fort Collins, Colorado, July, 1986a.

Salas, J.D., H.W.Shen, and J.T.B.Obeysekera, Operational and safety studies of the Valdesia Reservoir,Volume V:Organization and Functions for Operating the Valdesia Reservoir System, Colorado State University, Fort Collins, Colorado, July, 1986b.

Shen, H.W., J.D.Salas, J. Regenstreif, and C. Mendoza, Operational and safety stdies of the Valdesia Reservoir,Volume IV:Inspection, Maintenance, and Safety Studies, Colorado State University, Fort Collins, Colorado, July, 1986.

U.S. Weather Bureau, Generalized Estimates of Probable Maximum Precipitation and Rainfall Frequency for Puerto Rico and Virgin Islands, Technical Paper No. 42, Washington D.C.,1961.

A PROBABILISTIC METHOD FOR DETERMINATION OF
THE REQUIRED CAPACITY OF RESERVOIRS

BY: Dr. Hormoz Pazwash*, M.ASCE

ABSTRACT

Evaluation of water supply potential of multi-purpose dams generally involves determination of the required capacity of reservoirs for new dams and analysis of safe yield or reliable yield for existing reservoirs. These parameters are commonly computed by using a mass curve (a deterministic approach) or an stochastic method of generating traces of data (an indirect probability method).

This paper presents a practical method for determination of the required capacity-draft relation for reservoirs based on a probability analysis of the actual streamflow records. This method was employed for prefeasibility study of 15 dams, situated in different parts of Korea. Design of reservoir capacities was based on a 20 year, two year long low flows which were distributed among months of year according to a selected non-dimensional model. The method was not only useful in analyzing the required capacity of each reservoir for a wide range of demand, but provided to be a valuable key In ranking them in terms of their potential for satisfying the projected future water demands.

1. INTRODUCTION

Analysis of capacity-draft relation for reservoirs is the most important basis of storage - reservoir design. In practice this relation is generally obtained by one of the following methods:

1. Mass curve (also called Rippl diagram) plotting
2. Reservoir reliability analysis

The first method is a deterministic approach involving cumulative plotting of actual streamflow records during a dry period which may have lasted for one year, two years or longer (1, 3). The second method includes generating stochastrically a large number (500 to 1,000) traces of streamflow, each trace equal in length to the designed project life, calculating the required storage to meet a specified demand for each trace, and plotting the required storages on a Gumbel extreme-value probability paper (2).

---

*Project Engineer, Woodward-Clyde Consultants, Wayne, NJ 07470

This paper presents a simple method which was employed for prefeasibility study of 15 medium size, multipurpose dams to determine the required storage and water supply potential of the proposed reservoirs (5, 6). The method involved:

1. Selection of a non-dimensional model to describe the monthly distribution of streamflow.

2. Determination of the cumulative stream discharge over a typical dry period (which was found to be two consecutive years) for a given recurrence interval by a simple probability method.

3. Computation of reservoir capacity by analyzing non-dimensional monthly values of supply and demand.

4. Development of non-dimensional capacity - demand diagram, showing the required capacity for a feasible range of variation in demand.

Since no gaging station existed at the dam sites, streamflow records at each site were generated by a regional rainfall - runoff analysis at major river basin in the country. Stream data at existing gaging stations were obtained from annual reports entitled "Hydrological Annual Report in Kore" (4). This paper describes the application of the method to the NAK Dong River Basin, one of the seven major river basins in Korea.

2. SELECTION OF MONTHLY STREAMFLOW DISTRIBUTION MODEL

An initial study was done to determine the effects of monthly distribution of streamflow on the required storage of a reservoir. Analysis showed that the required storage for a given drought flow depends largely on:

a. The annual distribution of streamflow in the two year drought period, and
b. The amount of demand

The monthly flow distribution was found to have a much smaller effect on the required storage. Consequently monthly distribution of normal flow was applied to the two-year drought flow of the river basins. However, monthly flow distributions of all river basins were investigated regarding their application to a single model. Table 1 provides a comparison between the monthly distribution of streamflow in various river basins and presents a model fitted to the average values. Distribution of monthly streamflow as presented in this figure is non-dimensional and expressed in terms of percentages of the average annual streamflow for each river basin.

TABLE 1
MONTHLY DISTRIBUTION OF NORMAL RUNOFF
IN VARIOUS RIVER BASINS

Month of Year	Geum	Nak Dong	Yeong San	River Basins Seom Jin	Han	Ave.	Model
Oct.	4.7	4.5	3.9	3.8	3.7	4.1	3.0
Nov.	3.7	2.9	2.8	2.4	3.1	3.0	3.0
Dec.	3.2	2.1	2.4	1.8	1.8	2.3	3.0
Jan.	2.6	1.7	2.6	1.9	1.1	2.0	3.0
Feb.	3.9	2.2	3.4	2.7	1.2	2.7	3.0
Mar.	6.3	4.1	4.0	3.6	3.5	4.3	3.0
Apr.	8.1	7.6	8.0	7.3	9.5	8.1	7.7
May	6.6	6.5	7.2	7.0	7.0	6.8	7.7
June	5.7	7.4	10.2	9.0	6.9	7.8	7.7
July	22.2	25.9	26.8	28.2	22.6	25.1	22.5
Aug.	19.5	18.6	16.1	19.6	25.0	19.8	22.5
Sept.	13.5	16.5	12.6	12.7	14.6	14.0	13.9
TOTAL	100.0	100.0	100.0	100.0	100.0	100.0	100.0

3. LOW FLOW ANALYSIS

Examination of annual variation of precipitation and streamflow for the NAK Dong River Basin indicates that low flows have often occurred for a period of two consecutive years and that low flows were distributed according to approximately 40 percent and 60 percent of cumulative two year low flow among the two years. A probability analysis was then carried out to determine the cumulative two year streamflows with a return period of twenty years.

Figure 1 present the frequency curve for two year low flow for this basin on log-log paper. The probability of two year flow precipitation for this basin is also shown on the same figure. This figure shows that data for return periods of over two years (i.e. for discharges smaller than the average value) exhibit a log-log relationship approximately.

4. ANALYSIS OF NON-DIMENSIONAL RESERVOIR STORAGE

The change of storage of a reservoir during a time period, Δt, can be expressed by the equation:

$$\Delta S = \Delta t\,(\bar{I} - \bar{O}) - E_n - L_s + D_s \tag{1}$$

where $\overline{I}$ and $\overline{O}$ are the average inflow and outflow (draft) from the reservoir within the time period, respectively. E_n is the net evaporation from the reservoir, L_s is the seepage loss and D_s is the quantity of deposited sediments in the reservoir, all during the time period Δt. Net evaporation is given by:

$$E_n = E - (P-q) \tag{2}$$

where E is evaporation from the lake surface, P is precipitation on the lake and q is the surface runoff from the lake area prior to dam construction. Considering that the average annual values of precipitation and evaporation are approximately equal in Korea and assuming that seepage losses (a negative term) and sediment deposition (a positive term) are negligible, equation 3 simplifies as:

$$\frac{\Delta S}{\Delta t} = \overline{I} - \overline{O} \tag{3}$$

This equation may be expressed in the following non-dimensional summation form.

$$\frac{S_n - S_1}{Q} = \sum_{i=1}^{n} \frac{I_i}{Q} - \sum_{i=1}^{n} \frac{O_i}{Q} \tag{4}$$

where the right hand terms represent the cummulative values of non-dimensional streamflow and draft during the time period (at the end of n'th month of drought) and the left hand side is the change in storage for the same time period.

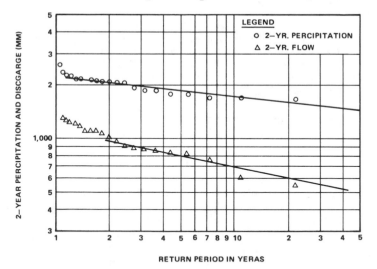

FIG. 1   ESTIMATION OF 2—YR. LOW FLOW
FOR THE NAK DONG RIVER BASIN

5.   DEVELOPMENT OF NON-DIMENSIONAL STORAGE-DRAFT DIAGRAM

The determination of the required storage capacities of the proposed reservoirs in the NAK Dong River basin was based on simulation of the reservoir's release during a two year drought. The cumulative two year drought flow for this basin was 115 percent of average annual flow for the basin. The selected model was employed to determine the monthly distribution of non-dimensional streamflow during two year drought during which the flow was allocated according to 40 percent and 60 percent proportions among the first and second year, as discussed already. The calculated non-dimensional monthly flows were used for determination of storage required to meet demands over a range of variation from 20 percent to 90 percent of average annual flow. Calculations of the required storage was done numerically by tabulating monthly streamflows and demands, calculating mass supply and mass demand and differentiating these. This numerical method provides a more detailed information about the condition of the reservoir (depleting, filing, spilling), than the graphical method and can be most readily adapted by the use of spread sheets.

Figure 2 shows the storage-yield relation for all dam sites in the NAK Dong River Basin. The actual storage required to meet a certain demand at each dam site were then obtained by multiplying the non-dimensional storage obtained from this figure by the average annual stream flow expressed in million cubic meters.

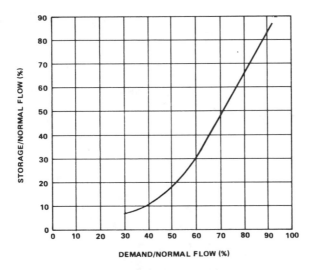

**FIG. 2   NONDIMENSIONAL STORE–DEMAND
RELATIONSHIP FOR THE NAK DONG
RIVER BASIN**

REFERENCES

1. Golze, A.R. (Editor), Handbook for Dam Engineering, Van Nostrand Reinhold Co., New York, N.Y., 1977.

2. Linsley, R.K. and J.B. Franzini, Water Resources Engineering, 3rd Ed. McGraw Hill, 1979.

3. Linsley, R.K., M.A. Kohler and J.B. Franzini, Hydrology for Engineers, 3rd Ed., McGraw Hill, 1982.

4. Ministry of Construction, The Republic of Korea, "Hydrological Annual Report in Korea," 1962 to 1984.

5. Ministry of Construction, The Republic of Korea, Final Report for Prefeasibility Study for Medium Size Multi-Purpose Dams, prepared by Woodward-Clyde Consultants in association with Korea Engineering Consultants Corp. and Dohwa associated Engineering Services Co. Ltd., Vol. 1, Main Text, September 1986.

6. ibid, Vol. II, Appendices, September 1986.

Flood Forecasting Using Local Resources

Richard C. Sorrell[*]

ABSTRACT: A dynamic wave flood routing computer program has been used to develop a flood forecasting model for a 5000 mi² (12,950 km²) watershed in lower Michigan. The modeled area covers over 200 miles (320 km) of the Grand River, including the lower portions of its six main tributaries, and incorporates the operation of 16 dams. The dynamic wave solution for the flood routing equations is used to more accurately represent the hydrodynamics of the basin which is characterized by flat channel slopes and numerous control sections. The model can forecast at virtually any point along the river rather than the limited number of locations given by the model currently used by the National Weather Service. The model has been constructed over a three year period with assistance from the National Weather Service River Forecast Center (NWSRFC) in Minneapolis, MN. Runoff hydrographs needed as input at the upstream boundaries of the model are obtained from the NWSRFC via their DATACOL system. Flow conditions are monitored during the flood through contact with the dam operators and calling eight stream flow gages equipped with Telemarks. Contact with the dam operators also allows us to revise the forecast when gate settings are changed during the course of the flood. Unit hydrographs are developed for 26 subbasins to represent local runoff. The NWSRFC supplies the surface runoff estimates needed to generate the local runoff hydrographs. Results from the model are telephoned back to the NWSRFC for their evaluation. The entire model package will eventually be given to the NWSRFC for their use.

The model is calibrated to six floods, including four snowmelt floods and two late summer thunderstorm events. The model has been used in real-time mode to forecast the last two major floods with highly successful results. The small errors were due to initial estimates of the input hydrographs at the upstream boundaries rather than errors in the routing itself. Hindcasts with the observed flows produces nearly exact results. The complete model package can be run on an IBM PC. Since hydrologists can estimate rainfall-runoff amounts from sources other than the National Weather Service, it is feasible for a local agency to provide flood warnings in areas not served by the NWSRFC forecasts.

[*] Hydrologic Engineer, Michigan Department of Natural Resources, Hydrologic Studies Section, P.O. Box 30028, Lansing, MI 48909

Introduction

The Grand River is a major river basin in Michigan that was formed in a geologic setting consisting of glacial till, end moraines, and glacial outwash. The average slope of the Grand and each of its six main tributaries ranges between two and four feet per mile. Most of the watershed is either forested or used for agriculture. Major floods in the basin usually result from the early spring snowmelt although thunderstorm induced floods are not uncommon. Most of the urban development in the basin is concentrated in Lansing and Grand Rapids and increased flooding due to urbanization effects is limited primarily to smaller streams in these areas.

The NWSRFC currently provides flood forecasts at the Telemark stream flow gages and a few other selected locations as shown in Figure 1. This project was conceived in 1983 when Amway Corporation, located in Ada, requested that the NWSRFC establish Ada as one of the forecast points along the Grand River. Amway is located at the confluence of the Grand and Thornapple Rivers and forecasts at Ada would help Amway better respond to impending floods. By this time, a number of dynamic wave flood routing programs were available which were capable of improving the accuracy and timeliness of flood forecasts that were currently provided by the NWSRFC. These improved computer programs were being used by staff of the Michigan Department of Natural Resources (MDNR).

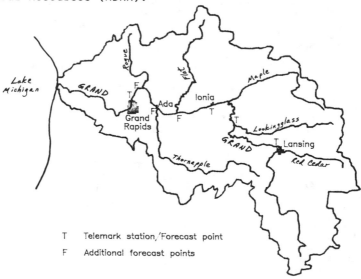

T	Telemark station,'Forecast point
F	Additional forecast points

Figure 1.   The Grand River Basin

MDNR, Amway, and NWSRFC discussed the feasibility of developing one of these programs into a forecasting model for the lower portion of the Grand River. After agreeing that a project like this would provide a significant benefit to all the communities in this area, Amway made a Public Service Grant to MDNR to develop a flood forecasting model for the Grand River from Ionia to Grand Rapids. This also included the lower portions of the Flat, Thornapple, and Rogue rivers. Meetings were also held with the owners and operators of the eight dams located on the three tributaries and they agreed to provide MDNR with data concerning impoundment levels and gate settings during a flood. This information can then be used to check the accuracy of the flood forecasts during the event and to update the forecast if the observed flows and stages are significantly different from the forecast.

After completing the model for this lower portion of the Grand, MDNR received a grant from the Federal Emergency Management Agency to extend the model upstream through Lansing. This portion of the model also includes the lower portions of the Maple, Lookingglass, and Red Cedar tributaries as well as seven additional dams located on the Grand River.

There are three objectives for this project. First is to develop a flood routing model that accurately depicts the hydraulic features of the river valleys and can reproduce past flood events. The degree of success in completing this objective indicates how reliable the model may be in forecasting floods in the future. The second objective is to try to optimize the operation of the dams during flood conditions. This objective determines if dam operations can be effective in controlling the severity of a flood. The last objective is to provide a sound model framework for the incorporation of data from automated remote data stations. Accomplishment of this objective ensures that real-time data can be utilized effectively as input to the forecasting process.

Model Development

Flood forecasting requires a means to generate surface runoff hydrographs and a means to combine the hydrographs and route them downstream. Computer program HEC-1, developed by the U.S. Army Corps of Engineers, is used to develop all of the local runoff hydrographs in the model that are not supplied by the NWSRFC. Unit hydrograph and infiltration loss parameters are developed during calibration of the model. Since the NWSRFC provides runoff volumes, it is only necessary to input unit hydrographs to HEC-1 although generating hydrographs directly from rainfall is possible using data from the calibration. Data are also calibrated to generate snowmelt hydrographs.

Computer program DAMBRK, developed by the National Weather Service, is used to perform the flood routing. This program routes by the same procedure as computer program DWOPER which is used for operational forecasting in other major watersheds around the country. DAMBRK is used in this model because of its ability to better define dam hydraulics. The major drawback is that DAMBRK cannot route floods in a river network. Consequently, seven separate computer runs are needed to determine flood profiles on the Grand and the six tributaries. The Grand River model is run first using the unrouted tributary hydrographs as lateral inflows. The six tributary models are then run using the stage hydrographs from the initial Grand run as a downstream boundary condition. The Grand model is rerun using the modified outflows from the tributaries. This process is repeated until the assumed and computed outflows from the tributaries are reasonably equal. In practice, this usually requires the seven initial runs and one final run of the Grand.

The cross section data used for all of the models come from readily available sources, including National Dam Inspection Reports, Flood Insurance Studies, and U.S. Army Corps of Engineers Flood Plain Information Reports. Additional data are used from hydrologic reports on file with the MDNR. Actual cross section data are input about every three to five miles unless the valley or the presence of control structures, such as bridges or dams, warrants finer detail. In the case of this project, much less cross section data are used than are actually available. Initial Manning roughness coefficients for each reach are available from the flood profile computations associated with the particular source of cross section data. The actual cross sections and roughness coefficients input to the model have been adjusted slightly during calibration to match observed high water elevations. Additionally, the roughness coefficients for summer floods (May to September) are approximately 20 per cent higher than during the rest of the year to reflect the increased resistance to flow caused by vegetation in the growing season.

Both HEC-1 and DAMBRK are used on an IBM Personal Computer. Although the computation time is significantly longer than running on a mainframe computer, forecasts can can be generated in approximately three hours. A modem is used to access the NWSRFC DATACOL system to receive input hydrographs and other related data. Communication software is available to interrogate and receive data from automated remote data stations if these are part of the flood warning system. The ability to base the model on a microcomputer makes it feasible and convenient to both produce the forecast and communicate the results to many people.

Model Results

The model is calibrated to the last six major floods in the Grand River basin. The most recent calibration flood in

1985 was the first real-time use of the model, occurring shortly after work began to extend the model to its present upstream boundaries. The initial stages forecasted by the model for Ionia and 26 miles (41.8 km) downstream at Ada were 22.8 ft (6.95 m) and 21.1 ft (6.43 m), respectively. The model predicted the Grand would crest at these stations in two days. Each succeeding day the forecast was rerun with the flows observed up to that point. The final run was made two days after the peak had passed with revised forecasts of 22.98 ft (7.00 m) and 21.66 ft (6.60 m), respectively, for Ionia and Ada. These stages compare well with the observed values of 23.15 ft (7.06 m) and 21.55 ft (6.57 m). The time of peak predicted by the initial forecast two days earlier was within 6 hours of what was observed. The accuracy of the predicted peak time provides confidence that the model is accurately representing the hydraulics of the Grand River. The minor error in the forecasted and observed stages could be attributed to a slight error in the initial runoff estimates.

The model has also been used to forecast the floods which occurred in the basin in late September 1986. This flooding followed 26 consecutive days of record rainfall through most of Michigan which led difficulties estimating runoff from the saturated basin. Tributaries in the north-ern portion of the watershed had experienced 100 to 500-year floods in the middle of the month while the southern areas had only minor flooding. By the end of the month heavy rains occurred primarily over the southern tributaries. Even with the complex hydrologic condition of the basin at the time, the model was again able to predict stages and peak times with very reasonable accuracy. The peak stages forecasted by the model were within 0.4 ft (0.12 m) and 0.2 ft (0.06 m) at Ada and Ionia, respective-ly. The forecasted peak time was within 1.5 hours and 6 hours at the two locations. Figure 2 illustrates the observed and forecasted stages at Ionia.

Summary

A computer model has been developed which predicts flood stages in a large watershed using a dynamic wave flood routing method. The use of a model like this accomplishes two important objectives in flood forecasting. First, since the model incorporates actual cross section data, additional forecast points can be added merely by inserting the cross section data at the appropriate point in the model. Since the routing program that is used has the ability to interpolate additional cross sections from those actually input, flood stages can be predicted at many locations without any additional data if the river valley is relatively uniform. Secondly, since the dynamic wave routing method calculates the travel time of the flood as part of its solution, it is not necessary to input a lag parameter to move the flood wave downstream. Routing methods which use a lag parameter like this may not produce accurate results if the flood is not similar to the past floods that were used to calibrate this input parameter. The forecast model that has been developed removes this uncertainty and increases the confidence that it can be used to predict a wider range of floods.

All of the data used to develop the model come from readily available sources. The only unknown input needed to generate a forecast is the rainfall-runoff for a storm. Although these data come primarily from the National Weather Service, other sources can be found to provide this information.

The ability to run the models and access the data with microcomputers makes forecasting feasible for anyone who has the expertise and resources needed to calibrate the model. A modem for the microcomputer can provide efficient communication for both receiving needed data and transmitting the forecast to emergency service officials and others who need this information.

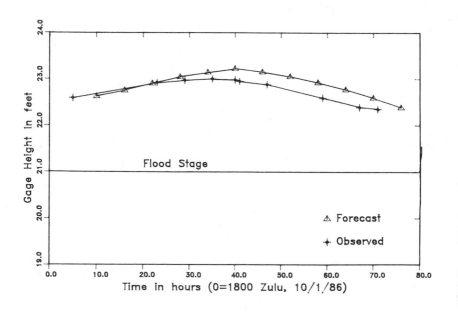

Figure 2.   Stages on the Grand River at Ionia

FLOOD FORECASTING WITH MINIMAL DATA

by Gary R. Dyhouse*, M. ASCE

Abstract. The Corps of Engineers' HEC-1 program has been used with a specialized pre-processor program to provide interim emergency flood forecasting for a 4000 sq mi (10,360 sq km) watershed in east-central Missouri. Results to date have been excellent despite inadequate real-time data, particularly rainfall information. Flood crests have been predicted within 1-2 ft (0.3-0.6 m) with 24-48 hr warning time on all but one occasion.

Introduction. The Meramec River drains a watershed of about 4000 sq mi (10,360 sq km) of Ozark Highlands country in east-central Missouri and empties into the Mississippi River, about 20 mi (32.2 km) downstream of St. Louis, Missouri. The basin is composed of the three main rivers shown schematically in Figure 1. The lower 6-8 mi (9.7-12.9 km) of the Meramec are greatly influenced by backwater from the Mississippi, with a lesser effect upstream. The initial 51 mi (82.1 km) reach of the Meramec (to Pacific) is referred to as the lower Meramec. During this century, flood plain use in the reach has gradually changed from totally agricultural to include much residential and commercial use. Much of this development occurred from the late 1940's into the 1970's, when modest summer cottages were gradually converted into permanent residences. During this period, only one significant flood occurred. Major communities lying partially in the flood plain include: Arnold, Fenton, Valley Park, Eureka-Times Beach and Pacific. An additional 15 small communities, each having at least 20 structures, are scattered throughout the reach.

The Corps of Engineers, St. Louis District (SLD), has been periodically directed to review the basin for flood control improvements since the 1930's. A system of reservoirs was proposed with construction started on the first one--Meramec Park Dam near Sullivan--in 1974. Construction was well underway when a referendum, held in 1978, voted down any further state support for this project, terminating further construction. It appears unlikely that any reservoirs will be constructed in the foreseeable future.

The last eight years have seen flooding along the Meramec unlike any similar period in this century. Major floods in 1979, 1982, 1983, and 1985, along with several smaller ones, have caused significant damages. Without reservoirs, other structural solutions to alleviate flooding are unlikely to be economically feasible. Future damage reductions would seem to rest solely in the non-structural arena, particularly with improved flood forecasting.

December 1982 Flood. The Meramec Basin is one of 26 tributary basins to the Mississippi River forecasted by the National Weather Service's (NWS) regional office in Minneapolis. During a major flood in the region, NWS resources are spread sufficiently thin that up-to-date forecasts for all 26 basins plus the Mississippi are extremely difficult. Such was the case during the December 1982 flood, when the SLD became unintentionally involved in

* Chief, Hydrologic Engineering Section, Hydro & Hydr Br, St Louis District, Corps of Engineers, 210 N Tucker Blvd, St Louis, MO 63101-1986

emergency flood forecasting in the Meramec Basin. Before December 1982,
most people living in the Meramec flood plain had not experienced a major
flood. The opinion often expressed at meetings concerning flood plain
management and the flood insurance program was that "flooding used to occur
but is no longer a problem". Many Meramec communities scoffed at the
hypothetical flood profiles published in various reports and resented the
zoning and building restrictions required to participate in the flood insurance
program. Consequently, early in the December 1982 flood, most residents
refused to believe the initial flood forecasts issued by the NWS, which turned
out to be far too low. Local officials' concern with the rapidly-rising
Meramec, which was already exceeding stages predicted for 24-48 hours later,
prompted a barrage of phone calls to the Corps' Emergency Operations
Center, then fighting a flood on the Mississippi. To provide additional
information, SLD hydraulic engineers gathered early on Sunday morning,
attempting to supplement the Meramec forecast. An existing hydrologic model
of the Meramec, using the HEC-1 computer program (1), was retrieved from
computer archives. Input consisted of six basin rainfall totals called in by
the general public after a radio plea from the Corps and a time distribution
pattern using the NWS gage at the St. Louis Airport (well out of the basin).
Model output was compared to one stage reading taken that afternoon at the
Byrnesville gage on the Big River. In spite of the sparsity of data, the model
indicated that stages along the lower Meramec would approach record levels
within 24-48 hours. This information was passed to the NWS, the Corps'
Emergency Operations Center and to Meramec floodfighters by early that

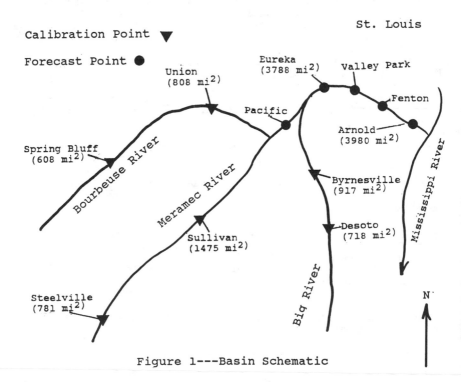

Figure 1---Basin Schematic

evening. With additional stage data obtained the next day, the model
satisfactorily predicted final crest stages within one foot (0.3 m) of actual.
Peak stages of the flood approximated the 100-year return interval event and
caused $40-50 million dollars in property damage and the loss of several lives
(2).

Forecast Model Development. The hydrologic model used for forecasting had
actually been developed several years earlier for plan formulation purposes.
The HEC-1 model consisted of 44 sub-areas and accompanying routing reaches.
The Soil Conservation Service loss rate technique (3) was used for each subarea, with Clark parameters (1) for unit hydrograph computations. Storage-
outflow methods, using modified Puls techniques (1), were incorporated for
routings. The model had previously been calibrated to several actual floods
and then used to reconstitute the long-term discharge-frequency relationships
at gages throughout the basin. Effects of various flood control measures
were also assessed.

Modifications for Forecasting. The Meramec outflow during the December
1982 flood had a major impact on Mississippi River flood levels, more than
had been previously experienced. This impact required the SLD to obtain
additional data for use during future Mississippi River flood situations.
Although the NWS provides forecasts for the Mississippi, the St Louis District
has a responsibility for obtaining sufficient runoff data to responsibly regulate
its projects, minimizing Mississippi flooding. To fulfill this responsibility,
improvements in hydrograph estimation and additional real-time data were
needed.

   a. Model Improvements. Several problems surfaced during the December
1982 flood that required rectification for easier future use : effect of
Mississippi River backwater on lower Meramec elevations, quicker calibration
of the model to existing stage data early in the rising limb of the
hydrograph, and easier methods to code the data in the correct format for
application by the HEC-1 program. These three problems were overcome
during 1983 by developing a pre-processor program called STORMER. This
program provided a quick and easy way to enter real time data required by
the HEC-1 model. STORMER arranges the data into the correct HEC-1
format and keeps track of the output from different calibration and
forecasting runs throughout the course of a flood. All Meramec River
recording gages had the stage-discharge relationship added to the HEC-1 input
to display stage-hydrographs for actual and predicted stage comparisons. The
effects of the Mississippi River were incorporated by adding its 3-day
forecast to STORMER and developing relationships at all affected gages for
any combination of Mississippi River backwater and Meramec River discharge.
Families of curves were computed at each gage site and damage center along
the first 37.5 mi (60.4 km) of the lower Meramec. The HEC-1 output is
accessed by STORMER to graphically display the results of each calibration or
forecasting run. A Tektronix 4014 with a hard copy printer has been used
exclusively for output display, although the program has recently been
converted to run and display output data on the Harris 1000 system. The
only problem unlikely to be overcome is the sparseness of the real-time data
used to prepare emergency forecasts.

   b. Real-Time Data. To obtain a minimum of real-time data, the SLD
contracted for eight rainfall observers in the Meramec to collect and forward
daily rainfall data. The observers data is needed only during floods, when

rainfall totals are phoned to the District office once or twice per day. To more accurately distribute the average rainfall in time, a 15-minute rainfall recorder (with satellite comlink) was added near the basin centroid. Six recording stage gages on the Meramec River and the two major tributaries were modified for satellite communications, providing hourly changes in river stage. During a flood, forecasts are made with as many as eight rainfall totals from throughout the basin with average basin rainfall distributed against data from one continuous recorder gage. Calibration is made against actual stages received from the six satellite gages and from stages that may be called in from observers at 1-4 other sites (non-recorders). The existing raingage network density is approximately one per 500 sq mi (1295 sq km), considerably more area/gage then optimal. While these gages are adequate to provide a Meramec outflow hydrograph to the Mississippi, the number of gages is less than satisfactory for lower Meramec forecasts.

Current Procedures. The SLD's Meramec model is used for forecasting only during periods of potential flooding (usually less than 3 times per year). Usage of the model has occurred about 10-12 times over the past several years. Each application is used to evaluate the potential flood situation, to calculate a discharge hydrograph for Mississippi forecasts, and to evaluate flood stages on the lower Meramec. The Meramec information generated is passed to the NWS for their use and information, and to Corps floodfighters. From past experience with the hydrologic model, 2-3 inches (51-76 mm) of rainfall basin-wide over 24-48 hours is typically necessary to potentially cause flooding along the lower Meramec. Rainfall information is critical only during a potential flood situation, although the data are recorded daily by each observer. Standing instructions for all raingage observers are to report, via phone, any 24-hour rainfall total exceeding 0.5 inch (13 mm). During an actual flood emergency, they may be requested to report rainfall amounts at 12-hour intervals and to report any rainfall at all. During a potential flood situation, SLD hydraulic engineers make initial preparations for running the forecast model. This process involves readying the STORMER package for the watershed conditions at the start of the flood. Initial streamflow is entered at the Sullivan, Byrnesville and Union gages. Rainfall and river stage data are encoded to STORMER, as received. Initial calibration runs are made and the SCS curve numbers in each of the three major watersheds may be ratioed to approximately match the rising limb of the hydrograph. This simplified calibration procedure through loss rate adjustments has been adequate for nearly all of the flood situations thus far encountered. The input to the model is updated as new information is received; several times daily early in the flood and once or twice a day with no additional rainfall data later in the event. The initial setup and run requires 2-4 hours. Updates and reruns take less than 30 minutes. In addition to evaluating flood potential from actual data, the possibilities involving "what if" situations are also reviewed. Each day, the SLD receives a 24- and 48-hour quantitative precipitation forecast from the Corps' Lower Mississippi Valley Division office at Vicksburg, Mississippi. This potential precipitation is routinely added to analyze a "worst case" flooding scenario, based on these rainfall predictions.

Model Output. Model output is primarily graphical plots of actual and forecasted stage hydrographs at key gages along the lower Meramec and an outflow discharge hydrograph for the Mississippi River. Comparisons at the calibration points upstream are also plotted. Figure 2 shows typical examples of actual and predicted stage hydrographs during a flood. The forecasted hydrographs displayed were computed during the morning of 24 February 1985,

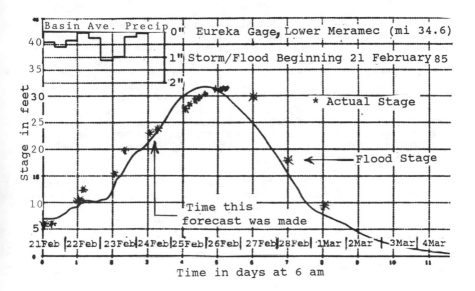

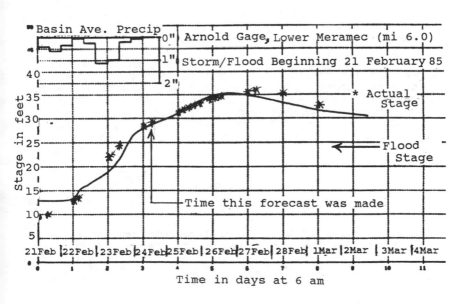

Figure 2---Typical Forecast Hydrographs

48-72 hours in advance of the actual peak. Accuracy and warning time has generally been excellent for nearly all past applications, in spite of the inadequate precipitation data. The only exception occurred from a series of isolated thunderstorms moving through the Meramec Basin over a several day period in early October 1986. Precipitation was not sufficiently widespread across the basin for the sparse raingage network to give accurate information. The model consistently over-predicted Meramec stages throughout this event. However, stages are usually within one foot (0.3 m) of actual, with 24-48 hours warning time, depending on the lower Meramec location being forecast. The model generally predicts the time of peak about 12 hours early, and the forecasted time of peak is adjusted accordingly by the responsible hydraulic engineer. This underprediction has not caused any difficulties to date.

Future Applications. In spite of its success, the continued use of the Meramec model is uncertain. This emergency operation was always envisioned as interim testing, to be eventually taken over by a responsible local entity or perhaps the NWS. The forecasting system would be a key component of one local protection project (levee) envisioned for Valley Park and would provide warning to the unprotected communities along the lower Meramec. As such, the operation would be greatly improved by incorporating an additional 12-23 real-time gages for rainfall or river stage throughout the Meramec. However, no local sponsor for the flood forecasting/warning system has been found and continued use of the model for anything other than providing a discharge hydrograph for Mississippi River forecasts is in doubt. As the HEC-1 model is not compatible with the NWS forecast model, that agency has not indicated interest in further use and development. A real possibility exists that flood forecasting applications of the Meramec model will decline or cease.

Summary. A hydrologic model of the Meramec River Basin has been adapted for real-time flood forecasting using the HEC-1 program and an SLD-developed, pre-processor program to facilitate the input/output of data. It has been successfully tested on 10-12 occasions and has satisfactorily forecasted peak stages with 24-48 hours warning time in all but one situation. Input data are less than adequate for long term use, particularly rainfall data. Improvements in the model and continued application seem in doubt unless a local entity comes forward to fund/operate the system.

This paper represents the findings and opinions of the author and not necessarily those of the Corps of Engineers.

References.

1. Hydrologic Engineering Center, HEC-1, Flood Hydrograph Package Users Manual, Corps of Engineers, Davis, CA, January 1985

2. St Louis District, Corps of Engineers, Lower Meramec River Flood Damage Reduction Project, General Design Memorandum, January 1987

3. Soil Conservation Service, National Engineering Handbook, Section 4, August 1972.

REAL-TIME HYDROLOGIC MODELING

Stanley M. Wisbith *, M. ASCE, and Karen V. Miller **

ABSTRACT

The Huntington District of the Corps of Engineers is currently using daily computer modeling for guidance in regulating all nine of its major river basins. Both pool elevation and flow forecasts are generated by using simulated run-off and streamflows combined with observed data. Observed hourly precipitation and streamflow values required for the models are automatically collected from remote data collection platforms (DCP's) via satellite.

Initial data screening and conversion is performed by a data conversion program. "Human" input is required for secondary screening and correction of input data and for estimating a schedule of releases from various dams. If current model results indicate a need, the schedule of releases can be modified, after which the forecasts are re generated to produce final estimates. Execution of all these processes and the programs necessary to generate the forecasts are coordinated by a model control program.

Real-time modeling has proven to be a useful tool for regulation of complex hydrologic systems. The increase of detail from the modeling output provides regulation guidance not possible through manual methods.

INTRODUCTION

The Huntington District of the U.S. Army Corps of Engineers straddles the central Ohio River basin, with its operational area spanning portions of Ohio, West Virginia, Kentucky, Virginia, and North Carolina. Topography within this area covers almost all possible types, ranging from the flat glaciated planes of central Ohio to the mountainous regions of West Virginia and Kentucky. The drainage patterns of the district, excluding the main stem Ohio River, form nine major river basins. Streamflows in these basins are regulated by 34 flood control dams.

The Huntington District's real-time numerical modeling effort is composed of nine sets of computer models, one for each of the major river basins. The modeling system used by the district is based on a blend of observed and computed values. This combination maximizes the amount of information extracted from observed data and improves the overall accuracy of the model's output.

The basin models produce runoff and stream flow forecasts for over 32,000 square miles of drainage area, as well as pool elevation forecasts for all 34 of the above mentioned flood control dams. In addition to producing forecasts, the models are used in an iterative

* Hydraulic Engineer, U.S. Army Corps of Engineers, Huntington District, Huntington, WV 25701
** Hydraulic Engineer, U.S. Army Corps of Engineers, Huntington District, Huntington, WV 25701

manner to determine and verify operational decisions.

Because of the use of observed values by the models, it is necessary to use the latest possible data available. The use of these
values, however, requires the implementation of several processes
before actual modeling is performed. These processes include data acquisition, screening, conversion, and distribution.

DATA ACQUISITION

Initial data acquisition from individual gages is accomplished
through approximately 225 satellite platforms distributed throughout
the district.   These platforms transmit a total of over 350 products
from the gages to the Geostationary Operational Environmental Satellite
(GOES).   The relayed data contains hourly readings for stream stages,
reservoir pool elevations, and cumulative precipitations. Additional
information in the form of water temperature, pH, and conductivity is
also transmitted from several locations.

Observed hourly data are transmitted at four hour intervals. Each
gage transmission includes data collected for the last four hours, and
in most cases, four hours of redundant data from the previous
transmission.   Although platforms transmit only once every four hours,
timeslots for transmission are arranged so that some data is being
received each hour.

The transmitted GOES values are received by a satellite downlink
located at the Ohio River Division (ORD) Headquarters at Cincinnati,
Ohio.   These values are passed through the division's computer system
before being sent by dedicated (X.25) phone lines to the district office at Huntington, West Virginia. As part of the system backup, an
alternate means of transmission is available by using remote job entry
(RJE) through normal commercial lines.

DATA SCREENING AND CONVERSION

Once received by the Huntington District, the observed values are
processed by decoding software. The raw values are converted from the
various alphanumeric transmission protocols used by different platform
manufacturers into engineering units (feet, inches, etc.) After decoding, the values are then entered into a single master data base for
further processing.

The data base system used by the water control program was
developed by the Corps of Engineer's Hydrological Engineering Center
(HEC), and is called "Data Storage System" (DSS). This system is
designed to efficiently store and retrieve time series data. In addition, HEC has developed a package of several supporting programs for
data manipulation, data viewing, and report generation using DSS data
bases.

Initial automated data screening and conversion is performed by
the specialized data conversion program DATCON developed by the
Huntington District.   Automated screening primarily attempts to identify and correct errors incurred during the transmission process. Once
identified, transmission spikes are replaced with values obtained from
interpolating between good points. Small blocks of missing data, up to
four hours, are also replaced by interpolated values. Data conversion
involves determining incremental precipitations, stream flows, and
reservoir conditions.

Incremental precipitations are computed by differencing the observed cumulative precipitation values. The resulting values from this

operation are precipitation rates in inches per hour. Stages are then used to compute streamflows by interpolating the gage rating table for each gage. The resulting values are flow rates in cubic feet per second (cfs).

The conversion process for a reservoir is more complex, and involves the computation of reservoir outflow, storage, and inflow values. Reservoir outflows are normally derived by interpolating outflow gage elevations in a rating table. However, when an outflow gage is not available, releases are generally computed from lake elevations, gate openings, and gate rating tables. The exception to this rule is when outflows are from an uncontrolled spillway. In this case, releases are computed directly from lake elevations and a spillway rating table. After outflows have been determined, reservoir storages are derived from lake elevations by interpolating an area-storage table. Finally, inflows are then computed from outflows and changes in storage.

Once preliminary screening and conversion is completed, the data values are transferred by DATCON to the individual basin data bases. These data bases are screened manually every morning by the basin operators. This screening is performed graphically using the DSS package's DISPLAY program.

Graphic viewing of the data allows for quick identification of erroneous data whether the errors are from an improperly operating gage or from transmission noise not corrected by the automated screening process. Once identified, erroneous values are corrected by the basin operator using the graphical editing capabilities of the DISPLAY program. It should be noted here that all data editing is performed on the data in the basin data bases. The master data base is never changed, and forms a backup for reconstructing the individual basin data bases.

SIMULATION PREPARATION

The actual modeling process is controlled by the basin operator through the interactive model control program MODCON developed by HEC. This program acts as an outer shell sequencing and controlling execution of the various programs required for the simulation. MODCON is also the interface between the basin operator and the models. This is where parameters are set and reservoir operational decisions are entered.

The flow simulation process itself is begun by first selecting a "Time of Forecast" (TOF) to be used as the time from which the basin's forecast is computed. Generally this is chosen to be the latest time that data is available from all gages in the basin. For a normal daily forecast a TOF from between 0400 hours and 0600 hours is used. This time is entered through MODCON to initialize the simulation.

The flow simulation runs are made during a time window that straddles the time of forecast. This window looks back from the TOF four to seven days and forward at least five days. Observed data within the look back period will be used by the model with forecasted data generated for the period after TOF.

The HEC developed program EXTRACT is then used to extract observed data from the basin data base into a local data base. EXTRACT transfers all data required to run the particular basin during the simulation time window. During this operation, the observed values are converted from one hour data to the time step used by the model. In most cases this requires a conversion from one hour to three hour data.

EXTRACT also aligns the values from all the extracted gages to fall on the same hour.

After extraction into the local data base, the subbasin average precipitation is computed from the observed incremental gage rainfall. This is accomplished by using a program called PRECIP developed by HEC. PRECIP computes area average precipitation for all of the subbasins within each model. The subarea precipitation is derived from the closest gage available in each of the four cardinal directions. A method based on Thiessen polygons is then used to compute average precipitation with adjustments made for the distance from each gage to the centroid of the subbasin.

FLOW SIMULATION

The core of the actual simulation is performed by the Corps of Engineers' HEC-1F computer program. Two successive runs of this program are required to complete the simulation. Adjustments of parameters and project operations are made as required between the two model runs.

The first, or "E-model," run is made to estimate hydrological parameters in headwater and selected intermediate subbasins. The unit hydrograph, loss rate, and base flow optimization capabilities of the HEC-1F program are used to determine the values for those parameters which produce hydrographs best fitting observed flows during the time window up to the TOF. The event specific unit hydrographs, derived from observed discharges in these subbasins, form a much better in-dication of hydrologic conditions at the time of forecast than would hydrographs derived from a few large scale events. Hydrographs gener-ated from these unitgraphs in this model are then saved to be used as input for the second step.

After the E-model has been run and the optimized parameters deter-mined, these parameters are then entered though MODCON. Also at this time decisions on project operations are made and entered if required. A default condition of releases equal to each dams outflow at TOF is automatically set for the forecast period. However if an alternate release schedule is desired it would now be entered through MODCON.

The second, or "F-model," run routes and combines flows from the E-model hydrographs, project releases, and hydrographs which it gener-ates for the remaining areas. These generated hydrographs are computed from the subbasin average precipitation and the parameters entered through MODCON. The routing and combining is performed by the model in a general downstream direction. As each control point is reached the observed and computed flows for this point are blended to form a single hydrograph which is then used for the next downstream routing step.

The process of blending is a key operation performed by the F-model and consists of the merging together of both observed and com-puted flows. This procedure maximizes the amount of information extracted from the observed data by using observed flows as part of the direct calculation within the model. A blended hydrograph consists of observed values up to time of forecast then a gradual transition into purely computed values within the forecast period. The transition from observed to computed flows occurs during the first five time periods after the TOF.

After the F-model run has been made, revised reservoir operational decisions may be made based on graphical displays of forecasted flows and pool elevations. These decisions are then reentered into the modeling process through MODCON and the F-model is rerun to confirm their validity. The optimized coefficients are also sometimes modified before final forecasts are made to improve the matching of observed and computed flows. Determination of the correlation between observed and computed flows is again made by graphical display of these values.

FORECAST REPORTS

During normal daily operations, the final steps of the forecasting procedure are the compilation and transmission of basin and district forecast reports. The basin reports are created by REPGEN, a program from the DSS package developed to generate reports from a data base. Each basin report contains observed and forecasted values of stage and flow for the basin's final downstream control point as well as pool elevation and outflow for all the basin's reservoirs. The observed values are composed of current 24 hour instantaneous readings, while forecasted values are 24 hour values for the next five days for downstream control points and the next three days for reservoirs. The basin reports are given a final screening by the basin operators and then certified. This procedure marks the basin modeling as being complete and the forecast as ready to be transmitted.
Once all the basins have been certified the district forecast report is automatically compiled. This is simply the collecting of the basin reports into a single report. The district report is once more scanned for accuracy by a team leader, then transmitted to the ORD Headquarters over the X.25 lines.
A schematic of the data flow path during the modeling process for three basins is shown on Figure 1. This figure shows the process from reception of raw data to transmission of the final district forecast.

RECOMMENDATIONS

Experience has shown that modeling events due to large area storms generally provide good results. The computation of subarea average precipitation for these events assumes uniform distribution between precipitation gages. However, more common events such as multiple cell thunderstorms are not modeled quite as well using the same assumption. These events are often the cause of flash flooding and their proper modeling is very important for flood control forecasting. A system based on radar or satellite imagery could possibly produce better precipitation distributions therefore better modeling of cellular events.
The existing models have at their core a program designed to simulate single large scale events using batch processing. Using this basic program to model constantly changing conditions requires the addition of an outer shell of programs to produce input decks with current data. This method results in both the redundant manipulation of data and a rigidly structured model. Incorporating the existing algorithms into a unified system, that could determine the best data and procedure to use for the varying conditions that may exist during real-time modeling, would provide a more fluid approach. Such a unified model could then function as an "Expert System."

CONCLUSIONS

The Huntington District's current modeling system, used in conjunction with real-time data aquisition, enables a more realistic calculation of hydrologic properties than have previously been

430 ENGINEERING HYDROLOGY

available. The use of event specific unit hydrographs, along with loss
rate estimates, routing of observed flows, and one or three hour time
increments produces forecasts that are more detailed, more timely, and
extract more information from the observed data than those produced by
manual methods. This improved forecasting capability has since
resulted in a corresponding improvement of the Huntington District's
water control operations.

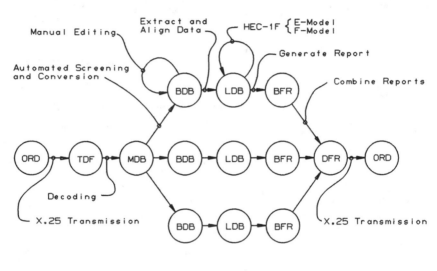

ORD	Ohio River Division
TDF	Transmitted Data File
MDB	Master Date Base
BDB	Basin Data Base
LDB	Local Data Base
BFR	Basin Forecast Report
DFR	District Forecast Report

Figure 1
Data Flow Schematic

HEPPNER, OREGON FLOOD WARNING SYSTEM
An Interagency Cooperative Effort

Jeffrey D. Hanson*

ABSTRACT

The Portland District, U.S. Army Corps of Engineers, operates an eastern Oregon project that is unique in the Corps -- the Willow Creek project at Heppner, Oregon, in Morrow County. All Corps projects serve authorized functions. Usually, flood control is the primary purpose with conservation uses secondary. A Flood Warning System, normally a responsibility of the National Weather Service, was later added to the Willow Creek project under discretionary authority by the Office of Chief of Engineers. This system's main purpose is to eliminate or minimize the potential for loss of life.

Cooperative agreements between Morrow County, the National Weather Service (NWS), and the Corps of Engineers (COE) outline agency responsibilities. Basically, the Corps designed, installed, and maintains the flood warning system, and funds its operation, communications, and maintenance. The NWS assisted in site selection, and provides software for data collection, the advisory system, and forecast capability for thunderstorm and rain-snowmelt runoff. NWS also notifies local officials of flash flood potential. Morrow County provides office space and utilities for the base station computer, round-the-clock monitoring and notification of flood events, and drew up an Emergency Warning and Evacuation Plan for the City of Heppner.

The flood warning system is a network of event-actuated precipitation and river level gages in two flash flood prone drainage basins of 44 and 6.8 square mile areas. These basins enter the City of Heppner below Willow Creek Dam. The radio telemetered gages report to the base station which processes the data and provides five levels of flood warnings based on certainty and severity of the event. The base station also receives project operational data such as lake elevation, inflow, and outflow, and combines that data with the forecasts from the two basins for a downstream routed forecast.

The base station's mini-computer/data logger monitors the various sensors, evaluates incoming data, and by using a continuous self-adjusting model, computes basin hydrographs. The forecasted hydrograph is converted to stages, levels are evaluated, and audio-visual alarms are given, if applicable. The numerical model, maintained by the NWS, is derived from portions of the NWS-Sacramento River Forecast Center small basin runoff model. The NWS-Portland River Forecast Center (NWS-RFC) wrote the evaluation and warning portion of the system.

INTRODUCTION

The flood of 1903, which killed 247 Heppner residents out of a total population of about 1,200, ranks as one of the worst natural flood disasters of its kind in the history of this nation. To alleviate the potential for a recurrence of such a disaster, local interests and agencies expressed a strong desire for full investigation of the flood control problem, and of other types of water resource development.

Since flash flooding from thunderstorm activity in the area is a significant cause of damage and loss of life, and since Heppner is affected by three flash flood prone drainages, a composite reservoir system including all three drainages was studied by COE.

*Hydrologist, Reservoir Regulation and Water Quality Section, U.S. Army Corps of Engineers, Portland District, 319 SW Pine Street, Portland, Oregon 97204

The result was the construction of Willow Creek Dam to protect Heppner from floods originating in Willow Creek and Balm Fork. Although flood retention structures on Shobe and Hinton Creeks were found to be infeasible, primarily because of economic reasons, a flood warning system was developed as a nonstructural solution to provide advanced warning of an impending flood and allow evacuation of residents in the flood path.

For a state-of-the-art automated flood warning system to be fully effective, 24-hour continuous monitoring is required. Although the NWS provides notices of flash flood potential, there is usually a wide-spread area of concern. Information pertaining to localized conditions such as storm duration and amounts, resultant runoff, and actual flash flooding cannot be comprehensively disseminated. To overcome this deficency, the NWS-RFC provided a "composite category score" method that allows non-technical personnel to determine the event's seriousness without sifting through masses of hydrologic data for which they are not adequately trained. Using this scoring method, local officials can initiate appropriate flood emergency actions.

BACKGROUND

Willow Creek Basin is an 890-square-mile area in north central Oregon. The basin is 60 miles in length with a maximum width of 23 miles. Willow Creek flows in a northerly direction from its headwaters in the Blue Mountains of Oregon to enter the Columbia River 250 miles from the Pacific Ocean. Basin elevations range from 5,000 feet in the headwaters to 195 feet at the mouth.

The City of Heppner, population 1,500, is situated at the confluence of three streams: Willow, Hinton, and Shobe Creeks. The drainage areas of these creeks are 96, 44, and 6.8 square miles, respectively. The small portion of the drainage area above Heppner lies on the rugged slopes of the Blue Mountains and is partially forested. The remaining major portion is undulating, treeless plains with shallow soils and separated by numerous small watercourses. The rolling prairie lands are used for grazing or raising grain by dryland farming methods.

The climate of the area can be described as semi-arid. It is characterized by low precipitation, wide variations in annual temperatures, low humidity, and high evaporation during the summer. Severe frontal systems are rare, but intense summer thunderstorms occur frequently. Occasionally they result in significant amounts of rainfall over limited areas within less than an hour.

Although other types of flooding such as winter rain-snowmelt and spring snowmelt can occur, the thunderstorm-generated flash flooding is by far the most damaging in regard to lives and property as such floods develop rapidly and without warning, as evidenced by historical events. The severity of the peak flow is caused by the multiple factors of concentrated rainfall over a small area, rugged terrain, and the almost total lack of vegetation in the watershed. As the rainfall sweeps down barren steep-sloped gulches, it combines with other stream courses while picking up mud and debris. The wave height is increased and usually travel time is detained, adding to the destructive effects.

As typical with the western United States, early history is relatively unknown. Incomplete accounts have been used to reconstruct known hydrologic events, the earliest occurring in 1883. The June 1903 flood is the worst known flood, particularly because of the amount of damage and loss of life. Approximately 1.5 inches of rain over a 20-minute period resulted in an estimated peak flow of 36,000 ft^3/s and a total volume of 1,100 acre-feet. The flood peak occurred almost simultaneously with the initial rise and the creek was reported to be out of its banks for less than 2 hours. Seven flash flood events causing various degrees of damage have occurred in the Heppner area, with two of the more recent originating in the Shobe Creek basin. The most recent, in May 1971, resulted in a peak flow estimated at 6,000 ft^3/s from the 6.8 square mile basin. The flow, from a 1-hour storm of nearly 2.5 inches, swept trees, cars, and bridges into the city, severely damaging several homes and businesses.

Because historical events had shown that flash-flooding from any of the three drainages could have an impact on the City of Heppner, a comprehensive study was made by the COE. Since structures on Hinton and Shobe Creeks were found to be economically infeasible, two separate components were authorized. The first, Willow Creek Dam, was constructed under Congressional authority to provide flood control on Willow Creek and Balm Fork. The second, developed under discretionary authority by the Chief of Engineers, is a state-of-the-art flood prediction and warning system will alert residents of impending floods originating in the other two basins. This warning system would protect lives, but not property.

PROJECT DESCRIPTION

The Heppner flood warning system consists of three major elements: Willow Creek Dam and associated project gages, the tributary precipitation and stream gaging configuration, and the base station equipment.

Willow Creek Lake Operational System. Willow Creek Dam is the world's first gravity dam to be built completely of roller compacted concrete (RCC). The dam, completed in September 1982, is 154 feet high, 1,780 feet long, and consists of 403,000 cubic yards of RCC. Within the 1.8-mile long lake, 9,765 acre-feet of storage is available for flood control. Two inflow stream gaging stations, a lake elevation gage, and a downstream outflow gaging station provide data required for the project operation. This data is available through telephone interrogation and is also transmitted to the base station minicomputer by event-activated (self-reporting) radio telemetry.

Tributary Gaging Facilities. The tributary gaging system consists of 17 precipitation gages, 7 in Shobe Creek and 10 in Hinton Creek, distributed throughout the drainage basins on NWS-RFC recommendation. These tipping bucket gages are event-activated and transmit when one millimeter of precipitation is registered. Each basin has a temperature sensor transmitting data every 12 hours for modeling runoff in a rain/snow environment. Hinton Creek has two stream gaging sites, at river mile 0.8 and 6.0, while Shobe Creek has one gage at river mile 1.0. Again, these are event-activated gages used to verify high water conditions following intense rainstorms, and to provide streamflow data for a composite flow in Heppner while evacuating Willow Creek Lake of stored flood waters.

Flood Warning System Base Station. The base station is located in the Morrow County Sheriff's Dispatch Center in the Morrow County courthouse. This office is manned 24 hours a day and is a "911" emergency response center. The base station consists of a Data General NOVA 4X with a CRT and printer, an uninterruptable power supply with backup batteries, and communications equipment. The CRT and printer are conveniently located in the dispatcher's work area where they have immediate access to the information. The base station's basic functions are to store system input data and software, monitor data transmissions from the various sensing equipment and process the data into information files, receive warning messages from NWS offices, compute and display flood warning categories for flood events, and dial-up the NWS office, alerting them of category changes while providing system output data for display. Communication facilities are also available for dial-in monitoring of the system and retrieval of data.

COOPERATIVE AGREEMENTS

Because of the flood warning interaction between the NWS and the COE, the cooperating interest of Morrow County officials, stream gaging responsibility by the U.S. Geological Survey (USGS), and overall project operation by the Corps, numerous Cooperative Agreements were drawn up identifying responsibilities of the various entities.

Basic organizational responsibilities for the installation, operation, and maintenance of the flood warning system are as follows:
   a. Corps of Engineers
       (1) Monitor the project gages and lake level.

    (2) Monitor tributary gages via the base station.
    (3) Provide a spare parts inventory for the system.
    (4) Perform major repair or parts replacement.
    (5) Provide funding to NWS for monitoring the system,
        maintaining software, computer and communications
        maintenance, and travel for software maintenance.
    (6) Provide funding to USGS for operation and
        maintenance of stream gaging equipment and
        publication of records.

b. U.S. Geological Survey
    (1) Operate, maintain, and publish permanent records
        of the project lake and stream gages.
    (2) Operate and maintain tributary stream gages.

c. National Weather Service
    (1) Update and maintain warning system software.
    (2) Monitor the warning system and adjust model
        parameters for seasonal conditions.
    (3) Issue flood warnings, flash flood warnings and
        watches, and any advisories.
    (4) Assist Morrow County with formulating and updating
        an emergency evacuation plan.

d. Morrow County
    (1) Provide electrical power and telephone service
        for the base station.
    (2) Provide 24-hour monitoring of the system.
    (3) Notify Corps of system component malfunction.
    (4) Provide routine nontechnical maintenance of
        system equipment.
    (5) Maintain a supply of Corps-furnished spare parts.
    (6) Formulate and update an emergency evacuation
        plan with State of Oregon and NWS assistance.
    (7) Coordinate emergency plans with City of Heppner.
    (8) Disseminate all warning and evacuation notices.

Early in the system operation, it became clear that it would be difficult for Morrow County to comply with their responsibility to provide routine nontechnical equipment maintenance. Because of manpower restrictions after the agreement was signed, the County no longer had staff personnel qualified to do the work. This responsibility has since been picked up by the Corps.

FLOOD WARNING SYSTEM

The Heppner Flood Warning System was developed to provide real-time warning to local officials in the event of a flash flood in Hinton or Shobe Creeks. In addition, it alerts local officials to rapid rises in Willow Creek Lake which may result in uncontrolled spillway discharges. Due to the quick response nature of the events, early recognition of a flood problem by dispatch center personnel was the primary objective. Two key elements, to provide real-time data and to present it in non-technical format for immediate problem recognition, were necessary. An event-reporting data network satisfied the first element and the "category scores" concept was developed by the NWS-RFC to satisfy the latter.

Data Collection. Within the two basins, NWS and COE cooperatively located 17 precipitation gages in areas felt to have topographical features conducive with thunderstorm-runoff conditions. These sensors consisted of a 10-foot high standpipe housing a tipping bucket precipitation gage, radio transmitter, battery, and antenna. Because the majority of the sites were located on rangeland, barriers were installed to prevent cattle from using the standpipe as a rubbing post. Two of the sites, one in each basin, have an air temperature sensor, and one s.te in the Hinton basin has a radio repeater to provide a reliable radio path from five sites in the upper Hinton watershed. In conjunction with these meteorological sensors, there are two stream gaging stations on Hinton Creek and one station on Shobe Creek. With the exception of the temperature data which is transmitted twice a day, the sensors are set up to transmit when a hydrologic "event" occurs. In the case of precipitation, it is when the bucket tips, 1 mm of precipitation measured. There were some difficulties in setting up the stream gage portion of the system. Due to the radio equipment's transmission rate, the stage rise increment had to be large enough to allow transmission to occur and yet small enough to provide data changes during non-flood events. A 0.15 foot change was selected. Float (trip) switches at the gaging stations are being planned to provide backup as they would more accurately detect a debris-laden flood wave. If the switches are tripped or washed away, a series of signals will be transmitted to the base station.

Flood Warning Software. The Flood Warning software was written by the NWS-RFC to provide non-technical information to local authorities by a method known as "Composite Category Score" (CCS). The category score computations reduce a myriad of information to a single CCS for each defined basin in the system. A CCS can range from 1 (no significant activity in the basin) to 5 (major flooding). The CCS for a specific basin is found by combining the influence of four individual category scores: (1) flash flood watches and warnings in effect; (2) observed precipitation; (3) observed river levels; and (4) forecast river levels (See Figure 1). By defining the 5 scores allows local emergency services personnel to easily tailor their response to the CCS by associating specific actions with each of the scores (See Figure 2).

```
                   HEPPNER FLASH FLOOD WARNING SYSTEM

        COMPOSITE CATEGORY SCORE SUMMARY FOR 4/20/87 @14:36 PLT
        .............................................................

                        BASIN #1         BASIN #2         BASIN #3

                       SHOBE CANYON     HINTON CREEK      CONFLUENCE
                       ............     ............      ..........

                          *****            *****            *****
        COMPOSITE         * 1 *            * 1 *            * 1 *
        CATEGORY SCORE    *****            *****            *****

        CATEGORIES (0=not eval.)
        ------------------------
        FF WATCHES/WARNINGS       1                1                1
        OBS. PRECIPITATION        1                1                1
        OBS. RIVER LEVELS         1                1                0
        FORECAST RIVER LEVELS     0                0                0

        .............................................................
                                Figure 1


        . . . . COMPOSITE CAREGORY SCORE DEFINITIONS . . . .
        .............................................................

        CATEGORY 1.  THERE IS NO SIGNIFICANT HYDROLOGIC ACTIVITY IN THE BASINS
                     ABOVE HEPPNER.  EVERYTHING IS OK.

        CATEGORY 2.  MINOR HYDROLOGIC ACTIVITY IN 1 OR MORE BASINS IS OR MAY BE
                     OCCURRING.  THIS SITUATION WARRANTS INCREASED MONITORING.
                     NO ACTION IS REQUIRED AT THIS TIME.

        CATEGORY 3.  MODERATE HYDROLOGIC ACTIVITY IN 1 OR MORE BASINS IS OR MAY
                     BE OCCURRING.  STREAM LEVELS MAY APPROACH FLOOD STAGE.
                     INCREASED MONITORING AND LIMITED ACTION ARE WARRANTED.
                     REFER TO "RED BOOK" FOR INSTRUCTIONS.

        CATEGORY 4.  SERIOUS HYDROLOGIC ACTIVITY IN 1 OR MORE BASINS IS OR MAY BE
                     OCCURRING.  FLOODING AND PROPERTY DAMAGE ARE LIKELY.  TAKE
                     PROMPT ACTION AND MONITOR FREQUENTLY.  REFER TO "RED BOOK"
                     FOR INSTRUCTIONS.

        CATEGORY 5.  MAJOR HYDROLOGIC ACTIVITY IN 1 OR MORE BASINS IS OR MAY BE
                     OCCURRING.  MAJOR FLOODING AND PROPERTY DAMAGE ARE LIKELY.
                     TAKE IMMEDIATE EMERGENCY ACTION.  REFER TO "RED BOOK' FOR
                     INSTRUCTIONS.
        .............................................................
                                Figure 2
```

Because of the flash flood potential in the area, the CCS is recalculated every 30 seconds for each basin in the system. Whenever a CCS increases, the system updates the numeric indicator on the terminal in the Sheriff's Dispatch Center, sends a category score summary to the line printer, and initiates an auto-dial to the Pendleton, Oregon, Weather Service Office.

Software was also developed to: receive and store data transmitted from the field sensors for later retrieval for post-event evaluations; provide various types of raw and processed data displays for detailed localized activities and evaluation of basin conditions; and report sensor outages and system diagnostics.

In the 3 years that the project has been operating, numerous situations have occurred that affected the system's reliability to some extent. Battery malfunctions have resulted in numerous sensor outages and replacement was delayed by nearly inaccessible precipitation gages. Uninterruptable power supply capacity was not large enough for the prolonged power outages incurred by the base station. Stream gage locations and sensing equipment are prone to siltation problems, thus rendering them ineffective during high water periods. On the positive side, however, whenever the above outages occur, the system adjusts the hydrologic modeling to compensate for the missing information.

SUMMARY

The Heppner Flood Warning System design process was fairly unique for the Corps of Engineers. The system integrates various agency responsibilities: Corps, flood control operation through use of structural facilities; National Weather Service, dissemination of flood watches and warnings; the U.S. Geological Survey, stream gage operation and maintenance; State of Oregon Emergency Services, formulation of warning and evacuation plans; and Morrow County, implementation of the warning and evacuation plans.

Because of the crossed responsibilities, Cooperative Agreements were written between the COE, NWS, and Morrow County, which outlined responsibilities to ensure optimal use of the system. Morrow County became a key component because of their isolation from immediate help in this sparsely populated area of the state.

Although an Agreement was signed, new County administrators have taken it lightly, and have been particularly lax in programming funding or personnel. Because of the County's size and widespread population areas, elected officials may not be aware of, or not choose to consider, the seriousness of non-compliance with their obligations. This factor is being acted upon by all parties, but progress is very slow. In the meantime, many of the agreed upon County responsibilities, particularly in the way of sensor maintenance, have been assumed by the Corps. Yet, local involvement still is the key element for successful operation of the system.

In spite of the problems that have occurred, the system is up and operating. There have not been any flood events to truly test the system, but personnel from all agencies are keeping in touch, and watching the system closely. Hopefully the Heppner Flood Warning System can prevent loss of life in the event of a flash flood.

REFERENCES

Hartman, R.K., Bissell, V.C., Halquist, J.B., 1986. "Category Scores For Local Flood Warning System". at 2nd Conference on Weather, Climate, and Hydrology of the Pacific Northwest. Oregon Branch, American Meteorological Society, 25 January 1986.

U.S. Army Corps of Engineers, 1981. Flood Warning System - Supplement 3 to General Design Memo 2, Phase II. Walla Wall District, Walla Walla, Washington

U.S. Army Corps of Engineers, 1984. Water Control Manual for Willow Creek Lake. Walla Walla District, Walla Walla, Washington.

MINIMIZING THE IMPACT OF SALINITY INTRUSION
DUE TO SEA LEVEL RISE

G. P. Lennon[1], B. L. Du[2], and G. M. Wisniewski[3]

Abstract

A rise in sea level will cause saline water to intrude
up estuaries including those adjacent to aquifers with major
pumping centers. During a drought a significant amount of
saline estuary water can recharge certain aquifer systems.
The objective of this study is to consider various response
options that are appropriate to minimize the impact of the
salinity intrusion into the Delaware Estuary and adjacent
Potomac-Raritan-Magothy aquifer system (PRM) during a
drought. Extraction wells and a combination of extraction
and injection wells are considered during the drought
period. The results indicate that the PRM can be adequately
protected during a drought with an adequate short term water
supply and delivery system.

Introduction

As a result of human activities, atmospheric concentrations of carbon dioxide and other gases are increasing,
causing global warming by a mechanism known as the greenhouse effect. If such a global warming occurs, the 2.4 ft
sea level rise scenario considered by Hull and Titus (1986)
is very likely to occur over the next 100 years.

The objective of this paper is to evaluate selected
response options for minimizing salinity intrusion into
aquifers as a result of a 2.4 ft sea level rise: an
extraction well barrier and a combined extraction-injection
well barrier. To carry out this objective,the Potomac-
Raritan-Magothy (PRM) aquifer system adjacent to the
Delaware River Estuary is considered.

The Delaware River Basin Commission (DRBC) monitors the
salinity levels in the Delaware Estuary including tracking
the location of the salt front, (250 mg/l isochlor). The
salt front is primarily controlled by river geometry,
flowrate and elevation of the ocean. During the worst

[1]Department of Civil Engineering, Lehigh University,
    Bethlehem, PA 18015
[2]Shanghai Geological Department, Shanghai, People's
    Republic of China
[3]Charles T. Main, Inc., Charlotte, NC

drought on record in late 1964 the salt front had migrated
upstream to River Mile 102, adjacent to Philadelphia and
Camden County (Hull and Titus, 1986). The DRBC conducted
simulations using the Delaware Estuary Simulation Model, a
one-dimensional, transient numerical model for the 1964-65
drought flows in conjunction with a sea level rise of 2.4
ft. (Hull and Titus, 1986). The duration and salinity
levels from the 2.4 ft sea level rise scenario were used as
the boundary condition along the river for the simulated
transport of saline water into the Potomac-Raritan-Magothy
(PRM) aquifer system.

Salinity Intrusion Simulations in the PRM

     The PRM is a seaward thickening wedge of unconsolidated
materials that become finer offshore. Luzier (1980), and
Walker (1983) provide hydrogeological descriptions of the
PRM system. A large cone of depression is centered in
Camden County as a result of extensive pumping from the PRM.
Vowinkel and Foster (1981) estimated that the flow from the
Delaware River into the PRM was 102.5 cfs in 1973 and 112.9
cfs in 1978. The greatest inflow occurred between River
Mile 101 to 106.5 adjacent to the Camden County cone of
depression.

     A finite-difference solute transport and dispersion
model (Konikow and Bredehoeft, 1978) is used to calculate
piezometric heads and chloride concentrations in the PRM in
vicinity of River Mile 103. A model area 5200 ft by 5200 ft
is composed of a grid of 26 by 26 cells, each 200 ft square.
The important parameters are:

     Thickness, b = 100 ft
     Hydraulic Conductivity, $K = 173$ ft/day
     Transmissivity = 17,200 ft^2/day
     Porosity, n = .30
     Longitudinal dispersivity, a_L = 100 ft
     Transverse dispersity, a_T = 30 ft

     The northeast and southwest boundaries are assumed
impermeable whereas the heads along northwest boundary were
set at 0 ft mean sea level (msl). The heads along the
southeast boundary were set at -30 ft msl creating a
gradient of 30 ft/mile and inducing a river flow of
approximately 7.6 cfs/mi, in the same range of 1978 flows
reported by Vowinkel and Foster (1981) for this area. A
more accurate representation can be obtained by extending
the grid or using a transport window in a larger flow grid
(see Lennon and Rumbaugh, 1987).

     The chloride concentration in the river and the aquifer
is assumed to be zero at the beginning of the simulation
period. A 3-month drought river flow superimposed on a 2.4
ft sea level rise creates an estuary chloride distribution

that varies from 300 ppm to 200 ppm along the northwest
portion (top) of the model area (see Fig. 1). At
the end of the 3-month drought period, an increase in river
flow causes the salt front to migrate down the river,
resulting in zero chlorides in the river adjacent to the
model area.

Response Options

    A variety of response options are available to minimize
possible salt water intrusion into an aquifer system such as
the PRM (Hull and Titus, 1986). The response options
considered here are extraction barriers and a combination of
extraction and injection barriers.

Case 1: Extraction Barrier - Three wells are included at
the locations shown in Fig. 1, creating a local groundwater
trough that provides an effective groundwater barrier
preventing flow out of the southeast boundary. Wells P_1 and
P_2 are operating at 6 cfs and Well 3 is operating at 4 cfs.
Figure 2 illustrates the chloride distribution in the PRM at
the end of the 3-month drought flow; Fig. 3 shows the
distribution 12 months later. The intrusion of saline water
is effectively terminated by the groundwater trough barrier.

Case 2: Combination of Extraction and Injection Barriers -
The model area and grid are the same as Case 1. Five
injection wells are located parallel to and 1000' from the
Delaware River, and a row of extraction wells located
another 1000' further away (see Fig. 4). The 0.5 cfs pumped
from each extraction well is returned through the injection
wells creating a low gradient from the river that slows the
rate of salinity intrusion. A higher net pumping rate will
reduce the gradient further, creating a more efficient
barrier. Successful long term groundwater barrier to
protect the groundwater adjacent to the West Coast Basin,
California from the intrusion of ocean salinity.

    Figure 4 shows the chloride distribution in the aquifer
for the operation of the combination barrier after the
3-months of salinity recharge. The combination barrier
(Fig. 4) significantly reduces the concentration compared to
Case 1, of pumping wells alone (Fig. 2). Increasing the
injection rate will lower the gradient even more, creating a
ridge barrier. Changing the injection flow rate from 0.5 to
1.0 cfs for each well lowers the chloride concentration even
more.

Case 3: Extraction/Injection Barrier during Drought,
Exclusively Extraction AFter Drought - After the 3-month
salinity intrusion period has ended, salinity levels are
still elevated above background levels but no additional
salinity is entering from the estuary. At this time, the

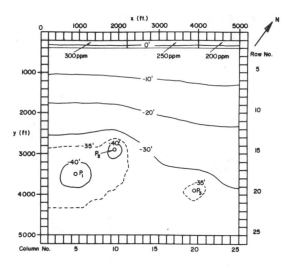

Figure 1   The grid, the steady-state piezometric head, and
           the chloride distribution in the river during the
           intrusion period.

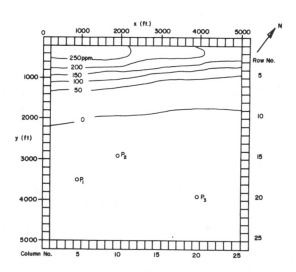

Figure 2   The contour of chloride concentration at the end
           of the 3-month drought flow.

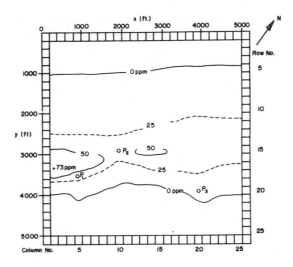

Figure 3   The contour of chloride concentration 12 months
            after the end of the drought flow period with an
            extraction barrier.

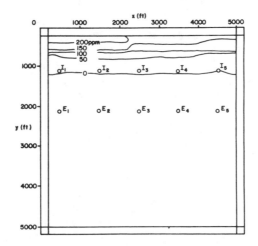

Figure 4   The distribution of chloride after a 3-month
            salinity intrusion period with a combination
            barrier.

injection wells are changed to withdrawal wells. The
pumping wells extract saline water from the river side and a
smaller volume of fresh water from the downgradient side.
After 3 months of pumping each well at 0.5 cfs, the
concentration in the aquifer is reduced further. Compared
with the natural river recharge rate, the extraction rate is
low, but high enough to accelerate the restoration of the
groundwater to background levels.

Conclusions

     A rise in sea level will cause saline water intrusion
into the PRM aquifer adjacent to the Delaware Estuary during
drought events. Minimizing salinity intrusion is very
important because once contaminated, elevated salinity
levels persist for decades. Of the numerous methods for
protecting groundwater from saline water intrusion a
comparison of the extraction and combination extraction/
injection barriers was performed. The combination
extraction/injection barrier has several advantages: 1) A
relatively small amount of groundwater is withdrawn and then
recharged back into the aquifer during the year or two after
a severe drought. 2) Very little fresh water is wasted. 3)
A flexible pumping rate can be used to match the degree of
salinity expected. 4) The aquifer is restored efficiently.

References

Barth, M. C., and Titus, J. G., Greenhouse Effect and Sea
     Level Rise, Van Nostrand Reinhold Company, New York, 1984.
Hull, C. H. J., and Titus, J. G., ed., Greenhouse Effect,
     Sea Level Rise, and Salinity in the Delaware Estuary,
     GPO, EPA 230-05-86-010, May, 1986.
Konikow, L. W., and Bredehoeft, J. D., "Computer Model of
     Two-Dimensional Solute Transport and Dispersion in
     Ground Water", Techniques of Water Resources Investi-
     gations of the U.S. Geological Survey, Chap. C2, Book 7,
     1978.
Lennon, G. P., and Rumbaugh, D., "Two Dimensional Ground-
     water Transport Subgrids in a 3-D Flow Model, Proceedings
     of the National Conference on Hydraulic Engineering,
     Williamsburg, VA, 1987 (in press).
Luzier, J. E., Digital-Simulation and Projection of Head
     Changes in the Potomac-Raritan-Magothy aquifer system,
     coastal Plain, New Jersey: U.S. Geological Survey
     Water-Resources Investigations No. 80-11, 1980.
Vowinkel, E. F., and Foster, W. K., "Hydrogeologic
     Conditions in the Coastal Plain of New Jersey", U.S.
     Geological Survey Water Resources Division, Open File
     Report 81-405, 1981.
Walker, R. L., 1983, "Evaluation of Water Levels in Major
     Aquifers of the New Jersey Coastal Plain, U.S. Geological
     Survey Water-Resources Investigations Report 82-4077.

Transport Model for Aquifer Reclamation Management

Tissa H. Illangasekare[1], A.M.ASCE and Yuen C. Yap[2]

Abstract: A contaminant transport model which is specifically designed for management applications is developed. A key feature is the transient velocity field needed in the solution of the advection-dispersion equation is computed using a linear system's technique. The kernel coefficients of aquifer drawdowns are generated by solving the groundwater flow equation. The advection-dispersion equation is solved using an improved particle tracking method.

Introduction

Once a contaminated aquifer is identified, decisions may have to be made to design appropriate remediation schemes to either reclamate or mitigate the effects of potential contamination. Mathematical models which are designed to simulate the movement of solutes in groundwater could be used for this purpose. The model is executed repeatedly under various scenarios which are designed to achieve a particular objective. Under most general conditions the use of the optimization approaches become restrictive due to the highly nonlinear transport processes. Also, when the decisions are not always made based on well rationalized and structured situations and under uncertainty of objectives the management problem could not be formulate as a problem in mathematical optimization. The simulation approach could still be used in these cases if the model is computationally efficient so that a large number of alternatives could be evaluated cost effectively.

The senior author has developed management oriented groundwater flow models in conjunctive use studies(Illangasekare and Morel-Seytoux, 1986 and Illangasekare et al.,1984). Few models for transient groundwater quality management have been reported (e.g. Willis 1979, Gorelick and Remson 1982) The models developed by the senior author used a special linear system's technique (discrete kernel method). In the model reported here this approach is used in conjunction with an improved particle tracking technique.

Model Formulation and Numerical Procedures

The movement of a solute in groundwater is governed by two equations; namely: (1) groundwater flow (2) the advection-dispersion.

---

[1] Associate Professor of Civil and Environmental Engineering, University of Colorado, Boulder, CO80309-0428
[2] Graduate Research Assistant

The equation describing the flow of water is the Bossinesq
equation:

$$\phi \frac{\partial s}{\partial t} - \frac{\partial}{\partial x} (T \frac{\partial s}{\partial x}) - \frac{\partial}{\partial y} (T \frac{\partial s}{\partial y}) = Q_p \tag{1}$$

where ϕ = storage coefficient, s = drawdown, t = time, x and y are
the cartesian coordinates, T = transmissivity and Q_p = pumping at
p.

It has been shown (Illangasekare and Morel-Seytoux, 1982) that
the solution to Eq. 1 is of the form:

$$\bar{s}_c(n) = s_c^o + \sum_{p=1}^{P} \sum_{\nu=1}^{n} \bar{\delta}_{ce}(n-\nu+1)\bar{Q}_e(\nu) + \sum_{p=1}^{P} \sum_{\nu=1}^{n} \tilde{\tilde{\delta}}_{ce}(n-\nu+1) \tilde{\tilde{Q}}_e(\nu)$$

$$- \sum_{\lambda=1}^{\Pi} \tau_{c\lambda}(n) \; s_\lambda^o \tag{2}$$

where $s_c(n)$ is the drawdown in cell "c" at time n, s_λ^o = initial
average drawdown in cell λ, $Q_e(\nu)$ = distributed excitation in a cell
"e" during period ν, $Q_p(\nu)$ a point pumping excitation in well "p",
E number distributed excitation cells, P is the number of wells and Π
= number of finite difference cells. The kernal coefficients are:
the cell-by-cell discrete kernels $\bar{\delta}_{ce}(\;\;)$, cell-by-point kernels
$\tilde{\tilde{\delta}}_{cp}(\;\;)$ and redistribution kernals $\tau_{c\lambda}(\;\;)$. A computer program to
generate these coefficients is given by Illangasekare and
Morel-Seytoux, 1983.

The model is based on the method of characteristics (MOC)
developed by Garder, et al.,(1964). In the MOC particle tracking
scheme, a set of uniformly distributed points are assigned to each
calculation cell. At the beginning of each time step, each particle
is assigned a concentration representative of the cell. The solution
to the average head is obtained using Eq. 2. The computed heads are
used to estimate the velocities of the particles. Using these
velocities the particles are moved to the new locations. The change
in concentration due to advection is determined by averaging the
concentrations of all particles within each cell. In the existing
MOC models(eg. Konikow and Bredehoeft,1978), these average
concentrations are obtained by using an arithmetic mean. This scheme
does not recognize the relative location of a particle within a
cell. A scheme which uses an area of influence associated with each
particle was adopted, The influence area of each particle is used as
a weighing factor in computing the average. The influence area of a
particle is defined as the portion of the representative area of each
 particle overlapping the calculation cell (Fig.1). The details of
the of the model (KERNMOC) and a user's manual are given in
Illangasekare (1986).

Model Testing and Application

The model accuracy and the efficiency were evaluated by applying the model to a set of test problems. The model simulations were compared with two existing models, namely: (1) Method of Characteristic model(MOC) by Konikow and Bredehoeft(1978), and (2) a Saturated/Unsaturated Transport Model (SUTRA) by Voss(1984).

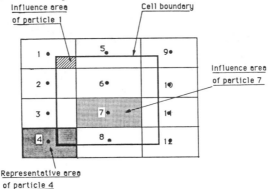

Figure 1: Definition of Influence Areas

In the first test, the simulation, results are compared with the analytical solution for transport through a confined aquifer with steady flow. The concentrations at the end of the simulation period are compared on Fig. 2. A comparison of CPU time in an IBM 3081 computer shows that the two MOC models are about ten times efficient as SUTRA.

In the second test problem, a case of transient flow in an unconfined aquifer with no-flow and constant head boundary conditions was considered. A recharge well representing a solute source is located within the aquifer. Fig. 3 shows a comparison of the simulations. A comparison of computing time shows that the KERNMOC model gave the highest efficiency with 1.6 secs followed by MOC with 2.1 secs and the SUTRA with 25.6 secs.

The third test problem for a transient case is based on a simple representation of Rocky Mountain Arsenal in Colorado (Voss,1984). A uniform square grid system was used in the finite difference formulation in MOC and KERNMOC. A slightly modified finite element grid mesh was used in SUTRA. A comparison of the concentrations at a representative cell is given on Fig. 4.

Discussion of Results

The comparison of the model results with analytical solution show that the KERNMOC and MOC agreed extremely well. The accuracy of SUTRA would have improved by decreasing the grid size but this will be at the expense of increased computing cost. The relative concentrations as computed by SUTRA and KERNMOV for the transient

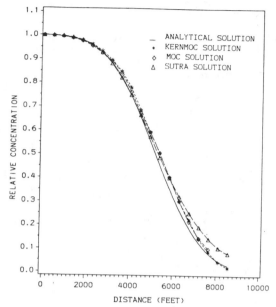

Figure 2: Comparison with Analytical Solution

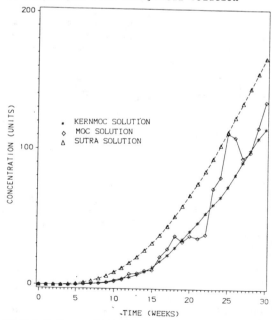

Figure 3: Model Comparison for Transient Case

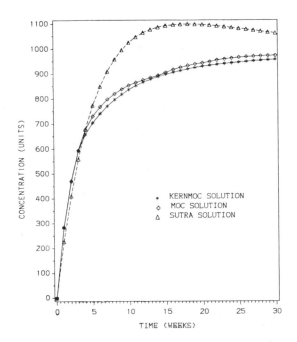

Figure 4: Model Comparison for a Transient Case Based on
          Rocky Mountain Arsenal Example

flow problems deviated at all the observation cells. MOC and KERNMOC
results were reasonably close. KERNMOC and SUTRA show smooth
variation of concentrations. Whereas the MOC shows discrete jumps in
the computed concentrations. The improvement in KERNMOC solution
could be attributed to the "influence area" averaging scheme.

    The cost comparison shows the efficiency of the method of
characteristics models. Even with the computationally cumbersome
"influence area" averaging scheme, KERNMOC seem to have a slight cost
advantage over MOC.

Conclusions

    The model comparisons show that KERNMOC could achieve comparable
accuracies(or even better in some cases) with very low computing
costs. The higher computer efficiency of KERNMOC do not in any way
make it superior for all applications. The USGS models are general
models which could be applied to a wide class of problems. In its
present form KERNMOC model does not incorporate all the capabilities
of the kernel approach. Using superposition it is possible to obtain
detailed velocity distributions in parts of the aquifer where a
higher resolution of the velocity is required. Research is currently
in progress to implement this and incorporate reactive processes.

References

Garder, A.O., D.W. Peaceman, A.L. Pozzi, 1964, Numerical calculation
of multidimensional miscible displacement by the method of
characteristics, Soc. Pet. Eng. J. 4(1), 26-36.

Gorelick, S.M. and I. Remson, 1982, Optimal management of waste
disposal facilities affecting groundwater quality. Water Resources
Research, 18(1), 43-51.

Illangasekare, T. and H.J. Morel-Seytoux, 1982. Stream-Aquifer
Influence Coefficients for Simulation and Management, Water Resources
Research, Vol. 18, No. 1, pp. 168-176,

Illangasekare, T.H. and H.J. Mortel-Seytoux, 1983, User's Mannual for
DELSCAN andd ANADEL: Computer Programs to Generation of Discrete
Kernels. HYDROWAR program report, Colorado State University, Fort
Collins, Colo., p.50.

Illangasekare, T.H. and H.J. Morel-Seytoux, 1986, "A Discrete Kernal
Model for Conjunctive Management of a Stream-Aquifer System," J. of
Hydrology,85,pp. 319-330.

Illangasekare, T.H. and H.J. Morel-Seytoux, and E.J. Koval,
1984. Application of a physically based distributed parameter model
for arid zone surface-groundwater management, J. of Hydrology,
74(3-4), pp. 233-257.

Illangasekare, T.H., 1986, An aquifer solute transport model for
pre-operational and post-operational management of lignite mining,
Completion report to the U.S Geological Survey, p. 119.

Konikow, L.F. and J.D. Bredehoeft, 1978, Computer model of two-
dimensional solute transport and dispersion: Techniques of water
resources investigation, U.S. Geological Survey, Book F, Chapter C2.

Voss, C.I., 1984, A finite element simulation model for
saturated-unsaturated fluid-density-dependent groundwater flow with
energy transport of chemically reactive single species solute
transport, U.S. Geological Survey, Water Resources Investigations
Report.

Willis, R., 1976, Optimal groundwater management: well injection of
wastewater, Water Resources Research, 12(1), 47-53.

MODELING COAL PILE LEACHATE - A CASE STUDY

by Gail G. Crosbie, P.E., A.M. ASCE* and
Gerard P. Lennon, Ph.D., M. ASCE**

Abstract

Ground water transport of coal leachate from an existing unlined coal storage pile was modeled using the USGS "Computer Model of Two-Dimensional Solute Transport and Dispersion in Ground Water."[1] Flow calibration was obtained by varying head, transmissivity and recharge. Solute transport modeling achieved a good match with some observed contaminant levels. Simulated results point to unlined runoff collection ponds as major potential sources of ground water contamination. The model proved useful in defining gaps and inconsistencies in the existing data base. More data from and near these sources would help determine the validity of the results.

Site Description

Approximately 1.3×10^9 kg (1.3 million tons) of coal are stockpiled 91-122 m (300-400 ft) along the western shore of the Susquehanna River (north-central Pennsylvania) at an industrial facility in a storage yard covering 117,000 m^2 (29 acres). Site geology consists of an alluvial sand/gravel layer, overlying a partially weathered, variably fractured and intact shale layer. The facility has operated since 1949. Since installation of an extensive ground water monitoring network (17 wells) in 1984, initial monitoring has revealed high levels of sulfate (SO_4), iron (Fe), manganese (Mn), nickel (Ni), and aluminum (Al), typical of acid mine drainage.

Data Analysis and Interpretation

Interpolation of well water levels showed typical contours, parallel to the river. From the observed chemical levels for twenty wells (17 new wells and 3 existing wells), isoconcentration contours were mapped solely by interpolation. Overall, the contaminant plumes did not follow the water table contours except for upgradient background levels. These unusual chemical patterns, with the usual ground water contours, suggested the need for further investigation (Figure 1a).

Two possible additional contaminant sources are the two unlined coal pile runoff collection ponds. These surface sources are directly upgradient of the wells where the highest terrain conductivity values were measured. The North Runoff Collection Pond (NRCP), located upgradient of the central portion of the pile, is in hydraulic

*Project Engineer, Pennsylvania Power & Light Company, Allentown, Pennsylvania, 18101-1179.
**Associate Professor, Lehigh University, Bethlehem, Pennsylvania, 18015

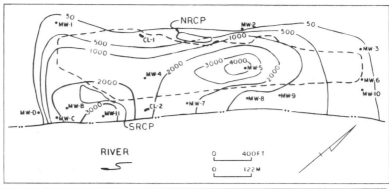

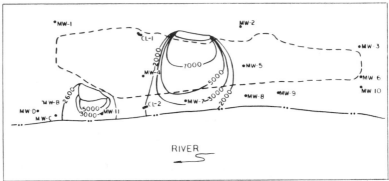

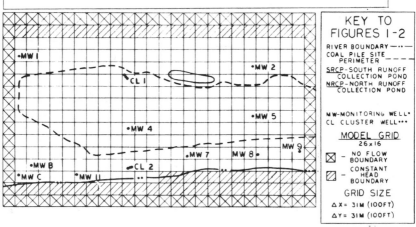

Figure 1 — Evaluation of Coal Pile Leachate Data (Plan View): (a) Sulfate Concentrations (mg/l) — Interpolated Contours (top), (b) Sulfate Concentrations (mg/l) — Contours Assuming a Strong Effect of Runoff Collections Ponds (RCPs) (middle), (c) Ground Water Computer Model Grid (bottom).

connection with the water table, while the South Runoff Collection
Pond (SRCP) is not.

A possible revision to the chemical contours postulated plumes
emanating from the ponds. Without violating the observed chemical
data from the wells, these plumes represent conditions that are more
reasonable, given the observed ground water flow patterns (see
Figure 1b). SRCP, was not studied further; what was learned from
modeling NRCP could be applied to SRCP.

Extensive review of well installation/sampling records provided
estimates for modeling: 1) core borings (for estimating saturated
thickness of soil/rock layers), 2) water levels (for estimating
hydraulic gradients--changes in piezometric head), and 3) slug,
recovery and packer tests (for estimating hydraulic conductivities).

Flow Modeling

Increasingly complex analytical and computer ground water flow models
were used to match actual ground water levels. Initially, gross
water quantities were estimated using a water budget model and
drainage area calculations. Then transmissivity, recharge and head
values were varied, within one- and two-dimensional modeling, to
simulate actual ground water contours.

The "Computer Model of Two-Dimensional Solute Transport and Dispersion in Ground Water"[1] was used one- and two-dimensionally to model
slices and plan views of the ground water below the pile and adjacent
areas. The model couples the ground water flow equation, solved by a
finite difference method and the solute transport equation, solved by
the method of characteristics, to obtain the governing equations and
boundary conditions. The resulting equations and boundary conditions
are solved using the finite difference method.

Two-Dimensional Flow Model Simulation

A 26 x 16 two-dimensional grid of cells, 100 ft on a side (equal to
929 m^2 or 10,000 ft^2), encompassed twelve of the twenty wells in or
near the coal pile (see Figure 1c). These wells showed the highest
contamination and represented a majority of the pile (its central and
southern portions). The northern portion of the pile showed little
contamination and was, therefore, eliminated from further study in
order to concentrate effort (and more detailed modeling) on the
contaminated area. The calibrated base case parameters are shown in
Table 1.

Aquifer transmissivity was represented by one combined geologic layer
(sand/gravel and weathered/fractured/intact shale) of varying saturated thickness. Transmissivity and recharge values were varied, by
blocks of cells, to achieve an excellent match of ground water levels
[within ±0.3 m (or ±1 ft) of observed levels for eleven of twelve
wells in the study area]. The coal pile recharge of 0.203 m/yr.
(8 in./yr.) and a NRCP recharge thirty times this amount were used to
achieve the match. The extra recharge at NRCP produced the correct

TABLE 1 - BASE CASE PARAMETERS

	SI Units	English Units
Storage Coefficient	0	0
Effective Porosity	0.25	0.25
Longitudinal Dispersivity	30.5 m	100 ft
Transverse to Longitudinal Dispersivity Ratio	0.3	0.3
Ratio of T_{yy} to T_{xx}	0.5	0.5
Transmissivity	$8.6 \times 10^{-5} - 3.8 \times 10^{-4}$ m²/sec	80-350 ft²/day
Constant Head Boundaries Above the River	7.3 m	24 ft
NRCP (Pond) Recharge Rate	0.0024 m³/sec	2300 ft³/day
Areal (Pile) Recharge Rate	20.3 cm/yr	8 in/yr
Leakage Rate	1.0	1.0
Initial Concentration, C_o	100 mg/l	100 mg/l
Number of Columns	16	16
Number of Rows	26	26
Grid X- or Y-length/cell	30.5 m	100 ft
Pumping Period	10, 20, 30 yrs	10, 20, 30 yrs
Aquifer Thickness	14.0-48.8 m	46-160 ft

TABLE 2 - COMPUTED VS. OBSERVED GROUND WATER LEVELS

Monitoring Well	Observed Water Level (10/18/85)	Base Case Computed Cell Value	Deviation from Observed
MW-2	438.44	438.41	-0.03
MW-5	N/A	427.50	N/A
MW-8	419.68	420.51	+0.83
MW-7	421.19	422.25	+1.06
CL-1	433.45	433.64	+0.19
MW-4	422.01	422.49	+0.48
CL-2	418.52	418.85	+0.33
MW-1	437.21	437.47	+0.26
MW-C	418.07	417.65	-0.42
		Average Deviation =	0.45

TABLE 3 - COMPUTED VS. OBSERVED GROUND WATER QUALITY - SULFATE

Monitoring Well	Observed Sulfate Values, C (11/11-19/84) (mg/l)	If C_o = 5000 mg/l, Then C/C_o (Column 2 $\div C_o$)	Base Case Assumed C_o = 100 mg/l; at 30 Yrs, C/C_o	Computed Value Was: High (H), OK, Low (L) OK = ±0.10
MW-2	45.3	0.01	0.09	OK
MW-5	4840	0.97	0.26	L
MW-8	560	0.11	0.29	H
MW-7	2820	0.56	0.67	OK
CL-1	65	0.01	0.25	H
MW-4	1940	0.39	0.38	OK
CL-2	1800	0.36	0.35	OK
MW-1	200	0.04	0.03	OK
MW-C	1910	0.38	0.18	L

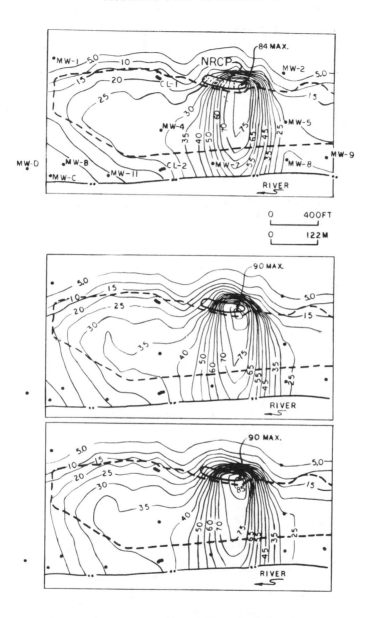

Figure 2 —Time Period Sensitivity-Simulated Solute Plume (C_o = 100 mg/l) at: (a) 10 years (top), (b) 20 years (middle), and (c) 30 years (bottom).

"bulge" in the ground water contours at the most highly contaminated well downgradient of the pile (MW-7). The overall transmissivity varied from 8.6 x 10^5 to 3.8 x 10^4 m^2/sec. (80 - 350 ft^2/day). While this combination of values may not mirror exact site conditions, they do represent reasonable values for this level of study and achieve a remarkably good match of water levels (see Table 2).

Solute Transport Modeling

Leachate rates, chemical concentrations, and dispersivities were assumed to attempt a match of chemical constituents measured within the area of interest. No attenuation was considered. With the strong assumption of the impact of NRCP, a good match was achieved with observed data (sulfate data shown in Table 3). Seven of the thirteen wells within the study area [two upgradient wells (MW-1, MW-2), one coal pile well (MW-4), and four downgradient wells (MW-7, CL-2--a cluster of three wells)] are within 10% of observed values. Two more downgradient wells (MW-8, MW-C) are within 20% of observed values. These values were achieved with an assumed NRCP value. Simultaneous sampling of NRCP and the wells should be performed to validate these results.

Time Period Study

Lengthy operation was modeled, resulting in simulations that were nearly identical for the 20- to 30-year time period. Slightly more lateral dispersion of contaminant occurred after 30 years than after 20 years (see Figure 2). This simulated contamination match indicates that a dynamic equilibrium has been reached. This simulated stable condition of dynamic equilibrium infers that the "clean" upgradient flow combines with the leachate and transports it downgradient without causing the ground water below the pile to ever reach the source chemical concentration. In essence, the modeled system reaches a chemical steady-state flux condition.

Conclusion

Ground water flow and solute transport modeling achieved a good match with observed water levels and some observed contaminant levels. Simulated results point to unlined runoff collection ponds as major potential sources of ground water contamination. The modeled system reaches a chemical steady-state flux in the twentieth year of simulated operation. The model proved useful in defining gaps and inconsistencies in the existing data base. More data from and near the ponds would help determine the validity of simulated results.

Reference

1. Konikow, L. F., and Bredehoeft, J. D., Computer Model of Two-Dimensional Solute Transport and Dispersion in Ground Water, U.S. Geological Survey Techniques of Water Resource Investigations, Book 7, C2, 1978.

GROUNDWATER QUALITY MODELING AT
REDLANDS/CRAFTON AREA

Shih-Huang Chieh[1], Mark J. Wildermuth[2], M.ASCE
P. Ravishanker[3]

ABSTRACT
    Organic groundwater contamination exists in the Redlands area of the
Bunker Hill Groundwater Basin which is of major concern to the health and
water agencies in the region. The contaminants found are trichlorocthylene
(TCE), 1,1-dichlorethylene (DCE), perchloroethylene (PCE) and dibromo-
chloropropane (DBCP). Groundwater elevations were measured in the area of
interest and limited water quality parameters were measured or obtained
from Santa Ana Regional Water Quality Control Board to establish initial
condition. A finite element groundwater flow model and a finite element
particle- tracking model was used to predict the movement of groundwater
and contaminants. The model was used to predict the movement of contaminants with time and also to evaluate the impacts in groundwater quality due
to the remedial alternatives considered. The alternatives include pumping
of contaminated groundwater at different pumping rates and locations.

INTRODUCTION
    TCE, DCE, PCE, and DBCP contamination has been observed in wells in the
Redlands area. The Santa Ana Watershed Project Authority (SAWPA) recognized the need for immediate action to protect down-gradient groundwater
producers and to recover the groundwater resource lost by contamination.

    The authors, under contract to SAWPA, conducted a reconnaissance level
study with the following objectives: determine the occurrence of TCE and
DBCP contamination in the study area; determine the extent and rate of
movement of TCE and DBCP in the study area; and determine the most effective method to control TCE and DBCP contamination in the study area. The
subject of this paper is a discussion of the contaminant modeling approach
that was used in our study, specifically as it relates to TCE.

    For purposes of this study, the study area consists of the eastern edge
of the Bunker Hill Basin bound by the Santa Ana River on the north, the
Redlands fault on the east, Mountain View Avenue on the west and Beaumont
Avenue and its easterly extension on the south.

    The study area overlies the eastern Bunker Hill Basin and Redlands
Subbasin in the Upper Santa Ana River Watershed. These groundwater basins
are separated by the Bryn Mawr barrier as shown in Figure 1. The Redlands
Subbasin is located between the Redlands Fault/Mentone Barrier and the Bryn
Mawr Barrier. The Mentone Subbasin is located between the Mentone Barrier
and the Redlands Fault. The main part of the Bunker Hill Basin lies west
of the Bryn Mawr Barrier.

---

[1] Senior Engineer, Camp Dresser & McKee Inc., Irvine, CA 92715
[2] Principal Engineer, James M. Montgomery Engineers, Irvine, CA 92715
[3] Project Manager, Santa Ana Watershed Project Authority, Riverside, CA,
    92503

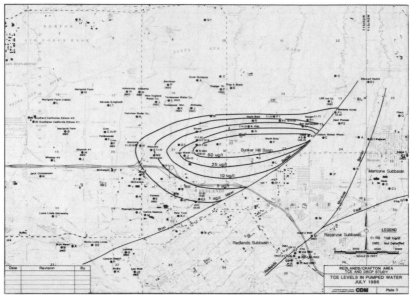

Figure 1   STUDY AREA AND ESTIMATED TCE PLUME

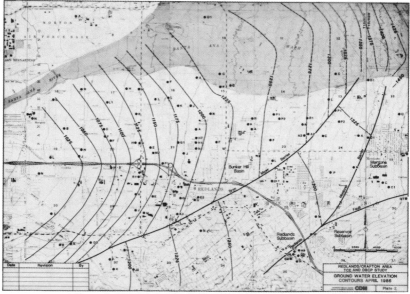

Figure 2   GROUND WATER ELEVATION CONTOURS APRIL 1986

AQUIFER CHARACTERISTICS

Groundwater in the study area is found in unconsolidated alluvial sediments which consist of interbedded clay, silt, sand, gravel, and boulders deposited in alluvial fans by the Santa Ana River and Mill Creek. The total thickness of the alluvial sediments varies from basin to basin, and ranges from over 800 feet in the Redlands Subbasin, to over 1,000 feet in the Bunker Hill Basin. Saturated thickness of the aquifers was estimated based on groundwater level data for spring 1986. The saturated thickness of the Bunker Hill Basin in the study area is over 800 feet. The saturated thickness in the Redlands Subbasin is over 600 feet. Wells in the study area do not produce from the full saturated thickness.

GROUNDWATER FLOW

The authors prepared a groundwater level contour map of the study area for spring 1986 conditions. This map is based on groundwater level data collected by the City of Redlands and the San Bernardino Valley Water Conservation District. This groundwater elevation map, shown in Figure 2, illustrates the influence of the Bryn Mawr and Redlands faults on groundwater flow. Along the Redlands Fault, the drop in groundwater level ranges between 25 feet in the north to about 225 feet along the southern end of the fault. The groundwater flow in the Redlands Subbasin is southwest with leakage through the Bryn Mawr fault west into the Bunker Hill Basin. The groundwater flow in the Bunker Hill Basin west of the Bryn Mawr barrier is west, following the direction of the Santa Ana River. Inflow to the Redlands Subbasin is primarily from the north side of the subbasin consisting of subsurface flows under the Santa Ana River and stormwater percolation in the Santa Ana River. Inflow to the study area part of the Bunker Hill Basin consists of subsurface inflows of groundwater beneath the Santa Ana River, percolation of stormflow in the Santa Ana River and subsurface flows through the Bryn Mawr fault. The groundwater level drop across the Bryn Mawr Barrier is between 25 and 50 feet. The leakage through the Bryn Mawr Barrier is substantial. Computer simulations of regional groundwater flow in the study area were done for the 1983 Santa Ana Basin Plan (CDM, 1982). The Basin Plan study estimated the flow across the Bryn Mawr Barrier averaged 12,000 af/yr. The groundwater elevation data presented in Figure 2 are relatively high levels but are consistent in flow direction and gradient from year to year. The regional flow conditions represented in Figure 2 were used in this study to represent average conditions.

CURRENT TCE LEVELS

The city of Redlands, the Regional Water Quality Control Board, and the authors conducted a sampling program to determine the current areal extent of TCE contamination in the study area. Forty-four wells were sampled.

The TCE data collected for this study were plotted to determine the regional extent of TCE contamination. This map (shown in Figure 1) shows the measured TCE levels in pumped water at wells sampled during the period June through July 1986. The map also shows estimated contours of TCE concentration in pumped water. Our interpretation of the well data indicates the existence of a plume approximately 1.5 miles wide and just over 3 miles long. Our interpretation of the TCE plume is conservative in the sense that it provides a "worst" case condition for a mitigation program. TCE concentration at the plume center are as high as 58 ppb and sharply decline to less than 1 ppb in the transverse direction. The leading edge of the plume has a more gradual change in TCE concentration than the north side of the plume. The north side of the plume is more compressed than the south side.

The impact of the Bryn Mawr Barrier can also be seen in Figure 1.   The TCE plume appears to enter the Bunker Hill Basin near the Bryn Mawr Barrier.   The TCE plume, as delineated on Figure 1 spreads out in a transverse fashion away from the centerline of the plume and along the Bryn Mawr Barrier.   The shape of the plume near the Bryn Mawr Barrier is speculative due to the paucity of TCE data in the Redlands Subbasin.

The source of this TCE plume has not yet been determined.   Two sites have been identified as possible TCE sources, however conclusive evidence has not yet been presented to indicate the responsible party and point(s) of contamination.

MODELING APPROACH
The movement of a contaminant plume is controlled by the velocity field in which the plume resides.   The groundwater velocity field in the Redlands area is complicated due to proximity of faults and a major recharge source along the study area boundaries.

A groundwater model is necessary in this study in order to determine how future groundwater pumping will influence the velocity field and hence the TCE movement.   Remedial alternatives that were identified and studied consisted of pumping at existing wells in the Texas Street well field at the leading edge of the plume.

The flow model is a finite element groundwater model developed by Geotrans called SEFTRAN.   The transport model was developed by CDM and is called DYNTRACK.   The study area was discretized into 167 triangular elements.   The hydrogeologic properties for the study area were, in part, based on the Santa Ana Basin Planning model calibrations done in 1981 (CDM, 1982).   The permeability of the Bunker Hill Basin and Redlands Subbasins are 11 and 17 feet per day respectively.   The specific yield in the Bunker Hill and Redlands Subbasins are 0.1 and 0.14 respectively.   Longitudinal and transverse dispersivity coefficients were estimated at 100 feet and 10 feet respectively for the entire study area.   The boundary conditions for this Study were taken as fixed groundwater elevations along the study area boundary.   The fixed groundwater-water elevations were taken from Figure 2 and were assumed constant for the study period.   The retardation coefficient was conservatively estimated at 1.0.

A steady state calibration was performed to adjust permeability and specific yield values at the faults.   The flow model reproduced the 1986 water levels very accurately.   Pumping within the confines of the study area was estimated by well to be equal to the annual average pumping of the previous 5 years for each well.   Wells shut down prior to June 1986 due to contamination were assumed to remain shut down during the study period. Wells which became contaminated during the simulations continued to pump during the study period.   The study period was arbitrarily set to 20 years, starting in 1986.

MODELING RESULTS
The SEFTRAN and DYNTRACK models were used to simulate a no-action alternative and several different pumping schemes.

Alternative 1
Under this alternative, the TCE plume moves west with the regional groundwater flow.   During the modeling period, the groundwater extraction remained the same.   Figure 3 is a computer-generated contour map showing

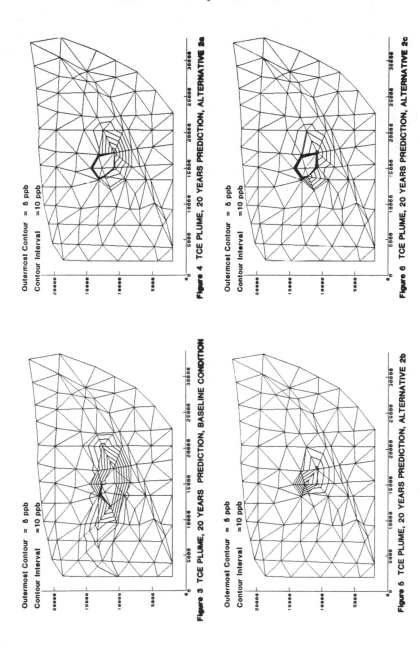

Outermost Contour = 5 ppb
Contour Interval = 10 ppb

Figure 3 TCE PLUME, 20 YEARS PREDICTION, BASELINE CONDITION

Outermost Contour = 5 ppb
Contour Interval = 10 ppb

Figure 4 TCE PLUME, 20 YEARS PREDICTION, ALTERNATIVE 2a

Outermost Contour = 5 ppb
Contour Interval = 10 ppb

Figure 5 TCE PLUME, 20 YEARS PREDICTION, ALTERNATIVE 2b

Outermost Contour = 5 ppb
Contour Interval = 10 ppb

Figure 6 TCE PLUME, 20 YEARS PREDICTION, ALTERNATIVE 2c

the movement of the TCE plume for the time period of 20 years. The outer-most contour is 5 parts per billion (ppb). The contour interval is 10 ppb. TCE concentrations range from 0 to 55 ppb all through the 20 year study period. The plume moved west approximately 2.5 miles contaminating every well along a path approximately 1 mile wide.

Alternative 2a
    This alternative is identical to the baseline conditions with the exception that Well 30A is run continuously at 1800 gpm for 20 years. Figure 4 is a 20-year prediction of the TCE plume.

Alternative 2b
    This alternative is identical to the baseline condition except that Well 31A is run continuously at 4200 gpm for 20 years. Figure 5 is a 20-year prediction of the TCE plume.

Alternative 2c
    This alternative is identical to alternative 2a with the addition of groundwater pumping from the North Brae Well, the Rees Well, Well 35 and Judson Well. Figure 6 is a 20-year prediction of the TCE plume.

ACKNOWLEDGMENT
    The authors would like to express appreciation to Dr. Brendan Harley of CDM's Boston office, and Mr. Don Schroeder of CDM's Ontario office, for their contribution to this project. The project was sponsored by the Santa Ana Watershed Project Authority.

BIBLIOGRAPHY
1. Geotrans, Inc. "SEFTRAN: A Simple and Efficient Two-Dimensional Groundwater Flow and Transport Model," Herndon, VA, 1986.

2. Camp Dresser & McKee Inc., "DYNTRACK, A Three-Dimensional Contaminant Transport Model for Groundwater Studies," Boston, MA, 1984.

3. Camp Dresser & McKee Inc., "Appraisal Level Investigation Redlands/ Crafton Area TCE and DBCP Study," Irvine, CA, 1986.

4. Camp Dresser & McKee Inc., Task 3, Memorandum Basin Plan Model Calibration, Irvine, CA, 1982.

IMPROVED SWMM APPLICATIONS
FOR STORMWATER MANAGEMENT MASTER PLANS:
A CASE STUDY OF VIRGINIA BEACH, VIRGINIA

by John A. Aldrich, A.M. ASCE, John E. Swanson, A.M. ASCE,
Kelly A. Cave, A.M. ASCE, and John P. Hartigan, M. ASCE*

A case study of a recent master plan for a 16.5 sq mi watershed in
Virginia Beach, Virginia is presented to illustrate stormwater master
planning approaches. The analysis showed that public perceptions of
increases in fluvial flooding potential were unfounded, and that coastal
storm surges cause more flooding in the impacted area than fluvial
flooding under an ultimate land use scenario. Final recommendations
were to prevent development and prohibit filling activities in the
floodplain. Approximately $1 million worth of improvements which were
originally proposed for the study area were found to be ineffective and
environmentally harmful by the master plan study.

The Master Plan Approach to Stormwater Management

Stormwater management is a significant concern in the City of
Virginia Beach, a rapidly growing municipality of 250 sq mi in
southeastern Virginia. Rapid development has caused the City to
question whether existing primary drainage facilities will be adequate
in the future. A City-wide stormwater master plan was prepared to
determine cost-effective solutions for both runoff-induced flooding of
drainage facilities and nonpoint pollution of environmentally sensitive
waterways. An appropriate mix of regional solutions to stormwater
problems are proposed, including regional detention/retention,
channelization, stream crossing improvements, and nonstructural
controls. The stormwater master plan also seeks to manage future
stormwater problems by setting a framework for subdivision drainage
design, recommending locational guidelines for major stormwater
management facilities and providing design tailwater elevations based
upon an ultimate land use scenario.

U.S. EPA's Stormwater Management Model (SWMM) is an ideal tool for
regional stormwater master planning. SWMM has the ability to produce
time histories of flow, water surface elevations, and backwater profiles
at various locations and can simultaneously analyze natural channels,
storm sewers and road culverts. Simulations of complete time histories
are critical to watershedwide master planning since sizing, location,
and overall effectiveness of proposed stormwater facilities is highly
dependent upon time variable hydraulic interactions.

Two alternative mechanisms for funding the improvements recommended
by the master plan are also explored. A stormwater utility involves
user-charges for both existing and new development based on the runoff

*Camp Dresser & McKee, 7630 Little River Turnpike, Suite 500,
Annandale, VA 22003

potential of each parcel. A fee-in-lieu of program involves charging
new developments for a portion of the capital cost of regional
stormwater facilities serving their sites.

Redwing Lake - A Case Study

Study Area Description

For the master plan study, Virginia Beach was divided into 25 major
watersheds (see Figure 1). The master plan for the Redwing Lake watershed (number 9 in Figure 1) demonstrates the key issues confronted by a
master planning study: the impacts of rapid growth on existing drainage
systems, the variety of alternatives available for stormwater management, and the environmental consequences of these alternatives. As
shown in Figure 1, the major features of the study area include: two
interconnected lakes (Redwing Lake and Lake Tecumseh), a drainage canal
(Canal-1 South) originally envisioned as a diversion channel to bypass
runoff around Redwing Lake; and a tidal boundary (Back Bay) at the mouth
of Canal-1 South. Wetlands which exist between the lakes and along
Canal-1 South below Lake Tecumseh represent a significant environmental
constraint to channel improvement projects. The Redwing Lake drainage
area is 5.5 sq mi, while the drainage area of the entire study area is
16.5 sq mi. Currently, Redwing Lake provides nearly 2,000 ac-ft of
flood storage during the 100-year storm.

Identifying Stormwater Problems

At the time the master plan study began, flooding around Redwing
Lake was reportedly increasing as a result of recent upstream development. Currently, 584 acres of the area around the Lake is slated for
residential development. One-third of this area lies within the
existing 100-year floodplain (5 ft msl) caused by wind-generated tide
levels in Back Bay. Under existing ordinances, this land could be
developed if filling does not significantly increase flood stages.
Other suspected stormwater problems in this watershed include development induced flooding along Canal-1 South and the performance of
existing culverts and detention facilities under future conditions.

Under existing City regulations, the design storm for new stormwater
facilities is based on the tributary drainage area (Table 1). Since the
Redwing Lake drainage area exceeds 500 acres, a 50-year design storm was

Table 1: Design Storms for Stormwater Master Planning Study

Return Period	Purpose
2-year	Stream bank erosion control
10-year	Drainage facilities with tributary area less than 300 acres
25-year	Drainage facilities with tributary area between 300 and 500 acres
50-year	Drainage facilities with tributary area greater than 500 acres
100-year	Flood insurance and minimum floor elevations

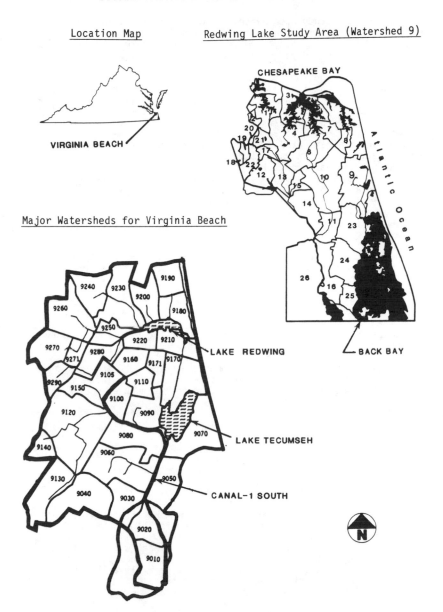

Figure 1. Study Area Location

required for the study area master plan. Additional stormwater management requirements limit peak flows from new development to the
downstream channel capacity, and require minimum floor elevations to be
one foot above the 100-year flood level.

To determine the increase in rainfall-runoff from urbanization,
three land use scenarios were analyzed: undeveloped, existing plus
committed development, and ultimate development under the City's comprehensive land use plan. The amount of urban development is 36% of the
drainage area under the "existing plus committed" scenario, while the
ultimate development scenario calls for urban development in 71% of the
Redwing Lake drainage area. In analyzing flooding potential for a
50-year design storm, an annual mean high wind tide of 2 ft msl was set
as the tidal boundary condition at Back Bay. Based upon SWMM model
results for the 50-yr and 100-yr design storms, the following conclusions were reached about the existing hydraulic system:

- Flooding around Redwing Lake for 50-year and 100-year design storms
  is not significantly worse now than when the watershed was
  undeveloped.

- The 50- and 100-year flooding around Redwing Lake under ultimate
  land use conditions will impact less than 450 acres of undeveloped
  and permanently open land areas.

- The City's present 100-year Redwing Lake flood levels from tidal
  events are not exceeded by the 100-year fluvial event.

- Canal-1 South is nearly at capacity under the existing land use
  pattern which is very close to buildout conditions.

- The unimproved channel which connects Redwing Lake and Lake Tecumseh
  significantly constrains Redwing Lake outflow, resulting in flood
  waters being retained in Redwing Lake for as much as 60 hours.

- The maximum 50- and 100-year flood levels in Redwing Lake are not
  significantly affected by development along Canal-1 South.

- Backflow from Canal-1 South is a significant inflow to Lake Tecumseh
  (see point A in Figure 2) because the lake level is not significantly raised by local drainage or outflows from Redwing Lake.

Master Plan Alternatives

Figure 2 shows seven stormwater management alternatives considered
for the Redwing Lake watershed stormwater master plan. The most expensive alternative (capital costs in excess of $40 million) involved ex-
tending Canal-1 South northward to either intersect with Redwing Lake
(alternative 4) or to divert upstream flow around the Lake (alternative
5). These alternatives lowered Redwing Lake peak elevations by up to 1
ft for a 50-yr design storm, but caused serious flooding along Canal-1
South. Thus, a major channelization project for Canal-1 would be
necessary as part of these alternatives. The total project cost for
alternatives 4 and 5 included significant expenditures to mitigate
wetlands impacts along Canal-1 South.

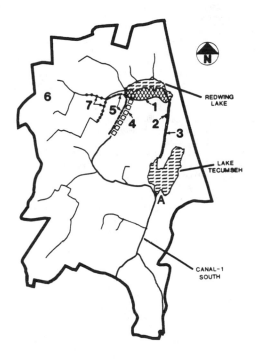

Figure 2.   Map of Redwing Lake Study Area Showing Key Features
and Stormwater Management Alternatives

Alternative 7, improvement to channels immediately upstream of
Redwing Lake, was proposed by the owner of the adjacent land.   This
channelization alternative failed to decrease either the 50-yr maximum
lake level (4.7 ft msl) or the flood levels in the upstream channels
because flooding was due entirely to backwater from Redwing Lake rather
than to insufficient channel capacity.

Peak Redwing Lake water levels for 50- and 100-yr design storms are
about 1.0 ft higher than Lake Tecumseh flood levels.   Thus improvement
of a major stream crossing on the channel connecting the two lakes
(Alternative 2) was considered a prime alternative.   However, the
unimproved channel between the lakes was also found to be a major constraint to Redwing Lake outflow.   Another alternative (no. 3) involved a
combination of alternative 2 (culvert improvement) and channelization
between Redwing Lake and Lake Tecumseh.   Flood levels in Redwing Lake
were only lowered by 0.3 ft for a design which minimized wetlands
destruction.   Even under this approach, the mitigation of wetlands
impacts increased capital costs of alternative 3 to $1.1 million.

Detention basins upstream of Redwing Lake were found to be ineffective at controlling flooding levels around the lake (Alternative 6). Because of the 60 hour retention time of Redwing Lake, these upstream facilities would need to retain significant runoff volumes for more than one day to significantly reduce lake levels. This would require the loss of a greater amount of developable land upstream of Redwing Lake than would be flooded around the lake with no controls. Onsite detention basins were recommended to prevent streambank erosion and local flooding of streams and culverts upstream of Redwing Lake.

In general, structural stormwater management measures would only produce a marginal reduction in Redwing Lake flooding, achieved at a relatively high capital cost with significant environmental and community impacts. Therefore, it was concluded that nonstructural alternatives offered the most cost-effective form of stormwater management for the entire Redwing Lake watershed. The master plan recommends that Redwing Lake be treated as an existing regional detention basin with sufficient capacity to control the impacts of upstream urban development and protect downstream areas. The recommended nonstructural alternative uses ordinances preventing filling or any development within Redwing Lake's 100-year floodplain to protect the Redwing Lake flood storage capacity and minimize the risk of flooding around the lake (alternative 1). Since the northern shoreline of Redwing Lake is primarily publicly owned open space, the recommended alternative addresses the need for development controls along the lake's southern shoreline. If southern shoreline development cannot be totally restricted, developable land could be purchased by the City for less than the cost of the most feasible structural alternative. Tax-base impacts should not be significant since only about 40%, or 190 acres, of the 100-year floodplain around Redwing is currently planned for land development, with potentially undevelopable wetlands comprising a significant portion of this area. Use of Redwing Lake as a regional stormwater detention facility also provides greater protection of Back Bay water quality from nonpoint pollution loadings than would structural alternatives involving the diversion of stormwater runoff around Redwing Lake.

Summary

The Redwing Lake watershed master plan is founded on the conclusion that development restrictions around Redwing Lake can maintain regional flood storage capacity for upstream areas and protect downstream areas as well. Since most of the floodplain around the lake is publicly owned or potentially undevelopable due to wetlands, the impact of strict floodplain zoning on developable land is minimal. The cost of protecting a small amount of lakefront property was significantly higher than the value of the land. In addition, severe environmental impacts can be expected if structural stormwater management measures are implemented. Wetlands areas would be permanently lost, and nonpoint pollution loadings currently removed by the Redwing Lake and Lake Tecumseh system would instead be transported to the environmentally sensitive Back Bay estuary. Original City plans for channelization projects both upstream and downstream of Redwing Lake have been replaced by the proposed nonstructural approach, at a cost savings of $1.1 million.

An Application of EPA EXTRAN Model to a
Flood Management Study in South Florida

By Steve Lin[1] and Richard S. Tomasello[2], M. ASCE

Abstract

The coastal floodplain of South Florida is generally
characterized by extremely flat terrain and heavy vegetation
interspersed with numerous ponds, and is subject to periods
of inundation during the wet season. The drainage in the
area is largely controlled and regulated by the South
Florida Water Management District (SFWMD). The SFWMD
operates the Central and Southern Florida Flood Control
Project canal system to manage water supply distribution and
drainage. A number of local private landowners, drainage
districts, and subdivisions operate secondary drainage
systems under District permit. Increasing population and
rapid land development in the area have necessitated the
updating of flood management studies to protect life,
property, and water resources from flood damage.
     In performing the flood management studies, somewhat
unique concerns in South Florida are the backwater
conditions due to tidal or non-tidal tailwater, flow
reversals from project canal to low flat lands, and flow
diversion by weir, culvert, and pumping facilities. The use
of storage in the EPA EXTRAN model minimizes these
concerns. This model can also handle several culvert flows
at each subbasin and provide hydraulic routing under
automatic gate operations of the coastal structures.
     A case study was performed for the C-18 canal basin,
which discharges into the tidewaters of the Loxahatchee
River near Jupiter, Palm Beach County. The results are
compared with the 1981 U. S. Army Corps of Engineers study
for the same basin.(Key words: Flood management; Backwater)

Introduction

     In performing flood analyses, river basins are normally
subdivided into a number of subbasins. The runoff hydrograph
for each subbasin is computed and routed to the outlet point
of the subbasin. A flood routing in the river begins at the
uppermost subbasin. The simulation proceeds downstream until

[1]Senior Engineer, Water Resources Div., Resources
Planning Dept., South Florida Water Management District,
West Palm Beach, Florida.

[2]Supervisor of Professional Engrs, Water Resources
Div., Resources Planning Dept., South Florida Water
Management District, West Palm Beach, Florida.

a confluence of two branches or the outflow point of the
next subbasin is reached. All flows above that confluence
are computed and routed to that confluence before they are
routed downstream. The process is repeated until the flood
wave reaches the outlet point of the river basin. In many
routing models the effect of backwater on flat low land
subbasins common to South Florida are not considered. These
subbasins may in fact serve as temporary stormwater storage
areas for the river basin. Routing models which lack direct
hydraulic interfacing between subbasin and river reaches can
give erroneous results when the available storage in the
flat terrain subbasins is not taken into account.

The coastal floodplain of South Florida is generally
characterized by extremely flat terrain and swampy wetlands.
Recent increasing population and rapid land development in
the areas has necessitated updates to flood management
studies to protect life, property, and water resources from
flood damage.

One of the objectives in the flood management study is to
provide flood protection criteria as guidelines for future
stormwater management. These criteria are based on the
evaluation of discharge limitations and degree of flood
protection of the existing system. The backwater flow due to
tidal and/or non-tidal conditions including flow reversals
due to automatic tidal gate operations, flow transfer by
weirs, culverts, and pumps are major concerns in defining
these criteria.

The SFWMD has performed a case study in C-18 basin, a
typical South Florida watershed, to illustrate how the
method presented in this paper can reduce these concerns. A
brief description of the basin, the methodology applied, and
simulation of a severe storm event are presented. Comparison
of the study results with the 1981 Corps of Engineers' study
in the same basin is also presented.

Study Area Description

The C-18 basin is located in the northeasterly portion of
Palm Beach County (Figure 1) and consists of about 100
square miles land tributary to the Southwest Fork of
Loxahatchee River.

The land elevation in the basin ranged from 14 ft NGVD
in the Loxahatchee Slough to 25 ft NGVD in the northwest
portion of the basin. The basin can be divided into 21
subbasins according to the land ownerships, highways, and
water control structures (Figure 1). The topographical
variation within each subbasin is normally one to two feet.
The original C-18 canal was about 3.5 miles until 1956 when
the Corps of Engineers extended and widened the canal. This
canal runs through the Loxahatchee Slough in a south-north
direction (east leg) with a connecting west leg draining the

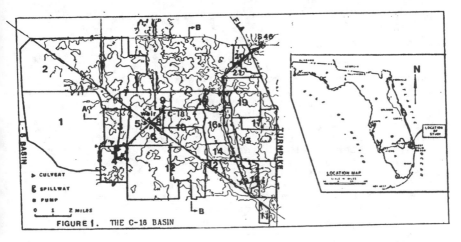

FIGURE 1. THE C-18 BASIN

western C-18 basin. A water control structure at the basin
outlet, S-46,and the steel sheet pile C-18 weir, located
about 200 feet downstream of State Road 710, were
constructed in 1956.

S-46 is a automatic vertical lift gate ogee spillway
structure, and located near the outlet of the C-18 canal to
maintain an optimum upstream stage of 14.8 ft NGVD. When the
upstream stage rises above 14.8 ft, the gate opens at a rate
of 0.4 inch per minute. The gate becomes stationary when the
upstream stage falls to 14.8 ft. The S-46 gate begins to
close when the upstream stage falls to 14.5 ft NGVD. During
major storm events, when the tailwater at the C-18 weir
rises above its crest elevation of 17.6 ft NGVD, the gate
will be operated manually to lower the upstream stage at S-
46 to 12.8 ft NGVD until tailwater at the weir drops below
its crest or the headwater at S-46 drops below 12.8 ft NGVD.

There are a number of private landowners and subdivisions
operating smaller pumps and secondary drainage systems.
Most of which are separated from C-18 by dikes. The drainage
connections to C-18 are via culverts with risers. The
culverts downstream of the weir were installed by the Corps
of Engineers as part of project design to prevent
overdrainage of the wetlands. These culverts with risers
were not maintained and the risers have deteriorated,
allowing adjacent wetlands to overdrain during dry months.
This condition also allows free exchange of stormwater in
the C-18 canal and the subbasin during floods. The developed
areas in these subbasins are generally protected by an
elevated outfall system with crest elevation above 18 ft
NGVD to prevent backwater inflow.

The present land use in the basin is 70 percent wetland,
10 percent urban, and remainder agriculture. However, more
than 50 percent of the basin is currently proposed to be

developed into residential and industrial parks.

Method of Approach

    The EPA EXTRAN model was chosen to analyse the C-18
basin. EXTRAN receives runoff hydrograph input at specific
nodal locations, and performs dynamic routing of stormwater
flows through the major storm drainage system to the outlet
point of the basin. The model can simulate branched or
looped networks, backwater, free-surface flow, pressure
flow, flow reversals, flow transfer by weir, orifice, and
pumps, and storage at on-line or off-line facilities. In the
application to C-18 basin, the following minor modifications
were made to the EXTRAN model:
    1. One inlet per catchment was changed to handle multiple
       pipes (culverts) to each inflow point due to several
       culverts existing in each subbasin.
    2. Automatic gate operational scheme at S-46 described
       previously was included.
    3. The number of printout cycles and storage junctions were
       increased.
    The use of the storage junctions provided a means to
handle flow exchange and hydraulic linking between subbasin
and the main canal. The generated runoff hydrographs were
input into the storage junctions and nodal points prior to
discharging into the C-18 canal via culverts. For subbasins
with elevated outfalls, the storage option was not used
unless its outlet became submerged during a more severe
storm event.
    A routing time step of 30 seconds was used. A culvert in
EXTRAN is treated as a conduit. An equivalent longer culvert
was used to meet the stability requirement of the model.
Figure 2 presents the schematic system of the C-18 basin
    Two approaches were used to generate subbasin runoff.
First, the Tracor's procedure adjusted with local data was
used to develop a 30-minute unit hydrograph. Second, the
HEC-1 stream network model was used to generate runoff
hydrographs for subbasins with elevated outfall systems.
    The composite hydrographs developed from the actual
rainfall event of September 21-24, 1983 (rainfall up to
12.55 inches), and the design rainfall conditions were
input into the storage junctions and nodal points prior to
discharging into the C-18 canal.

Results

    Figure 3 presents the computed and recorded discharge at
S-46 during Sept. 21-24, 1983 storm event. The gate at S-46
was operated manually at 8 a.m. Sept. 24, 1983 to 5 ft for 2
hours, then the gate was closed to 4.15, 3.65, 3.07, 2.42 ft
opennings for time intervals of 10 to 14 hours. The gate
was shut and put back on automatic setting at 10 a.m. Sept.

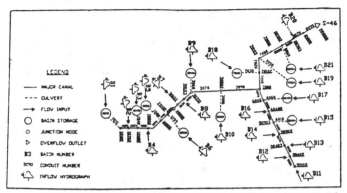

FIGURE 2. SCHEMATIC REPRESENTATION OF THE C-18 CANAL AND
DRAINAGE SYSTEM FOR USE IN THE EXTRAN MODEL.

26. Figure 4 compares the computed and recorded headwater
stage at S-46. The bi-hourly data were shown for both
computed hydrographs. In general, the stage hydrographs
agree well with the recorded values. The computed discharge
tends to be slightly higher. This difference may be
attributed to the manual operation of the gate during the
storm event.

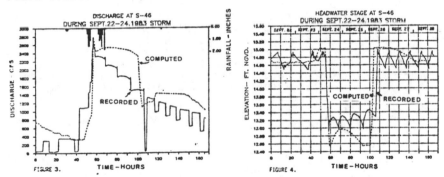

Figure 5 compares the discharge profiles along the east
leg of C-18 canal which resulted from this study (WMD), the
1956 General Design Memorandum (GDM) study, and the 1981
Corps of Engineers' study (COE-1981). The dam break version
of the Hydrologic Engineering Centers' "Flood Hydrograph
Package" (HEC-1), was applied in 1981 study, and the
hydrograph summation approach discussed in the introduction
was used in the 1956 GDM study. The discharge values
resulting from this study are comparable with the 1981
Corps' study within 200 cfs. Some differences are
attributable to the fact that outflow from subbasin 11 is
currently limited by a 10,000 GPM pump while 240 cfs was

estimated in the 1981 Corps' study, and a great portion of the subbasin 21 has been diverted downstream of S-46.

The result also indicated that the maximum discharge at S-46 occurred at the time when the headwater stage reached 14.8 ft NGVD instead of the 12.8 ft NGVD stage condition assumed by the Corps' study (Figure 6). Therefore, the design flood profile in the C-18 canal should be higher than the Corps' study.

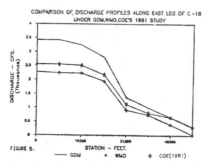

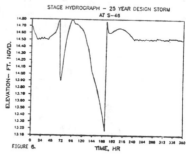

Conclusions

The results presented here indicated that the application of the EPA EXTRAN model to South Florida flood study is one step forward in the right direction to minimize the concern of backwater caused flow exchange, and hydraulic linking between subbasin and project canal. There are more improvements which can be made to the EXTRAN model to further alleviate these concerns and reduce the computer time. For example, a subroutine such as a direct computation of the flow through culverts with risers can be added to EXTRAN instead of treating the culvert as a conduit. This will provide more accurate result and reduce computer time.

Reference
1. Corps of Engineers, 1956. The Construction of Canal 18, General Design Memorandum, C-18 and S-46, Part V, Supp. 20. Dept. of the Army, Jacksonville District. Fl. 1956.
2. Corps of Engineers,1983. Feasibility Report and Environmental Assessment for Canal 18 Basin, Loxahatchee Slough, Central and Southern Florida, June 1983.
3. Roesner, Larry, Shubinski, R. P. Aldrick, J. A. 1983. Stormwater Management Model User's Manual, Version III, Addendum I. EXTRAN, Camp Dresser & Mckee, Inc, Jan. 1983.
4. Tracor, Inc. 1968. The Effects of Urbanization on Unit Hydrographs for Small Watersheds, Houston, Texas, 1964-67. Tracor, 6500 Tracor Lane, Austin,Texas 78721, OWRR Contract No. 14-02-001-1580. U. S. Dept. of the Interior, Washington D.C. 20240, Sept. 1968.

Shoal Creek Flood Control Plan
Austin, Texas

Duke G. Altman,[1] M. ASCE
Dr. William H. Espey, Jr.,[2] M. ASCE, and
Ramon F. Miguez,[3] M. ASCE

Abstract: Following the devastating 1981 Memorial Day flood, a comprehensive and innovative flood control plan was developed and implemented for Austin's Shoal Creek Watershed. The $12 million plan funded by City bond elections is comprised of a combination of property acquisition, bridge and channel improvements as well as regional-type detention ponds. Utilization of Northwest Park as an off-channel/high-flow detention pond, while enhancing its recreational function, provides an essential element of the watershed's overall flood control plan.

A watershed model developed with the HEC-1 program was utilized in the overall planning process. Model calibration was accomplished with 1981 Memorial Day rainfall and runoff data gaged by the U.S. Geological Survey. The calibrated hydrologic model produced storm hydrographs and peak discharges that closely matched those recorded at various locations along Shoal Creek for the extreme event. A HEC-2 model calibrated with storm high water marks was also extensively utilized.

To assess existing flood conditions as well as analyze various flood control alternatives, the HEC-1 and HEC-2 programs were applied in a variety of innovative ways. As an example, the HEC-1 program was used to develop flood hydrographs adjacent to Northwest Park. The park's design function to allow certain portions of high flows to enter the park over an inflow wall was considered in reducing design discharges in downstream areas. The HEC-2 split flow routine was used to divide creek flows between those entering the park and those continuing downstream. Up to 145 acre-feet and 175 acre-feet of stormwater storage may be achieved in the park for the 25- and 100-year design events, respectively.

The interactions of existing hydrologic conditions and proposed flood control projects were closely considered in developing a project sequencing plan that prevents potential increases in flooding along the watercourse during, and after, plan implementation.

Introduction

During the past six years the City of Austin has scoped, funded, designed and implemented a master flood control plan for the Shoal Creek Watershed in

[1] Associate, Espey, Huston & Assoc., Inc., P.O. Box 519, Austin, TX 78767
[2] President, Espey, Huston & Assoc., Inc., P.O. Box 519, Austin, TX 78767
[3] Asst. Dir., Trans. & Pub. Services, City of Austin, P.O. Box 1088, Austin, TX 78767

Austin, Texas. Like many other cities and governmental agencies across the country, Austin spent considerable time in the 1970s trying to match the advances in floodplain hydrology with needs created by rapid urbanization and drainage problems inherited from the past. To meet this challenge the City: 1) upgraded their drainage ordinances (to include, among other items, on-site stormwater detention); 2) developed a drainage criteria and policy manual; 3) joined the Federal Emergency Management Agency's Flood Insurance Program; 4) prepared floodplain studies for many of its major watersheds (including the Shoal Creek Watershed) and 5) constructed certain channel and detention pond improvements. However, it was obvious to the City that these efforts could not avert disaster if a major flood event occurred in the almost fully urbanized Shoal Creek Watershed.

On May 24 and 25, 1981, four to six inches of rain fell on the lower portions of the Shoal Creek Watershed and from seven to eleven inches of rain occurred in the upper watershed area. Almost all of the storm's rainfall occurred in a two to three hour period. Six of the City's 13 deaths and approximately $15 to $20 million in damages were incurred in the watershed as a result of the storm. The storm rainfall, peak discharges and high water marks closely matched the City's 100-year floodplain study performed by Espey, Huston & Associates, Inc. and URS/Forrest and Cotton, Inc. in 1974. This disastrous Memorial Day flood made it clear even to past nonbelievers that additional flood control was needed in the watershed. Table 1 provides an updated comparison of the 1981 Memorial Day and 100-year peak discharges.

Table 1. A Comparison of Memorial Day and 100-Year
Peak Discharges (cfs)* Along Shoal Creek

Location	Drainage Area (sq mi)**	1981 Memorial Day Flood USGS[1]/Model[2]	100-year[3]
Steck Avenue	3.19	5,100/6,300	6,800
Northwest Park	7.03	14,600/15,100	14,400
White Rock Drive	7.52	15,700/16,200	15,400
12th Street	12.80	16,000/15,800	16,900

Notes:

[1] Indirect measurement by U.S. Geological Survey
[2] Modified HEC-1 program incorporating empirical 10-min. unit hydrograph equations (Espey, Altman and Graves, 1977)
[3] Model as specified in (2) used with City of Austin 100-year, 3-hr design storm

*cfs = cubic feet per second
**sq mi = square miles

Master Plan Development

With the support of its citizens, the City of Austin developed and passed a $12 million bond program for flood control in the Shoal Creek Watershed in 1981, 1982 and 1984 bond elections. As a result, hundreds of homes and businesses have been totally removed from the creek's 100-year floodplain while hundreds more now experience a reduced flood hazard. This added protection is provided by a combination of flood control components including acquisition and removal of 22 homes from the floodplain, six regional-type detention ponds, as well as five new bridges and over two miles of channel improvements. Figure 1 presents the location and sequential implementation scheduling for the individual components of the overall master flood control plan.

The plan components were designed and built such that flooding potential would not increase in the future. More specifically, bridge and/or channel improvements were built only after compensatory detention storage was provided in appropriate areas.

A watershed model utilizing a modified version of the HEC-1 program was used for developing the plan following its calibration with data obtained from the 1981 Memorial Day event. To assess existing flood conditions as well as analyze various flood control alternatives, the HEC-1 and HEC-2 programs were applied in a variety of innovative ways. The accompanying loss of floodplain storage along the creek due to proposed bridge and/or channel improvements was quantified with the HEC-2 program by developing pre-improvement and post-improvement flow versus storage values for respective channel reaches. Stormwater detention facilities in selected locations, as shown in Figure 1, were systematically put into operation as the bridge and/or channel improvements were constructed to prevent peak discharge increases along Shoal Creek due to the projected loss of floodplain storage. Redevelopment of Northwest Park as a stormwater detention facility, while enhancing its recreational potential, provides a good example of how this type of planning process was utilized.

Northwest Park Redevelopment and Detention Facility

The $5 million Northwest Park project is an integral part of Shoal Creek's master flood control program. The park encompasses approximately 30 acres with over 20 acres being used for detention purposes. The project not only reduces downstream peak discharges for flood events but also allows construction of other channel and bridge improvement projects that might otherwise increase downstream peak discharges if not compensated for by the park detention. These channel and bridge projects remove local residences from the floodplain by increasing the carrying capacity of Shoal Creek.

For example, the new Greenlawn Parkway bridge and channel improvements extending upstream to Foster Lane will remove 92 houses from the 100-year floodplain. However, by significantly improving the creek-carrying capacity in this upper portion of the watershed, the potential exists for increasing downstream peak discharges. As shown in Figure 1, Northwest Park is immediately downstream of Greenlawn Parkway and provides more than enough detention storage to offset any potential downstream peak flow increases. The Northland Drive bridge and channel project also relies on the

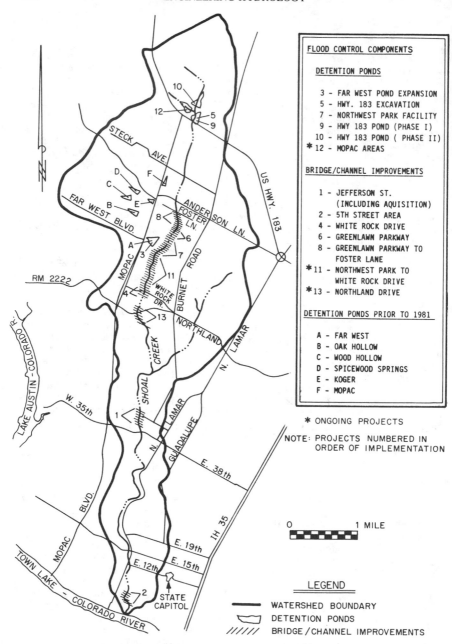

Figure 1. Shoal Creek Flood Control Plan

park's detention to offset any potential increases in downstream peak discharges due to increasing the creek's carrying capacity in that project area.

Being downstream of approximately half of the watershed's thirteen-square-mile drainage area and adjacent to the creek itself, the park is ideally located for an off-channel "peak shaving" detention pond. The facility's basic design function is to allow high flows to enter the park over an inflow wall located at the upper end of the park while allowing lower flows to pass by the park and continue downstream. In this manner, the park's detention storage capacity is reserved for high flows that cause the most damage in downstream areas. The park's detention temporarily traps a part of the flood and releases it slowly as flood flows and elevations in the creek subside following the storm.

To maximize the effectiveness of the available pond storage, and to maintain the park's recreational availability, stormwater inflow into the detention facility was designed to occur when the flow in Shoal Creek adjacent to the inflow wall reaches approximately 4,000 cubic feet per second (cfs) which approximates a 2- to 5-year frequency. This flow value was selected after reviewing downstream Shoal Creek channel capacities primarily, between the park and Northland Drive, along with the available storage of the pond system during the design 10-, 25-, and 100-year flood events.

To model the complex hydrologic conditions associated with this off-channel detention pond design, the modified HEC-1 and HEC-2 programs were extensively utilized. The following basic procedure was employed to assure the proper hydrologic design of the facility.

1. Obtain 10-, 25- and 100-year design hydrographs along the park's inflow wall from the modified HEC-1 watershed modeling.

2. Determine portions of design flood flows entering the park versus flows continuing downstream past the inflow wall (weir) using the HEC-2 split flow option.

3. Account for reduced inflows when the park's ponding elevations create tailwater conditions on the inflow wall (weir) which occurs for the 25- and (especially) 100-year events.

4. Route inflows into park storage and determine park outflow at the overflow spillway for the 100-year event.

5. Determine outflow exiting the park back into Shoal Creek over the inflow wall during the time period immediately following the park's maximum storage time until the park's ponding elevation reaches the top elevation of the inflow wall. This will occur as the creek elevation at the inflow wall falls below the park ponding level.

Figure 2 provides a graphical summary of the hydrologic inputs and outputs of the process.

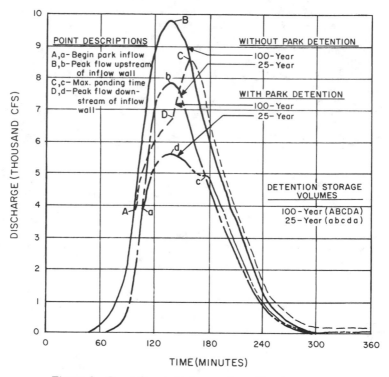

Figure 2. Shoal Creek Hydrographs at Northwest Park

The park's ponding elevation was designed to reach the overflow spillway elevation during a design 25-year flood event resulting in interior ponding volumes near 145 acre-feet. The 25-year peak flow continuing past the inflow wall was estimated to be reduced from 8,000 cfs to near 5,500 cfs which is a 31% decrease. During a design 100-year flood event, the overflow spillway will be utilized while the park achieves approximately 175 acre-feet of stormwater detention storage. The 100-year peak flow continuing past the inflow wall will be reduced from near 10,000 cfs to about 8,800 cfs. With the additional increased storage capacity recently built into the nearby Far West Detention Facility, this 12% reduction in 100-year peak flows increases to near 20% in some downstream areas due to the significant delay in runoff continuing downstream during the storm.

Appendix I. References

1.   Espey, W. H., Jr., Altman, D. G. and Graves, C. B. "Nomographs for Ten-Minute Unit Hydrographs for Small Urban Watersheds," ASCE Urban Water Resources Research Program, Tech Memo No. 32, 1977.

Multi-purpose Stormwater Detention Facility Completed
as First Phase of Englewood, Colorado, Flood Control Project

John M. Pflaum*, M. ASCE, Michael R. Galuzzi*
and William C. Taggart, M. ASCE*

Introduction

The City of Englewood has ambitious plans for redevelopment and revitalization of the central downtown area. However, nearly all of the downtown area lies within the 100-year flood plain. Studies have shown that the City of Englewood would suffer approximately $14 million in damages from a 100-year flood along Little Dry Creek. Consequently, in 1980 the City of Englewood Urban Renewal Authority proceeded with planning for flood control improvements along Little Dry Creek. The selected plan includes capacity improvements for an existing 3,500-foot (1,066.8 m) long box culvert and, at several bridges, widening of the existing narrow channel and construction of an off-stream detention facility near the upstream city limits. The detention facility would reduce peak flows and would economize the sizing of required improvements downstream.

Existing Conditions

The only site available for an off-stream detention area adjacent to the existing channel was an area near Englewood High School. The site was occupied by the school baseball field, wood/steel bleacher facilities, a basketball court and a child care playground. Two existing private residences also abutted the site. An existing irrigation ditch pipeline also traversed the area.

Design Procedure

As part of the initial data acquisition for the project, numerous soil borings were drilled and a complete soils report was prepared. Groundwater was encountered 12 to 23.5 feet (3.7 m to 7.2 m) below the ground surface and the groundwater levels appeared directly related to the surface water levels in Little Dry Creek. The Geotechnical Consultant concluded that groundwater levels in the area around the creek would, therefore, rise when levels in the creek rise. Since the excavated off-stream detention pond would have bottom elevations near the elevation of the creek, an underdrain system would need to be provided to maintain dry conditions for athletic fields during non-flood periods. The Geotechnical Consultant also noted that a significant stratum of generally clean sands existed in the area of the detention pond.

Meetings were held with representatives of the City and the School Board during which conceptual grading plans were reviewed and the following goals and criteria adopted:

*   Engineers, McLaughlin Water Engineers, Ltd., 2420 Alcott St., Denver, CO 80211.

1. Detention storage would be maximized by acquisition of the two adjacent private properties. This would enable construction of two soccer fields in addition to the reconstruction of the baseball field.

2. Surface grades for the athletic fields would not be less than 1 percent, with 2 percent desirable. Similarly, longitudinal slopes for perimeter swales would be a minimum of 0.5 percent, with 0.75 to 1.0 percent desirable.

3. A subdrain system would be required to maintain a stable, dry bottom for the detention area.

As the grading plan for the detention area was refined, hydrologic routing of flood flows for various design frequencies was conducted utilizing the Massachusetts Institute of Technology Catchment Model (MITCAT). Routing iterations enabled optimization of the key components of the structure. As Figure 1 illustrates, a combination drop structure and constriction structure in the creek channel provides the hydraulic control by which the upstream water surface is raised and flow is directed over a 320-foot-(97.5 m) long grassed side channel spillway into the detention area. To determine the hydraulic characteristics of the side channel spillway for the full range of design flows, backwater calculations were first performed for the improved channel downstream of the constriction structure to determine the tailwater conditions for the structure. Then iterative backwater and overflow weir spill calculations were performed upstream of the constriction structure to analyze the spatially varied flow conditions at the side channel spillway. As a result of the hydraulic refinements, it was determined that the spillway would operate at flows greater than the 20-year flood of 2,900 cfs (82 m³/s). Approximately 940 cfs (27 m³/s) of the total 100-year peak flow of 4,400 cfs (125 m³/s) would spill into the detention pond. Water contained within the detention area drains to the northwest corner where it is discharged by gravity via a 36-inch (91 cm) reinforced concrete pipeline. A flap valve at the outlet end of the pipe prevents creek flows from backing into the detention area.

A concrete maintenance path, which also serves as a recreational trail, was planned along the crest of the overflow spillway as a means of providing a fixed control of the spillway crest width and elevation. Hydraulic analyses conducted for various inflows at the spillway resulted in the conclusion that the 4:1 grassed slope would withstand velocities for the full range of design flows. While some minor erosion was likely at the toe of the spillway slope for severe flood events, the risk of such erosion was determined to be acceptable by the City in lieu of construction of structural protective measures, such as rock riprap.

Since the detention area would normally provide athletic fields for Englewood High School and community league play, it was extremely important that the fields be well drained. Sources of water which, under normal circumstances, could contribute to wet and unstable bottom conditions included groundwater, potential seepage from the adjacent irrigation ditch pipeline, surface flows from precipitation on the site, and surface flows from sprinkler irrigation. Because the bottom of the detention facility would be near the creek elevation and below the groundwater table, a complete underdrain system was designed to convey subsurface flows to the northwest corner of the site.

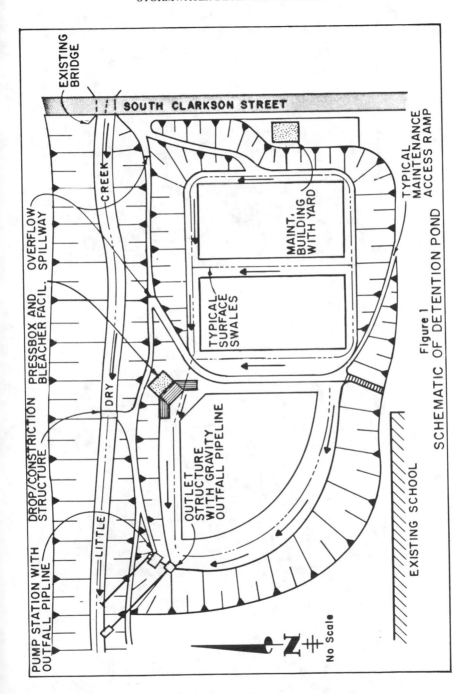

Figure 1
SCHEMATIC OF DETENTION POND

Figure 2.   Stormwater Detention Facility Under Construction

Figure 3.   Completed Stormwater Detention Facility

Manholes were designed at subdrain pipeline junctions to provide maintenance access to the system. A pump station was required to pump the subsurface drainage to the creek. The below-grade station was designed to have two wetwells. The underdrain system drains to one wetwell, where two 5-horsepower (5.1 hp) submersible pumps direct the water via a 16-inch (40.6 cm) ductile-iron pipeline to the creek. A second wetwell houses a 14-horsepower (14.2 hp) submersible pump that provides overflow or surcharge capacity for the subdrain pumps. As an added feature, a surface drain is connected to the overflow wetwell so that the pump can facilitate draining surface water from the detention area.

A free-draining sand layer was chosen as the ideal surface for the athletic fields to provide good drainage and stability. A layer of sand 2 feet (0.61 m) deep would provide a good hydraulic connection to the subsurface drainage system. Fortunately, excavation for the detention facility provided enough clean sandy material to form the lower 12-inch (30.5 cm) layer. Imported clean sand was planned for the upper 12-inch (30.5 cm) layer of material. For the athletic field surface, a special sand-grown sod was chosen, and a complete sprinkler irrigation system was designed for the facility. Extensive tree planting was also planned for the banks of the detention area. Landscaping and irrigation design was provided by EDAW, Inc., Fort Collins, Colorado.

Construction
The construction plans and specifications were completed and bid in April 1983. Construction of the facility was completed between April 1983 and October 1984 at a cost of $1.52 million. Figures 2 and 3 show the detention area during construction and as completed. To date all systems have performed well and the facility has provided the City with the multipurpose benefits of flood control and recreation. Construction of the adjacent and downstream channel improvements is currently underway.

Conclusions
1.   The project demonstrates that off-channel detention storage is an effective flood control measure, particularly in corrective projects.

2.   Subdrainage is very cost effective for maintaining a dry, stable bottom in an excavated detention facility. Proper planning and design for subsurface drainage can eliminate wet bottom conditions which prohibit use and inhibit maintenance.

3.   The design of the pump station provides the added benefit of supplementary pumping to facilitate removal of stormwater and reduce maintenance costs for silt removal. However, the detention facility is not dependent on the pump station for draining itself following a flood. Impounded stormwater is discharged by the gravity outlet pipeline.

4.   The Englewood detention facility is an excellent example of the secondary recreational benefits which can be achieved in conjunction with the planning and design of flood control improvements.

MULTIPLE PARAMETER RULE CURVE FOR RESERVOIR OPERATIONS

*Donald R. Jackson
M. ASCE

Abstract

A procedure for deriving operations rules for water supply storage in a multipurpose reservoir, based on physical principles and data analysis, is described. This procedure has several advantages compared to optimization procedures.

Introduction

The Susquehanna River Basin Commission has adopted a regulation which requires new consumptive uses beginning after the effective date of the Susquehanna Compact to replace their consumptive use (or cease operations) whenever the natural flow of the supply source is less than the 7-day 10-year low flow plus consumptive use (Q7-10+C). The Corps of Engineers has determined (1) that 25,600 ac. ft. ($31.5 \times 10^6 m^3$) of flood control storage in the existing Cowanesque reservoir can be reallocated for water supply purposes, to meet the consumptive use makeup needs for two power plants, the Susquehanna Steam Electric Station (SSES) located about 20 miles (32 km.) south of Wilkes-Barre, Pennsylvania, and the Three Mile Island Nuclear Station (TMI) located about 12 miles (19 km.) south of Harrisburg, Pennsylvania. The Cowanesque Reservoir is located on the Cowanesque River in north central Pennsylvania about 12 miles (19 km.) south of Corning, New York. The development of criteria for determining when to begin and end consumptive loss makeup releases is the subject of this paper.

Procedure

The distance from the Cowanesque Reservoir to SSES is about 175 stream miles (282 km.) and to TMI is about 285 stream miles (459 km.). The U.S. Geological Survey operates long record stream gages at Wilkes-Barre and Harrisburg. The wave travel time for releases from Cowanesque to reach the USGS gage at Wilkes-Barre is about 5 days, and the travel time to the Harrisburg gage is about 10 days. These travel times were based on separate flow routing studies. The Q7-10 at Wilkes-Barre is 814 cfs. (23 cu. m./sec.), based on records back to about 1900; the Q7-10 at Harrisburg is 2547 cfs. (72 cu. m./sec.) based on records back to 1892. The design consumptive use for SSES is 54 cfs. (1.5 cu. m./sec.) and for TMI is 39 cfs. (1.1 cu. m./sec.).

*Staff Hydrologist, Susquehanna River Basin Commission, 1721 North Front Street, Harrisburg, PA   17102

There were two basic conditions which the criteria had to meet. The first was that the Commission's consumptive loss makeup regulation had to be satisfied for the historical events available. The second was that impact of water supply operations on the recreation use had to be minimized to the extent feasible.

The procedure for developing the criteria involved using historical streamflow records to simulate beginning and ending dates in an iterative process.

The following assumptions were made in evaluating proposed operations criteria:

1. In order to satisfy the Commission's regulation, consumptive loss makeup must be provided at the beginning of the first day on which the average daily flow is less than or equal to the Q7-10+C. Therefore the release should begin at midnight T days, where T is the travel time, before the day on which the natural average daily flow becomes less than the Q7-10+C at the target location.

2. If the release is started less than T days before it is needed, it will arrive at the target location too late. In this case, the criteria is unconservative in that it doesn't meet the requirements of the regulation. If the release is made between T and T+3 days before it is needed, it will arrive on the day needed or within 3 days before it is needed. In this case, the criteria is considered satisfactory for that event. If the release is made more than 3 days before it is needed, it will arrive more than 3 days too early. In this case the criteria is too conservative, and therefore unsatisfactory.

3. In general the consumptive loss makeup release cannot be ended until the natural flow at the respective location rises above the highest level at which the release could have been initiated.

4. The decision to make the release should be made at least T+1 days before the natural average daily flow at the respective target location first becomes equal to or less than the Q7-10+C.

5. There should not be any events in which the decision to make the release is made less than T+1 days before the release is needed at the target location.

6. In order to minimize the effect of water supply operations on the recreation use, the period of time between beginning the release and the time it is needed should be as short as possible, subject to the limitations imposed by the travel time assumptions.

7. In order to minimize the effect of water supply operations on the recreation use, events where a consumptive loss makeup release is made but the natural flow did not go as low as the Q7-10+C should be minimized as much as possible.

There were four phases in the development of the criteria. The first phase was basically a preliminary investigation of the concept, which resulted in the conclusion that it appeared to be feasible. The second and third phases were attempts at refinement of the criteria, and some investigation of alternative assumptions regarding the design consumptive use. These phases were further divided into scenarios which basically represented alternative criteria.   In these early phases only data for the Wilkes-Barre and Harrisburg gages were used.

In order to carry out the evaluation of each scenario daily streamflow data were extracted from the USGS WATSTORE data files and reformatted in a table which included the following information:

a. Date
b. Streamflow at the target location (Wilkes-Barre or Harrisburg);
c. The daily rate of change of streamflow at the target location;
d. Streamflow at the upstream locations being used in the procedure;
e. The daily rate of change of streamflow at the upstream locations;
f. The sum of the streamflow at the upstream locations;
g. The daily rate of change of the sum of streamflow at the upstream locations;
h. The difference between the daily streamflow at the target site and the sum of the streamflows at the upstream site(s).

The upstream gages were selected to measure conditions on the major streams which affect future flows at Wilkes-Barre and Harrisburg.   The upstream locations used with Wilkes-Barre initially included Chemung River at Chemung, New York and Susquehanna River at Waverly, New York. Subsequently, it was determined that due to the large (about 25%) increase in drainage area downstream from Chemung and Waverly and upstream of Wilkes-Barre, the following additional streamgages in that reach should be included:

Susquehanna River at Towanda, PA.
Towanda Creek Near Monroeton, PA.
Tunkhannock Creek Near Tunkhannock, PA.
Lackawanna River At Old Forge, PA.

The upstream locations used with Harrisburg were the Susquehanna River at Sunbury and Juniata River at Newport.

Also, two tools were developed. The first is an electronic spread-sheet which included the following data:

a. Date consumptive use makeup needed;
b. Date consumptive loss makeup release began for the scenario being considered;
c. Number of days between the two dates;
d. Streamflow at the target location T+1 days prior to the date consumptive loss makeup was needed;
e. 1-day and T-day rates of decline of streamflow at the target location;
f. Streamflow at each upstream location on the date that the consumptive loss makeup release was started for that scenario;
g. Sum of the streamflow at the appropriate upstream locations on the date that consumptive loss makeup release was started;
h. Short description of behavior of the hydrograph at each upstream location and of the difference between the sum of the flows at the target location and the upstream sites, during the days immediately following the beginning of the release;

i. Date consumptive loss makeup release ended under the scenario.

This tool was very useful in the early phases but became less useful as the analyses became more sophisticated.

The T-day rate of decline is defined as the difference between the average daily flow T days prior to the present day and the present day's (average daily) flow, divided by T.   Initially it was expected that the T-day rate of decline would average out the day-to-day fluctuations in flow rates, and be a good indicator of future trends of the flow hydrograph. That was found not to be true in many cases. The 1 day rate of decline was found to be a better indicator of the future trend in some situations. A ratio of 1-day to 10-day rates of decline was found to be a useful indicator for determining when to make releases for the TMI power plant, because when the ratio is less than unity it means that the rate of decline is decreasing, and flow is being maintained by groundwater rather than surface runoff.

The second tool was a narrative which summarized each event for each scenario, in order to facilitate the detection of patterns.   Initially this tool was only a summary and was not very useful. However, this tool became more sophisticated and more useful as the analyses became more sophisticated.

These phases resulted in several key decisions. The first was that the average consumptive use during a repeat of the worst drought on record should be used as the design consumptive use.

The second decision was that it was necessary to make a detailed day-by-day simulation of operations for all events in order to evaluate the criteria.   The third decision was that upstream stations had to be included, and the fourth decision was that very specific criteria had to be developed for all the streamgages.   The practical effect of the latter decision was to reduce the period of record and number of events for which the criteria could be developed and/or tested since some of these other streamgages had records which began in the late 1930's.

The fifth decision was that the easiest way to eliminate the unconservative events is to set the initial criteria level at Wilkes-Barre or Harrisburg so that the release will always be made in time.   In other words the basic level of the hydrograph at the target location at which the release will be made should correspond to the highest level of flow T+1 days prior to the date on which the consumptive loss makeup is needed, in any historical event needing consumptive loss makeup. Variations among events then are handled by including additional information to determine whether to actually begin the release or not.

With the first three phases completed and the conclusions drawn, Phase IV was started. There were two scenarios, 7 and 8, considered in this phase. Only low flow events occurring after 1937 were considered in Scenario 7 or 8.

In Scenario 7 the release was started whenever the natural flow dropped below 1470 cfs. (42 cu. m./sec.) at Wilkes-Barre or 4360 cfs.

(123 cu. m./sec.) at Harrisburg.   These are the largest values of natural flow T+1 days before the date when the Q7-10+C is reached at the respective target location.

Each consumptive loss makeup event was simulated in detail to determine the date the release would be started under this criteria, and to determine characteristics of that event which might help to develop refinements to the criteria for use in Scenario 8.   Statistics for this Scenario are shown in Table 1.   For some events a simple exponential decay curve was fitted to parts of the hydrograph.   Then the recession factor, defined as $R = 10^{-kt}$, became part of the criteria for determining when to begin releases.

Certain events were difficult to handle because of previous rainfall which caused the hydrograph to rise above the level at which the release would cease, but then decline very rapidly.   These events were handled by raising the level at which the release would be stopped once it had been started, in effect combining two events into one.

Criteria for Scenario 8 at both locations were developed based on the evaluation of each event under Scenario 7.   Preliminary testing of the criteria, developed as a result of the detailed analysis of scenario 7, showed a need to modify some of the criteria.

Type of Event	Number of Events			
	Harrisburg		Wilkes-Barre	
	Scenario 7	Scenario 8	Scenario 7	Scenario 8
W/Consumptive Loss				
Makeup	9	8	12	12
Handled Well	3	5	1	5
Unconservative	2	0	0	1
Conservative	4	3	11	6
W/O Consumptive Loss				
Makeup	50	15	50	7

TABLE 1

STATISTICS OF SCENARIOS FOR COWANESQUE OPERATIONS FOR
PERIOD 1938-82

The actual criteria are too lengthy to repeat here, but are given in detail by Jackson (2).

Pertinent statistics for this scenario are shown in Table 1 for comparison with Scenario 7.

Note that the criteria is reasonably effective in reducing the number of events for which a release is made but the natural flow doesn't reach the level of Q7-10+C.   For Harrisburg the number of such events remaining in Scenario 8 is only 30% of the number of events in

Scenario 7. For Wilkes-Barre the number of such events has been reduced by 86% compared to Scenario 7. The Scenario 8 criteria are also reasonably effective in reducing the number of consumptive loss makeup events which are handled conservatively, and in increasing the number of consumptive loss makeup events handled well.

Based on the Corps of Engineers design and operations criteria for the recreation facilities, the proposed scheme for determining when to make releases results in closing the boating and swimming facilities during the recreation season (assumed to end September 7) in two years and four years, respectively, of the 31 years of record available for storage computations.

With the exception of one event at Wilkes-Barre, the Scenario 8 criteria is expected to ensure that the consumptive loss makeup release arrives in time under all conditions experienced in the historical record considered (1938-1982). The criteria reduces the number of events for which a release is made but the natural flow does not reach the level of the Q7-10+C, and minimizes the impact of the consumptive loss makeup release on the recreation pool. The Scenario 8 criteria provide essentially complete compliance with the consumptive loss makeup regulation and are effective in minimizing unnecessary releases and the consequent impact on the recreation pool. Scenario 8 protects the recreation pool almost as well as another scenario which was specifically designed to maximize protection of the recreation pool, but which did not satisfy the requirements of the consumptive loss makeup regulation. For these reasons, Scenario 8 is the recommended criteria for determining when to begin and end releases, subject to studies presently underway to further refine the criteria.

This method of deriving operating rules has a number of advantages over the optimization procedures. It is easier to understand, and to explain to non-technical people. It is hydrologically more intuitive, and avoids the mathematical and computational difficulties inherent in optimization procedures.

References

1. Cowanesque Lake Reformulation Study, Baltimore District, U.S. Army Corps of Engineers, March 1982.

2. Jackson, D. R., Engineering Studies, Report on Task F2, Cowanesque Lake Water Supply Storage Project Special Studies, Susquehanna River Basin Commission, February 1986.

A Riverflow-Water Quality Control Model

Bernard B. Hsieh*, A.M. ASCE

In order to implement how regulated riverflow (control variable) impacts the salinity (output variable) under a tidal dynamic environment, a multiple input control system is developed. The subtidal heights and local wind stress are considered as two uncontrollable inputs. This controller, which consists of variable coefficients by means of a self-tuning scheme and recursive computation algorithms, is tested by the desired salinity patterns after calibrating proper stabilizing factors and pure delays. The low estimate error and variable and controllable forgetting factor show that this system can perform a dynamic output function. A simulation test indicates that this controller can respond to the sudden change of the output variable even when a constant desired level of salinity is provided.

Introduction

The required riverflow from the Susquehanna River is discharged into the Chesapeake Bay after being released from a multiobjective reservior, the Conowingo Dam. The circulation patterns of the bay, as a typical estuary, are mainly dominated by tidal fluctuations, seasonal freshwater discharge and wind forcing on the water surface. Maintaining the water quality level in the bay, especially for salinity, is one of the major issues for the operation of the reservior. The new release strategies can be made if the necessary riverflow is obtained by giving the specified salinity levels for some critical locations. The recent research (Hsieh, 1986) indicated that the Susquehanna Riverflow, the Chesapeake Bay subtidal signals and the cross-bay component of the wind stress from the Delaware Bay mouth are the three most significant forcing functions, which are calculated daily, dominating the daily regime of the Elk River area, one target station of the upper bay.

Figure 1 shows the design for a tidal water quality control system. In this system, the system output, salinity, is the function of three inputs. The controller, combined with the cost function and system model, is used to calculate the required riverflow when the desired salinity or set-point salinity is proposed. The system parameters and coefficients of the controller are adjusted by a recursive algorithm for every time step after the initial conditions are given. It was found that this system output versus each system input shows a very strong non-linearity. Usually, a non-linear transfer function can be used to simulate the approximate system coefficients. However, in a noise water environment, the order selection is a very difficult task.

*Senior Energy Resources Officer, Maryland Department of Natural Resources, Annapolis, MD 21401

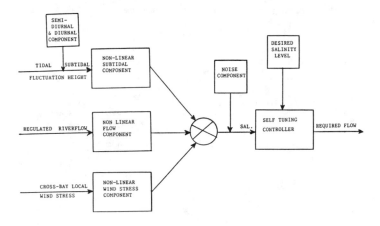

FIGURE 1. Tidal Water Quality Control System

The higher order system model is not practical for the calculation,
whereas the lower order model loses its accuracy. This problem can be
solved by recursive processes. The Kalman's estimation method with
variable forgetting factor (weighting factor) enables the parameter
estimates to follow both show and sudden changes in the dynamic system
(Fortescue, Kershenbaum and Ydstie, 1981). Furthermore, the use of a
variable forgetting factor with correct choice of information bound can
avoid one of the major difficulties which associated with constant exponential weighting of past data. The best advantages of this type of
the system no longer are the dominate factor for generating a good solution although the initial covariance still needs to be estimated correctly.

The results from this study could be linked to a regional control
center as one of the subsystems. All control loops with different purposes can be used to meet the water resources management needs of the
river basin.

Multivariate Control Model

Astrom (1970), Astrom and Wittenmark (1973) and Clarke and Gawthrop
(1975) introduced and developed a self-tuning regulator or controller
which uses a stochastic difference equation as a system identification
algorithm and a feedback control law to minimize the variance of the
output function. The basic self-tuning controller structure has been
extended for the multivariate case by Borisson (1975; 1979), Koivo
(1980) and Bayoumi, Wong and El-Bagoury (1981). Ganendra (1978) used a
self-tuning controller applied to the real-time control of release from
a river regulating reservior. In this paper, the self-tuning controller with characteristics of unknown parameters and variable forgetting
factor is used to define the estuary water quality when a multiple case
is applied.

The linear vector difference equation describes the system as:

$$A(q^{-1})y(t) = B(q^{-1})u(t-k_1) + H(q^{-1})T_c(t-k_c) + M(q^{-1})WS(t-k_w)$$
$$+ C(q^{-1})e(t) \tag{1}$$

where q^{-1} = backward shift operator
y = system output, salinity
u = system input/control variable, riverflow
T_c = subtidal heights from the Chesapeake Bay mouth
WS = wind stress from the Delaware Bay mouth
e = a sequence of independent random numbers or noise
$A = 1 + a_1 q^{-1} + \ldots + a_n q^{-n}$
$B = b_0 + b_1 q^{-1} + \ldots + b_n q^{-n}$
$C = 1 + c_1 q^{-1} + \ldots + c_n q^{-n}$
$H = h_0 + h_1 q^{-1} + \ldots + h_n q^{-n}$
$M = m_0 + m_1 q^{-1} + \ldots + m_n q^{-n}$
k_1 = pure time delay between riverflow and salinity
k_c = pure time delay between subtidal heights and salinity
k_w = pure time delay between wind stress and salinity

The controller has a set-point term $w(t)$ (a desired output function) included in its cost function given by

$$I = E[(y(t+k_1) - w(t+k_1))^2 + \lambda^*(u(t-1) - u(t))^2] \tag{2}$$

where $E(.)$ = expectation operator

λ^* = stabilizing factor of the controller

The identity $C = AF + q^{-k}1 G$ is introduced with the polynomials F and G defined as

$$F = 1 + f_1 q^{-1} + \ldots + f_{1-k_1} q^{1-k_1} \tag{3}$$

$$G = g_0 + g_1 q^{-1} + \ldots + g_{n-1} q^{1-n} \tag{4}$$

Using the above information and the concepts of minimum square error predictor and the minimum cost control, we can obtain the control variable as

$$u(t) = \frac{-1}{E}[Gy(t) - Cw(t+k_1) + HFT_c(t-k_c+k_1) + MFWS(t-k_w k_1)] \tag{5}$$

where $E = BF + C(1-q^{-1}) = e_0 + e_1 q^{-1} + \ldots + e_n q^{-n}$
$J1 = HF$
$J2 = MF \tag{6}$

In order to calculate the feedback control law, the coefficients of the polynomials E, G, C, J1 and J2 can be estimated by recursive least squares method and to simplify the notation, the data and parameter vectors are introduced.

$$\phi(t) = [u(t), \ldots, u(t-n), y(t), \ldots, y(t-n+1), w(t+k_1-1), \ldots,$$
$$T_c(t+k_1-k_c), \ldots, T_c(t+k_1-k_c-n), WS(t+k_1-k_w), \ldots,$$

$$WS(t+k_1-k_w-n)]\qquad\qquad(7)$$

$$\theta^T(t) = [e_0, \ldots, e_n, g_0, \ldots, g_{n-1}, -c_1, \ldots, -c_n, J1_0, \ldots, J1_n,$$
$$J2_0, \ldots, J2_n]\qquad\qquad(8)$$

Finally, a required riverflow can be computed by

$$u(t) = \frac{-1}{e_0}[\phi(t)\theta(t) - e_0 u(t) - w(t+k_1)]\qquad\qquad(9)$$

The computation algorithm is shown as Figure 2.

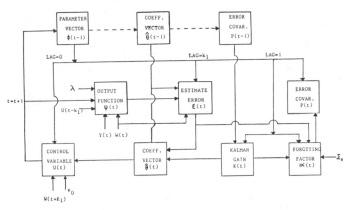

FIGURE2.   Self-Tuning Controller with Variable
           Forgetting Factor

Model Calibration

    The estimation of coefficients for the self-tuning controller can be
made by giving an initial coefficient vector, and initial error covariance, pure delays and the stabilizing factor. Since this system is assumed to be unknown, those values can be estimated roughly. For simplicity purpose, the lower order of the system is easier to handle.
The sum of square errors and minimum forgetting factor also can be estimated by the degree of noise. Usually, calibration is done by assuming the desired salinity equals the observed salinity, then by adjusting   and k_1, k_c and k_w until an optimal and stable condition is arrived at. The forgetting factor basically responds to the sensitivity of
estimate error for each time step. During the calibration process, a
minimum value of forgetting factor is needed to handle the most dynamic
condition. Under the best selection of these system parameters, the
required flow is nearly reproduced by the measurement flow. The final
selection for this study is k_1, k_c and k_w equal to 1 and λ equal to
0.20. (n=2)

    Figure 3(a)-3(b) indicate the behavior of the forgetting factor and
estimate errors. Under steady-state operation, the forgetting factor
is close to one and the estimator behaves very much like an unweighted
filter. The noisy output function and small forgetting factor occur
every year which explains the stochastic inputs in this system. The

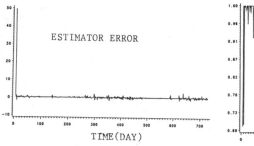

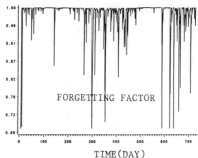

FIGURE 3(a).   Estimate Error of   FIGURE 3(b).   Forgetting Factor
               the Controller   of the Controller

poor fit between model and process usually occurs during the late part
of the year because all input and output have larger fluctuations.
This self-tuning controller can handle the process very well even when
poor initial system coefficients are provided as in Figure 3(a).

Simulation Experiments

   The simulation test is conducted by calculating the required river-
flow as the desired salinity level is given. The high flow season 1982
and low flow season 1983 are selected for this test. The desired salinity level is assumed to fix at some certain level in order to calculate how much flow change between the required riverflow and actual
measurement flow. They are shown on Figure 4(a) and 4(b). In 1982's
high flow season, two desired salinity levels (s=0.5 0/00 and s=1.0
0/OO) are used to simulate the required riverflow. It should be mentioned that even when a constant salinity level is provided, the required riverflow does not respond to a constant level due to two other
significant inputs which are included in this system. During the period from the end of May to the middle of June, 1.5 units of salinity
change causes the flow change to be about 500 CFS. This sensitive
change occurs partially because the flow is the most dominate factor
compared to the rest of the input variables. During the low flow season, the set-up salinity level is 4.0. This large change produces more
flow than we can provide during that season. It can be concluded that
the same amount of salinity change needs different flow change depending on its seasonality.

Conclusions

   For a dynamic environment as estuary, a multivariate control system
including variable forgetting factor is required to conduct the sudden
change of input and output functions. The pure delay between output
function and control variable is the most critical factor that dominates the stability of the system. From an application point of view,
this model can be extended to interpret the real case if the release
policy and outflow series from the dam and actual desired salinity level are proposed.

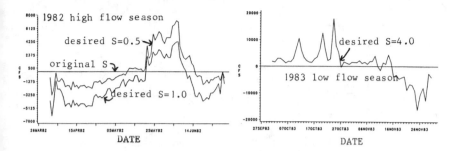

FIGURE 4(a). Flow Change Pattern   FIGURE 4(b). Flow Change Pattern
during 1982 High Flow Season   during 1983 Low Flow Season

Appendix - References

Astrom, K.J.   1970. Introduction to Stochastic Theory. Academic Press,
    N.Y.
Astrom, K.J. and B. Wittenmark.   1973.   "On Self-Tuning Regulators."
    Automatica, Vol 9: 185-199.
Bayoumi, M.M., K.Y. Wong and M.A. El-Bagoury. 1981. "A Self-Tuning Re-
    gulator for Multivariable Systems." Automatica 17, 575-592.
Borisson, U.   1975. "Self-Tuning Regulators - Industrial Application
    and Multivariable Theory." Report 7513, Dept. of Automatic Control,
    Lund Inst. of Tech., Lund, Sweden.
Borisson, U.   1979.   "Self-Tuning Regulators for a Class of Multivaria-
    ble Systems." Automatica 15, 209-215.
Clarke, D.W. and P.J. Gawthrop.   1975.   "Self-Tuning Controller." Proc,
    I.E.E., Vol 122, no. 9: 929-934.
Ganendra, T.   1978. "A Self-Tuning Controller Applied to River Regula-
    tion." Proceedings of Modeling, Identification and Control in Envi-
    ronment Systems, Vansteenskists, Editor.
Fortescue, T.R., L.S. Kershenbaum and B.E. Ydstie.   1981.   "Implementa-
    tion of Self-Tuning Regulators with Variable Forgetting Factors."
    Automatic 17, 831-835.
Hsieh, B.B.   1987. "A System Approach to Analysis of Water Quality due
    to Transport Mechanisms in the WAter Environment." Advances in Water
    Pollution Control, v3. Pergamon Journals Ltd., London. (in press)
Koivo, H.N.   1980. "A Multivariable Self-Tuning Controller." Automatica
    16, 351-366.

A REAL-TIME FLOOD MANAGEMENT MODEL FOR RIVER-RESERVOIR SYSTEMS

Larry W. Mays[1], M.ASCE and Olcay Unver[2], A.M.ASCE

Introduction

Large reservoir systems such as the Highland Lake System in the Lower Colorado River Basin in Texas (Figure 1), operated by the Lower Colorado River Authority (LCRA), are characterized by an integrated operation of multiple facilities for multiple objectives. The portion of the river basin operated by LCRA extends from the headwaters of Lake Buchanan 443 mi. downstream to the mouth of the Colorado River at the Gulf of Mexico. The Highland Lakes System consists of 7 man-made lakes that are serially connected: Lake Buchanan, Inks Lake, Lake Lyndon B. Johnson, Lake Marble Falls, Lake Travis, and Lake Austin, all operated by LCRA, as well as Town Lake, which is operated by the City of Austin. Development in the floodplains of the Highland Lakes System has caused severe problems in operation of the reservoirs under flooding conditions. Flood control operation of the Highland Lakes is further complicated because only two of the lakes, Buchanan and Travis, can store any significant flood volumes. Due to the potentially severe flooding conditions of the river-lake system, a flood forecasting model has been developed for the LCRA that can be used in a real-time framework to make decisions on reservoir operations during flooding. This real-time flood forecasting model is based upon state-of-the-art techniques for flood routing, rainfall-runoff modeling, and graphical display capability and is controlled by interactive software.

An optimization model has also been developed to optimally determine the real-time operation of reservoirs under flooding conditions. The optimization problem can be stated as to determine the operation of reservoirs in a multireservoir river system to minimize total flood damage, or deviations from target levels, or water surface elevations in flood areas, or spills from reservoirs, or to maximize storage in reservoirs subject to: a. unsteady flow equations and other relationships that describe the flow in the different components of a river-lake system; b. maximum and minimum allowable reservoir releases and flow rates at specified locations; c. maximum and minimum allowable water surface elevations at specified locations (including the reservoirs); and d. operating rules, targets, limitations, etc. The variables that are used in the mathematical model are: water surface elevations at all computational points; flow rates at all computational points; and gate operations at all gated reservoirs.

---

[1] Professor of Civil Engineering, The University of Texas at Austin, Austin, TX 78750.
[2] Engineer, Lower Colorado River Authority, P.O. Box 220, Austin, TX 78767.

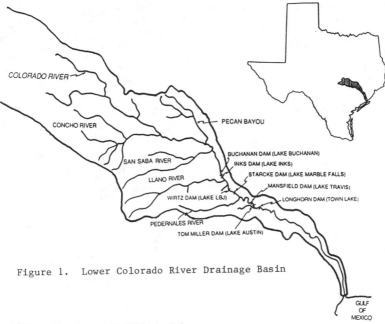

Figure 1. Lower Colorado River Drainage Basin

Real-Time Flood Forecasting Model

Overall structure of the real-time flood forecasting model for the Highland Lake System is shown in Figure 2. The data management module was developed by the LCRA. The real-time flood control module includes the following submodules:

1. DWOPER submodule – U.S. National Weather Service Dynamic Wave Operational Model for unsteady flow routing by Fread (1982).
2. GATES submodule – Computer program developed at The University of Texas at Austin to determine gate operation information for DWOPER.
3. RAINFALL-RUNOFF submodule – Rainfall-runoff model developed at The University of Texas at Austin for the ungaged drainage area surrounding the lakes for which streamflow data is not available.
4. DISPLAY submodule – Graphical display software originally developed by the U.S. Army Corps of Engineers and modified by The University of Texas at Austin.
5. OPERATIONS submodule and user control interface – The software developed by The University of Texas at Austin that basically operates the other submodules and provides the control interface with the user.

Data input for this flood forecasting model includes both the real-time data and the physical description of the system and system components that is stored and remains unchanged. The stored data include: 1) DWOPER data describing the physical system and including cross-section information, roughness relationships, etc.; 2) characteristics of reservoir spillway structures for GATES; and 3) drainage area description and hydrologic parameter estimates for RAINFALL-

RUNOFF. The entire river-lake system contains 871 cross-sections for DWOPER. This system is broken down into 5 subsystems because of the computational aspects and practicality of running the model. The real-time data include: 1) streamflow data at automated stations and head-water and tailwater elevations at each dam; 2) rainfall data at recording gages; 3) information (which subsystem) pertaining to which lakes and reservoirs will be considered in the routing; and 4) reservoir operations.

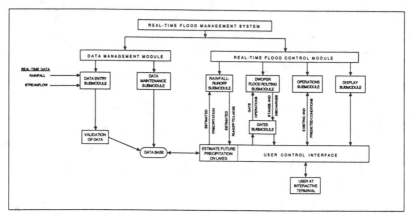

Figure 2. Structure of Real-Time Flood Management Model

The LCRA real-time data collection network referred to as the hydromet system, provides information of two types: (a) the stages at various locations (Figure 3) throughout the river-lake system, and (b) rainfall from a rain gage network for the ungaged drainage areas around the lakes. The LCRA hydrometeorological data acquisition system consists of (a) remote terminal unit (RTU) hydrometeorological data acquisition stations installed at USGS/LCRA river gage sites, (b) microwave terminal unit (MTU) microwave to UHF radio interface units located at LCRA microwave repeater sites, and (c) a central control station located at the LCRA System Operations Control Center (SOCC) in Austin. The system is designed to automatically acquire river level and meteorological data from each RTU; telemeter this data on request to the central station via an LCRA supplied UHF/microwave radio system; determine the flow rate at each site by using USGS tables stored in the central system memory; format and output the data for each site; and maintain a historical file of data for each site which may be accessed by the local operator, a local computer, and a remote dial-up telephone line terminal. The system also functions as a self-reporting flood alarm network.

Optimization Model

A mathematical statement of the optimization/operation model (Figure 4) is given. The objective function can be expressed in terms of the water surface elevation, h, and discharges, Q, as

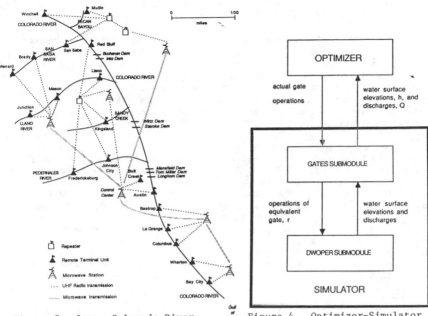

Figure 3. Lower Colorado River
Authority Hydrometeorological
Data Collection System (Hydromet)

Figure 4.   Optimizer-Simulator
Coordination

Minimize z = f(h,Q)

subject to

i. hydraulic constraints (Saint-Venant equations and internal
   and external boundary conditions)

$$g(h,Q,r) = 0 \qquad\qquad (1)$$

ii. bounds on water surface elevations and discharges

$$\underline{h} \leq h \leq \bar{h} \qquad\qquad (2)$$

$$\underline{Q} \leq Q \leq \bar{Q} \qquad\qquad (3)$$

iii. range for gate settings r (closed to full open)

$$0 \leq \underline{r} \leq r \leq \bar{r} \leq 1 \qquad\qquad (4)$$

iv. maximum rates of change of gate settings

$$W(r) \leq 0 \qquad\qquad\qquad (5)$$

The problem described above is a very large mathematical programming problem for most real-world situations. A typical 24-hour operation horizon with 1-hour time steps for a river system with 5 reservoirs and 150 computational points would give rise to a problem with
more than 7200 hydraulic constraint equations (1) and over 7200 flow
variables. This is far beyond the capacity of existing nonlinear
programming codes. The logical approach in solving a problem this big
is reduce the size.

The operations problem, hereafter called the General Operations
Model (GOM) has certain characteristics that can be used in reducing it
to a smaller problem. GOM has the general structure of a discrete time
control problem. It contains three basic groups of constraints: the
hydraulic constraints, the bound constraints, and the gate operation
constraints. The hydraulic constraints (Eq. 1) can be solved sequentially forward in time for the water surface elevations, h, and the
flow rates, Q, by using a simulation model such as DWOPER, once the
gate settings, r are specified. Then the problem can be solved more
efficiently by incorporating the simulation model into a procedure in
which a set of gate operations, r is chosen, the simulation model is
run with these to solve the hydraulic constraint set for the elevations
and discharges. Then the objective function is evaluated, the bound
constraints are checked for any violations and the procedure is repeated with an updated set of gate operations until a convergence criterion is satisfied and no bound constraints are violated.

Choosing the water surface elevations, h, and the dischares, Q, as
the state variables and the reservoir gate settings, r as the control
variables of the optimum control terminology, the GOM is converted into
a reduced problem by making use of the implicit function theorem. The
implicit function theorem is applied to the problem given by Eqs. (1-5)
in such a way that the hydraulic constraints (Eq. 3) are solved by the
simulation model. The simulation model computes the values of the
state variables, h and Q for given values of the control variables r
and the optimization model seeks the optimal values of r that will
minimize the objective function. The implicit function theorem states
that $h(r)$ and $Q(r)$ exist if, and only if, the basic matrix of the
system of equations given by Eq. 1 is nonsingular. This condition is
always warranted as the simulation model, DWOPER, uses the same matrix
for its Newton-Raphson iterations, so $h = h(r)$ and $Q = Q(r)$.

Then, the objective function, now called the reduced objective
function, is expressed similarly using the implicit functions $h(r)$ and
$Q(r)$ which can be evaluated once the state variables h and Q are found
for the given set of control variables, r. The reduced problem, which
is called the Reduced Operations Model (ROM), is now expressed as

$$\text{Minimize } z = F(r) = f[h(r), Q(r)] \qquad\qquad (6)$$

Subject to Eqs. (2) − (5).

The reduced problem can now be solved by a nonlinear programming algorithm. As the reduced problem still contains bound-type constraints on the state variables h and Q, the algorithm adopted should have provisions to assure the feasibility of the solutions as far as the state variables are concerned. In other words, the solution of the DWOPER model does not guarantee that the bounds, (2) and (5), on the state variables are satisfied. An augmented Lagrangian (AL) algorithm that incorporates the bounds on the state variables in the objective function is used for this purpose.

The reduced operation model with AL terms (ROMAL) becomes

$$\min L_A(r, \mu, \sigma) = F(r) + 0.5 \Sigma_i \sigma_i \min[0, (c_i - \mu_i/\sigma_i)]^2$$
$$+ 0.5 \Sigma_i \mu_i^2/\sigma_i \qquad (7)$$

Subject to Eqs. (4) and (5), where i denotes the constraint set which is formed of the bounds on the state variables, h and Q, and σ_i and μ_i are respectively the penalty weight and the Lagrange multiplier associated with the i-th bound. The term c_i is the violation term which is defined as:

$$c_i = \min [h_i - \underline{h}_i), (\overline{h}_i - h_i)] \text{ or } \min [(Q_i - \underline{Q}_i), (\overline{Q}_i - Q_i)] \qquad (8)$$

The constraints of the new problem are the bounds on the control variables and the operating constraints. The ROMAL is solved using the generalized reduced gradient method, GRG2 by Lasdon and Waren (1984).

Summary

The original problem, GOM with the water surface elevations and the discharges at the computational points, and the gate settings at the reservoirs as the decision variables. GOM is reduced to a smaller optimal control problem, ROM, using the implicit function theorem by solving Eq. (1) by DWOPER for h and Q. The number of the control variables in ROM, i.e., the reservoir gate settings, is only a fraction of those in GOM. An augmented Lagrangian (AL) algorithm is used to solve ROM, with the bounds on the water surface elevations and discharges incorporated into the objective function. The algorithm first solves the inner problem, Eq. 7 subjected to 4 and 5, using GRG2 for fixed values of the AL parameters. Then an outer cycle is iterated updating these fixed values. The overall optimization is attained when the AL parameters converge.

References

Fread, D.L. (1982). National Weather Service Operational Dynamic Wave Model, Hydrologic Research Laboratory, U.S. National Weather Service, NOAA, Silver Spring, Maryland.

Lasdon, L.S. and Waren, A.D. (1984). GRG2 User's Guide, Department of Business, The University of Texas, Austin, Texas.

ESOLIN: A Model to Optimize Water Resource Systems

Christian Guillaud* and Nick Grande**

Abstract

ESOLIN is a simulation model designed to study and optimize the
development and operation of large hydro resource schemes. It was
developed as a tool for the development of large hydro resource
complexes in Northern Canada. Experience shows that simulation models
like ESOLIN are preferred over Systems Analysis by most of the
designers and operators of hydro schemes, because of their flexibility
and versatility. The operation of the ESOLIN model features a
modified space rule called "Storage Factor" which enables one to
determine, for each time step, the priorities between each reservoir
of an integrated multireservoir system. The structure of the model is
such that it may be easily adaptated to the special features of any
project.

Introduction

Because of the ever growing role of computers in engineering, it
is now standard practice to solve problems through the use of
mathematical models. The design and operation of large hydro resource
networks is a typical application where simulation models are
particularly efficient because of the shear quantity of input data,
the type of calculations to be performed and the number of alternative
solutions which seem to increase during the course of the standard
study of a water resources project.

Simulation models were initially developed in Canada in the 60's,
to evaluate and harness Canada's remote northern hydro resources.
These models were initially based on the experience and intuition
developed by the engineers who developed Canada's southern hydro
resources in the 40's and 50's. The simulation models were then
refined in the 70's and made more versatile by applying them to
projects of all sizes in developing countries. The ESOLIN model
(Energy Simulation and Optimization of Large Interconnected Networks)
resulted from this process. It had originally been developed by SNC
to evaluate hydro projects in James Bay and on the Churchill river in
Labrador (Wiebe and Nguyen, 1979), and improved and generalized on
projects worldwide.

*   Principal Discipline Engineer, Hydraulics, Hydrology. SNC Inc.,
Montréal, Canada
** Specialist, Water Resources, SNC Inc., Montreal, Canada.

At first sight, it might appear contradictory that extensive simulation models such as ESOLIN are used for the design and operation of hydro resource systems given the large variety of Systems Analysis programs presented every year in the literature (Rogers and Fiering, 1986). A comprehensive and detailed evaluation of the state-of-the-art concerning reservoir management and operation models (Yeh, 1985) suggests that simulation models are the preferred engineering tool. The reasons for preference being given to simulation models over Systems Analysis are numerous, but a very important one, if not the major one, is that experience shows that the design of a hydro resource system is unique, and that the first task facing the designer is to correctly identify the controlling parameters; those parameters can be technical (dam design, hydrology etc...), environmental, political... It is therefore obvious that the flexibility of simulation models presents great advantages over Systems Analysis, which normally implies that the nature of the problem to be solved is well known. Another advantage of simulation models is that their operation and results can be easily understood and visualized by the managers and the decision makers, who in general do not possess the prerequisite theoretical knowledge necessary to evaluate the results of System Analysis programs.

Model Parameters

The rationale for developing a hydro resource project is to optimise the production of a certain commodity (electricity, irrigation, water supply), while respecting certain constraints (maintain or improve navigation, minimize impact on environment, maximize flood protection, etc...). The major inputs to the model are the natural inflows to each catchment reservoir and the system component characteristics. Therefore the reliability of the results strongly depend on the inflow series input to the model which are assumed to represent, with reasonable accuracy, the flows that will be experienced by the project under study. These flows must correctly reflect the long term mean and variability experienced historically. In most hydro resource projects, the criteria for guarantee of the output product (electricity, irrigation, etc.) is 95%, or 1/20 year. Therefore, the absolute minimum duration of the hydrologic series to be input is 20 years, however a duration of 40 years or more is preferable. A thorough investigation must be carried out to verify that the hydrologic sample does not contain outlying values with frequencies much beyond 1/20 years (drought cycles in particular). If natural inflow series are not readily available or are not of sufficient duration, the following two alternatives are possible:

- Correlation with nearby long term hydrometric or meteorologic stations.

- Reconstitution of flows by stockastic methods.

The operation of the model consists in setting, for each time period, a demand for each of the different products, (electricity, irrigation, water supply) the order of priority between them and their respective constraints (minimum flow for navigation, environmental protection, reservoir storage for flood protection, etc...). At the beginning of each time period, the inflows from each subbasin are stored in their respective reservoirs. The required minimum outflow, consumptive water usage, and evaporation are then subtracted. If the remaining inflow exceeds the reservoir capacity, then production demands are met (electricity, irrigation, etc.). If after meeting the production demands, the remaining inflow still exceeds the reservoir capacity, spillage occurs.

Once all inflows have been processed, a check is made to see if the demand for the given period is met and if all operating rules have been respected. If this is so, the simulation proceeds to the next time period.

If the demand is not met, the prescribed operating rules, for each reservoir, are applied. Typical reservoir operating rules are shown in Figure 1. Five zones are identified (A to E), separated by five curves (1 to 5).

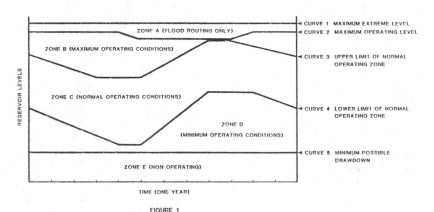

FIGURE 1

TYPICAL RESERVOIR RULE CURVES

- Zone A is used only in case of extreme flood. The spillway is operated at full capacity until the reservoir level is restored to curve 2.

- In Zone B, production is set to full capacity or to a preset percentage above normal demand until the reservoir level is restored to curve 3.

- In Zone C, production is set to meet the demand. In a well balanced system, the reservoir level should always stay within Zone C.

- In Zone D, the production is reduced by a preset percentage below normal until the reservoir level attains curve 4.

- In Zone E, no production is allowed.

Frequently, the operation of a hydro resource complex will be such that a global demand must be met for the entire complex, for each time period, irrespective of which individual plant supplied the production (this is particularly true for hydroelectric production). In such a case, it is important to determine the sequence and extent to which each of the reservoirs are to be drawn from. In order to do so, a modified space rule is introduced, in the model, which operates as follows:

- A variable called "Storage Ratio" is defined for each reservoir, which is the ratio of the storage currently available above curve 4 and the total storage between curve 3 and curve 4.

- For each reservoir, and for each time step, an empirical curve is defined relating the so called "Storage Factor" to the storage ratio. A typical set of these curves is shown in Figure 2.

- The ESOLIN model attempts to maintain the storage factors equal, at the end of each time period, by ordering the reservoirs in decreasing order of storage factors and drawing from the reservoirs with the largest storage factors first. Checks are made at each step to determine whether the demand is met for the time period, and whether drawings from some reservoirs cause spill at other downstream reservoirs.

- If the demand is met, the simulation proceeds to the next time step.

- If the demand is not met, curve 4 is crossed and water is drawn from reservoirs creating downstream spills.

- If all reservoirs are drawn downs to curve 5 and demand is still not met, a shortage occurs for the time period.

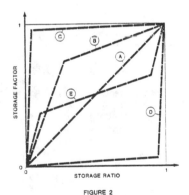

FIGURE 2

TYPICAL STORAGE FACTOR CURVES

CURVE A: RESERVOIR USED EVENLY
    B: RESERVOIR USED IN PRIORITY OVER OTHERS
    C: RESERVOIR STORAGE COMPLETELY USED BEFORE ANY
       OF THE OTHERS
    D: RESERVOIR STORAGE USED ONLY AS LAST RESORT
    E: RESERVOIR STORAGE INITIALLY SAVED, THEN FULLY
       USED AT MID POINT

Note that if curve 4 is crossed, a slight shortage is allowed to occur, because below this curve maximum allowable production is reduced by a predefined percentage.

The experience of applying ESOLIN to a large variety of projects all over the world, with different constraints and climatic conditions, has shown that such a tool must be very versatile. To achieve this goal the following two alternatives are possible:

- Identify all options that might be encountered during the study of a water resource system and include them in the model, allowing for easy access to each option in the input data.

- Isolate all key computational modules in the program, and program them in such a way that any qualified engineer/programmer will be able to introduce site specific options without destroying the structure of the model.

The second alternative is obviously more flexible than the first and results in a model tailored for each project under study. Therefore the selected procedure is to prepare a general version of the ESOLIN model applicable to any project with standard features. The model is then tailored to each particular project at the beginning of the project or during the course of the study, when specific operating rules are well defined.

Applications of the Model

Following is a list of some of the particular project features which have been successfully coped with during the course of utilizing the ESOLIN model.

Multiple demand patterns, with restricted connections

On the Lower Churchill River studies, the purpose was to optimize the production of two hydroelectric plants on the Lower Churchill River (in Labrador), which is almost 100% regulated by the existing Upper Churchill development. Part of the production of the Lower Churchill system was to be sent to the Island of Newfoundland, through a submarine cable of limited capacity, and the rest was to be sold to Quebec or to New England. Therefore, the optimization of the Lower Churchill system had to cope with two electrical demands with different patterns (Quebec and Newfoundland), had to balance two different hydrologies (Labrador and Newfoundland) and had to compute electrical exchanges between three networks (Quebec, Labrador and Newfoundland), with the limiting constraint at the submarine cable crossing connecting Labrador and Newfoundland.

Discharge capacity restriction in canals

For some projects in Northern Canada (LA-2 or Evans, in James Bay, Norman Wells in the Northwest Territories) the turbine flow in winter is supplied by drawing down distant reservoirs. In winter the canals conveying water from these reservoirs are partially blocked by ice cover and frazil which affect their discharge capacity. Moreover, there is also a significant time delay between the drawing down of the reservoirs and the actual turbining.

Seasonal or peaking operations

In the Island Falls project, in Northern Saskatchewan, the purpose was to produce base energy in summer, when water supply is plentifull, and to produce peaking energy during winter, when flows are at a minimum. Peaking in winter introduced two constraints; a significant difference in net head between peaking and non peaking periods during the day and the maximum rate of change of turbine discharge necessary to maintain a stable ice cover on the downstream reach.

The applications of ESOLIN, described above, relate to optimizing a scheme at the design stage, however, a simulation model such as ESOLIN can also be successfully used to monitor the operation of an existing scheme. One version of the ESOLIN model is currently being adapted to monitor the operation of an existing complex hydro system. The first application will be on the hydroelectric network of Kerala State, in Southwestern India. The instrinsic operation of the model is the same as for a design study, the main difference being that, instead of simulating long term hydrological inflows, the model computes what the response of the system would be for a given demand sequence for each of the individual years of hydrologic record. The results are then evaluated and the optimum mode of operation is recommended, for the next time period, for a given accepted level of risk.

References

Rogers P.P., Fiering M.B., 1986 - "Use of Systems Analysis in Water Management" (Paper G.W. 0189). Water Resources Research - Volume 22 - Number 9.

Wiebe P.A., Nguyen N.T., 1979 - "A Practical Optimization Technique for Hydroelectric Development Projects". Canadian Society for Civil Engineering Fourth Hydrotechnical Conference. Vancouver, B.C.

Yeh W.W.G., 1985 - "Reservoir Management and Operations Models: a State-of-the-art Review". Water Resources Research. Volume 21 - Number 12.

Overview--Committee Report
on Dam Safety Guidelines

Donald W. Newton, Fellow ASCE*

Abstract

In November 1984 the Surface Water Hydrology Committee appointed a
task committee with the assignment to suggest standards for the
hydrologic safety design of dams. A draft of the committee report
is currently receiving peer review. A final report is expected to
be published in late 1987 or early 1988. This paper provides an
overview of the committee report and its recommendations.

The committee report recommends procedures for selecting the
safety design flood for a dam assuming that flood and dam failure
probabilities have been defined. The report reviews the evolution
of design flood selection procedures, recommends selection
procedures, and discusses a number of technical issues including
probable maximum determinations, defining flood probabilities of
extreme events, dambreak flood routing, future projections,
evaluating the economic and social consequences of failure, risk
analysis, legal liability for dam failures, and computing
indemnification costs. The recommended procedures utilize a
risk-based analysis which compares the likelihood of dam failure
to the social, economic, and environmental consequences directly
attributable to dam failure. The financial analysis proposed for
incorporating into the design decision those failure consequences
which can be quantified in monetary terms is based upon the
estimated cost to fully indemnify possible victims for their
financial loss. A separate assessment is used to include into the
decision the social and environmental impacts not measurable in
monetary terms.

The current national interest in dam safety; the identification of
some 2,900 unsafe, non-Federal dams; the fact that inadequate
spillway capacity accounts for the deficiency in some 2,350 of
these dams; and the considerable cost to upgrade the deficient

---

*Supervisor, Hydrology Section, Tennessee Valley Authority,
 200 Liberty Building, Knoxville, Tennessee 37902

dams have focused attention on the selection of the safety design flood. Because ownership of unsafe dams is widespread including Federal, state, and local government and private individuals or groups, the selection of design floods is made by engineers in private practice as well as those in state, Federal, and local government and the state and Federal regulatory agencies. The diversity of interest and lack of agreement about selection of a safety design flood afforded the ASCE a unique opportunity to serve the profession by bringing together interested engineers and related professionals from all groups to address the controversial aspects of safety design flood selection.

In November 1984 the Surface Water Hydrology Committee formed a Task Committee on Spillway Design Flood Selection. The committee's assignment was to develop standards for the hydrologic safety design of dams which would include consideration of all impacts of failure--economic, social, and environmental--and which could be readily implemented and consistently applied nationwide. The committee was composed of engineers from state and Federal regulatory agencies, Federal and private practice and education, a lawyer, an engineer-economist, and an insurance executive.

There are two major aspects to the hydrologic safety design of a dam. One is the evaluation of the flood potential at the dam and the resulting likelihood of dam failure and the other is the selection of the safety design flood. Each is a critical but separate step. The efforts of the task committee were focused on selection of the safety design flood. The committee has used the term "safety design flood" throughout its report. It concluded that the term "spillway design flood" was misleading because a safety design includes consideration of flood storage as well as spillway capacity. Further, it clearly separates the safety design consideration from the design to achieve project purposes such as flood control.

The report starts with a review of the evolution of design practice and philosophy. It is instructive in that the philosophy and considerations involved in safety design have remained remarkably consistent. What has changed is the engineers' ability to define the flood potential at a site and consequently the likelihood of dam failure and to evaluate the economic and social consequences directly attributable to dam failure. The consequences have changed as dams have become larger and the extent of development which could be affected by their failure has increased. Current dissatisfaction with dam safety standards stems both from the arbitrary nature of the current safety design standard and the perception that current standards are too conservative. The concern, particularly for designing modifications to existing dams, is that the current standard represents a dedication to public interest to the exclusion of economic considerations.

The procedures proposed by the task force are based upon a
thorough identification of the consequences of failure for present
and anticipated future conditions and their quantification to the
extent necessary to make a decision. The loss of a dam affects
two different groups: (1) those who receive benefits from dam
operation such as water supply for irrigation or domestic use,
flood control, navigation, power, and recreation and (2) those who
are damaged by the dam failure floodwave. These two groups are
not mutually exclusive. While the potential hazards and costs may
vary significantly between the impacted parties, each has a
critical stake in the safety of the structure. Thus, the
recommended dam safety criteria are based upon consideration of
the total economic, social, and environmental impact of failure
including those that occur downstream of the dam and the loss of
the dam, the reservoir, and their benefits.

The downstream consequences attributable to dam failure are to be
determined based upon the extent of downstream flooding with and
without dam failure. Current technology permits defining the
differences downstream between the flood profiles and flooded
areas with and without dam failure. Consequently, this is a
prerequisite for making the safety design decision. The paper
presented by Arthur Miller in this series describes the committee
recommendations regarding the identification and quantification of
the social, economic, and environmental impacts attributable to
dam failure.

Once the consequences of failure have been identified, an
important practical question is to what extent they need to be
quantified to reach a sound conclusion. To simplify the design
selection process, three categories of dams are identified
depending upon the extent and nature of the damage associated with
failure. Category 1 is for those dams where the consequences of
failure--social including loss of life, environmental, and
economic--are clearly so extensive that use of the PMF as the
safety design flood is appropriate. Category 3 is for the smaller
dams where the cost of construction is relatively small and the
failure damage is low and confined to the owner. In this
situation selection of a safety design can be based upon simple
rules of thumb considering dam height and reservoir volume
although a risk-based analysis might prove prudent. Category 2 is
for dams where the social, economic, and environmental failure
consequences are not large enough to categorically require use of
the PMF or small enough to permit use of simple rules of thumb to
select a design flood. Dams in Category 2 will require a more
thorough quantification of the impacts of failure to select an
appropriate design flood. A fourth situation is identified in
which the consequences of failure are so severe that they dictate
against building or maintaining a dam.

The consequences suggested to identify when a dam falls into
Category 1 include situations where (1) the number of people
exposed to flooding by dam failure is unacceptable (guidelines are

suggested), (2) there is a potential for catastrophic economic loss, (3) loss of some unique social or environmental resources is anticipated which are not replaceable by money, or (4) the political consequences are considered unacceptable.

The selection of an appropriate safety design flood for a Category 2 dam is to be based on a comparison of the consequences and likelihood of dam failure to the cost of protecting against failure. The safety design flood for these dams is the PMF unless it can be shown that a smaller safety design flood results in avoided costs which justify the increased risk of failure. In such cases, the dam may be designed for the appropriate flood between the 100-year and the PMF.

The comparisons to be made in the selection of the safety design flood for a Category 2 dam include both a financial analysis of the dollar-denominated failure costs and an evaluation of the cost to avoid the nondollar-denominated failure costs. The financial analysis proposed differs significantly from the typical benefit-cost analysis in that the expected value of failure consequences is replaced by the estimated cost of fully indemnifying possible victims against financial loss. Design decisions based solely on expected values can impose significant risk on third parties, who may well evaluate those risks in a completely different way. In order that the dam owner account for costs to parties who do not benefit from a less expensive design, it is proposed that the cost of fully indemnifying those parties from any financial loss be included in the economic evaluation as the cost of failure consequences incurred by others. The paper presented by John Boland in this series describes the computation and use of indemnification costs proposed by the committee for use in reaching the dam safety decision.

The indemnification cost incorporates into the safety design decision only those consequences which can be reduced to monetary terms. But the nonmonetary consequences which include the social and environmental impacts of failure are crucial to any final decision. These decisions involve complex judgmental issues relating to moral values, social concerns, and quality of life considerations. The nondollar-denominated failure consequences are displayed for specific inclusion in the safety design decision based on the cost to avoid these nonmonetary consequences. In the actual event of a dam failure it is likely that lawsuits will be filled on a legal basis and will be found for compensating victims for the loss of life and personal injury. Such lawsuits and the cost of providing insurance to cover such losses, should it become available, would provide one means to put a dollar value on the social and environmental impact of failure. However, the impacts are believed more pervasive than could be measured by such costs. Thus, there would remain a judgment decision about the additional cost appropriate to incur to avoid these nonmonetary consequences.

The committee report includes two examples showing application of
the proposed procedures and providing tables for displaying the
results of the analysis for decisionmaking. The paper presented
by Joe Haugh in this series outlines the procedures recommended by
the committee for selecting a safety design flood and provides
examples.

The decisionmaker faces a number of important technical issues
when developing the information necessary to make a safety design
decision. A number of these are discussed in the report but full
treatment of these issues is beyond the scope of this report.

A determination of the probable maximum flood (PMF) is an
essential step to selecting a safety design flood for a dam as it
defines a practical upper limit to flooding at a site considering
both climatic and watershed variables. Current practice for
defining the PMF varies widely. The task committee recognizes
that an accurate and consistent method to determine both the PMF
and probable maximum headwater levels, is essential to developing
a nationally consistent procedure for the safety design of a dam.

The proposed procedures postulate that it is possible to define
flood probabilities throughout the full range in flood potential
with sufficient accuracy to permit a dam safety design decision.
The task committee recognizes that estimates of flood probability
throughout the full range of flood magnitudes will probably not be
satisfactory from a purely statistical standpoint. Currently
available procedures, however, permit approximations with
sufficient accuracy for making the safety design decision.

In order to define the economic and social consequences directly
attributable to dam failure, it is necessary to determine if a dam
will breach and, if so, the duration and size of breach and the
resulting downstream discharge and stage hydrographs. Current
techniques are outlined which permit a determination adequate for
the dam safety design.

Selecting a safety design flood for a dam requires an evaluation
of the consequences of failure at the time of failure which
requires a projection of land uses and project benefits over the
life of the project. Experience with such projections has been
that they are often wrong. Adopting land-use controls or
acquiring easements to keep land use compatible with design as an
alternative to depending upon projections are discussed.

The dam safety decision requires identifying, quantifying, and
evaluating a mix of economic, social, and environmental impacts.
A number of the important issues which are critical to the dam
safety decision are briefly discussed. The recommended approach
to reaching the dam safety decision is a risk-based analysis.
There is currently much controversy about risk analysis. A brief
discussion of risk analysis is included for the purpose of
preventing misunderstandings about the perceived role of risk
analysis in engineering decisions.

In today's litigious society it is safe to assume that in the case
of a catastrophic dam failure, extensive litigation will ensue.
Lawsuits would likely be filed against everyone remotely involved
including those owning, operating, designing, constructing, and
inspecting the dam. In assessing the potential liability for dam
failure it is important to recognize that the legal standards will
vary from state to state. Even so, should dam failure result in
loss of life, personal injury, or substantial property damage, it
is fairly certain that most jurisdictions will find a legal basis
for compensating victims. The report provides an outline of legal
liability issues that arise from dam failures, a critical factor
in today's professional and public decisionmaking.

The committee recommends a procedure for determining the
appropriate safety design flood for a dam based upon a risk-based
analysis which includes as a major element the cost to indemnify
victims should the dam fail. To be fully effective where a safety
design less than the PMF is adopted, this approach must become
more than a calculus used by the engineer, regulator, or administrator to select a safety design flood. It requires that there
be some regulatory mechanism, state or Federal, to make sure that
the dam owner maintains the capability throughout the project life
for fully compensating all victims should the dam fail.

Criteria for Spillway Design Floods
Arthur C. Miller, Member ASCE[*]

Abstract

The procedures outlined in this paper only address the hydrologic safety criteria for spillway design floods. The recommended procedures are based on a careful identification of the social, economic, and environmental consequences of failure and the necessary quantification to make a decision. The computations and procedures utilized to evaluate alternative designs should be carried only to an extent to facilitate decisions. The decision process includes the economic and nondollar-denominated consequences of failure throughout the life of the project.

The analysis required to select an appropriate safety design flood includes the identification and quantification of all impacts attributable to dam failure to the extent necessary to make a decision. Damages attributable to a dam failure include those that occur downstream of the dam, the loss of the dam, the reservoir, and its benefits. The analysis would always include a dam breach analysis and the flood routing necessary to define the depth and extent of downstream flooding. It would also include the identification of the anticipated consequences of failure under both present and future conditions.

The consequences attributable to a dam failure are to be compared based on the extent of downstream flooding with and without dam failure. This requires that a dam breach analysis be performed, which would include routing the flood hydrograph downstream, to define the water surface profiles and flooded areas for a given discharge with and without dam failure. The computed profiles should extend from the dam downstream to a point where there is no significant difference in adverse flooding conditions with and without failure. Figure 1 illustrates the failure impact zone.

One of the decisions required to estimate flood damages is to determine the depth and flow velocity at which flooding becomes a threat to life and structural safety. For the purposes of determining loss of life it is recommended that the critical flood plain be defined as any area where the product of flood depth in ft (m) and velocity in fps (m/s) exceeds 7 f^{2}ps (0.65 m^2/s) or the water depth exceeds 3 ft (0.91 m). When these values are exceeded loss of life would be considered certain. For structural safety it is

---

[*]Professor of Civil Engineering, The Pennsylvania State University, University Park, PA 16802

recommended that the critical flood plain be defined as any area where the product of flood depth in ft (m) and velocity in fps exceeds 20 f^2ps (1.86 m^2/s) or the water depth exceeds 12 ft (3.66 m). When these values are exceeded complete property destruction would be considered certain. Figure 2 is a cross section on which the critical flood plains have been diagrammed.

Selection of the safety design flood for a dam is based on the anticipated consequences of failure due to hydrologic causes. The consequences of dam failure may range from minor damages affecting only the owner of the dam to a major catastrophe affecting a large segment of society. The effort required to select a design flood will vary depending upon the extent and magnitude of failure consequences. A reconnaissance-level appraisal of the major social and economic consequences of failure may be adequate for the selection process. However, where the decision is less obvious a more thorough and detailed evaluation of failure consequences may be required.

To simplify the design flood selection process, three categories of dams are identified depending on the extent and nature of the damage associated with failure. Category 1 is for dams whose failure is likely to cause significant loss of life or where the other social and economic consequences of failure are so extensive that use of the PMF as the safety design flood is prudent. Category 2 is for dams where the social and economic consequence of failure are not large enough to categorically require use of the probable maximum flood as the design flood. For these dams a more complete analysis is necessary to select the safety design flood. Category 3 is for smaller dams where the cost of construction is relatively small and the failure damage is low and confined to the owner.

A fourth situation can occur in which the consequences of failure are so severe that they dictate against building or maintaining a dam. A dam should not be built; or for existing dams, the dam should be removed if the consequences of failure are truly "unacceptable". Such a decision should be based on the fact that dams can fail from a number of causes including structural and seismic, as well as hydrologic. Figure 3 indicates the safety design decision sequences.

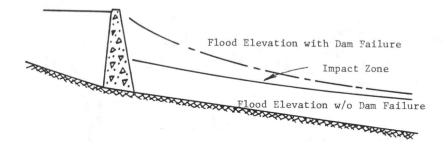

Figure 1   Definition of Failure Zone

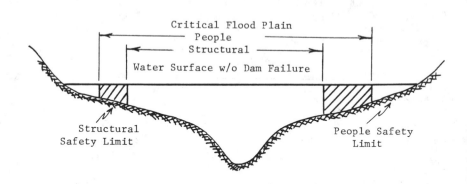

Figure 2   Definition of Critical Flood Plain

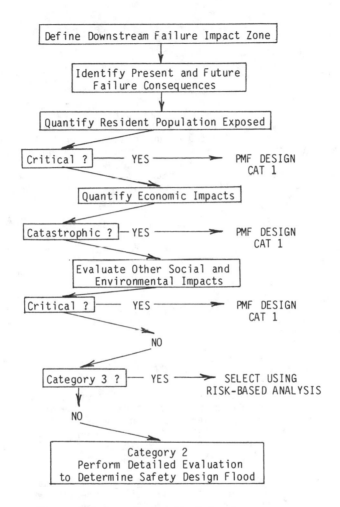

Figure 3 Safety Design Decision Sequence

Category 1 - Dams where the consequence of failure require a PMF design. This category includes dams where the consequences of failure are so large and adverse that the utmost conservatism and best engineering design efforts are required to make the dam safe against hydrologic failure. The design basis flood for such dams should be the PMF.

Category 2 - Intermediate class of dams. The selection of the appropriate design flood for a dam in this category should be based on a comparison of the consequences and likelihood of dam failure to the costs of protecting against failure. The prescribed safety design flood for these dams is the PMF unless it can be shown utilizing risk analysis that a smaller safety design flood is

justified. In such a case, the dam may be designed for the
appropriate flood between the 100-year and the PMF.
        The economic analysis proposed to be incorporated into the
safety design flood decision process differs significantly from the
economic analysis currently used in practice. Present day economic
analysis uses estimates of average annual flood damages to assess
alternate schemes. In the proposed procedure, average annual flood
damages are replaced by the estimated cost to fully indemnify
possible victims against financial loss. The economic evaluation is
broken into two parts: (1) those consequences incurred by persons and
organizations other than the dam owner and (2) the dam owner. The
indemnification cost is similar but not equal to insurance
costs. Insurance costs include administrative costs and profits as
well as the cost to settle lawsuits covering loss of life and other
social damages resulting from a dam failure.

        The indemnification cost incorporates into the safety design
decision only those consequences which can be reduced to monetary
terms. But nonmonetary consequences are crucial to any final
decision. These nonmonetary consequences include the many social,
environmental, and political impacts that would result from a dam
failure. The characteristics of these impacts vary from site to site
and require a separate assessment. In the proposed procedure the
nonmonetary failure costs are displayed for specific inclusion in the
safety design decision based on the cost to avoid these nonmonetary
consequences.

        The computational steps to quantify and apply the criteria for
comparing alternate designs are as follows:

  1. Select trial alternate hydrologic safety design discharges
      between the PMF and 100-year flows. These should include
      the PMF and sufficient additional alternatives to identify
      the appropriate safety design. For each design discharge,
      select the most economical design (spillway, flood storage,
      or other strategies) to accommodate the the flow.

  2. For each alternative, define the annual probability of dam
      failure and the failure probability over the project life.
      Define confidence limits for the probability estimates for
      use in a sensitivity analysis. Determination of dam failure
      from hydrologic causes requires estimation of both flood
      probabilities and, for a given flood, the probability of
      failure. A simplifying assumption would be to assume a dam
      fails when overtopped. The failure probability of a dam
      designed to safely pass the PMF, should a PMF occur, is
      assumed to be zero.

  3. Estimate the present value of the project cost and the
      equivalent annual operation and maintenance costs for each
      alternative.

4. Estimate the failure consequences for each alternative, both those that can be measured in monetary terms and those that cannot. No failure consequences are shown for the PMF design as the dam is assumed not to fail should the PMF occur.

5. Determine the indemnification cost for the economic consequences incurred by the owner and by others impacted by failure for each alternative.

6. Summarize the results of the analysis in a format to facilitate decision making.

The computations to develop the information necessary to select a safety design flood require the determination of a number of factors for which there is no exact value. These include the determination of the consequences of failure under both present and future conditions and the estimates of flood frequency and dam failure frequency. It includes assumptions which affect the economic analysis such as project life, discount rates, and cost of money. In the analysis, best estimates should be made and a sensitivity analysis performed to identify those factors where variations are critical to the decision. The final decision should be based upon conservative values for each factor affecting the decision.

Category 3 - Dams requiring minimum analysis. Dams in this category include dams where the failure is not expected to result in loss of life and no temporary or permanent residents are exposed under present or anticipated future conditions. There are no significant economic losses attributable to the failure of the dam incurred by persons or entities other than the owner, under present or anticipated future conditions. There are no significant or long-lasting environmental and social impacts from the loss of the dam under present or anticipated future conditions.

Indemnification Costs in the Dam Safety Decision

John J. Boland*

Introduction

The ASCE Task Committee on Spillway Design Floods has proposed safety criteria for three categories of dams. Category 1 dams are those with anticipated failure consequences so large and widely felt as to dictate, without further analysis, the safest design within the range of current practice. Category 1 dam spillways are to be capable of passing, without failure, the largest flow anticipated from a Probable Maximum Flood (PMF). At the other extreme, category 3 dams, where the likely consequences of failure are moderate to small and generally confined to impacts on the dam owner, are addressed by standard criteria similar to those used by the US Soil Conservation Service.

Between these widely different cases, the Committee has defined as category 2 all those dams with possible failure consequences too large or widespread to qualify as category 3, but not so severe as to confirm a PMF design regardless of cost (category 1). Criteria for category 2 dams require spillway capacity equal to the PMF unless a smaller design can be specifically justified. Justification takes account of the monetary impacts, whether sustained by dam owners, project beneficiaries, or downstream residents and activities, as well as non-monetary impacts including social, environmental, and other consequences.

In analyzing the monetary impacts of possible dam failure, problems arise when uncertain failure consequences, primarily affecting downstream populations and activities, are compared to the essentially certain avoided costs of dam construction, which accrue to the benefit of the dam owner. This paper discusses an innovative suggestion for treatment of this comparison: the use of indemnification costs.

Conventional Economic Analysis

The monetary consequences of dam failure, or of the measures necessary to reduce the risk of dam failure, are ordinarily evaluated by benefit-cost analysis. Identification and measurement of benefits and costs requires definition of a reference case, and of clearly defined alternatives to that reference case. For safety analyses for category 2 dams, the proposed criteria establish the reference case as a PMF design; the alternatives are a set of spillway designs for less than the PMF.

*Professor, Dept. of Geography and Environmental Engineering, The Johns Hopkins University, Baltimore, MD 21218.

Benefits associated with adopting one of the alternatives, then, are the construction and maintenance costs avoided by choosing a spillway design basis smaller than the PMF. Where smaller spillway capacity leads to other costs or affects project benefits during normal operation (shoreline changes, modified operating rule, reduced flood protection, etc.), benefits should be adjusted to incorporate those factors. Since benefits may occur at different times in the life of the dam, they are reduced to present value (or to annualized value).

Costs attributable to smaller spillway capacity are those associated with dam failure. Typically, these are very-high-magnitude, very-low-probability costs. In order to utilize the benefit-cost framework, the costs that would result from a dam failure in a particular year are multiplied by the probability of the failure occurring; this gives the expected failure cost for that year. If the PMF is assumed to have an exceedance probability of zero, then the probability of failure (for hydrologic causes) of a PMF design and the related expected failure cost are also zero. The added failure cost of an alternative design is, therefore, the present value of all expected failure costs throughout the life of the dam.

Benefit-cost analysis, as applied to monetary measures only, compares the benefits of a smaller spillway (present value of avoided construction costs) to the costs (present value of expected future failure costs). Neglecting the possible role of non-monetary factors (social, environmental, and other consequences), a smaller dam could be built if the benefits (costs avoided) exceed the expected costs (of failure), without regard to the degree of risk involved, or the identity of those who may bear that risk.

The Role of Risk

It is easy to imagine cases where benefit-cost analysis, applied in this manner, would result in a less-than-PMF design. The flaw in this procedure is simply stated: dam owners can, in certain cases, avoid costs by transferring both costs and risks to downstream residents and economic activities. It is argued that efficiency criteria are still met, since the dam owner could, in principle, pay downstream riparians an annual amount equal to the expected failure consequences, thus compensating them for the risk of failure and still leaving the owner better off with the smaller design. It is also argued that, where the dam owner is a public agency, benefits of smaller designs accrue to society as a whole, including potential victims, and that net benefits have been maximized.

Nevertheless, many within the economics and engineering professions remain profoundly uncomfortable with a procedure that trades off relatively certain benefits accruing to a dam owner against highly uncertain costs experienced by others. Failure consequences, while perhaps socially manageable, are often individually catastrophic. The mere possibility of such an event, even at very low probability, imposes a risk on those exposed. Little is known about how various individuals and organizations perceive such risks, or how they might measure them on a monetary scale, if in fact they do.

The only way to reduce the role of risk in such calculations is to arrange for actual compensation of failure consequences, rather than their expectation. (It should be remembered that this discussion omits consideration of any consequences not measured in dollar terms.) Such compensation of monetary damages does not change the magnitude of failure consequences, it merely shifts the associated risk from the victims to the dam owner, along with any increased cost associated with bearing that risk.

Indemnification

It is proposed, therefore, that the benefit-cost test of monetary consequences verify that the dam owner can bear the increased expectation of failure, along with the associated risk, and still prefer to build a spillway with less-than-PMF capacity. This is accomplished by replacing the conventional measure of cost--the expected present value of failure consequences--with a new measure-- the cost of indemnifying victims against the possible consequences of failure (indemnification cost).

Indemnification cost varies according to the assumed means of accomplishing potential compensation. At least three possible variants can be identified.

o   Self-insurance. Where the dam owner's resources and institutional permanency permit, it may be possible to guarantee payment of any foreseeable claim from current funds, no matter when in the future that claim might occur. The most likely example of this case is the US government, provided it can be assumed that Congress would always provide for prompt and full compensation in the event of failure.

    Indemnification cost, under a self-insurance assumption, is the expected present value of the opportunity cost of the funds needed to compensate victims in event of failure. If failure consequences are relatively moderate in size (opportunity cost equals dollars required), indemnification cost may converge on the conventional economic measure of cost. Where the consequences are large, the opportunity cost must be increased to reflect benefits lost from displaced programs or the cost of raising replacement revenue. In the latter situation, indemnification cost exceeds conventional economic cost.

o   Insurance. Where it is possible to obtain adequate no-fault liability insurance, so that all victims of dam failure will be risklessly compensated for all monetary losses associated with dam failure, the existence of such insurance constitutes indemnification. The Committee believes that policies of this kind are not generally available today. They may be, at least to a limited extent, in the future.

    Indemnification cost is the present value of the future

insurance premiums needed to keep adequate coverage in force throughout the life of the dam. This value will be substantially in excess of the conventional economic measure of cost.

o Escrow account. In the absence of suitable insurance or the capacity to self-insure, a dam-owner can indemnify by placing in escrow funds sufficient to compensate all potential victims. This requires that the funds be placed in a segregated account, with re-investments confined to relatively liquid, risk-free securities. The balance would be maintained at a level equal to the maximum damages anticipated. In the event of failure, the funds would be available to compensate victims; if no failure occurs, the account balance would be returned to the dam owner at the end of the dam's life. The calculation of indemnification cost associated with an escrow account is discussed in the following sections; the result is an amount substantially larger than conventional economic measure of cost.

Cost of Escrow Accounts

The cost of holding a sum in escrow for n years. then using it at the end of year n to compensate victims of a dam failure, consists of the present value of the opportunity cost of the necessary sum of money, reduced by any income from temporary re-investment, plus the present value of the sum paid out. This can be expressed as:

$$C_n = B * \left[\frac{1}{i} * [1 - \frac{1}{(1+i)^n}] * (r-s) + \frac{1}{(1+i)^n} \right] \tag{1}$$

Where: C_n = present value of net cost of escrow account, assuming dam failure at the end of year n ($)

B = estimated amount of monetary damage, in event of failure ($, at base year prices)

i = real discount rate (%/yr)

r = opportunity cost of money in fund (e.g., nominal borrowing cost) (%/yr)

s = nominal return on temporary investment of fund balance (%/yr)

The only exception to this occurs when the life of the dam ends without a failure. The cost of holding the fund for the full life of the dam is:

$$C_{N+1} = B * \left[\frac{1}{i} * [1 - \frac{1}{(1+i)^{N+1}}] * (r-s) \right] \tag{2}$$

Where: N = number of years of planned life for dam

Since a dam is only expected to fail once, the probability of a failure in year n is the joint probability of no failure for n-1 years, followed by failure in year n. If the probability of failure in any given year is equal to and independent of the probability of failure in any other year, the result is:

$$P_n = P*[(1-P)^{n-1}] \qquad\qquad (3)$$

Where: P_n = probability of first failure in year n

         P   = probability of failure in any single year

The probability that no failure occurs in the planning period is:

$$P_{N+1} = (1-P)^{N+1} \qquad\qquad (4)$$

There are, therefore, N+1 failure possibilities: the dam might fail in any of the N years of its life, or it may not fail at all. The total present value cost of maintaining an escrow account that would indemnify victims of failure is determined by calculating the present value cost of the fund for each of N+1 failure possibilities, multiplying each by the probability of its occurrence, and taking the sum of the products, as follows:

$$TC = P_1*C_1 + P_2*C_2 + \ldots + P_N*C_N + P_{N+1}*C_{N+1} \qquad\qquad (5)$$

Where: TC = value of total net cost of maintaining escrow account

By substituting expressions (1), (2), (3), and (4) into (5), an expression for the present value cost of the escrow account, stated in terms of B, i, P, N, and (r-s) can be obtained. The incremental cost of money, (r-s), reflects the net cost of money to the dam owner, considering the opportunities for re-investment. A large, credit-worthy government or institution may be characterized by a low (r-s), perhaps on the order of 1.0 %/year, but most private sector firms may experience substantially higher incremental costs. An illustration of escrow account costs for various parameter values is shown as Table 1. It should be noted that the conventional expected present value of damages, expressed as a fraction of total damages, would be 0.100, 0.050, and 0.010 for failure probabilities of 0.010, 0.005, and 0.001, respectively.

Table 1. Indemnification Costs as a Fraction of Total Damage*

(r-s)	Probability of failure (P)		
	0.010	0.005	0.001
0.005	0.136	0.095	0.059
0.01	0.182	0.143	0.109
0.02	0.273	0.238	0.208
0.03	0.364	0.333	0.307
0.05	0.545	0.524	0.505

*Values in table are $\frac{TC}{B}$, where i = 0.10 and N = 100 years

Conclusions

Conventional benefit-cost analysis measures the cost of providing smaller spillway capacity as the expected present value of those potential dam failure consequences which can be measured in monetary terms. This fails to account for the risk borne by possible victims, who may or may not be compensated for their losses, if they occur. Dam owners who may not be financially able to compensate victims in the future stand to reap the benefits of lower-capacity designs without proper consideration of the costs or risks borne by others.

Some allowance for risk and third-party losses can be incorporated into spillway safety design analysis by replacing the expected present value of potential failure costs with the present value cost of indemnifying victims. Indemnification cost may be calculated in several ways, depending on the funding options available to the dam owner, but all owners, with the possible exception of the federal government, will face indemnification costs substantially in excess of conventional measures of economic costs.

EFFECTS OF DATA EXTENSION ON DROUGHT PROBABILITIES

Dennis R. Horn, M. ASCE[1]

ABSTRACT

A condensed-parameter disaggregation model was applied to the monthly streamflow records on two rivers in Idaho, to investigate the occurrence and probabilities of drought periods. Using a 40,000 year sequence of generated flows, the cumulative probability density functions of maximum negative run-sums and run-lengths were obtained and the return periods of historical negative runs were estimated. A multi-variate stochastic model was then applied to the original streamflow sequences, to extend the historical data based on the longer term records of nearby streamflow gages. From these extended records, new model parameters were determined and the generation process repeated with the revised parameters. A comparison of the results, based on the extended and unextended station records, indicated significant differences in the probabilities assigned to historical drought periods. Although the changes in station statistics resulting from data extension were relatively minor, they affected the truncation levels used to define the negative run periods, and these periods were highly sensitive to truncation level. It is concluded that data extension, using multi-variate modeling, provides a better estimate of station statistics and hence an improved definition of both the long-term stochastic properties of the time-series and the probabilities assigned to observed drought events.

INTRODUCTION

Hydrologists and engineers involved in the planning and design of surface water projects have always had to deal with the problem of hydrologic uncertainty. This problem arises from either a total lack of critical streamflow information at or near the proposed development site, or a streamflow record which is far too short to adequately characterize the hydrologic regime. Even in cases where a relatively long streamflow record exists, a design based purely on the extreme event (such as a critical drought or low-flow sequence) in that historical record may imply an unquantifiable risk level associated with design failure. The use of historical records as the sole basis for design also introduces an element of inconsistency into the design approach, since different projects will have very different

---

[1]Asst. Prof., Dept. of Civil Engineering, University of Idaho, Moscow, ID 83843.

assumed risk levels. This may result in under-or-over-sized projects with economics far from the desired optimum.

For these and other reasons, the use of stochastic modeling methods has become a commonly accepted analysis procedure in the design of water resources projects. These methods generally involve the generation of long sequences of streamflow data from which relationships between probability and storage requirements can be developed. An analysis of these sequences also permits the assignment of probabilities or return periods to observed historical droughts.

As part of a broader study (Horn and Dieziger-Kim, 1985) of the use of condensed parameter disaggregation modeling (Lane, 1983) to generate monthly streamflow data, the issue of data extension prior to determining model parameters was explored. This paper presents the results and conclusions derived, as they impact the assignment of return periods to historical droughts.

MODELING PROCEDURE

Two Idaho rivers (the Coeur d'Alene and the South Fork of the Boise) were selected as study streams. Both have fairly long periods of good quality data, and have been proposed as potential sites of future water projects. Based on a review of the station statistics, and a comparison of various annual flow generation models, a pure probabilistic model was selected as the preferred generation method for annual flows at both locations. (Neither station exhibited any skewness to the annual data, or any time-dependency in terms of a serial correlation structure.)

Model parameters (the mean, u, and standard deviation, σ, of the annual flows) were determined from the observed streamflows, and a 40,000 year sequence of flows was generated. These sequences were divided into sample sizes, N, of 25, 50, and 100 years, and each resulting sample was examined using the theory of runs.

The use of runs as an objective definition of droughts has been extensively studied. (Yevjevich, 1967; Millan and Yevjevich, 1971). For negative runs, all annual flows below a defined truncation level, X_o, are examined for their length and magnitude (cumulative negative sum). These truncation levels are usually expressed as a function of the quantile, q_o:

$$q_o = P(X \leq X_o) \tag{1}$$

where X = streamflow. Probabilities are derived from an approximation to the gamma distribution. By application of selected truncation levels to each sample, the maximum negative run-lengths (equivalent to drought duration) and run-sums (a measure of the water deficit) can be determined, and cumulative probability distributions plotted for the array of samples within the sequence.

For the two study streams, these distributions of maximum run-sums and run-lengths were then used to assess the probabilities and return periods, T, of the maximum historical droughts observed at each site. Using procedures described by Millan and Yevjevich (1971), the median run-lengths, Lm, for each sample length and truncation level were determined from the cumulative probability distributions, and plotted versus N. Similarly, the median standardized deficits, Dm (obtained by dividing the median deficit, Sm, by the standard deviation, were obtained and plotted. These values of Lm and Dm may be considered to be "representative" droughts for the sample sizes, and an interpolation of the plots yields an estimate of the return periods for observed events (lengths or deficits).

The results of these analyses are summarized in Table 1, which presents the station statistics and the estimated return periods for drought lengths and deficits for the observed maximum events at both stations.

Table 1:   Return Periods Based on Unextended Records

1. Coeur d'Alene River:
        Mean annual flow   = 54.81 m³/s
        Standard deviation = 15.43 m³/s
        Record length = n = 44 years (1940-1983)

q_0	Lm (years)	T (years)	Dm	T (years)
0.50	3	23	3.71	33
0.35	3	70	2.55	45

2. South Fork Boise River:
        Mean annual flow   = 22.84 m³/s
        Standard deviation = 7.29 m³/s
        Record length = n = 38 years (1946-1983)

q_0	Lm (years)	T (years)	Dm	T (years)
0.50	6	>100	4.01	41
0.35	3	70	1.80	18

EFFECTS OF DATA EXTENSION

To test the sensitivity of these results and the modeling process to relatively small changes in the station statistics, a multivariate model was used to extend the records at both sites based on nearby longer-term streamflow gages. The performance of three multivariate models was compared, and a monthly model proposed by Yevjevich (1975) was selected for data extension. This model is designed to preserve all of the monthly statistics, including the lag-one cross correlations and the lag-one serial correlations of the two records, and, for the available data, had a model correlation coefficient in excess of 0.97 at both sites.

The Coeur d'Alene data were extended to cover the period 1921-1939, providing a total record length of 63 years; and the South Fork Boise data were extended to cover 1912-1945, for a total length of 72 years. The effects of these extensions were seen in relatively minor changes in the station means and standard deviations, with no significant impact on the skew coefficient or serial correlations (Table 2). Therefore, in standardized form, the results of generating a 40,000 year sequence at both stations were identical to the prior results, with the same plots of Lm and Dm.

Table 2: Extended Data Statistics

River	Statistic	Value (m^3/s)	% Change From Original
Coeur d'Alene	u	53.99	-1.5
Coeur d'Alene	σ	15.92	+3.2
Boise	u	21.07	-7.7
Boise	σ	7.42	+1.8

However, since the truncation levels, X_o, are a function of both the mean and standard deviation, the use of new truncation levels defined by the extended data statistics had a significant impact on the estimation of return periods associated with historical events, especially for the South Fork Boise River. To examine these impacts, the following comparison concentrates on the values of Dm, which, as a deficit, is particularly relevant to many water resources problems.

The period of data extension at both sites included two regional drought events, 1929-1931 and 1939-1942. Because of these occurrences, and the variability of the flows during this period, data extension resulted in the reduction in the means and the increase in the standard deviations shown in Table 2. The combined effect of these changes was to reduce the truncation levels, X_o, associated with each corresponding value of q_o. With a lower value of X_o, fewer negative runs appear during the run analyses.

For the Coeur d'Alene River, the historical maximum deficit period at both levels of q_o was 1940-1942. When the new truncation levels are applied to the historical data period, this period remains the deficit of record, but with an increased probability of occurrence (Table 3). If, however, the entire extended plus historical record is treated as a quasi-historical record for this site, the maximum deficits are shifted into the earlier years. For $q_o = 0.50$, the deficit of record includes the 1939 low-flow year, and the return period of this four-year drought event increases to 63 years. With $q_o = 0.35$, the maximum deficit occurs in the period 1929-1931, with a return period in excess of 100 years.

Table 3:   Comparison of Deficit Return Periods

1. Coeur d'Alene River with extended data statistics:

Data Period	q_o	Dm	T (Years)	Dates of Occurrence
Historical	0.50	3.44	27	1940-1942
(n=44)	0.35	2.88	33	1940-1942
Extended	0.50	4.55	63	1939-1942
(n=63)	0.35	3.39	>100	1929-1931

2. South Fork, Boise River with extended data statistics:

Data Period	q_o	Dm	T (Years)	Dates of Occurrence
Historical	0.50	2.15	10	1959-1961
(n=38)	0.35	1.44	12	1977
Extended	0.50	3.31	25	1939-1942
(n=72)	0.35	2.06	24	1929-1931

On the South Fork of the Boise River, the historical deficit of record occurred during the 1959-1964 period for both levels of q_o. With the revised truncation levels applied to the historical data, the maximum deficit for q_o=0.50 still occurs in this period, but with a greatly reduced return period. For q_o=0.35, the maximum deficit becomes the 1977 drought event, which has an estimated return period of slightly more than 10 years.

Again, by using the entire 72-year record as a quasi-historical record, the maximum drought events are shifted to the earlier years (1929-1931, 1939-1942) with return periods at both truncation levels of approximately 25 years.

SUMMARY AND CONCLUSIONS

By extending the records at both sites to include a period when several critical regional drought events were experienced, the station statistics appear to more adequately represent the long-term annual streamflow characteristics. Although the net changes in these statistics are relatively minor, and do not affect the underlying stochastic process model, they have a significant impact on the assignment of return periods to the observed historical events.

In the case of the Coeur d'Alene River, the original run analyses at both truncation levels would seem to indicate that the 1940-1942 deficits are the critical drought events at this location, with return periods representative of the sample length (n=44). However, the inclusion of the extended data alters this perception, since the extended data record now covers the 1939 drought year, as well as the 1929-1931 event. This shifts the critical events to a new time frame, and suggests that these maximum deficits, as defined by their return periods, are quite severe.

For the South Fork of the Boise River, the historical critical drought period of 1959-1964 becomes a relatively minor drought event after performing the data extension. Again, the analysis of the extended data record results in shifting the maximum deficits to the 1939-1942 and 1929-1931 time frame, and concludes that at this location these events were not particularly significant.

As a guide to future studies of drought events, as defined by negative run analyses, the following general conclusions are offered:

1. The determination of negative run events is highly sensitive to the truncation levels, X_0, which are primarily a function of the values of u and σ for the annual flow series.
2. By the use of data extension at a site, the increased data length can provide a larger sample size for estimating u and σ, as well as the other stochastic model parameters.
3. The extended data statistics, reflecting added periods of high, low or average flows, can result in significant revisions in the estimates of return periods associated with observed drought events.
4. The critical drought periods may be shifted into the extended data portion of the record, reflecting more accurately the occurrence of regional drought events.
5. Given the sensitivity of the run-analyses to the truncation level, the assignment of return periods to critical drought events is at best an approximate estimate.

REFERENCES

1. Horn, D. R., Dieziger-Kim, D., "Analysis and Generation of Low-Flow Sequences for Idaho Streams, Using Disaggregation Modeling," Research Technical Completion Report, IWRRI, University of Idaho, Moscow, ID, Sept. 1985.

2. Lane, W. L., Applied Stochastic Techniques, Bureau of Reclamation, Denver, CO, 1983.

3. Millan, J., Yevjevich, V., "Probabilities of Observed Droughts," Hydrology Papers, No. 50, Colorado State University, Fort Collins, CO, June 1971.

4. Yevjevich, V., "An Objective Approach to Definitions and Investigations of Continental Hydrologic Droughts," Hydrology Papers, No. 23, Colorado State University, Fort Collins, Co, Aug. 1967.

5. Yevjevich, V., "Generation of Hydrologic Samples, Case Study of the Great Lakes," Hydrology Papers, No. 72, Colorado State University, Fort Collins, CO, May 1975.

The GEV Distribution in Drought Frequency Analysis

Jose A. Raynal-Villaseñor**
M. ASCE
Jose C. Douriet-Cardenas**

Abstract

Three decades have passed since Jenkinson (1955) found the general solution to the Stability Postulate, which is the condition that all the
extremes must meet, and after him that solution has been called the ge
neral extreme value (GEV) distribution.

The GEV distribution has been widely used in flood frequency analysis,
but rarely has been applied to drought frequency analysis. This is the
topic of the paper. Furthermore, estimation procedures to obtain its -
parameters are included in the text of the paper. The methods depicted
are: moments, maximum likelihood and probability weighted moments.
Finally, the results of application of the GEV distribution to a region in Northwestern Mexico are reported, too.

Introduction

Among the most common distributions used to perform drought frequency
analysis are: 3 parameter Log-Normal, Gumbel, Extreme Value type III
and Pearson types III and V, (Gumbel, 1958 and Matalas, 1963). The -
usual methods for parameter estimation of such distributions are the
well-known methos of moments and maximum likelihood.

Recently, the so-called GEV distribution has been applied to flood -
frequency analysis successfully, estimating its parameters by the methods of moments and maximum likelihood, mainly. Due to the property
that such distribution can represent extreme value distribution types
II and III directly, and as a limiting condition when the shape parameter goes to zero, extreme value (EV) distribution type I can also be
represented, which it makes the distribution a good candidate among -
the possible distributions to model extreme values.

The method of probability weighted moments has been proposed in the -
literature a few years ago (Greenwood et al, 1979), and due to the --
straightforward expressions that usually are produced for the estimators of the parameters of directly invertible probability distribution
functions and the unbiased condition of the estimators in this method,
constitutes a powerful tool for parameter estimation.

** Water Resources Program, DEPFI, Universidad Nacional Autonoma de
   Mexico, Cd. Universitaria, 04510 Mexico, D.F., Mexico.

General Extreme Value Distribution for the Minima

The distribution function for the GEV distribution for the maxima is, (NERC, 1975):

$$F(X) = \exp\left[- \left(1 - \frac{(X - X_0)}{\alpha} \beta \right)^{1/\beta} \right] \tag{1}$$

where X_0, α and β are the location, scale and shape parameters, respectively. Now, using the simmetry principle, (Gumbel, 1958):

$$F_{min}(X) = 1 - F_{max}(-X) \tag{2}$$

the corresponding GEV distribution for the minima can be obtained as:

$$F(X) = 1 - \exp\left[- \left(1 - \frac{(w - x)}{\alpha} \beta \right)^{1/\beta} \right] \tag{3}$$

where w, α and β are the location, scale and shape parameters, respectively. The probability density function of eq. (3) is:

$$F(x) = \frac{1}{\alpha} \exp\left[- \left(1 - \left(\frac{w - x}{\alpha}\right)\beta\right)^{1/\beta} \right] \left(1 - \left(\frac{w - x}{\alpha}\right)\beta\right)^{1/\beta - 1} \tag{4}$$

The probability distribution function contained in eq. (3) has two - branches:

EV type II distribution $\beta < 0$; $-\infty < x < w - \dfrac{\alpha}{\beta}$

EV type III distribution $\beta > 0$; $w - \dfrac{\alpha}{\beta} \leq x < \infty$

and taking the limit when β goes to zero, the EV type I distribution is obtained and this distribution is unlimited in both sides.

Estimation Procedures for the Parameters of the GEV distribution for the Minima

Method of Moments

Using the well-known method fo moments for the GEV distribution for - the minima, the following relationship can be obtained:

$$\gamma = (-1)^i \frac{\Gamma(1+3\beta) - 3\Gamma(1+2\beta) \Gamma(1+\beta) + 2 \Gamma^3 (1+\beta)}{[\Gamma(1+2\beta) - \Gamma^2(1+\beta)]^{3/2}} \tag{5}$$

where γ is the skewness coefficient, i=1 for $\beta < 0$ and i=2 for $\beta > 0$, and Γ (.) is the complete gamma function for argument (.).

If eq. (5) is inverted by polynomial regression, the following expressions provide a direct estimation of the shape parameter:

$\beta > 0$; $-1.1396 < \hat{\gamma} < 11.35$

$$\hat{\beta} = 0.2794 \quad + 0.3335\ \hat{\gamma} + 0.0403\ \hat{\gamma}^2 - 0.0244\ \hat{\gamma}^3 + 0.0037\ \hat{\gamma}^4 \qquad (6)$$
$$-2.6316 \times 10^{-4}\ \hat{\gamma}^5 + 7.0135 \times 10^{-6}\ \hat{\gamma}^6$$

$\beta < 0$; $-19.04 < \hat{\gamma} \leq -1.1396$

$$\hat{\beta} = 0.2466 + 0.2866\ \hat{\gamma} + 0.0724\ \hat{\gamma}^2 + 0.0101\ \hat{\gamma}^3 + 0.0008\ \hat{\gamma}^4 \qquad (7)$$
$$+ 3.6385 \times 10^{-5}\ \hat{\gamma}^5 + 8.6489 \times 10^{-7}\ \hat{\gamma}^6 + 8.1446 \times 10^{-9}\ \hat{\gamma}^7$$

and the location and scale parameters are estimated as follows:

$$\hat{w} = \bar{x} + \frac{\hat{\alpha}}{\hat{\beta}}\ [\ 1 - \Gamma(1+\hat{\beta})\ ] \qquad (8)$$

$$\hat{\alpha} = \frac{\hat{\sigma}\ \hat{\beta}}{[\Gamma(1+2\hat{\beta}) - \Gamma^2(1+\hat{\beta})]^{1/2}} \qquad (9)$$

where $\hat{X}$ and $\hat{\sigma}$ are the estimated mean and standard deviation of the data.

Method of Maximum Likelihood

The likelihood function for the GEV distribution for the minima is:

$$L(x_i;\ w,\ \alpha,\ \beta) = \frac{1}{\alpha^N}\quad \exp\ [-\sum_{i=1}^{N} (1-(\frac{w-x_i}{\alpha})\beta)^{1/\beta}\ ] \prod_{i=1}^{N} (1-(\frac{w-x_i}{\alpha})\beta)^{1/\beta\ -1}$$

$$(10)$$

and the corresponding log-likelihood function is:

$$LL(X_i, w, \alpha,\ \beta) = - N\ Ln\alpha - \sum_{i=1}^{N} (1-\frac{(w-X)}{\alpha}\beta)^{1/\beta} + (\frac{1}{\beta} - 1) \sum_{i=1}^{N} Ln(1-\frac{(w-X_i)}{\alpha}\beta)$$

$$(11)$$

Now, using the approximation to maximum likelihood estimates provided by the method of scoring, the following iterative scheme is used to - obtain the maximum likelihood estimates:

$$w_{i+1} = w_i + \varepsilon_w \qquad (12)$$
$$\alpha_{i+1} = \alpha_i + \delta_\alpha \qquad (13)$$
$$\beta_{i+1} = \beta_i + \delta_\beta \qquad (14)$$

where δ_w, δ_α and δ_β are the deviations from the true maximum likelihood estimates at stage i. They are computed as:

$$\delta_{\dot{w}} = \frac{\alpha}{N}\ [\ bQ - \frac{h}{\beta}\ (P + Q) - \frac{f}{\beta}\ (R - (\frac{P+Q}{\beta}))\ ] \qquad (15)$$

$$\delta\alpha = \frac{\alpha}{N} \left(h \ Q - \frac{a}{\beta} (P + Q) - \frac{g}{\beta} \left(R - \frac{(P + Q)}{\beta} \right) \right) \tag{16}$$

$$\delta\beta = \frac{1}{N} \left[f \ Q - \frac{g}{\beta} (P + Q) - \frac{a}{\beta} \left(R - \frac{(P + Q)}{\beta} \right) \right] \tag{17}$$

where N is the sample size, a,b,c,f,g,h are teh coefficients of the -
variance-covariance of the parameters of the GEV distribution for the
minima, see table 1.

Table 1. Coefficients of the variance-covariance matrix of the parameters of the GEV distribution for the minima.

	a	b	c	f	g	h
0.10	0.2043	0.5109	-0.7818	0.9519	-0.7132	0.3231
0.20	0.1714	0.7273	-0.3862	0.5063	-0.4679	0.3531
0.30	0.1846	0.8461	-0.1998	0.2667	-0.2944	0.3952
0.40	0.2398	0.9298	-0.1147	0.1425	-0.1638	0.4722
0.50	0.3185	1.0109	-0.0731	0.0628	-0.1034	0.5674
0.60	0.4214	1.1004	-0.0484	0.0085	-0.0798	0.6810
0.70	0.2675	0.5012	-0.0299	0.0310	-0.0317	0.3661
0.80	0.0794	0.1575	-0.0577	-0.0094	-0.0419	0.1118

and P, Q and R are:

$$P = N - \sum_{i=1}^{N} e^{-y_i} \tag{18}$$

$$Q = \sum_{i=1}^{N} e^{(\beta-1)y_i} - (1-\beta) \sum_{i=1}^{N} e^{\beta y_i} \tag{19}$$

$$R = N - \sum_{i=1}^{N} y_i + \sum_{i=1}^{N} y_i \ e^{-y_i} \tag{20}$$

where:

$$y_i = -\frac{1}{\beta} \ Ln \ (1 - (\frac{w-X_i}{\alpha})\beta) \tag{21}$$

and the convergence criteria are:

$$(\frac{\partial LL}{\partial W})_i = \frac{Q}{\alpha} \approx 0 \tag{22}$$

$$(\frac{\partial LL}{\partial \alpha})_i = -(\frac{P + Q}{\beta}) \approx 0 \tag{23}$$

$$(\frac{\partial LL}{\partial \beta})_i = -\frac{1}{\beta}(R - (\frac{P + Q}{\beta})) \approx 0 \tag{24}$$

Method of Probability Weigthed Moments

The estimators of the parameters for the GEV distribution for the minima are, (Raynal-Villaseñor, 1987):

$$\hat{W} = M_0 - [1 - \frac{1}{\Gamma(1+\beta)}] \frac{[M_0 - 2M_1]^2}{[M_0 + 4M_3 - 4M_1]} \tag{25}$$

$$\hat{\alpha} = \frac{\hat{\beta} [M_0 - 2M_1]^2}{\Gamma(1 + \beta)[M_0 + 4M_3 - 4M_1]} \tag{26}$$

$$\hat{\beta} = Ln \left[\frac{M_0 - 2M_1}{2M_1 - 4M_3}\right] / Ln\, 2 \tag{27}$$

Examples of Application

The proposed GEV distribution for drought frequency analysis has been
applied to the data of the gauging stations contained in table 2, and
the computed parameters for the three methods depicted in the article
are contained in there.

Table 2. Parameter estimates for the methods of moments, maximum likelihood and probability weighted moments:

Station	Method								
	Moments			Maximum Likelihood			Probability Weighted Mom		
	$\hat{W}$	$\hat{\alpha}$	$\hat{\beta}$	$\hat{W}$	$\hat{\alpha}$	$\hat{\beta}$	$\hat{W}$	$\hat{\alpha}$	$\hat{\beta}$
La Huerta	1.74	0.78	0.42	1.74	0.74	0.40	1.78	0.81	0.49
Ixpalino	1.11	0.70	0.91	1.11	0.64	0.94	1.20	0.79	0.98
Huites	4.08	2.22	1.04	4.10	2.24	1.05	4.12	2.65	1.30

Conclusion

The GEV distribution for the minima has been presented and methods to
estimate its parameters have been provided. Due to its flexibility to
be adjusted to actual extreme value data its usage is recommended.

Acknowledgement

The authors wish to express their deepest gratitude to the Engineering
Graduate Studies Division (DEPFI), Universidad Nacional Autonoma de Mexico, for the support provided in the realization of this paper.

References

-Gumbel, E.J. (1958). "Statistics fo Extremes", Columbia University Press. New York.
-Greenwood, J.A. et al (1979). "Probability weighted moments: definition and relation to parameters of several distributions expressable in inverse form". J. Wat. Res. Res., 15 (5), 1049-1054.
-Jenkinson, A.F. (1955). "The frequency distribution of the annual maximum(minimum) values of meteorological elements". Quart. J. Roy. Met. - Soc., 81, 158-171.
-Natural Environment Research Council (1975). "Flood studies report". Vol. I: Hydrological studies. Whitefriars Press Ltd. London.
-Matalas, N.C. (1963). "Probability distribution of low flows". Statical studies in hydrology, professional paper 434-A, A1-A27.
-Raynal-Villaseñor, J.A. (1987) "Probability weighted moments estimators for the general extreme value distribution (maxima and minima)". Hydrological Science and Technology: Short Papers J. Accepted for publication.

KEYWORDS: droughts, extreme values, frequency analysis, maximum, likelihood, minima, moments, probability density function, probability distribution function, probability weighted moments.

ESTIMATING LOW-FLOW FREQUENCIES OF UNGAGED STREAMS IN NEW ENGLAND

S. William Wandle, Jr., M.ASCE*

Most locations where low-flow-frequency estimates are required do not coincide with locations of streamflow-gaging sites where flow statistics can be determined from the streamflow record. Estimates of these statistics are needed to design water supplies, protect aquatic habitat, allocate in-stream waste loading, and regulate ground-water development. Previous studies in Connecticut and New York demonstrated the importance of stratified drift, till, and bedrock; lake and swamp area; and mean runoff in explaining low-flow variations. Regional low-flow equations were developed from regression analysis of these and other basin characteristics and can be used to estimate low flows for ungaged sites.

Equations to estimate low flows were developed using multiple-regression analysis with a sample of 48 river basins, which were selected from the U. S. Geological Survey's network of gaged river basins in Massachusetts, New Hampshire, Rhode Island, Vermont, and southwestern Maine. This region of New England was chosen because of its diverse terrain and the availability of long-term streamflow data, surficial geologic maps and other geologic information. Independent variables used in the regression analysis included basin elevation, basin location, main-channel length and slope, precipitation, mean annual flow; and areas underlain by till and bedrock, alluvium, and coarse and fine-grained stratified-drift deposits; and the area of swamps and lakes underlain by each of these geologic units. Areas of the various geologic units were determined from geologic quadrangle maps, unpublished maps in the Geological Survey files, and reconnaissance maps of surficial geology prepared for this study. Low-flow characteristics are represented by the 7Q2 and 7Q10 (the annual minimum 7-day mean low flow at the 2- and 10-year recurrence intervals). These statistics for each of the 48 basins were determined from a low-flow frequency analysis of streamflow records for 1942-71, or from a graphical or mathematical relationship if the record did not cover this 30-year period. Low flows at short-term sites were adjusted to the 1942-71 period using the graphical relationship between annual 7-day mean low flows at an index and short-term sites. Estimators for the mean and variance of the 7-day low flows at the index and short-term sites were used for two stations where discharge measurements of base flow were available and for two sites where the graphical technique was unsatisfactory.

Equations are provided that yield estimates of the 7Q2 and 7Q10 low flows for ungaged, natural-flow streams. Mean basin elevation; areas underlain by till and bedrock, coarse materials, fine deposits; and the area of lakes, swamps, and alluvium were significant in explaining regional variations in low flow where the relief between the highest and lowest points in a basin exceeds 300 meters.

*Hydrologist, U.S. Geological Survey, 150 Causeway Street, Suite 1001, Boston, MA 02114-1384.

DERIVING STREAMFLOW DISTRIBUTION PROBABILITIES

Dapei Wang[1] and Barry J. Adams[2], M. ASCE

ABSTRACT

Streamflow distribution probabilities are often involved in the stochastic analysis of water resources systems. Limited by data availability, certain probability terms can hardly be abstracted directly from historical records with adequate precision to facilitate meaningful system analysis. This paper presents some methods for deriving the distribution probabilities of seasonal streamflows from historical records with moderate sample size.

INTRODUCTION

In the stochastic analyses of water resources systems, natural streamflows are usually regarded as periodic stochastic processes to recognize the hydrologic uncertainty and seasonality [Butcher, 1971; Loucks et al, 1981; Yakowitz, 1982; Yeh, 1985]. Streamflows, which exhibit annual cycles with seasonal (or monthly) variations, are often represented by random variates and described by discrete probability distributions.

Streamflows are thought to retain the same statistical characteristics as revealed by historical records. Historical streamflow records, which are usually of moderate lengths, provide most information for deriving streamflow distribution probabilities. Theoretically, these probabilities could be abstracted directly from historical records. However, when precise descriptions of distribution probabilities are required, the streamflow amounts should be discretized with reasonably small intervals. In such cases, the direct abstraction may not be practically applicable due to the limited data availability, i.e., the limited lengths of historical records.

Markovian transition probabilities of seasonal streamflows and the joint distribution probabilities of multi-site seasonal streamflows are often involved in the system analysis. For these terms, direct abstraction from historical records with moderate sample size (e.g., with lengths of 30 to 50 years), can hardly lead to precise or even meaningful results.

D. Wang Ph. D. Candidate, B. J. Adams Associate Professor, Department of Civil Engineering, University of Toronto, Toronto, Ontario, Canada, M5S 1A4.

Most of the previous studies involving these terms concentrated themselves on system analysis approaches, and only very coarse or hypothetical probabilities were used to demonstrate application of the approaches. It could even be said that the formal derivation of these probability terms was ignored. However, in practical applications of stochastic analysis to water resources systems, too coarse an estimation of the streamflow distribution probabilities not only impedes the accuracy but also may distort the reality of the analysis. Hence, it is necessary to develop methods for deriving streamflows distribution probabilities with satisfactory precision.

This paper presents some methods for precisely deriving distribution probabilities of seasonal streamflows from historical records with moderate sample size. The derivations for the joint distribution probabilities of multi-site seasonal streamflows, for the Markovian transition probabilities of seasonal streamflows and for the distribution probabilities conditional upon forecasted values are discussed separately. More detailed presentations of these methods can be found in another report of the authors [Wang and Adams, 1987].

These methods are based on fitting the recorded streamflows to theoretical distribution functions. The Normal, Log-Normal, Pearson type 3 and Log Pearson type 3 distributions are often used for the fitting of seasonal streamflows. The three parameter Log-Normal distribution function is employed herein based on its appliability to modelling seasonal streamflows, its flexibility for fitting and its convenience of mathematical manipulation.

In this paper, the boldfaced letters are used to mark random variables. Let $\mathbf{q}$ denote a random streamflow, which is fitted to a three parameter Log-Normal distribution function as follows:

$$\mathbf{q} \sim LN[A,\mu,\sigma] \tag{1}$$

where A is the lower bound parameter and μ and σ are the mean and standard deviation of the logarithmically transformed deviate, respectively. The standardized streamflow variate is then taken as:

$$\mathbf{y} = \{\ln[\mathbf{q}-A]-\mu\}/\sigma \tag{2}$$

DERIVATION OF JOINT DISTRIBUTION PROBABILITIES
FOR MULTI-SITE SEASONAL STREAMFLOWS

Let $\mathbf{q}(l)$ denote the streamflows at site l, l=1,2,...,L; where L is the number of sites in multi-site stream system. The deviates $\mathbf{q}(l)$ are fitted to Log-Normal distributions with parameters A(l), $\mu(l)$ and $\sigma(l)$. Let $\mathbf{y}(l)$ denote the respective standardized streamflow variates and let $q(i_l)$ denote the discrete streamflow values for site l. The marginal distribution probabilities of $\mathbf{q}(l)$ for each site l are specified as the empirical ones abstracted from historical records.

The lag-0 cross correlations between the individual streamflows guide the derivation of the joint probabilities. When the cross correlations are not significant, the joint probabilities can be simply derived as the factorial of the marginal probabilities. When the cross correlations are significant, the joint probabilities are expressed as follows:

$$p(i_1, i_2, \ldots, i_n) = p[i_n | (i_1, i_2, \ldots, i_{n-1})] \cdot p(i_1, i_2, \ldots, i_{n-1}) \qquad (3)$$

where the conditional probabilities are defined as follows:

$$p[i_n | (i_1, i_2, \ldots, i_{n-1})]$$

$$= \text{Prob.}\{q(n) = q(i_n) | [q(1) = q(i_1), q(2) = q(i_2), \ldots, q(n-1) = q(i_{n-1})]\} \qquad (4)$$

Suppose that the joint probabilities for a subset of the system with n-1 sites are available, those for the subset of n sites can be specified from Eq.(3) provided that the conditional probabilities are derived. In such a way, the problem is converted to the derivation of the conditional probabilities.

A linear regression model can be used to describe the dependence structure between the streamflow at site n and those at the n-1 sites as follows:

$$\mathbf{y}(n) = a + \sum_{l=1}^{n-1} b_l \cdot \mathbf{y}(l) + z \cdot \sqrt{1 - R^2} \qquad (5)$$

where z is the standard Normal deviate and R is the multiple correlation coefficient, the parameters a, b_l and R are determined from regression modelling with the historical records.

Corresponding to joint probability $p(i_1, i_2, \ldots, i_n)$, the values of $\mathbf{y}(l)$ and the lower and upper bounds taken by $\mathbf{y}(n)$ in Eq.(5), denoted respectively as $y^l(n)$ and $y^u(n)$, can be specified as follows:

$$y(l) = \{\ln[q(i_l) - A(l)] - \mu(l)\} / \sigma(l) \qquad l = 1, 2, \ldots, n-1.$$

$$y^l(n) = \{\ln[q(i_n) - \Delta/2 - A(n)] - \mu(n)\} / \sigma(n) \qquad (6)$$

$$y^u(n) = \{\ln[q(i_n) + \Delta/2 - A(n)] - \mu(n)\} / \sigma(n)$$

where Δ is the discrete interval for $\mathbf{q}(n)$.

Substituting the values of y(l) and $y^l(n)$, $y^u(n)$ into (5) for $\mathbf{y}(l)$ and $\mathbf{y}(n)$, respectively, the corresponding lower and upper bounds taken by the standard Normal deviate z in (5), denoted as z^l and z^u, can be calculated. It is noted that both the logarithmic function and Normal probability cumulative function are monodromic and strictly increasing. Hence, the respective conditional probability can then be specified as follows:

$$p[i_n | (i_1, i_2, \ldots, i_{n-1})] \qquad (7)$$

$$= \int_{y^l(n)}^{y^u(n)} f\{y | [y(l), l = 1, 2, \ldots, n-1]\} \, dy = \int_{z^l}^{z^u} \phi(z) \, dz = \Phi(z^u) - \Phi(z^l)$$

where $f(\cdot|\cdot)$ is the conditional probability density function, $\phi(\cdot)$ is the standard Normal probability density function (pdf) and $\Phi(\cdot)$ is the standard Normal cumulative distribution function (cdf).

With the conditional probabilities obtained, the joint distribution probabilities of the multi-site streamflows can then be derived recursively for n=2,3,...,L. For two-site streamflows, the expressions of the joint probabilities are simplified as follows:

$$p(i_1,i_2) = p(i_2|i_1) \cdot p(i_1) \tag{8}$$

where $p(i_1)$ is the marginal probability of $q(1)$. The dependence structure of two streamflows takes a simpler form as follows:

$$y(2) = r \cdot y(1) + z \cdot \sqrt{1-r^2} \tag{9}$$

where z is the standard Normal deviate and r is the cross correlation coefficient between $y(1)$ and $y(2)$.

DERIVATION OF MARKOVIAN TRANSITION
PROBABILITIES FOR SEASONAL STREAMFLOWS

Streamflow processes may be described as periodic Markov chains with annual cycles [Wang and Adams, 1986]. Denote $q(m)$ for streamflows in season m, which are fitted to Log-Normal distributions with parameters $A(m)$, $\mu(m)$ and $\sigma(m)$. Let $y(m)$ denote the respective standardized streamflow variates and let q_i and q_j denote the discrete streamflow values with increment Δ. The streamflow transition probabilities are defined as follows:

$$p_{ij}(m) = \text{Prob.}\{q(m)=q_j|q(m-1)=q_i\} \qquad \forall i,j \quad \forall m. \tag{10}$$

Let $\rho(m)$ denote the serial correlation coefficient between $y(m)$ and $y(m-1)$. The synthetic streamflow formula of Fiering and Jackson [1971] can be used to describe the dependence structure between stream flows in successive seasons as follows:

$$y(m) = \rho(m) \cdot y(m-1) + z \cdot \sqrt{1 - \rho^2(m)} \tag{11}$$

where z is the standard Normal deviate.

Corresponding to streamflow transition from q_i to q_j during season m, the value of $y(m-1)$ and the lower and upper bounds taken by $y(m)$, denoted as $y^l(m)$ and $y^u(m)$, respectively, can be specified in a similar way to Eq.(6). Substituting the values of $y(m-1)$ and $y^l(m)$, $y^u(m)$ into (11) for $y(m-1)$ and $y(m)$, respectively, the corresponding lower and upper bounds taken by the standard Normal deviate z in (11), denoted as z^l and z^u, can be calculated. Then, the streamflow transition probability $p_{ij}(m)$ can be derived similarly as follows:

$$p_{ij}(m) = \Phi(z^u) - \Phi(z^l) \tag{12}$$

where $\Phi(\cdot)$ is the standard Normal cdf.

DERIVATION OF STREAMFLOW DISTRIBUTION PROBABILITIES
CONDITIONAL UPON FORECASTED VALUES

In some cases, Markov models may not describe streamflow processes adequately and it is necessary to employ more sophisticated models
to utilize other information available for describing the upcoming
streamflows. These models are established from historical hydrologic
and meteorologic records. Time-series models and multivariate regression models are often used, where the random residual **e** is normally
distributed with 0 mean and standard variation of s_e, which is estimated in the modelling process. Let **q**(t) denote the streamflow in the
upcoming period t, the stochastic streamflow model usually takes the
following form:

$$q(t) = F(t) + z \cdot s_e \qquad (13)$$

where z is the standard Normal deviate and F(t) is the forecasted
value specified by the model.

The distribution probabilities conditional upon forecasted values
lead to more accurate description of upcoming streamflows than those
estimated from empirical distributions or simple Markov models. The
use of such probabilities will benifit the system analysis [Stedinger
et al, 1984]. Let q_i denote the discrete values of **q**(t) and $\hat{q}$(t)
denote the forecasted value of **q**(t), the distribution probabilities
conditional upon forecased values are defined as follows:

$$p_i[q(t)|F(t)] = Prob.\{q(t)=q_i|\hat{q}(t)=F(t)\} \qquad (14)$$

Corresponding to this conditional probability, the bounds of the
discrete interval, q_i^l and q_i^u, can be specified. Substituting these
values for **q**(t) into Eq.(13), the bounds taken by the standard Normal
deviate z in Eq.(13), denoted as z^l and z^u, respectively, can be
calculated. The conditional probability can then be similarly specified as follows:

$$p_i[q(t)|F(t)] = \Phi(z^u) - \Phi(z^l) \qquad (15)$$

where $\Phi(\cdot)$ is the standard Normal cdf.

SUMMARY

The methods for deriving distribution probabilities of seasonal
streamflows from historical records of moderate lengths are presented
in this paper. The distribution probabilities could be derived with
reasonably small discretization intervals for the streamflow amounts
so that the results are of satisfactory precision to facilitate accurate analyses of water resources systems. Such results cannot be
obtained by means of direct abstraction from historical records of the
same lengths.

The procedures for the joint distribution probabilities of multi-
site seasonal streamflows, for the Markovian transition probabilities

of seasonal streamflows and for the distribution probabilities conditional upon forecasted values are presented separately. Computing experiences indicated the effectiveness of these procedures [Wang and Adams, 1987]. Although the procedures are presented based on the Log-Normal distribution fitting of the streamflows, they can be easily revised to correspond to the fitting with other distribution functions.

REFERENCES

Butcher, W. S., Stochastic dynamic programming for optimal reservoir operation, Water Resour. Bull., 7(1), 115-123, 1971.

Fiering, M. B., and B. B. Jackson, Synthetic Streamflows, Water Resour. Monogr., vol.1, AGU, Washington, D.C., 1971.

Loucks, D. P., J. R. Stedinger, and D. A. Haith, Water Resources Systems Planning and Analysis, Prentice-Hall, New Jersey, 1981.

Stedinger, J. R., B. F. Sule, and D. P. Loucks, Stochastic dynamic programming models for reservoir operation optimization, Water Resour. Res., 20(11), 1499-1505, 1984.

Wang, D., and B. J. Adams, Optimization of real-time reservoir operations with Markov decision processes, Water Resour. Res., 22(3), 345-352, 1986.

Wang, D., and B. J. Adams, Derivation of streamflow distribution probabilities from historical records, Dept. of Civil Eng. Publ. 87-06, ISBN 0-7727-7092-1, Univ. of Toronto, Toronto, Canada, 1987.

Yakowitz, S., Dynamic programming applications in water resources, Water Resour. Res., 18(4), 673-697, 1982.

Yeh, W. W-G., Reservoir management and operation models: a state-of-the-art review, Water Resour. Res., 21(12), 1797-1818, 1985.

HYDROLOGIC RUNOFF MODELING OF SMALL WATERSHEDS:

THE TINFLOW MODEL

Andrew T. Silfer*, James M. Hassett*, and Gerald J. Kinn*

ABSTRACT

TINFLOW is a PC-based Geographic Information System (GIS) that utilizes the Triangulated Irregular Network (TIN) and associated data structures, together with a deterministic, finite difference hydrologic construct to model rainfall-runoff processes via overland flow and interflow. The Triangulated Irregular Network (TIN) is used to model a watershed as a series of triangular facets. The TINFLOW data structure allows the user to store or calculate the necessary physical information required by the hydrologic model. In addition attributes such as soil cover type may be specified and stored directly in the data structure. The attributes allow the user to specify, on a facet-by-facet basis, physical parameters that drive the hydrologic model. This type of analysis is particularly well suited to modeling of urban areas with its alternating areas of pavement and vegetation. It may offer the capability to predict the results of change within a watershed (for example, the effect of a clearcut in a forested watershed on runoff, water quality and soil erosion).

TINFLOW SYSTEM DESCRIPTION

TINFLOW is a geographic information system (GIS) written for a personal computer environment in Turbo Pascal. The GIS contains a hydrologic module that can predict a stream's response to storm events.
TINFLOW uses a discretized watershed, with physical attributes assigned to each subarea, as a basis for the associated hydrologic model. However, the method used to discretize the watershed is unique. TINFLOW employs the Triangulated Irregular Network (TIN) as a digital terrain model to represent the topography of the watershed (Figure 1). The TIN model is a series of triangular planes connected at the boundaries and is commonly used in photogrammetry and remote sensing as a means of modeling topography.
The TIN data structure for each facet contains the facet identification number, the three node id numbers and their relative locations in space, topologic information such as the id number of the facets opposite each node link of the facet of interest, and attribute information such as vegetative cover type and soil type.

*: State University of New York College of Environmental Science and Forestry, 312 Bray Hall, Syracuse, New York 13210. (315) 470 - 6633

TINFLOW consists of three major algorithms. PREPRO, the first algorithm, in addition to computing geometric information such as slope and area of the facet, determines the direction water would flow, or flow vectors, across the surface of each facet. This information is then stored in the database. The CHECKR algorithm is then applied to the database to search and label, by evaluating the flow vectors produced by PREPRO, stream and ridge segments in the watershed (Figure 2).

HYDROLOGIC PROCESSING

After the two preprocessors have prepared the database with the information and structure required by the hydrologic model, the model itself can be used. The hydrologic model component of the GIS simulates a watershed's response to storm events by modeling the two major components that contribute to storm runoff: overland flow and interflow.

TINFLOW uses a finite difference solution of the St. Venant equations with a kinematic cascade approximation to simulate overland flow (Hong and Eli, 1985). The hydrologic processor uses the continuity (1) and momentum (2) equations to model overland flow:

$$\partial h/\partial t + u(\partial h/\partial x) + h(\partial u/\partial x) = q - f$$

$$(1/g)(\partial u/\partial t) + (u/g)(\partial u/\partial x) + (\partial h/\partial x) = S - S_f - (q/gh)(u - u_x)$$

where h=depth of flow, u=velocity, x=distance along the flow vector, q=lateral inflow rate (i.e., rain input), f=lateral outflow rate, g=gravitational constant, S=slope of flow vector, S_f=friction slope. The overland flow velocity u is defined by Manning's equation:

$$u = (\Phi/n) \, h^{2/3} S^{1/2}$$

in which n is the roughness coefficient and Φ is 1 or 1.486 depending on the system of units of employed. Substituting (3) for u into (1) and (2) leads to:

$$h_{x,t} = (1/(1+\lambda a))[h_{x,t-1} + \lambda a h_{x-1,t} + q_{x,t}\Delta t]$$

where $a = 5/3 \; \alpha_x h_{x,t-1}^{2/3}$, and $\lambda = \Delta x/\Delta t$. In (4), $q_{x,t}$ represents rainfall excess, i.e., rainfall less infiltration.

Equation (4) is the basis for overland flow calculations. Each triangular facet is converted during the simulation process to a rectangle with equal length and aspect ratios and a backwards finite difference solution of equation (4) is used to model overland flow.

The interflow process is modeled with Darcy's Law. The interflow model uses an expandable interflow zone, so that as the storm progresses, the wetting front of the interflow zone expands downward. The TIN structure's capability of allowing attributes to be assigned on a facet-by-facet basis is vital to the hydrologic model. The attributes considered to have the greatest influence on simulating storm events are cover type and soil type. The cover type influences the amount of precipitation intercepted by vegetation before it reaches the ground. The soil type affects infiltration rates, roughness coefficients, and hydraulic conductivities.

Triangulated Irregular Network:

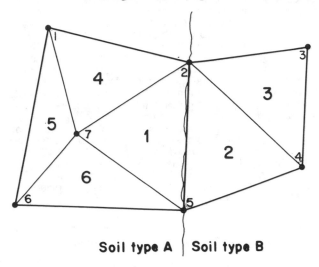

Soil type A | Soil type B

Figure One. Shematic representation of Triangular Irregular Network description of a segment of a watershed. The TIN structure is established by the TINFLOW model which then routes water (both overland and interflow) from the highest facet (5 in this schematic) to the stream boundary.

PREPRO

CHECKR

HYDROLOGIC PROCESSOR

Figure Two. Schematic description of TINFLOW model. The PREPRO algorithm performs a relational join of topologic and node coordinates, calculates physical parameters (slope, area, aspect) for each TIN facet, and determines how water should be routed. The CHECKR algorithm uses decision rules to locate ridge and stream segments. The HYDROLOGIC PROCESSOR calculates a mass balance for each facet given a rain input and calculates overland and interflow to the stream.

The hydrologic model simulates, for an individual TIN facet, the hydrologic process for that land parcel. A hydrologic mass balance is calculated to determine if excess water is present and runoff can occur. Inputs to the mass balance are precipitation and water flowing onto the current TIN from uphill neighbor(s) through overland flow and interflow. Outputs include the overland flow and interflow exiting to the TIN facet downhill. The data base is updated with these outputs so that they may be accessed by downgradient facets. This routing scheme requires that the processing begin with the most uphill facet in the watershed, (i.e., the facet that has no uphill neighbors).

For a given time period, precipitation, if it occurs, is assumed to occur for the entire duration. After one period of precipitation has occurred, the rain input from that time interval is routed through the basin by the sequential hydrologic processing of each facet. When a stream segment is encountered, as determined by CHECKR, the flow that would enter the stream is added cumulatively for the time period. The standard time period is one hour. It is assumed that, in a small watershed, any water that enters the stream during a given period will pass by the gage before the period ends.

MENU SYSTEM

After the completion of a hydrologic simulation, a postprocessing menu of graphic outputs is presented to the user. Possible choices include a stream runoff hydrograph, a precipitation hyetograph or bar chart, a mass balance summary of overland flow, interflow and infiltration, a cover type map, a soil type map, and a diagram of the watershed's stream network. Figures three, four, and five represent the storm hydrographs produced by TINFLOW with precipitation rates of one inch for one hour, one inch per hour for two hours and one inch for one hour, followed by no precipitation for one hour, followed by one inch in the third hour, respectively.

CONCLUSIONS

TINFLOW is a functioning GIS that performs hydrologic simulations on a personal computer. Although it has been tested on a synthetic watershed, the results shown in Figures three through five demonstrate the capability of the system to produce storm hydrographs which conform to the unit hydrograph theory (i.e. doubling the rain intensity that produced the unit hydrograph, doubles the height and runoff of the new hydrograph).

TINFLOW, with its unique method of discretizing the watershed, can be modified to model many discrete processes, including erosion, ground water flow, and agricultural runoff. Because it has the capability of assigning attributes to small parcels of land, TINFLOW could be used in suburban areas to predict the changes in stream flow patterns caused by changes in land use.

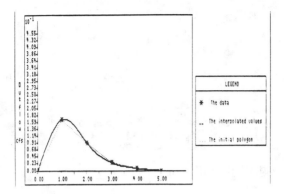

Figure Three. Outflow hydrograph (cfs versus hours) from wtershed as modeled by TINFLOW. Input storm is 1 inch for 1 hour.

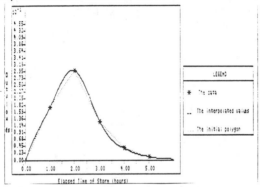

Figure Four. Outflow hydrograph for 1-hour 2-inch storm.

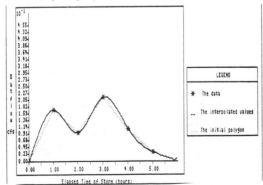

Figure Five. Outflow hydrograph for complex storm: one inch in first hour followed by one inch in third hour.

REFERENCES

Hong, H.M. and Eli, R.N. 1985, Accumulation of Streamflow in Complex Topography: Computer Applications in Water Resources, pp.196-205.

Monmonier, M.S. 1982, Computer Assisted Cartography Principles and Prospects, Prentice-Hall, Inc., Englewood Cliffs, New Jersey.

Peucker, T.K., Fowler, R.J., Little, J.J., and Mark, D.M., 1977, Digital Representation of Three-Dimensional Surfaces by Triangulated Irregular Networks (TIN). Technical Report 10, ONR Contract #N00014-75-c-0886, Dept. of Geography, Simon Fraser Univ., Burnaby, B.C., Canada.

Peucker, T.K. and Chrisman, N., "Cartographic Data Structures", American Cartographer,2, no.1 (April 1975), pp.55-69.

Silfer, A.T., Kinn, G.J., and Hassett, J.M.,"A Geographic Information System Utilizing the Triangulated Irregular Network as a Basis for Hydrologic Modeling", Eighth International Symposium on Computer-Assisted Cartography, Baltimore, Maryland, March 30 - April 3, 1987, pp.

Key Words:
 Geographic Information System
 Triangulated Irregular Network
 Digital Terrain Model

GEOGRAPHIC ESTIMATION OF RUNOFF-MODEL PARAMETERS

Arthur R. Schmidt[1], Linda S. Weiss[1], A.M. ASCE, and Kevin A. Oberg[1]

ABSTRACT

The U.S. Geological Survey is developing techniques to estimate and evaluate unit-hydrograph and loss-rate parameter values for rainfall-runoff models using Geographic Information System (GIS) procedures. The data base includes basin, soil, and climatological characteristics that will be stored in a GIS, and unit-hydrograph and loss-rate parameters obtained from calibration of a commonly used flood-hydrograph rainfall-runoff model for 616 storms in 98 gaged drainage basins. Development of unit-hydrograph and loss-rate parameter-estimation techniques includes statistical methods (exploratory data analysis, regression analysis, and categorical data analysis) to relate the model parameters to hydrologic characteristics.

The estimation techniques will be evaluated by use of error analysis of simulated hydrograph characteristics (peak discharge, flood volume, and time to peak discharge). The hydrographs will be simulated with parameters estimated by the techniques for (1) 102 storms occurring at 36 gaged basins; and (2) a large storm system (one which produced floods with a 50-to 100-year recurrence interval).

INTRODUCTION

Procedures to model the rainfall-runoff process vary from simple empirical methods to sophisticated deterministic or probabilistic approaches. However, present methods are inadequate to completely describe the complex hydrologic processes involved (Hagar, 1985, p. 499). Because of the difficulty and expense of using more sophisticated methods, water-resource managers commonly use simpler methods, such as lumped-parameter rainfall-runoff models, to compute discharge hydrographs at particular locations along a stream.

State and local water-resources planners in Illinois commonly use the U.S. Army Corps of Engineers' Hydrologic Engineering Center flood-hydrograph model (HEC-1) (1981a) to design and evaluate channels, culverts, storm drainage systems, reservoir spillways, and other hydraulic structures. The HEC-1 model is a commonly used lumped-parameter rainfall-runoff model. The model uses generalized functions

---

[1] Hydrologist, U.S. Geological Survey, WRD, Urbana, Illinois 61801

to estimate base flow, unit-hydrograph parameters, and the amount of precipitation lost due to infiltration and interception (loss rates) (U.S. Army Corps of Engineers Hydrologic Engineering Center, 1981b, p. 5). Because the physical processes modeled depend on factors such as land use and development, soil properties, and hydrologic and meteorologic conditions in the area, the parameters in the model differ geographically.

The U.S. Geological Survey, in cooperation with the Illinois Department of Transportation, Division of Water Resources, is conducting an investigation to develop and evaluate techniques to estimate unit-hydrograph and rainfall loss-rate parameters used in rainfall-runoff models. A technique based on the geographical distribution of hydrologic characteristics will be developed to increase the utility of the model for predictive purposes. The study incorporates computer-based Geographic Information System (GIS) procedures to relate basin, soil, and climatological characteristics to the unit-hydrograph and loss-rate parameters. This paper describes the approach of the study. The parameter-estimation techniques and evaluation of these techniques will be presented in future publications.

PREVIOUS STUDIES

The HEC-1 rainfall-runoff model requires time-area histograms, weighted precipitation hyetographs, and regionalized unit-hydrograph and loss-rate parameters to estimate discharge hydrographs on ungaged streams. Graf and others (1982a and 1982b) published unit-hydrograph parameters and a technique for estimating their values for ungaged basins in Illinois. In their investigation, the HEC-1 flood-hydrograph model was used to calculate two unit-hydrograph parameters, the time of concentration and storage coefficient, for 98 gaged drainage basins in Illinois.

As part of their investigation, Graf and others (1982a) used a four-parameter function to compute rainfall loss for the 98 gaged basins. This function--the Exponential Loss-Rate function--relates rainfall loss rate to rainfall intensity and accumulated losses (U.S. Army Corps of Engineers, 1981a).

Garklavs and Oberg (1986) used a different function--the Initial and Uniform Loss-Rate function--to compute rainfall losses for 209 storms at 32 of the 98 gaged basins. Weiss and Ishii (U.S. Geological Survey, written commun., 1986) calibrated the remaining 66 basins (of the 98) using the same loss-rate function as Garklavs and Oberg (1986), as a part of a separate investigation. The Initial and Uniform Loss-Rate function uses two parameters--the initial volume of rainfall to satisfy antecedent soil-moisture deficiency and a constant rate at which rainfall is lost after the initial loss is satisfied. The reader is referred to the HEC-1 users manual U.S. Army Corps of Engineers, (1981a) for a more detailed explanation of the two rainfall-loss functions. Weiss and Ishii (U.S. Geological Survey, written commun., 1986) have developed techniques to estimate the values of the rainfall-loss parameters for both rainfall-loss functions.

METHOD OF STUDY

The approach of the study consists of (1) data collection and analysis, (2) development of parameter-estimation techniques, and (3) evaluation of the parameter-estimation techniques. Data collection includes the acquisition and digitization of maps that indentify basin, soil, and climatological characteristics needed to describe the hydrologic processes. Parameter-estimating techniques will be developed by defining conceptual models of the basin, soil, and climatological characteristics that affect each parameter, and then using statistical methods to relate empirically unit-hydrograph and loss-rate parameters for both rainfall loss-rate functions to spatially and temporally variable hydrologic characteristics. Parameter-estimation techniques will be evaluated by error analysis of hydrograph characteristics produced by simulating (1) storms from 36 gaged basins not used in development of the estimation technique and (2) a large (producing flows of 50- to 100-year recurrence intervals) storm.

Drainage areas, soil types, land uses, and other characteristics of the 134 basins used in previous studies are being digitized. Tabular files of basin and other characteristics are being assembled for use with the digital data. These files include items such as calibrated unit-hydrograph and loss-rate parameters, drainage area, channel slope, and channel length.

Techniques to estimate storm-dependent rainfall-loss parameters will include the effect of antecedent soil moisture by including basin, soil, and climatologic characteristics that may be used to estimate soil moisture. These characteristics include areal temperature and precipitation trends for given dates or seasons, soil types, and land use.

Techniques for estimating basin-dependent unit-hydrograph and rainfall-loss function parameters will be defined by use of statistical methods, including exploratory data analysis, regression analysis, and categorical data analysis. Percent area covered by different spatially varying characteristics will be determined for each drainage basin by the GIS. Percent area covered and physical properties of the spatially varying characteristics, along with previously calibrated unit-hydrograph and rainfall-loss parameters for both loss-rate functions will be used to develop the parameter-estimation techniques. Techniques will be developed to relate the parameters (dependent variables) empirically to basin, soil, and climatological characteristics (independent variables).

For the evaluation phase, characteristics of hydrographs (peak discharge, flood volume, and time to peak discharge) simulated by use of the parameter-estimation techniques are compared to measured hydrograph characteristics for storms at 36 previously uncalibrated gaged basins. A large storm system that produced floods with a 50- to 100-year recurrence interval and for which precipitation and runoff data are available, will also be used to assess the applicability of parameter-estimation techniques to large storms. This part of the technique evaluation is planned, because the HEC-1 model commonly is used to simulate large storms for design and evaluation purposes.

SUMMARY

The U.S. Geological Survey is currently developing and evaluating techniques to estimate unit-hydrograph and rainfall-loss parameters for a rainfall-runoff model using GIS procedures. The data base consists of basin, soil, and climatological characteristics stored in a GIS, and unit-hydrograph and rainfall-loss function parameters obtained from calibration of the HEC-1 flood-hydrograph model for 616 storms at 98 gaged drainage basins. Techniques to estimate antecedent soil moisture from basin and climatological data are being considered. Statistical methods will be used to relate model parameters to hydrologic character- istics. Techniques will be evaluated by error analysis of hydrograph characteristics simulated on the basis of application of the techniques to (1) 102 storms at 36 gaged basins and (2) a large storm system.

REFERENCES CITED

Garklavs, George, and Oberg, K. A., 1986, Effect of rainfall excess calculations on modeled hydrograph accuracy and unit-hydrograph parameters: Water Resources Bulletin Paper No. 85063, Vol. 22, No. 4, p. 565-572.

Graf, J. B., Garklavs, George, and Oberg, K. A., 1982a, Time of con- centration and storage coefficient values for Illinois streams: U.S. Geological Survey Water-Resources Investigations 82-13, 35 p.

----- 1982b, A technique for estimating time of concentration and storage coefficient values for Illinois streams: U.S. Geological Survey Water-Resources Investigations 82-22, 16 p.

Hagar, W. H., 1985, A non-linear rainfall-runoff model: Paper from the 21st IAHR Congress, Melbourne, Australia, August 1985, pp. 498-503.

U.S. Army Corps of Engineers, 1981a, HEC-1 Flood hydrograph package users manual: Davis, Calif., Hydrologic Engineering Center, 190 p.

----- 1981b, Hydrologic analysis of ungaged watersheds with HEC-1: Davis, Calif., Hydrologic Engineering Center, 120 p.

A Flood Prediction Geographic Information System

Edward J. VanBlargan* and John C. Schaake*

ABSTRACT

The paper discusses a GIS that derives kinematic wave model
parameters using stream network, elevation, and basin boundary data.
Various approaches are being investigated for estimating several
parameters not directly obtainable from the GIS data. Additional work
is planned to determine the most feasible approach and to fully test the
system on a variety of basins.

INTRODUCTION

This paper describes a project dealing with the use of a geographic
information system (GIS) for hydrologic forecasting. This is an ongoing
project in the National Weather Service (NWS) that uses a kinematic wave
model. Discussion is given to the background of problems leading to the
project, objectives, system components including the model and data
bases, approach being taken, a sample case, and future directions.

The NWS has the mission of providing river forecasts nationwide.
For most large gaged rivers, hydrologic model parameters have been
derived and site specific forecasts are issued primarily using unit
hydrographs. However, for many small basins (i.e., 10 to several
hundred square miles) parameters have not been derived and generalized
area forecasts are issued (e.g., expect moderate rises on small creeks
in the county). In the future, the NWS would like to provide more site
specific forecasts for smaller basins and newly gaged areas. To do this
effectively for many basins requires sound and easy means of deriving
model parameters.

Parameter estimation methods often rely on time consuming fitting
with rainfall and streamflow data. Also, the required historical data
is often inadequate for many small basins (Clarke, 1973). Thus, an
alternative method was sought that would be more automated and rely more
on easily measurable basin characteristics. Also, three geographic data
bases were obtained by the NWS. GIS technology has found useful
application in hydrology and seemed to be a potentially feasible
alternative (Hill et.al., 1987; Ragan and White, 1985). Since
procedures using basin properties are not typically employed in the NWS
(with some exception, Sheridan, 1953), a model amenable to a GIS was
needed. While various models and synthetic techniques exist, a
kinematic wave model was chosen because of its physical basis and other
salient features (see MODEL).

* Hydrologists, National Weather Service, Silver Spring MD 20910

OBJECTIVES

The project goal is develop and test a GIS for deriving kinematic wave model parameters for use in small or newly gaged basins. The system is constrained to using digital data that will be readily available nationwide to the NWS. Specific objectives are: 1) Define the approach for deriving parameters directly from the GIS data; 2) Identify possible approaches for deriving information (e.g., basin area) that may have to be inferred indirectly from the GIS data; 3) Identify possible approaches for deriving any parameters (e.g., roughness) that cannot be obtained from the digital data; 4) Define which approach(es) is most feasible; and 5) Evaluate the GIS performance on small test basins. The performance evaluation will asses several criteria including accuracy, efficiency, ease of use, and amount of user input required.

THE SYSTEM: Model and Data Bases

The kinematic wave model (KWAVE) produces a hydrograph by routing of surface runoff to a watershed outlet (Schaake,1971). It was chosen because required parameter input can be tied to observable watershed properties. Also, it potentially may give improved forecast accuracy since it is non-linear and can handle spatially varied rainfall input For this project, the SCS curve number, which has been used in GIS's (Ragan and White, 1985), serves as the surface runoff component. However, the curve number is to be fitted for each basin in the project. KWAVE uses a finite difference solution and requires: 1) Basin segmentation into channel and overland flow segments; 2) Length of each segment,; 3) Two kinematic parameters for each segment derived from slope, roughness, and channel shape. The overland flow segments represent the myriad of small channels that cannot be defined but are the mechanism for conveying lateral inflow to the channel. This type of model has been applied on a variety of small basins (Schaake, 1973; Ross and Shanholtz, 1979). Further, the system could be used to estimate a unit hydrograph if desired (Hjelmfelt, 1984).

Geographic data containing gridded elevations, stream networks, and basin boundaries are obtained from the DMA half mile topography data base, the EPA River Reach data base, and the USGS hydrologic units basin boundary data base. The system relies almost totally on these data. The GIS reads the necessary geographic data and estimates the KWAVE parameters which are then filed (Figure 1). These parameters are then used in the real-time KWAVE model for any rainfall events that occur. Any parameter that cannot be defined from the geographic data are obtained from a file of regional values.

APPROACH

The basin is segmented at junctions in the river data. Each segment is idealized as a uniform channel receiving lateral inflow two symmetrical planes of overland flow (Figure 2). For each segment, the geographic data yields the channel length, drainage area, overland flow length, channel slope, and overland flow slope. Roughness and channel shape cannot be inferred from the geographic data so are assigned regionalized values.

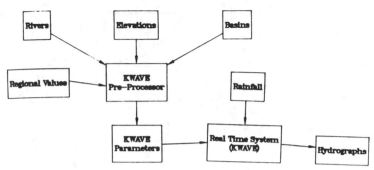

Figure 1.   System Architecture

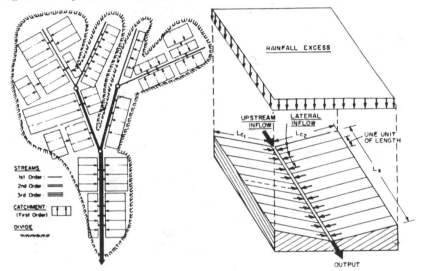

Figure 2.   Concept of segmentation and the idealized segment

The software for the basic approach is completed, however, several approaches are being investigated to address some existing questions, such as how to to regionalize values. Also, the basin boundary data do not always coincide with the user's outlet (e.g., Johnstown Figure 3). Another potential problem is that small areas may have a short digitized channel resulting in a poor overland flow length.

Terrain analysis is being investigated to enhance the resolution of the basin and river data for better drainage area and overland flow length estimates. It uses elevation data to "grow" a basin from a specified outlet (Ragan and Fellows, 1985). Alternatives are being considered that use terrain analysis but constrain it with a user specified drainage size or with the existing basin and river data.

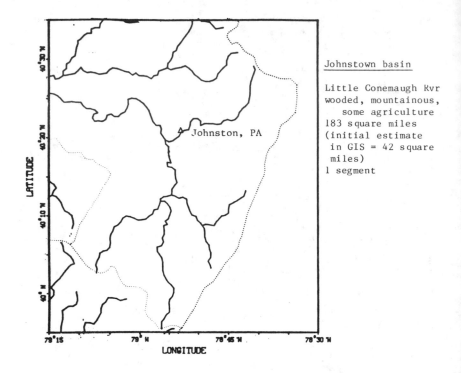

Figure 3. Map of Johnstown, PA area from the GIS

An approach is still needed for roughness and channel shape since
these cannot be obtained from terrain analysis. Several approaches
being considered are nominally assigned values (Arcement and Schneider,
1984), (i.e., the current approach), regional geomorphic drainage area
relationships for channel shape, and regionalized index watersheds.
Index watersheds would have fitted roughness and shape values which
would be used on any basin in the same physiographic region.

The test phase will use a variety of basins with fitted "optimum"
parameters. Differences between parameters estimated from each approach
and optimum values will be identified and used with parameter
sensitivity analysis to evaluate the accuracy potential of each
alternative. The most feasible approach(es) will be identified by
trying to maximize accuracy and minimize subjective user input.
Finally, the performance of the system will be tested on a variety of
storms in each basin using the most feasible approach.

EXAMPLE CASE

A test was done on the Little Conemaugh river near Johnstown, PA, using the July, 1977 event that averaged 10 inches of total rainfall. With only the outlet location specified the GIS estimated the initial parameters (Table 1). Drainage area was estimated from a regional geomorphic area-stream length relationship since no terrain analysis exists and the basin data was inadequate (see Figure 3). The drainage area estimate was very low so the actual size was input and a corresponding correction made to the overland flow length. The main channel presented no problem since it was digitized nearly to the basin divide. Roughness and channel shape were assigned nominal values.

The simulation using the non-fitted nominal roughness values was very close to the observed peak stage and time (Table 2). Sensitivity analysis of the variables not directly obtained from the GIS indicated that the overland flow length (or overland flow roughness) had the greatest effect on the results (Figure 4). No conclusions are extrapolated to other basins but further testing will be done to see how results compare.

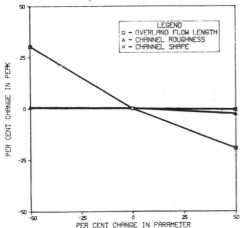

Table 1. Parameter Estimates

Parameter	Channel	Overland Flow
Length (mile)	19	4.8
Slope (feet/mile)	25	190
Roughness	.06	0.5
Shape	10	–

Table 2. Simulation Results for Peak

	Observed	Simulated
Stage (feet)	18.1	18.0
Time to Peak (hours)	6	6

Figure 4. Sensitivity Analysis

SUMMARY

The initial development phase of the GIS is complete. However, several questions remain that may require enhancement of the system. The current configuration uses nominal default values for roughness and channel shape, and an area versus stream length relationship to obtain drainage area when the basin data is inadequate. The next phase will investigate terrain analysis and regionalization approaches to establish how results can be improved. Once complete, the GIS will be tested on a variety of basins.

The potential value of the GIS is very high. It will give a sound
and easy means of deriving kinematic model parameters for many small
basins. Such basins are ungaged or lack adequate historical rainfall
and streamflow data to do traditional parameter fitting. Further, the
GIS has potential value in larger basins where it can be used to
subdivide and parameterize the basin into small segments giving better
results for spatially varied rainfall events.

Special acknowledgment is given to Bob Ragan, University of
Maryland, whose ideas and support are prime movers in this project.

REFERENCES

Arcement, G.J., and V.R. Schneider, 1984: Guide for Selecting
Manning's Roughness Coefficients for Natural Channels and Flood
Plains, USGS, Baton Rouge, LA, 62 pp.

Clarke, R.T., 1973: A Review of Some Mathematical Models Used in
Hydrology with Observations of Their Calibration and Use. Journal
of Hydrology, 19, pp. 1-20.

Hill, J.M., V.P. Singh, and H. Aminian, 1987: A Computerized Data
Base for Flood Prediction Modeling. Water Resources Bulletin,
23(1), pp. 21-27.

Hjelmfelt, A.T., 1984: Convolution and the Kinematic Wave Equations.
Journal of Hydrology, 75, pp. 301-309.

Ragan, R.M., and J.D. Fellows, 1985: The Role of Cell Size in Hydrology
Oriented Geographical Information Systems. IAHS Publication 160,
Proceedings, Hydrologic Applications of Space Technology,
Cocoa Beach, FL, pp. 453-460.

Ragan, R.M., and E.J. White, 1985: The Microcomputer as an Intelligent
Workstation in Large Scale Hydrologic Modeling. Proceedings,
Specialty Conference Hydraulics and Hydrology in the Small Computer
Age, ASCE, Lake Buena Vista, FL, pp. 504-509.

Ross, B.B., and V.O. Shanholtz, 1979: A One-Dimensional Finite Element
Structure for Modeling the Hydrology of Small Upland Watersheds.
Proceedings, Hydrologic Transport Modeling Symposium, ASAE, New
Orleans, LA, pp. 42-59.

Schaake, J.C., 1971: Deterministic Urban Runoff Model. Treatise on
Urban Water Systems, Colorado State University, Ft. Collins, CO,
pp. 357-383.

Schaake, J.C., G. Leclerc, and B.M. Harley, 1973: Evaluation and
Control of Urban Runoff. Preprints, Annual and National
Environmental Meetings, ASCE, New York, NY, 30 pp.

Sheridan, J.F., 1953: Methods for Developing Flash Flood Forecasting
Tables. MS Thesis, Oklahoma State Univ., Stillwater, OK, 87 pp.

MODELING SUPPORT FOR STORMWATER MANAGEMENT

D. W. Meier, Associate Member[1] and D. F. Lakatos, Member, ASCE[2]

ABSTRACT: An innovative "digitization" procedure has been developed and successfully utilized to accurately and economically prepare and manage the extensive amount of data required for watershed-wide stormwater management simulation computer models. Developed for use with the IBM-PC versions of the Penn State Runoff Model (PSRM) and TR-20, the digitization package is initially used to define physical watershed characteristics required as input into the models. Additional routines (which are available to the client for use on their personal computer after the initial model is prepared and calibrated) include the capability to digitally update the computer model as land use changes occur within an urbanizing watershed. An overlay analysis procedure superimposes the "digital map files" of different attributes to highlight areas having desired land characteristics. A case study illustrates the usefulness of the program.

Introduction

Increased awareness of the adverse impacts of urban runoff on flooding and water quality problems has recently led to the implementation of progressive stormwater management plans in many areas of the country. The State of Maryland is currently implementing a state-wide stormwater management program that emphasizes the management and control of adverse environmental impacts (e.g., flooding, water quality degradation, and stream erosion) from increased stormwater runoff resulting from new land development (State of Maryland, 1976; Prince George's County, Maryland, 1984). Similar programs have been enacted in Pennsylvania, New Jersey, and Florida (State of Pennsylvania, 1984; State of New Jersey, 1981; State of Florida, 1972).

Computer models for hydrologic and hydraulic simulation of watersheds, such as TR-20 and PSRM, have been proven to be useful technical support tools for these new types of water resources management programs. (USDA, Soil Conservation Service, 1965; Aron, G. and D. F. Lakatos,

1. Water Resources Engineer, Walter B. Satterthwaite Associates, Inc., 720 North Five Points Road, West Chester, PA 19380.
2. Principal, Water Resources Engineer, Walter B. Satterthwaite Associates, Inc., 720 North Five Points Road, West Chester, PA 19380.

1976). These models require watershed-wide information on
land use, soil types, and slopes as input or for
development of input parameters such as the SCS Runoff
Curve Number (CN). However, at a time of extreme financial
constraints on both the public and private sectors, it is
difficult to revise the models to quickly and cost-
effectively evaluate the impacts of land use changes or a
proposed development. The Digitization Program developed
by Satterthwaite Associates, Inc. provides the water
resources manager with the important capability to readily
and cost-effectively update the original planning tool as
land use and drainage pattern changes occur in an
urbanizing watershed, and to quickly revise the model to
analyze the impacts of proposed developments.

Digitization

      The development of watershed-wide information on land
use, soil types, and slopes is conventionally done by
preparing a set of transparent attribute overlays at the
same scale as a base map of the watershed. For most model
applications, a watershed is divided into numerous smaller
sub-basins, or "subareas." The modeler must then hand
planimeter the regions associated with a desired composite
attribute. This manual method quickly becomes unmanageable
for large watersheds in terms of record keeping and
computation, is prone to human error, and requires a large
expenditure of additional labor for each change to one of
the attribute maps.

      The Digitization Program provides an alternative
method for preparing PSRM and TR-20 input parameters which
avoids the aforementioned problems. With this computer-
assisted approach, the details of each attribute map are
translated by the computer into a digital map file. Once
the pertinent digital map files are generated, "overlaying"
of the individual attribute map files and measurement of
single attribute and composite attribute areas are handled
by the computer. Manipulation of the area data to
calculate the required input for simulation models (e.g.,
land area, SCS CN, percent impervious land area, average
land slope) is automatically done by the computer.

      Because of the flexibility that is inherent in digital
processing, the Digitization Program offers many
advantages, including:

a) Digital map files, created from original work maps that
    have been compiled at widely-varying scales, can be
    easily and effectively combined.

b) Checks are built into the Digitization Program to
    protect against the introduction of data errors.

c) Changes made to one or more digital map files (e.g., to reflect urbanization) do not require extensive additional labor to reanalyze. The computer quickly handles the recomputation of composite attribute areas and input parameters.

d) Digital map files created by different workers in adjacent study areas can, in most cases, be merged at some later date.

e) Since no physical overlays are involved, all of the map information and the results of computations are conveniently stored on personal computer diskettes.

The attribute information is coded into the computer by digitizing "breakpoints" between adjacent attributes along horizontal "strips" on each attribute map. The "strips" are then divided into "cells" by the program, as shown in Figure 1. Each cell is assigned a subarea designation, land use, soil type, and slope based on the digitized strip information. The program enables the user to readily "edit" any attribute in any cell to reflect changes in the watershed. These changes might include: a) land use changes, e.g., farmland to residential, b) subarea boundary changes due to installation of storm sewer systems, and c) slope changes due to land grading activities.

In addition to the above attributes, up to ten other attributes can be digitized, such as wetlands, flooding

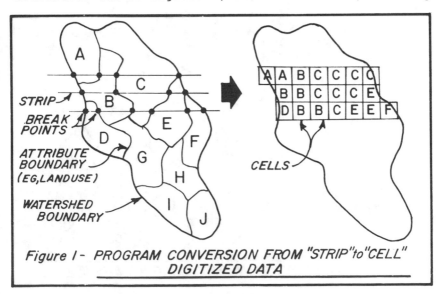

Figure 1 - PROGRAM CONVERSION FROM "STRIP" to "CELL" DIGITIZED DATA

problem areas, and groundwater recharge areas. The
Digitization Program can "overlay" up to ten different
digital map files to help the user look for areas within a
watershed with specific characteristics, such as optimum
areas for stormwater quality management facilities. The
graphic capabilities of the program enable the user to
print out maps of individual or composite attributes. The
user can also specify pollutant loading factors for
different land uses and quickly determine pollutant loads
from individual subareas, as well as changes in loads due
to proposed developments. Loading factors for up to ten
pollutants (e.g., lead, phosphorous, nitrogen, coliform,
etc.) can be specified for each land use.

Case Study

     The Digitization Program and PSRM were used to
evaluate the impact of a proposed 127-acre residential
development on 25- and 100-year flows for downstream areas
in a rapidly urbanizing watershed in Pennsylvania. The
watershed was originally digitized and modeled using PSRM
in 1982 as a pilot study in response to the Pennsylvania
Stormwater Management Act (State of Pennsylvania, 1984).
Several of the watershed subareas and the location of the
proposed development are shown in Figure 2. The land use
portion of the database was digitally updated to include

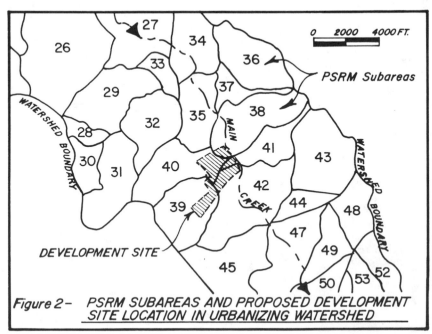

Figure 2 — PSRM SUBAREAS AND PROPOSED DEVELOPMENT
SITE LOCATION IN URBANIZING WATERSHED

all new developments since 1982 so that the proposed development's compliance with the Act could be analyzed based on current conditions within the watershed.

Since the proposed development site overlapped five subareas, the subarea boundaries were digitally revised to correspond with the site boundary along the main creek, as shown in Figure 3. Subareas 39 and 40 were digitized twice, once to reflect the existing predevelopment condition and then the future post-development condition. To develop flows for stream crossing design within the development, the two main creek subareas were divided into eleven "sub-subareas" based on the desired points of flow computation (Figure 3).

The Digitization Program automatically calculated the PSRM input parameters for the two scenarios, pre- and post-development, and the site was modeled using PSRM to compute the flows at the desired points both on-site and in the main creek downstream for the 25- and 100-year events. The results of this analysis showed that, although the peak flows from the proposed development increased slightly, the development would have no impact on the main creek peak flows due to differences in peak flow timing. This finding also showed that an on-site detention facility would serve no real purpose, thus providing additional land for

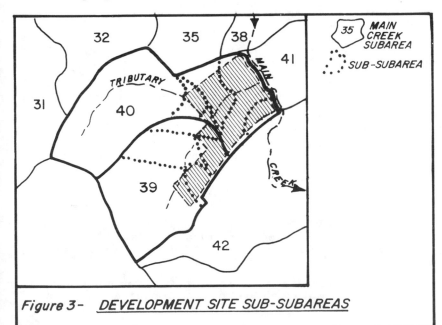

Figure 3- *DEVELOPMENT SITE SUB-SUBAREAS*

development or open-space dedication, and saving the expense of designing and building such a facility.

Summary

    A digitization method for the processing of map information used in watershed stormwater quantity and quality management studies has been developed. Digitization has resulted in a substantial savings in the cost and time required for data preparation, watershed modeling, and solution development in stormwater simulation projects. Digitization provides an efficient means of accomplishing the task of overlay map preparation and computation of hydrologic model input parameters which avoids the problems of tedious planimetering and the associated accumulation of human error. The inherent flexibility of the digital method allows for the development and graphic display of spatial information and composite map generation which would otherwise not be economically feasible. Furthermore, digitization allows the digital map files to be merged and updated without access to the master digitization equipment used to develop the original calibrated model.

References

Aron G., and Lakatos, D.F. 1976. "Penn State Runoff Model User's Manual," Institute for Resource and Land and Water Resources. The Pennsylvania State University, (Revised 1980, 1983, 1984.)

Prince Georges County, Maryland, 1984. Prince Georges County Maryland Stormwater Management Ordinance, CB-52-1984.

State of Florida, 1972. Florida Water Resources Act of 1972. Chapter 373, Florida Statutes.

State of Maryland, 1976. Maryland Flood Hazard Management Act of 1976, Natural Resources Article, Title 8 - Water and Water Resources, Subtitle 9A. Flood Control and Watershed Management.

State of New Jersey, 1981. New Jersey Stormwater Management Act, P.L. 1981, C.32, N.J.A.C.7:8-1.1 et. seq.

State of Pennsylvania, 1984. Pennsylvania Storm Water Management Act of October 4, 1978, P.L. 864, No. 167, 32 P.S. Sections 680.1-680.17, as amended by Act of May 24, 1984, No. 63.

USDA, Soil Conservation Service, 1965. Computer Program for Project Formulation Hydrology, Technical Release No. 20.

MUSKINGUM FLOW ROUTING USING PERSONAL COMPUTERS

G. Padmanabhan MASCE*
Ronald K. Williams MASCE**

Introduction

A linear storage–inflow–outflow relationship as in Eq. 1 is commonly used in the Muskingum method of flow routing. S_t, I_t and O_t

$$S_t = k[xI_t + (1-x)O_t] + A \qquad (1)$$

are the simultaneous storage, inflow and outflow. k, x and A are the parameters to be determined from past flow data through the river reach. Theoretically, x can be shown to have a range 0 – 0.5. The parameter 'A' can be omitted if absolute storage values are used. Although the linear form of storage–flow relationship in Eq. 1 is commonly used, two non-linear forms of the relationship are frequently discussed in the literature (2,3,4,11). It has been observed that the storage– inflow–outflow relationship may not be linear for all river reaches. In such cases, non-linear forms, as Eq. 2 or 3, should be preferred (4,11). k, x, m and A are the parameters to be estimated. In the actual routing process, the parameter 'A' can be shown to have no influence (4).

$$S_t = k[xI_t + (1-x)O_t]^m + A \qquad (2)$$

$$S_t = k[xI_t^m + (1-x)O_t^m] + A \qquad (3)$$

The routing equations (5–9) are derived from the continuity of flow through the reach and the assumed storage–inflow–outflow relationship, Eq. 1, 2 or 3. Continuity equation for the reach is given by Eq. 4 which reduces to Eq. 5 in a usable form. $\bar{I}$ and $\bar{O}$ are the average inflow and outflow during the time interval Δt, and ΔS is the incremental storage. Substituting Eq. 1, 2 or 3 in Eq. 5 we get the routing equations 6, 7, and 8 applicable to models of Eq. 1, 2, and 3 respectively.

$$\bar{I}\Delta t - \Delta S = \bar{O}\Delta t \qquad (4)$$

$$[\frac{I_t + I_{t+1}}{2} - \frac{O_t + O_{t+1}}{2}] \Delta t = S_{t+1} - S_t \qquad (5)$$

$$[\frac{I_t + I_{t+1}}{2} - \frac{O_t + O_{t+1}}{2}] \Delta t = k[\{xI_{t+1} + (1-x)O_{t+1}\}-\{xI_t+(1-x)O_t\}] \qquad (6)$$

*Associate Professor and **Lecturer respectively, Department of Civil Engineering, North Dakota State University, Fargo, ND 58105.

$$[\frac{(I_t+I_{t+1})}{2} - \frac{(O_t+O_{t+1})}{2}] \; \Delta t = k[\{xI_{t+1}+(1-x)O_{t+1}\}^m-\{xI_t+(1-x)O_t\}^m] \quad (7)$$

$$[\frac{(I_t+I_{t+1})}{2} - \frac{(O_t+O_{t+1})}{2}]\Delta t = k[\{(xI_{t+1}{}^m+(1-x)O_{t+1}{}^m)\}]-\{xI_t{}^m+(1-x)O_t{}^m\}] \quad (8)$$

In routing, typically I_t, O_t and I_{t+1} for any routing period Δt are known. Therefore, O_{t+1} can be solved for successive routing periods by using the estimated parameters in Eq. 6, 7 or 8. Eq. 6 can be shown to reduce to Eq. 9 in which the coefficients are functions of k, x and Δt, and the solution is relatively simple.

$$O_{t+1} = C_0 \; I_{t+1} + C_1 I_t + C_2 O_t \quad (9)$$

$$C_0 = - \frac{kx-0.5 \; \Delta t}{k-kx+0.5 \; \Delta t} \qquad C_1 = \frac{kx+0.5 \; \Delta t}{k-kx+0.5 \; \Delta t} \qquad C_2 = \frac{k-kx-0.5 \; \Delta t}{k-kx+0.5 \; \Delta t}$$

However, estimation of parameters of non-linear models in Eq. 2 or 3 and solution of the corresponding routing Eq. 7 or 8 are more complex resulting in an increase in computational burden. Since Eq. 7 and 8 are non-linear in O_{t+1}, routing requires the solution of a non-linear equation at every routing period.

Objective

Regardless of the model used – linear or non-linear, the computations involved in Muskingum flow routing consist of parameter estimation and the actual routing. Several methods of parameter estimation have been suggested by various authors for linear as well as non-linear models (1-5, 8, 10-12). The routing equations are solved differently for the two cases. The objective of this paper is to demonstrate the viability of carrying out these computations in a Personal Computer (PC) environment with ease and speed using commonly available softwares.

Use of Personal Computers

1. Linear Model

Traditionally, parameters of the linear model are estimated by plotting S vs. weighted sum of inflow and outflow for various values of x. The most appropriate value of x is the one for which the graph plots as a straight line instead of a loop and the value of k is the slope of that line. This method can be implemented readily on a PC and the graphs viewed on the screen with speed and ease. Other objective methods have been proposed by various authors (1-5,8). These include a multiple linear regression (MLR) (4) and a Linear Programming method (LP) (2,8). In the MLR method, Eq. 1 is rewritten as Eq. 10 in which $\alpha=kx$ and $\beta=k(1-x)$. Estimates of α, β and A can be obtained by using multiple regression and hence k and x. The authors

$$S_t = \alpha I_t + \beta O_t + A \tag{10}$$

used Lotus 1-2-3, a spreadsheet application software, for the computation of S_t as in Eq. 5 using the past inflow and outflow hydrographs and for the estimation of α and β. S_t, I_t and O_t, $t=1,\ldots,n$ are the data needed for multiple regression. In the LP formulation, the objective is to minimize the maximum absolute deviation between S determined by using Eq. 5 and by using Eq. 10, subject to certain constraints. Constraints 12 and 13 ensure that de-

Minimize: Y, the maximum absolute deviation   (11)

$$Y + [\alpha I_t + \beta O_t] \geq S_t, \quad t=1,2,\ldots,n \tag{12}$$

$$Y - [\alpha I_t + \beta O_t] \geq S_t, \quad t=1,2,\ldots,n \tag{13}$$

$$\alpha - \beta \leq 0 \tag{14}$$

$$Y, \alpha \text{ and } \beta \geq 0 \tag{15}$$

viations between storages computed by using Eq. 5 and 10 stay within Y. The usual condition that $0 \leq x \leq 0.5$ is equivalent to constraint 14. S_t is the storage computed from observed inflow and outflow using Eq. 5. A Fortran LP code was used to solve the LP problem. The code uses Simplex algorithm. The data file for the problem can be easily created with a text editor such as PC-Write. A Fortran compiler is necessary for linking the data file and executing this code. From the optimal values of α and β, k and x can be calculated. Any commercially available software for solving LP's in Personal Computers can be used.

With the estimated values of k and x, routing of inflow is carried out by using Eq. 9 for every time step. The routing calculations can be easily performed on a spreadsheet software such as Lotus 1-2-3. With the aid of an internal graphics package, all necessary plots can be viewed and reviewed on the screen.

2. Non-linear model

Gavilan et. al (2), Gill (4) and Wilson (11) have suggested methods for estimating parameters in the non-linear model, Eq. 2. A constrained optimization method was used by Gavilan et. al. Due to the complexity involved in estimating parameters k, x, m and A in Eq. 2 and 3, Gill used an x value obtained for the best straight line fit and then proceeded to obtain estimates of k, m and A. However, all parameters of these models can be easily estimated simultaneously by using non-linear regression (NLR). A PC version of statistical Analysis System (8) was used for estimating the parameters. There are other softwares also available for performing NLR.

The estimated parameters are then used to route given inflow hydrographs by solving the non-linear Eq. 7 or 8 successively for every time step. These computations can be easily done using any software for solving non-linear equations. The software 'TK Solver' was used by the authors.

Examples

The example problem chosen for linear model application is from Linsley et. al (7). The graphs of S vs. [xI + (1-x) 0) were generated for various values of x and viewed on the screen. The best value of x can be chosen by inspection of these plots. The parameters obtained from graphical analysis MLR and LP are given in Table 1. In the LP method, the maximum absolute deviation was minimum when α = 0.133 and β = 0.458. This corresponds to x = 0.225 and k = 0.591 days (Table 1). The spreadsheet software used by the authors has an internal graphics package. This was used to develop graphs of observed and predicted flows (plots not shown due to page limitations).

The examples 1 and 2 chosen for the application of non-linear models are from page 74 (6) and page 141 (11), respectively. The given inflow and outflow hydrographs were used in estimating the parameters of the linear as well as the non-linear models (Table 2). The estimated parameters were then used to route the same inflow hydrograph.

In Example 1, the non-linearity is not very strong. The parameter estimates also appear to indicate the same. Estimate of the exponent 'm' in the non-linear model is found to be close to unity (Table 2). It may be noted that for m=1, the model reduces to the linear case, k and x values obtained form linear as well as non-linear models for this example appear to be fairly close. Example 2 was chosen particularly because of the strong non-linear storage-weighted flow relationship. This can be easily verified by plotting storage versus weighted flows. This is also evident from the high values of exponent 'm' and widely differing values of 'k' for linear and non-linear models (Table 2). The outflow hydrographs obtained by using linear and non-linear models for Example 2 are shown along with the observed outflow hydrograph in Table 3.

Conclusions

The tedious computations involved in flow routing can be easily implemented on Personal Computers. Parameter estimation and routing calculations for both linear and non-linear models can be performed with ease and speed with the commonly available PC softwares. Although Lotus 1-2-3 and 'TK Solver' were used in this study, many other softwares are also commercially available for spreadsheet calculations, solving non-linear equations and Linear Programming problems. Graphing capability in PC's is of particular advantage for producing storage and flow plots.

References

1. Croley II, T.E., 1980, Hydrologic and Hydraulic Calculations in BASIC for Small Computers, Iowa Institute of Hydraulic Research, pp. 156-170.

2. Gavilan, G. and Houck, M.H., 1985, Optimal Muskingum River Routing, Proc. of the ASCE WRPMD Specialty Conference on Computer Applications in Water Resources, June 10-12 at Buffalo, New York, pp. 1294-1302.

3. Gavilan, G. and Houck, M.H., 1986, Optimal Parameter Estimation for Muskingum River Routing, TR CE-HSE-86, School of Civil Engineering, Purdue University, Indiana 47907.

4. Gill, M.A., 1978, Flood Routing By The Muskingum Method, Journal of Hydrology, Vol. 36, pp. 353-363.

5. Heggen, R.J., 1984, Univariate Least-Squares Muskingum Flood Routing, Water Resources Bulletin, Vol. 20, pp. 103-107.

6. Linsley, R.K. and Franzini, J.B., 1979, Water Resources Engineering, McGraw-Hill, New York, p. 74.

7. Linsley, R.K., Kohler, M.A. and Paulhus, J.L.H., 1975, Hydrology for Engineers, McGraw-Hill, New York, 2nd ed., pp. 316-317.

8. Padmanabhan, G., Williams, R.K., and Rogness, R.O., 1986, On Optimizing Coefficients of the Muskingum Flow Routing Model, Proc. of the North Dakota Academy of Science, 78th Annual Meeting, April 1986, p. 104.

9. SAS User's Guide: Statistics SAS Institute Inc., 1985, SAS Circle, P.O. Box 8000, Cary, NC. 27511-8000.

10. Singh, V.P. and McCann, R.C., 1980, Some Notes on Muskingum Method of Flood Routing, Journal of Hydrology, Vol. 48, pp. 343-361.

11. Wilson, E.M., 1974, Engineering Hydrology, Macmillan, London, p. 141.

12. Wu, J.S., Kilng, E.L., and Wang, M., 1985, Optimal Identification of Muskingum Routing Coefficients, Water Resources Bulletin, Vol. 21, pp. 417-421.

TABLE 1. Estimates of Parameters
Example used in linear modeling

Method	k (days)	x
Graphical (by the author)	0.780	0.250
Multiple Linear Regression	0.763	0.275
Linear Programming	0.591	0.225

TABLE 2.   Estimates of Parameters
Examples used in non-linear modeling

Model	Example	Eqn. No.	k	x	m	A
Linear	1	1	1.140	0.200	–	–
	2	1	6.000	0.254	–	–
Non-linear	1	2	1.554	0.203	0.937	-58.415
	1	3	1.571	0.209	0.935	-57.782
	2	2	0.005	0.275	2.495	5.967
	2	3	0.029	0.222	2.096	-10.156

TABLE 3.   Example 2[*]

Time	Inflow	Outflow			
		Actual	Linear Model	NL Model[**] (Eq. 2)	NL Model[**] (Eq. 3)
0	22	22	22.00	22.00	22.00
6	23	21	21.80	22.09	22.05
12	35	21	19.64	23.19	22.29
18	71	26	15.52	26.00	18.33
24	103	34	20.21	33.75	24.25
30	111	44	35.17	46.72	46.78
36	109	55	50.73	58.42	62.57
42	100	66	64.19	68.67	74.12
48	86	75	74.15	77.10	81.67
54	71	82	79.52	82.58	84.89
60	59	85	80.22	84.27	84.26
66	47	84	78.37	83.73	81.38
72	39	80	73.70	79.80	76.24
78	32	73	68.16	74.04	69.94
84	28	64	61.73	66.06	62.69
90	24	54	55.78	57.31	55.16
96	22	44	49.83	47.37	47.41
102	21	36	44.46	37.63	40.06
108	20	30	39.97	29.48	33.58
114	19	25	36.17	23.66	28.15
120	19	22	32.74	20.36	23.96
126	18	19	30.19	19.16	21.32

[*] Time in hours and flows in cumecs
[**] Parameter estimation by NL regression

A FILTERING APPROACH TO FLOOD ROUTING

Yun-Sheng Yu* and Wang Guang-Te**

INTRODUCTION

Hydrologic river routing and hydraulic channel routing are in general use for flood routing. Both approaches use deterministic mathematical models and treat inflow and outflow hydrographs for a given river reach as discrete time series. The former is based on the continuity equation and the latter solves the governing Saint-Venant equations numerically. Mathematical models are subject to errors in estimated model parameters, error in model structure, and error in numerical schemes used. In addition, errors in field measurements are virtually unavoidable. These errors may affect seriously the resulting flood hydrograph and should be dealt with in the solution. This paper combines a hydrologic river routing model and a filtering technique to provide optimal estimates of outflow hydrograph.

METHODOLOGY

Routing Equation

The hydrologic river routing may be considered as a special case of the more general input-output model. For a discrete, linear model, the general relationship between the time-dependent inflow $I(t)$ and outflow $Q(t)$ may be represented by the following equation (Box and Jenkins, 1976)

$$(1 - a_1 B - a_2 B^2 - \ldots - A_p B^p) Q(t)$$
$$= (b_0 + b_1 B + b_2 B^2 + \ldots + b_q B^q) I(t) \tag{1}$$

where B is a backward shift operator, for example, $B[Q(t) = Q(t-1)]$; a's and b's are time-invariant coefficients. Equation 1 represents an autoregressive and moving average process of orders p and q known as ARMA (p,q).

For flood routing application, an ARMA (3,2), which accounts for linear dependence of the current outflow on inflows and outflows of

*Professor, Department of Civil Engineering, University of Kansas, Lawrence, KS 66045
**Associate Scientist, Institute of Geography, Academia Sinica, Beijing, China

several time units back, is suggested. The equation is

$$Q(t) = a_1 \, Q(t-1) + a_2 \, Q(t-2) + a_3 \, Q(t-3)$$
$$+ \, b_0 \, I(t) + b_1 \, I(t-1) + b_2 \, I(t-2) \tag{2}$$

Equation 2 reduces to the Muskingum routing equation if b_2, a_2, and a_3 are set to zero. The six coefficients in Eq. 2 must be estimated based on available flood hydrograph data and will be presented later.

Filtering Equations

Filtering refers to estimating the state vector at the present time based on all past measurements. In flood routing, the state variable is the discharge at the downstream station. This paper uses the Kalman filter (Kalman, 1960) for estimating the state variable in a recursive manner. The state-space equations in augmented vector forms are:

System equation

$$\underline{Q}(t) = A \, \underline{Q}(t-1) + B(t) + \underline{W}(t) \tag{3}$$

and measurement equation

$$\underline{Z}(t) = C \, \underline{Q}(t) + \underline{V}(t) \tag{4}$$

where $\underline{W}(t)$ and $\underline{V}(t)$ are, respectively, model error and measurement error; $\underline{Z}(t)$ is observed state vector at time t; A, B, and C are matrices. Equation 3 is an augmented vector form of Eq. 2 with an additive noise term $\underline{W}(t)$. $\underline{W}(t)$ and $\underline{V}(t)$ are assumed to be white noises with zero mean.

The following prediction and updating equations (Gelb, 1974) are used for the state variable and its error-covariance matrix.

State prediction,

$$\underline{Q}(t+1 \mid t) = A \times \underline{Q}(t \mid t) + B \times \underline{I}(t+1). \tag{5}$$

Predicted error covariance matrix,

$$P(t+1 \mid t) = A \times P(t \mid t) \times A^T + R_w. \tag{6}$$

Kalman gain factor,

$$K(t+1) = P(t+1 \mid t) \times C^T \times [C \times P(t+1 \mid t) \times C^T + R_v]^{-1}. \tag{7}$$

State estimate update,

$$\underline{Q}(t+1 \mid t+1) = \underline{Q}(t+1 \mid t) + K(t+1) \times [\underline{Z}(t+1) - C \times \underline{Q}(t+1 \mid t)]. \tag{8}$$

Updated error covariance matrix,

$$P(t+1|t+1) = [I - K(t+1) \times C] \times P(t+1|t) \qquad (9)$$

where I is an identity matrix.

$Q(t+1|t)$ and $Q(t+1|t+1)$ are, respectively, predicted state based on the model and updated state after a measurement at time $(t+1)$ is made. The following example demonstrates the use of the algorithm.

APPLICATION

The 1974 Yangtze River Flood

The Yangtze River is the largest river in China known for its rampaging floods. For the 1974 flood, the river reach under consideration is between Wanxian, Sichuan province at the upstream end and Yichang, Hubei province at the downstream end. The drainage area at Yichang is about $1,000,500$ km^2. The flood lasted about 35 days from July 27 through August 30, 1974. The recorded daily inflows and outflows are shown in the second and third column of Table 1 (The Yangtze River Basin Planning Office, 1978). The inflow values also include the runoffs from the intervening drainage area between the two stations. The peak discharge at the downstream station is 61,000 m^3/s ranked the fifth among the highest peak discharges since flow records were kept in 1800. The estimated recurrence interval of the flood peak is about 37 years.

Determination of Model Parameters

The six parameters of ARMA $(3,2)$, Eq. 2, are estimated based on the data in Table 1 and by using the Lagrangean method. The Lagrangean method seeks estimators that, while satisfying prescribed constraints, minimize the sum of squared residuals between the observed outflows $\hat{Q}(t)$ and the outflows $Q(t)$ computed from Eq. 2. For uncorrelated residuals, the objective function may be written as

$$\text{minimize} \quad f(\underline{\beta}) = \underline{e}(t)^T \times \underline{e}(t) \qquad (10)$$

$$\text{subject to } a_1 + a_2 + a_3 + b_0 + b_1 + b_2 = 1 \qquad (11)$$

where $\underline{\bar{e}}(t) = \hat{Q}(t) - D\underline{\beta}$ and $\underline{\beta}^T = (a_1\ a_2\ a_3\ b_0\ b_1\ b_2)$. Equation 11 is obtained by summing all $Q(t)$ from ARMA $(3,2)$ model from $t = 1$ to m and using the continuity equation, Wang and Yu (1986). Equations 10 and 11 form a nonlinear optimization problem and can be solved conveniently by using the Lagrangean method. The resulting ARMA $(3,2)$ is

$$Q(t) = 0.740\ Q(t-1) - 0.182\ Q(t-2) + 0.0138\ Q(t-3) + 0.190\ I(t)$$

$$+ 0.350\ I(t-1) - 0.120\ I(t-2) \qquad (12)$$

The computed outflows for July 30 through August 30 from Eq. 12 are shown in the fourth column of Table 1.

Table 1.   The 1974 Flood of the Changjiang (Yangtze) River
(July 27-August 30) in 10^3 m^3/s

Date	Inflow	Outflow		
		Historical	Eq. 12	Kalman Filter
July 27	17.9	17.9		
28	18.2	18.0		
29	25.7	18.1		
30	36.5	23.7	24.1	24.1
31	38.7	32.7	31.9	31.5
August 1	35.7	36.3	35.4	36.1
2	30.5	35.4	34.4	34.9
3	28.4	31.8	31.3	31.8
4	35.3	30.2	30.4	30.6
5	39.1	35.1	33.6	33.4
6	39.2	38.3	36.7	37.8
7	38.3	38.4	37.8	38.7
8	36.8	37.0	37.5	37.6
9	36.0	36.3	36.5	36.1
10	48.3	38.8	38.1	38.0
11	65.4	47.9	47.1	47.6
12	67.7	58.4	58.4	58.9
13	56.8	61.0	61.8	61.7
14	47.5	56.3	56.6	56.0
15	37.8	47.8	48.4	48.4
16	29.6	38.9	39.6	39.2
17	23.9	31.5	31.6	31.3
18	23.2	26.0	26.1	26.1
19	33.1	24.5	25.7	25.6
20	45.2	30.3	32.8	31.2
21	49.5	39.5	40.7	39.6
22	42.9	43.8	44.7	44.1
23	36.3	41.2	42.1	41.6
24	35.3	38.6	37.9	37.3
25	38.2	36.1	36.3	36.9
26	35.5	36.3	36.4	36.2
27	31.3	34.9	34.7	34.6
28	27.5	31.8	31.5	31.7
29	25.0	28.7	28.1	28.3
30	25.0	26.8	25.8	26.1

Flood Routing by Kalman Filter

For comparison, ARMA (3,2) model will be used together with the Kalman filter to route the 1974 flood. The recursive calculation is initiated by setting July 29 as t = 0 and using the first three values of historical outflows in Table 1 as the initial state vector. The augmented initial state vector is

$$\underline{Q}(0|0) = (18.1 \quad 18.0 \quad 17.9)^T. \tag{13}$$

The estimated error covariance matrix is

$$P(0|0) = \begin{vmatrix} 0.01 & 0 & 0 \\ 0 & 0 & 0 \\ 0 & 0 & 0.01 \end{vmatrix}. \tag{14}$$

Each number in the above matrix is in $10^6 \ m^6/s^2$. The covariance matrix of the model error in Eq. 6 is

$$R_w = \begin{vmatrix} \sigma_w^2 & 0 & 0 \\ 0 & \sigma_w^2 & 0 \\ 0 & 0 & \sigma_w^2 \end{vmatrix} \tag{15}$$

where $\sigma_w^2 = 0.681 \times 10^3 \ m^6/s^2$ obtained from the following equation

$$\sigma_w^2 = \frac{1}{m} \sum_{t=1}^{m} e^2(t). \tag{16}$$

A similar expression is used for the covariance matrix of model error R_v in Eq. 7.

With $\underline{Q}(0|0)$, $P(0|0)$, R_w and R_v known, $\underline{Q}(1|0)$ can be computed from Eq. 5, $P(1|0)$ from Eq. 6, $K(1)$ from Eq. 7, the updated state estimate $\underline{Q}(1|1)$ from Eq. 8, and the updated covariance matrix $P(1|1)$ from Eq. 9. Recursive calculations give the filtered values of outflow from July 30 through August 30 shown in the last column of Table 1.

The computed flood hydrographs with and without the Kalman filter are shown in Fig. 1 together with the measured hydrograph. The filtered values (open circles) generally are closer to the observed values than the values (crosses) obtained from Eq. 7. Figure 1 shows good agreement between the observed and the filtered hydrographs. The error variances of computed outflows from Eq. 7 without filter and with filter are, respectively, $0.653 \times 10^6 \ m^6/s^2$ and 0.399×10^6 m^6/s^2. Use of the Kalman filter reduces the error variance of estimated flood hydrograph by 39%.

CONCLUSION

Combined use of the Kalman filter and an autoregressive and moving average model for hydrologic river routing provides optimal

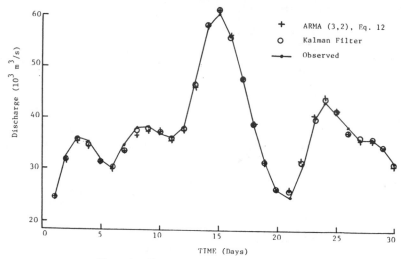

Figure 1.   Observed and Computed Flood Hydrographs for
ARMA (3.2) Model With and Without Kalman Filter

estimates of the flood hydrograph with the least error variance.
Excellent agreement between filtered hydrograph and observed
hydrograph has been achieved.   Conjunctive use of the ARMA (3,2)
routing equation and the Kalman filter reduces the error variance of
estimated flood hydrograph by about 39%.

APPENDIX - References

Box, G.E.P., and G.M. Jenkins, (1976).   Time-Series Analysis--
    Forecasting and Control. Holden-Day, San Francisco, CA.

Gelb, A., Editor, (1974). Applied Optimal Estimation. The MIT Press,
    Cambridge, MA.

Kalman, R.E., (1960).   "A New Approach to Linear Filtering and
    Prediction Problems."   Transactions, American Society of
    Mechanical Engineers, 82:   35-44.

Wang, Guang-Te, and Y.S. Yu, (1986).   "Estimation of Parameters of the
    Discrete, Linear, Input-Output Model." J. of Hydrology, 85: 15-
    30.

Yangtze River Basin Planning Office, Division of Hydrologic Data
    Analysis, (1978). "Flood Routing Techniques."

Utility Programs for DWOPER and DAMBRK

Philip A. Burns, Associate Member ASCE*

Abstract

Engineers and hydrologists that routinely use the National Weather Service (NWS) programs DWOPER (Operational Dynamic Wave Model) and DAMBRK
(Dam-break) may, as the author did, tire of struggling with their cumbersome data input formats. DAMBRK has distinct advantages over DWOPER
by having the cross section mileage, elevations, top-widths and storage
widths grouped together whereas with DWOPER this data is distributed to
four different areas of the input format, making it both difficult to
modify a cross section and to see the data for a cross section at a
glance. But even with DAMBRK, data for each reach such as lateral inflow hydrographs, interpolation distances, coefficients of expansion
and contraction, Manning coefficients, floodplain compartments, etc.
are in discrete areas of the format and are each linked to a designator
(an integer input) which instructs the program as to where features are
located. Adding or removing even one cross section requires time-
consuming and error-prone recoding of these designators and others. As
a result, time and computer resources greatly limit potential investigations and sensitivity analyses. In order to facilitate modeling with
theses programs, the author has written a series of utility programs
which reformat and/or manipulate the data to allow much easier, faster
and reliable editing. These programs also allow the user to make a
DWOPER deck from a DAMBRK deck or vice-versa so that the individual
capabilities of each NWS program are readily available for any study.
These utility programs are fast, running in several seconds on a
Harris-500 computer.

The Programs

The programs' functions are summarized below. Afterward, a more
thorough discussion is provided.

1a) DBFRNWS (DAMBRK From NWS): This program reformats the input data
of a DAMBRK deck into a sequential format. The sequential format organizes the data in an upstream to downstream sequence, allowing the
information to be seen as one might see it canoeing the stream.
Instead of different types of reach data being scattered throughout the
deck, they are organized into the following order: mile, elevations,
top-widths, storage-widths, Manning coefficients, expansion/contraction
coefficients and interpolation distances (all required input) and, if
they occur, lateral inflow hydrographs, floodplain compartments, pumps
and landslides.

*Hydraulic Engineer, Army Corps of Engineers, Baltimore District,
P.O. Box 1715, Baltimore, MD 31023-1715

1b) DBTONWS (DAMBRK To NWS): This program reformats the edited
sequential format back to the NWS format so that it may be submitted to
DAMBRK.

2a) DWFRNWS (DWOPER From NWS): This program has the same function as
DBFRNWS except it is for a DWOPER deck.

2b) DWTONWS (DWOPER To NWS): This program reformats the edited
sequential format back to the NWS format so that it may be submitted to
DWOPER.

3) DWPREP (DWOPER Preparation): This program will, after being provided the basic parameters for a DWOPER deck, provide a listing of all
input requirements, entry by entry, and will exclude all extraneous
input cards.

4) DWTODB (DWOPER To DAMBRK): This program will make a DAMBRK deck
from a DWOPER deck.

5) DBTODW (DAMBRK To DWOPER): This program will make a DWOPER deck
from a DAMBRK deck.

6) Miscellaneous Functions: The programs DBFRNWS and DWFRNWS identify
and read in all data types, which are assigned the same variable names
as used by the NWS programs. Consequently, equations may be inserted
directly into the programs so that the data is read, changed and then
"echoed" back. For example, assume that one wishes to add three cross
sections to the upstream end of a deck which already contains eighty
cross sections; rather than manually changing eighty cross section
mileages, one FORTRAN statement may be inserted to simply add an increment to each cross section mileage. If modified programs such as these
are preserved and if, as in this case, the increment is read as a variable, tedious and error-prone chores can be reduced to a minute's work.

Discussion

Programs DBFRNWS, DBTONWS, DWFRNWS and DWTONWS

The sequential formats are not intended for infrequent users of DWOPER
or DAMBRK. They are intended for persons who are already familiar with
the NWS formats and for those who are able to recognize the various
data at a glance. To the untrained eye, the sequential formats look
like a confusing and disorganized array of numbers; the organization is
apparent only if the user is knowledgeable of the input requirements
and is familiar with what each set of data looks like.

Frequent users of DAMBRK and DWOPER may take advantage of both the NWS
format and the sequential format. Both formats offer advantages over
the other. Consider, for instance, that one wishes to remove all
lateral inflow hydrographs from a deck rather than adding or removing
some of them. This modification is easier to do in the NWS format
since the data is grouped in one place. Adding or removing some of the
lateral inflow hydrographs, on the other hand, is more easily carried
out in the sequential format since it is simpler to identify the individual hydrographs and because location designators are unnecessary;

the programs automatically "know" where these and other features occur.

Besides having the Job Control Language (JCL) to go back and forth between the format types, it is handy to write JCL that will submit a sequential format directly to the NWS program to avoid a two-step procedure. This is easily accomplished by JCL which prompts for the sequential deck, executes DBTONWS or DWTONWS as appropriate, writes the NWS format to a scratch file, then submits the scratch file for execution in the NWS program.

Program DWPREP

The idea for this program was motivated by the awkward business of supplying correct data in the proper sequence to DWOPER. The documentation is plagued with notes such as the following: "If NCSS>0 (see card input no. 5), and NCSS1(J)>0 (see card input no. 9), read in BSS(K,I,J)"; "repeat card sequence 31, 32 and 33 for each I, I=1,NCSS(J) and repeat this sequence for each J, J=1,JN." The user is continually distracted by this type of note and spends more attention on how to input the data than on what is being input. First attempts to execute a newly input deck in DWOPER almost inevitable get aborted due to data input errors.

DWPREP provides an entry-by-entry listing for the particular model the user wishes to build. ALL DWPREP requires of the user is information such as the number of rivers being modeled, the number of cross sections for each, the type of boundary conditions which will be used, and so on. All the user then needs to do is answer each question from the DWPREP output. In this way, the user is allowed to concentrate on the actual data rather than on input procedure.

Programs DBTODW and DWTODB

These programs were written so that the individual capabilities of both DWOPER and DAMBRK would be easily and readily available for any study utilizing either of the NWS programs. The utility programs alleviate the necessity of typing all the data into the other format from scratch as well as providing protection from input errors and typos. Besides the input formats being different for the two NWS programs, the input requirements and options also have differences. It will, therefore, be necessary to make some modifications after converting one type of deck into another, but fifteen minutes work or so should provide a working model. Additional time will probably be required to calibrate it and to adjust storage volumes. The programs are designed to convert only the required data plus lateral inflow hydrographs; options such as levees and floodplain compartments are omitted.

Debug Status

These utility programs were written for the exclusive purpose of facilitating and expediting the use of NWS programs by the author and co-workers. Each program was designed to be simple to use and each is provided with a brief documentation so that others may use them without requiring personal instruction. The programs were written to have the capability of handling any and all options in DAMBRK and DWOPER;

however, in that the programming was justified only when it provided an overall savings to the office, the programs have been debugged only for the options actually used by the office. Untested options may need debugging and, undoubtedly, the programs could be improved upon with more time and effort.

Availability

At this time, there is no established procedure for the distribution of these programs. However, it is hoped that a procedure may be arranged through the Hydrologic Engineering Center (HEC) or that some other distribution procedure will be established. Those readers who wish to inquire as to program availability in the future may write to Philip Burns, 3420 Sylvan Lane, Ellicott City, MD 21043 or call (301)962-4840. Please include a self-addressed, stamped envelope with any correspondence which solicits a reply.

A HYBRID MODEL FOR STORAGE ROUTING THROUGH ROCK DUMPS
Anand Prakash*, F. ASCE

ABSTRACT

Construction of permeable structures across stream channels results in temporary impoundment of flood waters upstream of such structures. This happens because the hydraulic conveyance of the structure is considerably less than that of the natural channel. To develop the outflow hydrograph downstream of such structures, a hybrid model is developed which is made up of two main components. The first part performs storage routing computations through the impoundment and the second performs computations for unsteady one-dimensional turbulent flow through the rockfill. Like most storage routing programs, the model operates iteratively until the volume of outflow in each small time interval equals the difference between the incremental inflow and storage. The model can be used to predict the attenuated outflow hydrograph for storm hydrographs from watersheds upstream of a rockfill.

INTRODUCTION

Storm runoff entering stream channels obstructed by permeable check dams, debris basins, rockfill weirs, or rock dumps gets temporarily impounded upstream of such structures because their discharge capacity is generally lower than that of the upstream channel. As a result, the flood hydrographs get significantly attenuated and spread over longer time bases. The degree of attenuation is a function of the elevation-area-capacity characteristics of the impoundment upstream of the structure and the hydraulic characteristics of the structure itself. Flood routing computations for such cases require a method to route the inflow hydrograph through the impoundment and an algorithm to compute the discharging capacity of the rockfill. The objective of this paper is to present a hybrid model comprised of the Muskingum or step-by-step method for storage routing through the impoundment (USACE, 1981; Prakash, 1982) and an explicit finite-difference scheme to model unsteady flow through the permeable structure.

MATHEMATICAL FORMULATION

In principle, any of the commonly known methods of storage routing (USACE, 1981) can be used for the first component of the model. The Muskingum and step-by-step storage routing methods are arbitrarily selected as a matter of convenience. The Muskingum equations for flood routing are:

* Chief Water Resources Engineer, Dames & Moore, Golden, CO

$$O_2 = C_1 O_1 + C_2 I_1 + C_3 I_2$$

$$C_1 = \frac{k - kx - 0.5\Delta t}{k - kx + 0.5\Delta t} \quad ; \quad C_2 = \frac{kx + 0.5\Delta t}{k - kx + 0.5\Delta t} \quad ;$$

and $\quad C_3 = \frac{- kx + 0.5\Delta t}{k - kx + 0.5\Delta t}$

in which O_1, O_2 and I_1, I_2 = outflow and inflow rates at time steps 1 and 2, respectively; k = storage time constant; x = weighting factor; and Δt = time step of computations. The step-by-step method involves trial and error computations for each time step until the average inflow minus outflow volume matches the change in storage during that period.

The flow of water through a rockfill is generally in the transitional to turbulent regime. Empirical equations defining such flows have been proposed by Ward (1964), Leps (1973), Bear (1979), and others. Herein, the following equation given by Leps (1973) is used,

$$u = Cr^{0.5} \ i^{0.54} \tag{1}$$

in which u = velocity of flow through the voids; r = hydraulic radius of flow; and i = hydraulic gradient. The hydraulic gradient may be defined as,

$$i = S_o + \frac{dy}{dx} \tag{2}$$

in which S_o = bed slope of the rockfill; y = depth of flow through the rockfill; and x = longitudinal distance along the direction of flow. Leps (1973) has tabulated the values of the product, $Cr^{0.5}$, for mono-sized rocks of different sizes and has suggested that for graded rockfills with no more than 10 percent of the particles by weight being less than one inch (2.54 cm) in size, the dominant size may be taken to be the d_{50} of the rockfill.

The one-dimensional equation of mass conservation for unsteady flow of water through the rockfill is,

$$\frac{\partial}{\partial t} (Ap) + \frac{\partial}{\partial x} (uAp) = q \tag{3}$$

in which A = area of flow; p = porosity; t = time; and q = rate of infiltration per unit length of the rockfill. Using the quadratic upstream interpolation scheme given by Leonard (1979), Eq. 3 can be written in the following finite-difference form,

$$\phi_i^{n+1} = \phi_i^n + \frac{\Delta t}{\Delta x} \left[\left\{ \frac{U_L}{2} \left(\phi_{i-1}^n + \phi_i^n \right) - \frac{U_L^2}{2} \cdot \frac{\Delta t}{\Delta x} \left(\phi_i^n - \phi_{i-1}^n \right) \right\} \right.$$

$$\left. - \left\{ \frac{U_r}{2} \left(\phi_i^n + \phi_{i+1}^n \right) - \frac{U_r^2}{2} \cdot \frac{\Delta t}{\Delta x} \left(\phi_{i+1}^n - \phi_i^n \right) \right\} \right] + \Delta t . q \qquad (4)$$

in which ϕ = Ap; subscripts i-1, i, and i+1 refer to the cross-section numbers reckoned from the upstream boundary; superscripts n and n+1 refer to the time step of computations; Δx = distance between adjacent cross sections; Δt = time step; U_L = velocity of flow between cross sections i-1 and i; and U_r = velocity of flow between cross sections i and i+1.

For computational convenience, Eq. 4 any be written as,

$$\phi_i^{n+1} = \frac{1}{2} \left(- C_r + C_r^2 \right) \phi_{i+1}^n + \left[1 + \frac{1}{2} \left(C_L - C_L^2 - C_r - \right. \right.$$

$$\left. \left. C_r^2 \right) \phi_i^n \right] + \frac{1}{2} \left(C_L + C_L^2 \right) \phi_{i-1}^n + \Delta t . q \qquad (5)$$

in which C_r and C_L are Courant numbers given by $U_r \frac{\Delta t}{\Delta x}$ and $U_L \frac{\Delta t}{\Delta x}$, respectively. To ensure stability of this explicit formulation, the selected values of Δt and Δx should be such that the Courant number is always less than 0.5.

If the depth of flow at the exit end of the rockfill is equal to or less than that in the downstream channel, the hydraulic gradient at the downstream boundary is taken to be equal to the energy slope in the downstream channel. Otherwise, the hydraulic gradient at the downstream boundary is assumed to be equal to the slope of the downstream face of the rockfill. The upstream boundary is defined by the transient water surface in the impoundment upstream of the rockfill. The initial condition is taken to be a prescribed depth of steady flow through the rockfill. The inflow and outflow for the rockfill are given by the product, uA, for the upstream and downstream sections, respectively.

COMPUTATIONAL PROCEDURE

The computational steps required to perform flood routing through a system of impoundment and rockfill are:

(i)   Identify the inflow hydrograph from the upper part of the basin and divide it into small time segments Δt apart.

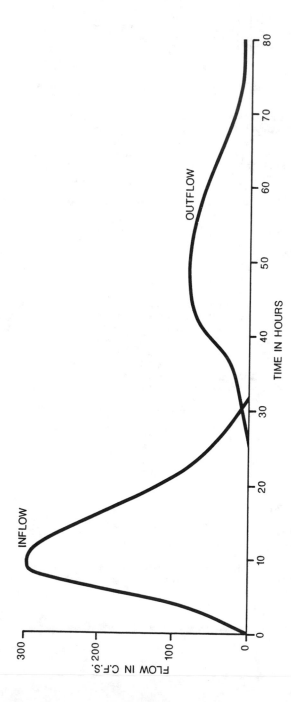

Figure 1. Inflow and Outflow Hydrographs for a Test Rockfill

(ii)   Identify I_1, I_2, and O_1 for the time step Δt and estimate the corresponding trial water surface elevation in the impoundment.

(iii)   Perform routing computations through the rockfill using the algorithm described previously and obtain O_2.

(iv)   Use the estimated O_2 to compute $\dfrac{O_1 + O_2}{2}$ and recompute the water surface elevation in the impoundment and repeat step (iii).

(v)   Repeat steps (ii), (iii), and (iv) until an acceptable match is obtained between the volumes of inflow and outflow and the change in storage during the period Δt.

(vi)   Start step (ii) for the next time step.

To illustrate the application of the methodology, the inflow hydrograph for a typical impoundment and the outflow hydrograph for the rockfill for a test case are shown in Figure 1.

CONCLUSION

Construction of permeable rockfill structures across natural stream channels results in restricting their hydraulic conveyance. As a result, storm runoff gets impounded upstream of such structures and the outflow hydrograph is greatly attenuated and extended over a longer period of time. To perform flood routing computations through such structures, a hybrid model is presented. It consists of an algorithm to perform storage routing through an impoundment and an explicit finite-difference scheme to perform computations for unsteady one-dimensional turbulent flow through a rockfill. The model is used to predict the outflow hydrograph corresponding to a given inflow hydrograph at the upstream edge of a rockfill.

APPENDIX

Bear, J. (1979). "Hydraulics of Groundwater," McGraw-Hill Book Company, New York, New York.

Leonard, B.P. (1979). "A Stable and Accurate Convective Modeling Procedure Based on Quadratic Upstream Interpolation," Computer Methods in Applied Mechanics and Engineering, 19 (1979) 59-98, North-Holland Publishing Company.

Leps, T.M. (1973). "Flow Through Rockfill" in Embankment Dam Engineering, Casagrande Volume, John Wiley & Sons, New York, New York.

Prakash, A. (1982). "Flood Routing with and without Dam-Break," ASCE Convention, Las Vegas, Nevada.

USACE (U.S. Army Corps of Engineers) (1981). "Flood Hydrograph Package," User's Manual, The Hydrologic Engineering Center, Davis, California.

Ward, J.C. (1964). "Turbulent Flow in Porous Media," Journal of Hydraulics Division, ASCE, HY5, No. 90.

SAFETY DESIGN FLOOD FOR TOO WIDE CREEK DAM

Joseph S. Haugh, M. ASCE*

Introduction. This paper presents an example application of
the procedure developed by the ASCE Task Committee on
Spillway Design Floods for a dam on the Too Wide Creek. The
example was developed from actual field data supplemented by
use of reasonable assumptions for ease in illustrating the
procedure. Although it is recognized that dams can fail due
to many causes, of which overtopping is only one, for
purposes of this paper it is assumed that all other modes of
possible failure have been adequately addressed. This paper
is therefore limited to a discussion of the design to prevent
an overtopping failure.

Background. In 1984, an ASCE task committee was appointed
to develop a procedure for determining the proper criteria
to be used for selecting the design flood for dams. The
committee has developed a recommended procedure and is in
the process of finalizing it for publication.

The procedure consists of a basic analysis of several
alternative design levels. The final design is selected
based on a trade-off of the various alternatives considered.
In that sense, the procedure is a "full disclosure" of
various alternatives whereby the decision makers can be made
fully aware of trade-offs from among the alternatives. In
addition, the public can be made aware of what it would cost
to design a dam to higher levels or, conversely, what is
sacrificed by designing to a lesser level.

Procedures. For any new dam, an early step is to evaluate
the impacts of dam failure. If the failure of the dam would
result in catastrophic damages and significant loss of life,
the answer is easy; design the dam for the most extreme
events, i.e., the probable maximum flood (PMF). The answer
is not so clear in the case of many dams, such as when the
failure would result in a loss but the loss is not
catastrophic. This is where an assessment of the risk comes
into play and various trade-offs are considered.

For those dams where the failure due to overtopping would
not result in a catastrophic loss, the first step is to make
an economic analysis comparing the costs for various
alternative design levels with the losses due to failure.

* Civil Engineer for Projects, Soil Conservation Service, US
Dept. of Agriculture, P.O. Box 2890, Washington, D.C. 20013.

One of the problems with the economic analysis is that when dealing with the less frequent events such as those used to size spillways and proportion a dam, the frequencies become very speculative. It is generally agreed that frequencies up to about the 100 year return level can be fairly well determined. However, floods normally used for design of dams are much less frequent and their frequency cannot be accurately determined. However, as will be pointed out later in the sensitivity analysis, this does not present a major problem for this example because it was determined that even changing the probabilities by several orders of magnitude did not appreciably affect the results. The biggest problem lies in the fact that many effects of dam failure are not quantifiable in monetary terms.

Therefore, it is often necessary to go to the next step of considering effects beyond economics so that the decision regarding proper design level considers all effects and not just those that can be expressed in monetary terms. For example, the value of a human life is not expressed in monetary terms although the decision maker is free to assign a value if he or she so chooses. It should be noted, in fact, that by selecting from among several alternative design levels when potential loss to human life is a factor, the decision maker is implicitly making these decisions by determining whether or not to go to higher design levels. For example, by selecting an alternative that costs $2 million more than another when one life is at risk, it is presumed that the one life is worth at least the $2 million.

Similarly, environmental and social effects are quantified to the extent possible and converted to economic values where possible. In addition, they are expressed in a separate account thus enabling the decision maker and the public to weigh potential effects to these items against the alternative design levels and their cost.

The procedure can be considered a form of risk assessment or risk analysis. The first step, i.e., based strictly on the economic analysis, can be considered a probabilistic risk analysis whereby all variables are expressed in common units (dollars) and the least costly alternative, considering all effects is determined. The succeeding steps then become the risk assessment in a more general sense whereby other factors are considered even though they are not measurable in monetary terms (threat to human life, environmental losses, social effects, etc). The procedure provides a system for display so that all alternatives and their effects can be compared and provide a rational basis for decision making.

It is important to note that each case is unique and the solution selected for one might not be appropriate for another. The example include in this paper is just that -

an example. If conditions are different from that presented
then the example may not apply.

Situation. The decision has been made to design a multiple
purpose dam on the Too Wide Creek for flood protection for
an agricultural area and municipal water supply for a small
rural community. The dam is located in a rural area which
is expected to remain rural with no potential for
development in the foreseeable future. See map of the area,
Figure 1.

The basic proportioning of the dam results in an initial
permanent pool for water supply and sediment allowances and
a temporary pool for flood storage. Given these basic
proportions, the problem is now to determine the appropriate
safety design flood and, considering the effect of reservoir
storage, proportion a spillway system to handle it.

The drainage area at the site is 1735 hectares. Two
occupied houses are located downstream. In each of the
various alternative designs considered in the example, these
houses are high enough above the floodplain that they would
not be affected by water levels during the passage of large
storms through the spillway (the non-failure condition) but
would likely be destroyed by the dam breach, resulting in
the probable loss of about four lives, based on current (and
projected future) conditions. For purpose of this example,
it is not considered a feasible alternative to relocate the
houses and occupants.

Alternatives. Four alternative designs were considered for
the dam with design precipitation levels ranging from a low
of 307 mm to a high of 752 mm, the Probable Maximum
Precipation (PMP). Each alternative design is characterized
by a unique inflow flood from the different amounts of
precipitation and resulting flood. For each alternative,
the dam is proportioned so that the resultant inflow flood

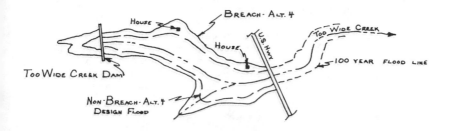

FIG. 1. DOWNSTREAM INUNDATION AREAS.

will not overtop the embankment. Dam heights varied from
15.6 meters to 18.6 meters. Initial installation cost
ranged from $1,460,000 to $3,280,000. The annual equivalent
operation and maintenance (O&M) costs ranged from $55,000 to
$70,000.

After determining the dam and spillway dimensions and cost
for each of the four alternatives, studies were made to
determine the effect of a dam breach at the top of dam
elevation for each alternative. The inundation information
for this condition for each alternative was compared to the
same hydrologic event routed downstream without a failure.
The difference between the "failure condition" and the "non-
failure condition" is the effect of the breach. This
information was used to develop cost data. Where the
effects were such that they could not be converted to
monetary terms, the information was quantified to the extent
possible in non-monetary terms.

Flood Frequencies. Precipitation for known frequencies at
the site were determined from TP-40. PMP was determined
from Hydromet Paper 51. This precipitation data was plotted
on semi-log paper and used as a basis for assuming
probabilities for Alternatives 2, 3, and 4.

Comparison of alternatives. Table I provides a summary of
significant data from the four alternative designs. For
purposes of comparing the economic effects, all costs must
be converted to a common time frame. Either average annual
costs or capitalized costs may be used but the time frame
must be consistent. In this case, capitalized costs (or
present values) were used. Indemnification cost is added as
another factor in Line 18. Consideration of indemnification
cost indicates the amount required by the owner to provide
financial relief to potential victims of the dam failure.
This cost is assumed to be zero for Alternative 1 since that
alternative is based on the PMF which is assumed to have a
zero probability of causing failure of the dam. For the
other alternatives, an assumed probability of occurrence is
used and can very easily be reviewed or reconsidered through
the use of a sensitivity analysis.

Sensitivity. Sensitivity studies can be conducted very
easily since the data was tabulated on a LOTUS 1-2-3 spread
sheet. This allowed changes in any one parameter to be
revised as desired and the results compared. Other changes
could also be easily made if additional studies are desired.

In the case of Too Wide Creek, Dam the final result was not
sensitive to the assumed probability of failure. It was
determined that the design flood for Alternative 4 would
need to have an annual probability of occurrence of about
one percent to begin affecting the results -- an unlikely
prospect since the design flood is based on a precipitation
amount of 307 mm, which is twice the size of the 100 year
precipitation. Significant changes in discount rate,
project life, and the incremental cost of money also did not
affect the ranking of the alternatives. The dam was most
sensitive to the consequences of failure. When the failure
consequences were doubled, Alternative 1 proved to be better
than 2 but was still more than Alternatives 3 or 4.

TABLE I. SUMMARY FOR DECISION MAKING - Too Wide Creek Dam.

		1 (PMF)	2	3	4
1.	ALT.NO.	1 (PMF)	2	3	4
2.	DESIGN PRECIP. (mm)	752	564	391	307
3.	DAM HEIGHT (meters)	18.6	18.0	16.8	15.6
4.	ANN PROB OF FAILURE	0	0.00001	0.0001	0.001

PROJECT COST ($ million)

5.	Initial Cost (Cap)	3.28	2.17	1.800	1.460
6.	Equiv Ann O&M	.070	.065	.060	.055

ONE TIME CONSEQUENCES OF FAILURE

7.	ECON DAMAGE ($m)	5.830	4.750	4.400	4.100

SOCIAL AND POLITICAL CONSEQUENCES

8.	RESIDENTS EXPOSED	4	4	4	4
9.	EXPCTD LIVES LOST	4	4	4	4
10.	TRAUMA	minor	minor	minor	minor
11.	WATER SUPPLIES	1	1	1	1
12.	ENV. IMPACTS	minor	minor	minor	minor
13.	POLITICAL CONSQ	minor	minor	minor	minor

ECONOMIC SUMMARY ($ MILLION)

14.	CAP PROJ COST	4.443	3.250	2.797	2.374
15.	EXP ANN FAIL CONSQ	0	0.000	0.000	0.004
16.	CAP FAIL CONSQ	0	0.001	0.007	0.068
17.	SUM OF COST & CONSQ	4.443	3.251	2.804	2.442
18.	INDEMN COST & CONSQ	4.443	4.023	3.576	3.199

Note: Project life = 100 yrs. Discount Rate = 6.0 %.
      Inc Cost of Money = 1.5 %.

From a strictly economic standpoint, the lowest level
design, Alternative 4, would be the selected alternative.
However, this ignores the threat to the four lives which
would be likely lost in event of failure. Looking at the
selection process, the question could be asked, "How much
should society (or the owner) pay to remove the threat to
the four lives?" From an analysis of the data summarized in
Table I, a move from Alternative 4 to Alternative 1, would
cost $1.244 million which would equate to $311,000 per life.
Even for the greatest changes made in the assumptions, the
cost equates to about $400,000 per life. If, in the
judgment of the decisionmaker, the human lives are worth at
least the $311,000, it would be rational to build this dam
for the most extreme conditions, or the PMP.

Summary. Selection of a safety design flood is not a
straightforward process except in those cases where the
consequences of failure are so severe that all reasonable
precautions should be taken to prevent failure. In such
cases, the proper design level is the PMF. For dams where
the consequences of failure are not considered catastrophic,
the safety design can be determined by a consideration of
several alternatives and displaying the results so that the
trade-offs can be readily compared.

In the case of Too Wide Creek, the economic analysis alone,
even when considering the indemnification cost to the owner
would have driven the decision to a flood smaller than the
PMF. However, when all other effects were displayed, the
decision can be based on all known facts especially the
threat to human life.

References

American Society of Civil Engineers, Flood Evaluation
Guidelines, Task Committee on Spillway Design Floods, Draft
Report, April 1, 1987.

U.S. Department of Commerce and US Department of the Army,
Hydrometeorlogical Report No. 51, Probable Maximum
Precipitation Estimates, United States, East of the 105th
Meridian, Washington, D.C., June 1978.

U.S. Department of Commerce, Technical Paper 40, Rainfall
Frequency Atlas of the United States, Washington, D.C., May
 1961.

Analysis of Drought Indicators
in Pennsylvania

D. F. Kibler, G. L. Shaffer, and E. L. White*

Abstract

The Pennsylvania Department of Environmental Resources currently monitors five parameters in its regional water supply management program. These are: (1) accumulated precipitation deficit; (2) groundwater level; (3) streamflow; (4) reservoir storage level; and (5) Palmer hydrologic drought index. This paper summarizes a recent research project to examine the historical record of each indicator and to evaluate the sensitivity and consistency of the indicators, together with their associated trigger points. Four phases of the project are identified as: (1) data collection and fill-in; (2) frequency analysis of raw indicators and adjustment of triggering criteria; (3) time-series modeling to establish lead-lag structure and box-plots to determine triggering consistency; (4) comparison of revised triggering criteria during periods of known drought. Although it was an extensive task, data fill-in procedures are not discussed here.

Background

At the height of the 1984-85 drought, more than 370 water suppliers in Pennsylvania alone experienced severe water shortages and were forced to use emergency water management procedures. By early summer 1985, conditions in the Pennsylvania portion of the Delaware River basin were so severe that Governor Thornburgh was forced to impose restrictions on all but the most essential uses of water. This experience was repeated in different degrees throughout much of the mid-Atlantic and southeastern regions of the country.

Experiences such as the 1980-81 and 1984-85 droughts have emphasized the need to adopt a set of indicators which can be used to measure the severity of regional drought and to provide an early warning system. The Pennsylvania Department of Environmental Resources (DER) currently monitors the following five indicators: (1) accumulated precipitation deficit; (2) groundwater level; (3) streamflow; (4) reservoir storage level; and (5) Palmer hydrologic drought index (PHDI). Each indicator is monitored weekly during dry periods throughout eleven regions and compared with a critical threshold or trigger point to determine whether conditions are normal,

*Respectively, Professor, Department of Civil Engineering, Penn State University and ASCE Member; Hydraulic Engineer, Earth Sciences Consultants, Exton, PA; Senior Research Associate, Department of Civil Engineering, Penn State University, University Park and ASCE Member.

drought watch, drought warning, or drought emergency. The more severe
the triggering, the more stringent the water supply conservation
measures must be.

Frequency Analysis of Drought Triggers

    A series of frequency analyses was performed on the raw indicator
data as well as on the indicator triggering occurrences. As an
example of this process, Table 1 is presented as a summary of raw
indicator frequency analysis carried out for the PHDI on a state-wide
basis. Reference is made to Palmer [1965] and Alley [1985] for
computational details of the PHDI. Monthly exceedence levels were
established in this way for each of the five indicators in all eleven
regions on an individual basis.

Table 1. Summary of regional PHDI excedence levels for the entire State

	JAN	FEB	MAR	APR	MAY	JUN	JUL	AUG	SEP	OCT	NOV	DEC	AVG
MEAN	-0.15	-0.16	-0.14	-0.14	-0.09	-0.06	-0.02	-0.02	-0.14	-0.05	-0.02	0.01	-0.08
STD.DEV.	0.30	0.31	0.28	0.28	0.27	0.28	0.29	0.29	0.30	0.31	0.31	0.32	0.29
MEDIAN	0.02	-0.08	-0.33	-0.06	-0.05	-0.01	0.23	0.25	-0.27	-0.25	0.08	0.25	-0.02

% OF TIME EXCEEDED	JAN	FEB	MAR	APR	MAY	JUN	JUL	AUG	SEP	OCT	NOV	DEC	AVG
0	4.51	3.99	3.85	3.82	3.63	4.35	4.01	4.15	4.34	4.21	3.96	4.03	4.07
5	3.28	3.10	2.59	2.51	2.69	2.82	2.80	2.68	3.12	3.34	3.31	3.21	2.95
10	2.71	2.50	2.25	2.20	2.28	2.34	2.30	2.31	2.67	2.76	2.88	2.86	2.50
25	1.71	1.61	1.51	1.53	1.45	1.55	1.59	1.65	1.60	1.82	1.82	1.82	1.64
40	0.82	0.68	0.73	0.61	0.70	0.95	0.97	1.00	0.79	0.78	0.95	1.01	0.83
50	0.02	-0.08	-0.33	-0.06	-0.05	-0.01	0.23	0.25	-0.27	-0.25	0.08	0.25	-0.02
60	-0.94	-1.01	-0.97	-0.83	-0.75	-0.82	-0.64	-0.78	-1.03	-1.02	-0.79	-0.67	-0.85
75	-1.80	-1.64	-1.56	-1.58	-1.50	-1.56	-1.62	-1.66	-1.77	-1.76	-1.81	-1.76	-1.67
90	-3.21	-2.82	-2.36	-2.48	-2.45	-2.55	-2.71	-2.66	-2.75	-2.83	-2.94	-2.98	-2.73
95	-3.79	-3.41	-3.02	-3.17	-3.07	-3.24	-3.35	-3.32	-3.36	-3.39	-3.48	-3.61	-3.35
100	-6.65	-6.58	-6.10	-5.62	-4.62	-4.70	-4.75	-4.33	-4.16	-4.12	-4.73	-4.90	-5.11

Adjustment of Drought Triggering Criteria

    A series of frequency analyses of indicator triggering was made
for all indicators operating under the original DER criteria. Using
region I as an example, we found that groundwater levels triggered
40.2 percent of the time, streamflows 24.6 percent, PHDI 20.6 percent,
reservoir levels 25.4 percent, 3-month precipitation deficit 18.5
percent, 6-month precipitation deficit 16.2 percent, 9-month
precipitation deficit 17.6 percent, and 12-month precipitation deficit
18.1 percent. Given these inconsistent triggering characteristics, it
was decided to adjust all trigger criteria to the 75, 90, and 95
percentiles for each indicator corresponding to drought watch, drought
warning, and drought emergency conditions, respectively. For the
groundwater indicator, this meant revising upward from the 60, 75, and

90 percent levels developed originally. Conversely, the PDHI criteria had to be adjusted downward as shown in Figure 1 to meet the revised baseline triggering targets set at the 75, 90, and 95 percentiles. The criteria for streamflow, storage, and precipitation deficits were similarly adjusted to the revised thresholds to establish the stages for drought watch, warning, and emergency.

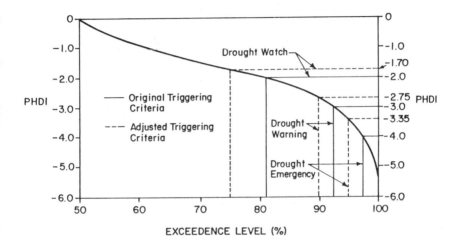

Figure 1. PDHI annual triggering criteria

Time Series Modeling to Establish Indicator Lead-Lag Structure

Before attempting to identify principal cross-correlations and dominant lags (months) between each of the indicators, it was necessary to "pre-whiten" the data series and thereby induce stationarity. Accordingly, an autoregressive, integrated, moving average model (ARIMA) was fitted to the time series for each indicator using methods described by Vandaele [1983]. The models for a given pair of indicators were found to be very consistent from region to region in terms of the lag, the order of differencing, and in the moving average component. The cross-correlations were also quite uniform from region to region. Table 2 summarizes the most significant lags between all indicators. It is noted that a positive lag means that the independent variable lags the dependent variable. In the case of multiple lags, the most significant is shown first. A dash (-) indicates that no significant lag was found.

Table 2. Summary of significant lags (months) between all indicators

Dependent Variable	Independent Variable	I.	II	III	IV	V	VI	VII	VIII	IX	X	XI
GWL	FLOW	0	0,-1	-1,0	0	0	+3	+3	0	0	0	-2
GWL	PHDI	+3	+2,+3	+4	+3	+3	+3	+5	+3	-	+3	+5
FLOW	GWL	0	0	0,-1	0	0	-3	-3	0	0	0	+2
FLOW	PHDI	+3	+3	+3	+3	+3	0	0	+3	+3	+3	+3
PHDI	GWL	-3	-3	-4,-3	-3	-3	-3	-3	-3	-3	-3	-3
PHDI	FLOWS	-3	-3	-3	-3	-3	-3	-3	-3	-3	-3	-3
DEF3	GWL	-3	-3	-4	-3	-3	-3	-3	-3	-3	0	-4
DEF3	FLOWS	-3	-3	-3,0	-3,0	-3	0,+3	0,+3	-3,0	-3,0	-3,0	-3,0
DEF3	PHDI	0,+3	0,+3	0,+3	+3,0	+3,0	+3,0	0,+3	+3,0	+3,0	0,+3	0,+3
DEF6	GWL	-3	-3	-3	-3	-3	-3	-3	+3,-3	-3	+3	-2
DEF6	FLOWS	-3	+3,-3	+3,-3	+3,-3	+3,-3	+6,0	+6,0	+3,-3	+3,-3	+3,-3	+3,-3
DEF6	PHDI	0,+6	+6,0	0,+6	+6,0	+6,0	+6,0	0,+6	+6,0	+6,0	+6,0	0,+6
DEF9	GWL	-3	-3	-4	+6,-3	-3	-3	-3	+6,-3	-3	-3	-3
DEF9	FLOWS	-3	-3,+6	-3,+6	-3,+6	-3,+6	0,+9	+9,0	-3,+6	+6,-3	-3,+6	-3,+6
DEF9	PHDI	0,+9	+9,0	+9,0	+9,0	0,+9	0,+9	+9,0	+9,0	+9,0	0,+9	0,+9
DEF12	GWL	-3	+9,-3	-3,+8	+7	-3	-3	-3	-3	-3	+9	-3
DEF12	FLOWS	-3	+9,-3	-3,+9	-3,+9	+9,-3	0,+12	0,+12	+9,-3	-3,+9	-3,+9	+9
DEF12	PHDI	0,+12	0,+12	0,+12	0,+12	0,+12	+12,0	+12,0	0,+12	0,+12	+12,0	0,+12
RES	GWL	-4	-	-2	-	-	-	-	-1	-	-	-
RES	FLOW	-2	-	-2	-	-	-	-	-3	-	-	-
RES	PHDI	+1	-	+1	-	-	-	-	-	-	-	-

Indicator Performance in the Period 1976-85

Two very significant droughts were observed in this period. They were especially severe from a water supply standpoint in regions II and III which lie in the Delaware River basin of central-eastern Pennsylvania. Individual and mass triggering diagrams were developed for this period using both the original DER and the adjusted PSU triggering thresholds. Mass triggering diagrams were compiled by simply adding the triggering occurrences in a given month, assigning 0 to normal condition, 1 to drought watch, 2 to drought warning, and 3 to drought emergency. Since there were eight indicators operating in this period (groundwater, streamflow, reservoir level, PDHI, and four precipitation deficits), a drought emergency would register as 24 units on the vertical scale of the mass diagram. This is shown in Figure 2 where the severity of the two droughts in 1980-81 and 1984-85 is quite evident. Overall, based on these and other comparisons, it is concluded that: (1) the revised triggering criteria are both more responsive and consistent than the original criteria; (2) the drought indicators themselves, despite well-known deficiencies, are useful and should be retained; and (3) the statistical properties of the drought indicators and their associated trigger points are now established and should provide a firm basis for future water supply management decision by the Pennsylvania DER. Further details of this study are available in publications by Shaffer [1987] and by Kibler et al. [1987].

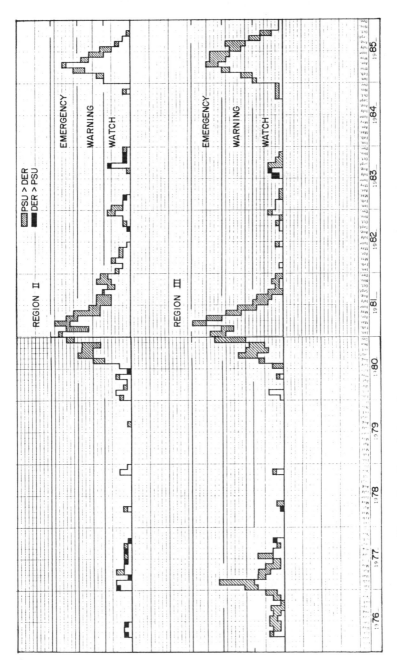

Figure 2. Mass triggering diagrams for 1976-85 in regions II and III

References

Alley, W. M., "The Palmer Drought Severity Index as a Measure of Hydrologic Drought," Water Resources Bulletin, Vol. 21, No. 1, pp. 105-114, February 1985.

Kibler, D. F., White, E. L., and Shaffer, G. L., "Investigation of the Sensitivity, Reliability and Consistency of Regional Drought Indicators in Pennsylvania," Vol. 1, Technical Report submitted to Pennsylvania Department of Environmental Resources, by Environmental Resources Research Institute, The Pennsylvania State University, May 1987.

Palmer, W. C., "Meteorological Drought," U.S. Weather Bureau, U.S. Department of Commerce, Research Paper No. 45, February 1965.

Shaffer, G. L., "Investigation of the Sensitivity, Reliability and Consistency of Regional Drought Indicators in Pennsylvania, M.S. Thesis in Civil Engineering, The Pennsylvania State University, May 1987.

Vandaele, W., "Applied Time Series and Box-Jenkins Models," Academic Press Inc., New York, 417 pp., 1983.

DROUGHT ANALYSIS IN THE OHIO RIVER BASIN

Tiao J. Chang[*], M. ASCE

ABSTRACT: A drought representing a shortage of water during a certain period may cause tremendous environmental impacts of the affected area if the intensity of the drought is substantially high. Since the development of new water resources becomes more costly, the drought problem is likely to be solved through the better management that requires a complete understanding of drought behavior. This study formulates the drought series, including drought occurrence time, duration, and deficit volume by the use of daily streamflows and a tracer level of one-half mean value. The drought occurrence rate function which expresses the intensity of drought, is developed based on the theory of the time-dependent nonhomogeneous Poisson process. The plots of drought occurrence rate function for the data from the Ohio River Basin reveal significant variations of drought intensity that need to be considered in the regional water management.

INTRODUCTION

Following the concept of the partial flood series, which was shown to be useful in analyzing extreme flow events (Chang, 1984, 1986), this paper attempts to study drought behaviors by the partial drought series. It is desirable to have a complete knowledge of a regional drought behavior so that the water management can be applied to reduce the hazardous impact caused by a prolonged drought.

Some of mathematical development for determining the distribution of flow extremes have been studied (Chang, 1986; North, 1980; Todorovic, 1978; Gupta, et.al. 1976; Todorovic and Zelenhasic, 1970; Shane and Lynn, 1964). Shane and Lynn (1964) used a probability model based on the time independent Poisson process in the analysis of base flow extreme data. Kirby (1969) defined the extreme as an exceedance in a sequence of randomly spaced Bernoulli trials. The time between exceedances, at sufficiently small exceedance probabilities, approaches those implied by trials from a Poisson process. Todorovic and Zelehasic (1970) developed the theory of extreme flow analyses based on partial flow series maintained that the magnitudes of exceedances are independent of exceedances. This implies that the occurrence times are independent of exceedance magnitudes. However the structure of time dependence among the occurrences of events were not addressed. Gupta el. al. (1976) constructed the expression for the joint distribution function of the largest flood flows and their occurrence times. North (1980) proposed the nonhomogeneous Poisson

* Assistant Professor, Department of Civil Engineering, Ohio University, Athens, Ohio 45701

distribution for flood occurrences with independent discharge volumes. Chang (1984) studied the behavior of the high-stage streamflow by the flood intensity function which was derived from the nonhomogeneous Poisson process. A flood simulation method, namely the Thinning method, was proposed to analyze the three-component flood series (Chang, 1986). Based on the assumption of the time-dependent Poisson theory, the formed partial drought series in this paper were investigated through the use of drought occurrence rate function.

THREE-COMPONENT DROUGHT SERIES

Considering a continuous streamflow hydrograph a drought event is constructed by taking a certain tracer level (e.g. Q-S; Q is the observation mean, S is the standard deviation) as shown in Figure 1. Given a starting time $\emptyset$, the occurrence time for a drought event is the period from the starting time $\emptyset$ to that of the down-crossing of the event. The difference between the occurrence time of a drought event and that of the following event is defined as the interval time of these two events expressed as

$$x_i = t_i - t_{i-1} \qquad (1)$$

where $t_o = \emptyset$. The time between the down-crossing and the up-crossing of a drought event is defined as the drought duration time, D_i. A drought duration series is formed by $\{D_i, i \geq 1\}$. The accumulated differences of discharge volumes between the tracer level and the observed discharges within the duration of an event is defined to be the deficit volume of the drought event, V_i. The deficit volume series is formed as $\{V_i, i \geq 1\}$. The three-component drought series are defined by all the drought events, associated with their occurrence times, durations, and deficit volumes, forming a discrete series $\{N(t), t \geq \emptyset\}$, which can possibly be modelled by the point process.

METHODOLOGY

Based on previous studies, extreme flow events can be interpreted by the Poisson process (North, 1980; Chang, 1984, 1986). A time-dependent nonhomogeneous Poisson process is developed for analyzing the formulated three-component drought series. There are many ways to generalize the time-dependent Poisson process (Cox and Lewis, 1966). Consider the number of events, N_t, occurring in an arbitrary interval of time length t. Then N_t has a Poisson distribution of mean λt, where λ is constant, and

$$P(N_t = r) = \frac{(\lambda t)^r e^{-\lambda t}}{r!} , \quad r = \emptyset, 1, 2 \ldots \qquad (2)$$

If we consider $\{N_{o,t}\}$ the number of events in $\{\emptyset, t\}$, as a stochastic process, i.e. as a random function of time, it has a property of having independent increments in non-overlapping time intervals and the corresponding probability generating function is

$$\Sigma \zeta^r p(N_t = r) = \exp[\lambda t(\zeta - 1)] \qquad (3)$$

Then if E and VAR denote, respectively, expectation and variance,

$$E(N_t) = \text{Var } (N_t) = \lambda t, \tag{4}$$

where N_t/t converges in probability to λ as $t \to \infty$ and λ is the rate of occurrence.

A second important group of properties of the Poisson process concern the intervals between events. Let X be the interval from the time origin to the first event. The distribution of X can be derived as follows: For no event occurs in $(\emptyset, x]$ if and only if X>x. Hence

$$p(X>x) = p(N_x=\emptyset) = e^{-\lambda x} \tag{5}$$

Thus $F_x(x)$, the distribution function and $f_x(x)$, the probability density function, are

$$F_x(x) = 1-e^{-\lambda x} , \quad x \geq \emptyset \tag{6}$$

$$f_x(x) = \lambda e^{-\lambda x} , \quad x \geq \emptyset \tag{7}$$

The Laplace-Stieltjes transform of the distribution is

$$E(e^{-sx}) = \int_0^\infty [\exp(-sx)] \lambda [\exp(-\lambda x)dx = \frac{\lambda}{\lambda+s} , \tag{8}$$

Therefore,

$$E(x) = \frac{1}{\lambda} , \tag{9}$$

$$\text{Var}(x) = \frac{1}{\lambda^2} \tag{10}$$

Further if X_1, X_2, ... are the intervals between the origin and the first event, between the first and second events, and so on, the random variables X_1, X_2, ... are mutually independent and each with the probability density function of λe^{-x}.

To generalize the Poisson process, consider the time-dependent Poisson process, i.e. λ is now a function of time. Let the occurrences in different time periods be independent and the $\psi(t';x)$ be the probability that there are no events in $(t', t' + x]$, i.e. the probability that, starting from t', the time to the next event is longer than x. Then

$$\psi(t';x+\Delta x) = \psi(t';x) \{1-\lambda(t'+x)\Delta x\} + o(\Delta x), \tag{11}$$

from which it results in

$$\lambda (t';x) = \exp \left\{ - \int_{t'}^{t'+x} \lambda (u) \, du \right\}. \qquad (12)$$

The probability density function of the interval to the next event is

$$\frac{\partial}{\partial x} \{1 - \psi(t';x)\} = \lambda(t'+x) \exp \left\{ - \int_{t'}^{t'+x} \lambda(u) \, du \right\}. \qquad (13)$$

The number of events occurring in the time interval $(\emptyset, t]$ has a Poisson distribution of mean

$$m(t) = \int_{o}^{t} \lambda(u) \, du \qquad (14)$$

The nonhomogeneous Poisson process proposed for the drought series is the time-dependent Poisson process i.e. its parameter is a function of time. Its occurrence rate function is expressed by $\mu(t)$ which is the derivative of $m(t)$.

To estimate the drought occurrence rate function, let the N_t be the number of events in $(t, t+\Delta t)$ and be considered in the limit as $\Delta t \to 0$, then

$$E(\Delta N_t) = \text{Var} \, (\Delta N_t) = \mu(t)\Delta t + o \, (\Delta t) \qquad (15)$$

Since the event at t is an arbitrary event in the stationary process, the drought occurrence rate of the process can be explained by

$$\mu(\tau) = \lim_{\Delta t \to o} \frac{p\{\text{event in } (t+\tau, \, t+\tau + \Delta t)/\text{event at } t\}}{\Delta t} \qquad (16)$$

The drought occurrence rate function, $\mu(t)$, can be interpreted as the expected number of droughts in an interval (o,t) starting at an arbitrary event.

APPLICATIONS

Assuming the time-dependent nonhomogeneous Poissonian behavior of a three-component drought series and that the observation of the drought series starts from some arbitrarily chosen time origin and last for a total time of T_o a histogram from all possible sums of contiguous times between events can be formed. Let n events occur at times measured from the time origin. Neglecting the time from the origin to the sums of contiguous times between events, t_2-t_1, t_3-t_2, ..., t_n-t_{n-1}, can be obtained. Dividing (o, T_o) into equal intervals of

length, p, the histogram can be formed by counting the number, q, of
the sums in each (sp, sp+p) and dividing np, where s=0, 1, 2, ...

Analytically it can be expressed by

$$\upsilon(t) = \frac{1}{n} \sum_{i=1}^{n-1} \sum_{j=1}^{n-1} \delta(t_{i+j} - t_i - t), \qquad (17)$$

where $\upsilon(t)$ is the estimated drought occurrence rate at time t, n is
the total number of observed events, $\delta(t)$ denotes the Dirac function
at time t.

Daily streamflows from four watersheds in the Ohio River Basin are
selected to form the drought series based on the tracer level of
one-half mean daily flow of the watershed. Table 1 gives the
watershed identification and the corresponding mean daily flow for
each watershed. From equation 1, the drought series include the
sequence of time intervals, $\{x_i\}$ which is associated with the
sequence of occurrence times $\{t_i\}$ that is used for the estimation of
drought occurrence rate. Additional drought associated components are
duration time, $\{D_i\}$, and deficit volume, $\{V_i\}$ which are not discussed
in this paper. Based on equation 17, the drought occurrence rates for
each drought series are estimated and plotted in Figures 2-5, which
show significant variations.

CONCLUSIONS

Assuming that the drought series are formed by taking a tracer
level of one-half mean daily flow and can be explained by the
time-dependent nonhomogeneous Poisson process, this study shows that
there are significant variations of drought occurrence rates for the
watersheds from the Ohio River Basin. This suggests that the regional
water resources management using annual drought series should consider
further modification in the estimation of design drought.

Table 1. Watershed Identification for Drought Analyses

Watershed Identification	Drainage Area (sq.km)	Mean Daily Flow (cms)
Mad River, Springfield, OH	1,270	13.5
Whitewater River, Alpine, IN	1,371	15.6
Salamone River, Dora, IN	1,443	14.2
Mississinewa River, Marion, IN	1,767	17.6

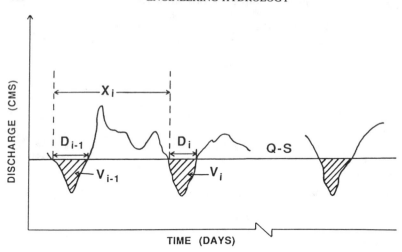

Figure 1. Schematic Drought

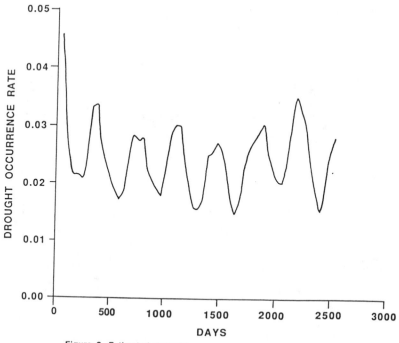

Figure 2. Estimated drought occurrence rates for Mad River Basin

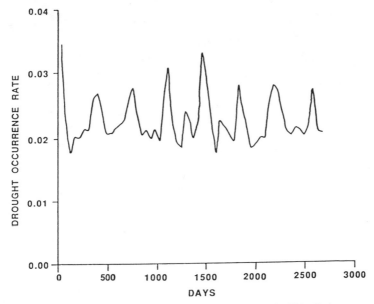

Figure 3. Estimated drought occurrence rates for White Water Basin

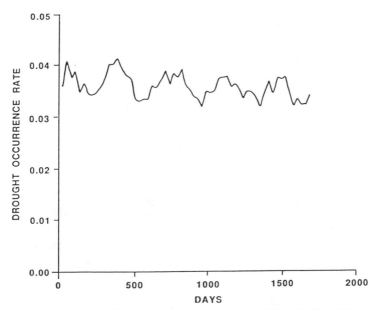

Figure 4. Estimated drought occurrence rates for Salamonie River Basin

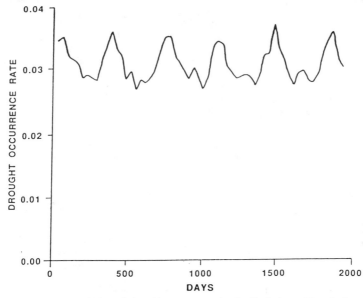

Figure 5. Estimated drought occurrence rates for Mississinewa River Basin

ACKNOWLEDGEMENTS

The author wishes to acknowledge his appreciation for the support
of the Stocker Research Fund at Ohio University. He also would like
to thank Pam Stettler for her assistance in typing this paper.

APPENDIX-REFERENCES

Chang, T.J., 1984, "High Streamflow Behaviors in the Ohio River
    Basin," Journal of Hydraulics Division, proc. ASCE, pp. 355-359.

Chang, T.J., 1986, "Flood Analysis in the Ohio River Basin,"
    proceedings of the International Symposium of Flood Frequency,
    Baton Rouge, Louisiana.

Cox, D.R. and P.A.W. Lewis, 1966, The Statistical Analysis of Series
    of Events, Methuen & Co. Ltd., London; John Wiley and Sons, Inc.,
    New York.

Gupta, V.K., L. Duckstein, and R.W. Peebles 1976, "On the Joint
    Distribution of the Largest Flood and Its Time of Occurrence,"
    Water Res. Res., Vol. 12, No. 2, pp. 295-304.

Kirby, W., 1969, "On the Random Occurrence of Major Floods," Water
    Res. Res., Vol. 5, No. 4, pp. 778-784.

North, M., 1980, "Time-Dependent Stochastic Model of Floods," J. of
    Hydrau. Division, ASCE, Vol. 10B, No. Hy5, pp. 649-665.

Shane, R.M., and W.R. Lynn, 1962, "Mathematical Model for Flood Risk
    Evaluation," J. of Hydrau. Division, ASCE, Vol. 90, No. Hy5, pp.
    1-20.

Todorovic, P. and Z. Zelenhensic, 1970, "A Stochastic Model for Flood
    Analysis," Water Res. Res., Vol. 6, No. 6, pp. 1641-1648.

Todorovic, P., 1978, "Stochastic Models of Floods," Water Res. Res.,
    Vol. 14, No. 2, pp. 345-356.

MAPPING OF DROUGHT STATUS IN SOUTH FLORIDA

George Shih*

INTRODUCTION

South Florida receives almost 54 inches of rainfall a year, but about 85% is lost to evapotranspiration (ET) and the ocean. These uncontrollable losses are equivalent to the entire rainfall amount in a 1-in-20 year drought. Even in a 1-in-5 year drought, with yearly rainfall being a little over 90% of normal, water supply is reduced to less than one half that of a normal year. In order to help residents cope with water shortages, the Water Management District (the District) sets regulations to guide water users in conservation of water use in four phases of drought severity. These phases are roughly divided by 1-in-10, 1-in-20, 1-in-50, and worse than 1-in-50 year drought conditions. Reductions of 15, 30, 45, and 60 of normal use for each phase respectively, are anticipated through voluntary conservation.

The District's Governing Board is responsible for water shortage management. Each month during a drought, a detailed report of hydro-meteorological parameters is prepared for evaluation by the Governing Board to decide whether or not a water shortage is in effect. An important parameter in this evaluation is the determination of the degree of drought at the time of report. In this paper, the determination of drought was attempted using data from rainfall, ground water levels, and lake stages to produce a map of drought recurrence intervals.

METHODS

Most drought frequency analyses specify a fixed averaging period. The fixed length of accumulation is not practical for determining the status of an on-going drought, for a drought may be: (1) initiated either earlier or later than the starting date of accumulation, and, (2) of longer or shorter duration than the fixed averaging period. Specifically, in this study, a model to transform monthly rainfall into hydrologic system soil moisture indicator is proposed. This indicator, and other parameters, were used in drought frequency analysis. With this approach, it is believed that severity of drought at any time can be determined regardless of the duration of the drought.

To report the drought status, the recurrence intervals determined at rainfall, lake stage and groundwater stations were integrated into a map. The point information was extended into a spatial distribution by kriging (Skrivan, 1980). The resultant map gives a snap-shot of the drought condition in south Florida.

* Senior Professional, South Florida Water Management District, P.O. Box 24680, West Palm Beach, Fl 33416-4680

RAINFALL AS A CURRENT DROUGHT INDICATOR

In most evaluations, drought severity was assessed at the end of the drought. During the progression of a drought, the severity was usually described by rainfall deficits referring to normal rainfall. The intention was to give an account of moisture status, which is a function of past rainfall and various losses. Let the moisture status, or drought status in times of moisture deficit, be denoted by S, then

$$S = a_0*(R_t - L_t) + a_1*(R_{t-1} - L_{t-1}) + a_2*(R_{t-2} - L_{t-2}) + \ldots \ldots \quad (1)$$

where R is the rainfall; L is the loss; t is the current time period, while t-1 is the previous time period; and a_0, a_1 are unknown coefficients. The known quantities in equation (1) are rainfall amounts. Also, if the quantities in the parentheses are considered as noise, then it is a moving average (MA) model of infinite order. To simplify the expression, a model for these coefficients is proposed. Thus,

$$a_{i-1} = q*a_i \quad (2)$$

where q is a decay coefficient to be calibrated later. Let $a_0 = 1$, then $a_i = q**i$; a_i is a power function of q. Thus, equation (1) becomes a first order auto-regressive (AR) model.

The losses, L, are known to be dependent on S, the system storage; which in turn makes equation (1) non-linear. If the losses are considered to be independent of S, and can be estimated from other historical data, then equation (1) becomes linear., Thus, S can be computed from historical rainfall when q is defined.

ESTIMATION OF LOSSES

In south Florida, ET is by far the most important mechanism of losing moisture from the hydrologic system. ET is assumed to be affected by temperature and humidity only. Energy available for ET is proportional to temperature, while it ET is hindered by increased humidity. A simple model for the loss is proposed

$$L_t = K * T_t/H_t \quad (3)$$

where K is a proportional constant; T is the temperature and H is the relative humidity. In south Florida, monthly temperature and humidity do not vary drastically from place to place and from year to year. Hence, yearly averages of T_m and H_m for m = 1,2...12, were considered sufficient for equation (3). The proportional constant K was adjusted so that the total yearly ET would sum up to be between 70 and 90 percent of the average yearly rainfall. mThe identification of q and K is explained in the next section.

CALIBRATION OF THE MODEL COEFFICIENTS

The decay coefficient, q, in equation (2) is a damping factor for the rainfall deficit incurred some time ago. Should a time series of S be available, it would be possible to identify the coefficient, q, through AR(1) modeling. In this study, however, S, was considered to be unmeasurable; and the task was to find a procedure that can properly construct such a series.

The purpose of calibrating q in this study was to construct samples of S to determine the drought in terms of recurrence year. More specifically, by using the annual series method, the minimum monthly values in each calendar year, from the created sample space S, was to be used to define a drought frequency distribution. Any method that can produce 12 values for S each year may satisfy this requirement. For example, if q = 0, S would simply become the monthly rainfall series, then the driest month of each year would be used for frequency analysis. Clearly, q should be between 0 and 1.

The concept used in this selection process was that various drought events were embedded in the historical rainfall data. Also, the worst drought in each year was to be captured for frequency analysis. A good criterion was expected to produce a uni-modal distribution with respect to q, so that no multiple values of q resulted in local maxima. Various criteria were tested, a successful one was to maximize the range of the constructed annual series of the drought indicator. The selection procedures were:

1) For each station, for each t, with given q and ET find

$$S_t = \Sigma_i(q**i) * [R_{t-i} - (K*T_{t-i}/H_{t-i})] \qquad (4)$$

K varies from 0.95 to 0.65; q varies from 0 to 1. T_{t-i} and H_{t-i} are cyclic between 1 and 12.

2) Let $D_t = S_t/(1. - 1./q)$ $\qquad (5)$

D_t is an equivalent monthly rainfall deficit, because

$$\Sigma_i (q**i) = (1. - 1./q)$$

3) Take the minimum in each 12 consecutive D_t; let it be denoted by A_y, y = 1, 2, ... Y, where Y is the number of years in the record. Therefore,

$$A_y = Min [D_t], \quad \begin{array}{ll} t = 1,............12; & y = 1 \\ t = 13,..........24; & y = 2 \\ .... \\ t = (12Y-11), ... 12Y; & y = Y \end{array} \qquad (6)$$

Thus the extreme values are obtained for the annual series, Ay, which is used as a drought indicator.

4) Take the range of A_y,

$$r_n(q, ET) = [Max A_y] - [Min A_y] \qquad (7)$$

The best value of q for drought analysis is when r_n is at the maximum. The largest range in the annual series will give the highest possible sensitivity in differentiating degrees of drought severity and will provide room for data variations. It was found that the decay factor, q, at the maximum value of r_n, was not sensitive to K.

5) To combine the r_n of all the stations, a normalization is required. Keeping K constant, $r_n(q, K = \text{const.})$ is a function of q only. Let the normalization scale for station n, SCL_n, be,

$$SCL_n(K = c) = [\text{Max } r_n(q, K = c)] - [\text{Min } r_n(q, K = c)] \qquad (8)$$

and

$$R_n(q,K = c) = \{ r_n(q,K = c) - [\text{Min } r_n(q,K = c)] \} / SCL_n(K = c) \qquad (9)$$

then, $[\text{Max } R_n] = 100\%$ and $[\text{Min } R_n] = 0\%$ for all stations and ET.

6) For multiple stations, R_n can be summed and averaged.

$$P(q,K = c) = \Sigma_n R_n(q,K = c) / N \qquad (10)$$

Obviously, P also falls between 100% and 0%. Because R_n does not peak at exactly the same q for all stations, P will be less than 100% but more than 0%.

7) The same procedure was applied for various values of q and K. These monthly reporting rainfall stations differed from other rainfall stations because their monthly rainfall totals were reported by phone at the end of month, so that these data were immediately available. The peaks of P were all above 80%, implying that over 80% of the widest ranges of droughts can be captured by using a single q factor for these 39 stations involved.

8) It was found that along the ridge, P was not sensitive to K. It was mentioned earlier that K was believed to be about 80% of the total rainfall. The steepest slope of P also occurred at K = 0.80 along the ridge, hence K was determined to be at that point. With K = 0.80, about 10.7 inches of rainfall was available to mitigate the moisture deficit in a year.

9) Along the curve of K = 0.80, the optimum q was 0.75 for P at the maximum; indicating an equivalent of four months of deficit accumulation was most sensible for drought analysis.

When the stations were grouped by the length of the records and the analyses repeated, it was found that K was still not critical in the determination of q; however, the optimum q tends to be lower for long-record stations. The optimum q for stations over 50-years old (13 stations) was about 0.65, with the P-maximum at about 90%; while the optimum q for stations less than 30-years old (16 stations) was 0.70, with the P-maximum at about 86%. The higher q value, implying longer drought memory. The higher P-maximum from over-50-years old stations also indicates that dry season rainfall in recent years were more erratic. For reporting the current drought status, it is more reasonable to adapt the statistics from recent years.

FREQUENCY ANALYSIS

Pearson type III distribution was used in this study because: (a). most of the District's frequency analyses used this distribution; and, (b) it fits a wide range of distributions with only three parameters. The procedure is given by (Haan, 1977).

$$Y_T = y_m + s_Y * K_T \tag{14}$$

Where Y_T is the magnitude of drought in T recurrence years; y_m is the mean of the sample; s_Y is the standard deviation of the sample; and K_T, available in a table form, is the frequency factor as a function of return year T, or occurrence probability in a year, and the sample coefficient of skewness, C_s.

The variance of Y_T as shown by Bobee (1973) is,

$$\text{Var } Y_T = s_Y^2/n * \{1 + K_T^2/2 * (1 + 3/4*C_s^2) + K_T*C_s$$
$$+ 6*(1 + 0.25C_s^2)(\partial K_T/\partial C_s) * [(\partial K_T/\partial C_s)(1 + 1.25C_s^2) + K_T C_s/2] \tag{16}$$

The statistical parameters, such as y_m, s_Y, and C_s, for each station are calculated from the historical data. In practice, Y_T is given as current status, while T, the recurrence years, is being sought. The transformation of Y_T to T also makes it possible to combine variables of different units, such as drought indicators, ground water levels, and lake stages, into a common variable, the year.

The procedure used to compute T was:

a) Compute $K'_T = -(y_m - Y_T)/s_Y$ (17)
b) Linearly interpolate T for K'_T with the given C_s from the table.

The variance of T was approximated by,
a) S_{YT} = sq. rt (Var Y_T) from equation (16)
b) Compute T_1 for $(Y_T - S_{YT})$ and T_2 for $(Y_T + S_{YT})$ by equations (17).
c) Var T = $[(T_1 + T_2)/2]^2$ (18)

To take care of the situation that different stations show contradictory wet and dry conditions at the same time, both wet and dry sides of the data were analyzed with the same procedure. These data are all combined in the mapping process.

MAPPING

If a set of rainfall, groundwater levels, and lake stage are expressed in terms of return year for the same instant of time, then it makes sense to combine all these pieces of information in a map. These data, however, are of different degrees of reliability, even conflicting with each other some times. Those reliability and conflicting problems were taken care of in a mapping method called kriging. A computer program of kriging is described by Skrivan (1980)

RESULTS

Numerical results and sample maps of this study are discussed in the slide presentation.

CONCLUSIONS

Drought severity can be estimated by several meteorologic and hydrologic parameters, such as rainfall, lake stage, and ground water level, at specific points in space for a given time. To use those parameters together, they had to be expressed in comparable units. Rainfall input was transformed into system moisture storage by an AR(1) model. The coefficient of the AR model was calibrated through the maximum range of the annual drought series. For south Florida, the decay coefficient for the AR model was identified to be 0.75, equivalent to four months of rainfall accumulation. All measures of system moisture were then expressed in terms of return year.

Drought is not a point phenomenon. To report drought status from the data, kriging techniques were incorporated in the mapping process. The reliability of point data and their relative locations are taken into consideration to create the contour map that can be readily used for making management decisions.

REFERENCES

Bobee, B. "Sample Error of T-Year Events Computed by Fitting a Pearson Type 3 distribution." Water Resource Research, Vol. 9:4, October 1973. pp 1264-1270.

Haan, C.T. "Statistical Methods in Hydrology." The Iowa State University Press. Ames, Iowa. 1977. 378pp.

Skrivan, J.A. and M.R. Karlinger, "Semi-Variogram Estimation and Universal Kriging Program." U.S.G.S WRD-WRI-80-064, Tacoma,Wash. 1980; NTIS: PB81-120560

SOUTHEASTERN DROUGHT OF 1986 - LESSONS LEARNED

C. Patrick Davis, F.ASCE and Albert G. Holler, Jr., M.ASCE

Abstract: The Southeastern drought of 1986 was one of the most severe in recorded history. Streamflows, in some areas, were twenty five percent of previous low flows of record. Because it followed so close to the 1981 drought, there was an excellent opportunity to test the previous lessons learned and the drought indicators that were developed. Much of the effort involved the Apalachicola-Chattahoochee-Flint (A-C-F) river basin in the States of Georgia, Alabama, and Florida and the Savannah River Basin in Georgia and South Carolina. Large cities, such as Atlanta, Georgia, and major industries, such as the Department of Energy's Savannah River Plant, are located near rivers within these basins that are regulated by Corps of Engineer multiple purpose reservoirs that form a hydropower system. An interim drought management plan for the A-C-F basin was completed in April, 1985, just in time for testing as the drought developed. Objectives of the plan include development of hydrologic factors which indicate the onset of a drought and a coordination program. This paper will describe the plan, the results, and needed changes. It will also discuss the hydrologic conditions of the drought, including precipitation, groundwater, and streamflow conditions. One of the lessons learned was the need for a more accurate and comprehensive system study which can account for the needs of hydropower, navigation, water supply, water quality, and recreation. This paper will describe the computer program used to develop the system study.

Introduction

    Water year 1986 produced the worst drought of modern record for many river basins within the South Atlantic Division area. However, those river basins with storage reservoirs were able to provide adequate water supply and water quality flows. Other reservoir project purposes, such as navigation, hydropower generation, and recreation, shared the adversity caused by the drought. Corps of Engineer water control projects in the drought area are shown on Figure 1.

Regional Rainfall

    Following a slightly wetter than normal fall of 1985, unusually dry conditions developed in December and continued through the following seven months. By July, eight month cumulative rainfall shortages of 20 inches had developed in northeast Alabama and North Georgia. Precipitation deficiencies of 10 to 15 inches also extended in a zone

----
1. Chief, Tech Engng Branch, Corps of Engineers, Atlanta, GA 30335-6801
2. Chief, Hydrology and Hydraulics Section, Corps of Engineers, Atlanta, GA 30335-6801

southwestward into Mississippi. In the December through July period, Atlanta, Georgia, received only 18.6 inches of rain and Birmingham, Alabama, only 19.4 inches, or about 50 percent of the normally expected amounts. Rainfall in the Savannah River basin above Augusta, Georgia, amounted to 42.5 inches for the water year, 9.1 inches below normal. Less than average rainfall amounts were recorded for 9 of the 12 months. Extremely dry conditions prevailed in the basin from November, 1985, through September, 1986.

In contrast, however, in the upper reaches of the Roanoke River basin in Virginia, high rainfall amounts occurred in November, 1985, causing severe flooding and record high lake levels. This was followed by record low rainfall in the summer of 1986 which produced a severe drought. New record low rainfall amounts were observed above John H. Kerr, Philpott, and W. Kerr Scott lakes during June, 1986. Reduced rainfall and high evaporation rates combined to make the summer of 1986 one of the driest periods ever in North Carolina and Virginia.

Inflows

By the summer of 1986, new record low flows were being recorded at most locations. The July inflow to Lake Lanier, near Atlanta, Georgia, fell to 287 cubic feet per second breaking the old record low of 717 cfs set in August, 1981. On the Etowah River at Canton, Georgia, the lowest stream flows in 58 years of record were observed. Further south on the Apalachicola River, the gage at Blountstown, Florida, fell to the lowest daily reading ever recorded (0.13 feet). On the Savannah River, inflows to Hartwell, Richard B. Russell, and Clarks Hill lakes were the lowest recorded this century. Inflow to Hartwell was 35 percent of normal for the year. In the Wilmington District area, inflows to John H. Kerr, Philpott, B. Everett Jordan, W. Kerr Scott, and Falls lakes were 70 percent, 88 percent, 59 percent, 68 percent and 64 percent of normal, respectively, for the year.

Lake Levels

The effect of low rainfall amounts and deficient inflows on the elevation of Lake Lanier (Buford Dam) is shown on Figure 2. Lake Lanier is the headwater reservoir in the Apalachicola, Chattahoochee, Flint (A-C-F) river basin. In addition, it is a major component of a hydropower system that produces electricity in amounts called for in a single large contract between the Southeastern Power Administration (SEPA) and the Georgia Power Company. Seven other Corps' multiple purpose projects are included in the contract. These are: Hartwell, Richard B. Russell, Clarks Hill, Allatoona, Carters, West Point, and Walter F. George. Also, Lake Lanier supplies water for most of the metropolitan Atlanta area and supports a downstream trout fishery. Lake Lanier is the most popular Corps Lake for recreation with over 15 million visitors annually.

In April, 1985, the Mobile District completed an Interim Drought Management plan for the A-C-F River basin. The purpose of the plan is to provide the States of Georgia, Alabama, and Florida, the Corps, SEPA, and others with a uniform strategy to effectively coordinate the management of the water resources within the basin during a severe

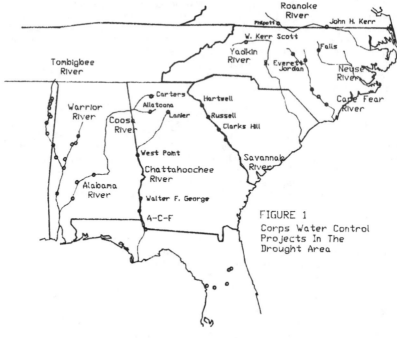

FIGURE 1
Corps Water Control
Projects In The
Drought Area

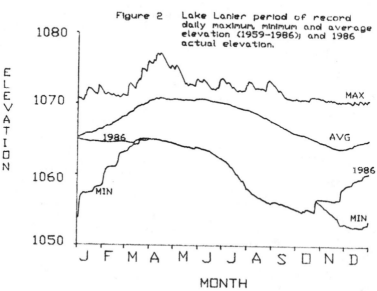

Figure 2   Lake Lanier period of record
daily maximum, minimum and average
elevation (1959-1986), and 1986
actual elevation.

drought. On March 21, 1986, a drought management committee consisting of representatives from the three states and the Corps met to discuss the dry conditions as indicated by a water availability index and other hydrometeorological data.

The water availability index for Lake Lanier was developed for the Mobile District office by personnel at the University of Alabama. It requires lake level and rainfall information and indicates appropriate management action. The index is defined by the equation:

$$I = R + \sum_{j=0}^{3} D_{-j}$$

in which:

I is the monthly index value;
R is the difference between the current lake elevation and the long term monthly mean; and
D is the difference between a given month's mean rainfall and the long term mean.

Negative values computed for the index indicate dry basin conditions. The following table is used to determine appropriate action to take:

INDEX	DESCRIPTION	ACTION
0 -1 -2	potential water supply is normal	none
-3 -4	water supply is below normal	alert
-5 -6	mild to moderate shortage	begin to conserve
-7 -8	moderate to severe water shortage	intensify conservation
-9 -10	extremely severe shortage	take every possible action

Another indication of the severity of dry conditions is the ability of Lake Lanier to satisfy the various water demands placed on its storage. To measure the lake's ability to meet the demands, a three zone storage delineation for the lake was developed. Additional information used by the committee included hydrologic information from the National Weather Service's Extended Streamflow Prediction (ESP) model, lake recovery probabilities, conditions at neighboring lakes, streamflows at key locations, and groundwater elevations.

Based on prevailing hydrologic conditions, the committee issued a

"water shortage alert" - meaning that conditions were dry enough to be
concerned. Accordingly, hydropower releases were restricted from the
lake, downstream lakes were held slightly higher than usual, and a
reduction in navigation depths was planned.

Lessons Learned

Perhaps the biggest lesson learned from the drought is the value of
having a functional drought contingency plan in hand before the onset
of a drought. The drought plan for the A-C-F basin grew out of the
1980-81 drought. In accordance with the plan, actions were taken to
conserve water in the Fall of 1985, well before the declaration of a
drought. Members of the drought committee were very effective in
balancing the water needs and minimizing disputes among project users.
Most of the water control actions in the basin during the drought were
based on the recommendations of the committee. However, additional
refinement of the water availability index equation is needed. The use
of the index during 1986 indicated that it is too sensitive to current
rainfall conditions which can produce overly optimistic index values.

Another lesson learned is that drought contingency plans are needed
for all of our water control projects. In some cases, these plans will
need to be linked together into basin plans and even into regional
plans. Key provisions of the plans will be utilized in hydropower
contracts.

Open and frank communications with the news media, river and lake
users, and the general public throughout the drought resulted in
excellent public relations. As many of the users as possible were
involved in decision making before priorities were set. The public was
kept fully informed well in advance of drought related actions which
affected them. Regular press briefings were held in connection with
drought committee meetings and were well received by media persons
throughout the region.

The drought also demonstrated the need for a simulation model of
our hydropower system which includes twelve multiple purpose reservoir
projects in three Corps' Districts. Projects in the system, besides
those previously listed in connection with the SEPA-Georgia Power
Company contract, include Jones Bluff and Millers Ferry (Alabama), and
John H. Kerr and Philpott (Virginia). The model is needed to study
various water control management strategies for real time hydrologic
situations, and it is needed to determine the long range hydropower
capabilties of the projects that can be made a part of the contracts
between SEPA and its customers. The model is being developed with the
assistance of the Hydrologic Engineering Center, Davis, California. The
system model begins with HEC 5 - "Simulation of Flood Control and
Conservation Systems." HEC 5 has been used for the individual river
basins which form the hydropower system - the Savannah, the A-C-F, and
the Alabama-Coosa basins. The basin models are being combined into a
system HEC 5 model and a period of record routing is being performed.
Upon completion of this first phase, a second phase to include
real-time optimization will be started possibly utilizing a recently
proposed "branch and bound" procedure.

Also, a more accurate inventory of water supply intake locations and elevations, water requirements and instream flow needs, and low flow river profiles is needed. This information is essential for effectively managing the projects and for determining the adequacy of existing and new facilities to function as intended during droughts without the need for costly intake modification after construction.

Although some of the impact costs of the drought on project purposes, such as hydropower, are known, the total impact has not been quantified. Even though 1986 recreational visitation figures for the lakes are surprisingly good, we did not have the "banner" year we might have experienced with full lakes. There was an important loss to recreation that has not been quantified. Also, with suspended navigation on some of the waterways because of inadequate depths, shippers incurred losses through the use of alternative, more expensive, transportation modes. Some of these impact costs are being obtained.

The drought has reaffirmed the importance of periodically restudying water control plans. In 1976, the top of the conservation pool at John H. Kerr Lake was modified to conserve water. Because of the modification, the lake was eight feet higher during 1986 than it would have been had not the change been made. Without the change, no supplemental releases from the lake would have been available in 1986 for the very important downstream striped bass fishery. At W. Kerr Scott lake (Yadkin-Pee Dee River basin, North Carolina), a revised low flow release schedule, which had previously been developed, prevented the lake from establishing a new record low pool elevation. The lake was 19 feet higher during the drought because of the revised release schedule.

Conclusion

By winter, 1986, heavy rainfall had raised the water levels at most of the lakes to acceptable levels and many of the project functions could be fully restored. Navigation depths were reestablished on the Alabama and Apalachicola rivers. Despite the drought, the seventh best tonnage year since 1960 was recorded on the Apalachicola River system. Also, we were then able to resume generating hydropower at our projects to the full amounts called for in the hydropower contracts because of improving water conditions. The A-C-F drought committee met formally for the last time on December 2, 1986. After the meeting a major press conference was held to update the improved hydrologic conditions.

The drought demonstrated the value of the Corps multiple purpose water control projects in the Southeast. Without these projects, many areas would not have survived this drought. Although there is always room for improvement, water control management activities during the drought, particularly coordination efforts with the states, with other federal agencies, and with the public were successfull. We will be striving to improve our coordination efforts even further and to fully respond to the lessons learned from the drought.

RESULTS OF A REGIONAL SYNTHETIC UNIT HYDROGRAPH STUDY

Michael E. Meadows*, M.ASCE

Abstract

Selected results of a regional synthetic unit hydrograph study are presented. Using the gamma function to approximate the SCS curvilinear unit hydrograph, peak rate factors were determined for three urban watersheds. The results indicate the peak rate value varies with urbanization and drainage system improvements. A geomorphologic method for estimating the gamma function shape parameter was tested to better understand the effect of the drainage network on the watershed hydrologic response.

Introduction

Synthetic unit hydrograph procedures are being developed for application in the Coastal Plain and Piedmont Provinces of South Carolina. One objective of the study is to test the standard SCS hydrograph methodology, and if the results are unacceptable, to recommend revised parameter values, eg., peak rate factors, or an alternative method. The SCS methodology is preferred because of its popular acceptance; many local ordinances specify it as the basis for drainage design; and it is available with computer models such as HEC-1 which run on micro-computers.

A second objective is to develop a better understanding of the fundamental processes underlying watershed hydrologic response and to incorporate this information into improved methods for estimating unit hydrograph parameters.

Background

Current synthetic unit hydrograph techniques have developed basically along two different approaches (Dooge 1973). One approach, embodied in time-area methods, is based on the assumption that every watershed has a unique unit hydrograph, and the other, typified by the SCS dimensionless (curvilinear) unit hydrograph, on the assumption that all unit hydrographs for all watersheds can

---

*Associate Professor, Department of Civil Engineering, University of South Carolina, Columbia, SC 29208

be represented by either a single curve, a family of curves or a single equation. According to Dooge (1973) the convergence of these two lines of development is represented by the two parameter gamma distribution or Nash model:

$$h(t) = [k\Gamma(n)]^{-1} (t/k)^{n-1} \exp(-t/k) \tag{1}$$

where h(t) is the instantaneous unit hydrograph (IUH), k is a scale parameter, n is a shape parameter and t is time.

Eq 1 can be rewritten in a more pragmatic form (Wu 1963) as

$$Q = Q_p [(t/t_p)\exp(1-t/t_p)]^{n-1} \tag{2}$$

where Q is the IUH ordinate, and Q_p and t_p are the IUH peak flowrate and time to peak, respectively. Peak flowrate is estimated as

$$Q_p = PRF \ A/t_p \tag{3}$$

where A is watershed area and PRF is the unit hydrograph peak rate factor, given by

$$PRF = 1290.67p \tag{4}$$

in which p is the proportion of the runoff hydrograph under the rising limb of the hydrograph. There is a unique relationship between p and n (Meadows and Blandford 1983); therefore, once n is specified, PRF can be determined.

Eq 1 serves as the basis for many unit hydrograph models; for instance, when n=4.7, p=37.5%, PRF=484, and Eq 1 and the SCS curvilinear unit hydrograph are essentially identical (Meadows and Blandford 1983).

McCuen and Bondelid (1983) used the relationships in Eqs 2, 3 and 4 to evaluate peak rate factors for the SCS curvilinear unit hydrograph and found the standard value of 484 is too large for Delmarva peninsula, which suggests that hydrographs at coastal watersheds should be simulated using smaller peak rate factors. In a different study, Meadows and Chestnut (1983) used data from a coastal watershed near Charleston, S.C., and found the standard SCS hydrograph methodology overpredicted peak runoff and underpredicted time to peak. They concluded the method does not work well at watersheds where the runoff is dominated volumetrically by subsurface flow, eg., coastal watersheds. The results of these two studies reinforce the opinion that the standard PRF=484 is not appropriate for all watersheds. Although vague guidelines are provided to modify the peak rate factor, a definitive and systematic means is needed for determining the value unique to any watershed.

In an effort to develop a unifying synthesis of the hydrologic response of a watershed, Rodriguez-Iturbe and Valdes (1979) linked hydrology with quantitative geomorphology. In their estimation of the IUH, they used geomorphologic concepts of stream order and the Horton watershed numbers: bifurcation ratio (R_B), stream drainage area ratio (R_A) and stream length ratio (R_L). Rosso (1984) extended their work by parameterizing the Nash model in terms of the Horton numbers and found that the shape parameter (n) depends only on the Horton numbers while the scale parameter (k) is a time-varying characteristic which depends on both geomorphology and average velocity along the stream network. Rosso's relationships for n and k are

$$n = 3.29(R_B/R_A)^{0.78}R_L^{0.07} \qquad (5)$$

and

$$k = 0.70[R_A/(R_B R_L)]^{0.48}v^{-1}L \qquad (6)$$

where v is the average streamflow velocity, and L is the mean length of the higher order stream. The significance of Eq 5 is that the shape of the IUH depends only on the Horton order ratios and therefore can be predicted from watershed geomorphology. This is consistent with the first approach to unit hydrograph developed discussed previously. According to Eq 6, the time scale of the IUH depends both on geomorphology and on streamflow velocity, where velocity is a surrogate for antecedent conditions and the intensity of the generating rainfall event.

Rosso (1994) also showed that the relationship in Eq 3 can be expressed in geomorphologic terms as

$$q_p t_p = 0.58(R_B/R_A)^{0.55}R_L^{0.05} \qquad (7)$$

where $q_p = Q_p/A$. Comparison of Eqs 3 and 7 provides a direct determination of PRF, which is merely the right hand side of Eq 7.

Approach

To test the proposed approach, peak rate factors were determined for three urban watersheds located along the fall line, that physiographic area formed by the intersection of the Piedmont and Coastal Plain. Using values estimated with Eq 7, several events at each watershed were simulated and the predicted and measured hydrographs compared. If the error in peak flowrate was less than 10 percent, the simulation was accepted; otherwise, the model was calibrated to the event by adjusting the PRF. For all events, the event curve number was determined and used to predict the rainfall excess pattern.

Results

The three watersheds studied are Marsh Creek, Senn Branch and Smith Branch. Marsh Creek is 9.4 sq mi (34.3 sq km) and is undergoing conversion to urban land use. The internal drainage system has not been altered significantly to affect the Horton order ratios, which were determined as 4.1, 5.2 and 2.2 for the area, bifurcation and length ratios, respectively. Senn Branch is 1.3 sq mi (8.5 sq km) and also is undergoing conversion to urban land use. Channel modifications are less than in Marsh Creek. The Senn Branch order ratios are 4.4, 3.5 and 2.2. Smith Branch is 5.6 sq mi (14.5 sq km) and is almost completely developed to a mix of commercial and residential land use. Extensive storm sewering and channel improvements have been made. The Horton order ratios based on the original drainage network are 4.1, 2.9 and 2.1.

The peak rate factors for Marsh Creek, Senn Branch and Smith Branch were estimated as 430, 341, and 324, respectively. Model calibration results for Marsh Creek resulted in PRF=430 for approximately one-half of the events and 250 for the rest. Those events with the higher PRF value also had event curve numbers greater than 90. When the event curve number was less than 90, PRF=250 provided the best agreement between the predicted and observed hydrographs. Similar performance was observed at Senn Branch. Both watersheds contain several ponds, a factor not considered by the Horton ratios and this might explain why a lower PRF provided better simulation results for the lower curve number events.

The results for Smith Branch were quite different but not unexpected due to the extensive modifications to the internal drainage system. Best results were obtained for all events using PRF=600 for the commercial area and 484 for the single family residential area. This watershed naturally is divided into lower and upper branches. The commercial land use is largely situated in the subwatershed draining to the lower branch and the residential land use is in the subwatershed draining to the upper branch.

This initial investigation has demonstrated that the standard peak rate value of 484 may be too large even for some urban watersheds, or at least watersheds undergoing conversion to urban land use. Clearly, there are other factors which influence the estimation of PRF such as the extent of ponds which the Horton numbers do not include. Alteration of the internal drainage network may have the greatest impact on the peak rate factor.

Discussion

Eq 7 provides a means of estimating a unique PRF for any watershed. One constraint on this estimate, however, is that the Horton ratios typically are measures for natural watershed conditions. No generalized relationship is available to include the effects of storm sewering on, for example, the bifucation ratio, R_B. Further, determination of the Horton order ratios is time consuming which is not desirable for regionalized parameter prediction equations.

Regardless of these criticisms, if regional equations can be developed which contain the same information as the Horton ratios and reflect the effect of drainage system improvements while requiring minimal measurements, then a worthwhile procedure will have been developed. The results reported herein are encouraging and seem to offer new directions in synthetic unit hydrograph theory.

Acknowledgements

This paper is based on research supported in part with funds provided by a cooperative study between the South Carolina Department of Highways and Public Transportation and the U.S. Geological Survey. The author especially wishes to acknowledge the assistance of Mr. Sean Roche who performed most of the watershed measurements.

Appendix I.--References

1. Dooge, J. C. I. (1973), "Linear Theory of Hydrologic Systems," USDA-ARS, Tech. Bull. No. 1468.
2. McCuen, R. H. and T. R. Bondelid (1983), "Estimating Unit Hydrograph Peak Rate Factors," J. Irr. and Drain., ASCE, 109(2), 238-25
3. Meadows, M. E. and G. E. Blandford (1983), "Improved Methods and Guidelines for Modeling Stormwater Runoff from Surface Coal Mined Lands," Report 147, Water Resources Res. Inst., Univ. of Kentucky, Lexington,
4. Meadows, M. E. and A. L. Chestnut (1983), "Advances in Irrigation and Drainage: Surviving External Pressures," Proc., ASCE Spec. Conf., held at Jackson, Wyoming.
5. Rodrigues-Iturbe, I. and J. B. Valdes (1979), "The Geomorphologic Structure of the Hydrologic Response," Water Resour. Res., 18(4), 877-886.
6. Rosso, R. (1984), "Nash Model Relation to Horton Order Ratios," Water Resour. Res., 20(7), 914-920.
7. Wu, I. P. (1963), "Design Hydrographs for Small Watersheds in Indiana, J. Hydr. Div., ASCE, 89(6), 35-66.

DEVELOPMENT, USE, AND SYNTHESIS OF S-GRAPHS

By George V. Sabol[1], M.ASCE

ABSTRACT
    A study to identify S-graphs from the southwestern United States
has been completed. This study has resulted in the compilation of 53
S-graphs, documentation of some of the watershed characteristics,
investigation into the development of an empirical relation for lag,
and an investigation into the synthesis of S-graphs from Clark unit-
hydrographs. Some of the results of that study are presented.

INTRODUCTION

    An S-graph is a form of unit-hydrograph and is often used in
performing flood studies. S-graphs are usually defined by the
reconstitution of recorded flood events and numerous S-graphs are
available from such reconstitutions. Existing S-graphs for the
southwestern United States have been compiled and reviewed (Sabol,
1987a). These and other S-graphs can be used (transposed) to other
watersheds for the purpose of defining a unit-hydrograph under certain
limiting conditions. The concept of the S-graph dates back to the
development of the unit-hydrograph itself, although the application of
the S-graph has not been as widely practiced as that of the unit-
hydrograph. The use of S-graphs has been practiced mainly by the U.S.
Army Corps of Engineers, particularly the Los Angeles District, and
the U.S. Bureau of Reclamation. Recently the S-graph has been adopted
as the unit-hydrograph procedure by several counties in southern
California and selected S-graphs have been presented in their
hydrology manuals. The S-graphs in those hydrology manuals have been
selected primarily from S-graphs that had previously been defined by
the Los Angeles District of the Corps of Engineers from a rather long
and extensive history of analyses of floods in California. Other
areas may not have the advantage of such an extensive data base.
    S-graphs may gain more popularity and usage with the release of
the Third Edition of Design of Small Dams by the USBR. This
engineering reference book is widely used as a guideline in flood
hydrology, particularly in the western United States, and the revised
Flood Hydrology Studies chapter will discuss and present the
application of S-graphs. Six regionalized S-graphs are to be
presented in the revised chapter of that book.
    Recently there has been an increase in interest in the use of S-
graphs for flood hydrology. The transposition of S-graphs from gaged
watersheds for use in ungaged watersheds needs to be carefully
evaluated. In addition there is a need for better empirical methods
for estimating the S-graph parameter, lag.

[1]Consulting Engineer, 1351 East 141st Ave., Brighton, Colorado 80601

S-GRAPHS

Definitions - An S-graph is a dimensionless form of a unit-hydrograph in which discharge is expressed in percent of ultimate discharge and time is expressed in percent of lag. An example of a typical S-graph is shown in Figure 1. In hydrology there are, regrettably, several different definitions of lag, and in every application requiring the use of a lag term the hydrologist should verify which definition of lag is to be applied. For an S-graph, lag is the elapsed time from the beginning of an assumed continuous series of unit rainfall excess increments over the entire basin to the instant when the rate of resulting runoff equals 50 percent of the ultimate discharge. An equivalent definition of this lag is the time for 50 percent of the total volume of runoff of a unit-hydrograph to occur. Ultimate discharge is the maximum discharge that will be achieved from a particular watershed when subjected to a continuous intensity of rainfall excess uniformly over the basin.

Development of S-Graphs - S-graphs are developed by summing a continuous series of unit-hydrographs, each lagged behind the previous unit-hydrograph by a time interval that is equal to the duration of rainfall excess for the unit-hydrograph. The resulting summation is a graphical distribution that resembles an S-graph except that the discharge scale is accumulated discharge and the time scale is in units of measured time. This graph is terminated when the accumulated discharge equals Q_{ult} which occurs at a time equal to the base time of the unit-hydrograph less one duration interval. The basin lag can be determined from this graph at the time at which the accumulated discharge equals 50 percent of Q_{ult}. This summation graph is then converted to a dimensionless S-graph by dividing the discharge scale by Q_{ult} and the time scale by Lag, the scales of the resulting S-graph are expressed as percent Q_{ult} and percent Lag, respectively.

In practice, S-graphs have generally been developed by reconstituting observed floods to define a representative unit-hydrograph and then converting this to an S-graph. Prior to the advent of computerized models, such as HEC-1, flood reconstitution was a laborious task of rainfall and hydrograph separation along with numerous hand-cranked simulations to define the representative unit-hydrograph. Modern S-graph development generally relies on use of optimization techniques, such as coded into HEC-1, to identify unit-hydrograph parameters that best reproduce the observed flood.

Parameter Estimation - Three parameters are used in the application of S-graphs; drainage area (A), selected duration of rainfall excess (D), and Lag. The size of the drainage area will be obtained from adequate watershed maps. The duration is selected after the lag is calculated and is the computational interval that will be used in convoluting the rainfall excess increments with the unit-hydrograph.

A general relationship for basin lag as a function of watershed characteristics is given by Equation 1.

$$\text{Lag} = C \left(\frac{L\, L_{ca}}{S^{\,p}} \right)^{m} \tag{1}$$

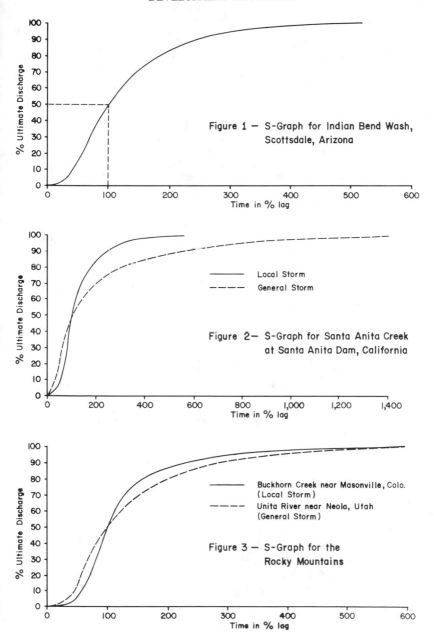

Figure 1 — S-Graph for Indian Bend Wash, Scottsdale, Arizona

Figure 2 — S-Graph for Santa Anita Creek at Santa Anita Dam, California

Figure 3 — S-Graph for the Rocky Mountains

where Lag is basin lag, in hours,
  L is length of longest watercourse, in miles,
  L_{ca} is length along the watercourse to a point opposite the
       centroid, in miles,
  S is watercourse slope in feet/mile,
  C is a coefficient, and
  m and p are exponents.
The Los Angeles District often uses C = 20n, where n is the estimated
mean Manning's n for all the channels within an area, and m = 0.38.
The USBR (1987) has recommended that C = 26n and m = 0.33. Both sets
of values in Equation 1 will often result in similar estimates for
Lag. Traditionally the exponent, p, on the slope is equal to 0.5.
    A major disadvantage of Equation 1 is that n must be selected
which is very subjective and introduces significant uncertainty into
the Lag prediction. Also, Equation 1 is not dimensionly homogeneous
when m and p as defined above are used. A modified Lag equation has
been developed (Sabol, 1987b) based on dimensional similitude, and in
this equation p = 2.0, m = 0.25, and C is determined from various
equations depending upon watershed location and characteristics.
    Data on watershed characteristics and lag as determined by S-
graph analysis was obtained from the USBR and was analyzed by stepwise
multiple regression in various groupings of data (Sabol, 1987b). For
example, the resulting prediction equation for C in the modified lag
equation for a region encompassing the Southwest Desert, Great Basin,
and Colorado Plateau is:
$$C = -21.72 + 1.79 \log A + 4.90 \log S \qquad (2)$$
where A is drainage area in square miles, S is watercourse slope in
feet per mile, and log indicates natural logarithm.

Limitations on Use of S-graphs - Although the S-graph is completely
dimensionless and does not have a duration of rainfall excess
associated with it as does an unit-hydrograph, its general shape and
the magnitude of Lag is influenced by the distribution of rainfall
over the watershed and the time distribution of the rainfall.
Therefore, the transposition of an S-graph from a gaged watershed to
application in another watershed must be done with consideration of
both the physiographic characteristics of the watersheds and the
hydrologic characteristics of the rainfalls for the two areas.
    Two examples are provided of the effect of rainfall on the shape
of S-graphs: S-graphs have been developed from two different types of
storms for Santa Anita Creek at Santa Anita Dam in California; one S-
graph is for a general storm and the other for a thunderstorm. These
two S-graphs are graphically compared in Figure 2. The USBR, in
preparing the Third Edition of Design of Small Dams, has identified
six S-graphs for application in generalized regional and physiographic
watersheds. Two of these S-graphs are for the Rocky Mountains; one is
for a general storm and the other is for a thunder-storm. These two S-
graphs are graphically compared in Figure 3.

SYNTHESIS OF S-GRAPHS
    An investigation has been performed (Sabol, 1987a) in which S-
graphs were synthesized from Clark unit-hydrographs. The magnitude of
T_c and R will vary with the size of drainage area, however the effects

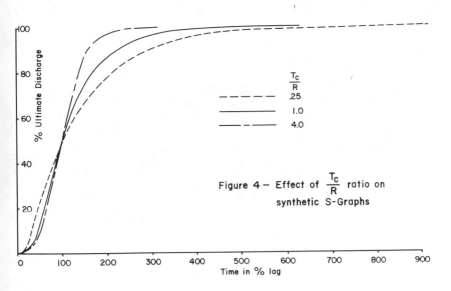

Figure 4 – Effect of $\frac{T_c}{R}$ ratio on synthetic S-Graphs

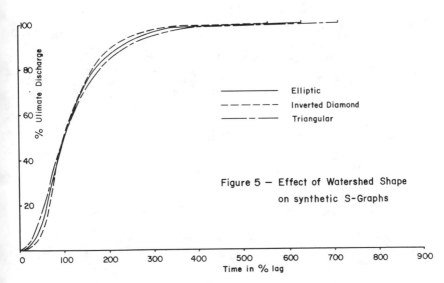

Figure 5 – Effect of Watershed Shape on synthetic S-Graphs

of the magnitudes of T_c and R can be eliminated if the ratio of T_c/R is held constant. That is, watersheds of any size and with any values of T_c and R will result in identical S-graphs if the ratio of T_c/R is the same. Therefore, a family of S-graphs can be developed with each S-graph being identified by a ratio T_c/R.   The results of the synthesized S-graphs are shown in Figure 4 for a range of T_c/R from 0.25 to 4.0.

The shape of a watershed should have an affect on the shape of the corresponding unit-hydrograph and S-graph. The effect of watershed shape on S-graphs was investigated by synthesizing S-graphs by the Clark unit-hydrograph for three different time-area relations. The first time-area relation is the default relation used in HEC-1, and is for a symmetric watershed of elliptic shape. The second relation is for a triangular shape with the largest contributing area being most removed from the watershed outfall location.   The third relation is for an inverted diamond shape with the largest contributing area being closest to the watershed outfall location.

The results of S-graph synthesis using the time-area relations are shown in Figure 5.   The three examples are for watersheds with T_c/R equal to 1.0 and for D equal to 20 percent of T_c. As shown in Figure 5 the S-graph is most affected in the range from 0 to 100 percent Lag, and the magnitude of this is about the same as the effect of a change in T_c/R over the range from 0.25 to 4.0 as illustrated in Figure 4.

CONCLUSIONS
1. The results of the compilation of 53 S-graphs from the south-
    western United States has recently been completed.
2. S-graphs are completely dimensionless forms of unit-hydrographs
    and S-graphs that have been developed for gaged watersheds can
    often be applied for hydrologically similar watersheds.
3. Storm characteristics can affect the shape of S-graphs. There-
    fore, S-graphs should only be transposed for areas that have
    similar rainfall characteristics.
4. S-graphs can be synthesized from Clark unit-hydrographs.
5. S-graphs can be defined by a T_c/R ratio.
6. Watershed shape has a significant effect on S-graph synthesis.
    Therefore, S-graph transposition should be performed with
    adequate consideration of the shape of the gaged and ungaged
    watersheds.

REFERENCES

Sabol, G.V., 1987a, S-Graph Study: preliminary report prepared for the Flood Control District of Maricopa County, Phoenix, Arizona, 30 pgs. plus appendices.

Sabol, G.V., 1987b, Inspection of Lag Relation: unpublished technical paper.

U.S. Bureau of Reclamation, 1987, Design of Small Dams, Third Edition: Unpublished manuscript.

TIME OF CONCENTRATION
IN SMALL RURAL WATERSHEDS

Dr. Constantine N. Papadakis *, Member ASCE
Dr. M. Nizar Kazan **, Member ASCE

INTRODUCTION

The determination of peak discharges for a given return period is necessary for the appropriate design of drainage structures. Peak discharges of a given frequency are related to rainfall intensity which in turn depends on rainfall duration. Since the maximum runoff for a given frequency occurs when the rainfall duration becomes equal to the time of concentration of the watershed, the time of concentration is the most significant variable in the computation of peak runoff.

Many empirical and a few theoretically founded equations used to compute the time of concentration were evaluated in this study. Some of these equations consider the time of concentration to be only a function of physical watershed parameters, such as length, slope, roughness and degree of imperviousness. Other equations also consider the characteristics of rainfall excess, such as rainfall intensity and duration. Times of concentration computed by these equations for a given watershed and for the same rainfall event were found to vary by more than 500%.

Data were gathered and analyzed from: (a) measurements from tests on three experimental watersheds conducted by the Corps of Engineers, Colorado State University, and the University of Illinois and from (b) measurements from 84 small rural watersheds from 22 states obtained by the USDA Agricultural Research Service for selected runoff events. From this data, two global regression equations were developed for the estimation of the time of concentration in small rural watersheds. One of these equations is based on four independent parameters and the second is based on only one independent parameter. These equations have general applicability and could be used in design with a predetermined degree of confidence.

REVIEW OF EXISTING METHODS

The first phase of this study involved an exhaustive literature search which revealed a plethora of methods developed to compute the time of concentration. These formulas share the general format:

$$T_c = k \, L^a \, n^b \, S^{-y} \, i^{-z} \tag{1}$$

where T_c = time of concentration, in minutes; L = length of flow path, in feet; n = roughness coefficient; S = average slope of flow path, in ft/ft; i = intensity of excess rainfall, in in/hr; k = constant; and a,b,y,z = exponents.

Eleven of the most frequently encountered formulas were reviewed [3]. These included the Kirpich, Izzard, Kerby/Hathaway, Carter, Eagleson, Kinematic Wave, Morgali and Linsley, Federal Aviation Agency (FAA); SCS Curve Number, SCS Velocity Charts, and the Singh's Kinematic Wave and Chezy Formulas. In some cases $b = 0$ and/or $z = 0$ which indicates that the time of concentration was considered to be independent of watershed surface roughness and/or excess rainfall intensity.

---

* Dean and Geier Professor of Engineering Education, College of Engineering, University of Cincinnati, Cincinnati, Ohio 45221
** Head, Hydraulic Engineering Department, College of Civil Engineering University of Aleppo, Syria

TIME OF CONCENTRATION OF NATURAL AND EXPERIMENTAL WATERSHEDS

A comprehensive data base was compiled for 84 natural rural watersheds from 22 states. Information was obtained from the US Department of Agriculture, Agricultural Research Service [1]. Watersheds were selected if they had an area of less than 500 acres, detailed basin topography and surface cover information was available, and a rainfall event had been isolated with good rainfall-runoff measurements. The length and the average slope of the flow path were measured for each watershed. Values of the other two independent parameters, the average surface roughness coefficient of the basin, and the excess rainfall intensity were estimated from the available data. Values of the time of concentration were also found from the available data.

From 1948 to 1952 the Corps of Engineers conducted simulated rainfall tests at the Santa Monica Municipal Airport. The tests were performed on airfield strips having flow-path lengths of 84 to 500 feet and slopes of 0.5, 1 and 2 percent. Simulated rainfall intensities of 0.25 to 10 inches per hour on concrete and simulated turf were utilized. The roughness coefficient for concrete surfaces was considered equal to n = 0.04 and for the turf covered flow surfaces equal to n = 0.20. The results of 162 of these tests as compiled by the Los Angeles District of the Corps of Engineers [2] were used in this study. The length of the flow path, the slope, the applied rainfall intensity, and the measured average time of equilibrium (equal to the time of concentration) were obtained for each of 89 cases involving concrete surface and for each of 73 cases involving simulated turf surface.

The experimental watershed constructed at the Engineering Research Center of Colorado State University consists of a conic sector which has an interior angle of 104 degrees and a radius of 116 feet with a slope of 0.05. Two 88-foot by 70-foot long intersecting plane surfaces join the edges of the conic sector with a maximum surface slope of 0.05 and a collecting channel slope of 0.03 [4]. The simulated rainfall tests were conducted in 1970-1971 and utilized different surface cover materials, such as gravel and butyl. Ninety three of these tests were utilized in this study including the watershed configuration, length of flowpath, slope and rainfall intensity. The actual times of concentration were estimated from the runoff hydrographs.

An experimental basin and a precipitator were used in an indoor laboratory at the University of Illinois at Urbana-Champaign. The size of the basin is 40 feet by 40 feet, its lateral slope is 0.01 and its longitudinal slope can be set at 0.005, 0.01, or 0.03. The length of the flow path is 60 feet and the roughness coefficient of the aluminum plate surface of the basin was estimated equal to n = 0.08. The results of the tests performed in 1974 at the University of Illinois were reported using relative nondimensional variables [5]. The necessary transformations were performed to obtain the values of the rainfall intensity and the time of concentration for 36 tests used in this study.

FOUR-PARAMETER TIME OF CONCENTRATION EQUATION

A four-parameter equation of the general format of Eq. 1 was chosen to fit the 375 data points obtained from natural and experimental watersheds. In Eq. 1 the time of concentration is the dependent variable and L, n, S and i are the independent variables. This equation exhibits a linear correlation of the logarithms of the variables involved.

A regression analysis was performed for each group of available data. A power model was used to regress the time of concentration on four predictor variables L, n, S and i. Only two predictor variables were used for the Colorado State University data because L and S were constant throughout the measurements, and similarly for the University of Illinois data where L and n were constant. Table 1 summarizes the results of the regression analysis for each data group and for the total data sample. The table includes estimates for the parameters k, a, b, y and z. The standard deviation of a sample of observations, σ, is also listed in Table 1 together with the coefficient of variation which is the ratio of the standard deviation to the mean, $\sigma/\log T_c$. The coefficient of determination is also included in Table 1. This coefficient is

TABLE 1
REGRESSION ANALYSIS RESULTS OF FOUR-PARAMETER T_c EQUATION 2

Data Source & Watershed Type	No.of Data	k	a	b	y	z	Standard Deviation σ	Coeff. of Variation $\sigma/\overline{\log T_c}$ (%)	R^2 (%)
USDA-ARS Natural	84	1.04	0.60	0.96	0.24	0.29	0.126	9.0	76.8
US Corps Experimental	162	0.95	0.41	0.49	0.33	0.47	0.035	3.7	98.6
CSU Experimental	93			0.43		0.30	0.089	12.4	75.2
Univ. of Illinois Experimental	36				0.30	0.24	0.044	8.4	76.4
All Experimental Corps + CSU + UI	291	0.75	0.42	0.48	0.26	0.42	0.062	7.7	95.5
All data Natural plus Experimental	375	0.66	0.50	0.52	0.31	0.38	0.092	9.8	94.1

TABLE 2
REGRESSION ANALYSIS RESULTS OF ONE-PARAMETER T_c EQUATION 4

Data Source & Watershed Type	No. of Data	k	x	Standard Deviation σ	Coef. of Variation $\sigma/\overline{\log T_c}$ (%)	R^2 (%)
USDA-ARS Natural	84	0.98	0.44	0.140	10.0	70.5
UC Corps Experimental	162	0.49	0.52	0.070	6.8	95.1
CSU Experimental	93	0.62	0.47	0.090	12.6	79.3
Univ. of Illinois Experimental	36	0.75	0.43	0.040	8.4	75.6
All Experimental Corps + CSU + UI	291	0.55	0.50	0.070	8.8	94.1
All data Natural plus Experimental	375	0.52	0.52	0.095	10.0	93.7

Table 3
TOLERANCE LIMITS

	P(%)	D	$\Delta\log T_c$	Tolerance Interval of T_c	Best Fit k	Tolerance Interval of k
Four Parameter Equation 2	75	1.23	± 0.113	$0.77 < T_c/\overline{T}_c < 1.30$		$0.51 < k < 0.86$
	90	1.75	± 0.161	$0.69 < T_c/\overline{T}_c < 1.45$	0.66	$0.45 < k < 0.96$
	95	2.09	± 0.192	$0.64 < T_c/\overline{T}_c < 1.56$		$0.42 < k < 1.03$
One Parameter Equation 4	75	1.23	± 0.117	$0.76 < T_c/\overline{T}_c < 1.31$		$0.40 < k < 0.68$
	90	1.75	± 0.166	$0.68 < T_c/\overline{T}_c < 1.47$	0.52	$0.35 < k < 0.76$
	95	2.09	± 0.199	$0.63 < T_c/\overline{T}_c < 1.58$		$0.33 < k < 0.82$

Table 4
COMPARISON OF EXPONENTS OF VARIOUS TIME OF CONCENTRATION EQUATIONS

Equations / Exponents	Kirpich	Izzard	Kerby	Carter	Eagleson	Kinematic Wave	Morgali	FAA	SCS Curve	SCS Velocity	Singh/Chezy	4-parameter Equation 2	1-parameter Equation 4
Exponent of L	0.77	0.33	0.467	0.60	0.76	0.60	0.593	0.50	0.80	1.0	0.667	0.50	0.52
Exponent of S	-0.385	-0.333	-0.233	-0.30	-0.19	-0.30	-0.38	-0.333	-0.50	-0.50	-0.333	-0.31	-0.35
Exponent of i	0	-0.667	0	0	0	-0.40	-0.388	0	0	0	-0.333	-0.38	-0.35

equal to the square of the correlation coefficient R and indicates the percentage of the variation in the variable that is explained by the regression equation.

Based on the total data sample available, the best-fit four-parameter time of concentration equation was found to be:

$$T_c = 0.66 \, L^{0.50} \, n^{0.52} \, S^{-0.31} \, i^{-0.38} \tag{2}$$

Values of T_c computed by Eq. 2 versus those measured are plotted in Figure 1 using logarithmic scales. Tolerance limits containing 75, 90 and 95% of the sample points are also enveloped in Figure 1. Statistical tolerance limits for a given population are limits within which a stated proportion of the population are expected to lie with respect to some measurable characteristic. The width of the two-sided tolerance limits is:

$$\Delta \log T_c = \pm \, D\sigma \tag{3}$$

where σ is the standard deviation computed from a sample of m observations. The factor D is such that the probability is γ that a proportion P(%) of the m observations will be included between the tolerance limits. The factor D is a function of γ, P and m and can be obtained from statistical tables.

The probability γ is called the level of confidence and is equal to (1-a) where a is the level of significance. There is an a% risk of error, for even if the null hypothesis does hold, there is an a% probability that it will be rejected. The value of a is often based on convention and the availability of statistical tables. A value of a = 0.05 is being selected frequently. The tolerance interval encloses P percent of the population with a given confidence γ.

For a level of confidence equal to $\gamma = 0.95$ and a sample size of m = 375, values of D for three selected values of P are listed in Table 3 together with the corresponding tolerance limits of the dependent variable T_c. These limits can be transformed to a tolerance interval of the constant k of Eq. 1 as shown in Table 3. It is of interest to note that as P increases from 75% to 95%, given the same level of confidence and sample size, the width of the two-sided tolerance limits also increases.

ONE PARAMETER TIME OF CONCENTRATION EQUATION

The exponents of L and n are almost identical in Eq. 2. Furthermore the exponents of i and S are also nearly equal. Therefore, Eq. 2 was simplified by combining the four independent parameters in one.

A linear regression analysis was performed for each group of available data and for the total data sample. Table 2 summarizes the results of the regression including the parameter k and the exponent, the standard deviation of the sample, the coefficient of variation and the coefficient of determination. Based on the total data sample available, the best-fit one-parameter time of concentration equation was found to be:

$$T_c = 0.52 \, [Ln \, (Si)^{-2/3}]^{0.52} \tag{4}$$

Observed values of Tc versus those computed by Eq. 4 are plotted in Figure 2 using logarithmic scales. Tolerance limits containing 75, 90 and 95% of the sample are enveloped in Figure 2.

For a level of confidence equal to $\gamma = 0.95$ and a sample size of m = 375, values of D for three selected values of P are listed in Table 3 together with the corresponding tolerance limits of the dependent variable Tc. These limits can be transformed to a tolerance interval of the constant k of Eq. 4 as shown in Table 3.

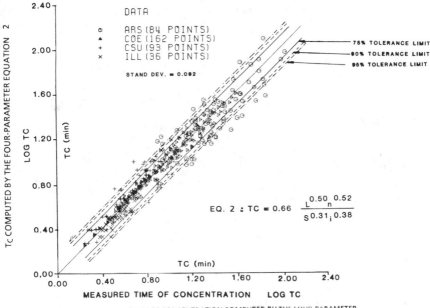

Figure 1.   TIMES OF CONCENTRATION COMPUTED BY THE FOUR-PARAMETER
EQUATION 2 VERSUS MEASURED VALUES.

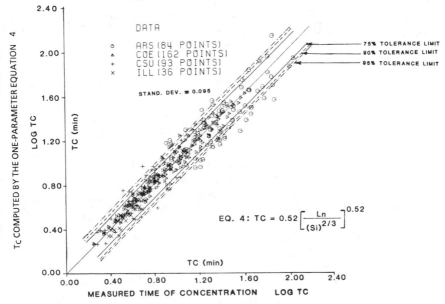

Figure 2. TIME OF CONCENTRATION COMPUTED BY THE ONE-PARAMETER
EQUATION 4 VERSUS MEASURED VALUES.

CONCLUSIONS

A comparison was performed between the exponents of the independent variables L, S and i of the four-parameter time of concentration Eq. 2 and eleven frequently encountered equations listed in Table 4. This comparison led to the following conclusions:

a) The exponents of L in the equations of Carter, Kinematic Wave, Morgali, FAA and Kerby agree within $\pm 20\%$ with the exponent a of Eq. 2.

b) The exponents of S in the equations of Izzard, FAA, Kerby, Carter, Kinematic Wave, Morgali and Singh agree within $\pm 25\%$ with the exponent y of Eq. 2.

c) The exponents of i in the equations of Kinematic Wave, Morgali and Singh agree within $\pm 15\%$ with the exponent z of Equation 2.

Three case studies were also used to compare the time of concentration computed from Eq. 2 with the values obtained from the other eleven equations of Table 4 and those measured [3]. Results from Eq. 2 closely agreed with the measured time of concentration. Almost the only other equation that showed good agreement with the measurements is the Kinematic Wave equation. However, the derived four-parameter Eq. 2 has more general applicability compared to the Kinematic Wave equation which is more appropriate for computing the time of concentration of very small rural watersheds where surface flow is predominant.

ACKNOWLEGEMENTS

The authors would like to thank Professor Vujica Yevjevich of George Washington University, Professor Tom Sanders of Colorado State University, Professor Ben Yen of the University of Illinois, Chief Hydrologist Andrew Sienkiewich of the Corps of Engineers Los Angeles District, and Research Hydrologist Ted Engman of the Beltsville USDA-ARS Hydrology Lab for making available the data that made this study possible. Appreciation is extended to the Civil Engineering Department of Aleppo University in Syria for having offered one year of leave to Professor M. N. Kazan to perform this investigation in collaboration with Dr. Papadakis at the Civil Engineering Departments of Colorado State University and the University of Cincinnati.

REFERENCES

1. Agricultural Research Service, "Selected Runoff Events for Small Agricultural Watersheds in the United States," Soil and Water Conservation Research Division, 1955-1975.

2. Corps of Engineers, U.S. Army "Data Report - Airfield Drainage Investigation", prepared by the Los Angeles District, October, 1954.

3. Papadakis, C. N. and Kazan. M. N., "Time of Concentration in Small Rural Watersheds", Technical Report 101/08/86/CEE, Civil Engineering Department, University of Cincinnati, Cincinnati, Ohio, August, 1986.

4. Schulz, E. F. and O. G. Lopez, "Determination of Urban Watershed Response Time", Colorado State University, Hydrology Paper No. 71, Fort Collins, CO, 1974.

5. Shen Y. Y., B. C. Yen and V. T. Chow, "Experimental Investigation of Watershed Surface Runoff", Hydraulic Engineering Series No. 29, University of Illinois, Urbana-Champaign, IL, 1974.

APPLICATION OF THE NEW SCS TIME OF CONCENTRATION METHOD

BY

C.K. Taur[1], A.M.ASCE and George E. Oswald[2]

INTRODUCTION

The second edition of the Soil Conservation Service TR-55, published in June 1986, introduces a new method for calculating the time of concentration for storm runoff flowing through the small watersheds. The method divides runoff into three flow conditions, i.e., sheet flow, shallow concentrated flow and channel flow. The travel time for sheet flow is estimated by a Manning's kinematic solution and is limited to a range of 300 feet. After 300 feet, sheet flow becomes shallow concentrated flow. Then, shallow concentrated flow travels into open channels.

Sheet flow replaces overland flow in the first edition of TR-55 and shallow concentrated flow replaces storm sewer and gutter flows. Shallow concentrated flow is divided in two categories: paved and unpaved. The two equations used in computing the average velocity for paved and unpaved shallow concentrated flows are exactly the same as those used for storm sewer and gutter flows in the first edition of TR-55. The Manning's equation is applied to calculate the average velocity of channel flow in both the first and the second editions.

The difficulty in application of the new method is in calculating the travel time for sheet flow. The Manning's kinematic solution uses the Manning's coefficient as the decision factor in computing the travel time. This results in a much longer travel time than that from the

---

[1]Staff Engineer and [2]Division Engineer, Watershed Management Division, City of Austin, 505 Barton Springs Road, Austin, Texas 78704.

639

overland flow method in the first edition of TR-55. The longer travel time creates a lower peak flow in hydrologic calculation and affects hydraulic design in urban drainage system.

This study is to make comparison between the two TR-55 time of concentration methods, especially for the travel time methods of overland flow and sheet flow. An example of calculating the travel time for four different land types with three different slopes and three different flow distances is used.

OVERLAND FLOW

Overland flow, storm sewer or road gutter flow, and channel flow are the three phases used in the first edition of TR-55 for computing the travel time for storm runoff through an urban watershed. The travel time for overland flow was defined as the time which it takes water to travel from the uppermost part of the watershed to a defined channel or inlet of the storm sewer system. The velocities of the overland flow vary differently with the surface covers and the land slopes. Figure 1 was provided to compute overland flow velocities of six different surface covers. There is no limit being set for the flow length.

After determining the average velocity from the Figure 1, the travel time can be calculated by

$$T_t = L/V \qquad\qquad\qquad \text{(Eq. 1)}$$

in which, L is the length of the reach in feet and T_t is the travel time in seconds.

SHEET FLOW

In the second edition of the SCS TR-55, the three phases for calculating the travel time are sheet flow, concentrated flow and channel flow. The sheet flow takes the place of the overland flow with a new limit that the flow length cannot exceed 300 feet. The following

Manning's Kinematic equation is introduced to calculate the travel time for the sheet flow.

$$T_s = 0.007(nL)^{0.8}/(P_2)^{0.5}S^{0.4} \qquad (Eq.\ 2)$$

in which

T_s = travel time of sheet flow (hr)
n = Manning's coefficients
L = flow length (ft)
P_2 = 2-year, 24 hour rainfall depth (in)
S = land slope (ft/ft)

The Manning's coefficient used in Eq. 2 represents very shallow flow depths of about 0.1 foot for various surface conditions. Table 1 is provided for roughness coefficients in different land surface covers. The 2-year, 24-hour rainfall depth can be obtained from the TR-55 or the National Weather Service TP-40.

Table 1 - Manning's Coefficients for Sheet Flow

Surface description	n
Smooth surfaces (concrete, asphalt, gravel, or bare soil)	0.011
Fallow (no residue)	0.05
Cultivated soils:	
Residue cover 20%	0.06
Residue cover 20%	0.17
Grass:	
Short grass prairie	0.15
Dense Grasses	0.24
Bermudagrass	0.41
Range (natural)	0.13
Woods:	
Light underbrush	0.40
Dense underbrush	0.80

The value of 2-year, 24 hour rainfall is 4.1 inches for the areas in Austin, Texas. Equation 2 is modified as follows:

$$T_s = 0.208 \ (nL)^{0.8}/S^{0.4} \qquad\qquad\qquad (Eq. \ 3)$$

in which T_s is travel time in minutes and other parameters are the same as those used in Eq. 2.

COMPARISON

An example calculating and comparing travel time for the overland flow and the sheet flow is shown in the following steps.

Step 1. Find the average velocity for four different overland flow from Figure 1. The land surface slopes are 2, 5, and 7%.

Velocities of Four Different Overland Flows (fps)

Slope (%)	Forest or Meadow	Fallow	Short Grass	Bare Ground
2	0.35	0.66	1.0	1.4
5	0.56	1.10	1.6	2.3
7	0.66	1.30	1.9	2.6

Step 2. Choose the corresponding Manning's coefficients from Table 1 for the above four different surface covers. The roughness coefficient are

Forest or Meadow	Fallow	Short Grass	Bare Ground
0.40	0.05	0.15	0.011

Step 3. Calculate and compare the travel time from two different methods. Equation 1 (the old method) is used to calculate travel time for overland flow and Equation 3 (the new method) is used for sheet flow. Since the flow length is limited in the range of 300 feet for sheet flow, three flow distances of 100, 200 and 300 feet are used in comparison. The results are shown in the following tables.

Travel Time (in minutes) for Flow Distances of 100 Feet

Slope (%)	Forest or Meadow Old	New	Fallow Old	New	Short Grass Old	New	Bare Ground Old	New
2	4.76	19.02	2.53	3.60	1.67	8.68	1.19	1.07
5	2.98	13.18	1.52	2.50	1.04	6.01	0.72	0.74
7	2.53	11.52	1.28	2.18	0.88	5.26	0.64	0.65

Travel Time (in minutes) for Flow Distance of 200 Feet

Slope (%)	Forest or Meadow Old	New	Fallow Old	New	Short Grass Old	New	Bare Ground Old	New
2	9.52	33.11	5.05	6.27	3.33	15.11	2.38	1.87
5	5.95	22.95	3.03	4.35	2.08	10.45	1.45	1.29
7	5.05	20.06	2.56	3.80	1.75	9.15	1.28	1.13

Travel Time (in minutes) for Flow Distance of 300 Feet

Slope (%)	Forest or Meadow Old	New	Fallow Old	New	Short Grass Old	New	Bare Ground Old	New
2	14.29	45.80	7.58	8.68	5.00	20.90	3.57	2.58
5	8.93	31.74	4.55	6.01	3.13	14.48	2.17	1.79
7	7.85	27.75	3.85	5.26	2.63	12.66	1.92	1.57

CONCLUSION

From the above comparison, it is easy to find that there are obviously different results in travel time from the two methods, especially for the land types of forest or meadow and short grass. The short grass areas include lawns, parks, greenbelts and golf courses which are usually encountered by a drainage engineer in the urban hydrologic analysis.

For a storm runoff flowing through a small watershed with flat slope (2%) and 300-ft flow length, the values of travel time from the old and the new methods are 5 minutes and 20.9 minutes. It means that rainfall intensities will be 12.0 and 7.8 in/hr, respectively (by the Austin Intensity-Duration-Frequency Curves). The peak flows from the Rational Method will be 1.54 times differently by application of the two travel time methods for the same watershed.

It seems that the second edition of the SCS TR-55 provides a more comprehensive approach to the travel time calculation than that in the first edition. However, a sound engineering judgement is required when this new time of concentration method is applied to the urban drainage study.

Appendix - References

1. City of Austin, Austin Drainage Criteria Manual, First Edition, Austin, Texas, 1977.

2. Soil Conservation Service, Urban Hydrology for Small Watersheds, First Edition, Technical Release 55, January 1975.

3. Soil Conservation Service, Urban Hydrology for Small Watersheds, Second Edition, Technical Release 55, June 1986.

Figure 1. Average Velocities for Estimating Travel Time for Overland Flow.

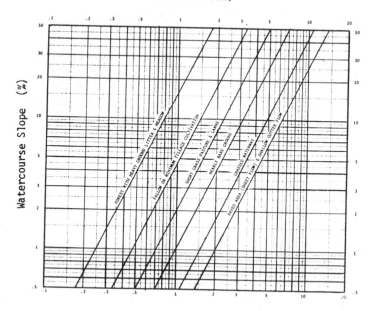

Velocity (fps)

Some New Snyder Type Coefficients For Small Watersheds

by

Victor L. Zitta[1], A. M. ASCE and Timothy J. Hubbard[2]

Abstract

Data collected on the Goodwin Creek watershed has produced a continuous hydrologic record from 37 rain gages and runoff record from 14 runoff stations within a drainage area of 8.26 sq mi (21.39 sq km). All of the data has been collected by electronic sensors monitored by radio telemetry and recorded in synchronous time by a mini computer. Thirteen simple storms have been identified at two of the 14 runoff stations. Basin #1 includes the entire drainage area of 8.26 sq mi (21.39 sq km) and basin #5 includes an area of 1.66 sq mi (4.30 sq km). For each of 13 simple storms for basin #1, time of excess precipitation, t_r; basin time lag, t_1; time base of the surface runoff, T; precipitation excess, P_e; peak discharge, Q_p; and the ordinates of the unit hydrograph were determined. Storms which may be considered simple for the entire basin were complex when considered over a sub-basin. Only 9 of the 13 storms could be considered simple for basin #5. Basin parameters of main channel length, L; and distance along the main channel L_{ca}, to a point opposite the centroid of the basin were determined. Values of Snyder's coefficients C_t and C_p were calculate from standard formulas.

For basin #1, Snyder's C_t values ranged from 0.710 to 1.228 with a mean value of 0.894. Values of C_p ranged from 0.559 to 1.128 with a mean of 0.843. For basin #5, Snyder's C_t values ranged from 0.276 to 2.173 with a mean of 9 values equal to 1.18. Values of C_p ranged from 0.174 to 1.900 with a mean value of 0.815. Published values for C_t range from 1.8 to 2.2 with no mean value specified. Published values for C_p range from 0.4 to 0.8. For both basins, values of C_t tend to be lower and values of C_p tend to be higher than published values.

Introduction

Goodwin Creek watershed is a well instrumented catchment on which continuous rainfall and runoff data have been collected since October 1, 1981. Rainfall data are collected from 37 rain gages distributed uniformly both within and surrounding the watershed. A total of 25 gages are within the basin, giving a raingage density of 1 raingage per 0.3308 sq mi (1.42 sq km). There are 14 runoff stations within the watershed covering a range of drainage areas from 8.26 sq mi (21.39 sq km) to 0.02 sq mi (0.06 sq km).

The watershed is described in Figure 1 with the basin boundaries, the drainage waterway network, the raingage and streamgage stations located. There is one complete climatological station located between streamflow stations 4 and 14. All raingage and streamgage stations have automatic sensors which are interrogated by a central computer

---

[1]Professor, Dept. of Civil Engineering, Mississippi State
  University, Mississippi State, MS 39762
[2]Hydrologic Technician, Vicksburg District, U.S. Army Corps of
  Engineers, Vicksburg, MS 39180

through radio telemetry signals. Data collection is event actuated with
data points stored in response to magnitude change. Data collection is
the responsibility of the USDA–ARS Sedimentation Laboratory, Oxford, MS.

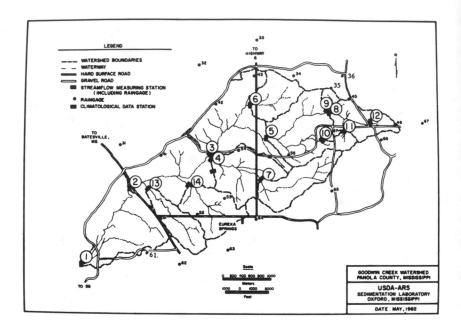

Figure 1. Goodwin Creek Watershed

The data record was scanned from 1981 to 1984 to identify simple
storm events with precipitation excess of approximately one inch (2.54
cm). No storms in the record met the ideal criteria for unit hydrograph
development (Linsley, et al, 1982; Viessman, et al, 1977), but 13
adequate storms were identified for basin #1 and 9 for basin #5.

Rainfall–Runoff Concepts

The purpose of the data analysis was to measure for each storm the
parameters necessary to calibrate the basin for the Snyder type
coefficients. Traditional unit hydrograph concepts as set forth by the
Corps of Engineers (1973), Viessman, et al (1977) and Linsley, et al
(1982) were used in the analysis. A sketch of a typical rainfall
hyetograph and runoff response is given in Figure 2. The base flow was
abstracted from the surface runoff by straight line techniques assuming
channel precipitation and interflow are negligible. Precipitation and
runoff data were analyzed for precipitation excess, P_e; time of
rainfall, t_r; basin lag, t_1; and the time of surface runoff, T.
Precipitation excess was found as the volume under the surface runoff
hydrograph reduced to depth over the drainage area. This value was used
to abstract losses from the rainfall hyetograph.

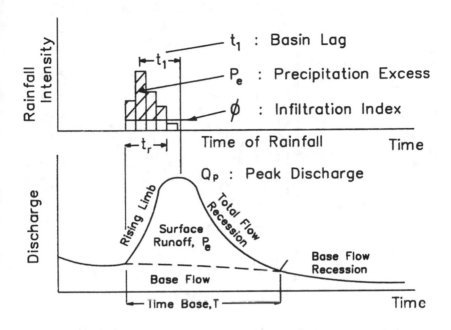

Figure 2.   Rainfall-Runoff Concepts

Precipitation Data Analysis

Precipitation data for each storm were collected in mass accumulation format. The data are reduced to a hyetograph, as for the storm of March 27, 1984 in Figure 3, with time increments that may be varied. For a time increment of 30 minutes, the time of rainfall is also 30 minutes. Reducing the time increment to 10 minutes did not change the 30 minute time of precipitation excess. The mass center of the precipitation excess was determined numerically. Basin lag, t_1, defined by Snyder (1933) as the time from the center of mass of the excess rainfall to the peak of the runoff hydrograph was determined from the synchronous rainfall-runoff data. All events were analyzed in a similar manner to determine the time of rainfall, t_r, aided by computer programs written to facilitate the numerical and graphical analysis.

Runoff Analysis

A typical runoff hydrograph for basin #1 is shown in Figure 3. Abstraction of the base flow was by straight line or linear technique. Precipitation excess was determined from the area under the surface runoff hydrograph using the trapezoidal rule of numerical integration. This value was used to abstract losses from the precipitation hyetograph to obtain the time of rainfall.

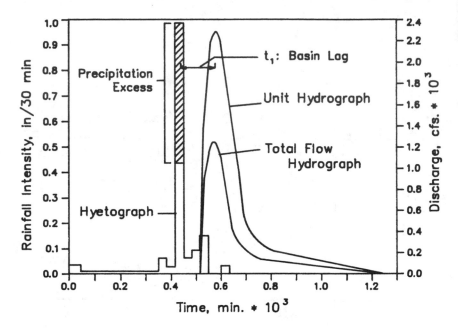

Figure 3.   Rainfall Hyetograph and Discharge Response for the
Event of March 27, 1984 - Basin #1

Precipitation excess was divided into the ordinates of the surface
runoff hydrograph to obtain the unit graph. The unit hydrograph is
given in Figure 3 for the event of March 27, 1984 on basin #1. The peak
discharge, Q_p, and the time lag, t_1, to the peak were determined
numerically and compared graphically.

Snyder Coefficients

Snyder coefficients C_t and C_p were calculated from the standard
relationships

$$C_t = t_1/(L*L_{ca})^{0.3} \tag{1}$$

$$C_p = Q_p*t_1/640*A_d \tag{2}$$

where t_1 = basin lag, hrs; L = length of main stream, mi; L_{ca} = length
of main stream to a point opposite centroid of area, mi; Q_p = peak
discharge of unit graph, cfs; A_d = drainage area, sq mi; and 640 =
constant. All geometric, length and area values were determined by
digital graphics and checked for judgment errors.

Results of Data Analysis

Data analyses for 13 storms on basins #1 and #5 are summarized in Tables 1 and 2 respectively. For basin #1, the time of rainfall for the storms varied from 14 minutes to 450 minutes. Basin lag varied from 115 minutes to 199 minutes. Total storm precipitation varied from 1.19 inches to 6.17 inches. The precipitation excess varied from a low of 0.3075 inches to a high of 1.9329 inches.

Values for the Snyder coefficient C_t on basin #1 range from a low of 0.710 to a high of 1.228 with a mean of 0.0894. Published values range 1.8 to 2.2. Thus, the C_t values tend to be lower than the standard. Values of the coefficient C_p ranged from a low of 0.559 to a high of 1.128 with a mean of 0.843. These values tend to be higher than published values.

Table 1. Summary Data Analysis–Basin #1–Goodwin Creek
Qbase = Constant

Storm	t_r min	t_1 min	T min	TQ_p cfs	Ppt in	UGQ_p cfs	Ppt Ex in	C_t	C_p	$640C_p$
1/02/82	18	137	698	688	1.85	2140	0.3154	0.864	0.924	592
5/28/82	14	121	403	841	1.92	2670	0.3075	0.747	1.019	652
8/27/82	50	142	668	1341	2.27	2021	0.6490	0.877	0.905	579
12/15/82	450	199	1592	1370	1.55	1258	1.0778	1.228	0.789	505
12/27/82	50	127	1050	1285	1.19	1396	0.8954	0.784	0.559	358
1/31/83	240	168	1484	1065	1.40	1283	0.8076	1.037	0.680	435
3/04/83	320	115	682	2423	1.90	1745	1.3074	0.710	0.633	405
5/03/83	140	146	955	1574	2.49	1472	1.0504	0.901	0.678	434
5/15/83	50	147	622	2516	2.66	2110	1.1544	0.907	0.978	626
9/20/83	120	139	786	3758	6.17	1935	1.9329	0.858	0.848	543
11/19/83	20	168	1050	1159	2.00	2129	0.5398	1.037	1.128	722
11/23/83	40	139	680	1416	1.24	1878	0.7327	0.858	0.823	527
3/27/84	30	134	819	1301	1.51	2340	0.5460	0.827	0.989	633

A_d = 8.26 sq mi, L = 7.33 mi, L_{ca} = 3.74 mi

Table 2. Summary Data Analysis–Basin #5–Goodwin Creek
Qbase = Variable Straight Line

Storm	t_r min	t_1 min	T min	TQ_p cfs	Ppt in	UGQ_p cfs	Ppt Ex in	C_t	C_p	$640C_p$
1/02/82	20	70	268	304	1.82	629	0.4749	0.841	0.691	442
5/25/82	20	57	200	377	1.92	944	0.3939	0.684	0.844	540
2/15/82	350	151	540	383	1.52	332	1.1191	1.813	0.786	503
12/27/82	200	23	378	379	1.10	483	0.7763	0.276	0.174	112
1/31/83	210	118	494	281	1.30	312	0.8774	2.173	0.578	370
5/15/83	60	64	244	970	2.49	763	1.2492	0.768	0.766	490
11/19/83	40	80	336	287	1.54	591	0.4804	0.961	0.742	475
11/23/83	60	187	315	593	1.47	646	0.9004	2.245	1.900	1213
3/27/84	40	73	285	528	1.01	744	0.6974	0.877	0.852	545

A_d = 1.66 sq mi, L = 2.54 mi, L_{ca} = 1.17 mi

Standard Snyder coefficients and observed coefficients for basins #1 and #5 are summarized in Table 3. Values for C_t tend to be lower than published values. Values for C_p tend to be high. For both basins outliers result in a wide range of values. As a general observation, there is more variability in the data from the smaller basin than from the larger basin.

Table 3.   Summary of Snyder Coefficients (* Mean Value)

Standard Values	$1.8 < C_t < 2.2$	$0.4 < C_p < 0.8$
Basin #1	$0.71 < C_t < 1.2$ (0.894)*	$0.56 < C_p < 1.1$(0.893)*
Basin #5	$0.28 < C_t < 2.2$ (1.18)*	$0.17 < C_p < 1.9$(0.815)*

Analysis of data from basin #5 shows the same trend as data from basin #1. The values of C_t are lower than standard values, and the values of C_p are high.

Conclusions

From analysis of 13 simple storm events on a well instrumented watershed of 8.26 sq mi (21.39 sq km) drainage area, the values of Snyder's C_t values ranged from 0.710 to 1.228 with a mean value of 0.894. Values of C_p ranged from 0.559 to 1.128 with a mean of 0.843. Values for C_t tend to be lower, and values for C_p tend to be higher than published values.

Analysis of data from the sub-area of 1.66 sq mi (4.30 sq km) within the larger basin gave C_t values from 0.276 to 2.173 with a mean of 9 values of 1.18. These data are smaller than published values. Values of C_p ranged from 0.174 to 1.900 with a mean value of 0.815. These values exceed published values.

References

Linsley, Ray K., Jr., Max A Kohler and Joseph L. H. Paulhus, Hydrology for Engineers, McGraw-Hill, 1982.

Snyder, F. F. "Synthetic Unit Graphics", Trans American Geological Union, 19, 1938.

U. S. Army Corps of Engineers, "Hydrograph Analysis", The Hydrologic Engineering Center, IHD, Hydrologic Engineering Methods for Water Resources Development, Vol. 4, October 1973.

Viessman, Warren, Jr., John W. Knapp, Gary L. Lewis and Terence E. Harbaugh, Introduction to Hydrology, IEP A Dun-Donnelley Publisher, New York, N.Y., 1977.

CHANGING HYDROLOGY - THE MIDDLE MISSISSIPPI RIVER

By Frederick R. Bader, M. ASCE[1]

Abstract: Observed daily flow and stage at major stream gages on the middle Mississippi River are numerically and graphically analyzed to quantify and illustrate the magnitude and seasonality of the changing hydrology of the river in recent years. The analysis is developed using desktop computer technology.

INTRODUCTION

The media has given considerable coverage to the extraordinary high levels of the Great Lakes and of the Great Salt Lake. These levels have been attributed to a climatological trend of several years duration in central North America. This same meteorological condition has also effected the watersheds encompassed by the upper and middle Mississippi River. A result of these meteorological trends has been rapidly increasing mean discharge and stage at major river gaging stations on the middle Mississippi as well as many new daily record stages in normally unseasonable months.

BACKGROUND

The changing hydrology of the middle Mississippi is more difficult to illustrate to the public and to decision makers than the more obvious conditions on lakes. Riverine problems related to flooding, erosion and sedimentation created by changing meteorology are more numerous and diffused than in the lake situation. Myriad flood emergency situations of the decade have provided a cause to investigate the magnitude of the observed conditions and lead to a discussion of implications of the present hydrologic conditions upon the above topics.

The middle Mississippi River is generally considered to include the reach of the Mississippi from it's confluence with the Missouri River to the confluence with the Ohio river. The basis for locations on the middle Mississippi is river mile 0 at Cairo, Illinois a short distance downstream of the mouth of the Ohio River. The reach extends to the confluence with the Missouri River at river mile 195.0 at a location 7.7 miles (12.4 km) south of Alton, Illinois. St. Louis gage is located at river mile 179.6, approximately 20 miles (32 km) south of Alton. The next major gage in the downstream direction is Chester, Illinois at river mile 109.5, approximately 70 miles (113 km) south of St. Louis. These locations are illustrated by Figure 1.

---

[1]Hydraulic Engineer, Hydrologic Engineering Section, St. Louis District, Corps of Engineers, 210 Tucker Blvd. North, St. Louis, MO 63101-1986

FIG. 1-Location Map.

DISCHARGE MASS CURVE - ST. LOUIS

   Streamflow measurements on the Mississippi River at St. Louis Mo. have been undertaken by two organizations since 1870 (4).   Systematic daily observations were effectively established by the Corps of Engineers by 1861 and continued by the United States Geological Survey (USGS) since 1933. This combined record of daily streamflow is undoubtedly one of the longest and most accurate in the United States.   It has been noted that the streamflow measurement techniques utilized by the Corps of Engineers in the earlier era (before 1920) were not as accurate as present techniques and may have introduced a slight bias towards overestimation (7).   The magnitude of this overestimate cannot be precisely determined, but it has been appraised to be nearly within the range of accuracy of more contemporary methods. This potential bias is significant to the present discussion only through comparisons with average annual trends of the last decade.
   One technique for examining long-term trends in river flows is mass curve analysis (1).   By simple summation of mean daily flow values for each calendar year, and for each year in the period of record, a graph may be plotted showing the cumulative runoff since the beginning of the period. This mass curve is plotted for the period from Jan 1, 1905 to Dec 31, 1986 as Figure 2.
   Inspection of Figure 2 reveals the long term trends of mass flow for the Mississippi River at St. Louis, a watershed totaling 697,000 square miles (1,805,000 km^2).   In chronological order, mass flow is fairly uniform from 1905 to about 1926, subject to the potential bias mentioned above, followed by a drought period in the 1930's, a moderate increase through the flood years of 1943-1951, a sharp decline in the drought of the mid 50's, a steady increase due to the major flood years of 1973-74, and an rapid increase since 1982 as a result of the many floods of the most recent 5 years.

The mean daily flow for this 82 year period of record is 183,000 cfs (5182 cms) a rate which is only typical of the 1905-25 period. As a result of the potential bias in flow measurement in that early period it is likely that the true mean for the period is somewhat lower and that the mean flow for the period is unprecedented by any interval in the early records. In any case, the mean flow for the 14-year period 1973-1986 (229,000 cfs - 6485 cms) and the mean flow for the 5-year period 1982-1986 (276,000 cfs - 7815 cms) are substantially beyond any expectation based upon observations in this century.

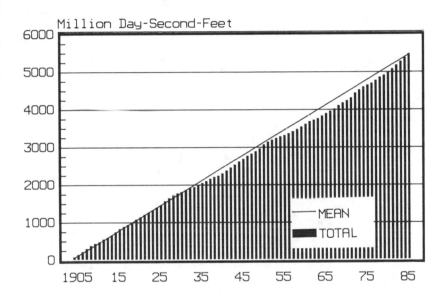

Fig. 2.-Mass Flow at St. Louis 1905-86

CHESTER STAGES 1892-1986

To extend the period of record for analysis and to confirm the findings from the analysis of flow records at St. Louis, the daily 8 a.m. stage readings at the Chester gage site on the Mississippi River were utilized. Systematic observations were available for the calendar years from 1892 to the present. Short intervals without records, generally a result of icing in the winter months, were estimated. The historical stage recordings were adjusted, when applicable, for the present conditions of low-flow regulation, primarily by the navigation releases from the Missouri River Basin projects, by increasing any negative stages during the navigation season to zero stage (approximately equal to navigation release). The watershed at Chester gage is 708,600 square miles (1,835,000 km^2) including two tributary rivers and numerous creeks downstream of the St. Louis gage.

Figure 3 presents the 5- ,10- ,20- , and 40-year means of the daily stage observations at Chester gage for the period from Dec 31, 1896 to Dec 31, 1986. Points plotted are the mean of all readings for the x-year period preceding the last day of the year. Trends of shorter duration are illustrated by the 5-year running means, etc. It is apparent that the mean stage for any duration reaches a historic high level at the end of 1986. Other stage trends are noted to be rapidly rising from as far back as the drought of the mid-60's.

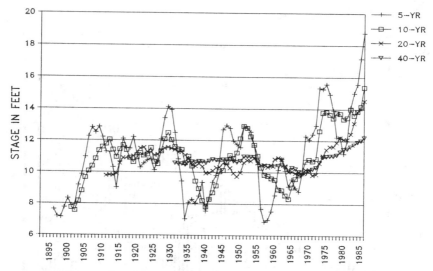

FIG. 3. — MEANS OF STAGE AT CHESTER, ILLINOIS 1892—1986

It is evident, from comparisons to the St. Louis flow analysis that the rising trends are primarily the result of rapidly increasing discharges rather than any impacts of sedimentation or changes in the hydraulics of the river reach at Chester. For comparison the mean stage at Chester for the 82-year period from 1905-1986 was computed to be 11.52 feet (3.51 m). In the more recent 14-year interval from 1973-1986 the mean stage is 15.34 feet (4.68 m), and for the most recent 5-years from 1982-1986 the mean stage is 18.77 feet (5.72 m). The middle Mississippi River at this location is generally considered to be in a stable configuration with regard to sedimentation and has not been significantly impacted by construction of hydraulic features since the early 1950's.

Figure 4 is a daily stage hydrograph for the period 1982-1986 at Chester with the daily maximum, mean, and minimum stages for each date of the year (period of record = 1900-1986) and the mean stage for the 5-year period. Flood stage at Chester, the level above which damage to property commences, is 27 feet on the gage. This figure shows that many floods occurred during the period and that 9 intervals of multiple days were new record high stages for that date in this century. The only spell significantly below the mean was in the late summer of 1985. The 5-year mean stages indicate a substantial deviation from seasonal patterns by the large increases in the winter months.

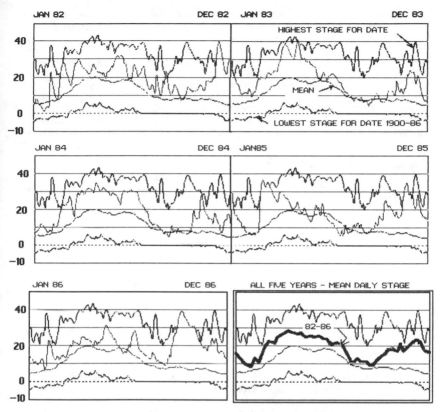

FIG. 4.—DAILY STAGES AT CHESTER

CONCLUSIONS

Mississippi River flows at St. Louis have increased 25% for the most recent 14-year term, compared with the 85-year mean. Likewise, stages at Chester have increased 33%. In the more recent 5-year interval the St. Louis flows have increased 51% and Chester stages are 63% higher. These increases signal a general increase in precipitation throughout central North America. Many floods have occurred in the fall and winter months, establishing record stages for these dates.

ACKNOWLEDGEMENTS

No endeavor of this type could be attempted without the cooperative efforts of many individuals of the Corps of Engineers and of the United States Geological Survey, Department of the Interior. This paper reflects the findings and opinions of the writer and not necessarily those of the Corps of Engineers.

This paper was written and illustrated entirely by desktop computers and general purpose commercial software and hardware. Many calculations were expedited by utilization of the spreadsheet "Supercalc4". The text was prepared and printed by the computer program "Wordperfect" on the Hewlett-Packard Laserjet printer. Figure 1 was developed using "Freelance" and printed on the laserjet. Figure 2 was developed using "Supercalc4" and plotted by "Graph-in-the-Box" on the H-P model 7550 graphics plotter. Figure 3 was developed and plotted by "Supercalc4" on the HP plotter. Figure 4 was developed by "Supercalc4", consolidated by "Show Partner", and printed on a NEC P5XL dot matrix printer. MS-DOS (Microsoft disk operating system) compatible computers manufactured by IBM, NEC, and ZENITH were utilized.

APPENDIX.--REFERENCES

1. Butler, Stanley S., "Engineering Hydrology", Prentice-Hall, Englewood Cliffs, N.J., 1957.

2. "Freelance", Lotus Development Corp., Cambridge, MA, 1986.

3. "Graph-in-the-Box", New England Software, 1986.

4. "Results of Discharge Observations, Mississippi River and its Tributaries and Outlets, 1838-1923", Mississippi River Commission, St. Louis, MO, 1925

5. "Show Partner", Brightbill Roberts and Co. Ltd., Syracuse, NY, 1986.

6. Smith, Roger H., "Engineering Evaluation of Corps Monitoring Efforts at Eight Selected Dikes in the Middle Mississippi River", SLD Potamology Study S-27, University of Missouri-Rolla, Rolla, MO, August 1986.

7. Stevens, Glendon T., Jr., SLD Potamology Study (S-3), The Institute of River Studies, University of Missouri-Rolla, Rolla, MO, 1979.

8. "Supercalc4", Computer Associates International Inc., San Jose, CA, 1986

Flood Hydrology Modeling: U.S. Virgin Islands

James E. Scholl, M. ASCE*

Abstract

The steep mountain slopes, a dominance of clay soils, the absence of perennial streamflow, and the potential for high volume and intense rainfall create a significant potential for flooding problems in the Caribbean Islands.   To develop a consistent basis for evaluating existing flooding problems and for planning activities to reduce future flood damage, a watershed modeling project was initiated by the Government of the U.S. Virgin Islands in 1979.   This paper summarizes the procedures utilized to model flood hydrology in the U.S. Virgin Islands, presents results of hydrograph analysis performed using observed streamflow data, and concludes by briefly summarizing the status of watershed modeling performed to date.   Flood hydrographs were developed for 10-, 25-, 50-, and 100-year design storm conditions using the HYMO computer program.   Thirteen major watersheds in the Islands were evaluated using HYMO, and conceptual solutions to flooding problems were developed.

HYMO Description

HYMO, derived from the words hydrologic model, is a computer language developed by the Agricultural Research Service, U.S. Department of Agriculture, in cooperation with the Texas Agricultural Experiment Station, Texas A&M University (Williams and Hann, 1973). The original program was developed for mainframe equipment, but has been adapted for use on microcomputers.   HYMO commands are simple to use and offer a great deal of flexibility.   For example, any design storm duration and distribution can be input along with site-specific unit hydrograph shape parameters.

Flood hydrographs are developed using unit hydrograph theory and the rainfall-runoff relationship developed by the U.S. Soil Conservation Service (SCS).   Channel routing is accomplished by the variable storage coefficient method, which accounts for changes in channel velocity or reach travel time with flood stage.   The storage-indication method as documented by Mockus (1972) is used for reservoir routing computations.   Details of rules for HYMO commands are presented in the Users Manual by Williams and Hann (1973) along with an example problem.

*Water Resources Engineer, CH2M HILL, P.O. Box 1647, Gainesville, FL 32602.

657

Recent enhancements to HYMO, developed at the University of Ottawa, include an urban hydrograph routine, a user-specified initial abstraction for the SCS rainfall-runoff equation, a Nash unit hydrograph option, and a kinematic storm sewer routing routine. Documentation for this program is by Wisner (1982) and is called OTTHYMO.

Hydrograph Analysis

To establish a site-specific basis for using HYMO in the U.S. Virgin Islands, published and unpublished streamflow data were obtained from the USGS (Robison et al., 1972 and McCoy, 1979). Because long unbroken periods of record were not available, observed data did not provide an adequate basis to develop an annual maximum flood series. The data did, however, contain several isolated single-event runoff hydrographs that could be analyzed to determine site-specific flood hydrograph parameters. Streamflow records from the four gaged watersheds on St. Croix considered in this study are summarized in Table 1.

Table 1. USGS Streamflow Data Available for St. Croix,
          U.S. Virgin Islands

Station Name	Station Number	Drainage Area (mi^2)	Record Examined
River Gut at River	3320	1.42	1963 - 1967 (5 years)
River Gut at Golden Grove	3330	5.12	1963 - 1971 (9 years)
Jolly Hill Gut at Jolly Hill	3450	2.10	1963 - 1968 (6 years)
Creque Gut above Mount Washington Reservoir	3470	0.50	1965 - 1967 (3 years)

($1 mi^2$ = 2.590 km^2)

The first step in the hydrograph analysis was to obtain the original stage hydrographs (strip charts) from the USGS files in Puerto Rico (McCoy, 1979). Strip charts were obtained for selected days for all four stream gaging stations listed in Table 1. These strip charts were then screened for single-peak events resulting from short duration rainstorms. Since synchronized rainfall records were not available at these gages, historic simulation and rainfall/runoff calibration was not possible.

Once the stage hydrographs were selected for analysis, they were converted to discharge hydrographs by application of the appropriate rating table, and the following hydrograph parameters were measured:

1. Time to peak (T_p), in hours, defined as the time from the beginning of rise to the peak of the hydrograph

2. Peak flow rate (q_p) of the runoff hydrograph, in cfs (m^3/sec)

3. Total volume of runoff (Q), in inches (mm)

Using these measured parameters, the hydrograph shape factor, B, was computed for each event as follows:

$$B = \frac{q_p T_p}{AQ} \qquad (1)$$

where:

B = Hydrograph shape factor
A = Watershed drainage area, in square miles (km²)
(all other terms are as previously defined)

These parameters are reported in Table 2 for each event analyzed.

Table 2. Hydrograph Analysis Results, Based on USGS Data

Location	Event Date	T_p (hours)	Q (inches)	q_p (cfs)	B
River Gut at	4/06/65	1.7	0.033	11.44	415
River No.	5/30/65	3.0	0.064	11.78	389
3320	11/09/65	1.0	0.023	18.91	579
Mean		1.9			461
River Gut at	8/01/63	4.00	0.240	144.0	467
Golden Grove,	11/02/63	1.30	0.100	123.0	311
No. 3330	11/21/63	1.25	0.021	28.1	325
	10/24/69	1.50	0.112	169.0	440
	10/30/69	5.50	0.062	26.5	457
	12/02/69	2.75	0.070	41.5	317
Mean		2.72			386
Jolly Hill at	1/03/63	0.60	0.024	39.5	470
Jolly Hill,	1/04/63	0.50	0.097	177.0	434
No. 3450	12/12/65	0.60	0.009	17.5	556
	12/12/65	0.70	0.023	39.5	572
Mean		0.60			508
Creque Gut	10/13/65	1.0	0.297	91.1	613
above Mt.	11/12/65	1.0	0.250	79.8	638
Washington	11/09/65	0.5	0.174	83.7	481
Reservoir,	5/17/65	0.5			462
No. 3470					
Mean		0.75			549

(1 inch = 25.4 mm, 1 cfs = 0.0283 m³/sec)

Equation 1 is an algebraic transformation of the hydrograph peak rate equation employed by the SCS (Mockus, 1972). The standard value for B used by the SCS in the majority of hydrologic design applications is 484. However, Mockus (1972) reported that the hydrograph shape factor has been known to vary from about 600 in steep terrain to 300 in very flat swampy country. From equation 1, it can be seen that the larger the hydrograph shape factor, B, the larger the peak rate of runoff generated by a given volume of runoff, Q. Thus, the hydrograph shape, which is generally related to topography and watershed storage, can be an important variable influencing flood hydrology.

The hydrograph shape factor is a watershed property that should be a constant. However, considerable variation is reported in Table 2 for individual event B values at each gaging station. These variations are primarily a result of all errors present in each measured hydrograph parameter (i.e., T_p, Q, and q_p) being combined in the computation of B. The time to peak, T_p, is a particularly difficult parameter to measure accurately, because of the time scale of the original strip charts (1 inch [25.4 mm] = 10 hours) and the short times of observation. Therefore, the mean of the event B values is probably the best estimate of the hydrograph shape factor for each watershed analyzed.

Plotting the results from Table 2 as shown in Figure 1, the relationship between the unit hydrograph shape factor, B, and drainage area provides a basis for modeling ungaged watersheds.

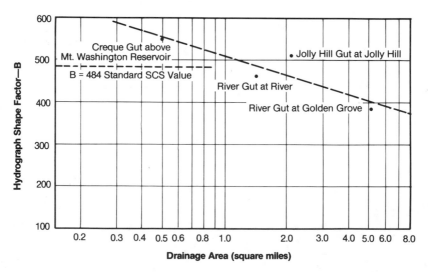

FIGURE 1. Hydrograph Analysis Results for the U.S. Virgin Islands.

Modeling Procedure

A common step-by-step procedure was used to evaluate flooding problems for each of the watersheds studied during this project. These steps are summarized as follows:

1. Compile watershed data and delineate major watershed boundaries and flow paths.

2. Conduct a site visit to confirm watershed boundaries and flow paths and to collect measurements on drainage structures such as culverts, channels, storm sewers, inlets, and storage ponds.

3. Delineate sub-watershed boundaries and compute parameters such as the runoff curve number (CN), time of concentration, drainage area, and unit hydrograph factors.

4. Use HYMO computer language to compute the 10-, 25-, 50-, and 100-year flood hydrographs for existing and zoned land use conditions.

5. Identify areas that may be subject to flood damage.

6. Develop conceptual plans for flood damage mitigation activities when technically feasible.

Thirteen watershed areas in the U.S. Virgin Islands ranging in size from 365 acres (1,478.25 ha) to 3,396 acres (13,753.8 ha) have been modeled using this procedure. Nine of these watersheds were on St. Croix and four on St. Thomas. The results of this modeling work are contained in three technical reports by CH2M HILL (1979, 1982a, and 1982b).

Conclusions

HYMO is a practical computer program for evaluating flood hydrology for a wide range of conditions and project requirements. The analysis of observed streamflow hydrographs, as demonstrated using data for the U.S. Virgin Islands, can provide a site-specific basis for assigning unit hydrograph parameters for ungaged watersheds.

References

CH2M HILL. (1979). " A Flood Damage Mitigation Plan for the U.S. Virgin Islands." Prepared for the Disaster Programs Office, Office of the Governor, Government of the U.S. Virgin Islands.

CH2M HILL. (1982a). "Planned Drainage Basin Studies for the Protection of Roads from Flood Damage in the U.S. Virgin Islands, Volume 1--St. Thomas." Prepared for the Public Works Department, Government of the U.S. Virgin Islands.

CH2M HILL. (1982b). "Planned Drainage Basin Studies for the Protection of Roads from Flood Damage in the U.S. Virgin Islands, Volume 2--St. Croix." Prepared for the Public Works Department, Government of the U.S. Virgin Islands.

McCoy, J. (1979). Personal Communications. U.S. Geological Survey, San Juan, Puerto Rico.

Mockus, V. (1972). Section 4, Hydrology. In: SCS National Engineering Handbook (NEH-4). U.S. Department of Agriculture, Soil Conservation Service, Washington, D.C.

Robison, T. M. et al. (1973). Water Records of the U.S. Virgin Islands, 1962-69. U.S. Geological Survey, San Juan, Puerto Rico.

Williams, J.R. and Hann, R.W. (1973). "HYMO: Problem-Oriented Computer Language for Hydrologic Modeling--Users Manual." ARS-S-9, Southern Region Agricultural Research Service, U.S. Department of Agriculture.

Wisner, P. (1982). "IMPSWM Urban Drainage Modeling Procedures, Part III, OTTHYMO: A Model for Master Drainage Plans." University of Ottawa, Dept. of Civil Engineering, Ottawa, Canada.

IMPACT OF TEMPERATURE ON COLUMBIA RIVER WINTER FLOODS

By Daniel J. Barcellos[1], M.,ASCE and George D. Holmes[1], M.,ASCE

ABSTRACT

During the course of an extensive study of the Columbia River drainage to evaluate a major levee system near Portland Oregon, it was noted that flood runoff produced by winter storms with large amounts of precipitation was not properly reconstituted. Analysis of the model output resulted in the conclusion that the determination of areal distribution of precipitation as rain or snow was the underlaying reason for failure to reconstitute runoff. The key parameter in determining this distribution of rain and snow is temperature as effected by elevation. Further investigation lead to a number of conclusions about the impacts of temperature on Columbia River winter flood runoff and the implications regarding future levee design floods.

INTRODUCTION

The purpose of this paper is to present study findings regarding the impact of air temperatures on winter flood runoff in the Columbia River. The paper begins with background information and a brief description of the computer program used in the study. Then discussion proceeds by using as successive examples a 278 sq mile watershed, a 37.3 sq mile zone within the watershed, and the runoff from a 237,000 sq mile portion of the Columbia River Basin. The intent of this development is to provide the reader insight into both regulation and design aspects of Columbia River winter flood runoff.

COLUMBIA RIVER BASIN HYDROLOGY

The Columbia River Basin, as shown on Figure 1, drains a total area of 259,000 sq miles and stretches 730 miles in an east-west direction and about 820 miles in a north-south direction. Topographic features have a dominating effect on the regional meteorology and hydrology. The Cascade Range separates the coastal from the interior portions of the area. West of the Cascades maximum precipitation occurs in the winter, and winter rain-produced runoff is the most significant characteristic of the annual discharge regime. East of the Cascades at higher elevations precipitation during the fall, winter and early spring generally occurs in the form of snow. Late spring and summer snowmelt-produced runoff then becomes the most significant characteristic of the annual discharge regime;

---

[1]Hydraulic Engineer, North Pacific Division, Corps of Engineers, P.O. Box 2870, Portland, Oregon 97208-2870.

however, major winter storms from the Pacific Ocean do occasionally bring relatively warm temperatures and copious amounts of rainfall to areas east of the Cascades and it is these events which are of concern in this paper. In the two largest such events in modern times, December 1964 and January 1974, 89% and 59% of the flow in the Columbia River at The Dalles Dam originated from the area downstream of Grand Coulee Dam. The locations of Grand Coulee and The Dalles Dams are shown below in Figure 1.

Figure 1:   Columbia River Basin

COMPUTER PROGRAM

The Streamflow Synthesis and Reservoir Regulation (SSARR) program was used for this study. This program has been in continual development and use by the Corps of Engineers North Pacific Division and the National Weather Service Northwest River Forecast Center since 1956, and has been widely distributed throughout the world. A reference listed in appendix I (Rockwood 1982) describes elements of the SSARR program which simulate watersheds, river networks, and reservoir systems. Watershed accounting is accomplished through the use of elevation bands (or zones) with independent accounting of precipitation, snow water equivalent, soil moisture and other characteristics for each band. References on the snow band option in the program include a description (Speers, et al. 1978) and initial use and testing (Kuehl 1979).

WINTER FLOOD SIMULATIONS

Reconstitutions. - A model was developed for the entire drainage basin above Portland. This included 70 watersheds, 56 routing reaches, and 30 reservoirs. The watersheds were initially calibrated on annual runoff volume and spring snowmelt hydrographs for a minimum continuous 5-year period; however, it was found that individual winter storm events such as occurred in December 1964 and January 1974 were being under-computed. In both the December 1964 and January 1974 events only about 75 percent of observed volume and estimated peaks were being generated by the model during the storm event. Investigation of individual watershed computations revealed that the most likely problem was the temperature prevailing during the time of intense rainfall. Most watershed temperatures were determined from relatively low elevation station data which were adjusted at a 3.3 $^\circ$F/1000 ft rate to obtain higher elevation temperatures. This is acceptable for spring melt computations, but does not describe the situation which normally occurs with the inflow of moisture-laden relatively warm marine air. In order to compute reasonable watershed temperatures, data from high and low elevation stations were used to compute daily temperature lapse rates. This refinement to the watershed simulation produced acceptable results.

Watershed Example. - Using data for a 278 sq mile watershed during the January 1974 storm, figure 2 is used to illustrate the effect of temperature on the percentage of this watershed which contributes to runoff, and the sensitivity of temperature, elevation, and runoff relationships. Starting with mean daily temperatures

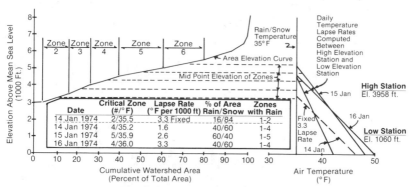

Figure 2: Percent of Watershed Receiving Precipitation as Rain

which are adjusted to mean sea level there are only 3 days when temperatures computed with variable lapse rates (not less than zero or greater than 3.3 $^\circ$F/1000 ft) are above 35 $^\circ$F at the lowest elevation in the watershed. Even on 14 and 15 January 40 to 60 percent of the watershed is subject to rainfall with snow accumulation occurring on the remainder. For comparison, if the fixed lapse rate of 3.3 $^\circ$F/1000 ft is used on 14 and 15 January the area subject to rainfall is reduced to 16 and 40 percent respectively. The dashed lapse rate line on figure 2 illustrates the effect of a 3.3 ^{o}F/1000 ft lapse rate on 14 January. Temperature at mean sea level is 46.3 ^{o}F and is decreased to 35.0 ^{o}F at elevation 3400 feet. This elevation is 100 feet below the average elevation of zone 3, therefore zones 3 and above are receiving snow and only zones 1 and 2 are receiving rainfall.

Figure 2 also is used to show temperature sensitivity in another manner; for example, the temperature at sea level is 2 ^{o}F colder on 15 January than on 16 January, however the small change in lapse rate results in 60 percent of the area with rainfall on 15 January as opposed to 40 percent on 16 January.

Zone Example. - Figure 3 illustrates what is occurring in zone 4 (elevation 4000 to 4500 feet) between 10 and 23 January 1974.

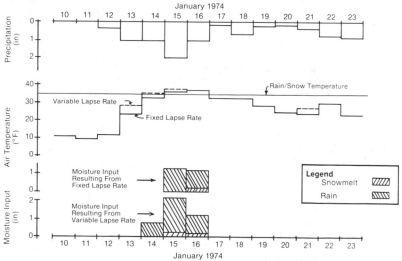

Figure 3:   Zone 4 Moisture Input to Watershed Model

Precipitation is falling for 12 of the 14 days and is in the form of snow for 9 of these days. Zone temperature ranges from 8.9 to 36.1 ^{o}F with a fixed lapse rate. Zone temperature computed with a variable lapse rate was above the rain/snow temperature on 14 January so that rainfall rather than snowfall was occurring. Just as importantly the zone was "primed" so that melt and a greater percentage of rainfall were available for moisture input on 15 January. Moisture input is the rain and melt accumulation before loss to soil moisture (evapotranspiration). This figure also shows that rainfall is the dominant contributor to runoff and snowmelt is relatively unimportant during this type of event. The rise in temperature during periods of increasing precipitation as shown here is typical of periods of high winter discharges; December 1964 and January 1974 being the largest such events, with December 1955, January 1970 and a number of other events demonstrating this sensitivity to temperature to a lesser extent.

Runoff. - The volume of runoff using a variable lapse rate as opposed to a fixed rate with the watershed discussed above was increased from 8,600 acre-ft to 12,900 acre-ft over a 7-day period. The maximum daily flow during this period increased over 50% from 1050 cfs to 1600 cfs. The sensitivity to temperature on a single watershed and a single zone is clearly illustrated by figures 2 and 3 and the effect on major tributary and main stem Columbia River maximum daily discharges is shown below in Table 1.

TABLE 1. - **MAXIMUM DAILY DISCHARGE IN 1000 CUBIC FEET PER SECOND**

Location	December 1964			January 1974		
	(a)	(b)	(c)	(a)	(b)	(c)
Snake River	204.0	320.0	413.0	97.0	129.0	247.0
Columbia River above Grand Coulee Dam	43.0	48.0	65.0	127.0	151.0	218.0
Columbia River at The Dalles Dam	323.0	445.0	574.0	306.0	366.0	533.0

a. - Observed temperature and fixed lapse rate
b. - Observed temperature and computed lapse rate
c. - Observed temperature + 5 $^{\circ}$F for 4 days
    and computed lapse rate

In Table 1, the Columbia River runoff above Grand Coulee Dam is less sensitive to temperature variation during winter floods because it is in the northern part of the basin where winter storm events from the Pacific have less influence on the colder before-storm temperatures. The major runoff response is from the area below Grand Coulee Dam and the lower part of the Snake River which comprises less than 50 percent of the area above The Dalles Dam. Use of the variable lapse rate computation, when compared to use of a fixed lapse rate, produces increased runoff response to winter precipitation; reasonably reproducing flow at The Dalles Dam for better model calibration, and further demonstrating the sensitivity of temperature adjustments in reconstituting winter flood runoff.

Increasing temperatures, as shown in Table 1, by 5 $^{\circ}$F at all stations for 4 days prior to the peak at The Dalles Dam results in even greater runoff response. The two events shown indicate that each degree increase in temperature will increase the maximum daily discharge at The Dalles Dam by an average of 30,000 cfs during severe storms.

CONCLUSIONS

Columbia Basin winter storm runoff is highly dependent upon temperature sequences experienced during individual events. Figures 2 and 3 illustrate the importance in watershed simulation of temperature and elevation in determining how much of a watershed is subject to rainfall and how much is subject to snowfall during winter storm events. It is the amount and areal distribution of rainfall which determines the magnitude of resultant runoff, snowmelt being relatively unimportant in this case.

The figures 2 and 3 also strongly suggest that winter storm lapse rates are less than the 3.3 °F per 1000 feet fixed rate which proved suitable for spring melt computations. Using a fixed lapse rate precluded runoff from much of the watershed area. Using data from the few available high elevation stations to compute lapse rates increased runoff from sufficient watersheds to produce the results shown in Table 1.

Data in Table 1 indicate that arbitrarily raising all temperatures by 5 °F for 4 days preceding the maximum daily flow at The Dalles Dam yielded about the same or greater results at The Dalles Dam as daily lapse rate computations. Concern for future events based on these data would indicate that discharge at The Dalles Dam will increase an average of 30,000 cfs for each 4 day increase in temperature of one °F during storms of this magnitude.

Implications of these study findings on levee design floods for the Portland area are that relatively minor changes in meteorologic conditions can result in significantly larger floods. As demonstrated by the December 1964 flood, the relatively rare winter runoff response from the eastern cascade basins combined with the backwater of typical western cascade runoff results in Columbia River stages in the Portland area approaching those experienced in spring with much higher Columbia River flow. Therefore, temperature is a very sensitive parameter which requires in-depth analysis for computation of future winter levee design floods.

APPENDIX I. - REFERENCES

Kuehl, D. W., "Volume Forecasts using the SSARR Model in a Zone Mode," Proceedings of the Western Snow Conference, 47th Annual Meeting, 18 Apr. 1979, pp. 38-47.

Rockwood, D. M., "Theory and Practice of the SSARR Model as Related to Analyzing and Forecasting the Response of Hydrologic Systems," Applied Modeling in Catchment Hydrology, Water Resources Publications, Littleton, Colorado, 1982, pp. 87-106.

Speers, D. D., Kuehl, D., Shermerhorn, V., "Development of the Operational Snow Band SSARR Model," Proceedings, Modeling of Snow Cover Runoff, U.S. Army Cold Regions Research and Engineering Laboratory, Hanover, New Hampshire, Sep. 1978.

APPENDIX II.- U.S. CUSTOMARY - S I CONVERSION FACTORS

To convert	To	Multiply by
Inches (in)	Millimeters	25.40
Feet (ft)	Meters	0.305
Miles (miles)	Kilometers (km)	1.61
Square miles (sq miles)	Square Kilometers (km^2)	2.59
Cubic feet/second (cfs)	Cubic meters/second (m^3/s)	0.0283
Acre feet (acre-ft)	Cubic meters (m^3)	1,233

		Use Equation
Degree Fahrenheit (°F)	Degree Celsius (°C)	$t_C = 5/9(t_F - 32)$

The Dallas Floodway Project

Gary M. Pettit, P.E.* and Nancy E. Begel**
Member and Associate Member
ASCE

This paper explores the history of the Dallas Floodway project, past performance of the project under various flood conditions, recently updated hydrologic and hydraulic evaluations, and ongoing engineering studies and design work in the floodway.

The Dallas Floodway project, as an integral part of the Corps of Engineers' flood protection plan for the City of Dallas, must be continually monitored, maintained and updated to provide the flood protection for which it was designed, and proven that it can provide.

Historical Background

The Dallas Floodway is often referred to as the most significant flood control and floodplain reclamation project in Dallas history. It was the calamitous, costly and tragic inundation of 1908 which spurred Dallas' leaders and engineers into action to free the City from future ravages of the Trinity River. The Trinity River, usually a calm stream 548 miles long and draining a total area of 17,845 square miles (6,106 square miles at Dallas) would periodically and spontaneously become a raging torrent, isolating the east and west parts of Dallas, and inundating all in between.

Historical records indicate major floods in 1844, 1866, 1871, 1890, and the most catastrophic flood of 1908: at 3:00 A.M. on May 25, 1908, the river was 41.5 feet and rising. The peak discharge of 184,000 cfs created a flood elevation of 420.7 ft msl at the Commerce Street Bridge. By 10:00 A.M., the river reached its maximum stage of 52.6 feet. The floodwaters extended two miles wide. Oak Cliff could be reached only by boat, and trips to Fort Worth had to be made via Terrell and Ennis, small towns approximately 30 miles east and south of Dallas, respectively.

It was a week before the Trinity River's floodwaters finally receded into its banks. Traces of the disaster lingered for months. Loss estimates reached $2.5 million, a considerable figure in 1908. Drainages from a similar flood today would reach into the billions. It is estimated that, without levees, all structures as far east as Akard Street in the Central Business District would be flooded, along with underpasses, roadways and other improvements.

*Associate, Water Resources Dept., Espey, Huston & Associates, Inc., 17811 Waterview Parkway, Dallas, TX 75252.

**Staff Engineer, Water Resources Dept., Espey, Huston & Associates, Inc., 17811 Waterview Parkway, Dallas, TX 75252.

Following the devastating 1908 flood, Dallas planners in 1910 began to masterplan the construction of levees to protect the City from a flood recurrence of such magnitude. George E. Kessler, a Dallas engineer who drafted the first city plan for the proper development of Dallas, and of the land adjacent to the Trinity, recognized as early as 1910 the potential growth possible to the City of Dallas if the Trinity River could be tamed and flood-controlled. He was persistent in his belief that a great city could spring up immediately upon construction of a flood control project. The river, instead of remaining as the western border of the City and a constant flood hazard, could become the center of a thriving city, unthreatened by imminent floodwaters.

During the decades following the 1908 flood, Dallas was expanding rapidly, but all growth stopped on its western edge due to the flood-scarred Trinity. In 1920, an extensive study of the Trinity River was begun by Myers, Noyes & Forrest Engineers. By 1926, plans were complete enough to permit the establishment of an assessment district. This district was to become the City and County of Dallas Levee Improvement District. Engineering plans showed that the cost of the flood control and reclamation work alone would exceed $6.5 million. The total cost of all related improvements (railroads, utilities, roads, bridges) within the district would total nearly $21 million. The construction of the flood control improvements, which took place from June 1928 to November 1931, included widening, straightening and diversion of the old channel, 35-ft high levees, and a floodway approximately 2,200 ft wide. In all, the project encompassed some 25 miles of levees, and the total length of the new diversion channel and floodway is approximately 14 miles.

In terms of benefits to real estate, land development and overall growth, the Trinity River flood control project immediately spurred the development of the Trinity Industrial District, 1,150 acres located immediately adjacent to downtown Dallas (although it was not confined to that one area). In terms of health, safety and welfare, thousands of acres of flood-prone land were protected from floodwaters, including a significant portion of the downtown Dallas Central Business District. There is doubtless no ONE project which has meant as much to Dallas, both in terms of the City's development into a major metropolis, and in dollars and cents, as the control of the Trinity and the development of lands outside of the levees which was thereby made possible.

As of 1986, the City of Dallas boasts a population of approximately one million, and the City and surrounding suburbs within the Trinity River watershed have experienced unprecedented growth and urbanization. The Dallas Floodway was designed to accommodate a peak flow which was computed based on a watershed largely undeveloped and agricultural in nature. Recent hydrologic and hydraulic studies by the U.S. Army Corps of Engineers indicate that the integrity of the levee system may now be threated as a result of increased flows. Additionally, siltation of the pilot channel has reduced its conveyance and resulted in problems in draining adjacent sump areas. Engineering studies are currently underway to restore the capacity of the pilot channel through approximately 10 miles of the floodway.

The 1926 Dallas Floodway plans called for a floodway with a maximum discharge capacity of 500,000 cfs, compared to Mr. Kessler's original figures which were based on 200,000 cfs (approximately the equivalent of the 1908 flood). Thus, the project was designed to provide for a discharge capacity roughly equivalent to that of the 1908 flood, according to hydrologic and hydraulic studies conducted by Myers, Noyes and Forrest Engineers. However, the applicability of these design flows is being reevaluated due to significant changes in the hydrologic conditions of the Trinity River Basin, particularly urbanization in the Dallas-Fort Worth area.

The effectiveness of the Dallas Floodway and other flood control works in the Trinity River Basin was demonstrated during the April 1942 flood, in which a discharge of 111,000 cfs was experienced with a crest elevation of 413.7 ft msl at the Commerce Street gage. This is the second largest discharge on record. The 1957 flood resulted in a discharge of 75,300 cfs and a maximum flood elevation of 409.8 ft msl at the Commerce Street gage. The April-June 1957 flood over the entire upper Trinity River Basin caused damages estimated at about $19,500,000. However, it is estimated that the flood control projects in operation in the basin at the time this flood occurred prevented an additional $85 million in damages. Based on the damage prevention estimates, these projects more than repaid their total initial cost by preventing the damages from this single flood.

The most recent major improvements to the Dallas Floodway were made in 1958 by the U.S. Army Corps of Engineers. The improvements included raising the levees to provide for conveyance of the Standard Project Flood (SPF) within the floodway with four ft of freeboard. To improve interior drainage, additional pump stations, including pumps and gravity sluices, were constructed and the pilot channel within the floodway was modified by excavating it to an average depth of 25 ft, with a 50-ft bottom width, to provide a design capacity of 13,000 cfs.

The capacity of the pilot channel within the floodway is currently reduced to approximately 6,500 cfs due to siltation and the growth of trees and brush in the channel, both within and downstream of the floodway. This loss of channel capacity, coupled with the irregular channel grade, causes higher water levels for low flows in the channel and greatly reduces the capacity of the existing gravity sluices. An engineering report for Pump Plant B in 1981 estimated that the level of protection for the Hampton-Oak Lawn area located adjacent to the floodway represented only a 37-year recurrence interval. Although the current level of flood protection has not been quantified for other sump areas, it has decreased significantly due to reduced gravity sluice capacity.

Ongoing Improvements

In 1986, the City of Dallas Department of Public Works acknowledged that improvements to the Dallas Floodway were a top priority. The Dallas Floodway, as an integral part of the Corps of Engineer's flood protection plan for the City of Dallas, must be continually monitored, maintained, and updated to provide the flood protection for which it was designed. In addition to restoring the integrity of the floodway by providing maximum flood protection through increased channel capacity and pump station and sump improvements, the City has recognized the opportunity to develop parks, off-channel water features and recreational facilities within the floodway for the benefit of the citizens of Dallas.

The engineering and surveying aspect of the Dallas Floodway Channel Improvements is divided into two phases, each representing approximately a five-mile length of channel. Phase 1 extends from the Hampton Street bridge (at the Delta Pump Station) to the Corinth Street Bridge. Phase 2 extends from the Corinth Street bridge downstream to U.S. Highway 75, downstream of the levee-protected floodway. Current survey data will be used to update the Corps' water surface profile model throughout the limits of the channel improvement project and to determine the extent of channel siltation. A single-benched channel configuration (a 70-ft bottom width, bench width of 70 ft, bench elevation 10 feet above the channel bottom, and 3:1 side-slopes) is envisioned for Phase 1.

An analysis of desiltation benefits will be performed on the lower five-mile segment (Phase 2), where it is believed that heavy vegetation and sedimentation is aggravating an already-existing backwater problem. Silt removal will be accomplished, if warranted based on the results of the analyses. In addition to the channel improvements and desiltation, bridge pier protection will be provided where necessary and improvements will be made to the Sylvan Avenue Bridge, a low water crossing which, as recently as June 7, 1986, was inundated when the Trinity River was flowing at 19,000 cfs. Other structural and engineering improvements currently under contract include improvements to the sump areas and gravity sluices.

Concurrent with engineering studies and design work, the City of Dallas Park Board formally accepted recommendations for a Trinity Park Development Plan on January 22, 1987. The Trinity Park Citizens Advisory Committee made several conceptual design recommendations to be incorporated into future planning and development, which will be compatible with the ongoing flood control improvements. These recommendations, based on the concept of developing a park which would encourage people to enjoy the beauty and tranquility of the Trinity River flood plain, include an environmental master plan, athletic complexes, off-channel lakes, a continuous park road, recreational walks and trails, equestrian facilities, preservation of wetland areas, art or environmental sculpture, major events area and civic area, and motorcycle motorcross area.

Conclusion

In conclusion, the Dallas Floodway has proven itself to be both an effective means of flood control and a significant catalyst to the growth of the City of Dallas, which would not have been possible due to the frequent inundation of the Trinity River flood plain. However, the level of protection afforded by the original levee and channel design is no longer being provided due to urbanization within the watershed, siltation of the channel and growth of vegetation where flood conveyance is needed. The dual goal of this improvement project is not only to restore the floodway to provide an adequate level of flood protection, but also to enhance an urban environment with both recreational and natural, environmentally-pleasing amenities.

REFERENCES

City and County of Dallas Levee Improvement District. 1931. Facts supporting the coordinated program involved in the Dallas County Flood Control and Reclamation Project. Dallas, Texas.

City of Dallas Department of Public Works. 1986. Scope of Work. Planning and Design of Channel Modification for the Dallas Floodway. Dallas, Texas.

_____. 1951. First Southwest Company. Dallas, Texas.

Dallas County Flood Control District. Publication date unknown. History. Industrial Properties, Inc. Dallas, Texas.

Trinity Improvement Association. 1968-1969. The Trinity River: New Vistas of Opportunity for Texas and the Great Southwest.

Trinity Park Citizens Advisory Committee. 1987. Recommendations for Trinity Park Development. Dallas, Texas.

U.S. Army Corps of Engineers. 1986. Draft Regional Environmental Impact Statement, Trinity River and Tributaries. Fort Worth District.

Probability Based Evaluation of Spillway Capacity

Thomas A. Fontaine and Kenneth W. Potter[1]

Spillway adequacy could be evaluated more objectively if estimates were available of the probabilities of discharges up to the Probable Maximum Flood (PMF). As a first step in making such estimates, we have developed a method to estimate the probability of rainfalls up to the Probable Maximum Precipitation (PMP). The method uses historic rainfall data from storms which have occurred in a meteorologically homogeneous region incorporating the drainage basin of interest. It is based on the assumption that these storms, subject to an appropriate adjustment of magnitude, have equal probability of occurring anywhere in the region. We have applied this method to a location in the upper midwestern United States.

Introduction

The Probable Maximum Flood (PMF) has traditionally been used to evaluate the adequacy of spillway capacity for significant and high hazard dams. The use of such a conservative standard has been questioned, especially for many existing dams for which expensive modifications would be required to safely pass the PMF. The evaluation of spillway capacity would be more objective if an estimate of the probability of the design event were available. We are developing a procedure to estimate probabilities of extreme floods using rainfall-runoff modeling. This paper describes an essential element of this procedure - a method for estimating the probability of storm rainfalls.

Description of Method

Consider a catchment with response time t in a meteorologically

[1] Department of Civil and Environmental Engineering, University of Wisconsin, Madison, WI 53706.

homogeneous region.   Define a " significant storm " to be a rainfall event, of duration t, which is sufficiently large to produce an average rainfall over the catchment, if optimally centered, which exceeds a critical value x_c. Assume that x_c is chosen sufficiently large so that the probability of two or more storms occurring in the region in a given year is much smaller than the probability of occurrence of a single storm.

Let X be a random variable representing the largest value of average rainfall occurring on the catchment over an interval t in any year.   The objective of the method presented here is to estimate the probability distribution of X, which subsequently can be used to estimate exceedance probablities of very large annual flood discharges.   Defining a storm to have a duration t implicitly assumes that an event of duration t " causes " the annual flood in each year in which the threshold x_c is exceeded.

We want to estimate p(x), where $p(x) = P\{ X > x \}$. This can be done by conditioning on the occurrence and location of significant storms in the region.

Let (y_0, z_0) be cartesian coordinates of an arbitrary point in the homogeneous region.   Assume that, given a significant storm has occurred in the homogeneous region, the probability that its center falls in any small region of area $\Delta y \Delta z$ centered at (y_0, z_0) is $\Delta y \Delta z / A_h$, where A_h is the area of the homogeneous region (This   is the first of two assumptions which define a meteorologically homogeneous region.)

Let p(y_0, z_0, x) be the probability that, given a storm is centered at (y_0, z_0) as shown in Figure 1, it will produce an average depth of rainfall on the catchment exceeding x (The second assumption which defines a meteorologically homogeneous region is that the only dependence of p(y_0, z_0, x) on (y_0, z_0) results from the geometry of the intersection of the storm and the catchment, and not from variations in the characteristics of storms with location.   This assumption can be relaxed if such variations can be identified and incorporated into the analysis.   For example, a simple deterministic uniform scaling can be applied to rainfall depths as a function of storm center location; or the storm pattern can be rotated to conform with observed orientations near the catchment.)

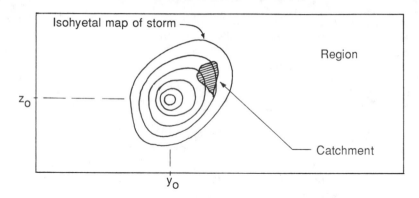

Figure 1. Location of a typical storm to estimate p(y_0, z_0, x)

Let P_h be the probability that a storm occurs in the homogeneous region in a given year. Then

$$p(x) = (P_h / A_h) \iint_{A_h} p(y, z, x) \, dy \, dz \qquad (1)$$

This relationship is approximate because it ignores the possibility that more than one significant storm will occur in a given year.

Estimation of p(x)

Assume that over some time period of length N years we have data on all significant storms which occurred in a meteorologically homogeneous region containing the catchment of interest. Let m be the number of significant storms. As a first approximation, P_h can be estimated by $P_h \approx m/N$.

The probability $(1 / A_h) \iint_{A_h} p(y, z, x) \, dy \, dz$ can be estimated numerically   by creating a square grid with nodes represented by (i, j) on the homogeneous region. Center each of the m storms on one of the grid nodes (i, j). Let n_{ij} be the number of storms producing an average depth of rainfall on the catchment exceeding x, when each storm is centered at node (i, j). Repeat for each node in the grid (i=1, K and j=1, L .)   Then

$$(1 / A_h) \iint_{A_h} p(y, z, x) \, dy \, dz \quad \approx \quad (1/mKL) \sum_{i=1}^{K} \sum_{j=1}^{L} n_{ij} \qquad (2)$$

Using $P_h \approx m/N$ and (2) in (1) yields

$$p(x) \approx (1/NKL) \sum_{i=1}^{K} \sum_{j=1}^{L} n_{ij} \qquad (3)$$

Preliminary results using this numerical approach in the upper Midwest indicate that probabilities can be estimated for rainfalls approaching the PMP.

This approach is based on two critical assumptions. First, it is assumed that the meteorologically homogeneous region is big enough to provide enough historical storms which exceed an average depth of rainfall x over the catchment after transposition to the site. This is possible for the upper midwestern states of the U.S. . In other situations the analysis may fail because the conditions causing the extreme rainfalls of interest change too abruptly with distance. The analysis may need to be modified or abandoned if the catchment is too large.

Second, the historical storm record for the region must be complete and continuous for a long enough period to provide enough storms for the analysis. The completeness and continuity of the record in the Midwest appears to be equal to or superior to that of records in other parts of the U.S..

If these two conditions are satisfied, the probabilities for various average rainfall depths can be estimated by the procedure given above. The resulting estimate can be used along with the probabilities of various antecedent conditions and time distributions of rainfall x within the duration t to estimate the probabilities of the floods simulated by a rainfall-runoff model.

Summary

The probabilities of extreme rainfalls can be estimated using rainfall data from historic storms which occurred in a meteorologically homogeneous region. An equation has been developed to estimate the probability of exceeding an average depth of rainfall x over the catchment.   The equation can be evaluated numerically provided that a sufficient number of historic storms have occurred in the region and that the historic rainfall data record is complete and continuous for a sufficiently long period.

The probabilities of extreme rainfalls for a specific catchment may be used in a rainfall-runoff model, along with probability distributions of the antecedent conditions and time distributions of rainfall, to estimate probabilities of discharges approaching the PMF.

Future work will combine the results of the rainfall analysis with antecedent conditions, time distribution of rainfall, and flood routing to produce estimates of extreme floods at specific dam sites.   An estimate of the uncertainty of the probabilities will be included in the analysis.   The results allow risk-based evaluation of spillway adequacy.

Monte Carlo Approach to
Spillway Design Floods

Jerson Kelman[1], Fernanda S. Costa[2], Jorge Machado Damazio[3]

ABSTRACT: This paper shows the use of a daily stochastic
streamflow model for the calculation of the hydrograph for
spillway design. Most of the literature related to the topic
deals with the problem of how to evaluate the peak of the
hydrograph with recurrence interval of T years, $x(T)$. The
usual approach is to fit a probability distribution $F(.)$ to
a set of m annual maxima $\underset{\sim}{x} = \{x_1, x_2, \ldots, x_m\}$ and get the estimate
$\hat{x}(T)$.

It is well known that the smaller m, the smaller the precision
of the estimate $\hat{x}(T)$. To be in the safe side, the engineer may
design for the upper limit of a confidence interval around $\hat{x}(T)$,
rather for $\hat{x}(T)$ itself. Confidence intervals may be calculated
through parametric and non-parametric methods. This paper shows
how a non-parametric method, the Bootstrap, can be used in two
different ways to estimate the standard deviation of $\hat{x}(T)$:
   a) re-sampling with replacement from the $\underset{\sim}{x}$ set
   b) postulating a stochastic model for daily flows and
      re-sampling with replacement from the corresponding
      noise set

A demonstration is made with both alternatives, using an
artificial process.

The availability of an extremely large number of synthetic
daily streamflow sequences produced by a stochastic model is
particularly helpful when one is designing jointly the spillway and the flood retention storage of a dam. In this
case a T flood hydrograph is required rather than just the peak
value. This paper describes how the Monte Carlo method was
used to define the 10,000 year flood hydrograph (peak of 17344
m^3/s and 20 days volume of 21.3 $\times 10^9 m^3$) for the Serra da Mesa
Dam, in the Tocantins River, Brazil.

[1]Prof., Engrg. Graduate School and Researcher, Electric Energy
Research Center-CEPEL, CP2754, Rio de Janeiro, 21910, Brazil
[2]Researcher, Electric Energy Research Center-CEPEL, CP2754, Rio de
Janeiro, 21910, Brazil
[3]Researcher, Electric Energy Research Center-CEPEL, CP2754, Rio de
Janeiro, 21910, Brazil

INTRODUCTION

There are several stochastic models for daily streamflow proposed in the literature which can be used in the selection of floods for spillway design. If there is no provision of retention storage the focus is on peak flows. In this case the classical approach aims to evaluate the peak flow with recurrence interval of T years, x(T), by fitting a probability distribution F(.) to a set of m annual maxima $\underset{\sim}{x}$ = { $x_1, x_2, \ldots, x_m$ }. The main disadvantage of this approach is that usually the sample size (m) is small (as compared with T) so that x(T) is difficult to estimate accurately. In these situations one can get a better design by incorporating the uncertainty in the analysis using the Bayesian approach (Donald et al, 1972) or the Bootstrap method (Efron,1979). The latter has the advantage of being distribution free. The re-sampling technique of the Bootstrap can be done either directly with the set x or with a set of independent noises of a stochastic model.

When the spillway is associated with some retention storage the design requires the calculation of a full hydrograph , rather than just the peak value. In this case the availability of an extremely large number of synthetic daily streamflow sequences produced by a stochastic model is particularly helpful. The last section of this paper describes how the Monte Carlo method can be used in this situation.

BOOTSTRAP ESTIMATES

The bootstrap method requires m independent observations and proceeds by sampling with replacement from the original observations to obtain a new sample of size m, from which a statistic is computed. The process of sampling and computation is repeated a large number of times (say B) in order to obtain an estimate of the distribution of the statistic.

Direct Method
As usual for an annual maximum series, we treat $\underset{\sim}{x}$ as a set of mutually independent values. We sample from $\underset{\sim}{x}$ to yield new sets $\underset{\sim}{x}_1, \ldots, \underset{\sim}{x}_B$, fit $F_i(.)$ to each $\underset{\sim}{x}_i$ and obtain the corresponding $\hat{x}_i(T)$.

The Bootstrap estimate of the standard deviation of $\hat{X}(T)$ is

$$STD^2 = \frac{1}{B} \sum_{i=1}^{B} (\hat{x}_i(T) - \hat{x}.(T))^2 \qquad (1)$$

where

$$x.(T) = \frac{1}{B} \sum_{i=1}^{B} \hat{x}_i(T) \qquad (2)$$

Use of Stochastic Daily Model
One may want to build a confidence interval for $\hat{x}(T)$ taking into account all the daily flows, rather than just the annual maxima. However the daily flows can not be considered independent, so it is necessary to employ a stochastic model to define a set of "noises" or "independent components" (Cover and Unny, 1986).

Consider the set of daily flows as a set of M "wet seasons" with N days each. With the MxN flows one builds a stochastic model and estimates a set of MxK noises (K < N). The re-sampling is done by the following algorithm:

    a) re-sample with replacement the MxK noises
    b) Separate the re-sampled noises in M sequences of K elements
    c) For each of the M sequences use the stochastic model to generate a "wet season" of N daily flows and calculate its maximum.

Repeat these steps B times to obtain new sets $x_1, ..., x_B$. Fit $F_i(.)$ to each x_i and compute the corresponding $\hat{x}_{\iota}(T)$. Equations (1) to (2) are used to estimate the standard deviation of $\hat{x}(T)$.

It should be noted that in the direct method only recorded maximum can be included in a re-sampled maximum series x_i. This is not true when the stochastic daily model is used, even if the daily flows are independent.

An Artificial Process

The simple AR(1) model was adopted to demonstrate the two approaches:

$$y_t = r\, y_{t-1} + \sqrt{1-r^2}\ a_t \qquad (3)$$

where y_t is the flow on day t, r is the lag-one autocorrelation and a_t is the noise on day t, which is modelled as a standard normal deviate. No claim is made that equation (3) can be used to model actual daily flows. It is adopted here for the sake of simplicity. When it comes to real cases, more sophisticated models need to be used, such as the DIANA model (Kelman et al, 1985), adopted in the example of the next section.

Also, for simplicity we have considered N=30 "days". If r=0., the T-year flow can be easily calculated using tables of the normal distribution. In other cases a simulation can provide near-perfect estimates. Simulation can also be used to obtain standard deviations of x(T). Table 1 shows the values of x(100).

The experiment included 100 sequences of M=10 "wet seasons" apiece. Each of these sequences was re-sampled one thousand times (B=1000). $x_i(T)$ was obtained using the Gumbel distribution with the parameters estimated by the method of moments.

Table 1 compares the known standard deviations of x(100) for different values of r with the mean of the 100 Bootstrap estimates. It can be seen that the method that uses the stochastic model is more accurate, for this particular case. Both methods underestimate the true value.

	r=0.00	r=0.50	r=0.95
x(100)	3.40	3.39	3.07
std	0.48	0.56	0.76
std_1	0.37	0.40	0.52
std_2	0.39	0.47	0.64

Table 1: Summary of the experiment.
std_1-direct method, std_2 -use of stochastic model

MONTE CARLO APPROACH TO DAM-SAFETY ANALYSIS

Usually a proposed spillway is tested through a routing calculation with design inflow hydrographs, assuming the maximum normal water level (MNWL) of the associated reservoir as the initial condition for the reservoir storage. From these simulations one gets the maximum water level (MWL), to which we add allowances to wave run-up due to wind speed. These two levels are them compared with the dam crest level and account is given to possible hazards. The eventual underdesign (or overdesign) should be corrected by changing either the crest level, or the MNWL or the spillway capacity.

Alternatively one may test the proposed spillway by calculating the required MNWL through a backward routing calculation with the design inflow hydrograph, assuming the MWL corrected by wave run-up as a boundary condition.

Following the latter alternative, it is necessary to calculate the minimum attenuation storage (which might be zero) sufficient to prevent overtopping the MWL. For an inflow hydrograph j, the attenuation storage s(j), can be calculated as:

$$s(j) = \max_t [s_t(j) = \max[0 \; ; \; s_{t+1}(j) - y_t(j) + d \; [\; s_t(j), s_{t+1}(j)] \;] \qquad (4)$$

where:

t is the day index, $t = h, h-1, \ldots, 1$
h is the last day of the flood season
j is the inflow hydrograph index
$s_h(j) = 0, \; \forall j$
$s_t(j) = 0$ implies that the water level is MWL
$y_t(j)$ is the inflow to the reservoir on day t
$d[s_t(j) \; ; \; s_{t+1}(j)]$ is the outflow from the reservoir through the spillway on day t

As the actual streamflow sequence is not known a priori, the attenuation storage is to be considered as a random variable. Its probability distribution can be inferred from a set $\{s(j), \; j=1,2,\ldots\}$ obtained from the use of equation (4) over thousands of synthetic sequences. The reliability of the design can be measured by the relative frequency of s(j) greater than the adopted value. This approach also can be easily used in the re-evaluation of operating constraints of existing dams (Kelman and Damazio,1983).

Table 2 shows the main characteristics of the Serra da Mesa project on the Tocantins River, central Brazil and data for testing a proposed spillway.

Total Reservoir Volume:	54.40 Km^3
Useable Volume:	43.25 Km^3
Drainage Area:	50,975 Km^2
Inundated area at MNWL:	1,784 Km^2
Installed Capacity:	1,200 Mw
Mean Inflow:	709 m^3/s
Proposed spillway characteristics	
MNWL:	460.00 m
MWL:	461.65 m
Attenuation storage:	2,91 Km^3
Number of gates: 5 (length of 15m, each)	

Table 2: Principal Characteristics of Serra da Mesa Dam and Reservoir.

In order to test the reliability of the proposed spillway 100000 synthetic "years" of 212 days each (October to April) were generated using the DIANA model (Kelman et al, 1985). Figure 1 shows a comparison between the sample accumulated distributions of maximum annual flow obtained from the historical information and from the 100,000 synthetic sequences. A comprehensive validation of these synthetic sequences is given by Damazio and Fuks (1985).

From the 100,000 synthetic sequences only 88 were considered for dam safety analysis. The adopted criteria was to select the hydrographs with peak flows greater than 15700 m^3/s which has a recurrence interval of 1000 years.

Table 3 gives the results. The required attenuation storage is only 0.98 Km^3, as compared to 2.91 Km^3, which is the value previously proposed for Serra da Mesa Dam. Because there may be some unexpected difficulties to operate the gates during a major flood, a sensitivity study was done assuming one of the gates to be inactive. In this case the attenuation storage should be 3.67 Km^3 and only 18 hydrographs out of 100,000 (recurrence interval of 5555 years) would possibly cause a dam break if the attenuation storage is selected as 2.91 Km^3, as originally planned. The 10th more adverse sequence, in terms of required attenuation storage, can be used as a 10,000-year flood.

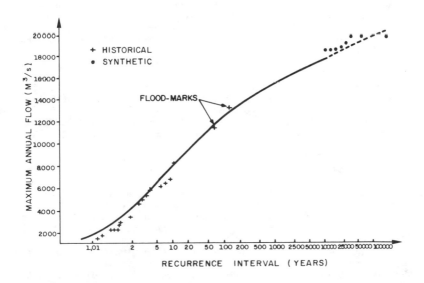

FIGURE 1 : ACUMULATED DISTRIBUTIONS OF ANNUAL MAXIMUM PEAK FLOWS AT SERRA DA MESA DAM, TOCANTINS RIVER, BRAZIL

Number of gates	Required attenuation storage for T=10000 (Km^3)	Risk for the originally proposed attenuation storage of 2.91 Km^3
5	0.98	none
4	3.67	18×10^{-5}

Table 3: Summary of the Monte-Carlo Study

CONCLUSIONS

Stochastic daily streamflow models can be very useful in the selection of floods for spillway design. Even if there is no space for retention storage the approach can be helpful in the determination of the peak flow with recurrence interval of T years, when the sample size is small as compared with T. When the spillway is associated with retention storage, the use of stochastic daily streamflow is a straightforward way to provide complete hydrographs.

ACKNOWLEDGMENTS

The Bootstrap section of this paper is part of the second author's M.Sc. research work at the Coordination of Engineering Graduate Programs of the Federal University of Rio de Janeiro (COPPE/UFRJ), Brazil. The comments of T.C. Hesterberg on Bootstrap enhanced the original draft. The authors thank ELETROBRAS for the funding of the research

APPENDIX I. - REFERENCES
1. Cover, K.A., Unny, J.E., "Application of Computer Intensive Statistics to Parameter Uncertainty in Streamflow Synthesis", Water Resources Bulletin, Vol. 22, No. 3, 1986.
2. Damazio, J.M., Fuks, E.; "Dam-Safety Analysis for Serra da Mesa" (in portuguese), Technical Report number 448/85, CEPEL, Rio de Janeiro, Brazil, 1985
3. Donald, R.D., Kisiel, C.C., Duckstein, L., "Bayesian Decision Theory Applied to Design in Hydrology", Water Resources Research, vol 8, number 1, 1972
4. Efron, B., "Computers and the Theory of Statistics: Thinking the Unthinkable", SIAM Review, vol. 21, number 4, 1979
5. Kelman, J., Damazio, J. M., "Synthetic Hydrology and Spillway Design",XX Congress of the International Association for Hydraulic Research, 1983
6. Kelman, J., Damazio, J.M., Costa,J.P., "A Multivariate Synthetic Daily Streamflow Generator", Fourth International Hydrology Symposium, Colorado State University, Fort Collins, Colorado, EUA, 1985

An Analysis and Comparison of Probable Maximum Floods
With 10,000-year Frequency Floods and
Maximum Observed Floods for Spillway Design
Cornelius L. Cooper *

ABSTRACT: The Federal Energy Regulatory Commission has under license
or exemption from licensing more than 2000 hydroelectric projects.
The Office of Hydropower Licensing published in September 1986, new
guidelines on selecting design floods for spillways. The guidelines
recommend the use of deterministic probable maximum floods as the
design floods for high hazard dams.

The use of deterministic probable maximum floods are supported
in part by the conclusions of an Interagency Work Group formed to
address the following questions: (1) Is it within the state of the
art to calculate the probability of the probable maximum flood
within definable confidence or error bands; and if not (2) what is
the minimum exceedance probability for which a flood probability
can be defined? The Work Group reviewed over 230 papers and reports,
then computed the estimated peak discharges of 10,000-year frequency
floods for some 3,000 sites which were compared with almost 900
observed extreme flood values and 550 Probable Maximum Flood computations. The Work Group concluded that it is not within the state of
the art to calculate the probability of the probable maximum or
other rare floods within definable confidence or error bands.

Introduction

The opinions expressed in this paper are those of the author and
not necessarily those of the Federal Energy Regulatory Commission.
The Federal Energy Regulatory Commission has under license or
exemption from licensing more than 2000 dams (Cooper, 1986). The
Office of Hydropower Licensing published in September 1986, guidelines for selecting and accomodating inflow design flood for
spillways. The guidelines recommend the use of the deterministic
Probable Maximum Flood (PMF) as the design flood for dams, if the
consequences of failure could include loss of life, damage to
national security installations and significant social, environmental, and economic impacts. The use of deterministic PMF's
instead of floods computed from flood frequency analysis for
spillway designs where failure of the dam would be unacceptable
is supported in part by the conclusions of a 2-year study conducted
by a Work Group of the Hydrology Subcommittee of the Interagency
Advisory Committee on Water Data (1986).

*Deputy Chief, Design Review Branch, Federal Energy Regulatory
Commission, Washington, D.C. 20426

The importance of the objectives of the work group are in part
described by the results of a study by Stedinger and Grygier (1985).
Their results indicate that the inability to assign a probability
to the PMF and to define the distribution between the 100-year flood
and the PMF impairs the ability to use risk-cost analysis.

The Work Group was formed in mid-1984 to investigate two
specific questions:

     °   Is it within the state of the art to calculate
         the probability of the PMF within definable
         confidence or error bounds?

     °   If the probability of the PMF cannot be defined,
         how far out on the probability scale can flood
         probability be determined within definable confidence
         or error bounds?

Discussion

The Work Group conducted a survey of over 230 papers and reports
from both technical journals and project reports of engineering
studies pertaining to the broad topic of large flood events. The
statistical literature and this study emphasize the peak discharge
of floods. However volume and duration estimates are sometimes of
interest, and the same considerations that prevent assessment of
the accuracy of peak flow frequency estimates also apply to volume
and duration estimates.

The Work Group's conclusions about defining the probability of
the PMF or other rare floods as based on a review of the literature
are summarized as follows:

     °   It is not within the state of the art to calculate
         the probability of PMF-scale floods within definable
         confidence or error bounds.

     °   Two types of errors affect the accuracy of extending
         the frequency curve from gaging station data without
         the benefit of additional information to define the
         flood: (1) random-sampling errors and (2) distributional
         errors. Although random-sampling errors may be large
         for rare hydrologic events, error or confidence bounds
         for these errors can be defined. Distribution errors
         which affect the correct tail of the frequency curve
         can also result in large errors. However these errors
         have not been described in terms of definable error
         or confidence bounds. Because this error source cannot
         be evaluated, it is not possible to define the overall
         accuracy of extrapolated flood estimates.

     °   There is no definable point on the probability scale
         at which it becomes impossible to define error bounds
         on flood magnitude and probability estimates. Rather

an analysis displays a gradual transition from common
place events, where estimation errors can be defined
by statistical random-sampling theory, to unprecedented
events where errors cannot be defined. Many professionals
believe this transition begins at recurrence intervals
of about twice the record length and is complete by
recurrence intervals in the general area of about 1000
years. The Work Group found no reason to contradict
these general perceptions.

Many of the papers suggested that the PMF may be assigned the
probability of the 10,000-year frequency flood. The Work Group found
no detailed studies to support this suggestion. The Work Group
questioned whether a recognizable correlation exist between the
10,000-year frequency flood and the PMF. The 10,000-year frequency
flood is based on extrapolation of flood frequency curves beyond
those defined by the length of observed floods, whereas the PMF is
based on storm transposition and maximization.

The Work Group addressed this question by comparing a sample
of computed PMF computations and maximum observed floods with
estimated 10,000-year frequency floods for comparable drainage
basins. A strict site-by-site comparison, however, was unnecessary.
It was taken for granted that the PMF could be computed for most
sites but the 10,000-year frequency flood could not. Records are
available for about 21,000 locations nationwide, with an average
of about 22 years of record at each site. Areal density of gages
ranges from over 20 per 1,000 square miles (2,590 square Kilometers)
in four or five states down to less than 1 per 1,000 square miles
(2,590 square Kilometers) in Alaska. About half the states have
densities between 5 and 10 per 1,000 square miles (2,590 square
Kilometers). The average density in the continental United States
is about seven per 1,000 square miles (2,590 square Kilometers).
Average record lengths for most states are close to the nationwide
average of 22 years.

The Nuclear Regulatory Commission (1977) has compiled a set
of over 550 PMF computations performed by various agencies for
drainage basins of various sizes in various regions of the country.
Crippen and Bue (1977) have compiled a similar set of almost 900
observations of extreme floods. The two samples were compiled
independently, and, although there may be a few sites common to
both, the samples are essentially disjointed. Because the two
samples should not be compared case by case, a generalized comparison
of the sample distributions was required.

Crippen and Bue (1977) and Crippen (1982) summarized their
extreme-flood data by drawing envelope curves of maximum discharge
as a function of drainage area for several hydrologically homogeneous
regions of the country. These curves represent the most extreme
upper limits of flood experience actually realized in the regions.

The envelope curves and hydrologic regions of Crippen and Bue
(1977) and Crippen (1982) were used to summarize the PMF computations

(Nuclear Regulatory Commission, 1977) as well as the 10,000-year
frequency floods. To describe these relationships in a concise and
informative way, it was first decided to express all discharges as
ratios to the corresponding envelope-curve discharges for the corresponding drainage area and hydrologic region. Next, the estimated
peak discharges of 10,000-year frequency floods for some 3000 sites
with moderate to long records of essentially unregulated annual peaks
were computed. The estimates were computed using the Bulletin 17B
guidelines (Hydrology Subcommittee, 1982) for basic preliminary
analysis of systematic records. Only the basic systematic-record
analysis was performed; no outlier adjustments or weighted skews were
used. Extreme high values of these estimates should be ignored,
because a complete Bulletin 17B analysis undoubtedly would reduce
them. In view of the areal density of gages, the short average
record lengths and the length of the extrapolation, the 10,000-year
frequency floods associated with these estimates are not as reliable
as the PMF computations and the curves. Despite their shortcomings,
these estimates have a rigorously defined and generally understood
conceptual meaning. Thus, they are a useful basis for interpreting
the envelope curve and PMF estimates. Finally, it was decided to
make the data sets more nearly comparable in terms of basin size by
considering only those sites with drainage areas greater than 100
square miles (259 square Km). Figure 1 (Hydrology Subcommittee,
1986) shows box-and-whisker plots (Tukey, 1977; Velleman and Hoaglins
1981) for all three data types (type E for envelope-curve data, K for
10,000-year flood estimates, and P for PMF estimates) for all 17
hydrologic regions in Crippen and Bue's (1977) study. The count of
the number of data values represented by each plot also is shown.

Conclusion

     Examination of Figure 1 shows that most of the PMF's are within
a factor of 2 of the envelope curve; this level of agreement is
generally considered acceptable in PMF estimation. In about half the
regions, the PMF estimates are generally above the envelope curve.
Occurrence of a few estimated PMF's below the envelope is not necessarily a problem because detailed examination may in fact show that
these basins are not capable of producing floods of the same magnitude
as the regional extremes. In most of the regions where significant
numbers of PMF's fall below the envelope curve, the envelope itself
is poorly supported by observed floods--there are large gaps between
the envelope curve and the observed extreme floods -- and the PMF's
are generally above the regional extreme values actually observed.

     The estimated 10,000-year frequency floods, except for the out-
side and far outside values, generally are of the same order of
magnitude as the observed regional extreme values, rather than the
PMF. Thus, Figure 1 provides the basis for a plausible argument that
the 10,000-year frequency flood has occurred during this century in
many of the hydrologic regions in Crippen and Bue's (1977) study.
Risk-based analysis, therefore, should not be used for selecting
inflow design flows for dam safety, unless more conclusive studies
involving more data and sophisticated analyses show a different
pattern.

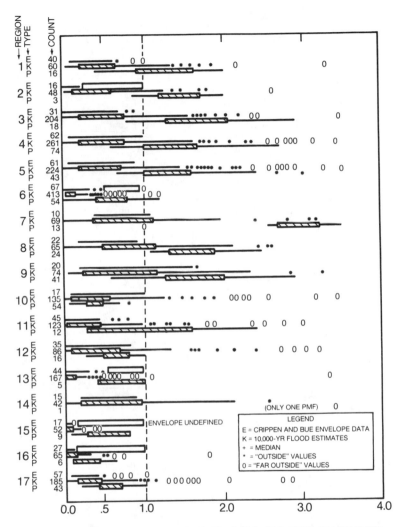

RATIO OF FLOOD MAGNITUDE TO ENVELOPE CURVE

Figure 1. —Comparison of extreme observed floods, estimated 10,000-year floods, and computed PMF's expressed as ratios to envelope-curve values, for drainage areas greater than 100 square miles in hydrologic regions 1-17 (Crippen and Bue, 1977).

Appendix. - References

Cooper, Cornelius L., 1986. Some Remarks on Practice of Flood
   Frequency and Risk Analysis by the Federal Energy Regulatory
   Commission, International Symposium on Flood Frequency and
   Risk Analyses, (May 14-17, 1986), at Louisiana State University,
   Baton Rouge, La.

Crippen, J.R., and Bue, C.D., 1977. Maximum Floodflows in
   the Conterminous United States. U.S. Geological Survey Water
   Supply Paper. 1887, Washington, D.C., 52 pp.

Crippen, J.R., 1982. Envelope Curves for Extreme Flood Events.
   Journal of the Hydraulics Division, ASCE, 108 (HY-10), 1208-1210.

Hydrology Subcommittee, 1982. Guidelines for Determining Flood Flow
   Frequency. Hydrology Subcommittee Bulletin 17B, with editorial
   corrections. Interagency Advisory Committee on Water Data, U.S.
   Geological Survey, Washington, D.C., 28 pp.

Hydrology Subcommittee, 1986. Feasibility of Assigning a Probability
   to the Probable Maximum Flood. Interagency Advisory Committee
   on Water Data, Office of Water Data Coordination, U.S. Geological
   Survey, Washington, D.C., 79pp.

Nuclear Regulatory Commission, 1977. Design Basis Floods for
   Nuclear Power Plants. Regulatory Guide 1.59 (Rev. 2),
   Washington, D.C., 66 pp.

Stedinger, J., and Grygier, J., 1985. Risk-Cost Analysis and
   Spillway Design. Conference on Computer Applications in Water
   Resources June 10-12, 1985. Buffalo N.Y., 1208-1217.

Tukey, J.W., 1977. Exploratory Data Analysis, Addison-Wesley,
   New York, 499 pp.

Velleman, P., and Hoaglin, D., 1981. Application, Basics,
   and Computing of Exploratory Data Analysis, Duxbury Press,
   Boston, Massachusetts, 354 pp.

Paleoflood Studies in Colorado

By Lyman R. Flook, Jr. [1] Life Member ASCE, and
Geoffrey M. Taylor, [2] M. ASCE

Abstract

Paleoflood hydrology is the study of botanic, sedimentologic and
geomorphic flood evidence remaining in a valley. Paleoflood studies in
the Colorado foothills and mountains have been used to determine maximum
flood flows back to the end of the ice age, some 10,000 years ago. Such
determinations can be made in other geographic locations. The results
are useful in dam spillway design and risk analysis, and can be used as
an independent means of confirming or correcting past accounts of
extreme flood and precipitation events.

Introduction

The development of sound engineering procedures for determining spillway
design floods for dams has remained stubbornly elusive. Early empirical
methods gave way to streamflow analysis as streamflow records became
available. Streamflow analysis gave way to precipitation and synthetic
hydrograph analysis. Synthetic hydrograph analysis has been superseded
by hydrometeorological analysis, in which storm processes are analyzed,
storm variables are maximized, and the probable maximum precipitation
(PMP) is estimated. The PMP estimates are used to develop a probable
maximum flood (PMF).

Generalized PMP estimates now cover the United States. These estimates
have gained wide acceptance for several reasons. First, they are
corroborated by a number of nearby extreme precipitation events that
approach the predicted PMP storm intensities. Second, in most areas
they represent a relatively modest extrapolation of a statistically
significant extreme storm data.

Such acceptance is not without serious consequences and legitimate
concerns. The hydrometeorological method is based on incomplete
knowledge of some important storm processes, undesirably brief and
uncorroborated basic data, and the need for subjective decisions. The
generalized estimates are subject to change in the near future as storm
processes become better understood and as additional storm data is
accumulated. The PMP estimates are issued without significant
independent review, and without the opportunity for public comment.

1. Vice President, McCall–Ellingson & Morrill, Inc., 1721 High St.,
   Denver, Colorado 80218
2. Project Manager, McCall–Ellingson & Morrill, Inc.

The automatic adoption of revised PMP estimates by regulatory authorities as a standard leaves dam owners and designers with continually changing spillway criteria. Recently developed paleoflood hydrologic studies of the foothills and mountainous areas of Colorado may be useful. They have convincingly shown that maximum flood flows can be estimated in these areas back as far as the end of the ice age, some 10,000 years ago.

Paleoflood Hydrology

Paleoflood hydrology is the study of botanic, sedimentologic, and geomorphic flood evidence remaining in a valley, Jarrett and Costa (1986). Paleoflood hydrology allows hydrologists to place the relatively short available streamflow and precipitation records in a much longer return period perspective, Costa (1978a). Such investigations can be used to differentiate between water flows, debris flows and mud flows, Costa and Jarrett (1981). They can also be used to confirm or correct flood discharge estimates based on slope-area flood measurements, and to confirm or correct questionable extreme precipitation depth and areal extent reports, Jarrett and Costa (1986) and Jarrett (1987).

Extremely large rare floods leave distinctive evidence of their occurrence in steep mountain valleys. If it is accepted that the frequency of extreme floods in the recent geologic past is the same as it will be in the near geologic future, paleoflood hydrology can be used to estimate the recurrence interval of such floods. Suitable areas for paleoflood investigations exist in the Colorado foothills and mountains. These areas are widespread in Colorado and suitable areas undoubtedly exist in many mountainous regions. Where paleoflood evidence of a recent extreme flood indicates that such a flood is the largest to have occurred in the past, say, 10,000 years it is probable that such evidence of its occurrence in a particular watershed is close to its true recurrence interval, Costa (1978a).

Paleoflood evidence has the advantage of indiciating past extreme flood characteristics directly. The evidence is permanent. It is unequivocal and can be readily confirmed by independent investigators. The evidence of past extreme floods strongly supports the existence and recurrence interval of such a flood. Conversely, the lack of such evidence in a particular watershed where it can be expected to be found, is conclusive evidence that such a flood has not occurred in the time period involved.

Colorado Paleoflood Data

Extensive and definitive paleoflood studies have recently been carried out in foothills and mountainous areas of Colorado. Much of the work has been done in the South Platte River basin, east of the continental divide. The area is included in the recently issued hydrometeorological report, Miller et al. (1984).

The paleoflood studies were accelerated as a result of the catastrophic Big Thompson Canyon flood of July 1976. The flood, which killed 139 people and destroyed 250 structures and more than 400 automobiles, is reported by Costa (1978b) to be a rare hydrologic event on the basis of paleoflood site evidence. This study effectively places the Big Thompson Canyon flood in a realistic 8,000 to 10,000-year return period perspective that is consistent with a reasonable statistical extrapolation of available short-term streamflow data.

Historical reports of small watershed floods in mountainous areas in Colorado were investigated and reported by Costa and Jarrett (1981). Their study shows that possible extreme precipitation amounts, derived indirectly from reported flood levels in small mountain watersheds, are not correct for watersheds above 7500 feet elevation (2300 meters). The paleoflood evidence at these high mountain sites shows that the reported flood flows were debris flows, not water flows. Since debris flows are 60 to 90 percent solids by weight, the reported floods do not support the indirect conclusion that extreme precipitation events occurred.

A multi-disciplinary study, including flood frequency analysis of rainfall peak and snow melt peak streamflow records together with paleoflood investigations, was carried out by Jarrett and Costa (1983). Their study shows that along the South Platte River Basin streams in Colorado snow melt floods dominate above about elevation 7500 feet (2300 meters), and that rainfall floods dominate below that elevation. The July 1982 dry weather Lawn Lake Dam break hydrology, geomorphology and dam break modelling were investigated as reported by Jarrett and Costa (1984). This dry weather dam failure produced profound geomorphic effects and a unique opportunity to observe the effects of debris ladened flood flows in a high mountain area above elevation 7500 feet (2,300 meters). The absence of such effects in other mountain valleys above that elevation confirms the lack of such flows since the ice age.

The Jarrett and Costa (1986) Big Thompson River Basin multi-disciplinary study demonstrates the importance of using paleoflood analysis in extreme flood frequency studies from a dam designer's point of view. The U.S. Bureau of Reclamation's Olympus Dam is located immediately below Estes Park, Colorado, on the Big Thompson River. The dam's spillway is designed for a flood of 22,500 cubic feet per second (637 cubic meters per second). The PMF for the dam is 84,000 cubic feet per second (2380 cubic meters per second), on the basis on the Miller et al. (1984) PMP estimates for the area. Preliminary streamflow and regional analysis, and paleoflood data indicate that the largest natural flood flow at Estes Park is about 5,000 cubic feet per second (142 cubic meters per second) during the past 10,000 years. Olympus Dam, there-fore, has a spillway capacity some 4.5 times the size of the largest flood since the ice age, the 10,000-year flood. Current PMP estimates result in a spillway requirement to pass a PMF nearly 17 times the size of the 10,000-year flood. Such gross differences produced by two acceptable methods require further investigation and reconciliation.

Jarrett (1987) makes use of hydrologic analysis methods and the extensive paleoflood data now available in Colorado to demonstrate convincingly that no extreme rainfall floods have occurred above about elevation 7500 feet (2300 meters) in Colorado since the ice age. It is also shown that flood peaks determined by use of the slope-area method on streams having slopes steeper than 0.002 foot per foot have been typically over estimated by 75 to 100 percent. The storm rainfall amounts indirectly derived to conform to such excessive flow determinations are also in error. All of the few extreme rainstorms and floods that have been reported above elevation 7500 feet (2300 meters) in Colorado have been found to be substantially in error, including the Leadville, Colorado extreme storm precipitation measurement of July 27, 1937, Jarrett (1987).

Although limited to about the 10,000-year flood in Colorado, paleoflood analysis promises to be a more permanent and convincing method of determining extreme flood size and frequency than other methods. Paleoflood analysis integrates all climatic and other variables that affect flood size. The method can be convincingly used to establish priorities for dam modifications to improve dam safety, by comparing the size of the spillway with the size of the 10,000-year flood to indicate degree of safety on a probability basis.

Conclusions

Paleoflood studies can be used to estimate the size of floods in the 10,000-year range, where suitable conditions exist. Such conditions exist in the extensive foothill and mountain areas of Colorado, and probably in many other such areas as well. Credibility of such estimates is enhanced by the clarity of the field evidence and the probability that independent investigators can confirm paleoflood study results. Such credible and reproducible extreme flood data will place PMP estimates in rational time perspective, and will help increase the validity and acceptability of risk analysis studies.

Paleoflood extreme flood flow estimates can be more accurate than estimates derived from slope-area measurements. They can also increase the accuracy of indirectly estimated extreme storm precipitation amounts and areal extent, thereby improving the accuracy of basic date required for hydrometeorological PMP estimates.

The paleoflood method inherently integrates all of the variables affecting extreme flood occurrence.

Recommendations

It is recommended that other suitable areas be identified and that paleoflood estimates of extreme flood flow amounts and frequency be made in such areas in order to rationalize dam safety risk analysis, and to improve basic data required for hydrometeorological PMP estimates.

Appendix - References

Costa, J.E. (1978a). "Holocene stratigraphy in flood-frequency analysis," Water Resources Research, 14:626-632.

Costa, J.E. (1978b). "Colorado Big Thompson flood, geologic evidence of a rare hydrologic event," Geology 6:617-620.

Costa, J.E., and Jarrett, R.D. (1981). "Debris flows in small mountain stream channels of Colorado and their hydrographic implications," Bulletin of the Association of Engineering Geologists 18:309-322.

Jarrett, R.D., and Costa, J.E. (1983). "Multidisciplinary approach to the flood hydrology of foothills streams in Colorado." In Johnson, A.I., and Clark, R.A., eds. International Symposium on Hydrometeorology, Bethesda, Maryland: American Water Resources Association, 565-569.

Jarrett, R.D., and Costa, J.E. (1984). "Hydrology, geomorphology, and dam-break modeling of the July 15, 1982 Lawn Lake Dam and Cascade Lake Dam failures, Larimer County, Colorado," U.S. Geological Survey, Professional Paper 1369.

Jarrett, R.D., and Costa, J.E. (1986). Evaluation of the flood hydrology in the Colorado front range using streamflow records and paleoflood data for the Big Thompson River Basin. U.S. Geological Survey, Water Resources Investigation Report   , in press.

Jarrett, R.D., (1987). "Flood hydrology of foothill and mountain streams in Colorado," dissertation presented to Colorado State University, Fort Collins, Colorado in partial fulfillment of the requirements for the degree of Doctor of Philosophy.

Miller, J.E., Hansen, E.M., Fenn, D.D., Schreiner, L.C., and Jensen, D.T. (1984). Probable maximum precipitation estimates - United States between the continental divide and the 103rd meridian. U.S. Department of Commerce Hydrometeorological Report No. 55, National Oceanic and and Atmospheric Administration, Silver Spring, Maryland.

Criteria for Selecting a Reservoir Routing Method

Raymond J. Kopsky Jr.[1], A.M. ASCE, and Roger H. Smith[2], M. ASCE

Abstract

Results of an investigation comparing dynamic and storage reservoir routing applied in dam safety analyses are presented. Outflow hydrographs computed by these reservoir routing methods are compared for spillway analysis and for dam failure analysis. A hypothetical reservoir, in the shape of a truncated pyramid, with variable dimensions is used to compare the routing methods. Test results, based upon the hypothetical reservoirs, are used to develop criteria for determining if dynamic routing should be used for a particular reservoir. Dam safety analysis case studies for several reservoirs in Missouri are performed to test the validity of the criteria developed. The National Weather Service computer model DAMBRK is used to perform all routing computations.

Introduction

The hydrologic study for a dam safety investigation includes the routing of a design flood through the reservoir. Basic to a dam safety investigation is a determination of the existing spillway capacity. In many cases, it is also necessary to perform a dam failure analysis. For both spillway analysis and dam failure analysis, a method of reservoir routing must be chosen in order to compute an outflow hydrograph from the reservoir. Two principal methods of reservoir routing exist. These methods are dynamic routing and storage routing. In order to compare outflow hydrographs computed by these reservoir routing methods, a hypothetical reservoir model was developed.

Hypothetical Reservoir Model

The hypothetical reservoir used in this study has the shape of a truncated pyramid with a trapezoidal base as shown in Figure 1. The basic reservoir dimensions are the reservoir length in miles (kilometers) (RL), the dam crest length in feet (meters) (BD), the upstream reservoir width in feet (meters) (BU), the dam height in feet (meters) (YD), the upstream reservoir depth in feet (meters) (YU) at the top-of-dam elevation, and the reservoir side slope (Z). This reservoir shape was chosen for investigation because previous channel routing experience published in the literature indicates that the greatest differences in hydrographs computed by hydraulic and hydrologic routing techniques occurred in cases where long, narrow and shallow river valleys were subjected to rapidly rising hydrographs.

---

[1]Hydraulic Engineer, Water Control Management Section, St. Louis District, U.S. Army Corps of Engineers, 210 Tucker Blvd. North, St. Louis, Missouri 63101

[2]Associate Professor of Civil Engineering, Memphis State University, Memphis, Tennessee 38152

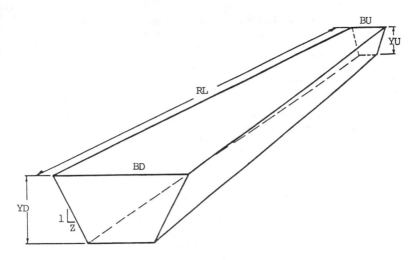

Figure 1. Hypothetical Reservoir Model

Reservoir parameters were based upon a statistical evaluation of data collected from the Missouri Dam and Reservoir Safety Council for 550 reservoirs under their jurisdiction. Based upon an analysis of dam height frequency in Missouri, hypothetical reservoirs with dam heights of 25, 50 and 100 feet (7.62, 15.2 and 30.5 meters) were evaluated. A similar analysis of dam crest lengths was made. The crest length selected for 25- and 50-foot (7.62- and 15.2-meter) high dams was 500 feet (152 meters) and the crest length selected for 100-foot (30.5-meter) high dams was 1000 feet (305 meters). Hypothetical reservoir lengths, which ranged from 0.5 mile (0.80 kilometers) to 27.0 miles (43.5 kilometers), were based upon typical reservoir bottom slopes in Missouri.

Scope of Dam Safety Analyses of the Hypothetical Reservoirs

In the first part of this study, differences in outflows computed by dynamic and storage reservoir routing were investigated for dams with uncontrolled rectangular spillways. Differences in spillways sized for a given reservoir with the two routing methods were analyzed. The spillways were sized to pass various ratios of the PMF without overtopping the dam. The initial reservoir water surface elevation was assumed equal to the spillway crest elevation. The depth of the rectangular spillway was equal to the depth of flood control storage. The amount of flood control storage was based upon data collected for 550 Missouri reservoirs. Reservoirs having flood control storages equal to 10, 20, 30 and 40 percent of the top-of-dam storage were selected for investigation.

In the second part of this study, differences in outflows computed by the two reservoir routing methods were investigated for static-pool dam failures. Static-pool failure analysis requires that the inflow to the reservoir be zero and that the initial reservoir water surface elevation be equal to the spillway crest elevation. This type of dam failure was studied because the criteria for delineation of the downstream environmental zone required by the State of Missouri states that the

zone must be identified by routing the outflow from a static-pool failure
downstream until flood wave depths are again within river banks.

For dam failure simulations, DAMBRK uses parameters to describe the size,
the shape and the temporal formation of the dam breach. Various breach shapes
may be used depending upon the breach side slope specified. The final breach
bottom width and the corresponding elevation complete the shape description of
the breach. The time-dependent nature of breach formation employed in DAMBRK
requires specification of the time from beginning of breach formation until the
maximum size is attained. Also, the water surface elevation at the dam when
failure commences must be specified.

This study attempted to simulate the full range of expected breach
parameters. Final breach bottom widths of 0.5, 2 and 4 times the dam height were
simulated. Dam failure times of 0.25, 0.5, 1, 2 and 4 hours were used. Smith (2)
reported that, of the three breach characteristics (i.e., size, shape and time of
formation), shape is the least important in the determination of peak outflow.
Therefore, an average breach side slope of 1:1 for earthfill dams was assumed.
The lowest elevation reached by the bottom of the breach was assumed equal to
the elevation of the bottom of the dam at the upstream face.

Reservoir distance steps for dynamic routing were based upon the method
given by Fread (1). He recommended that distance steps be computed from the
equation

$$DX = C * DT \qquad (1)$$

where   DX = distance step in miles (kilometers),
           C = dynamic wave celerity in miles (kilometers) per hour,
          DT = routing time step in hours.

The dynamic wave celerity was computed as

$$C = (G * D)^{0.5} \qquad (2)$$

where   C = dynamic wave celerity in feet (meters) per second,
          G = gravitational acceleration (32.2 feet (9.81 meters) per second
              per second),
          D = reservoir mean hydraulic depth in feet (meters).

In both types of dam safety analyses, the percentage difference in peak
outflows from a given reservoir computed by the two routing methods (relative to
storage routing) was calculated. This percentage difference was defined as the
outflow ratio.

Results of Spillway Analysis

Spillway analysis indicates that reservoir length, percentage of flood control
storage and the time to peak of the inflow hydrograph are parameters which affect
the outflow ratio. In general, a large outflow ratio occurred when a long reservoir
with small flood control storage was subjected to an inflow hydrograph having a
short time to peak.

In order to determine when dynamic reservoir routing should be used for
spillway analysis, a methodology to relate the outflow ratio to reservoir and inflow
hydrograph parameters was developed. Two time parameters were used to develop
this methodology. They are the time to peak of the inflow hydrograph and the
travel time of a dynamic wave through the reservoir. For each reservoir, three
values of time to peak of the inflow hydrograph were simulated (0.5, 1 and 2

hours). The dynamic wave travel time between adjacent reservoir cross section locations was computed by dividing the reach length by the celerity of a dynamic wave based upon the initial mean hydraulic depth in the reach. The total travel time through the reservoir was computed by summing travel times between adjacent cross section locations.

Knowing the two time parameters, a dimensionless time ratio was computed for each reservoir as

$$TR = TP / TT \qquad (3)$$

where   TR = time ratio,
           TP = time to peak of inflow hydrograph in hours,
           TT = travel time of a dynamic wave through the reservoir in hours.

Large outflow ratios occurred for hypothetical reservoirs with 10 percent flood control storage. Outflow ratios under 3 percent occurred for hypothetical reservoirs having 20 percent or more flood control storage. For all reservoirs, dynamic routing yielded higher peak outflow values than did storage routing.

This analysis indicates that dynamic reservoir routing should be seriously considered for spillway analysis of reservoirs having less than 20 percent flood control storage and having small time ratios (i.e., less than 3). This analysis shows that spillways dimensioned for hypothetical reservoirs with 10 percent flood control storage and with small time ratios may be undersized if storage reservoir routing is used. For each reservoir, the spillway sized with storage routing is shorter in length than that sized with dynamic routing. The use of dynamic routing resulted in a longer and, consequently, a more expensive spillway. However, from a dam safety viewpoint, there would be less risk of a dam failure from overtopping of the dam.

Results of Static-Pool Dam Failure Analysis

Static-pool dam failure analysis indicates that reservoir length, dam failure time and the ratio of final breach bottom width to dam crest length (i.e., the width ratio) are parameters which affect the outflow ratio. In general, a large outflow ratio occurred when a short failure time and a large width ratio were simulated for a long reservoir.

In order to determine when dynamic reservoir routing should be used for dam failure analysis, a methodology to relate the outflow ratio to reservoir length, failure time and the width ratio was developed. These parameters must be assumed or determined by an engineer prior to his dam failure analysis. Knowing these parameters, the engineer can use this methodolgy to compute an expected outflow ratio. Based upon the expected outflow ratio, he can then decide if storage routing should be superceded by dynamic routing.

Knowing the maximum breach dimensions, an estimate of the negative wave celerity in the reservoir was computed as

$$C2 = (G * A / T)^{0.5} \qquad (4)$$

where   C2 = negative wave celerity in feet (meters) per second,
          G = gravitational acceleration (32.2 feet (9.81 meters) per second per second),
          A = area of maximum breach size in square feet (square meters),
          T = top width of maximum breach size in feet (meters).

A dimensionless ratio composed of reservoir length, failure time and negative wave celerity was computed for each simulation. This dimensionless ratio, termed the failure ratio, was computed as

$$FR(\%) = 147 * (RL / C2) / FT \qquad (5)$$

where
- FR(%) = failure ratio in percent,
- RL = reservoir length at spillway crest elevation in miles (kilometers),
- FT = failure time in hours.

(If the International System of Units is used, the coefficient is 27.8 rather than 147.)

For each failure analysis, the failure ratio was plotted versus the outflow ratio. For a given width ratio, all values plotted on the same curve regardless of dam height, reservoir length or failure time. Least-squares best-fit curves for the data are shown in Figure 2 along with the equation for the outflow ratio (OR).

For the hypothetical reservoirs, an analysis was performed to determine the effect of computing the negative wave celerity from the initial reservoir mean hydraulic depth rather than from the maximum breach dimensions. For a given width ratio, computed values plotted on the same general curve regardless of the technique used to compute the negative wave celerity.

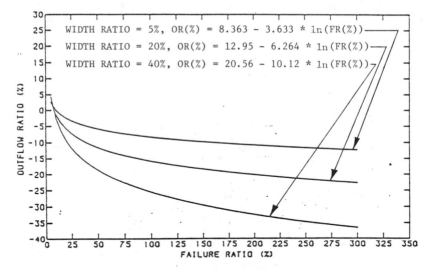

Figure 2. Best-Fit Curves for Hypothetical Dam Failures

Results of Case Studies

Case studies for reservoirs in Missouri were performed in order to test the criteria developed from studies of the hypothetical reservoirs. For the spillway analyses, outflow ratios less than 3 percent were found. This result was expected because all the reservoirs had over 20 percent flood control storage and had high time ratios (i.e., greater than 7).

For the dam failure analyses, expected outflow ratios were computed from both the maximum breach dimensions and the initial reservoir mean hydraulic depth. In general, expected outflow ratios based upon the initial reservoir mean hydraulic depth matched the actual outflow ratios more closely than those based upon the maximum breach dimensions. The areas of off-channel storage in actual reservoirs, which are not present in the hypothetical reservoirs, influence the negative wave celerity. For reservoirs with significant areas of off-channel storage, the initial reservoir mean hydraulic depth should be used in the computation of the negative wave celerity.

Conclusions and Recommendations

Dynamic reservoir routing should be seriously considered for use in dam safety analyses of certain reservoirs. For spillway analysis, dynamic routing should be considered for reservoirs with less than 20 percent flood control storage and with time ratios less than 3. For dam failure analysis, an expected outflow ratio can be computed with the methodology developed. The initial reservoir mean hydraulic depth should be used in the computation of the negative wave celerity for reservoirs with significant areas of off-channel storage. Based upon the expected outflow ratio, the engineer can decide if storage routing is acceptable.

Acknowledgement

The information in this paper was taken from the Master of Science theses of Raymond J. Kopsky Jr. and Mark L. Loethen. Both gentlemen did their graduate work at the University of Missouri - Rolla.

References

1. Fread, Danny L., "DAMBRK: The National Weather Service Dam-Break Flood Forecasting Model", Office of Hydrology, National Weather Service, January 1982.

2. Smith, Roger H., "Dam-Breach Wave Analysis", presented at Seminar on Dam Safety - Inspection and Design, held at Triton College, River Grove, Illinois, November 17-19, 1981.

USDA-Water Erosion Prediction Project (WEPP)

W. J. Rawls and G. R. Foster*
M. ASCE

Abstract

The USDA-Water Erosion Prediction Project (WEPP) will develop
improved erosion prediction technology based on modern hydrologic and
erosion science that will be process-oriented and conceptually a
significant improvement over the Universal Soil Loss Equation (USLE)
(USDA, 1978).

Introduction

The Universal Soil Loss Equation (USLE) (USDA, 1978) is by far the
most widely used method for predicting sheet and rill erosion. The
origin and form of the equation evolved from A. W. Zingg's 1940
equation for the effect of slope length and steepness on erosion. The
USLE is a lumped model since it does not define separate factor
relationships for the fundamental hydrologic processes of rainfall,
infiltration, and runoff and the fundamental erosional processes of
detachment by raindrop impact, detachment by flow, transport by rain
splash, transport by flow, and deposition by flow. Furthermore, the
USLE is empirical, depends on a large mass of data for its relationships, and is very much a "black box" model. Both its empirical
origin and lumped equation structure severely limit the potential for
increased accuracy and major improvement. Scientifically the USLE is
considered to be mature technology.

When the USLE was developed more than 20 years ago, it had to be a
simple mathematical expression so that it could be solved with the
computational equipment available at the time. However, small
computers are now readily available that can solve complex mathematical
relationships quickly and easily in the field, local office, and even
the farmer's home. Thus users now have the potential for using
erosion prediction technology based on modern science of hydrologic
and erosional processes that was impractical two decades ago.

The objective of this project is to develop improved erosion
prediction technology based on modern hydrologic and erosion science
that will be process-oriented and conceptually a significant improvement over the Universal Soil Loss Equation (USLE). The project is a
multiagency (USDA-Agricultural Research Service, USDA-Soil Conservation
Service, USDA-Forest Service, and USDI-Bureau of Land Management) and

*Hydrologist, USDA-ARS Hydrology Laboratory, Beltsville, MD 20705, and
Hydraulic Engineer, USDA-ARS National Soil Erosion Laboratory, W.
Lafayette, IN 47907.

multidiscipline effort lead by Dr. George Foster, USDA-ARS, National
Soil Erosion Laboratory, Purdue University, West Lafayette, Indiana
47907 (317-494-7748). The target group for the new technology consists
of all current USLE users, with most applications following within the
categories of conservation planning, project planning, and inventory
and assessment. The target date for completion of the project is
August 1989 with the output being a family of models and computer
programs. The user requirements developed for the project are given
in detail in USDA (1987).

Procedure Requirements

The erosion prediction procedure from this project is to apply to
"field-sized" areas or conservation treatment units. Although the size
of a particular field to which the procedure applies will vary with
degree of complexity within a field, the maximum size "field" is about
a section (260 ha) although an area as large as 810 ha is needed for
some rangeland applications. The procedure will not apply to agricultural fields or watersheds having incised, permanent channels such as
classical gullies and stream channels. The channels that the procedure
is to include are those farmed over and known as concentrated flow or
"cropland ephemeral gullies." Also, the procedure is to apply to
constructed waterways like terrace channels and grassed waterways. In
rangeland and forest applications, "fields" can include gullies up to
the size of typical concentrated flow gullies in 260-ha cropland
fields. These channels are on the order of about 1 to 2 m in width by
about 1 m deep. The land uses to be considered include: cropland,
vegetable land, hayland, pastureland, orchards, vineyards, nurseries,
rangeland, disturbed forest land, construction sites, road surfaces,
cuts and fills, surface mines, hazardous waste sites, recreational,
and other land uses where surface flow occurs over the entire area and
erosional processes are only slightly affected by "partial area"
hydrology. These lands may be non-irrigated, sprinkler irrigated, or
surface irrigated.

The procedure will compute erosion on the following basis: a) long
term, average annual, b) distribution by crop stages or seasons over an
average or typical year, c) frequency distribution of annual amounts,
d) frequency distribution by month, e) frequency distribution by event,
f) single design storm, and g) continuous simulation. These computations may assume time invariant soil, topographic and land use conditions except in cases where the cover conditions consistently change
over a period, like recovery following a disturbance in a forest.

Model Structure

The model is to be based on the fundamental erosion processes of:
a) interrill erosion (principally detachment by raindrop impact and
lateral transport by thin flow), b) rill and concentrated flow erosion
(detachment by flow), c) sediment transport by flow, d) deposition by
flow, e) deposition in impoundments and concentrated flow hydraulics.
The model will include major modules for 1) climate generation, 2) snow
accumulation, 3) snowmelt, 4) infiltration, 5) runoff, 6) soil
temperature, 7) erosion, 8) soil moisture, 9) crop growth, 10) plant

residue, and 11) tillage. Implicit in all of these modules except for
the climate module is the central role of soil and soil properties.
The model will have three basic versions (Figure 1): 1) a representative landscape profile version, 2) a watershed version, and 3) a grid
version that covers the entire field.

Profile Version

The profile version is applied to landscape profiles selected by
the user. The user will have two options to specify characteristics
of the landscape profile. The options are: a) to specify a slope
length, average steepness, location of the major inflection point,
degree of curvature of the upper slope and degree of curvature of the
lower slope, and b) locations and slope steepnesses along the profile.

The profile version is to compute: a) net erosion or deposition,
b) rill erosion or deposition by flow, c) interrill erosion, d) fineness of eroded or deposited sediment, e) sediment load, and f) fineness
of the sediment load. As a minumum, this version outputs average soil
loss for the slope length (sediment yield/unit of slope length).
Erosion rate for the erosional parts of the profile and deposition
rate for the depositional parts of the slope are to be computed for
slope length segments located along the slope length as specified by
user input.

Watershed Version

The user chooses a watershed within the field to apply the model
(Figure 1). Inputs for the profile portion of the watershed version
are to be the same as those for the profile version. Inputs for the
concentrated flow portion are to be: a) channel cross section properties, b) locations along the channel and grades at those locations,
c) information on the outlet control to the channel, and d) drainage
areas at the upper and lower ends of the channels. The channel outlet
control will consider backwater from: a) uniform flow in the last
reach of the channel or uniform flow in a specific outlet channel at
the end of the given channel, b) critical depth at the end of the
channel, or c) a natural or constructed outlet control with a known
rating curve. The input for a within-field impoundment will be the
sideslopes that form the basin or coefficients for an area-depth curve
and simple information on the rating function that describes flow out
of the impoundment.

The outputs for the slope length profile portion and the
concentrated flow portion are to be the same as that for the profile
version except that output is by channel reach rather than by slope
length segment.

Grid Version

The grid version describes erosion and sediment movement in grids
that can cover entire regular or irregularly shaped fields with
boundaries that may not coincide with watershed boundaries.

PROFILE VERSION

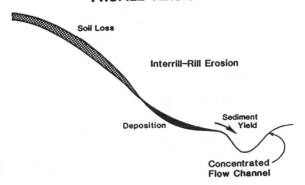

WATERSHED VERSION

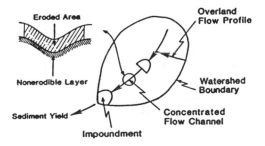

GRID VERSION

Figure 1.   Model structure.

The input to the grid model will be slope steepness and direction for each grid and the grade and direction of a channel reach within a grid. Properties of channel cross sections and channel outlet controls will be input like those in the watershed version.

The outputs will be the same as those for the watershed version except that the values are based on grid areas rather than on slope length or channel reach segments.

Operational Requirements

The major operation requirements that will be incorporated into the procedure are:

a) <u>Computational time</u>. The procedure will compute the frequency distribution of annual soil loss values for the profile version at the rate of one management practice per minute and one practice per two minutes for the watershed version running a single overland flow profile and a single concentrated flow channel. The rate can be proportionally slower for more complex systems.

b) <u>Ease of use</u>. The procedure will be easy to use, especially for the infrequent user, by accepting simple inputs that are commonly available and understood by personnel in the local field office. It should require little structured training or support. Also, it will be flexible and accept inputs on increasing detailed and complex levels if the user determines that more detail is needed or that default values need changing.

c) <u>Applicable to broad range of conditions</u>. The procedure will be applicable to all sheet-rill erosion and concentrated flow erosion situations.

d) <u>Robustness</u>. The procedure will tolerate out-of-range input data and combinations of inputs that might cause problems. However, the procedure will alert the user to loss of accuracy when inputs are over simplified, and for incorrect data entries.

e) <u>Validity</u>. The procedure will be sufficiently accurate to lead to the planning and assessment decision that would be made in the large majority of cases when full information is available. Since the procedure is composed of a number of modules, each major module will be individually validated, and the procedure will be validated as a package.

Supporting Research

An extensive set of rainfall simulation experiments will be conducted during 1987 and 1988 on more than 50 key agricultural and rangeland soils distributed over 26 states. The purpose of these experiments is to obtain data for relating the interrill and rill erodibility factors and infiltration parameters for a broad range of soil properties and management conditions.

References

USDA, Agricultural Research Service. 1987. User Requirements, USDA-Water Erosion Prediction Project (WEPP). USDA-ARS National Soil Erosion Research Laboratory, 43 pp.

USDA. 1978. Predicting rainfall erosion losses – A guide to conservation planning. Agricultural Handbook 537, 58 pp.

VERIFICATIONS AND APPLICATIONS OF
A NONPOINT SOURCE POLLUTION MODEL

by Ming T. Lee[1]

Abstract

To meet the needs of the Illinois Comprehensive Monitoring and Evaluation (CM&E) project of the Rural Clean Water Program (RCWP), the Agricultural Nonpoint Source Pollution Model (AGNPS) was selected for evaluating the effectiveness of Best Management Practices (BMPs). AGNPS is a distributed storm event model consisting of runoff, erosion, sediment yield, and agricultural related water quality components. Before the model was used to assess various watershed scenarios, 2 years and 10 months of field data were used to verify the model. The detailed model verification is discussed. The results of the model applications for the scenario analysis are presented.

Introduction

Watershed modeling is a tool for hydrologic system synthesis, prediction, and optimization. Nonpoint source pollution is a spatially related and multi-constituent water quality problem. Field data are not sufficient to evaluate all possible control schemes at the source and during transport and deposition processes. In order to overcome these problems and provide insights on the potential impacts of control schemes, watershed modeling in addition to field monitoring is considered a feasible approach. However, modeling requires simplification and abstraction, which may create deviations from the real processes. Hence verification on the basis of field monitoring data becomes a necessary step before the model can be used for analyzing the proposed alternatives.

This paper describes the case of a Rural Clean Water project at the Highland Silver Lake watershed in Illinois and the process of model verification and applications.

Model Selection

To meet the requirements of the nationwide RCWP, the Highland Silver Lake project (Illinois State Coordinating Committee, 1981 and 1986) was developed to: (1) determine the effectiveness of Best Management Practices in the water quality problem areas; and (2) project the possible impacts on water quality based on the limited implementation to date. To meet the first objective, extensive monitoring systems were installed at the lake, streams, and agriculture fields (Makowski and Lee, 1985, 1986). To meet the second objective, with the support of limited field data, the hydrologic modeling approach was adopted. Because of the spatial characteristics of nonpoint source pollution problems, a spatially distributed

---

[1]Hydrologist, Illinois State Water Survey, 2204 Griffith Drive, Champaign, IL 61820.

model is needed.   Through a team model-selection process, the Agricultural Nonpoint Source Pollution Model (AGNPS) (Young et al., 1985) was selected on the basis of its ability to describe BMPs and reflect their changes.

AGNPS is a grid-cell-oriented, single-storm-event model.   For modeling purposes, the watershed is divided into small square areas (cells) which are interconnected according to stream drainage patterns.   The AGNPS model predicts the runoff, eroded and transported sediment, and the nitrogen, phosphorus and chemical oxygen demand concentrations carried by the runoff and sediment for every cell in the watershed.   A detailed description can be found elsewhere (Young et al., 1985).

Model Verification

Before the model can be utilized for scenario analysis, validation on the basis of field monitoring data is essential.   For the Highland Silver Lake project, the AGNPS model was tested by (1) comparisons of different grid-cell sizes, (2) verification with observed storm event data, (3) verification with lake sedimentation survey data, (4) comparisons with observed annual water quality loadings at the outlet of watershed.   The AGNPS model was tested with two different grid-cell sizes.   The total watershed is 50 square miles.   Ten- and 40-acre grid-cell sizes were tested on a 300-acre sub-watershed.   The results indicated that the resolution of a 40-acre grid-cell is not sufficient to depict the land treatment boundaries, even though the runoff and sediment yield of the test storm at the outlet of the watershed did not show significant differences between two grid-cell sizes.

To verify modeling results with monitoring data for a single event, it would be necessary to prepare the input data for all cells in the AGNPS model so as to duplicate actual field conditions at the time the event occurred.   While this would be a worthwhile pursuit, time constraints and the lack of detailed land management data made such a comparison infeasible, considering that the main application of the model is to predict the average conditions of the watershed. Therefore, cells were set to depict 1984 watershed conditions which closely represent the condition during the data collection period.   The AGNPS model was tested by changing rainfall events ranging from 0.7 to 5.9 inches, which represent the range of five-times-a-year to once-in-50-years events.   In order to compare the model output and observed data, runoff volume and total suspended loads of a main stream gaging station were plotted (Figure 1).   The results showed the model output line which depicts the average condition passing through the scattered observed data points.   These discrepancies are fully expected due to seasonal variations of land use and soil antecedent moisture conditions for different storm events.   The trends of observed and predicted values are similar.

The second way to verify the model is to compare the predicted sediment deposition rate with the results of the lake sediment survey.   The 1984 lake sediment survey showed lake sedimentation is at the rate of 0.9 tons per acre per year, which is equivalent to 27,870 tons per year deposition in the lake as shown in Table 1.   The model predicted 19,100 tons per year, which is lower than the field data indicated.   It is worthwhile to note that in the period 1981-1984, rainfall was 14.5 percent higher than the long-term average.   Consequently, the field-measured lake sedimentation rate would be higher than the long-term average.

The third way to verify model performance is to compare average annual water quality loads with field monitoring data.   Results of field average annual

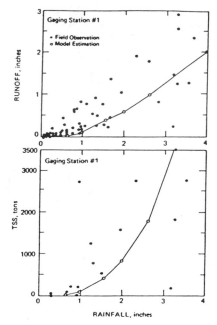

Figure 1.   Comparison of Observed and Predicted Runoff and Total Suspended
Solids at Stream Gage Site #1.

Table 1.   Comparison of Field-Observed Data with
Model-Estimated Data at Spillway Site

Parameter	Observed (tons/year)	Estimated (tons/year)
Sedimentation Rate	27,870	19,100
Total Suspended Solids	2,580	2,702
Total Kjeldahl Nitrogen	145	510
Total Phosphorus	20	106
Chemical Oxygen Demand	1,817	3,414

loading and model prediction at the spillway monitoring station are shown in
Table 1. The predicted annual total suspended solids load was 2702 tons per year,
compared to the monitored average annual load (2 years and 10 months data) of
2580 tons per year. The predicted total phosphorus load was 106 tons per year,
which exceeded the observed loads (20 tons per year) by a factor of five.
Predicted total Kjeldahl nitrogen and chemical oxygen demand exceeded
monitored results by factors of 3.5 and 2, respectively. While the actual and
predicted total suspended solids loading differed by less than 5 percent, the model
clearly overpredicted nutrient loadings. This may be due to inadequate
monitoring data, since not all significant events were captured, or it may be a
deficiency of the model.

In summary, while actual event-by-event verification of the model remains to be accomplished, comparison of model-predicted results with monitoring data for average watershed conditions indicated varying degrees of accuracy. The model provided especially good predictions of total suspended solids loading at the outlet of the watershed and did an adequate job of estimating lake sedimentation. Major discrepancies, however, existed between actual and predicted nutrient loadings at the outlet of the watershed. Since the main objective of this project is to assess sediment-related nonpoint pollution, the model was accepted for analyzing a series of scenarios.

Applications of Model

Four kinds of model applications were conducted: (1) incremental analysis of BMPs; (2) comparison of future conditions with and without the project; (3) critical-area analysis; and (4) particle size analysis. The incremental analysis is based on the concept of consecutively adding BMPs to the watershed to illustrate the relative effectiveness of various sets of BMPs. A comparison of future conditions with and without the project indicates net effects of BMPs attributable to RCWP. The critical-area analysis examines the operant critical-area definition and the model-selected critical area. The particle size analysis illustrates the variation of soil particle size due to transport process and shows the possibility of controlling fine particle sediment.

Incremental Analysis. To illustrate the effectiveness of various levels and kinds of BMP applications in reducing pollutants, the following scenarios were examined: first, application of contracted nonstructural BMPs such as conservation tillage on the watershed; second, adding grass waterways and impoundment structures; third, adding animal waste management systems (installing filter strips at feedlots); fourth, adding fertilizer treatment at the contracted farms. The results indicated that nonstructural BMPs were effective in reducing sediment yield and sediment-associated nitrogen and phosphorus. No significant changes in runoff, peak discharge, or soluble nitrogen and phosphorus were shown. The grass waterways and impoundments were effective in reducing peak discharge, sediment yield, and sediment-attached nitrogen and phosphorus at four field sites. Less reduction was evident at downstream gaging stations. The animal waste management system was effective in reducing soluble nitrogen and phosphorus at field sites with feedlots in their drainage areas. However, the downstream gaging stations show no significant reduction in most pollutants, mainly because there are only a few units of planned animal waste management systems in the watershed. By adjusting contracted cropland to a low fertilization level (50 lb nitrogen per acre, 20 lb phosphorus per acre) the reductions in soluble nitrogen and phosphorus ranged from about 1 to 22 percent at field sites and 3 to 9 percent at stream gaging stations.

Comparison of Future Conditions with and without Project. The net change due to RCWP showed that sediment yield had the greatest reduction, ranging from 0 to 54 percent. Generally, field sites exhibited larger variations than gaging stations. Peak discharge and runoff volume showed moderate reductions, ranging from 0 to 6 percent, mainly due to terracings and sediment retention basins.

Critical Area Analysis. The results showed that the project critical-area definition was fairly successful in designating the highest soil erosion and sediment yield area as compared with the model-selected high soil erosion and sediment yield areas (figure 2), but did not represent well the areas producing high amounts of soluble nitrogen and phosphorus.

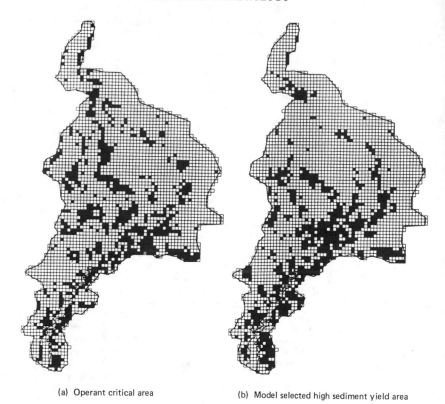

(a) Operant critical area   (b) Model selected high sediment yield area

Figure 2. High Sediment Yield Areas determined by the Model and by the Project
           Critical-Area Definition.

Sediment Particle Size Analysis. The results indicated that in small field sites
small aggregates are the dominant components. However, as drainage area
increases, silt and clay become the dominant particle size. The results also
showed that sediment deposited in the lake was predominantly clay-size.

Conclusions

Modeling provided insights into the relative effectiveness of BMPs in
improving water quality, overall impacts of RCWP, particle size transport, and
critical area selection. AGNPS has proven to be a useful tool in evaluation of the
Illinois RCWP project. Most of the model input data is physically-based, and very
little parameter calibration is required. However, it is strongly recommended that
model outputs be compared with field monitoring data and necessary adjustments
be made before the model is used for prediction or scenario analysis.

Acknowledgments

Funding for this project was partially provided by the U.S. Department of Agriculture through the Rural Clean Water Program. The State Office of the Soil Conservation Service in Illinois, Illinois EPA, Economic Research Service, USDA, Southwestern Illinois Metropolitan Area Planning Commission, and Illinois State Water Survey are participating members of the comprehensive monitoring and evaluation team.

Appendix - References

Illinois State Coordinating Committee (1981). "Comprehensive Monitoring and Evaluation Program for the Highland Silver Lake Watershed Rural Clean Water Program - Plan of Work." Illinois State Coordinating Committee, Springfield, Illinois.

Illinois State Coordinating Committee (1986). "Highland Silver Lake Rural Clean Water Project: Summary Report Fiscal Year 1986." Illinois State Coordinating Committee, Springfield, Illinois.

Makowski, P.B. and Ming T. Lee (1985). "Hydrologic Investigation of Highland Silver Lake Watershed: 1984 Progress Report." Illinois State Water Survey Contract Report 357, Champaign, Illinois.

Makowski, P.B. and Ming T. Lee (1986). "Hydrologic Investigation of Highland Silver Lake Watershed: 1985 Progress Report." Illinois State Water Survey Contract Report 380, Champaign, Illinois.

Young R.A., C.A. Onstatd, D.D. Bosch, and W.P. Anderson (1985). "Agricultural Nonpoint Source Pollution Model, a Large Watershed Analysis Tool, a Guide to Model Users." Agricultural Research Service, USDA, Morris, Minnesota.

LEGAL MISUSE OF URBAN HYDROLOGY CONCEPTS
AND REGULATIONS FOR RURAL AREAS

James E. Slosson,[1] (M. ASCE), Robert C. MacArthur[2] (A.M. ASCE)
and Gerard Shuirman[3] (F. ASCE)

This paper presents recent observations of the misuse of
technical definition, hydrologic concepts, sedimentation engineering
principles, and flood control regulations by the legal profession during litigation involving property damages caused by various kinds of
hydrologic events. Recommendations are offered so that prospective
expert witnesses and technical advisors can avoid the embarrassment
and possible professional discreditation associated with theatrical
courtroom procedures.

INTRODUCTION

At a time when more lawsuits are being filed for flood and
other natural hazards or natural/man-made hazard damage and losses,
more scientific and analytical procedures, explanations, and presentations are warranted.

Forensic engineering and the use of engineers and scientists
as expert witnesses by the court system is escalating rapidly as the
frequency of filings of natural hazard-related lawsuits increases.
Both the public and private sectors were once precluded from such lawsuits because of legal immunities. This situation has drastically
changed during the last decade. Cases appear to have motivated both
the attorneys and, to some extent, some of the experts to win "at all
costs" because of the enormously high monetary rewards to the professionals involved. In many cases obfuscation and deception seem to pre-
vail, contrasting to the scientific method of analysis that professional engineers and geologists are taught.

Part of the problem is the unknowing or deliberate misuse
and/or misapplication of codes, regulations, and policies. The authors
have witnessed opposing attorneys setting the tenor of the trial by
using codes, regulations, and/or policies unrelated to the facts
and/or the hydrologic and geographic setting of the area in question.

Examples of misuse appear herein. However, because the cases
are either in progress or pending appeal, a reference coding, rather

---

[1]Chief Engineering Geologist, Slosson and Associates,
 14046 Oxnard Street, Van Nuys, California 91401 USA

[2]Vice President, Simons, Li & Associates, Inc.,
 3901 Westerly Place, Suite 101, Newport Beach, California 92660 USA

[3]Civil Engineer, 15928 Ventura Boulevard, Suite 225, Encino,
 California, 91436 USA

than legal designation, will be utilized. These cases are all highly complex with many legal and technical issues. Summaries below are intended to stress only the theme of this paper, that is, the misuse of experts by attorneys.

CASE HISTORIES

CASE A: Geographically, this case involves a rural agricultural area in the southern desert portion of Imperial County in California. The affected farm land lies on the lower (or downslope) portion of the Pinto Wash alluvial fan. Approximately 161,500 acres of watershed drains from higher elevation mountain canyons and alluvial fans by way of the Pinto Wash to a lower fan where the channel of the Pinto Wash first bifurcates into two branches, then into multiple branches, and eventually turns to sheet wash plus many small distributary channels in the area of the farm land. Infrequent hurricane-type storms caused serious damage in 1976, 1977, 1982, and 1983, as well as in earlier years, when high intensity rainfall resulted in flood conditions. The 1976 storm caused flooding inundating agricultural land on the lower portion of the fan studied. Some erosion occurred but primary damage was due to deposition of sediment and ponded water. The upstream farmer releveled his field and re-established his irrigation ditches. Farmers downstream from him suffered some damage in the storm of 1976. In 1982, a storm of similar intensity occurred and land on the agriculturally developed portion of the fan suffered damage. Downstream farmers sued the upstream farmer contending a flood control dam should have been constructed by him upstream on Bureau of Land Management land where Pinto Wash was confined to a broad canyon prior to reaching the fan. They also contended that the dam should have been associated with a developed flood control channel extending from this hypothesized dam to a river a few miles below the farm land.

The attorney argued and expert witness for the Plaintiffs testified that in development of land, the parties involved are obligated to construct flood control facilities to protect downstream residences. While this is good practice where subdivisions are created for housing by subdividing the land into residential parcels (lots) with planned construction of roads and flood control facilities, it is not the rationale to adopt for large-scale agricultural land use. An additional misuse of technical data and expertise was adhered to by the attorney and witness for the Plaintiffs on introduction of a Flood Insurance Rate Map to indicate flood inundation based on a single channel typical to riverine areas of California rather than the methodology adopted by FEMA for alluvial fans.

Flooding on the lower Pinto Wash alluvial fan occurred as the FEMA methodology describes, not as described by the Plaintiffs' expert witness and attorney. The expert for the Plaintiffs indicated he was following instructions -- he was an advocate rather than an impartial judge of what happened, why it had happened, and what could realistically and legally be done to prevent such damage.

CASE B: A sequence of drought years in the central Mojave Desert during the 1950's caused somw ranchers to forget that desert flash floods can occur and cause severe damage from both erosion and sedimentation. After a storm in 1969, the U.S. Department of Agriculture Soil Conservation District assisted local farmers by providing plans for "flood proofing" their land(s). In the instance of one agricultural parcel severely flooded in the 1969 storm, plans and specifications for an earth channel following the old stream thread were prepared by the District as a means of protection. The plans were essentially adhered to by the owner in 1979. When severe storms occurred in 1980 and 1983, several of the downstream farmers suffered damage from sediment deposition. Their attorneys alleged the deposition was greater than it would have been without the improved channel construction.

In evaluating added damage, the Plaintiffs' attorneys asked their expert to estimate cost of restoring plaintiffs' properties to their pre-1980 levels. He complied by estimating the cost of truck removal and land-fill disposal of all the sediment, a cost at least 10 times the fair market value of the property. The expert, when deposed stated he really didn't anticipate anyone would take him seriously regarding his recommended corrective measure, but the attorney had asked him to estimate the cost of restoring the property to its original elevation. Under further interrogation by the upstream farmer's attorney, it was established that the expert agreed that farm land was usually restored by relevelling the land and resetting the irrigation system--costing about 1/25 the amount of the method calculated.

CASE C: In 1978, a high intensity storm impacted Southern California causing flooding and dramatic generation of debris flow. One of the most severely affected flood/debris flow areas was Hidden Springs located in the San Gabriel Mountains immediately north of Los Angeles. Rainfall during the winter of 1977-1978 was the third highest recorded during a 100-year sequence, exceeding fourteen inches during a 3-day storm sequence with approximately 9 inches falling the third day.

Flooding and debris flow were exacerbated by a brush fire which had burned approximately 2,440 acres of the upper watershed of Mill Creek in July of 1977 -- about seven months prior to the storm. This resulted in the creation of the extreme hazard of the fire-flood sequence. Additionally, all of the structures were located within the boundaries of the 50-year flood event.

The slopes laid bare by the fire were thoroughly saturated when the approximately 9"/24 hours high intensity storm occurred causing a flood height exceeding even the 100-year flood damage and destroying the structures in the flood-inundation area. Thirteen people living in the small community lost their lives.

The experts for the Plaintiffs argued that there was sufficient time, between fire and flood, to analyze, design, and construct a debris basin to prevent the flood. Analyses by Los Angeles County Flood Control District, Defense consultants and others determined that debris production exceeded 400,000 cubic yards of sediment and that a dam for debris would have been of such magnitude as to require approval by the State Division of Safety of Dams. Estimated time for analysis, design, approval of design, and construction exceeded the time between fire and flood. These data did not deter the attorney and experts for the Plaintiffs who argued the debris dam should have been constructed whether or not the procedures were adhered to. Information available suggests that the experts for the Plaintiff were instructed as to what to advocate knowing full well that their opinions would be contrary to good engineering practice. Further, residents of Hidden Springs were warned that they resided in a flood-prone area and on the day of the event were given warning of flood potential during the anticipated torrential downpour.

CASE D: This case alleged flood damage having international ramifications. The Tijuana River flows northerly through Tijuana, Mexico outletting into a broad alluvial, agricultural floodplain within the City of San Diego and extending to the Pacific Ocean, about ten miles from where the river crosses international boundary. In the late 1970's, after years of flooding within densely urbanized Tijuana, Mexico constructed a concrete-lined flood control channel terminating at the boundary, leaving San Diego with two basic methods of protecting its properties: to continue the Mexican channel to the ocean or to attempt to dissipate and spread the high velocity flow to minimize its erosive effect. The latter was chosen for economic and land-use consideration. Accordingly, an energy dissipator structure was designed and built by the U.S. Corps of Engineers in 1978, directing the flow westerly toward the ocean in line with general topography, but contrary to the northerly direction of the then existing low flow channel.

In both 1980 and 1983, storms of high intensity and long duration caused flood damage to many properties in the floodplain. Flooding was further exacerbated by release of large volumes of water from Rodriquez Reservoir, upstream from Tijuana, to protect the structural integrity of the dam.

Property owners joined to file a lawsuit alleging the dissipator structure should have been aligned with the low flow channel since that was the natural alignment. Plaintiffs experts were apparently instructed to analyze what the effect would have been on the subject properties if the dissipator had been aligned to direct the flow northerly, based on assumption that this was indeed the direction of natural flow. An investigation of the history of low flow channel locations showed this assumption to be false. Research indicated that for the previous 75 years there had been at least eight different low flow channel alignments meandering over nearly the entire width of the floodplain. This particular aspect of the case is a classic example of experts basing an analysis on unfounded assumption, leading to meaningless results. It also exemplifies the misapplication of ordinary diversion concepts and legalities to a floodplain where the low flow channel is constantly changing courses.

SUMMARY

Unfortunately, the experts who testify for these attorneys become willing, or unwilling, pawns by agreeing to confine investigation and analysis to limited scope of assignment, sometimes based on erroneous assumptions. As such, the experts do not necessarily lose their objectivity in terms of their narrow assignment; however, in not

examining the whole problem, they create confusion and misunderstand-
ing in the minds of judges and juries and frequently create geofanti-
sies.

It is gratifying to receive support of many practicing attorneys who express a belief in factual and authentic presentation and
documentation of scientific observations in courtroom testimony by
expert witnesses. We have worked with countless attorneys in developing our program who have endorsed the substance of our paper. One such
attorney was Jerry Ramsey (1987), who concluded:"I prefer to work in a
fashion that boils every case down to a simple, understandable, major
theory of how the case can be won on defense which is consistent with
the expert witness' opinions, with the law, and can be simply stated
for the jury".

The philosophy championed by the authors is finally emerging
in the world of scientific testimony. Noteworthy is an article
appearing in "The Sciences" by E.J. Imwinkelried (1986), professor of
law, in which he discusses witness participation. To further
strengthen our position, we quote directly points applicable to our
premise.

- "Virtually all of the tough restrictions that once governed the use
  of science in the courtroom have been softened in recent years,
  with the result that scientists have gained unprecedented power
  over the outcome of civil and criminal cases. Regrettably, it is
  a power that can be -- and has been -- abused".
- "The scientific witness typically testifies in a deductive fashion,
  applying general rules to the particular facts of the case".
- "With the gradual loosening of rules governing the legitimate bases
  for expert opinions, it is relatively commonplace for witnesses to
  consider evidence that would not be independently admissible".
- "Until quite recently, what was known as the 'ultimate fact' prohi-
  bition barred expert witnesses from trying to answer the questions
  that a trial was being held to resolve".
- "The net effect of these changes in evidence law is that courts now
  accept a wide range of scientific testimony that would have been
  patently inadmissable ten years ago. Experts now rely on hypoth-
  eses, consider information, and voice opinions that formerly would
  have been rejected out of hand".
- "The problem was that judges and lawyers were unfamiliar with
  scientific methodology.... Even a cursory reading of recent rul-
  ings...shows that courts are still largely ignorant of how scientific theories are validated".
- "It is not just by wandering outside their fields that experts
  generate unfounded conclusions. Problems can also arise when they
  couch conclusions about a particular case in general, statistical
  terms".
- "Today science *is* being distorted, and the legal system is suffering for it. Judges, juries, and lawyers too often accept unproved
  theories as gospel. Expert witnesses, unproved theories as gospel. Expert witensses, for their part, are often willing to analyze suspect or fragmentary information -- and to assert conclusions that exceed either their expertise or the proper limits of
  statistical inference".

●"Each liberalization of the evidentiary rules has been an act of
faith in the scientific community -- faith that responsible scientists will critically evaluate new theories, carefully review the
facts of cases in which they testify, and acknowledge the limitations of their techniques.... The new evidentiary rules presume
only that scientists will honor the same conventions in the courtroom that they do in the laboratory and clinic. So far the record
is not impressive".

CONCLUSIONS AND RECOMMENDATIONS

Messages from attorneys and others in the legal profession probe deeply into the basis of scientific concepts and how they
are researched, analyzed, and prepared for presentation. Those of us
in the engineering and scientific disciplines are constantly faced
with misuse and misapplication of codes, regulations, and policies by
attorneys and experts. Be it by naive interpretation or by deliberate
fabrication, it behooves leaders in expertise to correct questionable
practices found to exist in "legal misuse" and to establish standards
of practibility and careful interpretation of existing codes and regulations, followed by accurate guidance for the attorneys for whom you
testify. The most important chore is to educate the attorney - take
the lead and don't allow the attorney to dictate "your opinions" to
you. Recommendations for avoidance of legal misuse include:

● Inform the attorney(s) at the onset that you will only provide
  forensic study if allowed freedom to do a complete analysis and be
  provided with all available data. Never accept an assignment where
  or when the attorney(s) restricts your analysis to a limited
  scope.
● Never accept "assumptions" provided by the client/attorney without
  a separate analysis to determine validity.
● Be sure to keep in perspective the codes/regulations in effect at
  the time of land-use planning, design, and construction. Do not
  allow the attorney to cause you to cite the wrong codes, regulations, or standards of practice. Never agree to overlook or omit
  data related to cause and effect.
● Remember basic fundamentals of the scientific method of analysis -
  think as a scientist/engineer, not as an attorney. Define problem
  on own terms and make inquiries on your own analysis or problem.

It is basically the technical profession's responsibility to
police itself and redirect the legal industry in order that both disciplines do not abuse the powers that have evolved through changes in
evidentiary rules.

REFERENCES CITED

Imwinkelreid, E.J., Esq., 1986, "Science takes the Stand", *in* The
    Sciences, Nov.-Dec., pp. 20-25: New York Academy of Sciences,
    New York
Ramsey, J.A., Esq., 1987, Personal letter communication

Sedimentation Impacts On a Pumpback Storage Facility

By Wayne G. Dorough[1] and Tsong C. Wei,[2] Members ASCE

Abstract

Concern over future sedimentation impacts at a proposed pump storage hydroelectric facility at Lake Francis Case on the Missouri River led to use of a numerical model to analyze lake hydrodynamics. The results from this model along with field data on sediment bed critical shear stress, were then used in a second, more detailed model to design protective dike configurations.

Introduction

A depth-averaged two-dimensional hydrodynamic finite element model, RMA2 (3), was used to simulate Lake Francis Case flow patterns to study impacts of sediment inflow and remedial measures for proposed Gregory County pump storage hydropower project. To facilitate the investigation, a larger scaled lake model was developed to simulate unsteady lake flow and to provide the boundary flow conditions for a more detailed local model at the project. The local model, on the other hand, utilized the boundary conditions and the proposed project capacities to study the magnitude and patten of flow in the vicinity of the proposed pump storage facilities for various dike and levee alignment minimizing the impacts of sediment inflow. The lake model included a fine grid mesh near the project site, in a similar pattern of that used for the local model, so that hydraulic parameters determined from the model could be easily applied to the local model.

Three dike configurations were studied to minimize impacts of sedimentation on the project outlet, including the reciprocal impact of project operation on the lake. These included an L-shaped dike measuring 3000 feet by 7000 feet in length and oriented downstream of the outlet works; a streamlined dike of similar orientation and dimension; and a semicircular, submerged dike, which would surround the outlet works and have a radius of 3000 feet and a crest elevation of 17-1/2 feet below the minimum normal operating pool level. A no-dike condition was also modeled for comparative analysis.

Project Description

Lake Francis Case of Ft. Randall Dam is a 107-mile long, 2-mile wide reservoir located in south central South Dakota and constructed and operated by the U.S. Army Corps of Engineers. The dam is located 920

1. Chief, River & Reservoir Engineering Section, Omaha District, U.S. Army Corps of Engineers, Omaha, NE.
2. Hydraulic Engineer, Missouri River Division, U.S. Army Corps of Engineers, Omaha, NE.

miles upstream of the confluence of the Missouri with the Mississippi
River at St. Louis, Missouri. The outlet works for the Gregory County
facility were to be located upstream of Fort Randall Dam. The lake
extends upstream to Big Bend Dam, about 80 miles south of Pierre, South
Dakota. It has an average annual water inflow of 20,561,000 acre-feet
and an average annual sediment depletion rate of 22,000 acre-feet per
year. The lake pool operation varies about 20 feet seasonally from a
maximum normal pool elevation of 1360 feet msl to a minimum normal
operating pool level of 1337.5 feet msl. Typical release rates for the
Fort Randall and Big Bend Dams vary from about 29,000 cfs to 43,000 cfs
at Fort Randall to 0 to 104,000 cfs at Big Bend.

The only major tributary of any consequence is the White River
(located 30 miles upstream of the proposed project site). It has a mean
annual sediment load of 11.4 million tons and a mean discharge of 560
cubic feet per second. Annual sediment contributions vary from 5 to 40-
million tons, with the sediments composed mostly of fine clay with a
predominance of montmorillonite. Sediment deposits vary in density from
about 8 to 25 lbs/cu ft. The lake headwater delta extends from about
15-miles upstream of the White River to about 30 miles downstream, with
the foreset deposits ending about 10 miles upstream of the Gregory
County project site.

The lake cross-section is U-shaped at the project site and averages
5000 feet in width. Bed deposits average 12 feet in thickness under
existing conditions but are expected to rise another 30 to 50 feet by
Year 2050. The low friction to flow resulting from the clay bed
deposits, cause lake currents to vary from about 0.27 fps to 0.37 fps
for existing geometry, typical release and pool level conditions.
Similar variations for Year 2050 conditions were computed to be 0.74 to
2.22 fps.

Math Model Description

The two-dimensional (horizontal x and y coordinates), dynamic, finite
element model used was developed by the late Dr. Ranjan Ariathurai
(1,2). It is a generalized free surface hydrodynamic model that uses
the depth-averaged momentum equations with an eddy viscosity model for
turbulent diffusion. Bottom stress due to friction at the bed,
Coriolis acceleration, and wind shear at the free surface are accounted
for in the governing equations. These governing equations are solved
numerically by using the finite element method with a Galerkin weighted
residual that is minimized. Quadratic isoparametric triangles and
quadrilaterals are used to discretize the solution domain. Since the
equations are nonlinear, Newton-Raphson iteration is used to converge on
the solution. Time marching is accomplished by two point implicit
differencing.

Field Data Collection

Sediment depositions in Lake Francis Case (4) are too loose to obtain
samples by conventual methods. Therefore, grab samples had to be
collected using a large one-cubic yard scoop and dumped in to horse
feeding troughs, then placed in 55 gallon drums for shipment to the
laboratory for erodibility testing. Samples were collected within a
1/2-mile radius of the project site. Flow velocities from the power

generation and pump back operations were estimated to be small outside this radius. In addition to the grab samples, undisturbed core samples approximately 2-feet long, 1.5 inch diameter were also obtained at 19 locations within a 1-1/2 mile adjacent to the project site. These were used to determine the subsurface profile of sediment density, grain size distribution, and cation exchange capacity (CEC).

Laboratory Testing and Analysis

X-ray diffraction analysis of selected samples revealed predominance of montmorillonite clay in the sediments with lesser amounts of illite, chlorite and kaolinite. Total organic content varied from about 0.5 to 2.0 percent by weight.

Erodibility testing of the sediments were designed to gain as much information as possible on the depositional and erodibility characteristics of the sediments. The test were conducted in a rotating channel, a recirculating straight flume, a tilting flume, and a rotating cylinder. Each test relied on remodeled and redeposited samples including reconstituting the chemical properties of water in Lake Francis Case. The settling velocity of the sediments was measured by placing sediment suspensions in reconstituted lake water in glass cylinders and observing the rate of fall of sediment/water interface.

The critical shear stress for erosion of the surface deposits was found to range from 0.22 to 0.62 Newton/m2. A critical shear stress of 0.22 Newtons/m2 was generated by an average flow velocity of about 0.5 meters/sec in a 10 meter depth of flow and 0.62 Newtons/m2 corresponds to a velocity of about 0.9 meters/sec at the same depth of flow. Samples consolidated to a dry density of 800 gm/l did not erode at a shear stress of 13 Newton/m2 showing that there is a great increase in resistance to scour with increase in density.

Dry densities measured from core samples obtained from the upper 15-inches of the lake bed revealed dry densities in the 300 - 600 gm/l (19 - 37 lbs/cu ft) range. Consolidation test on the samples indicated that densities of these magnitudes can be achieved in a few days following initial deposition. Other test revealed that during periods of large sediment inflow from the White River, the deposited sediments are sometimes laid down at a lower range and is known as "static suspensions". It takes longer to consolidate due to the relatively large thickness that is deposited at once. Redistribution of this material, increased overburden pressure, drying when exposed by lower pool elevations accelerate consolidation.

Lake Hydrodynamics

The lake flow are mainly the result of regulated releases at Big Bend and Fort Randall Dams during power generation. Typical daily releases vary from 0 to 104,000 cfs at Big Bend, and from 29,000 to 43,000 cfs at Fort Randall. The White River provides the only significant tributary inflow with an annual peak flow of about 8000 cfs and a mean discharge of 560 cfs. Thus, it is considered negligible when analyzing local flows through Lake Francis Case. The stage varies about 20 feet over the year, with low pool levels in the winter and high pool levels in the summer.

Since the cross section of the lake at the Gregory County site is U-shaped, averaging about 5000 feet in width and 60 feet deep (at the minimum pool level), flow velocities are small under present day conditions. Even for an assumed steady 100,000 cfs discharge and a minimum normal pool level of 1337 feet, msl, the average velocity did not exceed 0.40 fps.

For future aggraded conditions, however, where the bed is expected to rise another 35 to 50 feet by end of the next 50-years, the velocity for a similar discharge and pool level was found to increase to about 2.0 fps. Such an increase is dramatic and of major importance when considering the erodibility of lake bed sediments and the consequential potential impact to the proposed project.

Flow Simulations to Establish Boundary Conditions

To effectively use the math model in simulating the spatially varied flow conditions, the model had to simulate lake boundary conditions, recognizing the differences in discharges at Big Bend and Fort Randall Dams, including White River inflows and variations in lake pool levels. Two simulations in unsteady flow were made. The first used the existing lake geometry and a starting pool elevation of 1337.5 feet, msl. The second took into account the projected bed elevation by year 2050, again with the same pool level. Each model run began with a steady 45,000 cfs flow through the lake. From this starting point, the releases at Big Bend were assumed to increase stepwise over a period of 45 minutes to a maximum of 104,000 cfs. The model used a finite element, approximately square-mesh grid consisting of three points in the cross (one each near the right and left banks and one in the center channel) and extending over the 150 mile reach from Big Bend fort Randall Dam.

For Year 2050, the flow simulations were based totally on geometrical extrapolations without regard to resuspensions of deposited materials. A bed elevation of 1330 feet, msl (50 feet above the original channel bed) and a pool elevation of 1337.5 feet, msl was chosen to be the equilibrium depth at the project site. This condition produced a velocity of about 0.6 m/sec (2.0 fps), and agreed with earlier determinations that a critical shear velocity range for a depth of 2 to 3 meters of flow (for existing deposits) would be about 0.5 - 0.8 m/sec (1.6 - 2.6 fps).

As to the effect of the proposed pumpback facility on changing lake flow velocity, the simulations revealed that velocities induced during generation and pumpback would be small under existing lake conditions. But for year 2050 the velocities were found to increase another six fold, thus warranting consideration of preventive measures to minimize the effect of sediment on project facilities, as well as the reciprocal impact of the project on the lake environment.

Alternative Dike Configurations

From the point of view of lake hydrodynamics and sediment transport, it was recognized that a dike configuration should (a) cause little or no head loss, (b) minimize resuspension during generation, (c) virtually eleminate resuspension during pumpback, and (d) minimize hydrodynamic disturbances in the lake.

Three dike configurations were considered: the first was an L-shaped dike extending 3000 feet in the lake and 7000 feet parallel to right bank; the second was similar to the first and of similar overall length, except streamlined into a shape best characterized as a one-half of an elongated light bulb; and the third a submerged semicircular dike of radius 3000 feet centered around the exit of the outlet channel. The submerged dike was designed to have a crest elevation 1310 feet, msl (27.5 feet below the minimum normal pool level). Each configuration was designed to prevent the sediment deposition in the immediate vicinity of the confluence of the outlet channel, and relied on the fact that generation of velocities from the pumpback facility are higher than those of the pumpback phase, thus keeping the flow path free of sediment resupensions during pumpback.

The finite element network of the local model was similar to that used in the lake model to establish boundary conditions. The mesh near the dikes was reduced drastically to permit a detailed study of direction and magnitude of localized flow velocities. In each case the mesh was made to conform as near as possible to the curvature and variation of the flow trajectories.

The flow simulations for all three dike configurations revealed that the streamlined dike, when properly configurated, could maintain internal velocities relatively unchanged over the project life. This offered definite flow advantages compared to the L-shaped configuration. The submerged semicircular dike was found to be of little if any advantage over the "no dike" condition once sediments reached the projected year 2050 elevation. The final streamlined configuration called for reversed curvature dike, inclosing an area 5000 feet in length, and varying in width from 1000 feet at the outlet to 550 feet at the throat, then gradually widening out to 800 feet at the dike outlet. The throat is located about 1500 feet from the outlet.

During maximum generation with Big Bend and Fort Randall Dams both undergoing full release, and for the worst pool condition (elevation of 1337.5 feet msl), velocities at the throat were found to be about 2.0 feet per second, sufficient to prevent significant sediment deposition and buildup of the bed within the diked area. Outside the dike, however, it was concluded that a scour hole would most likely develop outward and along the diked side of the lake as the lake bed elevation increased with time. For year 2050, in which the lake bed was projected to reach elevation 1328 feet msl (nearly 40 feet above the bed elevation within the diked area) velocities would be sufficient for the scour-hole to extend outward about 2000 feet. This would be the distance where outgoing velocities would defuse to less than the critical velocity required for erosion (i.e., 1.5 fps).

For the lake area adjacent to the diked area, the presence of the dike reduced the cross-sectional area available for flow in the lake thereby reducing the rate of deposition in the vicinity. It was concluded that material resuspended during generation would be directed downstream, producing less turbidity since less sediment would be resuspended by the slower flow and also because the discharge would not oppose the natural downstream flow of the lake.

Conclusion

The RMA2 numerical model demonstrated that the streamlined dike configuration offered certain advantages with respect to other configurations in minimizing sediment resuspension and turbidity problems. It also offered the added advantage of streamlining movement of flow velocities outside of the diked area and thus aiding in the movement of lake sediments past the project. Resuspension and deposition of sediments was of major concern throughout the study, however, the cohesive nature of the sediments precluded use of a conventual sediment transport model which deals with non-cohesive sediment materials. RMA2, on the other hand, provided a good simulation of lake flow hydrodynamics which could be used in conjunction with information on critical shear stress obtained from laboratory analysis of the sediments test to evaluate the effectiveness of the different dike configurations.

Acknowledgement

Acknowledgement is expressed to Dr. Ranjan C. Ariathuria who died in 1985 shortly upon completion of this study. The authors have summarized Dr. Ariathuria's technical report prepared under contract to the U.S. Army Corps of Engineers, Omaha District.

Appendix.-References

1.   Ariathurai, C. R. (1974), "A Finite Element Model for Sediment Transport in Estuaries," A Ph. D. thesis presented to the University of California, at Davis, Calif.
2. Aruathurai, C. R. and Krone, R. B. (1976), "Finite Element Model for Cohesive Sediment Transport," Journal of the Hydraulics Div. ASCE, Vol. 102, No.HY3, Proc. Paper 11987, pp. 323-338.
3.   Ariathurai, C. R. and Smith, D. J. (1981), "Prediction of Flows, Erosion, and Deposition as a Result of Proposed Hydropower Operations-Gregory County Project," report to Omaha District, Corps of Engineers.
4.   Livesey, R. H. (1955), "Deposition in Fort Randall Reservoir," M.R.D. Memorandum No.5, Missouri River Division, Corps of Engineers.

Optimization of Complex Hydrologic Models Using
Random Search Methods

Larry E. Brazil[1] and Witold F. Krajewski[2]

The primary objective of automated calibration of complex conceptual
hydrologic simulation models is to find the global optimum of a
specified response surface. While direct search techniques such as
gradient or Newton methods may be valuable tools for determining local
optimum points, they present many practical and theoretical difficulties in real applications and often are of limited utility for global
problems. As an alternative, four random search techniques have been
proposed and analyzed in this study. A comparison experiment was
performed on synthetic data using a state-space version of the
Sacramento Soil Moisture Accounting Model. Experiment results are
presented and implementation details are discussed.

Introduction

One important use of conceptual hydrologic simulation models is
hydrologic forecasting. Information obtained from forecasting models
is used as input to decisions concerning items such as water supply,
irrigation, power production, reservoir operation, and navigation.
Most importantly, hydrologic forecast models are used in the
preparation of river forecasts which include the issuance of flood
forecasts and warnings.

Forecast models must be calibrated for the specific area for which
they are to be used. A particular model's accuracy usually is
dependent on the accuracy of the calibration. The calibration process
generally consists of estimating the values for parameters which will
minimize the differences between observed historical streamflows and
streamflow values computed by the model. The actual procedures used
in calibrating hydrologic models vary considerably depending on the
form of the model; however, most calibration strategies include a
combination of manual and automatic fitting techniques.

A variety of automatic parameter identification procedures have been
developed and adapted to an assortment of models. Many of the models
have more than five parameters, making an exhaustive search of the
parameter space infeasible. As a result, most of the parameter
estimation algorithms are based on some type of directed search
procedure which attempts to find the global optimum on the objective
function response surface. These procedures typically have two major
drawbacks: 1) The final result is strongly influenced by the

[1],[2] Research Hydrologists, Hydrologic Research Laboratory, National
Weather Service, Silver Spring, MD. 20910.

parameter starting values. If the values are inaccurate, the search algorithm often converges to an unrealistic local optimum. 2) The search typically is driven by one objective function which may or may not be the best fitting criterion for the model and its application. The main advantages of the automatic techniques, though, are that they are computer rather than labor intensive and often can provide insight into modeling problems that may have been overlooked in manual calibration efforts.

Random search procedures offer a means of overcoming some of the major automatic technique disadvantages. Although they often are more computationally expensive, random procedures are less prone to local optima, since they search within an area rather than along a path determined by a starting point, and do not necessarily need to be driven by a single objective function. Random techniques are becoming particularly more attractive as computer hardware prices decrease and processing time becomes more available.

Application

The National Weather Service (NWS) is responsible for providing hydrologic forecasts for rivers and watersheds throughout the United States. Most of the forecasting is performed with the aid of computer simulation models such as the Sacramento Soil Moisture Accounting Model (Burnash et al., 1973). A major problem faced by the NWS is how to calibrate basins that are added to the forecast network and recalibrate existing modeled basins to reflect changing watershed conditions.

Ongoing research is being performed to improve the techniques used to calibrate NWS models. The purpose of this study was to determine the feasibility of using a random search optimization procedure with an NWS conceptual forecast model. The model used in the study was a modified version of the Sacramento model. The modifications resulted from the transformation of the model from its original FORTRAN algorithm into a set of state-space equations. The version being used in this study is a first order approximation of the integral of the nonlinear model developed by Georgakakos and Bras (1982). A detailed discussion of the model will be published soon.

Random Search Methods

Four random search optimization algorithms were used in this study. A brief description of the algorithms follows.

Algorithm 1

This is a uniform random (UR) search method. Parameter space Ω is searched through a random independent drawing and the best point $\underline{\alpha}_{OPT}$ is selected as

$$\underline{\alpha}_{OPT} = \left\{\underline{\alpha}_{OPT}: \ f(\underline{\alpha}_{OPT}) = \min f(\underline{\alpha}^j), \ j = 1,2,\ldots,K\right\}$$

where $\underline{\alpha}^j = \{\alpha_i^j: \ \alpha_i^j = U(\alpha_i^L, \ \alpha_i^U), \ i = 1,\dots,N\}$. Number of iterations K is predetermined and usually based on economic considerations. U(a,b) denotes uniform distribution on the range with lower bound a and upper bound b. This algorithm computes several objective functions at each iteration. The user selects the objective function(s) of interest after the program ends and determines the optimum point in a multi-objective analysis of the results.

Algorithm 2

This algorithm is based on an adaptive random search (ARS) algorithm described by Pronzato et al. (1984). The method is based on the following considerations: instead of uniformly searching the whole feasible region Ω, we can concentrate on those locations which show a potential for having a global optimum. This potential is based on the preceding search of the whole space Ω. The best point found is suspected of being within the vicinity of the optimum and its proximity is subject to an additional search. Since this can lead to a local optimum, a mechanism is provided to escape from such regions and continue the search in other parts of Ω. If, however, the search returns to the same vicinity a predetermined number of times, the best point in this vicinity is declared the global optimum. The algorithm consists of the following steps.

1. Select the criterion to be minimized $f(\underline{\alpha})$ and the admissible range of the parameters

$$\alpha_i^L \leq \alpha_i \leq \alpha_i^U \qquad \text{for } i = 1,\dots,N$$

$$R_i = \alpha_i^U - \alpha_i^L$$

2. Select the starting point as

$$\alpha_i^0 = \frac{1}{2} (\alpha_i^L - \alpha_i^U) \qquad \text{for } i = 1,\dots,N$$

3. Set MAX, LOC, K, L_{stop}, and k = 1, $L_{opt} = 0$

4. Compute $R_i^{(k)} = 10^{1-k} R_i$ \qquad for i = 1,\dots,N

5. Perform MAX iterations of uniform random search so that

$$\alpha_i^{j+1} = \alpha_j^j + U(\beta_i^L, \ \beta_i^U) \qquad \text{for } i = 1,\dots,N$$

where

$$\beta_i^L = \max\{\alpha_i^L, \ \alpha_i^j - \tfrac{1}{2}R_i^{(k)}\}$$

$$\beta_i^U = \min\{\alpha_i^U, \ \alpha_i^j + \tfrac{1}{2}R_i^{(k)}\}$$

Store the best found point and the corresponding k as $\underline{\alpha}^*(k)$.

6. Set k = k + 1.

    If k > K, go to 7, otherwise MAX = MAX/k, go to 4.

7. Select $\min\{\alpha^*(k): \ k = 1,...,K\}$. Record the "optimal" k as k*.

    If k* = K, then $L_{opt} = L_{opt} + 1$. If $L_{opt} = L_{stop}$, go to 10. If k* ≠ K, set $L_{opt} = 0$.

8. Perform LOC iterations of uniform random search around $\underline{\alpha}^*(k)$ within the neighborhood $\underline{R}^{(k^*)}$ corresponding to the optimal k*.

9. Reset the parameters MAX and k = 1. Go to 4.

10. Stop. The best point is $\underline{\alpha}_{OPT} = \underline{\alpha}^*(k)$.

The values of K, MAX, LOC and L_{stop} suggested by Pronzato et al. (1984) were 5, 100, 100, 5, respectively. We decided to use K = 3 and MAX = 200 for this problem. This decision was based on preliminary runs and economic considerations.

Algorithm 3

Algorithm 3 is the same as algorithm 1 except the Ω is now modified. This modification takes into account some functional relationships that are believed to exist between certain parameters of our model. Parametric relationships were used for three pairs of parameters in this study. For example, a quadratic relationship is believed to exist between the two percolation parameters REXP and ZPERC. This parametric relationship is used to restrict the search space to a band along the curve relating the parameters. The other two restrictions were based on relationships that are assumed between the interflow and baseflow components in two-dimensional space.

Algorithm 4

Algorithm 4 is the same as algorithm 2 with Ω modified as described above. Figure 1 presents the concept of the ARS method for both the algorithms 2 and 4.

Synthetic Data Experiment

In order to evaluate the performance of the above algorithms, a synthetic data simulation experiment was designed and conducted. Seven years of 6-hourly streamflow data were generated using the model described previously, and corresponding 7 years of actual rainfall record from the Bird Creek basin in Oklahoma. The original parameter values were obtained by manual calibration of the basin. The observations of streamflow are taken as error-free data; therefore, the difference between the parameter values used to generate the data (the optimal parameters) and the estimated parameters are due to sampling error and the estimation algorithm only. The resulting hydrograph is characterized by an average annual flow of 0.60 mm of runoff per day with flows ranging from 0.0 to 16.5 mm per 6 hours.

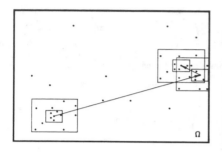 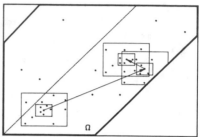

Figure 1. ARS concept representation. Left diagram (algorithm 2) shows search of a two-parameter space within upper and lower parameter bounds. Right diagram (algorithm 4) shows search being further restricted by the parametric relationship.

Results

The mean square error (MSE) was used as the objective function for all four algorithms. Five realizations were performed for each of the algorithms to evaluate bias introduced by the random seed choice. The results of the realization for each method are presented in Table 1. It shows that the best overall run was produced by algorithm 2 and that the mean of the MSE's for the ARS runs was only slightly improved by restricting the search space with parametric relationships. Relatively small improvement, due to parametric relationships among the model parameters, can be explained by the fact that, for this experiment, the implemented relationships only slightly modified the size of Ω space. The ARS algorithm was consistently better than the UR search. It should be pointed out, however, that each of the 20 runs found different solutions -- we did not find the global optimum -- and in the case of the UR search, we cannot even claim that the solutions are local optima.

Conclusions

The main conclusion resulting from the study is that the random search algorithms provide an attractive alternative to other nonrandom search techniques. ARS is more accurate than the UR search and also less expensive (about 3 times). Preliminary runs with real data confirm these findings. The random search methods soon will become a component of the system used to calibrate NWS hydrologic simulation models.

Table 1. Best Run for each Optimization Algorithm

Parameter	Algorithm 1	2	3	4	True Value	Lower Bound	Upper Bound
UZTWM	131.	121.	111.	117.	120.	100.	150.
UZFWM	15.1	15.1	14.8	15.6	15.0	10.	30.
LZTWM	200.	165.	193.	159.	160.	100.	200.
LZFPM	167.	168.	158.	150.	140.	100.	200.
LZFSM	14.	14.	15.	12.	14.	10.	60.
UZK	.280	.303	.382	.276	.300	.2	.4
LZPK	.0144	.0150	.0098	.0140	.0130	.001	.02
LZSK	.138	.102	.167	.130	.126	.02	.2
ZPERC	62.	53.	81.	49.	48.	10.	100.
REXP	2.90	2.70	2.57	2.29	2.10	1.5	4.
PFREE	.017	.053	.073	.037	.020	0.	.1
ADIMP	.135	.171	.175	.188	.170	.1	.2
PCTIM	.015	.000	.000	.000	.001	0.	.05
MSE (mm)	.0141	.0026	.0148	.0031			
Average MSE (mm)	.0158	.0060	.0166	.0058			

References

Burnash, R.J.C., Ferral, R. L., and McGuire, R. A., A Generalized Streamflow Simulation System: Conceptual Modeling for Digital Computers, Joint Federal-State River Forecast Center, Sacramento, CA (1973), 204 pages.

Georgakakos, K. P., and Bras, R. L., "A Precipitation Model and It's Use in Real-Time River Flow Forecasting," MIT TR286, Ralph M. Parsons Laboratory, Cambridge, MA. (1982), 302 pages.

Pronzato, L., Walter, E., Venot, A., and Lebruchec, J. F., "A General-Purpose Global Optimizer: Implementation and Applications," Mathematics and Computers in Simulation, 26 (1984), 412-422.

Automatic Calibration of Conceptual Rainfall-Runoff Models

Garry R. Willgoose[*] and Rafael L. Bras, Member, ASCE[*]

Abstract
An automatic calibration strategy applied to land and channel components of the National Weather Service River Forecast System is outlined. Aspects discussed include the objective function to be used, nonlinear optimization technique used, and reparameterization issues. An example of the use of this strategy on Bird Creek, Oklahoma, is presented and results discussed. A new optimization technique based on linear model approximations is presented which explicitly incorporates parameter interaction effects. The interaction between the complex model structure and the optimization technique is addressed.

I. Introduction
This paper is about the calibration of conceptual rainfall-runoff models. Specifically it is about the automatic calibration, by a computer program of the land (i.e., runoff) and channel components of the National Weather Service River Forecast System. The selected procedure attempts to optimize a given objective function. This objective function is generally a measure of the difference between the observed and estimated flows over a selected period of data.

The model components being considered (hereafter the NWS model) are basically those of the Sacramento Model, a commonly used conceptual rainfall-runoff model. This model is very complex, with between 21 and 27 parameters, depending upon the formulation. It is well known that the NWS model is difficult to calibrate even by experienced hydrologists. Thus the availability of an automatic, objective means of calibration would circumvent many problems.

The soil component of the NWS model is basically divided into two relatively independent components (Figure 1); the upper and lower soil zones. The upper zone models behavior in the top few meters of the soil horizon. These processes include surface runoff, interflow and soil saturation. The lower zone models long term groundwater processes. The two zones are connected by a percolation process. The percolation is driven by a complicated interaction within and between the upper and lower zones. It is commonly viewed as "the heart" of the long term behavior of the model.

The proposed algorithm was tested on data from the Bird Creek near Sperry, Oklahoma. Catchment area is 2344 km^2 and flows ranged from 0.2 m^3/s to 2520 m^3/s.

[*]Ralph M. Parsons Laboratory for Water Resources and Hydrodynamics, Civil Engineering Department, Massachusetts Institute of Technology, Cambridge, MA 02139

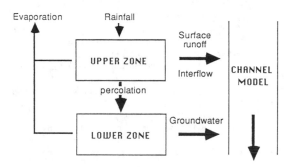

Figure 1   Schematic of the NWS Model

II. Proposed Algorithm

After testing a number of techniques (Willgoose, 1987) some recommendations can be made:

1. The Davidon-Fletcher-Powell procedure should be used for the nonlinear optimization procedure. The optimization process is one of repeated choice of search direction in the parameters space followed by optimization of the objective along this direction. For initial parameters far from the optimum it appears that a new procedure, described later, based on Fourier theory should be adopted.

2. The objective function optimized should be the sum of absolute values of the difference between the observed and simulated discharges. This objective is justified by both practical and theoretical evidence. Theoretically the distribution of residuals observed was shown to comply with the maximum likelihood estimator corresponding to the sum of absolute values. Practically it proved to be the most reliable estimator for our data set. A relatively high incidence of low flows in the flow-duration curve of the data, characteristic of semi-arid climates, was postulated as the reason for this good behavior.

3. The number of parameters to be estimated, at any given stage in the optimization, was reduced by breaking the full parameter set into groups of related parameters and optimizing on these groups separately. Parameter interaction was reduced, while not sacrificing parameter realism, so that the optimization procedure operated more

effectively. The best method of choosing the groupings of parameters is to group together those parameters that are relatively dependent upon each other. While apparently counter intuitive, this grouping strategy avoids situations where two dependent parameters in different groupings, take on unrealistic values that, because they compensate each other, produce reasonable looking simulated flows.

4.   The initial conditions were normalized by the capacities of their reservoirs. This lowered the interaction observed between the initial conditions and state capacities.

5.   The initial conditions were calibrated on the initial part of the data and the other parameters on the rest of the data. It should be noted that the traditional technique of ignoring a model warm up period is unsatisfactory since lower zone initial conditions may propagate up to two years into the future.

6.   At each stage, when a new search direction is found for the following parameter search, the parameter values should be rescaled to be equal to 1.0. The avoids serious round off errors in the search step resulting from the wide range of values the different parameters of the NWS model may take.

7.   Gupta and Sorooshian (1983) suggested a parameterization for the percolation function which is different to the standard one used in the NWS model. It was suggested that interaction of the percolation parameter was reduced. While this technique did reduce interaction within the percolation function, the interactions of the percolation parameters with other parameters in the model were increased. This negated much of the improvement postulated. Despite this, modest improvements were observed.

III. Final Results

As an example of the results obtained by the techniques discussed above, observed and estimated flows for an independent verification period, January to March 1961, are provided in Figure 2. The final parameters obtained are listed in Table 1. This result is compared to that using parameters determined by hydrologists at the National Weather Service by manual methods. The peaks of the automatic parameter set are lower and, for higher flows, in better agreement with reality. The recession of flood hydrographs for the automatic parameters is also in better agreement. For low flows (< 1 m^3/s) the automatic parameters gave lower values than the manual parameters which consistently overestimated these flows.

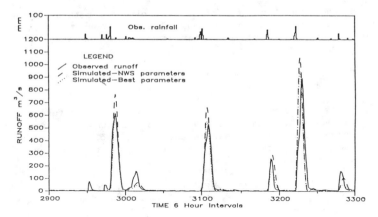

Figure 2   Manual and Automatically Fitted Parameters

	Manual	Best	Fourier		Manual	Best	Fourier
x_1^o	120	129.5	119.9	d_1''	0.126	0.209	0.133
x_2^o	15	27.2	15	d_1'	0.013	0.045	0.051
d_u	0.3	0	0.305	P_f	0.02	0.017	0.093
a_2	0.001	0.001	0.007	μ	3.55	5.41	3.46
a_1	0.17	0.188	0.172	$x_1'(0)$	0.05	0.718	0.904
SARVA	0.001	0.001	0.011	$x_2'(0)$	0	0	0.081
A	0.275	0.217	0.274	$x_3'(0)$	0.875	0.760	0.964
α	2.1	3.32	2.109	$x_4'(0)$	0	0	0.083
x_3^o	160	271.6	160	$x_5'(0)$	0.07	0.071	0.159
x_4^o	14	26.6	14	$x_6'(0)$	221	556	221
x_5^o	140	20.0	139.9				

Table 1   Calibrated Parameters for the NWS Model

These conclusions are general trends over both the calibration and
verification period. A monthly runoff mass balance did not suggest
either of the parameter sets to be superior.

IV.   Parameter Interaction and the Search Procedure

Parameter interaction is a major cause of poor performance in
nonlinear optimization procedures. As already noted, the NWS model

exhibits severe parameter interaction. To circumvent this problem a new search procedure, based on generalized Fourier theory, and which explicitly allows for parameter interaction, was developed. The steps in this procedure are:

1. Define an objective function. Weighted least squares works best.

2. Using the current best estimate of the parameters each parameter is perturbed a small amount and the effect on the simulated times series noted. If two parameters are independent, then their effect on the time series is different and vice versa.

3. Using the objective function and these perturbations, a new set of parameters, that minimize the objective function can be found by the use of Fourier theory. These new parameters, however, are only the best if an unconstrained, linear model is used. Since neither of these conditions are satisfied with the NWS model, then a direction of search is defined such that it passes through the current best estimate of the parameter and the new estimate determined above.

4. This direction is used for an univariate optimization.

5. This procedure is repeated until some termination criteria is satisfied.

The explicit incorporation of the interaction takes place in Step 3 through the use of Fourier theory. It is important not to confuse Fourier theory with Fourier sine and cosine series. Fourier theory provides the justification for the approximation of a vector (in this case the observed data series) by collections of orthogonal vectors (derived here by the Gram-Schmidt algorithm from the perturbations of parameters) where the difference between two vectors is given by an inner product (here weighted least squares). Fourier series, it should be noted, are one example of an application of this theory.

Results of application of this procedure are given in Figure 3 and in Table 1. Though the objective function value at the optimum is not as good as that found by the Davidon-Fletcher-Powell procedure, the greater physical feasibility of the parameters and simulated flows suggest a superiority under some circumstances.

IV. Conclusions

The preceding discussion has outlined a successful strategy for automatic calibration of a complex conceptual rainfall-runoff model. While many problems remain to be resolved progress has been made towards an operational computer code for parameter estimation. Most significant of the remaining problems is which objective function to adopt. This is intrinsically related to the error

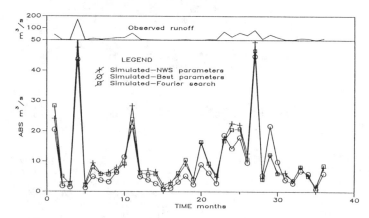

Figure 3   The Fourier Optimization Technique

structure of both the model and the data, which are both difficult
to determine. It is believed the authors' choice of sum of absolute values will be broadly applicable.

The question remains as to how feasible this approach is for
models as complex as the NWS model. One major problem is that
inordinate amounts of computer time are consumed. For parameter
estimation using three years of data 30 to 40 cpu hours on a
MicroVAX II were required. In addition problems of parameter
interaction, as noted above, are quite severe. Under normal flow
conditions many parameters are poorly activated (i.e., non-
observable) making reliable calibration extremely difficult. While
these parameters may be activated in particularly severe storms, it
is questionable whether the simulated flows will be realistic given
the difficulty calibrating them under normal flow conditions. This
leads to a contradiction. Conceptual rainfall-runoff models are
recommended because of their apparent capability to simulate flows
under conditions dramatically different from those that they were
calibrated for. However, the very parameters that govern behavior
in extreme events cannot be reliably calibrated in normal conditions due to interaction and observability problems. It is apparent that more thought needs to be given to the applicability of
complex conceptual models and the limitations that available data
place on their ability to reliably simulate flows.

V. References

Gupta, V. K. and Sorooshian, S.(1983), "Uniqueness and Observabil-
    ity of Conceptual Rainfall-Runoff Model Parameters: The Per-
    colation Process Examined," Water Resources Research, 19(1):
    269-276.
Willgoose, G. R. (1987), "Automatic Calibration Strategies for Con-
    ceptual Rainfall-Runoff Models," thesis submitted in partial
    fulfillment of requirement for S.M., Massachusetts Institute
    of Technology, Cambridge, MA.

A NEW METHOD FOR SYNTHESIZING
A SIX HOUR UNIT HYDROGRAPH

by

William H. Salesky, P.E., M. ASCE (1)

INTRODUCTION:

The unit hydrograph is one of the primary tools used by a
hydrologic engineer in performing a dam safety evaluation. This
single piece of information is used to predict what will happen
as a result of many different hydrologic and meteorologic
events. Obtaining a good unit hydrograph is not always easy
and at times is near impossible. Where stream gages exist,
records can be used to construct an actual unit graph. However,
as is more often the case, no records are available at the
point of interest and thus the engineer must derive a synthetic
graph based on available factors. There are several synthetic
methods available. Many methods involve extensive regional
analyses. Others which are developed using various basin physical parameters are much simpler. However, experience has
shown that simpler methods tend to be conservative with computed peaks two to four times as great as the actual peaks.

In order to provide an accurate and easily obtainable six
hour synthetic unit hydrograph, a new methodology was devised
that would use readily available basin physical parameters for
the determination of unit graphs in undeveloped and ungaged
areas. Rather than using small regional analyses, this analysis
was performed for the entire State of Ohio. Existing unit
hydrographs for gaged basins throughout the State were used to
derive regression equations for various components of the unit
hydrograph. These components consist of the peak flow, the
time of peak and a newly developed component, the width in
hours of the unit hydrograph at 1/2 of the peak flow. The
width component provides two additional points for use in
shaping the unit hydrograph. From the results of the regressions, the high correlations indicated that this method would
produce unit graphs with a high degree of confidence.

The major advantage of this method of synthetic unit hydrograph reproduction lies in its accuracy while still allowing a
relatively quick and easy application. By use of topographic
maps and a few simple equations, accurate unit hydrographs for
use in computing design flood flows can be obtained in a short
time period.

----

(1) Hydraulic Engineer, U.S. Army Engineer District, Pittsburgh
    1000 Liberty Avenue, Pittsburgh, Pennsylvania 15222

ANALYSIS:

The first step in the analysis was to obtain all available
unit hydrograghs for unregulated streams in the State of Ohio.
A limit of 250 square miles was established for two reasons:
records are usually available for larger streams; and normally,
a more sophisticated analysis is required for larger basins.

Certain features of a unit hydrograph must be defined. The
peak flow (Pk) and the time to peak (Tp) are obvious. Additionally, the width in hours of the unit graph at 1/2 the peak flow
(Wpk) was also developed as a feature. An inspection of the
developed unit hydrographs indicated that the points for the
width at 1/2 peak flow are located at a point .4 times this
width before the peak and .6 times this width after the peak
(see Figure 1). Finally, knowing the drainage area and the
associated unit runoff volume along with the three points, the
hydrograph can be drawn and the volume adjusted by trial and
error.

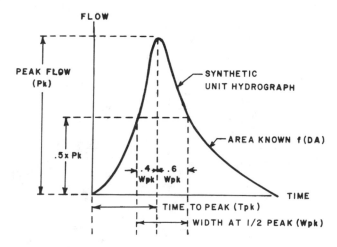

EXAMPLE SYNTHETIC UNIT HYDROGRAPH

FIGURE I

The next step was to determine which basin factors could be
used to develop the equations necessary to define the peak and
half peak points. The variety of physical basin parameters
that affect the runoff are many and varied, but in order to
simplify the procedure, the parameters had to be easily obtainable. Since many of the hydrologic features of State of Ohio
are similar, many factors were not directly addressed because
of the small variances throughout the State.

The factors that were selected included drainage area (DA), main stream length (.7L), slope of the stream (S) and width to length ratio (w/l). The need to use the drainage area is obvious. The length of stream was chosen as the distance from the point of interest to its flowing perennial end times a factor of 0.7. It was felt that this stretch of stream was most instrumental in the formation of the hydrograph. The slope was chosen by definition to be the difference in elevation from the point of interest to the 0.7 point divided by the 0.7 length. This factor has a significant influence on the peak and the time to peak. The width to length ratio is the width of the basin divided by the length of the basin. Note that the length of the basin is not the length of the stream. The width to length ratio is a determining factor in the runoff pattern and has an effect on the peak.

Twenty-one individual locations were used in this analysis. They were selected not only for their various geographic locations, but also for a variety of DA, .7L, S and w/l factors. The range and average of these physical parameters is shown in Table 1.

TABLE 1

RANGE OF PHYSICAL PARAMETERS

Parameter	Units	Low	Average	High
DA	sq. miles	5.44	80.1	251
.7L	miles	2.55	11.86	30.17
S	feet/mile	2.96	12.83	43.11
w/l	-----	.220	.513	1.232

RESULTS:

Values of peak flow, time to peak and width at the 1/2 peak flow were measured from the developed unit hydrographs for the 21 basins. The Hydrologic Engineering Center (U S Army Corps of Engineers) multiple linear regression program was used to determine if any relationship existed between the four physical parameters and the actual unit hydrograph values. Using log transforms, the program computed the regression coefficients, residuals, ratios of observed to calculated along with the statistically significant factors of R square (correlation), R bar square and standard error of estimate.

The final synthetic six hour unit graph equations are as follows:

$$Pk = 10.691 \times (DA)^{1.244} \times (w/l)^{-.229} \times (.7L)^{-.499} \times (S)^{.524} \qquad (1)$$

$$Wpk = 540.213 \times (Pk)^{-.985} \times (DA)^{.972} \qquad (2)$$

$$Tp = 66.229 \times (Pk)^{-.170} \times (.7L)^{.302} \times (S)^{-.424} \qquad (3)$$

Where:

Pk = Peak flow in cubic feet per second

Wpk = Width of unit graph in hours at 1/2 peak flow

Tp = Time to peak in hours from initial runoff

   (NOTE: If calculated Tp < 6 hours but > 4 hours, check
          basin characteristics and if calculated Tp is
          still in the range of 4 to 6 hours, set Tp equal
          to 6 hours for the unit graph plot. If Tp < 4
          hours use some other accepted method.)

DA = Drainage area in square miles

w/l = Ratio of width to length of basin (aligned with predom-
      inant stem)

.7L = Seven-tenths of the length of the predominant stem from
      the point of interest to the end in miles

S = Slope of stream measured from the point of interest to a
    point seven-tenths of the length of the stream in feet
    per mile

These equations were developed for use within the State of Ohio
for basins that are 250 square miles or less and that are within the general range of the other physical parameters. Basins
that exceed the physical parameter range should be analyzed by
other techniques and comparisons made. Additionally, the
effects of extensive urbanization have not been considered in
this analysis. The equations should be applicable to most Ohio
streams, but their application still requires good sound engineering judgment.

Analyses of these equations and their regression coefficients
indicate that they are statistically significant and make sense
hydrologically. For the peak analysis (eq. 1), a correlation (R
square) of .9803 was achieved between the calculated and actual
peaks. The average percentage difference of the calculated versus the observed peak was 9.0 percent. As would be expected,
the exponents for DA and S are positive causing the peak to
increase with an increase in these values. In the case of 0.7L
and w/l, the negative coefficients indicate that as the basin
is elongated, the runoff is spread out and the peak is
decreased.

For the width of the unit graph in hours at 1/2 peak flow

(eq. 2), the regression showed that using only physical characteristics would not produce adequate results. Therefore, the peak flow along with the DA were used with a resultant correlation of .9457. The average difference between the calculated and actual values was 1.4 hours as compared to the average width of 1/2 peak flow of 16.6 hours. With essentially a direct relationship to DA, it follows that the larger the DA, the greater the volume of runoff and hence a larger and wider hydrograph. The inverse relationship to the peak flow is reasonable since for higher peaks with a fixed volume, the unit graphs would necessarily be narrower at 1/2 peak.

The final equation (eq. 3), time to peak, also required the use of the peak flow to produce good results. Using the additional parameters of .7L and S, the regression achieved a correlation of .8945 . The average difference between the observed and the calculated peak time was 1.7 hours. Negative exponents for the Pk and S, indicating an inverse relationship, and a positive relationship for .7L are consistent with normal unit hydrographs. As the slope increases, the time to peak is earlier and as the length of stream increases, the time to peak is later. And for a fixed volume of runoff, the time to peak becomes earlier with an increasing peak flow.

CONCLUSIONS:

Based on the correlations and average deviations from actual data, the presented equations provide a good basis for developing unit hydrographs. An additional effort was made to verify these equations. One of the verifications involved a double peaked unit hydrograph that was not used in the equation development. If the equations could accurately represent this unit graph, it would further substantiate the validity of the equations. The stream used was Yellow Creek, a tributary to the Ohio River. The parameters were developed and the equations were used to synthetically produce a unit hydrograph. It was then compared to the unit graph developed from actual records as shown in Figure 2. By comparison, the calculated peak is within 11.4 percent of the actual, the time for the width at 1/2 peak flow is within 10.0 percent and the calculated time to peak falls between the two actual peaks.

From the results of the regressions and verifications, the correlations and excellent reproductions indicate that synthetic development of unit hydrographs can be performed with a high degree of confidence. All components produced by the equations were indicative of what should be expected for most hydrographs. With proper application, usable and realistic six hour unit hydrographs can be obtained for analyzing ungaged basins.

The major advantage of this unit hydrograph synthetic reproduction is its accuracy combined with relative quickness and ease of application. The use of the heretofore undeveloped width at 1/2 peak flow provides a well defined hydrograph that can be used with confidence.

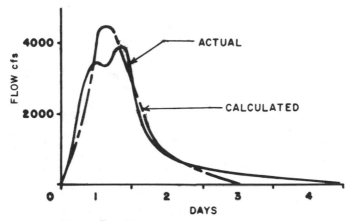

YELLOW CREEK NEAR HAMMONDSVILLE
ACTUAL & SYNTHETIC UNIT HYDROGAPHS

FIGURE 2

APPENDIX A

CONVERSION FACTORS, U. S. CUSTOMARY TO METRIC (SI)
UNITS OF MEASUREMENT

U. S. customary units of measurement used in this paper can be
converted to metric (SI) units as follows:

Multiply	By	To Obtain
square miles	2.589988	square kilometers
miles (U. S. statute)	1.609344	kilometers
feet	0.3048	meters
cubic feet per second	0.02831685	cubic meters per second

Time Base of a Hydrograph

Nazeer Ahmed[*] M. ASCE

Abstract

For design projects in water resources, a hydrograph is commonly used to determine the required flood volumes. The time base of a hydrograph is a crucial parameter in such studies, and its speedy and accurate determination is essential for both economic and hydrologic reasons. Based on the actual measurements of a large number of hydrographs, a technique is presented whereby the length of the time base can be computed expeditiously. The only variable required for this computation is the drainage area easily available from published sources. The technique developed is applicable for drainage areas from 1 mi^2 (or 1 km^2) to 2 000 mi^2 (or 2 500 km^2), and larger.

Introduction

The time base of a hydrograph is directly dependent on the storm as well as basin properties (1, 2, 3, 4). The physiographic characteristics of a basin such as size, shape, slope, drainage pattern, topography, and geology essentially remain constant, and their effects on the time base are accurately predictable. Similarly, the storm parameters consisting of duration, rainfall intensity, runoff volume, as well as, temporal and spatial distributions of intensity and runoff have definite influence on the time base and can be determined precisely.

The concept of ground water recession curve is the main ingredient that is normally used in different techniques for the division of total flow into direct runoff and base flow. Such a process usually provides a time base which is too long. Moreover, the separation of direct runoff and base flow by this method is cumbersome and time consuming. It is therefore desirable to devise a technique which is convenient to apply and would provide a satisfactory answer for the length of the time base.

Constancy of Time Base

The time base of a hydrograph is made up of two parts: one from the center of mass of rainfall excess to the peak flow, the other from the peak flow until the end of direct runoff. The first part is dependent upon duration, intensity, runoff, temporal and spatial

* Assistant Professor, Department of Civil Engineering, University of Louisville, Louisville, Kentucky, 40292

distributions of intensity and runoff, antecedent conditions, and time of concentration. The last factor indicates that watershed conditions have minor effects on the first part of the time base. The second part is heavily dependent on the ground conditions of the basin, and slightly dependent on temporal and spatial distributions of rainfall intensity and runoff volume.

According to the theory of unit hydrograph, rainfall excess of uniform intensity produces runoff of varying volumes such that the hydrograph time base essentially remains constant for all rainfall intensities.

From a practical stand point, the method of separation for direct runoff and base flow should be such that the length of time base remains relatively constant from storm to storm. In other words, it means that there should be a characteristic property pertaining to either the drainage basin or the storm that can provide a constant measure of the time base.

Since the second part of time base is primarily dependent upon the ground conditions, the drainage area is certainly an appropriate as well as a constant parameter that can be used to determine the time base of the hydrograph in some definite manner.

Variation of Base Flow

The total runoff represented by a hydrograph is the sum of direct runoff and the base flow. Therefore, it is necessary to determine independently the base flow distribution throughout the runoff cycle for separation purposes. In general, base flow obtained from a given drainage basin is governed by a large number of factors, and the determination of their individual effects is rather a formidable job. However, a description of their integrated influences can be appreciated through the following qualitative analysis.

Three possible paths followed by the base flow in a runoff cycle are shown in Fig. 1. The path designated by curve I represents the upper extreme case and indicates that base flow in the beginning increases in response to the increase in direct runoff, and then decreases as the runoff is reduced after the peak. Finally, it merges with the recession limb asymptotically at the point B where direct runoff is assumed to cease completely. Theoretically, maximum base flow should occur below the peak point, but usually it is delayed somewhat due to the time lag involved in ground water flow.

The flow path designated by curve II is the lower extreme case, where base flow decreases as the direct runoff increases. The reverse is true after the peak, i.e., base flow increases while the direct runoff diminishes. Again, minimum base flow occurs after the peak flow due to the time lag, and the base flow curve merges with the falling limb asymptotically at the point D where direct runoff is thought to be stopped. In this case however an exception of flow reversal from the stream to the banks should be noted: it is represented by the crosshatched area and is commonly known as the negative base flow or bank storage.

In general the base flow curve would be situated between these two extremes and usually above the horizontal axis. Curve III represents such a situation of base flow. The horizontal distance between the point of rise A of the hydrograph and the point of merger of base flow

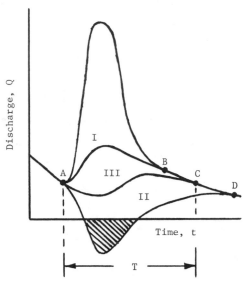

Fig. 1   Three possible divisions of total
         flow hydrograph into direct runoff
         and base flow.

with the falling limb, such as point C, is the required time base T.
The locus of the base flow curve would always be variable dependent
upon the storm and basin properties and the time lag involved.

Cessation of Direct Runoff

       To accurately determine the volume of direct runoff, it is
necessary to determine the precise plot of the base flow curve. The
exact location of the merger point, where direct runoff stops, is
determined by a plot of Q_0 versus Q_1 as shown in Fig. 2. Such plots
were prepared for all the sampled hydrgraphs used in the study. The
discharge Q_0 is considered at any time whereas Q_1 is the discharge a
unit time later depending upon the time scale used. The break in slope
at the point C represents the location of the merger point on the
hydrograph as shown in Fig. 3. The slope of the line OC provides the
value of the recession constant K.

Actual Volume of Direct Runoff

       Based upon the value of the constant K, the recession limb is
extended backward from the merger point C up to peak flow as shown in
Fig. 3. An arbitrary line is sketched in starting from the point of
rise A and joining it asymptotically to the extended recession curve
to accurately portray the storm properties, antecedent conditions, and

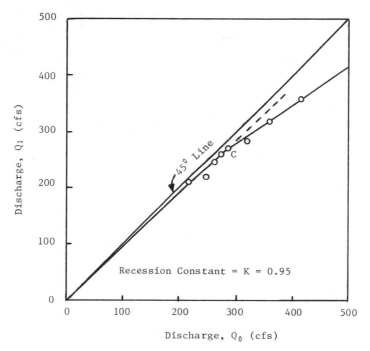

Fig. 2   Graphical method for the determination of the
         recession constant K.

         (U.S.G.S. Water Resources Data – Maryland and
         Delaware, Water Year 1985, April–May 1985,
         North Branch Potomac River at Kitzmiller, Maryland)

basin characteristics. The base flow curve thus developed represents
the division of total flow into base flow and direct runoff. The area
between the base flow curve and the hydrograph within the limits of
points A and C is planimetered and recognized as the volume of direct
runoff. By following this procedure, the volume of direct runoff for
each sampled hydrograph was computed. This is known as the actual
volume of direct runoff.

Empirical Volume of Direct Runoff

    It is a well known fact that the determination of the locus of
base flow is not only a complex process due to a large number of

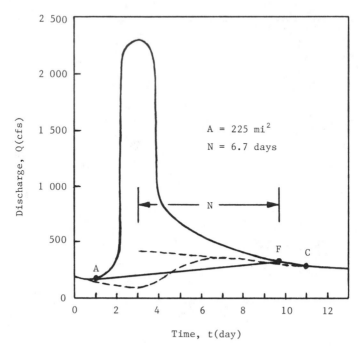

Fig. 3   Division of total flow hydrograph into base flow
         and direct runoff: (i) by the conventional
         technique, (ii) by the straight line technique
         of the present study.

         (U.S.G.S. Water Resources Data - Maryland and
         Delaware, Water Year 1985, April-May 1985, North
         Branch Potomac River at Kitzmiller, Maryland)

variables involved but also cumbersome and time consuming. To overcome
this type of difficulty, the area A of the drainage basin was chosen
as the crucial parameter and related to the number of days N after the
peak by the following formula as

$$N = A^n \qquad\qquad\qquad\qquad (1)$$

where n is an exponent. Eq. 1 was tested for various values of n to
yield a volume equal to the actual volume of direct runoff as explained

in the last section. The following technique was used to extract the empirical volume of direct runoff. The value of N yielded by Eq. 1 for a given value of the exponent n and the area A of the drainage basin was employed to draw a straight line starting from the point of rise A and continuing N days after peak intersecting the recession limb of the hydrograph at the point F as shown in Fig. 3. The area enclosed between the straight line and the hydrograph within the limits of the points A and F was planimetered and recognized as the empirical volume of the direct runoff. The values of the exponent n = 0.35 for areas in square miles and n = 0.40 for areas in square kilometers provided the empirical volume of direct runoff as an accurate estimate of the actual volume of direct runoff. This conclusion was arrived at by analyzing a large sample of hydrographs chosen from a variety of hydrologic areas from the 48 contiguous states. The hydrographs were selected from both regulated and unregulated flows.

Conclusions

It is shown that the time base of a hydrograph is composed of two parts. The first part can be computed from the hydrograph plot as the time interval between the point of rise and the peak flow; the second part can be determined by using the formula given by Eq. 1. The sum of these two parts is equal to the total time interval of the hydrograph time base. It is used in the form of a straight line as the locus of base flow curve to determine the empirical volume of direct runoff. The values of the exponent n = 0.35 and n = 0.40 are used in Eq. 1 for areas in square miles and square kilometers, respectively. Also, it is shown that the empirical volume of direct runoff computed by the technique detailed in this paper is equal to the actual volume of direct runoff determined through conventional means. Moreover, application of the present technique is analytical in nature and provides savings in time and results in economic benefits.

Acknowledgement

The author wishes to acknowledge and thank Suruyya Ahmed for her assistance in the preparation of this paper.

Appendix: References

1. Chow, V.T., Editor-in-Chief, Handbook of Hydrology, McGraw-Hill Book Co., New York, 1964.
2. Fetter, C.W., Jr., Applied Hydrogeology, Charles E. Merrill Publishing Co., Columbus, Ohio, 1980.
3. Linsley, R.K., Jr., M.A. Kohler, and J.L.H. Paulhus, Hydrology for Engineers, 3rd Ed., McGraw-Hill Book Co., New York, 1982.
4. Wilson, E.M., Engineering Hydrology, 2nd Ed., John Wiley & Sons, New York, 1974.

Fault-Tolerant Design
for
DATA ACQUISITION AND FLOOD FORECAST SYSTEMS

by David C. Curtis*

INTRODUCTION

Standing shoulder to shoulder in northern New Jersey, 132 communities face the constant threat of flooding from the Passaic River and its tributaries. More than two million people are affected. A recurrence of one turn-of-the-century flood would now reach 22,500 buildings and leave damages totaling more than $1.5 billion.

Federally sponsored flood control measures were proposed many times for the Passaic basin dating back to 1936. Local political opposition and economics prevented the construction of detention reservoirs, diversions, channel modifications, tunnels, and a variety of other local protection works. Then in 1976, congress mandated consideration of a full range of non-structural alternatives to alleviate flood risk in the Passaic. Finally, a 1984 report by the U.S. Army Corps of Engineers recommended enhancements to existing local and regional flood warning systems to help mitigate flood losses. Army engineers met with flood warning experts from the National Weather Service, and other interested agencies including the New Jersey Department of Environmental Protection, the New Jersey State Police and the U.S. Geological Survey to consider the recommended flood warning enhancements. A request-for-proposals was issued by the Army's New York District in July 1986 and in April 1987 a contract was awarded to Sierra/Misco, a Berkeley, CA firm, to implement the Passaic River Basin Flood Warning System.

Flood warnings in the Passaic are complex. Warning activities in three states, ten counties, and 132 communities must be coordinated along with four federal agencies, and at least two state agencies. The government design team recognized that reliable communication was essential to the success of the flood warning system. Several specifications imposed on the communications system by the government to promote reliability were extended by Sierra/Misco to create an innovative design that is tolerant of faults in critical communication links. The result is the most advanced local flood warning system design to date.

SYSTEM REQUIREMENTS

Thirty-one raingauges, including 23 new event-reporting gages and the automation of eight existing Fisher-Porter gages, were specified for the Passaic Basin. Data from the raingauge network were required at ten computerized receiving sites in three states:

* President, International Hydrological Services,
  1900 Point West Way, Suite 161, Sacramento, CA 95815.

1)	Bergen County Police Dept	NJ
2)	Essex County Police Dept	NJ
3)	Morris County Police Dept	NJ
4)	Passaic County Police Dept	NJ
5)	State Police HQ	Trenton, NJ
6)	NJ Dept of Environ. Protection	Trenton, NJ
7)	NWS Office	Newark, NJ
8)	NWS Forecast Office	New York, NJ
9)	NWS Forecast Office	Philadelphia, PA
10)	NWS River Forecast Center	Harrisburg, PA

The four county sites are critical 'need-to-know' sites within the Passaic Basin. The two sites in Trenton support state level emergency operations and the four sites in NWS offices support a variety of National Weather Service warning responsibilities in Passaic Basin.

In addition to acquiring data, the ten computerized receiving sites were required to pass text messages between computers. Messages could be National Weather Service forecasts, watches, and warnings, or simply administrative information passed from one site to another.

For reliability and rapid response, the government design team specified a VHF radio-based event reporting system for data acquisition. VHF systems have a proven track record of operation during severe weather and event reporting systems respond immediately to changing field conditions.

(An event reporting rain gage, for example, transmits a new data message immediately after receiving an additional increment--say 1mm-- of rainfall. The harder it rains, the faster each increment accumulates, and the more frequent the gage transmits data. The reporting rate of the network is directly proportional to the intensity of the storm. The highest intensity parts of the storm--those most likely to produce flash flooding--are exactly identified.)

The government design-team was especially concerned about over all system reliability. With $72 million dollars in annual damages and the potential for loss-of-life, system reliability was an obvious concern. The design team looked for opportunities to eliminate potential weak links in the communication systems. Recognizing that radio repeaters used in data acquisition created communication 'bottlenecks', a novel, dual data path feature was added to the system specifications. Data from each remote raingauge was required to have two alternate pathways to its target base station. (See Figure 1)
One pathway would be used routinely. Should the primary pathway fail (e.g. repeater failure due to lightning), the system operator could easily switch the radio receiver to the alternate data path frequency. When the primary data pathway returned to normal, the system operator could switch the radio receiver back to the primary frequency.

The remaining communication system design (inter-computer communications between each of the ten base stations) was left to the potential vendors to suggest this most appropriate design.

FIGURE 1

Dual Data Path

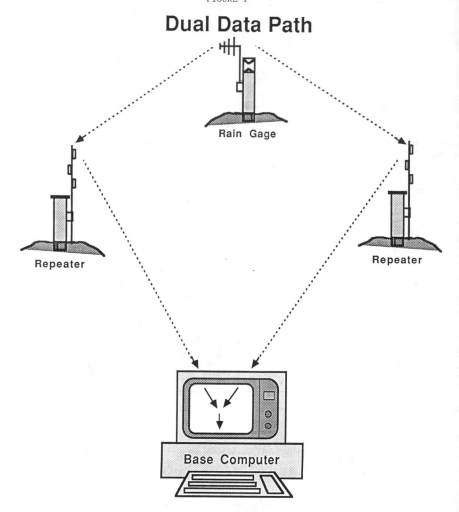

COMMUNICATIONS SYSTEM DESIGN

The final system design (Figure 2) reflected the three different communications problems presented by the Passaic system (i.e. data acquisition, communication in and near the Passaic basin, and regional communications to state and federal offices). First, the data acquisition component used VHF radio and the dual data path concept as specified by the government design team. For the second component, 2400 baud, multi drop, VHF radio links were chosen for communicating between computers (IBM PC AT compatible) in the Passaic area. The computer sites in the Passaic basin area are relatively close to one another making VHF a logical choice for economy and reliability.

The third communications system represented a more difficult problem. Communications between computers in the Passaic basin and regional centers at Trenton, Philadelphia, and Harrisburg cover several hundred miles. VHF and microwave systems required numerous repeater sites in series to complete the necessary communication pathways. The additional repeaters increase costs and raise system reliability questions since any single failure in the repeater series would cause the entire link to fail.

Telephone communications were considered but rejected as the regional communications component. Telephone communications were considered to be too vulnerable during severe weather - just the time they're needed most.

To economically cover the large distances required of the regional system, satellite communications were chosen. A single commercial geostationary communications satellite will be used to 'bounce' data and messages to offices outside the Passaic area. Alternate satellites as well as alternate transponders are available to ensure continued communications should the primary channel fail.

FAULT TOLERANT DESIGN

The combination of technologies used in the Passaic communications design enhances reliability by tolerating significant subsystem failures or faults. Major communications systems were isolated from one another, permitting operation of key elements despite the failure of a major component. For example, all rain gage data will be transmitted in real-time to two base station computers (Newark and Essex) by the data acquisition system using dual data paths. (See Figure 2) Should one repeater fail, a second repeater is available to takeover. Should one of the base stations fail, the data will still be picked up by the other. In either case, all the data gets to at least one base station in the Passaic basin in real-time.

(Real-time data will be collected at all sites within the Passaic basin but only Essex and Newark will collect data from all rain gages through the VHF data acquisition system. However, all computer base stations will receive data summaries every 15 minutes from the other computers to complete their rain data base.)

FIGURE 2

Passaic System

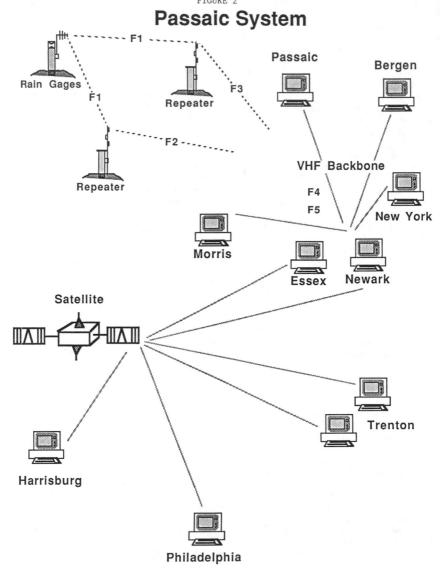

The VHF inter-computer communications with-in the Passaic area uses a VHF circuit that is separate and isolated from the data acquisition system. If any portion of the VHF inter-computer or backbone communications link fails, data reception by Newark and Essex remains unaffected. Should excessive rainfall amounts occur while the VHF backbone communications link is down, the Newark and Essex offices can initiate warnings by conventional means using existing voice radio and/or telephones.

A failure of the satellite would only affect the regional communications portion of the system. The data acquisition and VHF backbone communications in the Passaic area would continue uninterrupted. Communities most directly affected by potential flooding would still get timely data as well as on-line support from NWS offices in Newark and New York City.

Just as the VHF data acquisition system interfaces with two base stations, the satellite system is designed to interface with the Newark and Essex base stations. One base station (most likely Newark) will be the primary interface between the satellite and the VHF systems. However, should the primary satellite interface fail, the alternate interface is available to go on-line immediately.

Perhaps the most interesting fault-tolerant feature incorporated into the Passaic design by Sierra/Misco is an extension of the dual path concept for data acquisition. The government design team required the second data path to enable a base station operator to switch to an alternate reception path should the primary path fail. However, manual intervention by the operator is required to detect the path failure and to switch reception paths. Valuable data and time could be lost before the operator realized that the failure occurred.

The improved design calls for simultaneous collection of data from each data path. (See Figure 3) This means that for each data message transmitted by a remote rain gage, two data messages arrive at the base station computer. (The original data message is received by two different repeaters and retransmitted on separate radio frequencies.) The computer software receives the two data messages on separate communication ports, determines that the two messages are identical, and files only one, disregarding the second message as redundant.

By keeping both data paths on-line, the data acquisition system becomes an automatic fault-tolerant system. Should one of the repeaters fail, the alternate path is still on-line and single message data acquisition begins without operator intervention. When the failed repeater returns on-line, dual message reception resumes automatically without operator intervention.

To alert the operator that one of the data paths has failed, the software monitors the data reporting frequency of each repeater. If no messages are received from a repeater within an operator specified period of time, the computer sounds an alarm and notifies the operator that the repeater did not report as expected. The operator simply notifies a repair technician to fix the problem.

FIGURE 3

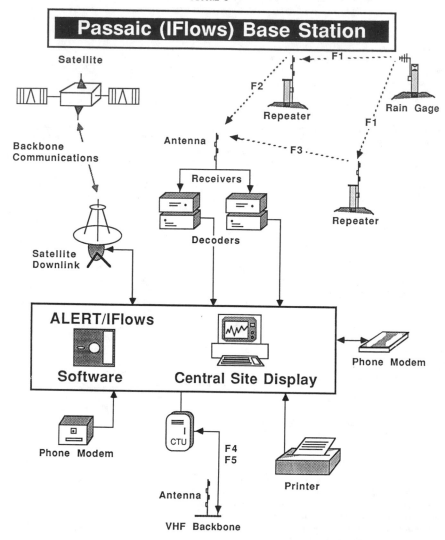

Passaic (IFlows) Base Station

IMPROVED DATA ACQUISITION

As data speeds on its way from a remote rain gage to the base station computer, many opportunities are encountered that can corrupt the data message. Atmospheric conditions, radio interference, and vegetation can disrupt the flow of data. The dual data path concept increases data acquisition performance since each data message has two opportunities to reach the base station computer safely.

For each data message sent from a remote rain gage, four different results can occur at the base station. First, both messages could arrive successfully. Second, both messages could fail. And, in two cases, one message could arrive successfully and one could fail. In three out of the four possible combinations, at least one message arrives successfully. Since only one message is ultimately filed in the data base no matter how many are successfully received, three out of the four combinations are a success from a data acquisition point of view. Table 1 summarizes the possible combinations.

Table 1. Possible Data Message Combinations.

Data Received		Net Result
Path 1	Path 2	
Success	Success	Success
Success	Failure	Success
Failure	Success	Success
Failure	Failure	Failure

To measure quantitatively how the dual data path improves acquisition performance, assume that the probability of success for each data path alone is 90%. This means that for a single path a data message has a 90% chance of reaching the base station safely and 10% chance of failure. Table 2 shows how the chance for success increases with the dual path approach.

Table 2. Probabilities of Message Combinations.

Data Received		Net Result	
Path 1	Path 2		
Success (0.90)	x Success (0.90)	= Success (0.81)	Success => 99%
Success (0.90)	x Failure (0.10)	= Success (0.09)	
Failure (0.10)	x Success (0.90)	= Success (0.09)	
Failure (0.10)	x Failure (0.10)	= Failure (0.01)	Failure => 1%

The dual data path raises the level of data acquisition to 99% as opposed to the 90% for a single path opportunity. In fact, to achieve a 90% probability of success for the dual path, only a 68% chance of success is required of each single path.

SUMMARY

An innovative fault tolerant flood warning system is currently being installed in the Passaic River Basin. The new system is designed to automatically tolerate the failure of major components without disrupting key elements of the flood warning system.

In addition, the dual data path concept used in the fault tolerant design significantly improves overall data acquisition performance. More data and more reliable data acquisition will support improved flood assessment.

The latest in available VHF and satellite communications technologies were combined with state-of-the-art computers and software in a system designed to reliably mitigate potential flood damages. When completed in late 1987, the Passaic River Basin Flood Warning System will be the most advanced local flood warning system in the world.

The Passaic Basin suffers an almost unbelievable $72 million in flood damages annually. Saving just one percent of the damages will more than pay for the new warning system in the first year of operation.

Modified CIMIS Irrigation Scheduling Program

by Ahmed N. Alajaji and Otto J. Helweg[1], F.ASCE

Abstract: CIMIS, California Irrigation Management Information System, is a program that collects weather data around the state to assist irrigators in planning and scheduling their irrigations. It also consists of a computer program that gives irrigation scheduling utilizing historic weather data averages. The program was not developed to be a real time irrigation scheduling program but was easily modified to include this capability. The main improvement reported on this paper is adding specific scheduling capabilities for center pivots.

Introduction

Real-time irrigation scheduling is fairly standard in the semi-arid west and south of the United States, especially for larger agribusinesses. A number of private consultants have proprietary programs and provide irrigation scheduling services for a fee.

All of the necessary techniques and data for real-time irrigation scheduling are easily available and the interested irrigator can do his own if he has the time and interest. This is not to imply that using a consultant might not be cost effective, but only that there are alternatives.

Real-time irrigation scheduling programs normally keep track of soil moisture so the irrigator then decides at which level of soil moisture depletion he desires to irrigate. Soil moisture depletion may be monitored by direct sampling of soil moisture or by calculating the amount of water evapotranspired (ET) from the soil using actual ET values. The CIMIS program utilizes the later method.

Any of a number of ET calculating methods may be used, but the most popular is the modified Penman equation (Doorenbos and Pruitt, 1984). Though this method has mostly been used for weekly or longer time frames, it has more recently been applied to daily ET estimates. The main requirements are finding "k" values for local conditions and various crops. The modified Penman equation can be written:

$$ET_c = k\ ET_o \qquad (1)$$

in which ET_c is the actual crop evaportanspiration, k is a correction coefficient, and ET_o is a reference ET (usually grass or alfalfa).

---

[1] Research Engineer and Professor, Water Studies Center, King Faisal University, PO Box 380, Al-Ahsa 31982, Saudi Arabia

ET_O can be found by:

$$ET_O = c[W\ Rn + (1-W)\ f(u)\ (ea - ed)] \qquad (2)$$

in which c is an adjustment factor for day and night weather conditions, W is a temperature related weighting factor, Rn is the net radiation in equivalent evaporation (mm/day), f(u) is a wind related function, ea is the saturated vapor pressure at mean air temperature in mbar, and ed is the mean actual vapor pressure in mbar.

CIMIS Model

The inputs to the CIMIS program may be classified as fixed site specific parameters, seasonal parameters, variable input.

The fixed site parameters are the soil characteristics

The seasonal parameters are:
1. crop type for k value
2. type of irrigation system
3. zones of root depth
4. desired minimum available soil moisture

The variable inputs (changes during the season) are:
1. correcting soil moisture content
2. changing well & pump efficiency
3. changing irrigation efficiency

Modified CIMIS

Because most of the agribusiness interested in irrigation in Saudi Arabia utilize center pivots, there was a need for real time irrigation scheduling for these systems. Though the CIMIS program covers a number of irrigation systems, including sprinkler, it did not specifically include center pivot systems.

The items of interest to a center pivot operator are: 1, when to irrigation; and 2, at what speed to run the pivot. Recall that the speed of the pivot determines the depth of irrigation. There are other items of information the center pivot operator should know such as:

1. when run off will occur
2. whether the time required for one irrigation is greater than the interval between irrigations
3. when to irrigate in order to keep the same pivot speed.

This latter item is important because many center pivots apply fertilizer and pesticides through the pivot so if the pivot speed is changed, the applicators need to be readjusted; consequently, they prefer to change the irrigation interval rather than the pivot speed.

The Modified CIMIS then, adds a subroutine which relates depth of applied water to pivot speed, time for one rotation at any speed, and maximum application rate to avoid runoff. During the season, the operator has the option of exceeding field capacity, going beyond the pre determined soil moisture minimum, allowing runoff, etc.

Any precipitation event may be input during the growing season and its effect will be incorporated into the soil water balance subroutine.

Fig. 1 shows a flow diagram of the modified CIMIS program. The first question is whether the program will be used for real time scheduling. The second question asks if it is the first run; if so, certain initialization subroutines are run. After that, weather data is entered. This may be done by hand or directly from a remote sensing unit. The field of interest is selected, the soil characteristics, other parameters (such as pump efficiency, etc.) and after that the irrigation schedule is given. If another field is to be examined, the program returns to the weather subroutine, if not, it stops.

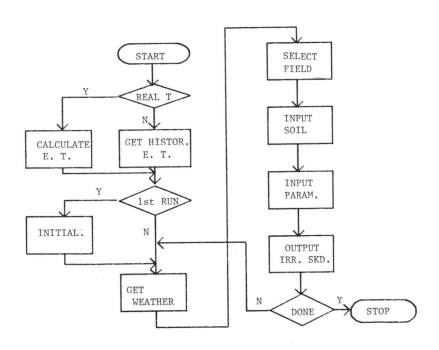

FIG. 1 - Flow Chart of the Modified CIMIS Program

The program is completely menu driven and fairly user friendly; however, it is not simple and does require some learning. Fig. 2 shows the main menu from which the major calculations are chosen. Fig. 3 shows the scheduling menu and Fig. 4 shows the real time irrigation scheduling output.

MODIFIED CALIFORNIA IRRIGATION MANAGEMENT INFORMATION SYSTEM

MAIN MENU

1   Develop or change a schedule

2   View or copy stored field data

3   View or copy a stored schedule

4   View or adjust crop coefficients

5   Input historical average ETo data

6   Calculate or input real time ETo

7   EXIT from program

Select

FIG. 2 - Modified CIMIS Main Menu

Results and Conclusions

The modified CIMIS was tested at the Hail Agricultural Development Company, HADCO, during the growing season of 1986-87. Two 50 ha center pivots were scheduled with the modified CIMIS and two adjacent pivots were scheduled by the normal method (usually a subjective determination). Though the harvest was not completed at the time this paper was submitted, it appeared that the CIMIS controlled pivots were doing better than the control ones.

Yield and applied water were calculated at the of the season to evaluate the effectiveness of the CIMIS program.

SCHEDULING MENU

Input or	....	1	Enter a field number —— 000
change	.	2	Select a crop ————
field or	.	3	Describe irrigation system
crop	.	4	Determine yield threshold depletion
data	....	5	Set development dates

| Run | .... | 6 | Generate a normal schedule |

| EXIT | .... | 7 | Return to Main Menu |

Select

FIG. 3 - Modified CIMIS Scheduling Menu

W S C   I R R I G A T I O N   S C H E D U L E

Date   Apr   14, 1987   Field 100   Crop WHEAT

Application Number	Date	Net Application inches	Gross Application inches	Center Pivot Speed (%)	Effective Irrigation (%)
		***** Pump discharge = 65 (L/S) *****			
1	Jan 29	1.08	1.35	25	80 *
2	Feb 08	1.16	1.45	23	80 *
3	Feb 18	1.33	1.66	20	80 *
RAIN	Feb 28	1.00			
		***** Pump discharge = 60 (L/S) *****			
4	Mar 10	0.39	0.52	59	75
5	Mar 20	2.05	2.73	11	75 *

TOTALS   6.01   7.72
FINAL WATER CONTENT - Desired 4.00 inches Actual -11.4 inches

Not possible to irrigate this day; last irrigation not yet over
* Surface runoff will occur at the outer part of the circle

FIG. 4 - Modified CIMIS Real Time Irrigation Schedule

Appendix I — References

1. Doorenbos, J. and W. O. Pruitt, **Guidelines for Predicting Crop Water Requirements,** FAO Irrigation and Drainage Paper 24, 1984

2. Snyder, R., D. W. Henderson, W. O Pruitt, and A. Dong, **California Irrigation Management System, Final Report, Vol II,** Department of Land, Air, and Water Resources, University of California, Davis, California, 1985

SUBJECT INDEX
Page number refers to first page of paper.

AUTHOR INDEX
Page number refers to first page of paper.